# Molecular Biology
# and Pathogenicity
# of Mycoplasmas

# Molecular Biology and Pathogenicity of Mycoplasmas

Edited by

## Shmuel Razin

*The Hebrew University-Hadassah Medical School*
*Jerusalem, Israel*

and

## Richard Herrmann

*University of Heidelberg*
*Heidelberg, Germany*

Kluwer Academic / Plenum Publishers
New York, Boston, Dordrecht, London, Moscow

ISBN 0-306-47287-2

©2002 Kluwer Academic / Plenum Publishers, New York
233 Spring Street, New York, New York 10013

http://www.wkap.nl/

10  9  8  7  6  5  4  3  2  1

A C.I.P. record for this book is available from the Library of Congress

# Foreword: From the Enigmatic Pleuropneumonia-like Organisms to the Paradigmatic Mycoplasma

Over a century ago, Edmond Nocard and Pierre Roux were engaged in the study of infectious bovine pleuropneumonia. The etiological agents were filterable, like the recently discovered viruses but could be cultured in sterile growth media similar to bacterial culture.[1] When related taxa were discovered, they were designated pleuropneumonia-like organisms (PPLO). The driving force for the first half of the century or more of PPLO research was the interest in a number of animal diseases associated with these organisms and the possibility of their involvement in human diseases.

In 1960, Volume 79 Article 10 of the Annals of the New York Academy of Sciences[2] presented the proceedings of a previous year's meeting entitled, "Biology of the Pleuropneumonia-like Organisms." Among the authors were several of the pioneers in the description of these still incompletely characterized microbes. The papers in this volume reveal great uncertainty about the relation of the PPLO to bacteria, newly discovered bacterial L forms, viruses, and other infectious agents such as the rickettsia. Of course, 1959 was less than a century after the founding of microbiology by Koch and Pasteur. Most of the papers in the Academy volume focus on pathogenicity with particular reference to animal diseases of importance to agriculture. A few papers began to probe the biochemical characterization of these organisms.

In the late 1950s, efforts were underway, from the perspective of biophysics, to seek the lower limit of life, the smallest, autonomous, self-replicating organisms. This began as a search for microbial oddities and seemed to lead relentlessly toward the pleuropneumonia-like organisms, which might therefore have a special role to play in molecular biology.

This second domain of interest in the PPLO resulted in the "Conference on the Molecular Biology of the Pleuropneumonia-like Organisms" held June 14-16, 1962, at the University of Connecticut. The site of the meeting

v

was the result of the efforts of Robert Cleverdon. The rapidly developing discipline of molecular biology and the rapidly expanding knowledge of the PPLO were brought together at this meeting. In addition to the PPLO specialists, the conference invited Julius Marmur to compare PPLO DNA to DNA of other organisms; David Garfinkel, who was one of the first to develop computer models of metabolism; Cyrus Levinthal to talk about coding; and Henry Quastler to discuss information theory constraints on very small cells. The conference was an announcement of the role of PPLO in the fundamental understanding of molecular biology.

Looking back 40-some years to the Connecticut meeting, it was a rather bold enterprise. The meeting was international and inter-disciplinary and began a series of important collaborations with influences resonating down to the present. If I may be allowed a personal remark, it was where I first met Shmuel Razin, who has been a leading figure in the emerging mycoplasma research and a good friend. This present volume is in some ways the fulfillment of the promise of that early meeting. It is an example of the collaborative work of scientists in building an understanding of fundamental aspects of biology.

In the three years between the 1959 New York Academy meeting and the University of Connecticut meeting, the problem of establishing the structural nature of PPLO cellularity had been approached in many ways. The first Academy meeting had left uncertainty about whether we were dealing with normal cells, syncitia, or some other form of biological organization, perhaps a novel method of structure. In retrospect the uncertainties resulted from small size and absence of a rigid cell wall. From a physiological point of view, we regarded a cell as an aqueous core surrounded by a membrane with limited conductivity for polar molecules and ions, but this required proof.

This type of problem could be studied by a biophysical method going back to J. Clerk Maxwell in 1873.[3] It involved dielectric dispersion measurements on a packed suspension of cells between two platinum plates. We solicited the collaboration of Herman Schwan of the University of Pennsylvania, who was a specialist in dielectric dispersion studies. This led to the experimental conclusion[4] that the organization was indeed cellular with membranes having a capacitance of 1.3 μfarad/cm., which is a normal value for other living cells. These results allowed us to speak more definitely of the smallest living *cells*. In subsequent years, S. Razin and others have carried out detailed characterization of the membrane that gave rise to the electrical properties. Although the dielectric work is seldom cited, the dielectric dispersion studies were to me of enormous significance. They established the nature of the mycoplasma. Robert Cleverdon and I in 1959 had posited another feature of the organism, the small amount of DNA per cell or minimum genome.[5] Our experimental value was misleadingly small, which I believe was due to errors in our method of determining cell number. In 1962 Mark Tourtellotte and I explored the question, "What are the smallest dimensions compatible with life?"[6] We were driven by the assumption that mycoplasma were primitive organisms. Another mea culpa: Carl Woese, Jack Maniloff, and L. B. Zablen later showed from molecular taxonomic arguments that, rather than being primitive, mycoplasma are the ultimate parasites and saprophytes.[7] I would now argue that primitive organisms must be autotrophs, and mycoplasma are the ultimate heterotrophs. Nonetheless, I believe that the taxon is of importance in the basic study of all living cells.

In May 1966, a second meeting of the New York Academy of Sciences brought together a considerably larger group to discuss "Biology of the Mycoplasma".[8] The impressive volume is 824 pages. In the opening remarks, Leonard Hayflick notes, "Since the first conference dedicated to them seven years ago, taxonomic dignity has been obtained by replacing the name PPLO with proper Linnean terminology."

The major subject of the 1966 volume is microbiology and characterization of mycoplasma. Much of what we now call molecular biology was in the section called physiology and pathogenicity. It divided according to the host taxon. In any case, from our point of view, the fundamentals of the material discussed in this volume had been laid down.

The 1966 publication also included a report on Hans Bode's work on unambiguously determining the genome size and configuration of mycoplasma DNA.[9] This was an important normalizing of this taxon among the prokaryotes. In the absence of methods for determining the sequence of DNA nucleotides, other methods were then undertaken, such as John Ryan's analysis of t-RNA and r-RNA coding regions.[10] This ultimately led to the studies on gene organization of *Mycoplasma capricolum* by Akira Muto and his coworkers and by Shmuel Razin and his coworkers.

In the 60s and 70s, mycoplasma also turned out to be a taxon of choice for certain basic membrane studies for two reasons. First, the small size resulted in a high surface-to-volume ratio and a large fraction of the cell's mass as membrane. Secondly, the absence of a cell wall facilitated the preparation of membrane. In addition, varying the fatty acid composition of the growth medium permitted considerable control in the membrane fatty acid composition. As a result, a series of physical chemical studies of bilayer-phase transitions was carried out on purified membrane and whole cells.[11],[12] It is doubtful that these fundamental membrane experiments could have been done with any other taxon. In a sense, all of this was serendipity. Cells that were being studied for other reasons turned out to be ideal for basic biophysical characterizations of cellularity and fundamentals of membrane structure and function.

The science that is the subject of this volume has covered about a century from the discovery of the etiological agent of bovine pleuropneumonia to the sequencing of the genome of *Mycoplasma genitalium* and other species. It is rooted in the bacteriology of Pasteur and Koch, expands in the biochemistry of Watson and Crick, matures into present day genomics and looks ahead to proteomics and physiomics.

Starting as a sideshow of early microbiology, mycoplasma have become central to modern computational and theoretical biology and the understanding of infectious disease. To look ahead, I'm confident that the first decade of the 21st century will lead to a complete computer model of mycoplasma cell function. As the biology of the 21st century unfolds, I suspect that the minimal cell concept as embodied in the mycoplasma will continue to be central to the understanding of life. The publishers and editors have chosen an appropriate moment in time for this volume.

HAROLD J. MOROWITZ
*Robinson Professor,*
*Krasnow Institute for Advanced Studies,*
*George Mason University,*
*Fairfax, Virginia 22030, USA*

[1] Nocard, E., E. R. Roux, Mm. Borrel, Salimbeni & Dujardin-Beaumetz. 1989. *Le microbe de la péripneumonie.* Ann. Inst. Pasteur **12**: 240.

[2] Biology of the Pleuropneumonia-like Organisms. Volume 79, Art. 10, Pages 305-758. *Annals of the New York Academy of Sciences,* 1960.

[3] Maxwell, J. C., *A Treatise of Electricity and Magnetism,* 1973, Oxford University Press.

[4] Schwan, H. P. and Morowitz, J. H. Electrical Properties of the Membranes of Pleuropneumonia-like Organisms, A5969, 1962, *Biophysical Journal* **2**, 295-407.

[5] Morowitz, H. J. and Cleverdon, R. C., An Extreme Example of the Coding Problem, 1959, Avian PPLO 5969. *Biochim. Biophys. Acta* **34**, 578-579.

[6] Morowitz, H. J. and Tourtellotte, M. E., The Smallest Living Cells, 1962, *Scientific American* **206**, 117-126.

[7] Woese, C. R., J. Maniloff and L. B. Zablen, 1980, Phylogenetic Analysis of the Mycoplasmas. *Proc. Nat. Acad. Sci.,* USA, **77**, 494-498.

[8] Biology of the Mycoplasma, 1967, Volume **143**, Art 1, Pages 1-824, *Annals of the New York Academy of Sciences.*

[9] Bode, H. R. and Morowitz, H. J., 1967, Size and Structure of the *Mycoplasma hominis* H39 Chromosome, *J. Mol. Biol.,* **23**, 191-199.

[10] Ryan, J. L. and Morowitz, H. J., 1969, Partial Purification of Native r-RNA and t-RNA Cistrons from Mycoplasma SP. (Kid), *Proc. Nat Acad. Sci.* 73, 1282-1289.

[11] Stein, J. M., Tourtellotte, M. E., Reinhert, J. C., McElhaney, R. N. and Rader, R. L., 1969, Calorimetric Evidence of Liquid-Crystalline State of Lipids in Biomembrane, *Proc. Nat. Acad. Sci.* USA, **63**, 104-109.

[12] Melchior, D. L., Morowitz, H. J., Sturtevant, J. M. and Tsong, T Y., 1970, Phase Transitions in Membrane Lipids, *Biochim. Biophys. Acta,* **219**, 114-122.

# Preface

The mycoplasmas (class *Mollicutes*) represent a wide spectrum of phylogenetically related parasitic bacteria. They have already served in the past as models for basic research but in many cases these studies were hindered by the lack of efficient methods for genetic transformation and by the fastidious growth of many mollicute species. The newly developed or improved methods of molecular biology and bioinformatics helped to overcome these problems to some extent. High through-put DNA sequencing, handling of large data sets, the PCR technology with the possibility to mutate DNA with relative ease and express mycoplasmal genes in foreign hosts have contributed to the success of many research projects summarized in the present book. The recent sequencing of the entire genomes of *Mycoplasma genitalium*, *M. pneumoniae*, *M. pulmonis* and *Ureaplasma urealyticum* has marked a turning point in the molecular genetic analysis of these organisms. Studying gene expression with complete cells at the level of transcription (transcriptome analysis) and at the level of translation (proteome analysis) and relating the products to genes or ORFs defined by total genome sequences promises to provide us with the definition of the total protein complement of a cell. The Mollicutes group includes the smallest known self-replicating organisms carrying the smallest number of genes. There is no wonder, therefore, that mycoplasmas have a special appeal to those interested in the definition of the minimal set of genes essential for life considering this as an important step on the way of reaching the goal of defining in molecular terms the entire machinery of a self-replicating cell.

The application of molecular markers has also pushed forward our understanding of the phylogeny of Mollicutes, placing their taxonomy on a sound molecular basis. The use of molecular markers and comparative genomics in taxonomy has extended the scope of mycoplasmology by enabling the classification of uncultivable mycoplasmas, such as the plant pathogenic phytoplasmas, and the recent inclusion in Mollicutes of the *Eperythrozoon* and *Haemobartonella* species, classified previously as rickettsia.

Considerable advances were also made towards better understanding of mycoplasma pathogenesis. Most impressive are the findings concerning the interaction of mycoplasmas with the immune system, macrophage activation, cytokine induction, mycoplasma cell components acting as superantigens, and autoimmune manifestations. Evasion of the host immune system by antigenic variation of mycoplasmal surface components is another subject that has gained much attention recently, as well as the molecular definition of mycoplasmal adhesins. The recent demonstration of the ability of mycoplasmas to enter host cells, cause fusogenic, apoptotic and oncogenic effects, as well as the possible association of mycoplasmas with activation of arthritis, and several other human diseases of unknown aetiology had their share in intensifying research on mycoplasma pathogenesis, bringing more researchers into the circle of those interested in this group of organisms.

The last multi-authored treatise on mycoplasmas: "Mycoplasmas: Molecular Biology and Pathogenesis" (J. Maniloff, R.N. McElhaney, L.R. Finch, and J.B. Baseman, eds.) was published in 1992. Large parts of this book are now out of date. Several reviews covering different aspects of mycoplasmology have been published during the last decade. Clearly, these reviews could not fill the gap created by the lack of a comprehensive, up-to-date multi-authored treatise including the new advances in the molecular aspects of mycoplasma research. The need for such a book has been felt for quite a while, not only by mycoplasmologists, but also by molecular biologists and the many researchers newly attracted to the study of Mollicutes as excellent models in genomics and proteomics.

Some comments as to the nomenclature used in the book: While the trivial terms "mycoplasmas" or "mollicutes" are used interchangeably to denote any species included in *Mollicutes*, the names ureaplasmas, entomoplasmas, mesoplasmas, spiroplasmas, acholeplasmas, asteroleplasmas, and anaeroplasmas are routinely used for members of the corresponding genera, and the term phytoplasmas is reserved for the uncultivable plant mycoplasmas.

Considering the relatively large number of chapters and contributors, keeping to the deadline set by the publisher is an achievement by itself. Obviously, this could not be accomplished without the cooperation of the many contributors. We express our gratitude and appreciation for their friendly collaboration in this endeavor. We thank also the Senior Publishing Editor (Biosciences) Joanna Lawrence, for her prompt and most efficient help in facilitating the fast publication of this book.

Shmuel Razin
Richard Herrmann

# Contents

# Contents

# Chapter 1

# Taxonomy of *Mollicutes*

KARL-ERIK JOHANSSON[*] and BERTIL PETTERSSON[#]
[*]*Department of Bacteriology, National Veterinary Institute, SE-751 89 Uppsala, Sweden;*
[#]*Division of Molecular Biotechnology, Department of Biotechnology, Stockholm Centre for Physics, Astronomy and Biotechnology (SCFAB), Royal Institute of Technology (KTH), Roslagstullsbacken 21, SE-106 91 Stockholm, Sweden*

## 1.    INTRODUCTION

Many commonly used bacteriological terms are far from being well defined and in scientific papers these terms may be used with different meanings[15,89]. One of the purposes of bacterial taxonomy is to avoid confusion by creating a database with a common language for bacteriologists. The word taxonomy is derived from the Greek words *taxis* and *nomos*, which mean arrangement or order and law, respectively. Microbial taxonomy includes three different, but related areas, namely classification, nomenclature and identification[2]. Taxonomy can be regarded as a complete system for organisation of information about organisms and classification is defined as the theories and principles for arranging organisms into hierarchic groups of the above database. Taxon (pl. taxa) is defined as a group of organisms of any rank in the taxonomic hierarchy[62]. For instance, species, family and domain are taxa and these taxa are ranked according to their level of inclusiveness. Nomenclature refers to the process of naming taxa on the basis of similarities or relations, which for bacteria is regulated by the so-called Bacteriological Code[72]. Identification is the practical application of a classification scheme to establish the identity of an isolate. Systematics is sometimes used as a synonym for taxonomy, but phylogenetic aspects are often also included in systematics[74]. Scientists have always been interested in classifying organisms and for higher animals and

*Molecular Biology and Pathogenicity of Mycoplasmas*, Edited by Razin and Herrmann, Kluwer Academic/Plenum Publishers, New York, 2002

1

plants this has been a comparatively easy and straightforward procedure, because of the useful and easily observable morphological characters. To avoid confusion, taxonomy should be as stable as possible to be useful, but when the amount of information about an organism has increased to a certain level, it may be necessary to accept a revised taxonomy. Bacterial taxonomy has traditionally been regarded as a rather conservative branch of bacteriology[12], but after the introduction of molecular techniques into taxonomy, the situation has changed and taxonomy has been transformed into an exciting and rapidly developing field in bacteriology[27]. Phylogeny plays a pivotal role in modern bacterial taxonomy, evident in the new (2nd) edition of Bergey's Manual on Systematic Bacteriology[23]. Phylogeny is, therefore, also discussed in this chapter, but at a more detailed and specific level than in the chapter "Phylogeny and Evolution", by J. Maniloff.

## 2.      TAXONOMY

Classification can be based on many different principles, but a useful system should have a predicative potential. In other words, if the taxon to which an organism belongs is known, it should be possible to predict many properties of that organism. A classification system based on the evolutionary history (phylogeny) of the organisms will have the potential to be very predicative. Classification of animals and plants have traditionally been based on morphological characters, which sometimes also reflects their evolutionary history. For unicellular organisms it is much more difficult to find common morphological characters, which mirror their phylogeny and which are easy to measure. Introduction of molecular methods has, therefore, resulted in a revolution in the fields of bacterial taxonomy.

## 2.1     General concepts

Traditional classification can be divided into special purpose classification and natural classification. Special purpose classification is in general only useful within particular areas and is often based on a few characteristics only. Special purpose classification can be regarded as artificial (in contrast to natural) and a classical example is the placement of *Escherichia coli* and *Shigella dysenteriae* in different genera, because of the more serious disease caused by the latter organism. Strains within these two taxa are phenotypically very similar, they show high DNA-DNA hybridisation values and they would normally be regarded as belonging to the same species. In natural classification the aim is to arrange organisms in a system which should be useful within as many areas as possible. For bacteria such a system should involve all described organisms and the belonging to a certain taxon should reflect their overall similarity as known at present. Such classification is called phenetic (in contrast to phylogenetic) and evolutionary relations are not taken into account. In phylogenetic classification, which is also natural, the grouping of organisms is based on their evolutionary history. Phenetic and phylogenetic classifications often give similar results unless convergent evolution or horizontal gene transfer has taken place. These two mechanisms can result in organisms with similar properties in spite of the fact that they are only distantly related.

The term polyphasic taxonomy was introduced by Colwell[10]. In this approach, phenotypic, genotypic and phylogenetic information are all considered[80]. The idea is to use all these types of information about an organism to create a picture, which illustrates the relations to other organisms as faithfully as possible. Polyphasic taxonomy is particularly useful for recently evolved taxa with small sequence differences in the 16S rRNA sequences[74]. Unfortunately, only few laboratories in the world have the capacity to apply this strategy for characterization and identification of new isolates.

## 2.2 The species concept

In spite of the fact that the species is the basic category in taxonomy, there is no general definition of what a species really is[15,67]. For higher organisms, there is, in general, no problem to define the species because the definition can be based on their sexual reproduction and the generation of fertile offspring. For bacteria, species definition is more complicated, because sexual reproduction is not the normal mode of multiplication. The definition of a bacterial species is, therefore, comparatively vague and it is important to keep in mind that the bacterial species concept is artificial, because nature tends to be continuous as was pointed out by Cowan[12]. Most bacteriologists probably accept that a bacterial species is a collection of strains, which are more similar to each other than they are to strains representing other species. However, if a sufficiently large number of strains representing closely related species are analyzed, strains that have intermediate properties will probably be found. Bacterial strains can be regarded as dots in the multidimensional space of life[12]. When a new species is described, it is therefore important to include analysis of several strains and it has been recommended that 25 strains, or at least 10 strains, should be included in such a study [62].

About 5000 bacterial species have been validly described so far and after 1980 approximately 200 new species are added each year to that list. To avoid confusion and repetition of work, a standardized nomenclature is important. Naming of new species is, however, not always an easy task and for researchers, which are inexperienced in the field, it is therefore recommended to contact someone at the editorial board of Int. J. Syst. Evol. Microbiol.[14]. In the List of Bacterial names with Standing in Nomenclature, LBSN (URL: http://www.bacterio. cict.fr/) some useful information on this matter can also be found[17] as well as in the new edition of Bergey's Manual[23]. A species may be divided to subspecies, a step which is also regulated by the Bacteriological Code. However for mollicutes, the use of subspecies has been discouraged[35], and there are only a few species within the *Mycoplasma mycoides* cluster that have been further divided into subspecies (see below).

### 2.2.1    The species genome concept

The possibility of introducing a species genome concept for bacteria has been discussed[41]. It has been estimated that about 20% of the DNA of one bacterial strain may be absent in another bacterial strain of the same species. It was therefore suggested that the species genome should comprise all genes found in all strains of that species and therefore the genome of an individual bacterium will comprise of a core set of genes (present in 95% or more of the strains) and auxiliary genes (present in 1 to 95% of the strains). Genes present in less than 1% of the strains may be regarded as foreign genes. The core genes are those which should be used to determine the characteristics of the species. When the presence of core genes has been more systematically studied for different bacterial taxa in whole genome sequencing projects, it may form a framework for a better way to define a bacterial species. If a species genome concept is applied it may well prove that taxa of mollicutes, which have been classified as different species because they are different by some criteria and similar by others, are in fact the same. Examples of such mollicutes, which are very similar as judged by 16S rRNA sequence data, but different as judged by other criteria are *Mycoplasma agalactiae* and *Mycoplasma bovis*, *Mycoplasma gallisepticum* and *Mycoplasma imitans*, *Mycoplasma cottewii* and *Mycoplasma yeatsii*, *Mycoplasma pneumoniae* and *Mycoplasma genitalium*, as well as certain members of the *M. mycoides* cluster.

### 2.2.2    Designation of taxa below the rank of subspecies

Rules and recommendations for designating taxa below the rank of subspecies (infrasubspecific subdivision) are not included in the Bacteriological Code, but advice on their nomenclature can be found in Appendix 10[72]. Large colony (LC) and small colony (SC) type of

*Mycoplasma mycoides* subsp. *mycoides* and biovars 1-2 and serovars 1-14 of *Ureaplasma urealyticum* are examples of infrasubspecific subdivision. An infrasubspecific taxon is one strain or a set of strains exhibiting the same or similar properties according to certain criteria. A strain is according to the Bacteriological Code defined as the descendants of a single isolation in pure culture. Most bacterial strains are not known to be clones, where a clone is defined as a bacterial population derived from a single parent cell. For valid description of a new species of the class *Mollicutes* according to the Minimum Standard Document (see below), cloning is, however, required. The necessity of cloning of mollicutes has been discussed by the Subcommittee on the Taxonomy of *Mollicutes* and for taxonomic and phylogenetic purposes cloning has been established as compulsory because of the great risk of working with mixed cultures when cloning is not performed.

## 2.3    The *Candidatus* concept

For a valid description of a bacterial species, it has to be cultivated. Most bacteria in nature cannot apparently be cultivated, but such bacteria can still be characterized by a number of molecular methods, for instance 16S rRNA gene sequencing. The *Candidatus* concept has, therefore, been introduced to describe prokaryotic taxa for which molecular data are available but for which characteristics required for description according to the Bacteriological Code are lacking[48,49]. The following mollicutes have, so far, not been cultivated and the *Candidatus* concept was, therefore, applied: *Candidatus* M. ravipulmonis[51], *Candidatus* Phytoplasma sp.[47] and the haemotrophic mollicutes (haemoplasmas)[19,46,50].

## 3.    PHYLOGENY

Phylogeny is defined as the evolutionary history of a group of taxa or the evolutionary history of a gene from a group of taxa. The aim of classification has been to introduce a system, which reflects the natural relation (or phylogeny) of organisms. When Zuckerkandl and Pauling[90] realized that the sequences of the building blocks of the macromolecules of the cells reflect the evolutionary history (or phylogeny) of the cells, it became possible to establish phylogenies in a scientific way. It was suggested that the amino acid sequence or nucleotide sequence of a protein or a gene, respectively, may change in a regular (clocklike) fashion. This idea stimulated Kimura to develop the neutral theory of molecular evolution in which it is stated that most mutations that occur are neutral and will not change the phenotype[40].

Neutral mutations are insensitive to natural selection, but can still be fixed in a population by random drift.

The phylogeny of some genes also reflects the phylogeny of the organisms, but it is important to keep in mind that this is far from being always the case. For instance, a high resolution of the branches in a phylogenetic tree of a certain gene does not automatically imply that the tree represents the true phylogeny of the organisms from which the gene originated. It was Carl R. Woese who first took advantage of the properties of 16S rRNA and used it as a tool for phylogenetic studies[53,85-88]. He started to use oligonucleotide mapping, but soon turned to complete sequence analysis of 16S rRNA or the 16S rRNA gene following the introduction by Sanger of the dideoxynucleotide sequencing method[69], which made it possible to easily determine the nucleotide sequence of 16S rRNA or its gene.

## 3.1    Some basic definitions

A group of taxa in a phylogenetic cluster is said to be monophyletic if the taxa share a common ancestor which is not shared by any other taxa outside the cluster. A clade is a monophyletic group of organisms. Sister lineages are two clades resulting from the splitting of a single lineage. A group of taxa is paraphyletic if they share an ancestor, which is also shared by other taxa outside of the group. If a group of taxa are descendants from different ancestors, which are not included in the group, they are said to be polyphyletic. Ideally, taxa should represent monophyletic groups at different levels. Phylogeny is based on homology and two features are regarded as homologous when they are inherited from the nearest common ancestor. Thus, homology is an all or none character, whereas similarity can be quantified. Synapomorphies, which constitute the basis for establishing phylogenetic relations are shared, derived characters, which are not present in distant ancestors, in contrast to primitive characters.

## 3.2    The 16S rRNA gene as a phylogenetic tool

Sequence data of the 16S rRNA molecule or its genes have proved to be very useful in studying the phylogeny of microorganisms[53,85]. All self-replicating organisms have ribosomes and rRNA and, therefore, it is possible to use 16S rRNA sequence data to construct universal phylogenetic trees in which any species may be included. The function of the 16S rRNA molecule is essential in the translation machinery of the cell and it has remained unchanged during evolution rendering this molecule a useful phylogenetic tool. A great span in phylogenetic differences can be studied because the molecule contains regions of different evolutionary variability. Furthermore, the universal regions and the secondary structure of 16S rRNA facilitate the process of sequence alignments. The size of about 1500 nucleotides is convenient for rapid sequence analysis and the presence of universal regions enables its amplification by PCR for subsequent sequence analysis[36,55]. PCR and sequencing primers can easily be designed to target universal regions of the 16S rRNA genes also from organisms which have not earlier been described or for organisms which cannot be cultivated[1,34]. This possibility has proved very useful for classification of phytoplasmas[26], haemoplasmas[19,46,50] and *Candidatus* M. ravipulmonis[51], which are all uncultivable. Some years ago there was still no evidence for horizontal gene transfer of rRNA genes and this was used as one of the arguments for 16S rRNA as a useful phylogenetic tool. However, it has recently been reported that even rRNA genes may under certain conditions be subject to horizontal gene transfer[81], although it seems to be an unusual event.

### 3.2.1    Signature nucleotides

Signature nucleotide and unique nucleotide positions in the 16S rRNA molecule are considered helpful in justifying a certain clade or ancestral state despite the fact that the actual node can be affected by a low statistical support due to e.g. weak bootstrap values. A signature nucleotide in this context is a nucleotide residue that is explicitly found in a position within the sequence where the base present differs from that found in the majority of the other bacteria. The characterization of a signature nucleotide feature is restricted to a certain phylogenetic group of the mycoplasmas and/or the class *Mollicutes*. Generally, a nucleotide residue at a certain position is said to be unique when present in the molecule of all strains within a cluster or subcluster, and with only a few exceptions among the strains of any other cluster or group. Similarly, analysis of elongations or truncations or stems and loops can be used as an independent method to phylogenetic trees to verify a certain clade, especially because gaps have to be introduced in these regions to obtain a proper alignment. These so-called gapped positions are normally removed from the alignment prior to the computation of trees and are, therefore, not taken into account in the resulting cladogram.

### 3.2.2    Intraspecific variation in the 16S rRNA genes

When the 16S rRNA gene sequences of the same bacterial species, deposited in GenBank, were compared, an unexpected large number of sequence differences were found[9]. The differences were too many to be explained by sequencing errors only, and it was postulated that these differences were due to sequence differences between 16S rRNA genes of different rRNA operons (polymorphisms) and sequence differences in homologous genes, but from different strains of the same species. These differences were collectively termed intraspecific variation. Intraspecific variation of the 16S rRNA genes has been reported for mycoplasmas[57,58,61,66] and it has been found to be quite significant for certain species and has, therefore, to be considered in phylogenetic analyses and when the 16S rRNA gene is used as a diagnostic tool. *M. agalactiae*[31], *M. bovis*[31], *M. capricolum* subsp. *capripneumoniae*[30,56] and *M. mycoides* subsp. *mycoides* type SC[54,57] have all been found to have a comparatively high intraspecific variation. Other members of the hominis group and the *M. mycoides* cluster have only a few or no polymorphisms. Phytoplasmas have also been found to have polymorphisms in their 16S rRNA genes [42].

## 3.3    Construction of phylogenetic trees

Diagrams which illustrate the relations between groups of organisms are called cladograms and an evolutionary (phylogenetic) tree is a cladogram that reflects the hypothetical phylogenetic relations between the organisms under considerations. The construction of phylogenetic trees was recently reviewed by Ludwig *et al.*[44]. In general, the alignment procedure is the most crucial part when calculating evolutionary relationships between organisms. By limiting the discussion to 16S rRNA gene sequences one should i) only consider almost complete sequences, since there is no reason for using partial data that can have severe effects on the branching orders; ii) perform the alignment of sequences by using the secondary structure of the 16S rRNA molecule as a guideline for proper identification of stems and loops. The best match is not necessarily the best way to achieve a good alignment; iii) select a reasonable set of outgroups to be included in the alignment. An outgroup is an organism which for good reasons can be assumed to be less related to and branching early off the other taxa under consideration than any other of these taxa; iv) several data sets should be derived from the alignment for subsequent computation of cladograms. The sets should differ with respect to e.g. choice of outgroups, selective use of nucleotide information such as, removing gapped positions and hypervariable regions. Moreover, one should select subsets with nucleotide position having certain percentage values of consistency, e.g. 50%, 75% etc. We do not recommend the use of automatic alignment procedures such as the program CLUSTAL. Instead, manual alignment of new data to prealigned sequences should be the method of choice. Prealigned 16S rRNA sequences can be downloaded from the Ribosomal RNA Database Project (RDP), the rRNA www server or the ARB-project at http://rdp.cme.msu.edu/html/, http://www-rrna.uia.ac.be/ or http://www.mpi-bremen.de/molecol/arb/, respectively.

All these different alignments should be judged carefully with regard to the resulting topology as obtained from distance based methods and character based algorithms for calculation of trees. Moreover, confidence tests (bootstrap, jack knifing etc.) of the trees should be performed to find week nodes. If certain weak nodes are important for the study, one can create new subsets from the original alignment in order to verify which conditions do affect the node stability. Weak nodes can sometimes be judged as the true nodes if they can be verified by the existence of signature nucleotides and/or structural attributes as discussed above.

## 3.4 Taxonomy and phylogeny based on genes other than 16S rRNA

The usefulness of the 16S rRNA gene as a phylogenetic tool has resulted in a large number of sequences deposited in public databases. More than 16 000 sequences of 16S (or 16S-like) rRNA (or rDNA) molecules are available for phylogenetic analysis. About half of these sequences are of bacterial origin[34]. The properties of the molecule (see above) has greatly contributed to its common use as a phylogenetic tool and it is definitely the first choice if the phylogeny of a new organism is going to be investigated[43,44, 53]. However, there are organisms (or taxa) for which the phylogenetic resolution of a 16S rRNA-based evolutionary tree is not sufficient. Other genes have, therefore, been used for taxonomic and phylogenetic studies of mollicutes as well as for other organisms. The gene for 23S rRNA, which contains more information than that of 16S rRNA, has been used for phylogenetic analysis[44]. However, rather as a support than evaluation of the results obtained by the 16S rRNA molecule because of the relatedness of the molecules. The genes of the elongation factors EF-Tu and EF-G have also been used as phylogenetic markers and these genes are more independent of the 16S rRNA genes although they represent components of the translation machinery. A gene, which is completely independent on the 16S rRNA gene and has been used for phylogentic analysis of bacteria is the gene of the β-subunit of the $F_1F_0$-type ATPase. The use of these and other genes have been discussed in recent reviews[44, 63, 64].

## 4. TAXONOMY AND PHYLOGENY OF THE *MOLLICUTES*

The members of the class *Mollicutes* are characterized by their small genome size (0.58 – 2.2 Mbp), a low G+C content (23 – 40 mol%) of the genome and a permanent lack of a cell wall. About 200 species of the class

*Mollicutes* have been validly described. This means that most of them have been described according to the rules of the Minimum Standard Document[35], the first version of which was published in 1972. The trivial name mollicutes will in this review be used to include all members of the class *Mollicutes*, whereas the trivial name mycoplasma will be reserved for the members of the genus *Mycoplasma* only. Haemoplasma will be used as trivial name for the haemotrophic mollicutes that were earlier classified as rickettsias of the genera *Eperythrozoon* or *Haemobartonella*[19,46,50]. According to the new taxonomy of prokaryotes, the mollicutes belong to the phylum *Firmicutes*[24]. This phylum is reserved for gram-positive *Bacteria* with low G+C content of the genome and it also comprises the classes *Bacilli* and *Clostridia*. The phylum *Firmicutes* belongs to the domain *Bacteria*. The domain concept, which was introduced by Woese and co-workers[88], is based on evolutionary trees derived from 16S (or 16S-like) rRNA sequence analysis and this concept has now also been accepted in the 2nd edition of Bergey's Manual[23]. All prokaryotic organisms are now arranged within one of the two domains *Archaea* or *Bacteria*[2]. Phylum, Class, Order, Family, Genus, Species and Subspecies are successive subsets with decreasing inclusiveness of the respective Domain. The names of these categories have "standing in nomenclature", which means that they are given formal recognition. The categories Kingdom and Division are not used any more. The lower taxa of the class *Mollicutes* are listed in Table 1. Because of their uncertain taxonomic affiliation, the genera *Erysipelothrix*, *Bulleidia*, *Holdemania* and *Solobacterium* will be included as *Incertae sedis* under the family *Erysipelothrichaceae* of Order V of the mollicutes in the 2nd edition of Bergey's Manual. For more information, see the link to "Taxonomic outline" at the home page of Bergey's Manual Trust (http://www.cme.msu.edu/Bergeys/).

Before 1989, only a few 16S rRNA sequences of mollicutes were available from GenBank. The first extensive phylogenetic analysis based on the 16S rRNA gene of the mollicutes was done by Woese and co-workers[82]. They defined five phylogenetic groups (the anaeroplasma group, the asteroleplasma group, the hominis group, the pneumoniae group, and the spiroplasma group) and 15 different clusters. Additional clusters as well as subclusters were defined when more sequence data became available. The phylogenetic groups of the mollicutes are listed in Table 2. The taxonomy of the mollicutes does not always reflect their phylogeny and vice versa. These discrepancies are shown in Fig. 1 and further discussed below. In 1993 a revised taxonomy was introduced for the mollicutes[75]. This classification was based on a polyphasic approach and in addition to the six previously described genera, two new genera (*Entomoplasma* and *Mesoplasma*) were introduced (Table 1). Biochemical characteristics are important for

Table 1. Taxonomy and characteristics of the genera of the class *Mollicutes*.

| Order[a] | Family | Genus[b] | No. of taxa[c] | CE req.[d] | Phenotypic characteristcs | Habitat |
|---|---|---|---|---|---|---|
| *Mycoplasmatales* | { *Mycoplasmataceae* | *Mycoplasma* | 107 (11) | Yes | Opt. growth: 37°C[e] | Humans, animals |
| | | *Ureaplasma* | 7 | Yes | Urease positive | Humans, animals |
| | { *Entomoplasmataceae* | *Entomoplasma* | 6 | Yes | Opt. growth: 30°C | Insects, plants |
| *Entomoplasmatales* | { | *Mesoplasma* | 12 | No | Opt. growth: 30°C | Insects, plants |
| | *Spiroplasmataceae* | *Spiroplasma* | 34 | Yes | Helical morphology | Insects, plants |
| *Acholeplasmatales* | { *Acholeplasmataceae* | *Acholeplasma* | 14 | No | Opt. growth: 30-37°C | Animals, plant surfaces |
| | | "*Phytoplasma*"[,f] | (6) | n.d. | Uncultured in vitro | Insects, plants |
| *Anaeroplasmatales* | { *Anaeroplasmataceae* | *Anaeroplasma* | 4 | Yes | Obligate anaerobes | Bov. and ov. rumen |
| | | *Asteroleplasma* | 1 | No | Obligate anaerobes | Bov. and ov. rumen |

[a] A fifth order including the family *Erysipelothrichaceae* will possibly be included as *Incertae sedis* in the 2nd edition of Bergey's Manual.

[b] The genera *Eperythrozoon* and *Haemobartonella* (haemoplasmas) were included in the family *Mycoplasmatacea* in the 2nd edition of Bergey's Manual[24]. This nomenclature may, however, be changed when Vol. III, which comprises the phylum *Firmicutes*, is published.

[c] Current number of validly described species and subspecies of the respective genus. The number of *Candidatus* species is given within parentheses and includes haemoplasmas under genus *Mycoplasma*.

[d] Cholesterol requirement for growth.

[e] Except some species isolated from cold-blooded vertebrates[3, 5, 39]

[f] Phytoplasmas have an uncertain taxonomic affiliation because they cannot be cultivated. However, phylogenetically they are closely related to the acholeplasmas and are, therefore, tentatively arranged among the members of the family *Acholeplasmataceae* in this table.

identification of mollicutes and the ability to ferment glucose and/or hydrolyse arginine are features, which also have to be established when a new species is described[35] (see chapter "Central carbohydrate pathways: metabolic flexibility and the extra role of some 'housekeeping' enzymes"). However, as many other biochemical features, these properties tend to be volatile and do not always reflect the phylogeny of the organisms. For instance, of the larger clusters, it is only the *M. hominis* cluster of the hominis group, which is comprised of 21 species that have an identical glucose arginine profile $(-/+)$[32, 59].

## 4.1     Taxonomy and phylogeny of the genus *Mycoplasma*

The members of the genus *Mycoplasma* belong to the family *Mycoplasmataceae* together with members of the genus *Ureaplasma* (Table 1). Today 107 species and subspecies of the genus *Mycoplasma* have been validly described, which makes this genus the largest one of the mollicutes. Many important pathogens for animals and human are found among the mycoplasmas. "*Mycoplasma monodon*" is still a remaining illegitimate name within genus *Mycoplasma*. "*M. monodon*" has been assigned to an isolate from prawns and its 16S rRNA sequence is available from GenBank. The sequence data show that this isolate is phylogenetically related to members of the spiroplasma group and particularly to the entomoplasmas and the mesoplasmas[22].

### 4.1.1     Discrepancies between taxonomy and phylogeny

One of the major problems concerning the relations between taxonomy and phylogeny in *Mollicutes* is that the genus *Mycoplasma* is distributed in four of the five phylogenetic groups and three of these groups also contain mollicutes of other genera (Table 3). In other words, the genus *Mycoplasma* is not monophyletic. One species of the genus *Mycoplasma*, *M. feliminutum*, belongs to the anaeroplasma group[4], the members of the phylogenetic *M. mycoides* cluster belong to the spiroplasma group and the pneumoniae group contains 13 *Mycoplasma* spp. and several *Candidatus* Mycoplasma spp. as well as 7 *Ureaplasma* spp. It is only the hominis group that solely contains members of the genus *Mycoplasma*. Thus, there is no phylogenetic support for the genus *Mycoplasma* as a whole. It would be possible to solve the problem by retaining the genus name *Mycoplasma* for only one of the four groups and a natural choice would be the phylogenetic *M. mycoides* cluster because it contains the type species of the mollicutes (see below). An alternative would be to retain the genus name *Mycoplasma* for the members of the hominis group only, which would involve the smallest number of name changes. The problem of the nomenclature of the mollicutes has been discussed in a recent review[83], but a final decision in this matter has to be taken by the Subcommittee on the Taxonomy of *Mollicutes*.

*Table 2.* Phylogenetic groups, clusters and subclusters of the mollicutes. The references refer to articles where the respective groups were introduced or redefined. The number of species or subspecies within a group is given within parentheses. Note that all mollicutes are not included because 16S rRNA sequence data are lacking.

| Group | Cluster or s. sp. l.[a] | Subcluster or s. sp. l.[a] |
|---|---|---|
| anaeroplasma[82] (17) | *Anaeroplasma* sp.[82] (4) | |
| | *A. laidlawii* (4) | |
| | *A. axanthum* (3) | |
| | *Cand.* Phytoplasma sp. (6) | |
| asteroleplasma[82] (1) | | |
| hominis[82] (86) | *M. hominis*[32, 59, 82] (21) | *M. alkalescens*[59] (6) |
| | | *M. hominis*[59] (15) |
| | *M. bovis*[60] (21) | *M. bovis*[60] (3) |
| | | *M. bovigenitalium*[60] (4) |
| | | *M. felifaucium*[60] (2) |
| | | *M. fermentans*[60] (2) |
| | | *M. iners*[60] (5) |
| | | *M. leopharyngis*[60] (2) |
| | | *M. lipofaciens*[60] (1) |
| | | *M. opalescens*[60] (1) |
| | | *M. spermatophilum*[60] (1) |
| | *M. equigenitalium*[60] (2) | |
| | *M. gypis*[59] (1) | |
| | *M. lipophilum*[60, 82] (2) | |
| | *M. neurolyticum*[82] (11) | |
| | *M. pulmonis*[82] (2) | |
| | *M. sualvi*[82] (3) | |
| | *M. synoviae*[61, 82] (22) | |
| | Cand. *M. ravipulmonis*[59] (1) | |
| spiroplasma[82] (17) | *M. mycoides*[34, 57, 82] (7) | |
| | *S. apis*[82] (6) | |
| | *S. citri*[82] (3) | |
| | *S. ixodetis*[b, 82] (1) | |
| pneumoniae[82] (29) | Haemotrophic mollicutes[38] (11) | |
| | *M. fastidiosum*[37] (2) | |
| | *M. muris*[82] (4) | |
| | *M. pneumoniae*[61, 82] (7) | |
| | *U. urealyticum*[82] (7) | |

[a] s. sp. l. = single species line.
[b] Originally termed the *Spiroplasma* strain Y-32 cluster[82].

*Table 3.* Phylogenetic belonging of members of the genus *Mycoplasma*. Species of single species lines can be found in Table 2.

| Pg[a] | Representative sp.[b] | Other sp. or subsp. of the cluster or subcluster |
|---|---|---|
| an | *Acholeplasma* sp. (cl) | *M. feliminutum* |
| ho | *M. equigenitalum* (cl) | *M. elephantis* |
| ho | *M. lipophilum* (cl) | *M. hyopharyngis* |
| ho | *M. neurolyticum* (cl)[c] | *M. bovoculi, M. collis, M. conjunctivae, M. cricetuli, M. dispar, M. flocculare, M. hyopneumoniae, M. molare, M. lagogenitalium, M. ovipneumoniae* |
| ho | *M. pulmonis* (cl) | *M. agassizii* |
| ho | *M. sualvi* (cl) | *M. moatsii, M. mobile* |
| ho | *M. synoviae* (cl)[c] | *M. alligatoris* (= "*M. lacerti*"), *M. anatis, M. bovirhinis, M. buteonis, M. canis, M. citelli, M. columborale, M. crocodyli, M. cynos, M. edwardii, M. corogypsi, M. felis, M. gallinaceum, M. gallopavonis, M. glycophilum, M. leonicaptivi, M. mustelae, M. oxoniensis, M. pullorum, M. sturni, M. verecundum* |
| ho | *M. alkalescens* (sc-ho)[d] | *M. arginini, M. auris, M. canadense, M. gateae, M. phocacerebrale* |
| ho | *M. hominis* (sc-ho)[d] | *M. anseris, M. arthritidis, M. buccale, M. cloacale, M. equirhinis, M. falconis, M. faucium, M. hyosynoviae, M. indiense, M. orale, M. phocidae, M. salivarium, M. spumans, M. subdolum* |
| ho | *M. bovigenitalium* (sc-bo)[e] | *M. californicum, M. simbae, M. phocarhinis* |
| ho | *M. bovis* (sc-bo)[e] | *M. agalactiae, M. primatum* |
| ho | *M. felifaucium* (sc-bo)[e] | *M. adleri* |
| ho | *M. fermentans* (sc-bo)[e] | *M. caviae* |
| ho | *M. iners* (sc-bo)[e] | *M. columbinasale, M. columbinum, M. gallinarum, M. meleagridis* |
| ho | *M. leopharyngis* (sc-bo)[e] | *M. maculosum* |
| pn | Haemotrophic mollicutes (cl) | "*M. ovis*", *Candidatus* sp.: M. haemodidelphis, M. haemocanis, "M. haemofelis", M. haemolama, M. haemominutum, "M. haemomuris", "M. haemosuis", "M. kahanii", "M. wenyonii" |
| pn | *M. fastidiosum* (cl) | *M. cavipharyngis* |
| pn | *M. muris* (cl) | *M. iowae, M. microti* (="*M. volis*"), *M. penetrans* |
| pn | *M. pneumoniae* (cl) | *M. alvi, M. gallisepticum, M. genitalium, M. imitans, M. pirum, M. testudinis* |
| sp | *M. mycoides* (cl) | *M. capricolum* subsp. *capricolum, M. capricolum* subsp. *capripneumoniae, M. cottewii, M. mycoides* subsp. *capri, M. mycoides* subsp. *mycoides, M. putrefaciens, M. yeatsii* |

[a] Phylogenetic group: an = anaeroplasma group; ho = hominis group; pn = pneumoniae group; sp = spiroplasma group[82].

[b] Species, which were used to name the cluster (cl) or subcluster (sc) or cluster name.

[c] The *M. neurolyticum* cluster and the *M. synoviae* cluster will probably be further divided into subclusters (Johansson and Pettersson, unpublished; Pettersson and Johansson, unpublished).

[d] Subcluster of the *M. hominis* cluster[59].

[e] Subcluster of the *M. bovis* cluster[60].

### 4.1.2    Taxonomy and phylogeny of the *Mycoplasma mycoides* cluster

The classical *M. mycoides* cluster[11] is composed of the following very closely related taxa. *Mycoplasma capricolum* subsp. *capricolum*, *Mycoplasma capricolum* subsp. *capripneumoniae*, *Mycoplasma mycoides* subsp. *capri*, *Mycoplasma mycoides* subsp. *mycoides* LC (large colony type), *Mycoplasma mycoides* subsp. *mycoides* SC (small colony type) and *Mycoplasma* sp. bovine serogroup 7. All these species cause diseases of great concern in ruminants. *M. mycoides* subsp. *mycoides* SC, the agent of contagious bovine pleuropneumonia (CBPP) is also the type species of the mollicutes. Phylogenetic analysis shows that *Mycoplasma cottewii*, *Mycoplasma putrefaciens* and *Mycoplasma yeatsii* are closely related to the members of the *M. mycoides* cluster[29,57,82] and all these species will collectively be referred to as members of the phylogenetic *M. mycoides* cluster. The members of the phylogenetic *M. mycoides* cluster form a clade within the spiroplasma group (Fig. 1).

### 4.1.3    Evolution of strains belonging to the taxon *M. capricolum* subsp. *capripneumoniae*

The members of the *M. mycoides* cluster have two rRNA operons designated *rrnA* and *rrnB*[7]. The 16S rRNA genes of the two rRNA operons belong to the same gene family and they have, like other genes of a gene family at some stage of the evolution, arisen by gene duplication. The evolution of genes belonging to the same gene family does not occur at random, but is controlled by a mechanism, which is termed gene conversion resulting in concerted evolution of the genes[18]. This means that if a mutation occurs in one of the genes, it can be repaired by homologous recombination with the other gene. Most mycoplasmas with two rRNA operons have only a small number of polymorphisms in their 16S rRNA genes. However, it has been found that the type strain (F38[T]) of *M. capricolum* subsp. *capripneumoniae* has an unusually large number of sequence differences between the two 16S rRNA genes[57,66]. When several strains of *M. capricolum* subsp. *capripneumoniae* were analysed for intraspecific variation, it was found that this species has an unusually large intraspecific variation, both in the form of sequence differences between the two 16S rRNA genes and sequence differences between homologous genes of different strains of *M. capricolum* subsp. *capripneumoniae*[30,56]. Interestingly, it proved possible to construct evolutionary trees by comparative analysis of these data, which indicates that *M. capricolum* subsp. *capripneumoniae* has a deficient system for gene conversion.

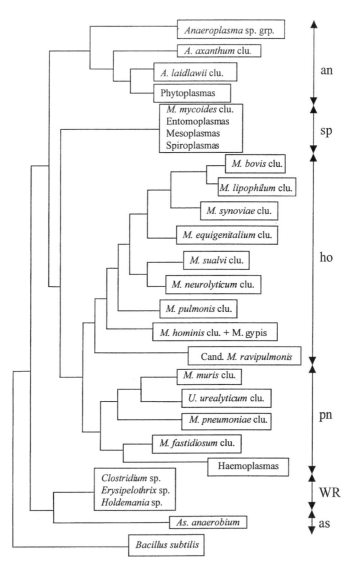

*Figure 1.* Schematic representation of the phylogenetic relations of the *Mollicutes* and some walled relatives (WR) based on 16S rRNA sequences. The phylogenetic groups are indicated with vertical arrows. Phylogenetic group abbreviations: an = anaeroplasma; as = asteroleplasma; ho = hominis; pn = pneumoniae; sp = spiroplasma.

## 4.2 Taxonomy and phylogeny of the genera *Entomoplasma* and *Mesoplasma*

The members of the genera *Entomoplasma* and *Mesoplasma* belong to the family *Entomoplasmataceae*, which in turn belongs to the order *Entomoplasmatales* together with the family *Spiroplasmataceae* (Table 1). Entomoplasmas, mesoplasmas and spiroplasmas, have insects or plants as hosts, but the two former genera lack the helical morphology of the spiroplasmas. Entomoplasmas and mesoplasmas were earlier classified as mycoplasmas and acholeplasmas, respectively, because the latter can be grown without sterols in the medium. Phylogenetic analysis clearly demonstrated that these species are closely related to members of the spiroplasma group[82] and a revised taxonomy with two new genera was, therefore, introduced for these mollicutes[75]. Since then, several new species have been described[78,79] and a total of six species of the genus *Entomoplasma* and 12 species of the genus *Mesoplasma* have now been reported. Many authors outside the mollicutes field are unfortunately not aware of the new classification and these species still appear as mycoplasmas or acholeplasmas in many publications.

### 4.2.1 Discrepancies between taxonomy and phylogeny

The members of the two genera *Entomoplasma* and *Mesoplasma* can be distinguished by their growth requirements. Entomoplasmas require serum in the growth medium, whereas mesoplasmas can be grown in the presence of Tween 80 instead of serum. Phylogenetic analysis shows that entomoplasmas and mesoplasmas form a paraphyletic group and share a common ancestor with members of the *M. mycoides* cluster[79]. Mesoplasmas, entomoplasmas and members of the *M. mycoides* cluster constitute a clade where *E. melaleucae*, *E. ellychniae* and *Me. entomophilum* form a sister lineage with the members of the *M. mycoides* cluster. There is no phylogenetic support, based on the 16S rRNA sequences, for the classification of the family *Entomoplasma* into the two genera *Entomoplasma* and *Mesoplasma*, because they do not form two coherent clusters but are rather intermixed in one paraphyletic group[79]. It was, therefore suggested that entomoplasmas and mesoplasmas should be combined into a single genus.

## 4.3 Taxonomy and phylogeny of the genus *Ureaplasma*

Members of the genus *Ureaplasma* belong to the family *Mycoplasmataceae* together with members of the genus *Mycoplasma* (Table 1).

Ureaplasmas can be distinguished from mycoplasmas by their ability to hydrolyse urea and by their very small colonies (about 10 μm in diameter). They were, therefore, earlier referred to as T- (tiny) strain mycoplasmas. Ureaplasmas were first isolated by Shepard in 1954 and *Ureaplasma urealyticum* was subsequently reported as a new species[71]. *U. urealyticum* strains have been divided into biovar 1 and biovar 2, and these biovars have been further subdivided into 4 and 10 serovars, respectively. Based on phylogenetic analysis of the 16S rRNA genes, the 16S-23S rRNA spacer regions, the genes of the urease subunits and the multiple-banded antigen genes, it was recently suggested that biovar 1 of *Ureaplasma urealyticum* should be classified as a new species, namely *Ureaplasma parvum*[18]. This reclassification has now also been formally proposed[65]. Ureaplasmas have been isolated from human, birds, cats, cattle and dogs and five other species of the genus *Ureaplasma* have been described. Based on 16S rRNA sequence data, the ureaplasmas form a monophyletic cluster within the pneumoniae group (Fig. 1). Thus, there is phylogenetic support for the genus *Ureaplasma*.

## 4.4    Taxonomy and phylogeny of the genera *Acholeplasma*, *Anaeroplasma* and *Asteroleplasma*

The members of the genus *Acholeplasma* belong to the family *Acholeplasmataceae* and the members of the genus *Anaeroplasma* belong to the family *Anaeroplasmataceae* together with the only member (*Asteroleplasma anaerobium*) of the genus *Asteroleplasma* (Table 1). So far, 14 acholeplasmas and 4 anaeroplasmas have been described. The acholeplasmas form two non-coherent clades within the anaeroplasma group (Fig. 1), which will be referred to as the *A. axanthum* and the *A. laidlawii* clusters. The *A. laidlawii* cluster, which also contains *A. oculi*, *A. palmae* and *A. vituli*, forms a sister lineage to the phytoplasmas. The *A. axanthum* cluster, which also comprises *A. modicum* and *M. feliminutum*, constitute a sister lineage to the cluster composed of the phytoplasmas and the members of the *A. laidlawii* cluster. Relatively few 16S rRNA sequences are available for the acholeplasmas and the topology of the phylogenetic tree may well change when more species will be included. All described anaeroplasmas form a clade within the anaeroplasma group (Fig. 1).

### 4.4.1    Discrepancies between taxonomy and phylogeny

A phylogenetic tree constructed with representatives of all phylogenetic groups and clusters of the mollicutes and some representatives of other members of the phylum *Firmicutes* suggests that *As. anaerobium* is more closely related to some members of the genera *Bulleidia*, *Clostridia*, *Erysipelothrix*, *Holdemania* and *Streptococcus* (Fig. 1). When *As. anaerobium* is excluded from such a tree, the mollicutes form a monophyletic group. It may, therefore, be a pertinent question to ask whether *As. anaerobium* should be regarded as a mollicute. On the other hand, it has been suggested that the genera *Erysipelothrix* and *Holdemania* should be included among the mollicutes[24].

## 4.5    Taxonomy and phylogeny of the phytoplasmas

Viruses were for many years believed to be the causal agents of so-called yellows-type diseases in plants. This conclusion was based on symptoms, routes of transmission by insect vectors and mode of multiplication in insects and plants (see the chapter "Mycoplasmas of Plants and Insects"). In 1967 it was shown that organisms, which were termed mycoplasma-like organisms, or MLOs, were able to cause some of these diseases in plants[16]. These organisms cannot be cultivated in cell free systems and their taxonomic affiliations were, therefore, for a long time uncertain. Several hundred diseases in different plants have been shown to be caused by phytoplasmas. By molecular techniques (sequence analysis of 16S rRNA and ribosomal protein genes) it has been shown that phytoplasmas are mollicutes and that they form a clade in the phylogenetic anaeroplasma group (Fig. 1). The taxonomy and phylogeny of the phytoplasmas have recently been reviewed[28,42]. By using general 16S rDNA primers it has been possible to detect a large number of phytoplasmas and a classification system based on RFLP of the 16S rRNA genes has been introduced. It was recently suggested that the phytoplasmas should be divided into 20 distinct groups[70] and 6 representatives have been assigned as *Candidatus* Phytoplasma sp.

## 4.6    Taxonomy and phylogeny of the genus *Spiroplasma*

Spiroplasmas were discovered in 1972 by phase contrast microscopy of sap from plants that were affected with corn stunt disease[13] and one year later, the first species, *Spiroplasma citri*, was described[68]. Spiroplasmas are characterised by their helical morphology and having insects and plants as their hosts. Many spiroplasmas cause diseases in insects and plants (see the chapter "Mycoplasmas of Plants and Insects"). Spiroplasmas have been classified by serological methods and 34 groups and 14 subgroups were introduced[77,83,84]. The groups were introduced to designate candidates for new species and the subgroups to designate possible candidates for new species. Thirty-four representative members of these groups or subgroups have been described as species. The members of the genus *Spiroplasma* belong to the family *Spiroplasmataceae*, which in turn belongs to the order *Entomoplasmatales* together with the family *Entomoplasmataceae* (Table 1). The phylogenetic spiroplasma group was divided into the *M. mycoides* cluster, the *S. apis* cluster, the *S. citri* cluster and the *Spiroplasma* sp. strain Y-32 cluster[82]. The latter strain has been named *S. ixodetis*, which should also be used as cluster name. The *M. mycoides* cluster is discussed above, the *S. apis* cluster also comprises *S. clarkii*, *S. diabroticae*, *S. gladiatoris*, *S. monobiae* and *S. taiwanense* and the *S. citri* cluster also comprises *S. mirum* and *S. poulsonii*. *S. ixodetis* was first described as a single species line[82], but has now been shown to contain several *Spiroplasma* sp. strains[22]. It may well be necessary to redefine the phylogenetic clusters when 16S rRNA sequences from all species have been reported. Today, 16S rRNA sequences from only 10 of the 34 validly described species are available from GenBank.

The spiroplasmas form a paraphyletic group and they share a common ancestor with members of the genera *Entomoplasma* and *Mesoplasma* as well as with members of the phylogenetic *M. mycoides* cluster (Fig. 1).

## 5.    THE USE OF 16S rRNA SEQUENCE DATA FOR DESCRIPTION OF NEW SPECIES

The procedures for describing new species within the class *Mollicutes*, are outlined in the so-called Minimum Standard Document[35]. When the family to which a potentially new species has been established, it has to be analysed by serological methods with antisera against all other previously described species within that particular genus. This method has become laborious for some of the genera, which contain many species, for instance *Mycoplasma* and *Spiroplasma*. The description of new species within these genera has probably been hampered by this requirement. There are very few laboratories in the world that keep all necessary antisera in stock. By using 16S rRNA sequence data as a complementary tool, the procedure can be facilitated. It has been suggested that the genera of the mollicutes should be arranged into a number of well-defined phylogenetic clusters and sub-clusters[29,32,36,37,59,60]. When the phylogenetic belonging of a potentially new species has been established, it should then be necessary to perform the serological analyses with antisera to members of the particular cluster or sub-cluster only. A necessary prerequisite for this approach is that the 16S rRNA sequences have been determined for all members of the family for which this strategy is to be applied. This goal has been reached for genus *Mycoplasma*, and all sequences are now available from GenBank.

The accession numbers to the most recently described species are listed in Table 4.

*Table 4.* Members of genera *Acholeplasma* and *Mycoplasma* for which the 16S rDNA sequences were recently deposited in GenBank (Pettersson and Johansson, unpublished; Johansson and Pettersson, unpublished).

| Species | Strain | Major host | Acc. no.[a] |
|---|---|---|---|
| *A. axanthum* | S-743[T] | Animals, plants | AF412968 |
| *M. alligatoris* | A21JP2[T] | Alligator | AF412969 |
| *M. anatis* | 1340[T] | Duck | AF412970 |
| *M. buteonis* | BbT2g[T] | Buzzard | AF412971 |
| *M. canis* | PG14[T] | Dog, cattle | AF412972 |
| *M. citelli* | RG-2C[T] | Rodent | AF412973 |
| *M. collis* | 58B[T] | Dog | AF412974 |
| *M. columborale* | MMP-4[T] | Pigeon | AF412975 |
| *M. cricetuli* | CH[T] | Rodent | AF412976 |
| *M. crocodyli* | MP145[T] | Crocodile | AF412977 |
| *M. cynos* | H831[T] | Dog | AF412978 |
| *M. dispar* | 462/2[T] | Cattle | AF412979 |
| *M. gallopavonis* | WR1[T] | Turkey | AF412980 |
| *M. glycophilum* | 486[T] | Chicken | AF412981 |
| *M. hyorhinis* | BTS7[T] | Pig | AF412982 |
| *M. lagogenitalium* | 12MS[T] | Afghan pika | AF412983 |
| *M. moatsii* | MK405[T] | Primate, rodent | AF412984 |
| *M. molare* | H542[T] | Dog | AF412985 |
| *M. mustelae* | MX9[T] | Rodent | AF412986 |
| *M. oxoniensis* | 128[T] | Rodent | AF412987 |
| *M. sualvi* | Mayfield B[T] | Pig | AF412988 |
| *M. verecundum* | 107[T] | Cattle | AF412989 |

[a]Accession number to the 16S rDNA sequence in GenBank.

## 5.1 The relationship between 16S rRNA sequence data and DNA-DNA reassociation values

One very useful technique to compare micro-organisms at the molecular level is to determine DNA-DNA reassociation values by DNA-DNA hybridisation experiments with whole genomic DNA[20,73]. In this technique, ssDNA from one organism is allowed to hybridise with ssDNA from a number of other organisms. If two organisms have a sequence similarity of at least 80% of their genomes, DNA preparations can form hybrids and the degree of reassociation can be determined on a scale where 0% represents a sequence similarity of less than 80% and a value of 100% represents 100% identity. This means that a scale from 80-100% is expanded to 0-100%. One of the advantages of this technique is that the reassociation values represent an overall similarity of the whole genome. A drawback is that the results are highly dependent on the experimental conditions, which are not always easy to control, which in turn means that the reproducibility of the method could be better.

DNA-DNA reassociation values have been used to define bacterial species and a value of 70% or higher indicates that the two strains belong to the same species. Note, however, that this corresponds to a sequence similarity of approximately 95% in general, although the correlation between sequence similarity and DNA-DNA reassociation values is not linear. As a rule of thumb a DNA-DNA reassociation value of 70% corresponds to a 16S rRNA sequence similarity of 97%[38,73]. However, bacterial isolates classified as belonging to different species, but showing 16S rRNA sequence similarity of above 97% have been found. Thus, 16S rRNA sequence similarity values are only indicative.

## 6.    THE IMPACT OF WHOLE GENOME SEQUENCING ON TAXONOMY AND PHYLOGENY

In 1995 microbiology went into the new era of whole genome sequencing and today more than 50 whole microbial genome sequences (including *M. genitalium*[21], *M. pneumoniae*[33], *M. pulmonis*[6] and *U. parvum*[25]) are available. Note that the strain of *U. parvum*, which was used for genome sequencing, was earlier referred to as *Ureaplasma urealyticum*[18,65]. The sequence of the genome of *M. hyopneumoniae* has also been reported as completed at the link TIGR Databases of the home page of The Institute for Genome Research (TIGR, URL: http:// www.tigr.org/). More than 150 whole genome sequencing projects of microorganisms (including *M. mycoides* subsp. *mycoides* SC) are known to be in progress. For updated information on this matter, see the home page of TIGR.

When it proved feasible to decipher whole genome sequences of microorganisms in the 1990s, it was believed that this would solve all taxonomic and phylogenetic problems in the future. However, to use whole genome sequences for this purpose may be more difficult than first anticipated. One of the problems will be to know which genes can be used to infer a meaningful phylogeny i.e. a phylogeny which reflects the phylogeny of the organisms. For instance, genes which are prone to horizontal gene transfer, are not suitable for this purpose. When the high extent of horizontal gene transfer was evident from different whole genome sequencing projects, it was also suggested that a phylogenetic tree is not the best way to illustrate the relations between microorganisms, but rather net-like patterns connecting organisms between which horizontal gene transfer has taken place[52]. Another approach that has been suggested for taxonomic and diagnostic purposes is the multilocus sequence typing (MLST) in which about 500 nt of a number of housekeeping genes are sequenced[45].

## 7.    CONCLUSIONS

Phylogeny is naturally dynamic, because the relation between organisms is dependant upon the species that have been included in the study and inclusion of a new species in a phylogenetic tree or selection of another outgroup, may well change the topology. A stable taxonomy is of course highly desirable, but when phylogenetic data are in complete contradiction to the taxonomy, it seems that it is time to change the classification. Classification must in the future be a more dynamic process than it was in the past[2,74].

Description of new species within the genus *Mycoplasma* has at least partly been hampered by the requirements of performing serological tests with antisera against all previously validly described species. The 16S rRNA (or rDNA) sequences of all members of this genus are now available from GenBank and, therefore, we suggest that the Minimum Standard Document[35] should be revised to include the use of 16S rRNA sequence data for classification of new isolates. Sequence analysis of the 16S rRNA genes of a bacterial isolate will, first tell if it is a mycoplasma or not and, second tell to which pylogenetic cluster or subcluster (according to the definitions in Tables 2 and 3) it belongs. Then, it should only be necessary to perform serological tests with antisera against the members of the cluster or subcluster. This strategy would decrease the number of serological tests from more than 100 to at most about 20 for the genus *Mycoplasma*. Phylogenetic clusters have been defined to facilitate the use of 16S rRNA sequence data for delineation of new species within the genus *Mycoplasma*. We strongly recommend authors to use the nomenclature for phylogenetic clusters, which was introduced by Woese *et al.*[82], and extended in this review to include all members of the genus *Mycoplasma*. A consistent terminology is essential to avoid confusion in the future.

## ACKNOWLEDGMENTS

We thank Göran Bölske, Janet M. Bradbury, Joseph G. Tully, Robert F. Whitcomb and the late Jan B. Ursing for many valuable discussions on mollicutes and on taxonomy.

# REFERENCES

1.   **Amann, R. I., W. Ludwig, and K.-H. Schleifer.** 1995. Phylogenetic identification and in situ detection of individual microbial cells without cultivation. Microbiol. Rev. **59**:143-169.

2.   **Brenner, D. J., J. T. Staley, and N. R. Krieg.** 2001. Classification of prokaryotic organisms and the concept of bacterial speciation, p. 27-31. *In* G. Garrity (ed.), Bergey's manual of systematic bacteriology. Vol. 1. 2$^{nd}$ edition. Springer-Verlag, N.Y., New York.

3.   **Brown, D. R., B. C. Crenshaw, G. S. McLaughlin, I. M. Schumacher, C. E. McKenna, P. A. Klein, E. R. Jacobson, and M. B. Brown.** 1995. Taxonomic analysis of the tortoise mycoplasmas *Mycoplasma agassizi*, and *Mycoplasma testudinis* by 16S rRNA gene sequence comparison. Int. J. Syst. Bacteriol. **45**:348-350.

4.   **Brown, D. R., G. S. McLaughlin, and M. B. Brown.** 1995. Taxonomy of the feline mycoplasmas *Mycoplasma felifaucium*, *Mycoplasma feliminutum*, *Mycoplasma felis*, *Mycoplasma gateae*, *Mycoplasma leocaptivus*, *Mycoplasma leopharyngis*, and *Mycoplasma simbae* by 16S rRNA gene sequence comparisons. Int. J. Syst. Bacteriol. **45**:560-564.

5.   **Brown, D. R., J. M. Farley, L. A. Zacher, J. M.-R. Carlton, T. L. Clippinger, J. G. Tully, and M. B. Brown.** 2001. *Mycoplasma alligatoris* sp. nov., from American alligators. Int. J. Syst. Evol. Microbiol. **51**:419-424.

6.   **Chambaud, I., R. Heilig, S. Ferris, V. Barbe, D. Samson, F. Galisson, I. Moszer, K. Dybvig, H. Wróblewski, A. Viari, E. P. Rocha, and A. Blanchard.** 2001.The complete genome sequence of the murine respiratory pathogen *Mycoplasma pulmonis*. Nucleic Acids Res. 29:2145-2153.

7.   **Christiansen, G., and H. Ernø.** 1990. RFLP in rRNA genes of *Mycoplasma capricolum*, the caprine F38-like group and the bovine serogroup 7. Zentralbl. Bacteriol. Hyg. Suppl. **20**:479-488.

8.   **Cilia, V., B. Lafay, and R. Christen.** 1996. Sequence heterogeneities among 16S ribosomal RNA sequences, and their effect on phylogenetic analysis at the species level. Mol. Biol. Evol. **13**:451-461.

9.   **Clayton, R. A., G. Sutton, P. S. Hinkle, Jr., C. Bult, and C. Fields.** 1995. Intraspecific variation in small-subunit rRNA sequences in GenBank: why single sequences may not adequately represent prokaryotic taxa. Int. J. Syst. Bacteriol. **45**:595-599.

10.  **Colwell, R. R.** 1970. Polyphasic taxonomy of the genus *Vibrio*: numerical taxonomy of *Vibrio cholerae*, *Vibrio parahaemolyticus*, and related *Vibrio* species. J. Bacteriol. **104**:410-433.

11.  **Cottew, G. S., A. Breard, A. J. DaMassa, H. Ernø, R. H. Leach, P. C. Lefevre, A. W. Rodwell, and G. R. Smith.** 1987. Taxonomy of the *Mycoplasma mycoides* cluster. Israel J. Med. Sci. **23**:632-635.

12.  **Cowan, T. S.** 1971. Sense and nonsense in bacterial taxonomy. J. Gen. Microbiol. **67**:1-8.

13.  **Davis, R. E., R. F. Whitcomb, T. A. Chen, and R. R. Granados.** 1972. Current status of the aetiology of corn stunt disease, p. 205-225. *In* K. Elliott and J. Birch (ed.), Pathogenic mycoplasmas. Ciba Foundation Sympoium. Elsevier-Excerpta Medica-North-Holland, Amsterdam.

14.  **De Vos, P., and H. G. Trüper.** 2000. Judicial Commission of the International Committee on Systematic Bacteriology. IXth International (IUMS) Congress of Bacteriology and Applied Microbiology. Int. J. Syst. Evol. Microbiol. **50**:2239-2244.

15.  **Dijkshoorn, L., B. M. Ursing, and J. B. Ursing.** 2000. Strain, clone and species: comments on three basic concepts of bacteriology. J. Med. Microbiol. **49**:397-401.

16. **Doi, Y., M. Terenaka, K. Yora, and H. Asuyama.** 1967. Mycoplasma or PLT-group-like microorganisms found in the phloem elements of plants infected with mulberry dwarf, potato witches-broom, aster yellows, paulownia witches-broom. Ann. Phytopathol. Soc. Japan **33**:259-266.

17. **Euzéby, J. P.** 1997. List of bacterial names with standing in nomenclature: a folder available on the internet. Int. J. Syst. Evol. Microbiol. **47**:590-592.

18. **Fanrong, K., G. James, M. Zhenfang, S. Gordon, B. Wang, and G. L. Gilbert.** 1999. Phylogenetic analysis of *Ureaplasma urealyticum* – support for the establishment of a new species, *Ureaplasma parvum*. Int. J. Syst. Bacteriol. **49**:1879-89.

19. **Foley, J. E., and N. C. Pedersen.** 2001. *"Candidatus* Mycoplasma haemominutum", a low-virulence epierythrocytic parasite of cats. Int. J. Syst. Evol. Microbiol. **51**:815-817.

20. **Fox, G. E., J. D. Wisotzkey and P. Jurtshuk, Jr.** 1992. How close is close: 16S rRNA sequence identity may not be sufficient to guarantee species identity. Int. J. Syst. Bacteriol. **42**:166-170.

21. **Fraser, C. M., J. D. Gocayne, O. White,** *et al.* 1995. The minimal gene complement of *Mycoplasma genitalium.* Science **270**:397-403.

22. **Fukatsu,T., T. Tsuchida, N. Nikoh, and R. Koga.** 2001. *Spiroplasma* symbiont of the pea aphid, *Acyrthosiphon pisum* (Insecta: Homoptera). Int. J. Syst. Evol. Microbiol. **67**:1284-1291.

23. **Garrity, G. M.** (ed.). 2001. Bergey's manual of systematic bacteriology. Vol. 1. 2nd edition. Springer-Verlag, N.Y., New York.

24. **Garrity, G. M., and J. G. Holt.** 2001. The road map to the manual, p. 119-166. *In* G. Garrity (ed.), Bergey's manual of systematic bacteriology. Vol. 1. 2nd edition. Springer-Verlag, N.Y., New York.

25. **Glass, J. I., E. J. Lefkowitz, J. S. Glass, C. R. Heiner, E. Y. Chen, and G. H. Cassell.** 2000. The complete sequence of the mucosal pathogen *Ureaplasma urealyticum.* Nature **407**:757-762.

26. **Gunderson, D. E., I.-M. Lee, S. A. Rehner, R. A. Davis, and D. T. Kingsbury.** 1994. Phylogeny of mycoplasmalike organisms (phytoplasmas): a basis for their classification. J. Bacteriol. **176**:5244-5254.

27. **Gürtler, V., and B. C. Mayall.** 2001. Genomic approaches to typing, taxonomy and evolution of bacterial isolates. Int. J. Syst. Evol. Microbiol. **50**:3-16.

28. **Harrison, N. A.** 1999. Phytoplasma taxonomy. Proceedings of the First Internet Conference on Phytopathogenic Mollicutes. (URL: http://www.uniud.it/phytoplasma/conf. html).

29. **Heldtander, M., B. Pettersson, J. G. Tully, and K.-E. Johansson.** 1998. Sequences of the 16S rRNA genes and phylogeny of the goat mycoplasmas; *Mycoplasma adleri, Mycoplasma auris, Mycoplasma cottewii,* and *Mycoplasma yeatsii.* Int. J. Syst. Bacteriol., **48**:263-268.

30. **Heldtander, M., H. Wesonga, G. Bölske, B. Pettersson, and K.-E. Johansson.** 2001. Genetic diversity and evolution of *Mycoplasma capricolum* subsp. *capripneumoniae* strains from eastern Africa assessed by 16S rDNA sequence analysis. Vet. Microbiol. **78**:13-28.

31. **Heldtander Königsson, M.** 2001. Ph.D. thesis. Phylogeny, diversity, detection: multiple uses of 16S rRNA genes in veterinary bacteriology. The Swedish University of Agricultural Sciences. Uppsala, Sweden

32. **Heldtander Königsson, M., B. Pettersson, and K.-E. Johansson.** 1998. Phylogeny of the seal mycoplasmas; *Mycoplasma phocae* corrig., *Mycoplasma phocicerebrale* corrig., and *Mycoplasma phocirhinis* corrig. based on sequence analysis of 16S rDNA. Int. J. Syst. Evol. Microbiol., **51**:1389-1393.

33.   **Himmelreich, R., H. Hilbert, H. Plagens, E. Pirkl, B. C. Li, and R. Herrmann.** 1996. Complete sequence analysis of the genome of the bacterium *Mycoplasma pneumoniae*. Nucleic Acids Res. **24**:4420-4449.

34.   **Hugenholtz, P., B. M. Goebel, and N. R. Pace.** 1998. Impact of culture-independent studies on the emerging phylogenetic view of bacterial diversity. J. Bacteriol. **180**:4765-4774.

35.   **International Committee on Systematic Bacteriology Subcommittee on the Taxonomy of *Mollicutes*.** 1995. Revised minimum standards for description of new species of the class *Mollicutes* (division *Tenericutes*). Int. J. Syst. Bacteriol. **45**:605-612.

36.   **Johansson, K.-E., M. U. K. Heldtander, and B. Pettersson.** 1998. Characterization of mycoplasmas by PCR and sequence analysis with universal 16S rDNA primers, p. 145-165. *In* R. J. Miles and R. A. J. Nicholas (ed.), Methods in molecular biology. Vol. 104. Mycoplasma protocols. Humana Press Inc., Totowa, N.J.

37.   **Johansson, K.-E., J. G. Tully, G. Bölske, and B. Pettersson.** 1999. *Mycoplasma cavipharyngis* and *Mycoplasma fastidiosum*, the closest relatives to *Eperythrozoon* spp. and *Haemobartonella* spp. FEMS Microbiol. Lett. **174**:321-326.

38.   **Keswani, J., and W. B. Whitman.** 2001. Relationship of 16S rRNA sequence similarity to DNA hybridization in prokaryotes. Int. J. Syst. Evol. Microbiol. **51**: 667-678.

39.   **Kirchhoff, H., P. Beyene, J. Flossdorf, J Heitmann, B. Khattab, D. Llopatta, R. Rosengarten, G. Seidel, and C. Yousef.** 1987. *Mycoplasma mobile* sp. nov., a new species from fish. Int. J. Syst. Bacteriol. **37**:192-197.

40.   **Kimura, M.** 1983. The neutral theory of molecular evolution. Cambridge University Press, N.Y. New York.

41.   **Lan, R., and P. R. Reeves.** 2000. Intraspecies variation in bacterial genomes: the need for a species genome concept. Trends Microbiol. **8**:396-401.

42.   **Lee, I.-M., R. E. Davis, and D. E. Gundersen-Rindal.** 2000. Phytoplasma: phytopathogenic mollicutes. Annu. Rev. Microbiol. **54**:221-255.

43.   **Ludwig, W., and Schleifer. K. H.** 1999. Phylogeny of bacteria beyond the 16S rRNA standard. ASM News **65**:752-757.

44.   **Ludwig, W., O. Strunk, S. Klugbauer, N. Klugbauer, M. Weizenegger, J. Neumaier, M. Bachleitner, and K. H. Schleifer.** 1998. Bacterial phylogeny based on comparative sequence analysis. Electrophoresis **19**:554-568.

45.   **Maiden, M. C., J. A. Bygraves, E. Feil, G. Morelli, J. E. Russell, R. Urwin, Q. Zhang, J. Zhou, K. Zurth, D. A. Caugant, I. M. Feavers, M. Achtman, and B. G. Spratt.** (1998). Multilocus sequence typing: a portable approach to the identification of clones within populations of pathogenic microorganisms. Proc. Natl. Acad. Sci. USA 95:3140-3145.

46.   **Messick, J. B., P. G. Walker, W. Raphael, L. Berent, and X. Shi.** 2002. "*Candidatus* Mycoplasma haemodidelphis" sp. nov., "*Candidatus* Mycoplasma haemolama" sp. nov. and "*Candidatus* Mycoplasma haemocanis" comb. nov., haemotropic parasites from a naturally infected opossum (*Didelphis virginiana*), alpaca (*Lama pocos*) and dog (*Canis familiaris*): phylogenetic and secondary structural relatedness of their 16S rRNA genes to other mycoplasmas. Int. J. Syst. Evol. Microbiol. In press.

47.   **Montano, H. G., R. E. Davis, E. L. Dally, S. Hogenhout, J. P. Pimentel, and P. S. Brioso.** 2001. '*Candidatus* Phytoplasma brasiliense', a new phytoplasma taxon associated with hibiscus witches' broom disease. Int. J. Syst. Evol. Microbiol. **51**:1109-1118.

48.   **Murray, R. G., and Schleifer, K.-H.** 1994. Taxonomic note: a proposal for recording the properties pf putative taxa of procaryotes. Int. J. Syst. Bacteriol. **44**:174-176.

49.  **Murray, R. G., and E Stackebrandt.** 1995. Taxonomic note: implementation of the provisional status *Candidatus* for incompletely described procaryotes. Int. J. Syst. Bacteriol. **45:**186-187.

50.  **Neimark, H., K.-E. Johansson, Y. Rikihisa, and J. G. Tully.** 2001. Proposal to transfer some members of the genera *Haemobartonella* and *Eperythrozoon* to the genus *Mycoplasma* with description of "*Candidatus* Mycoplasma haemofelis", "*Candidatus* Mycoplasma haemomuris", "*Candidatus* Mycoplasma haemosuis" and "*Candidatus* Mycoplasma wenyonii". Int. J. Syst. Evol. Microbiol. **51:**891-899.

51.  **Neimark, H., D. Mitchelmore, and R. H. Leach.** 1998. An approach to characterizing uncultivated prokaryotes: the grey lung agent and proposal of a "*Candidatus* Mycoplasma ravipulmonis". Int. J. Syst. Bacteriol. **48:**389-394.

52.  **Nelson, K. E., I. T. Paulsen, and C. M. Fraser.** 2001. Microbial genome sequencing: a window into evolution and physiology. ASM News **67:**310-317.

53.  **Olsen, G. J., and C. R. Woese.** 1993. Ribosomal RNA: a key to phylogeny. FASEB J. **7:**113-123.

54.  **Persson, A., B. Pettersson, G. Bölske, and K.-E. Johansson.** 1999. Diagnosis of contagious bovine pleuropneumonia by PCR-LIF and PCR-REA based on the 16S rRNA genes of *Mycoplasma mycoides* subsp. *mycoides* SC. J. Clin. Microbiol. **37:**3815-3821.

55.  **Pettersson, B.** 1997. Ph.D. thesis. Direct solid-phase 16S rDNA sequencing: a tool in bacterial phylogeny. Royal Institute of Technology, Stockholm, Sweden.

56.  **Pettersson, B., G. Bölske, F. Thiaucourt, M. Uhlén, and K.-E. Johansson.** 1998. Molecular evolution of *Mycoplasma capricolum* subsp. *capripneumoniae* strains, based on polymorphisms in the 16S rRNA genes. J. Bacteriol. **180:**2350-2358.

57.  **Pettersson, B., T. Leitner, M. Ronaghi, G. Bölske, M. Uhlén, and K.-E. Johansson.** 1996. Phylogeny of the *Mycoplasma mycoides* cluster as determined by sequence analysis of the 16S rRNA genes from two rRNA operons. J. Bacteriol. **178:**4131-4142.

58.  **Pettersson, B., K.-E. Johansson, and M. Uhlén.** 1994. Sequence analysis of 16S rRNA from mycoplasmas by direct solid phase DNA sequencing. Appl. Environ. Microbiol. **60:**2456-2461.

59.  **Pettersson, B., J. G. Tully, G. Bölske, and K.-E. Johansson.** 2000. Updated phylogenetic description of the *Mycoplasma hominis* cluster (Weisburg *et al.*, 1989) based on 16S rDNA sequences. Int. J. Syst. Evol. Microbiol. **50:**291-301.

60.  **Pettersson, B., J. G. Tully, G. Bölske, and K.-E. Johansson.** 2001. Re-evaluation of the classical *Mycoplasma lipophilum* cluster (Weisburg *et al.*, 1989) and description of two new clusters in the hominis group based on 16S rDNA sequences. Int. J. Syst. Evol. Microbiol. **51:**633-643.

61.  **Pettersson, B., M. Uhlén, and K.-E. Johansson.** 1996. Phylogeny of some mycoplasmas from ruminants based on 16S rRNA sequences and definition of a new cluster within the hominis group. Int. J. Syst. Bacteriol. **46:**1093-198.

62.  **Priest, F., and B. Austin.** 1993. Modern Bacterial Taxonomy. Chapman and Hall, London, UK.

63.  **Razin, S.** 2000. The genus *Mycoplasma*, and related genera (class *Mollicutes*). *In* M. Dworkin, S. Falkow, E. Rosenberg, K.-H. Schleifer, and E. Stackebrandt (eds.), The prokaryotes. An evolving electronic resource for the microbiological community, 3rd edition, Springer-Verlag, New York. (URL: http//:www.prokaryotes.com).

64.  **Razin, S., D. Yogev, and Y. Naot.** 1998. Molecular biology and pathogenicity of Mycoplasmas. Microbiol. Mol. Biol. Rev. **62:**1094-1156.

65.  **Robertson, J. A., G. W. Stemke, J. W. Davis Jr, R. Harasawa, D. Thirkell, F. Kong, M. C. Shepard and D. K. Ford.** 2001. Proposal of *Ureaplasma parvum* sp. nov. and amended description of *Ureaplasma urealyticum* (Shepard *et al.* 1974) Robertson *et al.* 2002. Int. J. Syst. Evol. Microbiol. **52:**587-597.

66.  **Ros Bascuñana, C., J. G. Mattsson, G. Bölske, and K.-E. Johansson.** 1994. Characterization of the 16S rRNA gene from *Mycoplasma* sp. strain F38 and development of an identification system based on the polymerase chain reaction. J. Bacteriol. **176:**2577-2586.

67.  **Roselló-Mora, R., and R. Amann.** 2001. The species concept for prokaryotes. FEMS Microbiol. Rev. **25:**39-67.

68.  **Saglio, P., M. L'Hospital, D. Laflèche, G. Dupont, J. M. Bové, J. G. Tully, and E. A. Freundt.** 1973. *Spiroplasma citri* gen. and sp. n.: a new mycoplasma-like organism associated with "stubborn" disease of citrus. Int. J. Syst. Bacteriol. **23:**191-204.

69.  **Sanger, F., S. Niklen, and R. Coulson.** 1977. DNA sequencing with chain terminating inhibitors. Proc. Natl. Acad. Sci. USA **74:**5463-5467.

70.  **Seemüller, E., C. Marcone, U. Lauer, A. Ragazzoni, and M. Göschl.** 1998. Current status of molecular classification of the phytoplasmas. J. Plant Pathol. **80:**3-26.

71.  **Shepard, M. C., C. D. Lunceford, D. K. Ford, R. H. Purcell,, D. Taylor-Robinson, S. Razin, and F. T. Black.** 1974. *Ureaplasma urealyticum* gen. nov., sp. nov: proposed nomenclature for the human T (T-strain) mycoplasma. Int. J. Syst. Bacteriol. **24:**160-171.

72.  **Sneath, P. H. A. (ed.).** 1992. International Code of Nomenclature of Bacteria. American Society for Microbiology, Washington, D.C.

73.  **Stackebrandt, E., and B. M. Goebel.** 1994. Taxonomic note: a place for DNA-DNA reassociation and 16S rRNA sequence analysis in the present species definition in bacteriology. Int. J. Syst. Bacteriol. **44:**846-849.

74.  **Stackebrandt, E., and P. Schümann.** 2000. Defining taxonomic ranks. *In* M. Dwozkin, S. Falkow, E. Rosenberg, K. H. Schleifer, and E. Stackebrandt (eds.), The prokaryotes. An evolving electronic resource for the microbiological community, 3rd edition, Springer-Verlag, New York. (URL: http//:www. prokaryotes.com).

75.  **Tully, J. G., J. M. Bové, F. Laigret, and R. F. Whitcomb.** 1993. Revised taxonomy of the class *Mollicutes*: proposed elevation of a monophyletic cluster of arthropod-associated mollicutes to ordinal rank (*Entomoplasmatales* ord. nov.), with provision for familial rank to separate species with nonhelical morphology (*Entomoplasmataceae* fam. nov.) from helical species (*Spiroplasmataceae*), and emended descriptions of the order *Mycoplasmatales*, family *Mycoplasmataceae*. Int. J. Syst. Bacteriol. **43:**378-385.

76.  **Tully, J. G., and S. Razin. (ed.).** 1996. Appendix in Molecular and Diagnostic Procedures in Mycoplasmology. Vol. II. Diagnostic Procedures. Academic Press, San Diego, Calif. USA.

77.  **Tully, J. G., and R. F. Whitcomb.** 2000. The genus *Spiroplasma*. *In* M. Dwozkin, S. Falkow, E. Rosenberg, K. H. Schleifer, and E. Stackebrandt (eds.), The prokaryotes. An evolving electronic resource for the microbiological community, 3rd edition, Springer-Verlag, New York. (URL: http//: www.prokaryotes.com).

78.  **Tully, J. G., R. F. Whitcomb, K. J. Hackett, D. L. Rose, R. B. Henegar, J. M. Bové, P. Carle, D. L. Williamson, and T. B. Clark.** 1994. Taxonomic descriptions of eight new non-sterol-requiring mollicutes assigned to the genus *Mesoplasma*. Int. J. Syst. Bacteriol. **44:**685-693.

79.  **Tully, J. G., R. F. Whitcomb, K. J. Hackett, D. L. Williamson, F. Laigret, P. Carle, J. M. Bové, R. B. Henegar, N. M. Ellis, D. E. Dodge, and J. Adams.** 1998. *Entomoplasma freundtii* sp. nov., a new species from a green tiger beetle (Coleoptera: Cicindelidae). Int. J. Syst. Bacteriol. **48:**1197-1204.

80.  **Vandamme, P., B. Pot, M. Gillis, P. De Vos, K. Kersters, and J. Swings.** 1996. Polyphasic taxonomy, a consensus approach to bacterial systematics. Microbiol. Rev. **60:**407-438.

81.  **Wang, Y., and Z. Zhang.** 2000. Comparative sequence analyses reveal frequent occurrence of short segments containing an abnormally high number of non-random base variations in bacterial rRNA genes. Microbiol. **146:**2845-2854

82.  **Weisburg, W. G., J. G. Tully, D. L. Rose, J. P. Petzel, H. Oyaizu, D. Young, L. Mandelco, J. Sechrest, T. G. Lawrence, J. Van Etten, J. Maniloff, and C. R. Woese.** 1989. A phylogenetic analysis of mycoplasmas: basis for their classifycation. J. Bacteriol. **171:**6455-6467.

83.  **Whitcomb, R. F., D. L. Williamson, G. E. Gasparich, J. G. Tully, and F. E. French.** 1999. Spiroplasma taxonomy. Proceedings of the First Internet Conference on Phytopathogenic Mollicutes. (URL: http://www.uniud.it/phyto plasma/conf.html).

84.  **Williamson, D. L., R. F. Whitcomb, J. G. Tully, G. E. Gasparich, D. L. Rose, P. Carle, J. M. Bové, K. J. Hackett, J. R. Adams, R. B. Henegar, M. Konai, C. Chastel, F. E. French.** 1998. Revised group classification of the genus *Spiroplasma*. Int. J. Bacteriol. **48:**1-12.

85.  **Woese, C. R.** 1987. Bacterial evolution. Microbiol. Rev. **51:**221-271.

86.  **Woese, C. R.** 2000. Interpreting the universal phylogenetic tree. Proc. Natl. Acad. Sci. USA **97:**8392-8396.

87.  **Woese, C. R.** 2000. Prokaryote systematics: The evolution of a science. *In* M. Dwozkin, S. Falkow, E. Rosenberg, K.-H. Schleifer, and E. Stackebrandt (eds.), The prokaryotes. An evolving electronic resource for the microbiological community, 3$^{rd}$ edition, Springer-Verlag, New York. (URL: http//:www. prokaryotes.com)

88.  **Woese, C. R., O. Kandler, and M. L. Wheelis.** 1990. Towards a natural system of organisms: proposal for the domains Archaea, Bacteria, and Eucarya. Proc. Natl. Acad. Sci. **87:**4576-4579.

89.  **Young, J. M.** 2001. Implications of alternative classifications and horizontal gene transfer for bacterial taxonomy. Int. J. Syst. Evol. Microbiol. **51:**945-953.

90.  **Zuckerkandl, E., and L. Pauling.** 1965. Molecules as documents of evolutionary history. J. Theor. Biol. **8:**357-366.

# Chapter 2

# Phylogeny and Evolution

JACK MANILOFF

*Department of Microbiology and Immunology, School of Medicine and Dentistry, University of Rochester, Rochester, NY 14642, USA*

## 1. INTRODUCTION

Beginning in the late 1970s, results from a variety of studies enabled an increasingly detailed *Mollicutes* phylogenetic tree to be reconstructed [4]. The overall tree is based on 16S rRNA sequence analyses and provides a framework for results on other relationships between *Mollicutes* strains and between *Mollicutes* and Gram-positive bacteria; e.g., other macromolecule sequences, lipid composition, metabolic enzymes and pathways, gene organization, and antibiotic sensitivity.

The *Mollicutes* are a single branch of the phylogenetic tree of Gram-positive bacteria with low genomic G+C contents [4]. The particular Gram-positive bacteria branch from which *Mollicutes* arose is that of the lactobacilli, bacilli, and streptococci. Details of the origin and divergence of *Mollicutes* species, the nature of the Earth's evolving flora and fauna in providing ecological niches for new *Mollicutes* lineages, and the role of rapid evolution in *Mollicutes* adaptation have been elaborated by further phylogenetic tree analyses.

Parenthetically, 16S rRNA sequence comparisons showed that the genus *Thermoplasma* (wall-less, thermophilic, acidophilic microorganisms), which had been grouped with the *Mollicutes*, are members of the *Archaea*.

*Molecular Biology and Pathogenicity of Mycoplasmas,* Edited by Razin and Herrmann, Kluwer Academic/Plenum Publishers, New York, 2002

## 2.    ANALYSIS OF 16S rRNA PHYLOGENETIC TREES

The distance from a phylogenetic tree node to a group of extant organisms has been calculated as the average of all branch lengths from the node to the organisms (J. Maniloff, submitted for publication). Using RDP (Ribosomal Database Project, release 7.0) 16S rRNA phylogenetic trees [3], averaging and normalization of branch lengths produced a tree with nodes as a function of relative time.

This analysis required each branch to represent an independent phylogenetic unit (i.e., a species). However, some 16S rRNA branches consist of clusters of minimally separated branches arising from either multiple strains of a single species or a single strain of one species in the midst of a number of strains of another species. For these types of branches, consisting of organisms with 16S rRNA similarity differences less than 1.5%, individual branch lengths were averaged to produce a single phylogenetic branch. Therefore, branches, each representing an independent phylogenetic unit, are referred to as lineages in this analysis.

The phylogenetic tree abscissa was calibrated by correlating the nodes corresponding to the origins of facultative and obligate aerobic bacteria with geological times for the appearance and increase in atmospheric oxygen to levels necessary for facultative and obligate aerobic growth, to produce a tree with nodes as a function of geological time (J. Maniloff, submitted for publication).

From these data, the number of lineages in a branch was plotted as a function of time. Such lineages-vs-time plots can be used to analyze molecular phylogeny [6,7]. For bacterial phylogenetic branches, the number of lineages increases exponentially with time, allowing lineage doubling times (i.e., rates of evolution) to be calculated from slopes of lineages-vs-time curves.

## 3.    ORIGIN AND EVOLUTION OF THE MOLLICUTES

Analysis of 16S rRNA phylogenetic trees as described above has enabled *Mollicutes* evolution to be examined in terms of geological and paleontological changes during the Earth's history (Table 1) (J. Maniloff, submitted for publication). These data show that *Mollicutes* diverged from the Streptococcus branch of Gram-positive bacteria with low G+C contents in the Late Proterozoic, at about 605 Myr (million years, with time measured back from the present). This *Mollicutes* branch, designated the AAP branch, consists of the extant families *Acholeplasmataceae* (genus *Acholeplasma*) and *Anaeroplasmataceae* (genera *Anaeroplasma* and *Asteroleplasma*) and the Phytoplasma (Fig. 1).

*Table 1.* Selected events in evolution of bacteria and *Mollicutes* during the Earth's geological history[a]

| Geological intervals | | | Start of interval | Selected events |
|---|---|---|---|---|
| Eon | Era | Period | | |
| Precambrian | Hadean | | 4600 Myr | Formation of Earth. Chemical evolution. Origin and evolution of cells. |
| | Archaean | | 3800 Myr | Evolution of oxygenic phototrophs. Increasing marine $O_2$. Evolution of facultative aerobes. |
| | Proterozoic | | 2500 Myr | Oceans become aerobic. Atmospheric $O_2$ reaches 1% present level. Evolution of aerobes. Evolution of *Mollicutes* AAP branch: loss of cell wall synthesis genes and some biosynthesis and rRNA genes. |
| Phanerozoic | Paleozoic | Cambrian | 544 Myr | |
| | | Ordovician | 505 Myr | Evolution of SEM branch: origin of sterol requirement, conversion of UGA a trp to a stop codon. |
| | | Silurian | 440 Myr | |
| | | Devonian | 410 Myr | Split of SEM branch into *Spiroplasmataceae-Entomoplasmataceae* and *Mycoplasmataceae* branches. |
| | | Carboniferous | 360 Myr | |
| | | Permian | 286 Myr | |
| | Mesozoic | Triassic | 245 Myr | |
| | | Jurassic | 208 Myr | Transition in *Mycoplasmataceae* phylogeny to rapid rate of evolution. Divergence of Phytoplasma from *Acholeplasma* branch. |
| | | Cretaceous | 146 Myr | Transition in *Spiroplasmataceae-Entomoplasmataceae* phylogeny to rapid rate of evolution. |
| | Cenozoic | Tertiary | 65 Myr | |
| | | Quaternary | 2 Myr | |

[a] References in Schopf (9, 10) and text.

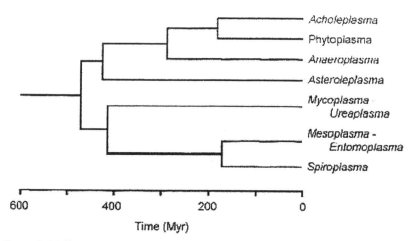

*Figure 1. Mollicutes* phylogenetic tree, showing divergence of the *Mollicutes* into two major branches, the AAP branch (with *Acholeplasma, Anaeroplasma, Asteroleplasma,* and Phytoplasma branches) and the SEM branch (with *Spiroplasma, Entomoplasma, Mesoplasma, Mycoplasma,* and *Ureaplasma* branches).

The ancestral *Mollicutes* was probably similar to acholeplasma (i.e., wall-less facultative aerobes metabolically similar to streptococci), with obligate anaerobic anaeroplasmas evolving later from the acholeplasma branch. Since *Mollicutes* arose from facultative aerobes on an aerobic Earth, this scenario is more parsimonious than one in which anaerobic *Mollicutes* arose from facultative aerobic streptococci and subsequently diverged at least twice to produce branches of facultative aerobic *Mollicutes*.

From the genome sizes of extant microorganisms, the *Mollicutes* arose from a Streptococcus branch containing species with genome sizes of 1700-2600 kb. Genome reductions must have accompanied establishment of the AAP branch and presumably involved loss of cell wall synthesis genes and possibly some biosynthesis and rRNA genes, to produce AAP branch species with 1500-1700 kb genomes.

A branch, designated the SEM branch, diverged from the AAP in the Middle Ordovician, at about 470 Myr, and consists of the extant families *Spiroplasmataceae* (genus *Spiroplasma*), *Entomoplasmataceae* (genera *Mesoplasma* and *Entomoplasma*), and *Mycoplasmataceae* (genera *Mycoplasma* and *Ureaplasma*) (Fig. 1). The SEM branch split at the Silurian-Devonian boundary, at about 410 Myr, to a *Spiroplasmatacae-Entomoplasmataceae* branch and a *Mycoplasmataceae* branch. The *Spiroplasmatacae* and *Entomoplasmataceae* diverged in the Middle Jurassic, at about 170 Myr.

From the phenotypes of extant *Mollicutes*, early events in evolution of the SEM branch must have included the origin of a sterol requirement and conversion of UGA from a stop codon to a tryptophan codon. There were also further genome reductions in some *Mycoplasmataceae* lineages.

The Phytoplasma branch arose on the AAP branch in the Middle Jurassic, at about 180 Myr, and is phylogenetically distant from SEM branch plant and insect *Mollicutes*. Some Phytoplasma lineages have also undergone further genome reductions.

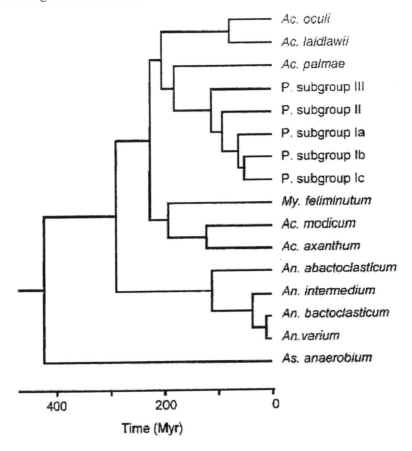

*Figure 2.* AAP phylogenetic tree. A lineage designated *Ac.* sp. strain J233 in the RDP is now designated *Ac. palmae* (K.-E. Johansson, personal communication), as shown here. The RDP phylogenetic tree shows a Mycoplasma-like organisms (i.e., Phytoplasma) branch with three branches (designated subgroups I, II, and III), and subgroup I with three branches (designated subgroups Ia, Ib, and Ic here).

## 4.       THE AAP PHYLOGENETIC BRANCH

The AAP branch diverged from the Streptococcus phylogenetic tree at about 605 Myr. This was the time (late Proterozoic) of the first major expansion of marine animals and during the early phylogenetic events of the Cambrian explosion of animal life[12]. The nature of the selective pressure for *Mollicutes* evolution at this time is not known.

The phylogenetic tree of AAP lineages is shown in Figure 2. The AAP lineage doubling time, calculated from the slope of a lineages-vs-time plot for these data, was about 100 Myr, compared to the Streptococcus branch lineage doubling time of about 65 Myr (J. Maniloff, submitted for publication).

The rate of evolution of an organism is the product of its mutation rate and fixation probability (i.e., the probability that an organism that has had a mutation will be viable and produce progeny). The slow rate of AAP evolution compared to the Streptococcus branch, as measured by a larger AAP lineage doubling time, may reflect a decrease in fixation probability of the AAP branch relative to the Streptococcus branch due to the smaller genome size and genetic complexity of AAP lineages.

## 5.       THE SEM PHYLOGENETIC BRANCH

The SEM branch diverged from the AAP at about 470 Myr, the time (Middle Ordovician) of the first land plants[1]. The most parsimonious model for SEM branch phylogeny is that a non-sterol-requiring *Mesoplasma* was the earliest SEM branch, with later divergences to sterol-requiring *Spiroplasma*, *Mycoplasmataceae*, and *Entomoplasma* branches.

The *Mycoplasmataceae* and *Spiroplasmatacae-Entomoplasmataceae* branches diverged at about 410 Myr, a time (the Silurian-Devonian boundary) of major expansion of marine life and the first land animals[2]. Phylogenetic trees of the *Spiroplasmataceae-Entomoplasmataceae* and *Mycoplasmataceae* lineages are shown in Figures 3 and 4-7, respectively. To facilitate presentation of *Mycoplasmataceae* lineages, Figure 4 shows the overall structure of the *Mycoplasmataceae* phylogenetic tree, with its four densely populated branches designated the $\alpha$-, $\beta$-, $\gamma$-, and $\delta$-mycoplasma branches, shown in detail in Figures 5-7. All *Mycoplasma* species are on the *Mycoplasmataceae* branch except the My. mycoides subgroup, which forms a branch within the *Spiroplasmataceae-Entomoplasmataceae* tree (Fig. 3).

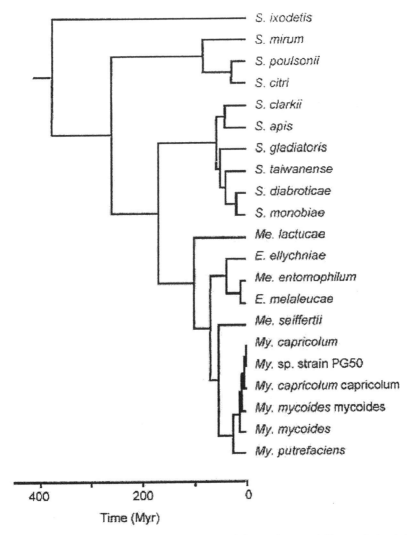

*Figure 3. Spiroplasmataceae-Entomoplasmataceae* phylogenetic tree. A lineage designated *S.* sp. strain DW1 in the RDP is now designated *S. poulsonii* (K.-E. Johansson, personal communication), as shown here.

*Jack Maniloff*

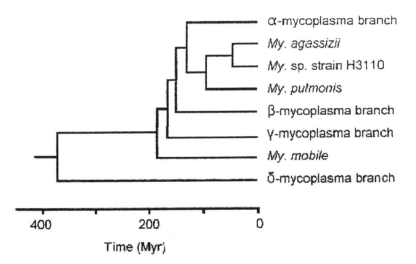

*Figure 4. Mycoplasmataceae* phylogenetic tree. The α-, β -, γ -, and δ-mycoplasma branches are described in the text and shown in Figs. 5-7.

For about the first 280 Myr after the origin of the SEM branch, SEM lineages were slowly evolving, with a lineage doubling time similar to the AAP branch from which it arose. However, a lineages-vs-time plot shows that, more recently, the rate of evolution of the two major SEM sub-branches each underwent a transition to rapid evolution: (1) at about 191 Myr, the rate of *Mycoplasmataceae* evolution increased several-fold to a lineage doubling time of about 41 Myr, and (2) the *Spiroplasmataceae-Entomoplasmataceae* branch continued its slow rate of evolution for about another 90 Myr and, at about 100 Myr, its rate of evolution also increased several-fold to a lineage doubling time of about 43 Myr (J. Maniloff, submitted for publication).

The times at which the transition to rapid evolution occurred for the *Spiroplasmataceae-Entomoplasmataceae* and *Mycoplasmataceae* branches correlate with major paleontological events in the evolution of the flora and fauna that formed ecosystems for lineages of each branch. The *Mycoplasmataceae* transition to rapid evolution at about 191 Myr was soon after the appearance of the first mammals and other vertebrate groups[13]. The *Spiroplasmataceae-Entomoplasmataceae* transition to rapid evolution at about 100 Myr occurred soon after the appearance of the first flowering plants and their associated insects1. In each case, the increase in potential hosts presumably provided niches for the selection of new lineages, although in the latter case it is not known whether new plant or new insect hosts were more important in providing a selective advantage for nascent *Spiroplasmataceae-Entomoplasmataceae* lineages.

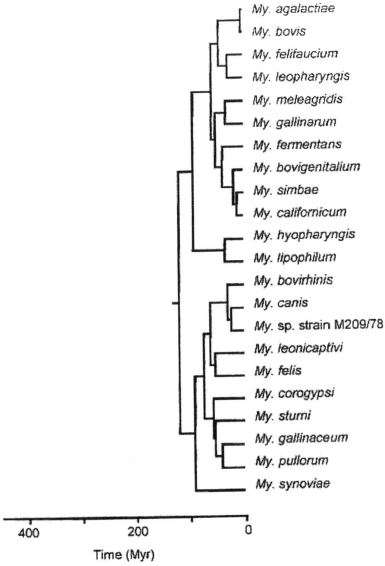

*Figure 5.* Phylogenetic tree of the α-mycoplasma branch. Lineages designated *My. leocaptivus* and *My. sturnidae* in the RDP are now designated *My. leonicaptivi* and *My. sturni*, respectively (K.-E. Johansson, personal communication), as shown here.

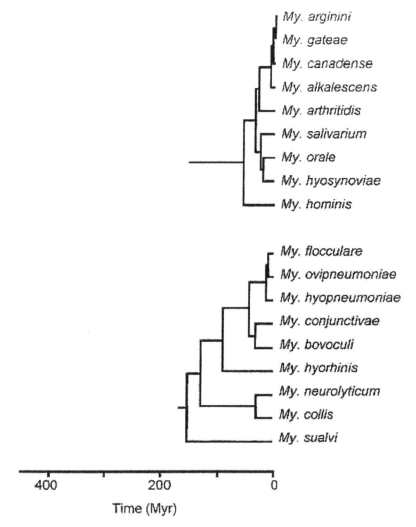

*Figure 6.* Phylogenetic trees of the β-mycoplasma (*Top*) and γ-mycoplasma (*Bottom*) branches.

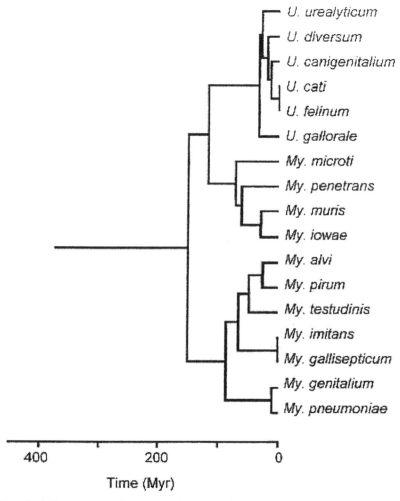

*Figure 7.* Phylogenetic tree of the δ-mycoplasma branch.    Lineages designated *My. volis* and *My. sualvi* in the RDP are now designated *My. microti* and *My. alvi*, respectively (K.-E. Johansson, personal communication), as shown here.

There is an interesting concurrence in the origin of some major nodes of plant and insect *Mollicutes*. Although the Phytoplasma and *Spiroplasmataceae-Entomoplasmataceae* branches are phylogenetically distant, the Phytoplasma arose at about 180 Myr, close to the 170 Myr time at which the *Spiroplasmataceae-Entomoplasmataceae* split into

Spiroplasmataceae and Entomoplasmataceae branches. This may be a coincidence or may reflect some currently unknown change in the Earth's flora and fauna about this time to provide a selective advantage for new plant and insect Mollicutes.

## 6.      EVOLUTION OF *MOLLICUTES* GENOME SIZE

Assuming genome sizes of ancestral microorganisms were comparable to those of extant lineages, the *Mollicutes* arose from Streptococcus lineages with relatively small bacterial genomes. The genome reductions during evolution of the AAP branch need only have involved 10-20 percent of the genome, to produce 1500-1700 kb *Acholeplasma*, *Asteroleplasma*, and *Anaeroplasma* genomes. The other *Mollicutes* (*Spiroplasma*, *Mesoplasma*, *Entomoplasma*, *Mycoplasma*, *Ureaplasma*, and Phytoplasma) contain lineages with a variety of genomes sizes, ranging from genomes slightly smaller (i.e., 1100-1400 kb) than AAP genomes to the smallest known genomes (i.e., 600-800 kb)[8]. These smallest genomes are the result of genome reductions of about 50%. Hence, on several different phylogenetic branches, *Mollicutes* lineages have undergone extensive genome reductions and, in each case, produced genomes in a relatively small size range. This suggests 600-800 kb is the lower limit of genetic complexity for a self-replicating organism on the Earth[5], even though, under laboratory conditions, some of these genes are not essential (see Mycoplasmas and the minimal cell concept chapter). The nature of the mechanism and selective pressure that led to genome reductions on independent phylogenetic branches is not known, nor is the reason genome reductions occurred in some lineages but not in others in the same genus and, in some cases, in some strains of a species but not in other strains of that species.

It is interesting that the obligate intracellular bacteria that have been studied have genomes of 1000 kb or larger[8], except for a *Buchnera* species that was recently reported to have a genome size of 641 kb[11]. So, genome size is not a simple function of life style.

## ACKNOWLEDGMENTS

I thank Karl-Erik Johansson for his careful reading of this manuscript and many useful suggestions.

# REFERENCES

1.  **Friedman, W.E., and S.K. Floyd.** 2001. Perspective: the origin of flowering plants and their reproductive biology–a tale of two phylogenies. *Evolution* **55**:217-231.
2.  **Gray, J., and W. Shear.** 1992. Early life on land. *Am. Sci.* **80**:444-456.
3.  **Maidak, B.L., J.R. Cole, T.G. Lilburn, C.T. Parker Jr., P.R. Saxman, R.J. Farris, G.M. Garrity, G.J. Olsen, T.M. Schmidt, and J.M. Tiedje.** 2001. The RDP-II (Ribosomal Database Project). *Nucl. Acids. Res.* **29**:173-174.
4.  **Maniloff, J.** 1992. Phylogeny of mycoplasmas. In *Mycoplasmas: molecular biology and pathogenesis* (J. Maniloff, J., R.N. McElhaney, L.R. Finch, and J.B. Baseman, eds), pp. 549-559. American Society for Microbiology: Washington, DC.
5.  **Maniloff, J.** 1996. The minimal cell genome: "On being the right size." *Proc. Natl. Acad. Sci. USA* **93**:10004-10006.
6.  **Nee, S., E.C. Holmes, R.M. May, and P.H. Harvey.** 1994. Extinction rates can be estimated from molecular phylogenies. *Phil. Trans. R. Soc. Lond. B* **344**:77-82.
7.  **Nee, S., E.C. Holmes, A. Rambaut, and P.H. Harvey.** 1995. Inferring population history from molecular phylogenies. *Phil. Trans. R. Soc. Lond. B* **349**:25-31.
8.  **Razin, S., D. Yogev, and Y. Naot.** 1998. Molecular biology and pathogenicity of mycoplasmas. *Microbiol. Mol. Biol. Rev.* **62**:1094-1156.
9.  **Schopf, J.W.** 1992. Times of origin and earliest evidence of major biologic groups. In *The Proterozoic Biosphere* (J.W. Schopf and C. Klein, eds), pp. 587-600. Cambridge University Press: New York.
10. **Schopf, J.W.** 1999. Deep divisions in the tree of life–what does the fossil record reveal? *Biol. Bull.* **196**:351-355.
11. **Shigenobu, S., H. Watanabe, M. Hattori, Y. Sakaki, and H. Ishikawa.** 2000. Genome sequence of the endocellular bacterial symbiont of aphids *Buchnera* sp. APS. *Nature* **407**:81-86.
12. **Stanley, S.M.** 1992. Exploring earth and life through time. W.H. Freeman: New York.
13. **Wyss, A.** 2001. Digging up fresh clues about the origin of mammals. *Science* **292**:1496-1497.

# Chapter 3

# Mycoplasmas of Humans

ALAIN BLANCHARD[*] and CÉCILE M. BÉBÉAR[#]
[*]*INRA, IBVM, 71 avenue Edouard Bourleaux, BP 81, 33883 Villenave D'Ornon, France;*
[#]*Université Victor Segalen Bordeaux 2, Laboratoire de Bactériologie, 146 rue Léo Saignat, 33076 Bordeaux cedex, France.*

## 1. INTRODUCTION

Although the mollicute responsible for contagious bovine peripneumonia, was isolated in 1898 (for an historical perspective see (13)), the first mycoplasma species (*M. hominis*) from humans was only recovered in 1937. In the 1940's, Monroe Eaton identified the agent (later named *M. pneumoniae*) of primary atypical pneumonia. In 1981, *M. genitalium*, now recognized as a common cause of nongonoccocal urethritis (NGU), was isolated from urethral swabs collected from men with this clinical condition (144) and in the last decade, *M. penetrans* was isolated in urine samples from HIV-infected patients (75). This progressive but slow discovery of the human mycoplasma flora is due to the fastidious growth of these bacteria which require special media and which, in many cases, are particularly difficult to culture from clinical samples. Consequently, although their detection greatly benefited from the development of PCR-based methods (see "Diagnosis of mycoplasmal infections", S. Razin), some of the aspects of mycoplasma pathogenesis in humans are still a matter of debate.

This chapter briefly reviews the clinical significance of the mycoplasmas of humans and the new developments concerning the biological features of these bacteria relevant to pathogenicity. Antibiotic treatment of these mycoplasma infections are reviewed in another chapter ("Antimycoplasmal agents", C. M. Bébéar and C. Bébéar).

*Molecular Biology and Pathogenicity of Mycoplasmas*, Edited by Razin and Herrmann, Kluwer Academic/Plenum Publishers, New York, 2002

## 2.      CLINICAL SIGNIFICANCE

### 2.1      Many species are commensals

Sixteen species of mycoplasmas have been described in humans (Table 1). They colonize mucosal surfaces of the respiratory and urogenital tracts. Most species reside extracellularly, but some like *M. pneumoniae*, *M. genitalium*, *M. penetrans* may localize and survive within the cells (22).

Many mycoplasmal species exist as commensals of the oropharynx, *M. salivarium* and *M. orale* being the most commonly found species. *M. pneumoniae* colonizes the lower respiratory tract and is clearly pathogenic. This mycoplasma is usually not detected in a carrier state, except during an outbreak period (see "*Mycoplasma pneumoniae* disease manifestations and epidemiology", E. Jacobs).

Of the seven mycoplasmas that have been detected in the genitourinary tract, only *M. hominis*, *U. urealyticum*, and *M. genitalium* are clearly associated with disease. However, *M. hominis* and *U. urealyticum* are frequently isolated from the lower urogenital tract of healthy adults, men and women. Colonization varies in relation with several parameters including age, race, hormonal status, and the lifetime number of sexual partners, and is greater among women, especially during pregnancy (137). *U. urealyticum* could be found in the vagina of 50% of healthy women and *M. hominis* in less than 10% of women. During pregnancy, *Ureaplasma* colonization could reach 50% to 70% (1, 103). *U. urealyticum* is an heterogeneous species which has been proposed for division into two separate species, *U. urealyticum* and *U. parvum*, corresponding to the two former biovars (64, 115). They will be considered together as *Ureaplasma* spp.. The presence of *M. genitalium* in the genital tract of healthy people is not documented. *M. fermentans* and *M. penetrans*, are rarely detected because of their extremely fastidious nature. Molecular diagnostic methods should enhance our knowledge of their possible pathogenic roles. The other species shown in Table 1 have been described only rarely.

It should be noticed that mycoplasmas are frequent cell culture contaminants (27). Two of the most frequent species involved in contamination are found in humans (*M. orale*, *M. fermentans*).

*Table 1.* Mycoplasmas isolated in humans (adapted from (138, 149)).

| Species | Primary site of colonization | | Metabolism of | | Pathogenic role[a] |
|---|---|---|---|---|---|
| | Oro-pharynx | Genitourinary tract | Glucose | Arginine | |
| *M. pneumoniae*[b] | + | - | + | - | + |
| *M. salivarium* | + | - | - | + | - |
| *M. orale* | + | - | - | + | - |
| *M. buccale* | + | - | - | + | - |
| *M. faucium* | + | - | - | + | - |
| *M. lipophilum* | + | - | - | + | - |
| *M. hominis* | - | + | - | + | + |
| *M. genitalium* | - | + | + | - | + |
| *M. fermentans* | + | + | + | + | ?+ |
| *M. penetrans* | - | + | + | + | ? |
| *M. primatum* | - | + | - | + | - |
| *M. spermatophilum* | - | + | - | + | - |
| *M. pirum* | ? | ? | + | + | - |
| *Ureaplasma* spp.[c] | - | + | - | - | + |
| *A. laidlawii* | + | - | + | - | - |
| *A. oculi* | ? | ? | + | - | - |

[a]In immunocompetent patients.
[b]Underlined species fulfil Koch's postulates.
[c]Metabolizes urea.
+, proven role; ?+, some evidence in favor; ?, questionable; -, non pathogenic.

## 2.2 Respiratory diseases

*M. pneumoniae* is an agent of atypical pneumonia and other respiratory diseases. Clinical manifestations and epidemiology of *M. pneumoniae* diseases are described in another chapter (see "*Mycoplasma pneumoniae* disease manifestations and epidemiology*", E. Jacobs).

Commensal mycoplasmas from the oropharynx do not spread usually to the lower respiratory tract. However *M. fermentans* has been detected in this site in a small number of immunocompetent adults developing fatal respiratory distress syndromes (77). It has also been recovered from the throats of children with pneumonia from whom no other respiratory pathogen was isolated, and from bronchoalveolar lavages from AIDS patients with pneumonia (2, 135). Thus, *M. fermentans* may have a possible pathogenic role both in immunocompetent and immunosuppressed patients (see below).

Exceptionally, *M. genitalium* was recovered from respiratory tract specimens, associated with *M. pneumoniae* (135).

## 2.3    Urogenital diseases

### 2.3.1    In men

*Ureaplasma* spp. and *M. genitalium* may cause nonchlamydial non gonococcal urethritis (NCNGU). *Ureaplasma* spp. have long been implicated as a cause of acute NCNGU on the basis of isolation studies including quantification of the organisms, controlled therapeutic and serological studies, and human intraurethral inoculation (137). However the proportion of cases due to the ureaplasmas, is unclear and controversial (135, 138). The frequent isolation of *Ureaplasma* spp. in the urethra of healthy men ranked them behind *Chlamydia trachomatis* and *M. genitalium* as a cause of acute NGU. Recently, *Ureaplasma* spp. have been implicated in chronic NGU with symptoms or signs re-emerging after treatment (49).

The molecular detection by PCR of *M. genitalium* has significantly enhanced our knowledge on its pathogenic role. *M. genitalium* appears to be strongly associated with acute NGU independent of the presence of *C. trachomatis* and ureaplasmas, as it has been described by several studies in Denmark (8, 56), Sweden (8), United Kingdom (49, 50), France (55), Italy (37), Japan (23), West Africa (97), and USA (143). This mycoplasma might be responsible for 15% to 25% of acute NGU. A recent study showed also its association with chronic NGU after treatment (49), confirming previous findings which detected by PCR *M. genitalium* in the urethra of men with persistent or recurrent disease following acute NGU.

Several studies have implicated *M. hominis*, *Ureaplasma* spp., and *M. genitalium* in prostatitis. However, there is still considerable controversy whether these genital mycoplasmas play a role in prostatitis. Krieger et al. detected *M. genitalium* in 4% of prostatic biopsies from men with chronic idiopathic prostatitis (67). There are also a few reports implicating ureaplasmas or *M. hominis* as causes of epididymitis (137). Nevertheless the role of mycoplasmas in prostatitis or epididymitis seems to be minimal.

### 2.3.2    In women

Mycoplasmas do not cause vaginitis, but proliferate in women with bacterial vaginosis (BV), together with other microorganisms like *Gardnerella vaginalis* and anaerobes. There is no doubt that *M. hominis* is strongly associated with BV, but the mechanism of its contribution in this pathology is still unknown (117). Indeed, *M. hominis* is found in the vagina of two-thirds of women with BV in high numbers, but in less than 10% of healthy women. *Ureaplasma* spp. have been associated with BV to a lesser extent (61, 117). Several studies suggested that *Ureaplasma* spp. could be responsible for some cases of urethral syndrome in women (102), but their role has never been proven.

To date, *M. genitalium* seems to have no role in BV. However, it has been detected in the lower genital tract of 3.5 to 20% of women visiting STD clinics (57, 96), and has been associated with cervicitis in one study (145).

*M. hominis* has been isolated from the endometrium and fallopian tubes of 10% of women with salpingitis, accompanied by a significant antibody response (135). However pelvic inflammatory disease (PID) is a disease of multifactorial etiology, and whether *M. hominis* is a primary agent of PID or acts in association with other bacteria in a context of BV, is still controversial. In contrast there is little evidence that *Ureaplasma* spp. play a similar role in PID. Several studies have reported direct isolation of ureaplasmas from the fallopian tubes of women with PID (137), but their occurrence is rare and always in association with other microorganisms. Furthermore experimental data with primates did not support a causal relationship. *M. genitalium*, however, could be an agent of PID owing to some serological evidence and primate inoculation studies (137). Further studies, including notably PCR detection of this mycoplasma in specimens from the upper genital tract, are needed to resolve this issue.

### 2.3.3    Reproduction disorders and infections during pregnancy

Evidence to incriminate *Ureaplasma* spp. in male and female infertility is sparse. *Ureaplasma* spp. have been reported to decrease sperm motility and alter the spermatozoa morphology (137). Animal models showed that ureaplasmas could adhere or be internalized in spermatozoa. However, the results obtained with antibiotic treatment of infertile couples have been contradictory (137), and whether it is necessary to detect and treat genital mycoplasmas in this case, is still debatable.

Doubts also remain about the role of genital mycoplasmas in recurrent spontaneous abortion, stillbirth, prematurity, or low birth weight (19). The presence of *Ureaplasma* spp. and *M. hominis* in the lower genital tract of women and their association with bacterial proliferation in BV complicate the data interpretation (26, 46). Whether genital mycoplasmas have a specific role or any role among the complex flora of BV is still an enigma.

*M. hominis* and *Ureaplasma* spp. were shown to be responsible for some cases of chorioamnionitis and endometritis. They have also been incriminated in post-partum or post-abortum fevers accompagnied by isolation of both organisms from the blood (89). In some cases it could represent the transient invasion of bloodstream following vaginal delivery, or it could be a real septicemia with isolation of the mycoplasma in both the mother blood culture and the endotracheal specimen of her newborn (89). It should be noticed that *M. hominis* has been implicated more frequently than *Ureaplasma* spp. in post-partum fever.

### 2.3.4   Disorders of the urinary tract

Several animal models and human studies have shown a possible role of *Ureaplasma* spp. in the development of infectious stones (141). This pathology is related to the urease activity of ureaplasmas with crystallization of struvite and calcium phosphates in urines.

*M. hominis* appears to cause a small number of cases of acute pyelonephritis in immunocompetent patients as indicated by the isolation of the microorganism from the upper urinary tract accompanied by a specific antibody response (142). Obstruction or instrumentation of the urinary tract could be a predisposing factor.

## 2.4   Neonatal infections

Colonization of neonates by genital mycoplasmas could occur *in utero*, but more frequently at the time of delivery by contact with mycoplasmas from the lower genital tract of the mother. The frequency of colonized newborns could be up to 50% when the mother is colonized. The highest rates concern the low-birth weight neonates (18). Colonization of infants tends to decrease rapidly after three months.

Several clinical features like bacteremia, septicemia, respiratory tract and central nervous system infections have been described for both *Ureaplasma* spp. and *M. hominis* (18). Respiratory tract infections are manifested most frequently as pneumonia and in some rare cases as respiratory distress syndromes (17). Since 1988, several studies have reported the association of colonization of the respiratory tract by *Ureaplasma* spp. with bronchopulmonary dysplasia in low-birth weight infants (<1,000 g), (17, 151). This association has still to be confirmed since bronchopulmonary dysplasia is a complex multifactorial pathology. However if a causality link is confirmed, it could lead to a distinct change in the therapeutic management of this disease (119).

Both the two common genital mycoplasma species could invade the cerebrospinal fluid of neonates causing some cases of meningo-encephalitis with neurological damage or death (3, 150). However, most of the infected newborns had a subclinical meningitis without sequelae (150).

## 2.5    Arthritis and systemic infections in the immunocompromised patient

Mycoplasmas can infect other organs than the respiratory or urogenital tracts. These infections are probably underestimated since in many cases mycoplasmas are usually considered late in diagnosis, when negative results have been observed for other microorganisms, or in case of treatment failure. Extrapulmonary or extragenital infections occur frequently in an immunodeficiency state as will be described below. The implication of mycoplasmas in arthritis will be discussed firstly since it is has been a matter of debate for many years, and then the other systemic infections will be reviewed.

### 2.5.1    Arthritis

Mycoplasmas are a major cause of arthritis in many animal species (see chapter "*Mycoplasma arthritidis* pathogenicity: membranes, MAM, and MAV1" by L. Washburn and B. C. Cole). Their ability to cause experimentally induced acute and chronic arthritis in animals led several authors to look for a similar potential in humans. To date, three types of causal links between mycoplasmas and human arthritis can be proposed (121, 139) as described below.

#### 2.5.1.1    Mycoplasma septic arthritis in hypogammaglobulinemia, a documented link

Mycoplasmas have been identified, either by cultural or non cultural methods, in about 40% of cases of arthritis in patients with hypogammaglobulinemia (34, 36). These arthritis cases are septic with generally a strong polymorphonuclear response in the joint and in some cases, isolation of a large number of mycoplasmas. They are generally cured successfully by antibiotics, especially when given early but only with a concomitant treatment of the immunosuppression. *Ureaplasma* spp. are the most frequently isolated species (104), followed by *M. hominis* (78, 84, 122). *M. pneumoniae* has been incriminated in a few cases (139). *M. salivarium* (34) and *M. fermentans* (122) were rarely isolated from synovial specimens of hypogammaglobulinemic patients.

Mycoplasmas, mainly *Ureaplasma* spp. and *M. hominis*, have been implicated in a few cases of septic arthritis in other types of immunodeficiency including systemic lupus, renal transplantation, lymphoma, or following treatment with immunosuppressive drugs (78, 121).

**2.5.1.2    Mycoplasmas and sexually-acquired reactive arthritis, a probable link**

*Ureaplasma* spp. have been detected by molecular methods in a few cases of reactive arthritis (72, 147). In one case, it has been cultured from the synovial fluid of a woman who developed arthritis following a cervicovaginitis episode. In this case *M. fermentans* was also detected in the joint by PCR (128). In addition, the fact that the synovial fluid lymphocytes, more than the peripheral blood ones, specifically proliferated in response to ureaplasmal antigens, argues also for a causal link between ureaplasmas and reactive arthritis (52). *M. genitalium* has been detected by PCR in a joint of a patient with reactive arthritis (139), but further observations are needed to envisage a possible role of this genital mycoplasma in this pathology.

In reactive arthritis the causal link with mycoplasmas is less evident than in septic arthritis, as *Ureaplasma* spp. are common commensals of the urogenital tract and ureaplasmal genital infections are often asymptomatic. However the data described above suggest strongly that ureaplasmas could induce sexually-acquired reactive arthritis. Further studies need to confirm the strength of this relationship.

**2.5.1.3    Mycoplasmas in chronic inflammatory arthritis, an hypothesis**

The role of mycoplasmas, namely *M. fermentans*, in chronic inflammatory arthritis has been discussed since the early 1970's (121, 139). More recently *M. fermentans* has been detected, mostly by PCR but also by culture in some cases, in at least 20% of synovial fluids or biopsies from patients with different chronic inflammatory arthritides including rheumatoid arthritis (RA), but not from patients with gout, osteoarthritis or chondrocalcinosis (51, 58, 124, 125, 127). However others did not find such an association (48, 105). An open question remains: is *M. fermentans* present as a bystander in the joint since it has been detected by PCR in the peripheral blood mononuclear cells (PBMCs) of about 10% of HIV negative patients (139), or does it play a role in initiating or perpetuating the chronic arthritis? Recently Horowitz et al. (51) found a prevalence of anti-*M. fermentans* antibodies in synovial fluids of patients with RA significantly higher than in those of patients with other arthritides. Furthermore for patients with RA, the level of anti-*M. fermentans* antibodies in synovial fluids was significantly higher than in sera. Such a difference was not found for patients with other arthritides. The hypothesis of a link between mycoplasmas and chronic inflammatory arthritis is reinforced by the finding of higher levels of antibodies to *M. arthritidis* mitogen (MAM), which is the superantigen of *M.*

*arthritidis*, a mycoplasma implicated in murine arthritis, in patients with RA compared to those with other rheumatic diseases or healthy controls (120).

*Ureaplasma* DNA has been detected in synovial specimens from 7% to 13% of patients with various chronic inflammatory arthritides, including RA, versus none of the patients with osteoarthritis or gout (126). The detection of other mycoplasmas either by culture (*M. hominis*, *M. orale*) or PCR (*M. genitalium*) is still too infrequent to be interpreted (139).

In summary, the implication of human mycoplasmas in chronic inflammatory arthritis is still a matter of debate and further investigations are needed to resolve this issue and to lead eventually to antibiotic therapeutic trials.

## 2.5.2    Other systemic infections

Other infections by mycoplasmas have often been discovered fortuitously. The ability of *M. hominis* to grow on blood agar used in routine bacteriological cultures, explains why it is much more frequently isolated than the other species. Several reviews related to *M. hominis* extragenital infections have been published (68, 84). Thus, this mycoplasma has been demonstrated to be responsible for a number of cases including septicemia (32), retroperitoneal abscesses and peritonitis (15), hematoma infection (71), vascular and catheter-related infections (31, 130), sternal wound infections associated with mediastinitis after thoracic surgery (81), prosthetic valve endocarditis (11), brain abscesses (157), and pneumonia (79) mainly by hematogenous spread. In most of the cases, these infections occurred in immunocompromised, or transplanted patients, or patients with major disruptions of anatomic barriers or polytraumatisms (84).

Exceptional cases of human infections due to animal mycoplasmas, *M. arginini* (156) and *M. felis* (12) have been reported in immunocompromised patients.

The possible role of mycoplasmas in HIV disease has been a great matter of debate and was fuelled by the ability of these bacteria to promote HIV replication *in vitro* and/or HIV-associated cytopathic effects (10). The role of *M. fermentans* as a HIV cofactor has been suggested since 1989 with the detection of this mycoplasma in tissues and PBMCs from AIDS patients (5, 73), but later also in HIV-negative patients (60). *M. penetrans*, a species recently isolated from the urine of HIV-positive homosexual male (75), has been associated by serological studies with HIV disease and Kaposi'sarcoma in some reports (153). However for at least the last ten years, no conclusive evidence has demonstrated none other than an opportunistic role for mycoplasmas in HIV infection.

*M. penetrans* has been recently isolated from the blood and the throat of a HIV-negative patient with primary antiphospholipid syndrome (155).

Recently, another matter of debate concerned the association of *M. fermentans* infection, detected either serologically or by PCR, and chronic fatigue syndrome or gulf war illness (88, 148). However others did not find such an association (42, 76), and to date, no evidence suggest that infection by *M. fermentans* is associated with the development of such diseases.

## 3.    BIOLOGICAL FEATURES OF HUMAN MYCOPLASMAS RELEVANT TO PATHOGENICITY

Biological features related to the *M. pneumoniae* pathogenesis are reviewed in another chapter (see "Cytadherence and the cytoskeleton" by M. F. Balish and D. C. Krause).

### 3.1     *Ureaplasma* species

#### 3.1.1     *U. urealyticum vs U. parvum*, biovars and serovars

The *U. urealyticum* species is divided into two biovars (64, 115) and 14 serovars (114) on the basis of discriminative features including manganese susceptibilities, protein gel electrophoresis, enzyme profiles, DNA hybridization studies, restriction fragment length polymorphism, genome size and 16S rRNA sequencing. *U. urealyticum* biovar 1 or the parvo biovar includes serovars 1, 3, 6 and 14 while biovar 2 or the T960 biovar encompasses serovars 2, 4, 5 and 7 to 13. Recently several authors (64, 65, 115) presented evidence that the species *U. urealyticum* should be separated into two new species, namely, *U. parvum* (previously *U. urealyticum* biovar 1) and *U. urealyticum* (previously *U. urealyticum* biovar 2).

Some *Ureaplasma* serovars identified by serotyping, especially serovars 4, 6 and 8, have been associated more frequently with disease in several studies (86) but not in others (159). Data are limited because of technical problems with serotyping. Indeed, serotyping by conventional methods remains difficult to use because of lack of commercial antisera and the presence of cross reactions even when monoclonal antibodies are used (85). Recently, a new ELISA assay has been developed using monoclonal antibodies, but with still some cross-reactions observed (28).

Both biovars can be differentiated by several PCR-based methods targeting the urease subunit genes (9), the multiple-banded antigen (MBA) gene (140), the 16S rRNA (115) and the 16S-23S rRNA intergenic spacer region (43). Arbitrarily primed PCR (40) and PCR-single strand conformation polymorphism analysis (100) have been also developed. Biovar 1 is the most commonly isolated from clinical specimens, especially in pregnant women (1, 41, 63, 65, 86, 103, 159). Biovar 2 was isolated significantly more often from women who delivered preterm with the clinical diagnosis of BV (103), women with miscarriage, and women with PID (1). Some genital specimens could contain strains belonging to both biovars (1, 65, 103).

Kong et al (65) described an algorithm for biovar identification and subtyping. This algorithm includes nine primer pairs targeting the 16S rRNA gene, the 16S-23S rRNA intergenic spacer region, the urease subunit genes, and the 5' end of the mba gene, used in combination to identify biovars and serovars. Genotyping methods based on several target PCRs to differentiate both *Ureaplasma* spp. could replace efficiently the 14-member-serotyping scheme established with polyvalent sera. Furthermore the new molecular typing systems developed would facilitated investigation of the pathogenic potential of different biovars or serovars of *Ureaplasma* spp.

### 3.1.2    What do we learn from the genome analysis?

Although *U. urealyticum* is one of the most significant pathogenic mollicutes in humans, its study was greatly hampered by the difficulty of obtaining high yields of ureaplasma cells in culture. Five ureaplasma proteins were initially identified as being putative virulence factors: urease, immunoglobulin-$\alpha$ (IgA) protease, the MBA antigen, and phospholipases C and A (for review see (109)). However, demonstration that these proteins play a role in the ureaplasma virulence is yet to be achieved because there are no genetic tools available to specifically disrupt or complement ureaplasma genes, and a good animal model for studying *U. urealyticum* urogenital infections is missing (14).

Among mollicutes, the 14 *U. urealyticum* serovars are characterized by heterogeneous genome size, ranging from 760 kpb for the serovars belonging to the biovar 1, to 840-1140 kpb for strains of the other biovar (113). Serovar 3 was chosen for complete sequencing of the ureaplasma genome as it is the most prevalent serovar among clinical isolates ((39); http://genome.microbio.uab.edu/uu/uugen. htm). The *U. urealyticum* genome has only a mean G+C content of 25.5% which is the lowest percentage among all sequenced bacterial genomes. It was hypothesized that this biased composition could be due to a decreased capacity to remove uracil from DNA resulting from misincorporation during genome replication or by spontaneous deamination of deoxycytidine residues. Indeed, some of the enzymes necessary to remove uracil from DNA were not found on genome annotation or the corresponding enzyme activities could not be detected. As for other bacterial sequenced genomes, there are a number of discrepancies between sequence annotation and identification of enzyme activities (101). The complete picture for metabolism can only be approached through a combination of *in silico* analysis from genomic data such as the one provided by the KEGG database (http://star.scl.genome.ad.jp/kegg/kegg2.html) with "*in aqua*" enzymatic findings (101), (see chapter "Central carbohydrate pathways: metabolic flexibility and the extra role of some housekeeping enzymes " by J. D. Pollack).

One of the most exciting perspectives resulting from the whole sequencing of *U. urealyticum* genome is the possibility to decipher the unique coupling between urease and ATP synthase ($F_0F_1$-ATPase) activities. Indeed, in *U. urealyticum,* the urease activity generates an ammonia chemiosmotic gradient which in turn activates the ATP-generating system (116, 131). This mechanism which is unique among living organisms is supposed to generate 95% of the cellular ATP. The key role of urease in the *U. urealyticum* metabolism has been demonstrated by the use of potent urease inhibitors such as flurofamide which blocks ureaplasma growth (62, 87). Although the *U. urealyticum* urease genes were identified and sequenced before the complete sequencing of its genome (9, 91, 154), the *in silico* analysis offers the opportunity to search for all the molecular components which are supposed to be required for this coupled metabolism between urease activity and ATP production. This is the case for membrane proteins involved in ammonium and $Ni^{2+}$ transport, for which three putative candidates have been indicated by genome annotation (39).

A striking feature of the ureaplasma genome is the limited amount of genetic information dedicated to DNA repair systems (see chapter "Comparative genome analysis of the mollicutes" by T. Dandekar *et al.*). In particular, the *mut* gene homologues, which play an important role in the methyl-directed mismatch repair system, are missing, a property also shared by the 3 other sequenced mollicute genomes (20, 35, 47). In enterobacteria, the lack of such a system enhances mutation rates by up to 1,000 fold (70) resulting in a mutator phenotype. This phenotype, which could be highly detrimental in a species fitted to survival in an environment with limited selection pressure, could be neutral or even advantageous for parasitic bacteria (133). For example, it has been shown recently that over 30% of the *Pseudomonas aeruginosa* isolates that chronically infect the lungs of cystic fibrosis patients are mutators (92). A link was also found in this study between high mutation rates and the fast evolution of antibiotic resistance among *P. aeruginosa* isolates. In addition, *Neisseria meningitidis* surface antigens are known to vary with a high frequency and this phase variation is mostly due to expansion-contraction of repeated nucleotide tracts located within or near coding regions, a mechanism similar to that described in mycoplasmas (see chapter "Antigenic variation" by D. Yogev, K. S. Wise, and G. Browning). For this major human pathogen, an association between high rates of phase variation of haemoglobin receptor and the mutations in the *mutS* or *mutL* genes was recently described (112).

In *U. urealyticum,* the main antigen recognized by antibodies of patients during infection is the MBA (158). The MBA gene encodes serovar specificity and its 3' moiety is composed of identical tandem repeats. In the reference strain of serovar 3, there are 44 such repeats which are 18 nt long with a deduced amino acid sequence of GKEQPA (39, 158). The number of repeats is variable among different clinical isolates and also varies at a high frequency during subculture of a single isolate (158). This variation of repeated elements is a common theme among variable surface antigens of mycoplasmas (see chapter "Antigenic variation" by D. Yogev, K. S. Wise, and G. Browning). Interestingly, sequence analysis of the *U. urealyticum* genome revealed 5 MBA paralogs which do not share a 3' repeated structure (39). Whether these paralogs are expressed is not known, but we found a repeat ($AT_{9x}$) upstream one of them (unpublished data), suggesting its possible role in phase variation.

## 3.2    *Mycoplasma hominis*

Compared to *Ureaplasma* spp., *M. genitalium,* and *M. pneumoniae,* little is known about the genome of *M. hominis,* since its complete sequence is not yet available.

Data based on serological and polyacrylamide gel electrophoresis studies suggest marked antigenic heterogeneity of the species, concerning especially membrane antigens (21). However no serovar or subspecies could be clearly distinguished among the *M. hominis* strains (68). Remarkable differences between strains of *M. hominis* have been also documented by DNA hybridization studies, restriction fragment length polymorphism and genome size (for a review see reference (68)).

Adaptative variation of surface protein expression and structure is a common strategy used by mycoplasmas to adapt to change in environment (109). In *M. hominis* several surface-exposed membrane lipoproteins belonging to the P120, Lmp, and Vaa adhesin families have been shown to be subject to size, phase, and antigenic variations as recently reviewed in reference (68), (see chapter "Antigenic variation" by D. Yogev, K. S. Wise, and G. Browning).

Depletion of arginine, production of ammonia, and activity of the membrane-bound enzymes phospholipase and aminopeptidase, have been previously brought up as classical factors of pathogenicity (66). Arginine deiminase produced by *M. hominis* has been shown to inhibit cell proliferation in vitro and act like an antitumor factor (44, 134).

Recently several studies have reported interdependence between *M. hominis* and *Trichomonas vaginalis* during infection (107, 136). *T. vaginalis* isolates cultured in vitro could be infected by *M. hominis* and infected *T. vaginalis* isolates were able to transmit bacteria not only to other protozoa, but also to human-derived epithelial cells (107). From these data, it has been suggested that *T. vaginalis* could be a carrier of the *M. hominis* infection in vivo. Such an association is not seen for *Ureaplasma* spp. (146).

## 3.3    *Mycoplasma genitalium*

*M. genitalium*, as indicated in this short review, has recently gained more recognition as a significant human pathogen. The ability to cause disease and hence challenge the sophisticated immune response of humans is fascinating considering that this bacterium possesses the smallest genome among living cells capable of autonomous replication. Indeed, with a genome size of only 580 kpb, this mycoplasma encodes a mere 480 proteins (99). Although a minimalist for many metabolic pathways and regulatory mechanisms, *M. genitalium* cells are characterized by a complex terminal structure, or tip, which mediates attachment to host cells, resembling *M. pneumoniae* (see chapter "Cytadherence and the cytoskeleton" by M. F. Balish and D. C. Krause).

### 3.3.1    Cytadherence and virulence

The molecular components of *M. genitalium* which play a role in cytadherence include two adhesins and accessory proteins and were mostly identified by homology with their counterparts in *M. pneumoniae* (54, 111). The *M. pneumoniae* P1 and P30 adhesins correspond to the P140 (MgPa) and P32 in *M. genitalium*, respectively. In addition to these two adhesins, homologs were found to the high molecular weight polypeptides HMW1, HMW2 and HMW3 (35). Evidence that the *M. genitalium* P140 acts as an adhesin came from studies showing that P140-specific monoclonal antibodies, inhibited attachment of the mycoplasma to sheep erythrocytes (93). In addition, spontaneous *M. genitalium* mutants that lack or are unable to correctly localize P140 lose the ability to cause hemagglutination or cellular adhesion (83). It has been demonstrated, at least for *M. pneumoniae,* that mutants that are unable to cytadhere are not pathogenic (for review see chapter "Cytadherence and the cytoskeleton" by M. F. Balish and D. C. Krause). Sequences homologous to MgPa which are distributed in the genome are thought to serve as a reservoir for generating antigenic variation by homologous recombination (98). The implication of the HMW2-homologue (MG218) in cytadherence was demonstrated by disrupting its genes by homologous recombination (25).

### 3.3.2    Gene inactivation and virulence

Functional genomics requires genetic tools to inactivate and complement specific genes. Although these tools are still in infancy for mollicutes, it was possible to obtain transposition with Tn*4001*-based constructs (see chapter "Extrachromosomal elements and gene transfer" by J. Renaudin). The use of this transposon allowed to identify molecular components potentially involved in cytadherence (110) but also to develop an ambitious strategy for identifying non-essential genes, with the ultimate goal of defining a set of genes necessary for a minimal cell (53). More recently, specific gene disruptions were also achieved by homologous recombination following electroporation with contructs including the Tn*4001*-derived gentamicin resistance gene (24, 25). It allowed to demonstrate the crucial role played by the peptide methionine reductase (MsrA) to resist oxidative stress for mycoplasmas which lack both catalase and superoxide-dismutase. Indeed, this enzyme has an activity directly linked to the repair of damages due to oxidative stress because it catalyses the reduction of methionine sulfoxide residues in proteins to methionine. Evaluation of the cytotoxicity of the mycoplasma following experimental infection in hamsters, and of the capacity of the *msrA* mutant to cytadhere suggested that this gene encodes a virulence determinant (24).

## 3.4    Other mycoplasmas

### 3.4.1    *Mycoplasma penetrans,* a species armed for pathogenicity

*Mycoplasma penetrans* is a mycoplasma recently isolated from urine samples collected from HIV-infected patients (75). Although the pathogenic role of this mycoplasma is still questionable, its unique biological properties suggest a pathogenic potential.

Its name came from its capacity to penetrate eucaryotic cells *in vitro* (4, 38). Furthermore, the detection of this mycoplasma by electron microscopy in the cytoplasm of uroepithelial cells collected from infected patients indicated that this is probably also true *in vivo* (74). This mycoplasma also induces cytopathic effects *in vitro* and kills a large proportion of chicken embryos experimentally infected via the yolk sac (45). In fact, the results indicating that both *M. fermentans* and *M. penetrans* are able to penetrate eucaryotic cells stimulated new studies aiming at re-evaluating the dogma according to which mycoplasmas were exclusively extracellular parasites. Co-culture systems of mycoplasmas with eucaryotic cells were used to evaluate the invasive properties of a number of mycoplasma species. It became clear that a number of mycoplasmas including *M. penetrans, M. fermentans, M. genitalium* and the poultry pathogen *M. gallisepticum*, have the ability to invade host cells (see chapter "Invasion of mycoplasmas into and fusion with host cells" by S. Rottem). However, it is not known whether this intracellular location can be envisaged as a strategy for escaping the host's immune reponse and has implications for chemotherapy. The interaction of *M. penetrans* with cultured cells results in a complex interplay in which the phospholipases produced by the mycoplasma were proposed to play a major role (118, 129). Furthermore, *M. penetrans* displays hemolytic and hemoxidative activities towards red blood cells of different origins (59). Hemoxidation was directly linked to the production of $H_2O_2$ but catalase treatment did not abolish completely the *M. penetrans* hemolytic activity. This result is interesting because the production of $H_2O_2$ is thought to be the sole *M. pneumoniae* hemolysin. However, the fact that catalase did not abolish completely the *M. penetrans* hemolytic activity suggests the secretion of a polypeptide acting as a hemolysin (59).

Similarly to other mycoplasmas, *M. penetrans* main surface antigens are highly variable (90). The major antigen recognized during natural and experimental infections is an abundant P35 lipoprotein (33), which is the basis of *M. penetrans*-specific serological assays (152). This P35 antigen undergoes spontaneous and reversible phase variation. The mechanism underlying this variation is unknown. Whether this antigenic diversity helps the mycoplasma to evade immune surveillance, as suggested for other similar systems in mycoplasmas remains to be determined.

### Nucleases as a new type of mycoplasma toxins?

Mycoplasmas are deficient in many metabolic pathways, in particular they lack the pathways for *de novo* synthesis of their nucleic acids, and adenine, uracil, guanine and thymine are required for growth (see chapter "Central carbohydrate pathways: metabolic flexibility and the extra role of some housekeeping enzymes " by J. D. Pollack). Production of nucleases has been proposed as a mechanism by which mycoplasmas could acquire the precursors for the synthesis of nucleic acids (108). Furthermore, early *in vitro* studies indicated that these nucleases could alter the nucleic acid metabolism of the host cells parasitized by mycoplasmas (82). Numerous effects of these nucleases were reported for mycoplasma-contaminated cultured cells including the apparent inhibition of HIV reverse transcriptase activity (80) and the induction of cell apoptosis (94, 95, 132). It was shown that the apparent inhibition of HIV reverse transcriptase was due to the degradation of the reverse transcriptase products by $Ca^{2+}$-dependent nucleases produced by contaminating mycoplasmas (29, 106). Among five different species of mycoplasmas, it was found that *M. fermentans* and *M. penetrans* produced the most potent nuclease activity. This led to a more detailed characterization of the nuclease(s) produced by *M. penetrans* (6). The main nuclease was purified to homogeneity as a 40 kDa polypeptide and required both $Mg^{2+}$ and $Ca^{2+}$ for maximum activity. The purified enzyme exhibited both a nicking activity of double-stranded DNA and an endonuclease activity of RNA and single-stranded DNA. Interestingly, a cytotoxic effect of this purified nuclease was detected when added to cultured cells with cell alterations suggestive of apoptosis (7). The 40 kDa nuclease was able to bind specifically in a dose-dependent manner to the cell membrane of a lymphocytic cell line. These results support a role of mycoplasma nucleases as potential virulence determinants. Interestingly, a toxin, termed cytolethal distending toxin, which is encoded by many bacterial pathogens has recently been shown to exhibit a DNase activity which by itself is responsible for the toxin-induced cell cycle arrest (30, 69). Whether the *M. penetrans* nuclease acts similarly to those cytolethal distending toxins remains to be determined.

### 3.4.2    *Mycoplasma fermentans*

Features concerning antigenic variation, modulation of the host immune system by *M. fermentans* or its clastogenic or oncogenic effects related to pathogenicity, are reviewed in other chapters of this book.

Recently two studies tried to assign some molecular characteristics to some strains of *M. fermentans* associated with either a specific disease or site of isolation. Seven strains of *M. fermentans* isolated from the joint of arthritic patients were characterized by genome size determination, arbitrarily-primed PCR, pulse-field gel electrophoresis and Southern blot analysis (123). No single profile could be constructed on the basis of these data. Campo et al (16) found a correlation between the expression of the *M. fermentans* surface antigen named Pra (proteinase-resistant antigen) and the site of isolation from the infected host. Strains which expressed Pra were most frequently associated with bone marrow cells while strains which lacked it were mostly isolated from epithelial cell surfaces. The Pra+ phenotype was shown to be associated with a specific insertion sequence (IS) genotype, hypothesizing a link beween the IS insertion site and the expression of Pra.

# REFERENCES

1.  **Abele-Horn, M., C. Wolff, P. Dressel, F. Pfaff, and A. Zimmermann.** 1997. Association of *Ureaplasma urealyticum* biovars with clinical outcome for neonates, obstetric patients, and gynecological patients with pelvic inflammatory disease. J. Clin. Microbiol. **35:**1199-1202.
2.  **Ainsworth, J. G., J. Clarke, R. Goldin, and D. Taylor-Robinson.** 2000. Disseminated *Mycoplasma fermentans* in AIDS patients: several case reports. Int. J. STD AIDS **11:**751-755.
3.  **Alonso-Vega, C., N. Wauters, D. Vermeylen, M. F. Muller, and E. Serruys.** 1997. A fatal case of *Mycoplasma hominis* meningoencephalitis in a full-term newborn. J. Clin. Microbiol. **35:**286-287.
4.  **Andreev, J., Z. Borovsky, I. Rosenshine, and S. Rottem.** 1995. Invasion of HeLa cells by *Mycoplasma penetrans* and the induction of tyrosine phosphorylation of a 145-kDa host cell protein. FEMS Microbiol. Lett. **132:**189-94.
5.  **Bébéar, C., B. de Barbeyrac, M. T. Clerc, H. Renaudin, H. J. Fleury, M. Dupon, J. M. Ragnaud, and P. Morlat.** 1993. Mycoplasmas in HIV-1 seropositive patients. Lancet **341:**758-759.
6.  **Bendjennat, M., A. Blanchard, M. Loutfi, L. Montagnier, and E. Bahraoui.** 1997. Purification and characterization of *Mycoplasma penetrans* Ca2+/Mg2+-dependent endonuclease. J. Bacteriol. **179:**2210-20.
7.  **Bendjennat, M., A. Blanchard, M. Loutfi, L. Montagnier, and E. Bahraoui.** 1999. Role of *Mycoplasma penetrans* endonuclease P40 as a potential pathogenic determinant. Infect. Immun. **67:**4456-62.
8.  **Bjornelius, E., P. Lidbrink, and J. S. Jensen.** 2000. *Mycoplasma genitalium* in non-gonococcal urethritis--a study in Swedish male STD patients. Int. J. STD. AIDS **11:**292-296.
9.  **Blanchard, A.** 1990. *Ureaplasma urealyticum* urease genes; use of a UGA tryptophan codon. Mol. Microbiol. **4:**669-676.
10. **Blanchard, A., and L. Montagnier.** 1994. AIDS-associated mycoplasmas. Annu. Rev. Microbiol. **48:**687-712.
11. **Blasco, M., L. Torres, M. L. Marco, B. Moles, M. C. Villuendas, and J. B. Garcia Moya.** 2000. Prosthetic valve endocarditis caused by *Mycoplasma hominis*. Eur. J. Clin. Microbiol. Infect. Dis. **19:**638-640.
12. **Bonilla, H. F., C. E. Chenoweth, J. G. Tully, L. K. Blythe, J. A. Robertson, V. M. Ognenovski, and C. A. Kauffman.** 1997. *Mycoplasma felis* septic arthritis in a patient with hypogammaglobulinemia. Clin. Infect. Dis. **24:**222-223.
13. **Bove, J. M.** 1999. The one-hundredth anniversary of the first culture of a mollicute, the contagious bovine peripneumonia microbe, by Nocard and Roux, with the collaboration of Borrel, Salimbeni, and Dujardin-Baumetz. Res. Microbiol. **150:**239-45.
14. **Brunner, H.** 1997. Models of mycoplasma respiratory and genital tract infections. Wien. Klin. Wochenschr. **109:**569-73.
15. **Brunner, S., P. Frey-Rindova, M. Altwegg, and R. Zbinden.** 2000. Retroperitoneal abscess and bacteremia due to *Mycoplasma hominis* in a polytraumatized man. Infection **28:**46-48.
16. **Campo, L., P. Larocque, T. La Malfa, W. D. Blackburn, and H. L. Watson.** 1998. Genotypic and phenotypic analysis of *Mycoplasma fermentans* strains isolated from different host tissues. J. Clin. Microbiol. **36:**1371-1317.
17. **Cassell, G. H., D. T. Crouse, K. B. Waites, P. T. Rudd, and J. K. Davis.** 1988. Does *Ureaplasma urealyticum* cause respiratory disease in newborns? Pediatr. Infect. Dis. J. **7:**535-541.

18. **Cassell, G. H., K. B. Waites, and D. T. Crouse.** 1994. Mycoplasmal infections, p. 619-656. *In* J. S. Remington and J. O. Klein (ed.), Infectious diseases of the fetus and newborn infant, 4th ed. The W. B. Saunders Co., Philadelphia, Pa.
19. **Cassell, G. H., K. B. Waites, and H. L. Watson.** 1993. *Ureaplasma urealyticum* intrauterine infection: role in prematurity and disease in newborns. Clin. Microb. Rev.:69-87.
20. **Chambaud, I., R. Heilig, S. Ferris, V. Barbe, D. Samson, F. Galisson, I. Moszer, K. Dybvig, H. Wroblewski, A. Viari, E. P. Rocha, and A. Blanchard.** 2001. The complete genome sequence of the murine respiratory pathogen *Mycoplasma pulmonis*. Nucleic Acids Res. **29:**2145-2153.
21. **Christiansen, G.** 1992. Genetic variation in natural populations, p. 561-573. *In* J. Maniloff, R. N. McElhaney, L. R. Finch, and J. B. Baseman (ed.), Mycoplasmas: molecular biology and pathogenesis. American Society for Microbiology, Washington, D. C.
22. **Dallo, S. F., and J. B. Baseman.** 2000. Intracellular DNA replication and long-term survival of pathogenic mycoplasmas. Microb. Pathogen. **29:**301-309.
23. **Deguchi, T., H. Komeda, M. Yasuda, K. Tada, H. Iwata, M. Asano, T. Ezaki, and Y. Kawada.** 1995. *Mycoplasma genitalium* in non-gonococcal urethritis. Int. J. STD. AIDS **6:**144-145.
24. **Dhandayuthapani, S., M. Blaylock, C. M. Bebear, W. G. Rasmussen, and J. B. Baseman.** 2001. Peptide methionine reductase (MsrA) is a virulence determinant in *Mycoplasma genitalium*. J. Bacteriol. **183:**5645-5650.
25. **Dhandayuthapani, S., W. G. Rasmussen, and J. B. Baseman.** 1999. Disruption of gene mg218 of *Mycoplasma genitalium* through homologous recombination leads to an adherence-deficient phenotype. Proc. Natl Acad. Sci. U S A **96:**5227-32.
26. **Donders, G. G. G., B. Vanbulck, J. Caudron, L. Londers, A. Vereecken, and B. Spitz.** 2000. Relationship of bacterial vaginosis and mycoplasmas to the risk of spontaneous abortion. Am. J. Obstet. Gynecol. **183:**431-437.
27. **Drexler, H. G., and C. C. Uphoff.** 2000. Contamination of cell culture, *Mycoplasma*, p. 609-627. *In* H. G. Drexler (ed.), The leukemia lymphoma cell lines Factsbook. Academic Press, San Diego.
28. **Echahidi, F., G. Muyldermans, S. Lauwers, and A. Naessens.** 2001. Development of an enzyme-linked immunosorbent assay for serotyping *Ureaplasma urealyticum* strains using monoclonal antibodies. Clin. Diagn. Lab. Immunol. **8:**52-57.
29. **el-Farrash, M. A., M. Kannagi, M. J. Kuroda, T. Yoshida, and S. Harada.** 1994. The mycoplasma-related inhibitor of HIV-1 reverse transcriptase has a DNase activity and is present in the particle-free supernatants of contaminated cultures. J. Virol. Methods **47:**73-82.
30. **Elwell, C. A., and L. A. Dreyfus.** 2000. DNase I homologous residues in CdtB are critical for cytolethal distending toxin-mediated cell cycle arrest. Mol. Microbiol. **37:**952-63.
31. **Fenollar, F., J. P. Casalta, H. Lepidi, P. Piquet, and D. Raoult.** 1999. Mycoplasma infections of aneurysms or vascular grafts. Clin. Infect. Dis. **28:**694-695.
32. **Fernandez Guerrero, M. L., J. Manuel Ramos, and F. Soriano.** 1999. *Mycoplasma hominis* bacteraemia not associated with genital infections. J. Infect. **39:**91-94.
33. **Ferris, S., H. L. Watson, O. Neyrolles, L. Montagnier, and A. Blanchard.** 1995. Characterization of a major *Mycoplasma penetrans* lipoprotein and of its gene. FEMS Microbiol. Lett. **130:**313-320.
34. **Franz, A., A. D. Webster, P. M. Furr, and D. Taylor-Robinson.** 1997. Mycoplasmal arthritis in patients with primary immunoglobulin deficiency: clinical features and outcome in 18 patients. Br. J. Rheumatol. **36:**661-668.

35.   Fraser, C. M., J. D. Gocayne, O. White, M. D. Adams, R. A. Clayton, R. D.
      Fleischmann, C. J. Bult, A. R. Kerlavage, G. Sutton, J. M. Kelley, and et al. 1995.
      The minimal gene complement of *Mycoplasma genitalium*. Science **270**:397-403.
36.   Furr, P. M., D. Taylor-Robinson, and A. D. Webster. 1994. Mycoplasmas and
      ureaplasmas in patients with hypogammaglobulinaemia and their role in arthritis:
      microbiological observations over twenty years. Ann. Rheum. Dis. **53**:183-187.
37.   Gambini, D., I. Decleva, L. Lupica, M. Ghislanzoni, M. Cusini, and E. Alessi.
      2000. *Mycoplasma genitalium* in males with nongonococcal urethritis: prevalence and
      clinical efficacy of eradication. Sex. Transm. Dis. **27**:226-229.
38.   Giron, J. A., M. Lange, and J. B. Baseman. 1996. Adherence, fibronectin binding,
      and induction of cytoskeleton reorganization in cultured human cells by *Mycoplasma
      penetrans*. Infect. Immun. **64**:197-208.
39.   Glass, J. I., E. J. Lefkowitz, J. S. Glass, C. R. Heiner, E. Y. Chen, and G. H.
      Cassell. 2000. The complete sequence of the mucosal pathogen *Ureaplasma
      urealyticum*. Nature **407**:757-762.
40.   Grattard, F., B. Pozzetto, B. de Barbeyrac, H. Renaudin, M. Clerc, O. G.
      Gaudin, and C. Bébéar. 1995. Arbitrarily-primed PCR confirms the differentiation
      of strains of *Ureaplasma urealyticum* into two biovars. Mol. Cell. Probes **9**:383-389.
41.   Grattard, F., B. Soleihac, B. de Barbeyrac, C. Bébéar, P. Seffert, and B. Pozzetto.
      1995. Epidemiologic and molecular investigations of genital mycoplasmas from
      women and neonates at delivery. Pediatr. Infect. Dis. J. **14**:853-858.
42.   Gray, G. C., K. S. Kaiser, A. W. Hawksworth, and H. L. Watson. 1999. No
      serologic evidence of an association found between Gulf War service and
      *Mycoplasma fermentans* infection. Am. J. Trop. Med. Hyg. **60**:752-757.
43.   Harasawa, R., and Y. Kanamoto. 1999. Differentiation of two biovars of
      *Ureaplasma urealyticum* based on the 16S-23S rRNA intergenic spacer region. J.
      Clin. Microbiol. **37**:4135-4138.
44.   Harasawa, R., K. Koshimizu, M. Kitagawa, K. Asada, and I. Kato. 1992.
      Nucleotide sequence of the arginine deiminase gene of *Mycoplasma hominis*.
      Microbiol. Immunol. **36**:661-665.
45.   Hayes, M. M., B. J. Li, D. J. Wear, and S. C. Lo. 1996. Pathogenicity of
      *Mycoplasma fermentans* and *Mycoplasma penetrans* in experimentally infected
      chicken embryos. Infect. Immun. **64**:3419-24.
46.   Hillier, S. L., R. P. Nugent, D. A. Eschenbach, M. A. Krohn, R. S. Gibbs, D. H.
      Martin, M. F. Cotch, R. Edelman, J. G. Pastorek, 2nd, A. V. Rao, and et al. 1995.
      Association between bacterial vaginosis and preterm delivery of a low-birth-weight
      infant. The Vaginal Infections and Prematurity Study Group. N. Engl. J. Med.
      **333**:1737-1742.
47.   Himmelreich, R., H. Hilbert, H. Plagens, E. Pirkl, B. C. Li, and R. Herrmann.
      1996. Complete sequence analysis of the genome of the bacterium *Mycoplasma
      pneumoniae*. Nucleic Acids Res. **24**:4420-4449.
48.   Hoffman, R. W., F. X. O'Sullivan, K. R. Schafermeyer, T. L. Moore, D. Roussell,
      R. Watson-McKown, M. F. Kim, and K. S. Wise. 1997. Mycoplasma infection and
      rheumatoid arthritis: analysis of their relationship using immunoblotting and an
      ultrasensitive polymerase chain reaction detection method. Arthritis Rheum. **40**:1219-
      1228.
49.   Horner, P., B. Thomas, C. B. Gilroy, M. Egger, and D. Taylor-Robinson. 2001.
      Role of *Mycoplasma genitalium* and *Ureaplasma urealyticum* in acute and chronic
      nongonococcal urethritis. Clin. Infect. Dis. **32**:995-1003.
50.   Horner, P. J., C. B. Gilroy, B. J. Thomas, R. O. Naidoo, and D. Taylor-Robinson.
      1993. Association of *Mycoplasma genitalium* with acute non-gonococcal urethritis.
      Lancet **342**:582-585.

51.  Horowitz, S., B. Evinson, A. Borer, and J. Horowitz. 2000. *Mycoplasma fermentans* in rheumatoid arthritis and other inflammatory arthritides. J. Rheumatol. **27**:2747-2753.

52.  Horowitz, S., J. Horowitz, D. Taylor-Robinson, S. Sukenik, R. N. Apte, J. Bar-David, B. Thomas, and C. Gilroy. 1994. *Ureaplasma urealyticum* in Reiter's syndrome. J. Rheumatol. **21**:877-882.

53.  Hutchison, C. A., S. N. Peterson, S. R. Gill, R. T. Cline, O. White, C. M. Fraser, H. O. Smith, and J. C. Venter. 1999. Global transposon mutagenesis and a minimal mycoplasma genome. Science **286**:2165-2169.

54.  Inamine, J. M., S. Loechel, A. M. Collier, M. F. Barile, and P. C. Hu. 1989. Nucleotide sequence of the MgPa (mgp) operon of *Mycoplasma genitalium* and comparison to the P1 (mpp) operon of *Mycoplasma pneumoniae*. Gene **82**:259-67.

55.  Janier, M., F. Lassau, I. Casin, P. Grillot, C. Scieux, A. Zavaro, C. Chastang, A. Bianchi, and P. Morel. 1995. Male urethritis with and without discharge: a clinical and microbiological study. Sex. Transm. Dis. **22**:244-252.

56.  Jensen, J. S., H. T. Hansen, and K. Lind. 1996. Isolation of *Mycoplasma genitalium* strains from the male urethra. J. Clin. Microbiol. **34**:286-291.

57.  Johannisson, G., Y. Enstrom, G. B. Lowhagen, V. Nagy, K. Ryberg, S. Seeberg, and C. Welinder-Olsson. 2000. Occurrence and treatment of *Mycoplasma genitalium* in patients visiting STD clinics in Sweden. Int. J. STD. AIDS **11**:324-326.

58.  Johnson, S., D. Sidebottom, F. Bruckner, and D. Collins. 2000. Identification of *Mycoplasma fermentans* in synovial fluid samples from arthritis patients with inflammatory disease. J. Clin. Microbiol. **38**:90-93.

59.  Kannan, T. R., and J. B. Baseman. 2000. Hemolytic and hemoxidative activities in *Mycoplasma penetrans*. Infect. Immun. **68**:6419-22.

60.  Katseni, V. L., C. B. Gilroy, B. K. Ryait, K. Ariyoshi, P. D. Bieniasz, J. N. Weber, and D. Taylor-Robinson. 1993. *Mycoplasma fermentans* in individuals seropositive and seronegative for HIV-1. Lancet **341**:271-273.

61.  Keane, F. E., B. J. Thomas, C. B. Gilroy, A. Renton, and D. Taylor-Robinson. 2000. The association of *Mycoplasma hominis*, *Ureaplasma urealyticum* and *Mycoplasma genitalium* with bacterial vaginosis: observations on heterosexual women and their male partners. Int. J. STD. AIDS **11**:356-360.

62.  Kenny, G. E. 1983. Inhibition of the growth of *Ureaplasma urealyticum* by a new urease inhibitor, flurofamide. Yale J. Biol. Med. **56**:717-22.

63.  Knox, C. L., and P. L. Timms. 1998. Comparison of PCR, nested-PCR, and random amplified polymorphic DNA PCR for detection and typing of *Ureaplasma urealyticum* in specimens from pregnant women. J Clin Microbiol. **36**:3032-3039.

64.  Kong, F. R., C. James, Z. F. Ma, S. Gordon, W. Bin, and G. L. Gilbert. 1999. Phylogenetic analysis of *Ureaplasma urealyticum* - support for the establishment of a new species, *Ureaplasma parvum*. Int. J. Syst. Bacteriol. **49**:1879-1889.

65.  Kong, F. R., Z. F. Ma, G. James, S. Gordon, and G. L. Gilbert. 2000. Species identification and subtyping of *Ureaplasma parvum* and *Ureaplasma urealyticum* using PCR-based assays. J. Clin. Microbiol. **38**:1175-1179.

66.  Krause, D. C., and D. Taylor-Robinson. 1992. Mycoplasmas which infect humans, p. 417-444. *In* J. Maniloff, R. N. McElhaney, L. R. Finch, and J. B. Baseman (ed.), Mycoplasmas: molecular biology and pathogenesis. American society for Microbiology, Washington, D. C.

67.  Krieger, J. N., D. E. Riley, M. C. Roberts, and R. E. Berger. 1996. Prokaryotic DNA sequences in patients with chronic idiopathic prostatitis. J. Clin. Microbiol. **34**:3120-3128.

68.  Ladefoged, S. A. 2000. Molecular dissection of *Mycoplasma hominis*. Apmis **108**:5-45.

69.    **Lara-Tejero, M., and J. E. Galan.** 2000. A bacterial toxin that controls cell cycle progression as a deoxyribonuclease I-like protein. Science **290:**354-7.
70.    **LeClerc, J. E., B. Li, W. L. Payne, and T. A. Cebula.** 1996. High mutation frequencies among *Escherichia coli* and *Salmonella* pathogens. Science **274:**1208-11.
71.    **Legg, J. M., T. T. Titus, I. Chambers, R. Wilkinson, R. J. Koerner, and F. K. Gould.** 2000. Hematoma infection with *Mycoplasma hominis* following transplant nephrectomy. Clin. Microbiol. Infect. **6:**619-621.
72.    **Li, F., R. Bulbul, H. R. Schumacher, Jr., T. Kieber-Emmons, P. E. Callegari, J. M. Von Feldt, D. Norden, B. Freundlich, B. Wang, V. Imonitie, C. P. Chang, I. Nachamkin, D. B. Weiner, and W. V. Williams.** 1996. Molecular detection of bacterial DNA in venereal-associated arthritis. Arthritis Rheum. **39:**950-958.
73.    **Lo, S. C., M. S. Dawson, D. M. Wong, P. B. Newton III, M. A. Sonoda, W. F. Engler, R. Y. H. Wang, J. W. K. Shih, H. J. Alter, and D. J. Wear.** 1989. Identification of *Mycoplasma incognitus* infection in patients with AIDS: an immunohistochemical in situ hybridization and ultrastructural study. Am. J. Trop. Med. Hyg. **41:**601-616.
74.    **Lo, S. C., M. M. Hayes, H. Kotani, P. F. Pierce, D. J. Wear, P. B. Newton, 3rd, J. G. Tully, and J. W. Shih.** 1993. Adhesion onto and invasion into mammalian cells by *Mycoplasma penetrans*: a newly isolated mycoplasma from patients with AIDS. Mod. Pathol. **6:**276-80.
75.    **Lo, S. C., M. M. Hayes, R. Y. Wang, P. F. Pierce, H. Kotani, and J. W. Shih.** 1991. Newly discovered mycoplasma isolated from patients infected with HIV. Lancet **338:**1415-1418.
76.    **Lo, S. C., L. Levin, J. Ribas, R. Chung, R. Y. H. Wang, D. Wear, and J. W. K. Shih.** 2000. Lack of serological evidence for *Mycoplasma fermentans* infection in army Gulf War veterans: a large scale case-control study. Epidemiol. Infect. **125:**609-616.
77.    **Lo, S. C., D. J. Wear, S. L. Green, P. G. Jones, and J. F. Legier.** 1993. Adult respiratory distress syndrome with or without systemic disease associated with infections due to *Mycoplasma fermentans*. Clin. Infect. Dis. **17 Suppl 1:**S259-S263.
78.    **Luttrell, L. M., S. S. Kanj, G. R. Corey, R. E. Lins, R. J. Spinner, W. J. Mallon, and D. J. Sexton.** 1994. *Mycoplasma hominis* septic arthritis: two case reports and review. Clin. Infect. Dis. **19:**1067-1070.
79.    **Lyon, G. M., J. A. Alspaugh, F. T. Meredith, L. J. Harrell, V. Tapson, R. D. Davis, and S. S. Kanj.** 1997. *Mycoplasma hominis* pneumonia complicating bilateral lung transplantation: case report and review of the literature. Chest **112:**1428-1432.
80.    **Lyon, M., and J. Huppert.** 1983. Depression of reverse transcriptase activity by hybridoma supernatants: a potential problem in screening for retroviral contamination. Biochem Biophys Res Commun **112:**265-72.
81.    **Mattila, P. S., P. Carlson, A. Sivonen, J. Savola, R. Luosto, J. Salo, and M. Valtonen.** 1999. Life-threatening *Mycoplasma hominis* mediastinitis. Clin. Infect. Dis. **29:**1529-1537.
82.    **McGarrity, G. J., V. Vanaman, and J. Sarama.** 1984. Cytogenetic effects of mycoplasmal infection of cell cultures: a review. In Vitro **20:**1-18.
83.    **Mernaugh, G. R., S. F. Dallo, S. C. Holt, and J. B. Baseman.** 1993. Properties of adhering and nonadhering populations of *Mycoplasma genitalium*. Clin. Infect. Dis. **17:**S69-78.
84.    **Meyer, R. D., and W. Clough.** 1993. Extragenital *Mycoplasma hominis* infections in adults: emphasis on immunosuppression. Clin. Infect. Dis. **17 Suppl. 1:**S243-S249.
85.    **Naessens, A., X. Cheng, S. Lauwers, and J. A. Robertson.** 1998. Development of a monoclonal antibody to a *Ureaplasma urealyticum* serotype 9 antigen. J. Clin. Microbiol. **36:**1125-1127.

86. **Naessens, A., W. Foulon, J. Breynaert, and S. Lauwers.** 1988. Serotypes of *Ureaplasma urealyticum* isolated from normal pregnant women and patients with pregnancy complications. J. Clin. Microbiol. **26:**319-322.

87. **Nagata, K., E. Takagi, H. Satoh, H. Okamura, and T. Tamura.** 1995. Growth inhibition of *Ureaplasma urealyticum* by the proton pump inhibitor lansoprazole: direct attribution to inhibition by lansoprazole of urease activity and urea-induced ATP synthesis in *U. urealyticum.* Antimicrob. Agents Chemother. **39:**2187-92.

88. **Nasralla, M., J. Haier, and G. L. Nicolson.** 1999. Multiple mycoplasmal infections detected in blood of patients with chronic fatigue syndrome and/or fibromyalgia syndrome. Eur. J. Clin. Microbiol. Infect. Dis. **18:**859-865.

89. **Neman-Simha, V., H. Renaudin, B. de Barbeyrac, J. J. Leng, J. Horovitz, D. Dallay, C. Billeaud, and C. Bébéar.** 1992. Isolation of genital mycoplasmas from blood of febrile obstetrical-gynecologic patients and neonates. Scand. J. Infect. Dis. **24:**317-321.

90. **Neyrolles, O., I. Chambaud, S. Ferris, M. C. Prevost, T. Sasaki, L. Montagnier, and A. Blanchard.** 1999. Phase variations of the *Mycoplasma penetrans* main surface lipoprotein increase antigenic diversity. Infect.Immun. **67:**1569-78.

91. **Neyrolles, O., S. Ferris, N. Behbahani, L. Montagnier, and A. Blanchard.** 1996. Organization of *Ureaplasma urealyticum* urease gene cluster and expression in a suppressor strain of *Escherichia coli.* J. Bacteriol. **178:**647-55.

92. **Oliver, A., R. Canton, P. Campo, F. Baquero, and J. Blazquez.** 2000. High frequency of hypermutable *Pseudomonas aeruginosa* in cystic fibrosis lung infection. Science **288:**1251-4.

93. **Opitz, O., and E. Jacobs.** 1992. Adherence epitopes of *Mycoplasma genitalium* adhesin. J. Gen. Microbiol. **138:**1785-90.

94. **Paddenberg, R., A. Weber, S. Wulf, and H. G. Mannherz.** 1998. Mycoplasma nucleases able to induce internucleosomal DNA degradation in cultured cells possess many characteristics of eukaryotic apoptotic nucleases. Cell Death Differ. **5:**517-28.

95. **Paddenberg, R., S. Wulf, A. Weber, P. Heimann, L. A. Beck, and H. G. Mannherz.** 1996. Internucleosomal DNA fragmentation in cultured cells under conditions reported to induce apoptosis may be caused by mycoplasma endonucleases. Eur. J. Cell. Biol. **71:**105-19.

96. **Palmer, H. M., C. B. Gilroy, E. J. Claydon, and D. Taylor-Robinson.** 1991. Detection of *Mycoplasma genitalium* in the genitourinary tract of women by the polymerase chain reaction. Int. J. STD. AIDS **2:**261-263.

97. **Pepin, J., F. Sobela, S. Deslandes, M. Alary, K. Wegner, N. Khonde, F. Kintin, A. Kamuragiye, M. Sylla, P. J. Zerbo, E. Baganizi, A. Kone, F. Kane, B. Masse, P. Viens, and E. Frost.** 2001. Etiology of urethral discharge in West Africa: the role of *Mycoplasma genitalium* and *Trichomonas vaginalis.* Bulletin of the World Health Organization **79:**118-126.

98. **Peterson, S. N., C. C. Bailey, J. S. Jensen, M. B. Borre, E. S. King, K. F. Bott, and C. A. Hutchison, 3rd.** 1995. Characterization of repetitive DNA in the *Mycoplasma genitalium* genome: possible role in the generation of antigenic variation. Proc. Natl Acad. Sci. U S A **92:**11829-33.

99. **Peterson, S. N., and C. M. Fraser.** 2001. The complexity of simplicity. Genome Biol **2:**COMMENT2002.1-2002.8 http://genomebiology.com/2001/2/2/comment/2002/

100. **Pitcher, D., M. Sillis, and J. A. Robertson.** 2001. Simple method for determining biovar and serovar types of *Ureaplasma urealyticum* clinical isolates using PCR-single-strand conformation polymorphism analysis. J. Clin. Microbiol. **39:**1840-1844.

101. **Pollack, J. D.** 2001. *Ureaplasma urealyticum:* an opportunity for combinatorial genomics. Trends Microbiol. **9:**169-75.

102. **Potts, J. M., A. M. Ward, and R. R. Rackley.** 2000. Association of chronic urinary symptoms in women and *Ureaplasma urealyticum.* Urology **55:**486-489.

103.   **Povlsen, K., P. Thorsen, and I. Lind.** 2001. Relationship of *Ureaplasma urealyticum* biovars to the presence or absence of bacterial vaginosis in pregnant women and to the time of delivery. Eur. J. Clin. Microbiol. Infect. Dis. **20**:65-67.

104.   **Puéchal, X., P. Hilliquin, M. Renoux, C. J. Menkes, H. Renaudin, and C. Bébéar.** 1995. *Ureaplasma urealyticum* destructive septic polyarthritis revealing a common variable immunodeficiency. Arthritis Rheum. **38**:1524-1526.

105.   **Puéchal, X., D. Roulland-Dussoix, and C. J. Menkes.** 1997. Rheumatoid arthritis: failure to isolate or detect mycoplasmas. J. Rheumatol. **24**:1445-1446.

106.   **Quillent, C., O. Grau, F. Clavel, L. Montagnier, and A. Blanchard.** 1994. Inhibition of HIV type 1 reverse transcriptase assay by nucleases produced by contaminating mycoplasmas. AIDS Res. Hum. Retroviruses **10**:1251-7.

107.   **Rappelli, P., F. Carta, G. Delogu, M. F. Addis, D. Dessi, P. Cappuccinelli, and P. L. Fiori.** 2001. *Mycoplasma hominis* and *Trichomonas vaginalis* symbiosis: multiplicity of infection and transmissibility of *M. hominis* to human cells. Arch. Microbiol. **175**:70-74.

108.   **Razin, S.** 1978. The mycoplasmas. Microbiol Rev **42**:414-70.

109.   **Razin, S., D. Yogev, and Y. Naot.** 1998. Molecular biology and pathogenicity of mycoplasmas. Microbiol. Mol. Biol. Rev. **62**:1094-1156.

110.   **Reddy, S. P., W. G. Rasmussen, and J. B. Baseman.** 1996. Isolation and characterization of transposon Tn*4001*-generated, cytadherence-deficient transformants of *Mycoplasma pneumoniae* and *Mycoplasma genitalium*. FEMS Immunol. Med. Microbiol. **15**:199-211.

111.   **Reddy, S. P., W. G. Rasmussen, and J. B. Baseman.** 1995. Molecular cloning and characterization of an adherence-related operon of *Mycoplasma genitalium*. J. Bacteriol. **177**:5943-5951.

112.   **Richardson, A. R., and I. Stojiljkovic.** 2001. Mismatch repair and the regulation of phase variation in *Neisseria meningitidis*. Mol. Microbiol. **40**:645-55.

113.   **Robertson, J. A., L. E. Pyle, G. W. Stemke, and L. R. Finch.** 1990. Human ureaplasmas show diverse genome sizes by pulsed-field electrophoresis. Nucleic Acids Res **18**:1451-5.

114.   **Robertson, J. A., and G. W. Stemke.** 1982. Expanded serotyping scheme for *Ureaplasma urealyticum* strains isolated from humans. J. Clin. Microbiol. **15**:873-878.

115.   **Robertson, J. A., A. Vekris, C. Bébéar, and G. W. Stemke.** 1993. Polymerase chain reaction using 16S rRNA gene sequences distinguishes the two biovars of *Ureaplasma urealyticum*. J. Clin. Microbiol. **31**:824-830.

116.   **Romano, N., R. La Licata, and D. Russo Alesi.** 1986. Energy production in *Ureaplasma urealyticum*. Pediatr. Infect. Dis. **5**:S308-12.

117.   **Rosenstein, I. J., D. J. Morgan, M. Sheehan, R. F. Lamont, and D. Taylor-Robinson.** 1996. Bacterial vaginosis in pregnancy: distribution of bacterial species in different gram-stain categories of the vaginal flora. J. Med. Microbiol. **45**:120-126.

118.   **Salman, M., and S. Rottem.** 1995. The cell membrane of *Mycoplasma penetrans*: lipid composition and phospholipase A1 activity. Biochim. Biophys. Acta **1235**:369-77.

119.   **Sarlangue, J., and C. Bébéar.** 1999. Infections néonatales à mycoplasmes. Médecine Thérapeutique Pédiatrie **2**:105-109.

120.   **Sawitzke, A., D. Joyner, K. Knudtson, H. H. Mu, and B. Cole.** 2000. Anti-MAM antibodies in rheumatic disease: evidence for a MAM-like superantigen in rheumatoid arthritis? J. Rheumatol. **27**:358-364.

121.   **Schaeverbeke, T., C. M. Bébéar, M. Clerc, L. Lequen, C. Bebear, and J. Dehais.** 1999. What is the role of mycoplasmas in human inflammatory rheumatic disorders? Rev. Rhum. Engl. Ed. **66**:23S-27S.

122. Schaeverbeke, T., M. Clerc, L. Lequen, C. M. Bébéar, Y. Morrier, B. Bannwarth, C. Bébéar, and J. Dehais. 1997. Isolation of *Mycoplasma hominis* and molecular detection by PCR assays of *M. hominis* and *M. fermentans* from synovial specimens of a hypogammaglobulinemic patient with non erosive polyarthritis. Rev. Rhum. Engl. Ed. **64**:688.

123. Schaeverbeke, T., M. Clerc, L. Lequen, A. Charron, C. M. Bébéar, B. de Barbeyrac, B. Bannwarth, J. Dehais, and C. Bébéar. 1998. Genotypic characterization of seven strains of *Mycoplasma fermentans* isolated from synovial fluids of patients with arthritis. J. Clin. Microbiol. **36**:1226-1231.

124. Schaeverbeke, T., C. B. Gilroy, C. Bébéar, J. Dehais, and D. Taylor-Robinson. 1996. *Mycoplasma fermentans* in joints of patients with rheumatoid arthritis and other joint disorders. Lancet **347**:1418.

125. Schaeverbeke, T., C. B. Gilroy, C. Bébéar, J. Dehais, and D. Taylor-Robinson. 1996. *Mycoplasma fermentans*, but not *M penetrans*, detected by PCR assays in synovium from patients with rheumatoid arthritis and other rheumatic disorders. J. Clin. Pathol. **49**:824-828.

126. Schaeverbeke, T., C. B. Gilroy, J. P. Vernhes, C. M. Bébéar, B. Bannwarth, C. Bébéar, and J. Dehais. 1996. Detection by PCR assay of *Mycoplasma fermentans*, *Ureaplasma urealyticum*, but not *M penetrans*, in synovial samples from patients with rheumatoid arthritis and other rheumatic disorders. Arthritis Rheum. **39** (Suppl.):S182.

127. Schaeverbeke, T., H. Renaudin, M. Clerc, L. Lequen, J. P. Vernhes, B. De Barbeyrac, B. Bannwarth, C. Bébéar, and J. Dehais. 1997. Systematic detection of mycoplasmas by culture and polymerase chain reaction (PCR) procedures in 209 synovial fluid samples. Br. J. Rheumatol. **36**:310-314.

128. Schaeverbeke, T., J. P. Vernhes, M. Clerc, H. Renaudin, B. Bannwarth, C. Bébéar, and J. Dehais. 1996. Isolement de *Mycoplasma fermentans* (Mf) et de *Ureaplasma urealyticum* (Uu) à partir du liquide synovial au cours d'une arthrite réactionnelle. Rev. Rhum. **63**:438.

129. Shibata, K., T. Sasaki, and T. Watanabe. 1995. AIDS-associated mycoplasmas possess phospholipases C in the membrane. Infect. Immun. **63**:4174-7.

130. Sielaff, T. D., J. E. Everett, S. J. Shumway, D. C. Wahoff, R. M. Bolman, 3rd, and D. L. Dunn. 1996. *Mycoplasma hominis* infections occurring in cardiovascular surgical patients. Ann. Thorac. Surg. **61**:99-103.

131. Smith, D. G., W. C. Russell, W. J. Ingledew, and D. Thirkell. 1993. Hydrolysis of urea by *Ureaplasma urealyticum* generates a transmembrane potential with resultant ATP synthesis. J. Bacteriol. **175**:3253-8.

132. Stolzenberg, I., S. Wulf, H. G. Mannherz, and R. Paddenberg. 2000. Different sublines of Jurkat cells respond with varying susceptibility of internucleosomal DNA degradation to different mediators of apoptosis. Cell Tissue Res. **301**:273-82.

133. Taddei, F., I. Matic, B. Godelle, and M. Radman. 1997. To be a mutator, or how pathogenic and commensal bacteria can evolve rapidly. Trends Microbiol. **5**:427-8.

134. Takaku, H., M. Takase, S. Abe, H. Hayashi, and K. Miyazaki. 1992. In vivo anti-tumor activity of arginine deiminase purified from *Mycoplasma arginini*. Int. J. Cancer **51**:244-249.

135. Taylor-Robinson, D. 1996. Infections due to species of *Mycoplasma* and *Ureaplasma*: an update. Clin. Infect. Dis. **23**:671-684.

136. Taylor-Robinson, D. 1998. *Mycoplasma hominis* parasitism of *Trichomonas vaginalis*. Lancet **352**:2022-2023.

137. Taylor-Robinson, D., J. G. Ainsworth, and W. M. McCormack. 1999. Genital mycoplasmas, p. 533-548. *In* K. K. Holmes, P. F. Sparling, P. A. Mardh, S. M. Lemon, W. E. Stamm, P. P., and J. N. Wasserheit (ed.), Sexually transmitted diseases, 3rd ed. McGraw Hill, New York.

138.   **Taylor-Robinson, D., and P. M. Furr.** 1998. Update on sexually transmitted mycoplasmas. Lancet **351:**12-15.

139.   **Taylor-Robinson, D., and A. Keat.** 2001. How can a causal role for small bacteria in chronic inflammatory arthritides be established or refuted? Ann. Rheum. Dis. **60:**177-184.

140.   **Teng, L. J., X. Zheng, J. I. Glass, H. L. Watson, J. Tsai, and G. H. Cassell.** 1994. *Ureaplasma urealyticum* biovar specificity and diversity are encoded in multiple-banded antigen gene. J. Clin. Microbiol. **32:**1464-1469.

141.   **Texier-Maugein, J., M. Clerc, A. Vekris, and C. Bebear.** 1987. *Ureaplasma-urealyticum*-induced bladder stones in rats and their prevention by flurofamide and doxycycline. Isr. J. Med. Sci. **23:**565-567.

142.   **Thomsen, A. C.** 1983. Occurrence and pathogenicity of *Mycoplasma hominis* in the upper urinary tract: a review. Sex. Transm. Dis. **10:**323-326.

143.   **Totten, P. A., M. A. Schwartz, K. E. Sjostrom, G. E. Kenny, H. H. Handsfield, J. B. Weiss, and W. L. H. Whittington.** 2001. Association of *Mycoplasma genitalium* with nongonococcal urethritis in heterosexual men. J. Infect. Dis. **183:**269-276.

144.   **Tully, J. G., D. Taylor-Robinson, R. M. Cole, and D. L. Rose.** 1981. A newly discovered mycoplasma in the human urogenital tract. Lancet **1:**1288-91.

145.   **Uno, M., T. Deguchi, H. Komeda, M. Hayasaki, M. Iida, M. Nagatani, and Y. Kawada.** 1997. *Mycoplasma genitalium* in the cervices of Japanese women. Sex. Transm. Dis. **24:**284-286.

146.   **van der Schee, C., H. J. F. Sluiters, W. van der Meijden, P. van Beek, P. Peerbooms, H. Verbrugh, and A. van Belkum.** 2001. Host and pathogen interaction during vaginal infection by *Trichomonas vaginalis* and *Mycoplasma hominis* or *Ureaplasma urealyticum*. J. Microbiol. Methods **45:**61-67.

147.   **Vittecoq, O., T. Schaeverbeke, S. Favre, A. Daragon, N. Biga, C. Cambon-Michot, C. Bébéar, and X. Le Loet.** 1997. Molecular diagnosis of *Ureaplasma urealyticum* in an immunocompetent patient with destructive reactive polyarthritis. Arthritis Rheum. **40:**2084-2089.

148.   **Vojdani, A., P. C. Choppa, C. Tagle, R. Andrin, B. Samimi, and C. W. Lapp.** 1998. Detection of *Mycoplasma* genus and *Mycoplasma fermentans* by PCR in patients with Chronic Fatigue Syndrome. FEMS Immunol. Med. Microbiol. **22:**355-365.

149.   **Waites, K. B., C. M. Bébéar, J. A. Roberston, D. F. Talkington, and G. E. Kenny.** 2001. Cumitech 34, Laboratory diagnosis of mycoplasmal infections. *In* F. S. Nolte (ed.). American Society for Microbiology, Washington D. C.

150.   **Waites, K. B., P. T. Rudd, D. T. Crouse, K. C. Canupp, K. G. Nelson, C. Ramsey, and G. H. Cassell.** 1988. Chronic *Ureaplasma urealyticum* and *Mycoplasma hominis* infections of central nervous system in preterm infants. Lancet **1:**17-21.

151.   **Wang, E. E., G. H. Cassell, P. J. Sanchez, J. A. Regan, N. R. Payne, and P. P. Liu.** 1993. *Ureaplasma urealyticum* and chronic lung disease of prematurity: critical appraisal of the literature on causation. Clin. Infect. Dis. **17 Suppl. 1:**S112-116.

152.   **Wang, R. Y., J. W. Shih, T. Grandinetti, P. F. Pierce, M. M. Hayes, D. J. Wear, H. J. Alter, and S. C. Lo.** 1992. High frequency of antibodies to *Mycoplasma penetrans* in HIV-infected patients. Lancet **340:**1312-6.

153.   **Wang, R. Y., J. W. Shih, S. H. Weiss, T. Grandinetti, P. F. Pierce, M. Lange, H. J. Alter, D. J. Wear, C. L. Davies, R. K. Mayur, and et al.** 1993. *Mycoplasma penetrans* infection in male homosexuals with AIDS: high seroprevalence and association with Kaposi's sarcoma. Clin. Infect. Dis. **17:**724-729.

154.   **Willoughby, J. J., W. C. Russell, D. Thirkell, and M. G. Burdon.** 1991. Isolation and detection of urease genes in *Ureaplasma urealyticum*. Infect. Immun. **59:**2463-9.

155. **Yanez, A., L. Cedillo, O. Neyrolles, E. Alonso, M. C. Prevost, J. Rojas, H. L. Watson, A. Blanchard, and G. H. Cassell.** 1999. *Mycoplasma penetrans* bacteremia and primary antiphospholipid syndrome. Emerg. Infect. Dis. **5:**164-167.

156. **Yechouron, A., J. Lefebvre, H. G. Robson, D. L. Rose, and J. G. Tully.** 1992. Fatal septicemia due to *Mycoplasma arginini*: a new human zoonosis. Clin. Infect. Dis. **15:**434-438.

157. **Zheng, X., D. A. Olson, J. G. Tully, H. L. Watson, G. H. Cassell, D. R. Gustafson, K. A. Svien, and T. F. Smith.** 1997. Isolation of *Mycoplasma hominis* from a brain abscess. J. Clin. Microbiol. **35:**992-994.

158. **Zheng, X., L. J. Teng, H. L. Watson, J. I. Glass, A. Blanchard, and G. H. Cassell.** 1995. Small repeating units within the *Ureaplasma urealyticum* MB antigen gene encode serovar specificity and are associated with antigen size variation. Infect. Immun. **63:**891-8.

159. **Zheng, X., H. L. Watson, K. B. Waites, and G. H. Cassell.** 1992. Serotype diversity and antigen variation among invasive isolates of *Ureaplasma urealyticum* from neonates. Infect. Immun. **60:**3472-3474.

# Chapter 4

# Mycoplasmas of Animals

JOACHIM FREY
*Institute for Veterinary Bacteriology, University of Bern, Länggassstrasse 122, CH-3012 Bern, Switzerland*

## 1. INTRODUCTION

Among the approximately 200 known different species of *Mollicutes* from animals, only a small number, mainly *Mycoplasma* species, are known as pathogens (Table 1). While pathogenic mycoplasmas play an important role in lifestock production, possibly with massive losses, commensal mycoplasmas which frequently inhabit mucosal tissues of animals without causing a disease, have also to be considered in veterinary medicine. Commensal non-pathogenic mycoplasmas often exhibit marked antigenic similarity with pathogenic mycoplasmas, thus causing serological cross-reactions strongly hampering diagnostic procedures that would be essential for the control and eradication of mycoplasmal diseases.

Pathogenic mycoplasmas have a pronounced affinity for mucous tissues and consequently show a predilection for the respiratory system, mammary gland, serous membranes and the urogenital tract. Most animal mycoplasmoses are chronic diseases with high morbidity and relatively low lethality. The incubation period often varies considerably and can last several weeks or months. Infected animals which do not show any clinical signs of disease, play an important role as carriers and are often the main obstacle in the control and eradication of mycoplasmoses. Pathogenic mycoplasmas are generally adapted to a specific host where they exhibit virulence and cause disease. Colonisation of secondary (atypical) hosts was observed only in a few cases, where mycoplasmas caused merely weak or no apparent symptoms at all. Nevertheless, considering its importance in control and eradication strategies of epidemics, the impact of secondary hosts as

*Molecular Biology and Pathogenicity of Mycoplasmas*, Edited by Razin and Herrmann, Kluwer Academic/Plenum Publishers, New York, 2002

*Table 1.* Major pathogenic Mollicutes of animals

| Animal host / Mollicutes species | Disease |
|---|---|
| **Bovine** | |
| *M. mycoides* subsp. *mycoides* SC | Contagious bovine pleuropneumonia, CBPP |
| *Mycoplasma* sp. bovine group 7 | Pneumonia and arthritis |
| *M. bovis* | Mastitis, pneumonia (calf), polyarthritis (calf), metritis, abortion, sterility |
| *M. dispar* | Pneumonia (calf) |
| *M. californicum* | Mastitis |
| *M. canadense* | Mastitis |
| *M. bovigenitalium* | Mastitis and genital disease (infrequently) |
| *M. bovocculi* | Conjunctivitis |
| *Ureaplasma diversum* | Metritis, sterility, abortion |
| *Eperythrozoon wengonii* | Anemia |
| **Sheep and goat** | |
| *M. capricolum* subsp. *capripneumoniae* | Contagious caprine pleuropneumonia |
| *M. capricolum* subsp. *capricolum* | Mastitis, arthritis |
| *M. mycoides* subsp. *capri* | Pneumonia, mastitis, arthritis, septicemia (goat) |
| *M. mycoides* subsp. *mycoides* LC | Pneumonia, mastitis, arthritis, septicemia (goat) |
| *M. agalactiae* | Infectious agalactia |
| *M. ovipneumoniae* | Pneumonia (lamb) |
| *M. conjunctivae* | Infectious keratoconjunctivitis (IKC) (sheep) |
| **Wild caprinae** | |
| *M. conjunctivae* | Infectious keratoconjunctivitis (ibex, chamois) |
| **Poultry** | |
| *M. gallisepticum* | Chronic respiratory disease (CRD) (chicken), sinusitis, infectious air sacculitis (turkey), |
| *M. synoviae* | Air sacculitis, arthritis, tendosynovitis |
| *M. meleagridis* | Air sacculitis, sinusitis, arthritis (turkey) |
| *M. anseris* | Respiratory tract infections (geese) |
| *Acholeplasma axanthum* | Air sacculitis (geese) |
| **Swine** | |
| *M. hyopneumoniae* | Enzootic pneumonia |
| *M. hyorhinis* | Pneumonia, arthritis |
| *M. hyosynoviae* | Arthritis |
| *Eperythrozoon suis* | Anemia |
| **Horse** | |
| *M. felis* | Pleuritis |
| *M. equirhinis* | |
| *M. equipharyngis* | |
| **Dog and cat** | |
| *M. cynos* | Pneumonia |
| *M. felis* | Conjunctivitis, pneumonia (cat) |
| *Haemobartonella canis* | Anemia (dog) |
| *Haemobartonella felis* | Anemia (cat) |
| **Small rodents** | |
| *M. arthritidis* | Arthritis (rat) |
| *M. pulmonis* | Respiratory & genital tract infections (rat, mouse) |

potential reservoirs of certain pathogenic mycoplasmas remains to be further investigated.

Therapeutic approaches (e.g. use of antibiotics) are mostly limited to temporary improvement of the clinical situation, since they cannot control the disease and eliminate infections in herds. Hence early detection of infected animals and identification of the infectious agent is a prerequisite for successful control and elimination of mycoplasmal diseases. During the last decennium, intensive research efforts have been directed at the molecular characterisation of many important animal mycoplasmas and their phylogenetic relationships (see chapters on Taxonomy and on Phylogeny and Evolution). They have led to the developments of new diagnostic tools and contributed to a significantly improved control of animal mycoplasmoses.

This chapter briefly reviews the major pathogenic mycoplasmas of animals, including the recent progress in research on basic molecular mechanisms of pathogenicity and new developments of molecular tools for diagnosis of mycoplasmal infections in animals.

## 2. MYCOPLASMAS OF RUMINANTS

### 2.1 Bovine mycoplasmas

*Mycoplasma mycoides* subsp. *mycoides* SC (small colony type), the etiological agent of contagious bovine pleuropneumonia (CBPP) affecting cattle and water buffaloes is both of historical importance and high actuality. It was the first Mycoplasma (or Mollicute) described by Nocard and Roux in 1898[18], when it caused severe losses of cattle in Europe. Today, CBPP is considered by the Food and Agriculture Organisation of the United Nations (FAO) as the most important threat to the cattle industry in all parts of Africa, with the exception of Northern Africa. CBPP is considered as the economically most important mycoplasmal disease in the world and is listed among the diseases to be eradicated (List A) by the International Organisation of Animal Diseases (OIE). CBPP is endemic in Africa. In Europe, CBPP re-emerged during the last two decades of the 20th century in Portugal and Italy, with minor outbreaks also in France and Spain, which entailed a significant research activity mainly with the purpose to improve diagnostic methods for the detection of *M. mycoides* subsp. *mycoides* SC directly in live animals or clinical material[7,32,80,88]. Molecular epidemiological studies, based on typing with insertion sequences IS*1296* as probes, revealed that the *M. mycoides* subsp. *mycoides* SC isolated from the re-emerging outbreaks in Europe represented a particular clone that was different from African isolates[27]. However, the origin of this particular European clone, which is characterised by the lack of a specific gene locus of 8.4 kb containing lipoprotein B (LppB) and the genes for uptake of glycerol[49,111], could never be traced back. The discovery of a novel insertion element IS*1634* [113] also allowed to distinguish *M. mycoides* subsp. *mycoides* SC strains of East Africa from those of West Africa[73], indicating two origins of African epidemics of CBPP. Insertion element typing furthermore offered the first reliable method to identify vaccination strains[27]. Recently, a PCR based method was developed for the rapid identification of the most currently used vaccine strain T1[71]. In spite of the high economical importance of CBPP, there is relatively little knowledge on the effectors of virulence of *M. mycoides* subsp. *mycoides* SC. Massive release of $H_2O_2$ by the potentially more virulent African strains, compared to the European strains which are virtually devoid of $H_2O_2$ production[57] is due to the presence of a highly active glycerol uptake system in the African strains[112]. In virulent African strains, glycerol uptake is mediated by a glycerol transport mechanism based on specific ATP-binding cassette (ABC) transporter proteins[112]. This finding led to the interesting hypothesis that mycoplasmas might have adopted sophisticated and highly efficient transport systems and metabolic pathways, allowing them to release highly active or toxic metabolic intermediates to cause inflammation and disease of the host. An interesting observation is the presence of an extraordinarily high number of insertion elements IS*1296* and IS*1634*[113,39] representing up to 7 % of the total genomic DNA of *M. mycoides* subsp. *mycoides* SC, a situation that was never found among other, less pathogenic mycoplasmas. Although *M. mycoides* subsp. *mycoides* SC is specific to bovine, it has also been isolated, as confirmed by molecular methods, from caprine and ovine hosts[99,19] which might act as reservoirs and must be taken into consideration in eradication programmes.

   ***Mycoplasma bovis*** is a relatively new emerging pathogen of cattle with increasing importance particularly as an etiological agent of endemic outbreaks of mastitis and calf pneumonia in industrialized countries[23,43,86]. New molecular tools have consequently been developed during the last decade for the identification and detection of *M. bovis* as well as for epidemiological investigations[22,55,56,67,101]. Virulence of *M. bovis* is assumed to be caused by its potential to induce TNF-α[63], by specific cytadhesive structures[93-95] and by its pronounced capability to outwit the host's immune system by adaptive surface antigen variation[8,38,70,72]. Phenotypic switching of variable surface lipoproteins in *M. bovis* involves high frequency chromosomal rearrangements[72], which are subject to extensive research efforts (see chapter on Antigenic Variation).

*Mycoplasma* **sp. bovine group 7** is a yet unnamed mycoplasma that in general is sporadically isolated from cattle with pneumonia and arthritis. This mycoplasma shares its major surface lipoprotein, LppA, with *M. mycoides* subsp. *mycoides* SC, showing 91 % identical amino acids and, hence, being a source of serological cross-reactions[40]. However, phylogenetic analysis based on sequence determinations of the *rrs* (16S rRNA) gene[90] or an intergenic *rrs-rrl* spacer region[51] as well as a silent potential membrane protein gene[103] clusters *Mycoplasma* sp. bovine group 7 closest to *Mycoplasma capricolum*. Recent outbreaks of mastitis, polyarthritis and abortion caused by *Mycoplasma* sp. bovine group 7 in diary cattle[61] attracted further attention to the importance of this mycoplasma as a pathogen of economical importance.

*Mycoplasma dispar* is the etiological agent of bronchopneumonia of calves. However, its role as a primary pathogen has been questioned in a recent study which showed that healthy calves could not be infected by *M. dispar* alone, suggesting that co-infection with a secondary agent such as *Haemophilus somnus* or *M. bovis* is necessary for *M. dispar* to act as a pathogen[58, 10].

*Eperythrozoon (Haemobartonella) wenyonii* causes anemia in cattle and is indirectly responsible for infertility of bulls and reduced milk production of milk cows. This haemotrophic organism originally classified as *Rickettsia* was shown recently to belong to the mycoplasmas based on phylogenetic analysis of its *rrs* (16S rRNA) genes[62,85]. It is now proposed as '*Candidatus* Mycoplasma wenyonii'[85a].

## 2.2 Mycoplasmas of sheep, goats and wild caprinae

Mycoplasmal diseases of small ruminants are spread world-wide and constitute an important socio-economical problem, particularly in regions where small ruminants represent a substantial source of milk and meat provision for the population.

*Mycoplasma capricolum subsp. capripneumoniae*, formerly known as the F38-group of mycoplasmas[69], is the agent of contagious caprine pleuropneumonia (CCPP). It causes severe infections with high morbidity and mortality of goats particularly in Africa, the Near East and Asia. Isolation and cultivation of *Mycoplasma capricolum* subsp. *capripneumoniae* is most difficult. The recently developed molecular genetic methods have significantly improved detection and identification[6,16] as well as evolutionary and epidemiological investigations[52, 89, 103].

*Mycoplasma mycoides* **subsp.** *capri* and *Mycoplasma mycoides* **subsp.** *mycoides* **LC** mainly cause pneumonia in sheep and goats and are also associated with mastitis and arthritis of sheep and goats as well as septicemia in goats. Both subspecies are indistinguishable by *rrs* (16S rRNA) sequence analysis[90] or an intergenic *rrs-rrl* spacer region[51] or a silent potential membrane gene[103]. Moreover, both *M. mycoides* subsp. *capri and M. mycoides* subsp. *mycoides* LC share the same major lipoprotein A (LppA)[82]. These two mycoplasmas being almost identical, a working group of the European COST (cooperation in science and technology) Action 826 on "Ruminants' Mycoplasmoses" recommended to refer to both of them as *M. mycoides* subsp. *capri*.

*Mycoplasma agalactiae*, the agent of infectious agalactia of sheep and goat is mainly spread in the Mediterranean area. Infections with *M. agalactiae* cause a sudden decrease in milk production and mastitis. *M. agalactiae* strains were, however, also isolated from bulk milk of healthy flocks. These apparently "avirulent" strains were shown to be fully infectious, indicating the importance of strictly controlling also apparently healthy flocks[96] , since the disease is transmitted by direct contact between animals as well as by milking. Diagnosis is based on detection of *M. agalactiae* in milk, particularly bulk milk, for which several powerful PCR methods have been developed recently[33,104]. Serological diagnosis of *M. agalactiae* is currently done with specific extracts of *M. agalactiae* antigens ELISA[68] as well as with monoclonal antibodies[91]. Due to the high antigenic variability of *M. agalactiae*, however, serological methods based on specific stable antigens are currently being developed[92]. *M. agalactiae* is phylogenetically very closely related to *M. bovis*[3,75]. The two species cannot be separated based on their *rrs* (16S rRNA) gene sequences and require methods based on other housekeeping genes (e.g. *uvrC*) for genetic identification[101]. Similar to *M. bovis*, *M. agalactiae* shows a high potential for antigenic variability of surface lipoproteins due to high-frequency DNA rearrangements of genes coding for variable surface proteins[11,37,47,98] which is discussed in detail in the chapter on antigenic variation. The genetic locus *vspA* of the variable antigen A has been mapped on the physical map of *M. agalactiae* strain PG2[105]. By using monoclonal antibodies recognizing permanently expressed epitopes, it was shown that the antigenic variability of *M. agalactiae* is not only based on frequently acting surface antigen switching but also upon differences of stable epitopes which can be related partially to the geographic origin[11]. Vaccination of sheep with cultured *M. agalactiae* inactivated by different methods showed variable success[21,106,114]. A new generation of vaccines, however, requires a detailed basic knowledge of the functions of the various antigenic determinants to define ideal targets for the immune defence against *M. agalactiae* infections and the ideal sub-units for the vaccine respectively.

*Mycoplasma conjunctivae* causes infectious keratoconjunctivitis (IKC) in sheep with mild to moderate symptoms[84], and in wild caprinae (chamois and ibex) in the alps where the disease takes a severe progression often leading to death of the animals[50,76]. Direct detection of *M. conjunctivae* in the lachrymal fluid by PCR[44] and serological methods[34,9] allowed broad epidemiological investigations of IKC and indicate transmission of IKC to wild caprinae by sheep where the disease is widely spread.

## 3. MYCOPLASMAS OF POULTRY

Various species of mycoplasmas are involved in respiratory diseases of domestic poultry, often in interaction with viruses and bacteria colonising the respiratory tract. *Mycoplasma gallisepticum* and *Mycoplasma meleagridis* are the primary respiratory mycoplasmal pathogens of poultry, causing significant losses in poultry production. Furthermore *Mycoplasma synoviae* causing contagious arthritis, which leads to lameness of chicken and turkey, and to some extent *Mycoplasma iowae*, which is responsible for embryo mortality of chicken and turkey, belong also to economically important poultry mycoplasmas. Detection and identification of these most important avian mycoplasmas are currently possible by means of specific PCR methods[36, 81, 115].

*Mycoplasma gallisepticum* is the etiological agent of chronic respiratory diseases of chicken and infectious sinusitis in turkeys, spread world-wide and causing most severe losses in poultry breeding. Infection with *M. gallisepticum* occurs laterally from infected chickens or by infected eggs. The infection can remain without symptoms in chicken and be induced by stress, followed by rapid spread throughout the flock through aerosol transmission. Considering the economical importance of this pathogen, most research on avian mycoplasmas was focused on *M. gallisepticum* during the last ten years. Surface located proteins involved in cytadherence, particularly the cytadhesin GapA,[48,54,81,87] and hemagglutinin pMGA[74] play an important role in pathogenicity and tenacity of *M. gallisepticum*. Rapid switching of the expression of hemagglutinin pMGA[45,46], and antigenic variation of cytadhesins[4,15,41] allow *M. gallisepticum* to escape the host immune system. A central virulence attribute of *M. gallisepticum* is its capacity to invade eukaryotic cells by a regulated mechanism and to survive within invaded cells, thus providing *M. gallisepticum* the opportunity to resist the host's immune defence and antibiotic therapy[117]. These new findings render *M. gallisepticum* a valuable model to study mycoplasmal cell invasion and intracellular parasitism. Antibiotic therapy is inefficient in controlling the disease, presumably due to the intracellular state of *M. gallisepticum*. Vaccination using colonizing live vaccine strains showed that virulent *M. gallisepticum* could be repressed in flocks[1,65,107]. However, in the view of the possibility of the intracellular state of some *M. gallisepticum* organisms, it remains to be determined how far full eradication of virulent strains is possible by vaccination.

## 4.    PORCINE MYCOPLASMAS

The most frequently encountered *Mollicutes* in pigs are *Mycoplasma hyopneumoniae*, the etiological agent of porcine enzootic pneumonia, *Mycoplasma flocculare*, *Mycoplasma hyorhinis*, *Mycoplasma hyosynoviae* and *Eperythrozoon suis* (formerly classified as *Rickettsia* and now assigned to the *Mollicutes* based on phylogenetic criteria[62,85]). Among these *Mollicutes*, *M. hyopneumoniae* is of highest importance as a pathogen, while *M. flocculare* is considered to be of low epidemiological importance but might interfere in diagnosis of enzootic pneumonia due to its close antigenic and phylogenetic relationship with *M. hyopneumoniae*. *M. hyorhinis* and *M. hyosynoviae* are only occasionally found in arthritis.

***Eperythrozoon suis*** is spread world-wide and causes fever and anemia in young animals and, sometimes, latent anemia in older animals[53]. New PCR-based methods are currently available for the detection and identification of *E. suis*[78]. *E. suis*, originally classified as *Rickettsia*, was recently shown to belong to the mycoplasmas based on phylogenetic analysis of its *rrs* (16S rRNA) genes and is now proposed as '*Candidatus* Mycoplasma haemosuis'[85a].

***Mycoplasma hyopneumoniae*** causes porcine enzootic pneumonia world-wide in industrialized livestock husbandry. Infections with *M. hyopneumoniae* often remain inapparent over long periods and become suddenly apparent upon secondary bacterial or viral infections. The virulence of *M. hyopneumoniae* seems to be rather low, as demonstrated by experimental infections. However, economical losses due to *M. hyopneumoniae* infections are estimated to be significant. This led to intensive research and development of vaccines[20,35,64,66] and sensitive methods for detection of the organism in clinical material and samples taken from live animals and from the environment[2,13,100,109,110]. These latter developments enabled significant improvement in the control and eradication programmes of enzootic pneumonia. The most striking effects observed upon infection with *M. hyopneumoniae* were ciliostatis and the marked loss of cilia in the mid-trachea and bronchi surfaces[14,31]. Attachment of *M. hyopneumoniae* to the tips of cilia, which is prerequisite of ciliostasis and an effacement of cilia, is due to the P97 adhesin in combination with other specific factors encoded by the cilium adhesin operon[59,60,79,116]. Besides ciliar adhesion, *M. hyopneumoniae* seems to require transport systems based on ATP-binding cassette (ABC) transporter proteins in order to be virulent. Spontaneous loss of a gene of an *M. hyopneumoniae* ABC transporter was correlated with the loss of virulence in experimental infections[13]. Although the function of this ABC transporter and its role in virulence is yet unknown, this observation is of high interest in view of the impact of ABC transporters for glycerol and fructose on the virulence of *M. mycoides* subsp. *mycoides* SC[112] and *Spiroplasma citri*[42], respectively, and may lead to the detection of a new branch of effectors for pathogenicity of *M. hyopneumoniae*.

## 5.    EQUINE MYCOPLASMAS

*Mycoplasma felis* is occasionally isolated from horses with severe pleuritis and lower respiratory tract diseases. *M. felis* was isolated and also serologically tested in the course of a severe outbreak of lower respiratory tract disease in horses, where it was found to be the main etiological agent of the disease[118].

*Mycoplasma equirhinis* and *Mycoplasma equipharyngis* frequently isolated from the respiratory tract of horses with non-specific respiratory diseases[24] do, in contrast to *M. felis,* not seem to be primary pathogens, since they are often isolated from healthy horses too.

## 6.    MYCOPLASMAS OF DOGS AND CATS

Most of the mycoplasmas isolated from dogs and cats seem to be normal inhabitants of the respiratory and genital tracts. However, *Mycoplasma cynos* seems to be associated with pneumonia of dogs, and *M. felis* is assumed to cause conjunctivitis and respiratory infections in cats. *M. felis* was also found in immuno-deficient humans with septic arthritis, and was supposed to be transmitted from cats[17].

*Haemobartonella canis* and *Haemobartonella felis* (formerly classified as *Rickettsia* and now, based on phylogenetic criteria, assigned to the *Mollicutes*[62,85]) are haemotrophic organisms of dogs and cats respectively, causing fever and persistent anemia, particularly in cats[10]. A PCR method based on the *rrs* (16S rRNA) gene of *H. felis* was developed for the detection and identification of *H. felis* in blood samples from cats[77]. Based on phylogenetic studies, *H. felis* is now proposed as '*Candidatus* Mycoplasma haemofelis' [85a].

## 7.    MYCOPLASMAS OF SMALL RODENTS

*Mycoplasma pulmonis* is the etiological agent of murine respiratory mycoplasmosis (MRM), a disease of laboratory rats and mice. Specific pathogen-free (SPF) laboratory animals must be free of MRM. *M. pulmonis* is considered as a particularly powerful model for studying mycoplasmal respiratory infections. Since it is a natural pathogen of mice and rats, it allows to study the role of innate immunity and humoral immunity in the defence of the lungs against colonization and against dissemination of the infection[25] and to search for basic principles of vaccination against mycoplasmas causing respiratory infections[28]. Most recently, the complete genome sequence of *M. pulmonis* strain UAB CT has been established[26], which is a significant step forward in the research on molecular mechanisms of pathogenicity of mycoplasmas. Genome sequence analysis revealed genes for potential virulence factors of *M. pulmonis*, including a hemolysin similar to *hlyA* of *Brachyspira (Sperulina) hyodysenteriae* and certain mycobacteria, secreted nucleases, glycoproteases and a thiol peroxidase[26]. This is of particular interest since the production of hydrogen peroxide is suggested to be a virulence factor of mycoplasmas (see also *M. mycoides* subsp. *mycoides* SC in this chapter). Furthermore, *M. pulmonis* serves as an ideal model to study antigenic variation, which is considered as a key factor in the ability of mycoplasmas to establish chronic infections[97].

*Mycoplasma arthritidis* causes acute polyarthritis in rats experimentally infected with large doses of *M. arthritidis*. Its role as a natural pathogen is not established since *M. arthritidis* is often found in colonies of rats without arthritic symptoms. The experimental infection of rats with *M. arthritidis,* however, serves as an important model to study the trigger molecules such as superantigens involved in the induction of inflammatory mediators and their role in inflammation and immune pathologies[5,12,29,30,83,108] (see chapter *Mycoplasma arthritidis* Pathogenicity: Membranes, MAM and MAV1).

## 8.    CONCLUSIONS

Intensive molecular genetic and biochemical research on animal mycoplasmas carried out during the last 10 years has significantly improved the knowledge of their phylogenetic relationships, the characterization of many of their major antigens and the mechanism of their variability. Furthermore, several basic biological mechanisms have been unravelled and led to a better understanding of the pathogenicity of several mycoplasmas. These perceptions allowed the development of new efficient methods for the detection and identification of important mycoplasmal pathogens and for the serological diagnosis of important animal diseases. The resulting new knowledge from genomic and proteomic analyses of animal mycoplasmas will be essential for a thorough understanding of their molecular mechanisms of pathogenicity and the development of novel targeted vaccines.

## ACKNOWLEDGMENTS

I am grateful to Pia Wyssenbach and Edy Vilei for their valuable help in preparation of this manuscript. This study is part of the European COST Action 826 "ruminants' mycoplasmoses" and was supported by grant no. C96.0073 of the Swiss Ministry of Education and Science and by the Swiss Federal Veterinary Office.

## REFERENCES

1.  **Abd-el-Motelib, T. Y. and S. H. Kleven.** 1993. A comparative study of *Mycoplasma gallisepticum* vaccines in young chickens. *Avian Dis.* **37**:981-987
2.  **Artiushin, S., L. Stipkovits, and F. C. Minion.** 1993. Development of polymerase chain reaction primers to detect *Mycoplasma hyopneumoniae. Mol. Cell. Probes* **7**:381-385
3.  **Askaa, G. and H. Erno.** 1976. Elevation of *Mycoplasma agalactiae* subsp. *bovis* to species rank: *Mycoplasma bovis* (Hale et al.) comp. nov. *Int. J. Syst. Bact.* **26**:323-325
4.  **Athamna, A., R. Rosengarten, S. Levisohn, I. Kahane, and D. Yogev.** 1997. Adherence of *Mycoplasma gallisepticum* involves variable surface membrane proteins. *Infect. Immun.* **65**:2468-2471
5.  **Atkin, C. L., S. Wei, and B. C. Cole.** 1994. The *Mycoplasma arthritidis* superantigen MAM: purification and identification of an active peptide. *Infect. Immun.* **62**:5367-5375
6.  **Bascunana, C. R., J. G. Mattsson, G. Bolske, and K. E. Johansson.** 1994. Characterization of the 16S rRNA Genes from *Mycoplasma* sp Strain F38 and development of an identification system based on PCR. *J. Bacteriol.* **176**:2577-2586
7.  **Bashiruddin, J. B., P. DeSantis, A. Vacciana, and F. G. Santini.** 1999. Detection of *Mycoplasma mycoides* subspecies *mycoides* SC in clinical material by a rapid colorimetric PCR. *Mol. Cell. Probes* **13**:23-28
8.  **Behrens, A., M. Heller, H. Kirchhoff, D. Yogev, and R. Rosengarten.** 1994. A family of phase- and size-variant membrane surface lipoprotein antigens (Vsps) of *Mycoplasma bovis. Infect. Immun.* **62**:5075-5084
9.  **Belloy, L., M. Giacometti, E.-M. Abdo, J. Nicolet, M. Krawinkler, M. Janovsky, U. Bruderer, and J. Frey.** 2001. Detection of specific *Mycoplasma conjunctivae* antibodies in the sera of sheep with infectious keratoconjunctivitis. *Vet. Res.* **32**:155-164
10.  **Berent, L. M., J. B. Messick, and S. K. Cooper.** 1998. Detection of *Haemobartonella felis* in cats with experimentally induced acute and chronic infections, using a polymerase chain reaction assay. *Amer. J. Vet. Res.* **59**:1215-1220

11.  **Bergonier, D., F. De Simone, P. Russo, M. Solsona, M. Lambert, and F. Poumarat.** 1996. Variable expression and geographic distribution of *Mycoplasma agalactiae* surface epitopes demonstrated with monoclonal antibodies. *FEMS Microbiol. Lett.* **143**:159-165

12.  **Bhardwaj, N., A. S. Hodtsev, A. Nisanian, S. Kabak, S. M. Friedman, B. C. Cole, and D. N. Posnett.** 1994. Human T-cell responses to *Mycoplasma arthritidis*-derived superantigen. *Infect. Immun.* **62**:135-144

13.  **Blanchard, B., M. Kobisch, J. M. Bove, and C. Saillard.** 1996. Polymerase chain reaction for *Mycoplasma hyopneumoniae* detection in tracheobronchiolar washings from pigs. *Mol. Cell. Probes* **10**:15-22

14.  **Blanchard, B., M. M. Vena, A. Cavalier, J. Le Lannic, J. Gouranton, and M. Kobisch.** 1992. Electron microscopic observation of the respiratory tract of SPF piglets inoculated with *Mycoplasma hyopneumoniae*. *Vet. Microbiol.* **30**:329-341

15.  **Boguslavsky, S., D. Menaker, I. Lysnyansky, T. Liu, S. Levisohn, R. Rosengarten, M. Garcia, and D. Yogev.** 2000. Molecular characterization of the *Mycoplasma gallisepticum pvpA* gene which encodes a putative variable cytadhesin protein. *Infect. Immun.* **68**:3956-3964

16.  **Bolske, G., J. G. Mattsson, C. R. Bascunana, K. Bergstrom, H. Wesonga, and K. E. Johansson.** 1996. Diagnosis of contagious caprine pleuropneumonia by detection and identification of *Mycoplasma capricolum* subsp. *capripneumoniae* by PCR and restriction enzyme analysis. *J. Clin. Microbiol.* **34**:785-791

17.  **Bonilla, H. F., C. E. Chenoweth, J. G. Tully, L. K. Blythe, J. A. Robertson, V. M. Ognenovski, and C. A. Kauffman.** 1997. *Mycoplasma felis* septic arthritis in a patient with hypogammaglobulinemia. *Clin. Infect. Dis.* **24**:222-225

18.  **Bové, J. M.** 1999. The one-hundredth anniversary of the first culture of a mollicute, the contagious bovine peripneumonia microbe, by Nocard and Roux, with the collaboration of Borrel, Salimbeni, and Dujardin-Baumetz. *Res. Microbiol.* **150**:239-245

19.  **Brandao, E.** 1995. Isolation and identification of *Mycoplasma mycoides* subspecies *mycoides* SC strains in sheep and goats. *Vet. Rec.* **136**:98-99

20.  **Brooks, E. and D. Faulds.** 1989. The *Mycoplasma hyopneumoniae* 74.5-kD antigen elicits neutralizing antibodies and shares sequence similarity with heat-shock proteins. In: *Vaccines89 – Modern Approaches to New Vaccines Including Prevention of AIDS* (R.A. Lerner, H. Ginsberg, R.M. Chanock and F. Brown, eds.), Cold Spring Harbor Laboratory, Cold Spring Harbor N.Y., pp. 265-267

21.  **Buonavoglia, D., A. Fasanella, P. Sagazio, M. Tempesta, G. Iovane, and C. Buonavoglia.** 1998. Persistence of antibodies to *Mycoplasma agalactiae* in vaccinated sheep. *Microbiologica* **21**:209-212

22.  **Butler, J. A., C. C. Pinnow, J. U. Thomson, S. Levisohn, and R. F. Rosenbusch.** 2001. Use of arbitrarily primed polymerase chain reaction to investigate *Mycoplasma bovis* outbreaks. *Vet. Microbiol.* **78**:175-181

23.  **Byrne, W. J., H. J. Hall, R. McCormack, and N. Brice.** 1998. Elimination of *Mycoplasma bovis* mastitis from an Irish dairy herd. *Vet. Rec.* **142**:516-517

24.  **Carman, S., S. Rosendal, L. Huber, C. Gyles, S. McKee, R. A. Willoughby, E. Dubovi, J. Thorsen, and D. Lein.** 1997. Infectious agents in acute respiratory disease in horses in Ontario. *J. Vet. Diagn. Invest.* **9**:17-23

25.  **Cartner, S. C., J. R. Lindsey, J. Gibbs-Erwin, G. H. Cassell, and J. W. Simecka.** 1998. Roles of innate and adaptive immunity in respiratory mycoplasmosis. *Infect. Immun.* **66**:3485-3491

26.  **Chambaud, I., R. Heilig, S. Ferris, V. Barbe, D. Samson, F. Galisson, I. Moszer, K. Dybvig, H. Wróblewski, A. Viari, E. P. C. Rocha, and A. Blanchard.** 2001. The complete genome sequence of the murine respiratory pathogen *Mycoplasma pulmonis*. Nucleic Acids Research **29**, 2145-2153.

27. **Cheng, X., J. Nicolet, F. Poumarat, J. Regalla, F. Thiaucourt, and J. Frey.** 1995. Insertion element IS*1296* in *Mycoplasma mycoides* subsp. *mycoides* small colony identifies a European clonal line distinct from African and Australian strains. *Microbiology* **141**:3221-3228

28. **Cimolai, N., D. G. Mah, G. P. Taylor, and B. J. Morrison.** 1995. Bases for the early immune response after rechallenge or component vaccination in an animal model of acute *Mycoplasma pneumoniae* pneumonitis. *Vaccine* **13**:305-309

29. **Cole, B. C. and C. L. Atkin.** 1995. Identification, characterization, and purification of mycoplasmal superantigens. In: *Molecular and Diagnostic Procedures in Mycoplasmology* (S. Razin and J.G. Tully, eds.), Academic Press, San Diego, pp. 439-449

30. **Cole, B. C. and A. Sawitzke.** 1995. Mycoplasmas, superantigens and autoimmune arthritis. In: *Mechanisms and Models in Rheumatoid Arthritis,* Academic Press, San Diego, pp. 47-66

31. **Debey, M. C. and R. F. Ross.** 1994. Ciliostasis and loss of cilia induced by *Mycoplasma hyopneumoniae* in porcine tracheal organ cultures. *Infect. Immun.* **62**:5312-5318

32. **Dedieu, L., V. Mady, and P. C. Lefevre.** 1994. Development of a selective polymerase chain reaction assay for the detection of *Mycoplasma mycoides* subsp *mycoides* SC (contagious bovine pleuropneumonia agent). *Vet. Microbiol.* **42**:327-339

33. **Dedieu, L., V. Mady, and P. C. Lefevre.** 1995. Development of two PCR assays for the identification of mycoplasmas causing contagious agalactia. *FEMS Microbiol. Lett.* **129**:243-249

34. **Degiorgis, M. P., E.-M. Abdo, J. Nicolet, J. Frey, D. Mayer, and M. Giacometti.** 2000. Immune responses to *Mycoplasma conjunctivae* in Alpine ibex, Alpine chamois, and domestic sheep in Switzerland. *J. Wildl. Dis.* **36**:265-271

35. **Djordjevic, S. P., G. J. Eamens, L. F. Romalis, P. J. Nicholls, V. Taylor, and J. Chin.** 1997. Serum and mucosal antibody responses and protection in pigs vaccinated against *Mycoplasma hyopneumoniae* with vaccines containing a denatured membrane antigen pool and adjuvant. *Aust. Vet. J.* **75**:504-511

36. **Fan, H. H., S. H. Kleven, and M. W. Jackwood.** 1995. Application of polymerase chain reaction with arbitrary primers to strain identification of *Mycoplasma gallisepticum. Avian Dis.* **39**:729-735

37. **Flitman-Tene, R., S. Levisohn, I. Lysnyansky, E. Rapoport, and D. Yogev.** 2000. A chromosomal region of *Mycoplasma agalactiae* containing *vsp*-related genes undergoes in vivo rearrangement in naturally infected animals. *FEMS Microbiol. Lett.* **191**:205-212

38. **Flitman-Tene, R., T. R., S. Levisohn, R. Rosenbusch, E. Rapoport, and D. Yogev.** 1997. Genetic variation among *Mycoplasma agalactiae* isolates detected by the variant surface lipoprotein gene (*vspA*) of *Mycoplasma bovis*. *FEMS Microbiol. Lett.* **156**:123-128

39. **Frey, J., X. Cheng, P. Kuhnert, and J. Nicolet.** 1995. Identification and characterization of IS*1296* in *Mycoplasma mycoides* subsp. *mycoides* SC and presence in related mycoplasmas. *Gene* **160**:95-100

40. **Frey, J., X. Cheng, M. P. Monnerat, E.-M. Abdo, M. Krawinkler, G. Bolske, and J. Nicolet.** 1998. Genetic and serological analysis of the immunogenic 67-kDa lipoprotein of *Mycoplasma* sp. bovine group 7. *Res. Microbiol.* **149**:55-64

41. **Garcia, M., M. G. Elfaki, and S. H. Kleven.** 1994. Analysis of the variability in expression of *Mycoplasma gallisepticum* surface antigens. *Vet. Microbiol.* **42**:147-158

42. **Gaurivaud, P., J. L. Danet, F. Laigret, M. Garnier, and J. M. Bove.** 2000. Fructose utilization and phytopathogenicity of *Spiroplasma citri*. *Mol. Plant Microbe Interact.* **13**:1145-1155

43.    Ghadersohi, A., R. G. Hirst, J. ForbesFaulkener, and R. J. Coelen. 1999.
       Preliminary studies on the prevalence of *Mycoplasma bovis* mastitis in dairy cattle in
       Australia. *Vet. Microbiol.* **65**:185-194
44.    Giacometti, M., J. Nicolet, K. E. Johansson, T. Naglic, M. P. Degiorgis, and J.
       Frey. 1999. Detection and identification of *Mycoplasma conjunctivae* in infectious
       keratoconjunctivitis by PCR based on the 16S rRNA gene. *Zentralbl. Veterinarmed.*
       *[B]* **46**:173-180
45.    Glew, M. D., N. Baseggio, P. F. Markham, G. F. Browning, and I. D. Walker.
       1998. Expression of the *pMGA* genes of *Mycoplasma gallisepticum* is controlled by
       variation in the GAA trinucleotide repeat lengths within the 5' noncoding regions.
       *Infect. Immun.* **66**:5833-5841
46.    Glew, M. D., G. F. Browning, P. F. Markham, and I. D. Walker. 2000. *pMGA*
       phenotypic variation in *Mycoplasma gallisepticum* occurs in vivo and is mediated by
       trinucleotide repeat length variation. *Infect. Immun.* **68**:6027-6033
47.    Glew, M. D., L. Papazisi, F. Poumarat, D. Bergonier, R. Rosengarten, and C.
       Citti. 2000. Characterization of a multigene family undergoing high-frequency DNA
       rearrangements and coding for abundant variable surface proteins in *Mycoplasma
       agalactiae*. *Infect. Immun.* **68**:4539-4548
48.    Goh, M. S., T. S. Gorton, M. H. Forsyth, K. E. Troy, and S. J. Geary. 1998.
       Molecular and biochemical analysis of a 105 kDa *Mycoplasma gallisepticum*
       cytadhesin (GapA). *Microbiology* **144**:2971-2978
49.    Gonçalves, R., J. Regalla, J. Nicolet, J. Frey, R. Nicholas, J. Bashiruddin, P.
       DeSantis, and A. P. Gonçalves. 1998. Antigen heterogeneity among *Mycoplasma
       mycoides* subsp. *mycoides* SC isolates: discrimination of major surface proteins. *Vet.
       Microbiol.* **63**:13-28
50.    Grattarola, C., J. Frey, E.-M. Abdo, R. Orusa, J. Nicolet, and M. Giacometti.
       1999. *Mycoplasma conjunctivae* infections in chamois and ibexes affected with
       infectious keratoconjunctivitis in the Italian Alps. *Vet. Rec.* **145**:588-589
51.    Harasawa, R., H. Hotzel, and K. Sachse. 2000. Comparison of the 16S-23S rRNA
       intergenic spacer regions among strains of the *Mycoplasma mycoides* cluster, and
       reassessment of the taxonomic position of *Mycoplasma* sp. bovine group 7. *Int. J. Syst.
       Evol. Microbiol.* **50** Pt 3:**1325-9**:1325-1329
52.    Heldtander, M., H. Wesonga, G. Bolske, B. Pettersson, and K. Johansson. 2001.
       Genetic diversity and evolution of *Mycoplasma capricolum* subsp. *capripneumoniae*
       strains from eastern Africa assessed by 16S rDNA sequence analysis. *Vet. Microbiol.*
       **78**:13-28
53.    Henderson, J. P., J. O'Hagan, S. M. Hawe, and M. C. Pratt. 1997. Anaemia and
       low viability in piglets infected with *Eperythrozoon suis*. *Vet. Rec.* **140**:144-146
54.    Hnatow, L. L., C. L. Keeler, Jr., L. L. Tessmer, K. Czymmek, and J. E. Dohms.
       1998. Characterization of MGC2, a *Mycoplasma gallisepticum* cytadhesin with
       homology to the *Mycoplasma pneumoniae* 30-kilodalton protein P30 and *Mycoplasma
       genitalium* P32. *Infect. Immun.* **66**:3436-3442
55.    Hotzel, H., M. Heller, and K. Sachse. 1999. Enhancement of *Mycoplasma bovis*
       detection in milk samples by antigen capture prior to PCR. *Mol. Cell. Probes* **13**:175-
       178
56.    Hotzel, H., K. Sachse, and H. Pfutzner. 1996. Rapid detection of *Mycoplasma bovis*
       in milk samples and nasal swabs using the polymerase chain reaction. *J. Appl.
       Bacteriol.* **80**:505-510
57.    Houshaymi, B. M., R. J. Miles, and R. A. Nicholas. 1997. Oxidation of glycerol
       differentiates African from European isolates of *Mycoplasma mycoides* subspecies
       *mycoides* SC (small colony). *Vet. Rec.* **140**:182-183
58.    Howard C.J., Taylor G., J. Collins, and Gourlay R.N. 1976. Interaction of
       *Mycoplasma dispar* and *Mycoplasma agalactiae* subsp. *bovis* with bovine alveolar

macrophages and bovine lacteal polymorphonuclear leukocytes. *Infect. Immun.* **14**:11-17

59. **Hsu, T. and F. C. Minion.** 1998. Identification of the cilium binding epitope of the *Mycoplasma hyopneumoniae* P97 adhesin. *Infect. Immun.* **66**:4762-4766

60. **Hsu, T. and F. C. Minion.** 1998. Molecular analysis of the P97 cilium adhesin operon of *Mycoplasma hyopneumoniae*. *Gene* **214**:13-23

61. **Hum, S., A. Kessell, S. Djordjevic, R. Rheinberger, M. Hornitzky, W. Forbes, and J. Gonsalves.** 2000. Mastitis, polyarthritis and abortion caused by *Mycoplasma* species *bovine group* 7 in dairy cattle. *Aust. Vet. J.* **78**:744-750

62. **Johansson, K. E., J. G. Tully, G. Bolske, and B. Pettersson.** 1999. *Mycoplasma cavipharyngis* and *Mycoplasma fastidiosum*, the closest relatives to *Eperythrozoon* spp. and *Haemobartonella* spp. *FEMS Microbiol. Lett.* **174**:321-326

63. **Jungi, T. W., M. Krampe, M. Sileghem, C. Griot, and J. Nicolet.** 1996. Differential and strain-specific triggering of bovine alveolar macrophage effector functions by mycoplasmas. *Microb. Pathog.* **21**:487-498

64. **King, K. W., D. H. Faulds, E. L. Rosey, and R. J. Yancey, Jr.** 1997. Characterization of the gene encoding Mhp1 from *Mycoplasma hyopneumoniae* and examination of Mhp1's vaccine potential. *Vaccine* **15**:25-35

65. **Kleven, S. H., H. H. Fan, and K. S. Turner.** 1998. Pen trial studies on the use of live vaccines to displace virulent *Mycoplasma gallisepticum* in chickens. *Avian Dis.* **42**:300-306

66. **Kobisch, M., L. Quillien, J. P. Tillon, and H. Wroblewski.** 1987. The *Mycoplasma hyopneumoniae* plasma membrane as a vaccine against porcine enzootic pneumonia. *Ann. Inst. Pasteur* **138**:693-705

67. **Kusiluka, L. J., B. Kokotovic, B. Ojeniyi, N. F. Friis, and P. Ahrens.** 2000. Genetic variations among *Mycoplasma bovis* strains isolated from danish cattle. *FEMS Microbiol. Lett.* **192**:113-118

68. **Lambert, M., M. Calamel, P. Dufour, E. Cabasse, C. Vitu, and V. Pepin.** 1998. Detection of false-positive sera in contagious agalactia with a multiantigen ELISA and their elimination with a protein G conjugate. *J. Vet. Diagn. Invest.* **10**:326-330

69. **Leach, R. H., H. Erno, and K. J. Macowan.** 1993. Proposal for designation of F38-type caprine mycoplasmas as *Mycoplasma capricolum* subsp *capripneumoniae* subsp. nov and consequent obligatory relegation of strains currently classified as *M. capricolum* (Tully, Barile, Edward, Theodore, and Erno 1974) to an additional new subspecies, *M. capricolum* subsp. *capricolum* subsp. nov. *Int. J. Syst. Bact.* **43**:603-605

70. **LeGrand, D. L., M. Solsona, R. Rosengarten, and F. Poumarat.** 1996. Adaptive surface antigen variation in *Mycoplasma bovis* to the host immune response. *FEMS Microbiol. Lett.* **144**:267-275

71. **Lorenzon, S., A. David, M. Nadew, H. Wesonga, and F. Thiaucourt.** 2000. Specific PCR identification of the T1 vaccine strains for contagious bovine pleuropneumonia. *Mol. Cell. Probes* **14**:205-210

72. **Lysnyansky, I., R. Rosengarten, and D. Yogev.** 1996. Phenotypic switching of variable surface lipoproteins in *Mycoplasma bovis* involves high-frequency chromosomal rearrangements. *J. Bacteriol.* **178**:5395-5401

73. **March, J. B., J. Clark, and M. Brodlie.** 2000. Characterization of strains of *Mycoplasma mycoides* subsp. *mycoides* small colony type isolated from recent outbreaks of contagious bovine pleuropneumonia in Botswana and Tanzania: evidence for a new biotype. *J. Clin. Microbiol.* **38**:1419-1425

74. **Markham, P. F., M. D. Glew, M. R. Brandon, I. D. Walker, and K. G. Whithear.** 1992. Characterization of a major hemagglutinin protein from *Mycoplasma gallisepticum*. *Infect. Immun.* **60**:3885-3891

75.  **Mattsson, J. G., B. Guss, and K. E. Johansson.** 1994. The phylogeny of *Mycoplasma bovis* as determined by sequence analysis of the 16S rRNA gene. *FEMS Microbiol. Lett.* **115:**325-328

76.  **Mayer, D., M. P. Degiorgis, W. Meier, J. Nicolet, and M. Giacometti.** 1997. Lesions associated with infectious keratoconjunctivitis in Alpine ibex. *J. Wildlife Dis.* **33:**413-419

77.  **Messick, J. B., L. M. Berent, and S. K. Cooper.** 1998. Development and evaluation of a PCR-based assay for detection of *Haemobartonella felis* in cats and differentiation of *H. felis* from related bacteria by restriction fragment length polymorphism analysis. *J. Clin. Microbiol.* **36:**462-466

78.  **Messick, J. B., S. K. Cooper, and M. Huntley.** 1999. Development and evaluation of a polymerase chain reaction assay using the 16S rRNA gene for detection of *Eperythrozoon suis* infection. *J. Vet. Diagn. Invest.* **11:**229-236

79.  **Minion, F. C., C. Adams, and T. Hsu.** 2000. R1 region of P97 mediates adherence of *Mycoplasma hyopneumoniae* to swine cilia. *Infect. Immun.* **68:**3056-3060

80.  **Miserez, R., T. Pilloud, X. Cheng, J. Nicolet, C. Griot, and J. Frey.** 1997. Development of a sensitive nested PCR method for the specific detection of *Mycoplasma mycoides* subsp *mycoides* SC. *Mol. Cell. Probes* **11:**103-111

81.  **Moalic, P. Y., F. Gesbert, and I. Kempf.** 1998. Utility of an internal control for evaluation of a *Mycoplasma meleagridis* PCR test. *Vet. Microbiol.* **61:**41-49

82.  **Monnerat, M. P., F. Thiaucourt, J. B. Poveda, J. Nicolet, and J. Frey.** 1999. Genetic and serological analysis of lipoprotein LppA in *Mycopasma mycoides* subsp *mycoides* LC and *Mycoplasma mycoides* subsp *capri. Clin. Diagn. Lab. Immunol.* **6:**224-230

83.  **Mu, H. H., A. D. Sawitzke, and B. C. Cole.** 2000. Modulation of cytokine profiles by the Mycoplasma superantigen *Mycoplasma arthritidis* mitogen parallels susceptibility to arthritis induced by *M. arthritidis. Infect. Immun.* **68:**1142-1149

84.  **Naglic, T., D. Hajsig, J. Frey, B. Seol, K. Busch, and M. Lojkic.** 2000. Epidemiological and microbiological study of an outbreak of infectious keratoconjunctivitis in sheep. *Vet. Rec.* **147:**72-75

85.  **Neimark, H. and K. M. Kocan.** 1997. The cell wall-less rickettsia *Eperythrozoon wenyonii* is a Mycoplasma. *FEMS Microbiol. Lett.* **156:**287-291

85a.  **Neimark, H., Johansson, K.E., Rikihisa, Y., Tully, J.G.** 2001. Proposal to transfer some members of the genera *Haemobartonella* and *Eperythrozoon* to the genus *Mycoplasma* with descriptions of 'Candidatus Mycoplasma haemofelis', 'Candidatus Mycoplasma haemomuris', 'Candidatus Mycoplasma haemosuis' and 'Candidatus Mycoplasma wenyonii'. *Int. J. Syst. Evol. Microbiol.* **51:**891-899

86.  **Nicolet, J.** 1994. *Mycoplasma bovis* - Extension of a new pathogen germ in the cattle population of Switzerland. *Schweiz. Arch. Tierheilkd.* **136:**81-82

87.  **Papazisi, L., K. E. Troy, T. S. Gorton, X. Liao, and S. J. Geary.** 2000. Analysis of cytadherence-deficient, GapA-negative *Mycoplasma gallisepticum* strain R. *Infect. Immun.* **68:**6643-6649

88.  **Persson, A., B. Pettersson, G. Bolske, and K. E. Johansson.** 1999. Diagnosis of contagious bovine pleuropneumonia by PCR-laser- induced fluorescence and PCR-restriction endonuclease analysis based on the 16S rRNA genes of *Mycoplasma mycoides* subsp. *mycoides* SC. *J. Clin. Microbiol.* **37:**3815-3821

89.  **Pettersson, B., G. Bolske, F. Thiaucourt, M. Uhlen, and K. E. Johansson.** 1998. Molecular evolution of *Mycoplasma capricolum* subsp. *capripneumoniae* strains, based on polymorphisms in the 16S rRNA genes. *J. Bacteriol.* **180:**2350-2358

90.  **Pettersson, B., T. Leitner, M. Ronaghi, G. Bolske, M. Uhlen, and K. E. Johansson.** 1996. Phylogeny of the *Mycoplasma mycoides* cluster as determined by sequence analysis of the 16S rRNA genes from the two rRNA operons. *J. Bacteriol.* **178:**4131-4142

91. **Rasberry, U. and R. F. Rosenbusch.** 1995. Membrane-associated and cytosolic species-specific antigens of *Mycoplasma bovis* recognized by monoclonal antibodies. *Hybridoma* **14:**481-485

92. **Rosati, S., S. Pozzi, P. Robino, B. Montinaro, A. Conti, M. Fadda, and M. Pittau.** 1999. P48 major surface antigen of *Mycoplasma agalactiae* is homologous to a *malp* product of *Mycoplasma fermentans* and belongs to a selected family of bacterial lipoproteins. *Infect. Immun.* **67:**6213-6216

93. **Sachse, K., C. Grajetzki, R. Rosengarten, I. Hanel, M. Heller, and H. Pfutzner.** 1996. Mechanisms and factors involved in *Mycoplasma bovis* adhesion to host cells. *Zentralbl. Bakteriol.* **284:**80-92

94. **Sachse, K., J. H. Helbig, I. Lysnyansky, C. Grajetzki, W. Muller, E. Jacobs, and D. Yogev.** 2000. Epitope mapping of immunogenic and adhesive structures in repetitive domains of *Mycoplasma bovis* variable surface lipoproteins. *Infect. Immun.* **68:**680-687

95. **Sachse, K., H. Pfutzner, M. Heller, and I. Hanel.** 1993. Inhibition of *Mycoplasma bovis* cytadherence by a monoclonal antibody and various carbohydrate substances. *Vet. Microbiol.* **36:** 307-316

96. **Sanchis, R., G. Abadie, M. Lambert, E. Cabasse, P. Dufour, J. M. Guibert, and M. Pepin.** 2000. Inoculation of lactating ewes by the intramammary route with *Mycoplasma agalactiae*: comparative pathogenicity of six field strains. *Vet. Res.* **31:**329-337

97. **Shen, X., J. Gumulak, H. Yu, C. T. French, N. Zou, and K. Dybvig.** 2000. Gene rearrangements in the *vsa* locus of *Mycoplasma pulmonis*. *J. Bacteriol.* **182:**2900-2908

98. **Solsona, M., M. Lambert, and F. Poumarat.** 1996. Genomic, protein homogeneity and antigenic variability of *Mycoplasma agalactiae*. *Vet. Microbiol.* **50:**45-58

99. **Srivastava, N. C., F. Thiaucourt, V. P. Singh, and J. Sunder.** 2000. Isolation of *Mycoplasma mycoides* small colony type from contagious caprine pleuropneumonia in India. *Vet. Rec.* **147:**520-521

100. **Stark, K. D. C., J. Nicolet, and J. Frey.** 1998. Detection of *Mycoplasma hyopneumoniae* by air sampling with a nested PCR assay. *Appl. Environ. Microbiol.* **64:**543-548

101. **Subramaniam, S., D. Bergonier, F. Poumarat, S. Capaul, Y. Schlatter, J. Nicolet, and J. Frey.** 1998. Species identification of *Mycoplasma bovis* and *Mycoplasma agalactiae* based on the *uvrC* genes by PCR. *Mol. Cell. Probes* **12:**161-169

102. **Tegtmeier, C., O. Angen, S. N. Grell, U. Riber, and N. F. Friis.** 2000. Aerosol challenge of calves with *Haemophilus somnus* and *Mycoplasma dispar*. *Vet. Microbiol.* **72:**229-239

103. **Thiaucourt, F., S. Lorenzon, A. David, and A. Breard.** 2000. Phylogeny of the *Mycoplasma mycoides* cluster as shown by sequencing of a putative membrane protein gene. *Vet Microbiol.* **72:**251-268

104. **Tola, S., A. Angioi, A. M. Rocchigiani, G. Idini, D. Manunta, G. Galleri, and G. Leori.** 1997. Detection of *Mycoplasma agalactiae* in sheep milk samples by polymerase chain reaction. *Vet. Microbiol.* **54:**17-22

105. **Tola, S., G. Idini, A. M. Rocchigiani, S. Rocca, D. Manunta, and G. Leori.** 2001. A physical map of the *Mycoplasma agalactiae* strain PG2 genome. *Vet. Microbiol.* **80:**121-130

106. **Tola, S., D. Manunta, S. Rocca, A. M. Rocchigiani, G. Idini, P. P. Angioi, and G. Leori.** 1999. Experimental vaccination against *Mycoplasma agalactiae* using different inactivated vaccines. *Vaccine* **17:**2764-2768

107. **Turner, K. S. and S. H. Kleven.** 1998. Eradication of live F strain *Mycoplasma gallisepticum* vaccine using live ts-11 on a multiage commercial layer farm. *Avian Dis.* **42:**404-407

108.  **Ulrich, R. G., S. Bavari, and M. A. Olson.** 1995. Bacterial superantigens in human disease: structure, function and diversity. *Trends Microbiol.* **3**:463-468

109.  **Verdin, E., M. Kobisch, J. M. Bove, M. Garnier, and C. Saillard.** 2000. Use of an internal control in a nested-PCR assay for *Mycoplasma hyopneumoniae* detection and quantification in tracheobronchiolar washings from pigs. *Mol. Cell. Probes* **14**:365-372

110.  **Verdin, E., C. Saillard, A. Labbe, J. M. Bove, and M. Kobisch.** 2000. A nested PCR assay for the detection of *Mycoplasma hyopneumoniae* in tracheobronchiolar washings from pigs. *Vet. Microbiol.* **76**:31-40

111.  **Vilei, E. M., E.-M. Abdo, J. Nicolet, A. Botelho, R. Gonçalves, and J. Frey.** 2000. Genomic and antigenic differences between the European and African/Australian clusters of *Mycoplasma mycoides* subsp. *mycoides* SC. *Microbiology* **146**:477-486

112.  **Vilei, E. M. and J. Frey.** 2001. Genetic and biochemical characterization of glycerol uptake in *Mycoplasma mycoides* subsp. *mycoides* SC: Its impact on $H_2O_2$ production and virulence. *Clin Diagn. Lab Immunol.* **8**:85-92

113.  **Vilei, E. M., J. Nicolet, and J. Frey.** 1999. IS*1634*, a novel insertion element creating long, variable-length direct repeats which is specific for *Mycoplasma mycoides* subsp. *mycoides* small-colony type. *J. Bacteriol.* **181**:1319-1323

114.  **Vizcaino, L. L., E. G. Abellan, M. J. C. Pablo, and A. Perales.** 1995. Immunoprophylaxis of caprine contagious agalactia due to *Mycoplasma agalactiae* with an inactivated vaccine. *Vet. Rec.* **137**:266-269

115.  **Wang, H., A. A. Fadl, and M. I. Khan.** 1997. Multiplex PCR for avian pathogenic mycoplasmas. *Mol. Cell. Probes* **11**:211-216

116.  **Wilton, J. L., A. L. Scarman, M. J. Walker, and S. P. Djordjevic.** 1998. Reiterated repeat region variability in the ciliary adhesin gene of *Mycoplasma hyopneumoniae*. *Microbiology* **144**: 1931-1943

117.  **Winner, F., R. Rosengarten, and C. Citti.** 2000. In vitro cell invasion of *Mycoplasma gallisepticum*. *Infect. Immun.* **68**:4238-4244

118.  **Wood, J. L., N. Chanter, J. R. Newton, M. H. Burrell, D. Dugdale, H. M. Windsor, G. D. Windsor, S. Rosendal, and H. G. Townsend.** 1997. An outbreak of respiratory disease in horses associated with *Mycoplasma felis* infection. *Vet. Rec.* **140**:388-391

Chapter 5

# Mycoplasmas of Plants and Insects

ERICH SEEMÜLLER*, MONIQUE GARNIER#, AND BERND
SCHNEIDER*
*Biologische Bundesanstalt für Land- und Forstwirtschaft, Institut für Pflanzenschutz im
Obstbau, D-69221 Dossenheim, Germany; #Laboratoire de Biologie Cellulaire et Moleculaire,
IBVM, INRA, F-33883 Villenave d'Ornon, France

## 1.     INTRODUCTION

Plants and insects are habitats of several mollicute genera including
*Acholeplasma, Entomoplasma, Mesoplasma, Spiroplasma*, and the
provisionally classified phytoplasmas. Plant- and insect-associated
acholeplasmas, entomoplasmas and mesoplasmas occur as saprophytes on
plant surfaces including flowers, and in insects probably as commensals or
symbionts (90, 106). They are pleomorphic, nonhelical in shape and
culturable in vitro and consist of relatively few species. There is no
indication that they are of economic importance, and only their phylogenetic
positions will be discussed in this chapter. For more information, the reader
is referred to references 89, 107, 109.

The helical spiroplasmas comprise apparently one of the most abundant
group of microbes. They have been identified in many insect species and
have also been isolated from ticks. In addition, spiroplasmas occur as
saprophytes on plant surfaces and in flowers, and three species are plant
pathogens (88, 108). While many spiroplasmas can be cultivated in complex
or defined media (16), some fastidious species from insects could only be
cultured in cell-free media conditioned in insect cell cultures (31, 32, 54).
However, many insect spiroplasmas have resisted cultivation (88).

Phytoplasmas comprise a large group of nonhelical mollicutes associated
with diseases collectively referred to as yellows diseases. Many of them are
of great economic importance. They were thought to be caused by viruses

until wall-less, pleomorphic bodies were detected in diseased plants, resembling the morphology and ultrastructure of known mycoplasmas (21). Due to these similarities these bodies were called mycoplasma-like organisms (MLOs). The failure to culture them under axenic conditions prevented their proper classification until DNA-based methods became available about a decade ago. In particular sequence analysis of the 16S rRNA gene confirmed their mycoplasma nature and their relationship to other mollicutes (29, 64, 102). As a consequence of these studies, the International Committee of Systematic Bacteriology, Subcommittee on the Taxonomy of Mollicutes has agreed to replace the trivial name MLO by the term phytoplasma (36, 38). Under the provisional taxonomic status '*Candidatus*' (79), this name became also the genus name for this group of mycoplasmas. Six *Candidatus* species names have been published to date (19, 28, 78, 93, 112, 122; see also Fig. 2), whereas traditional common names are still used to designate the majority of the phytoplasmas.

## 2.     PHYLOGENETIC DIVERSITY OF PLANT AND INSECT MYCOPLASMAS

Plant and insect mycoplasmas are located on two major branches of the mycoplasma phylogenetic tree (Fig. 1). The phytoplasmas cluster with the acholeplasmas in an early mycoplasma branch. The closest known relative of the phytoplasmas is *Acholeplasma palmae* which shares between 85.6 and 90.7% 16S rDNA sequence similarity with them. Most closely related to *A. palmae* and other acholeplasmas are the aster yellows (AY) and stolbur-type phytoplasmas which share between 89.3 and 90.7% 16S rDNA sequence similarity with *A. palmae* and form the first major phytoplasma branch arising from the acholeplasma clade (Figs. 1 and 2). Thus, they appear to be the earliest phytoplasmas. Due to the close phylogenetic relatedness, the phytoplasmas are considered to be a subline of the acholeplasmas (70). However, a formal taxonomic position of the phytoplasmas in relation to the acholeplasmas, which form the order *Acholeplasmatales*, has not been defined. The second branch comprises the spiroplasmas, entomoplasmas, and mesoplasmas, together with the mycoplasmas of the *Mycoplasma mycoides* group. The spiroplasmas, entomoplasmas, and mesoplasmas are assigned to the families *Spiroplasmataceae* and *Entomoplasmatacea*, respectively, which compose the order *Entomoplasmatales* (107).

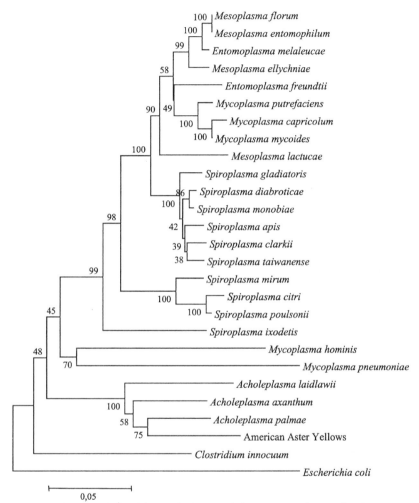

*Figure 1.* Phylogenetic positions of plant and insect mycoplasmas in the genera *Acholeplasma, Spiroplasma, Entomoplasma, Mesoplasma,* and the provisionally classified phytoplasmas (represented by the American aster yellows phytoplasma). Mollicutes of the *Anaeroplasma, Ureaplasma, Mycoplasma fermentans* and *M. hyorhinis* groups according to Maniloff (70) are not included. Bar represents a phylogenetic distance of 5%.

## 3. DIVERSITY OF PHYTOPLASMAS AND SPIROPLASMAS

## 3.1 Biological and pathological diversity

### 3.1.1 Plant association and symptomatology

Phytoplasmas and plant-pathogenic spiroplasmas are phloem parasites which reside almost exclusively in the sieve elements. Phytoplasmas affect a great number of plants. The most comprehensive list of phytoplasma hosts was published in 1989, at the end of the 'premolecular era' of phytoplasmology. This list contains about 750, mostly naturally infected species belonging to 98 families (77). Since then, many additional host plants have been identified, mostly by using molecular techniques, in particular PCR. The majority of the newly discovered and many of the previously known phytoplasmas have been taxonomically characterised, mainly by restriction fragment length polymorphism (RFLP) and sequence analysis of their 16S rDNA. Several lists arisen from such work have been published (20, 58, 59, 71, 94, 96, 101). From these papers and other work it can be estimated that about a thousand of plant species are affected by phytoplasmal infection. The potential host range is even greater. For example, North American aster yellows phytoplasmas were transmitted experimentally by polyphagous leafhopper vectors to 191 plant species belonging to 42 families (77). Most of them were not known as natural phytoplasma hosts.

Most of the phytoplasma host plants are dicotyledons in which a wide array of specific and nonspecific symptoms are induced (76, 77). Characteristic symptoms include floral discoloration and distortions such as virescence, phyllody, big bud, flower proliferation and other flower deformations. Other specific symptoms are witches' brooms and other abnormal proliferation of shoots and roots, enlarged stipules, undersized fruits, internode elongation and etiolation, shortened internodes, premature shooting and flowering in winter or early spring, and necrosis of vascular cambium and phloem. Less specific and nonspecific symptoms are often expressed by woody plants and include foliar yellowing and reddening, reduced leaf size, leaf roll, leaf deformation, premature leaf drop, vein clearing, vein enlargement, vein necroses, stunting and decline (reviewed in 77).

Fewer phytoplasmas have been detected in monocotyledons of which most hosts are from *Palmae* and *Poacea* families. Infections usually result in white leaf symptoms, foliar yellowing, witches' brooms, stunting, and decline (77). This may indicate that the physiological reaction of monocotyledons to phytoplasmal infection is more limited than that of dicotyledons. After several unconfirmed reports on the occurrence of phytoplasmas in gymnosperms, an hitherto unknown type of phytoplasma (PinP in Fig. 2) has recently been identified in *Pinus sylvestris* (B. Schneider and E. Seemüller, unpublished results). Because the sieve pores in gymnosperms are much narrower than in angiosperms (21a) it was thought that they would not allow phytoplasma spread.

Due to missing genotypic and physiological criteria, the phytoplasmas were at first differentiated and designated according to plant association, symptomatology and geographic occurrence. One such classification approach mainly based on symptoms expressed in *Catharanthus roseus* (periwinkle), to which many phytoplasmas have been experimentally transmitted, resulted in the differentiation of 6 primary groups in which 50 disease agents or groups of pathogens were distinguished (74). This grouping shares some similarities to the current phylogenetic classification (see below) but exhibits also significant differences. Some of the differences are due to the fact that genetically different phytoplasmas, such as grapevine yellows agents, may induce similar symptoms in the same host (69), whereas closely related organisms such AY group phytoplasmas may cause very different symptoms in periwinkle (71, 94). Also, the same pathogen may induce distinctly different symptoms in closely related plants, such as the apple proliferation agent in *Malus* species (44, 45). In this case, symptoms range from nonspecific yellowing to highly specific witches' brooms. Symptom expression in apple was clearly related to phytoplasma concentration. In trees with specific symptoms the titer was as high as 5.3 x $10^8$ phytoplasma cells/g tissue, whereas in trees with nonspecific symptoms the numbers were as low as 6.5 x $10^2$ phytoplasma cells/g tissue. This phenomenon was observed in other phytoplasma-infected plants, too (8).

The three plant pathogenic species among the spiroplasmas are *S. citri*, *S. kunkelii*, and *S. phoeniceum*. *S. citri* is the causal agent of citrus stubborn disease, which affects all *Citrus* species, and of horseradish brittle root. In addition, 35 plant species in 12 families have been reported as natural hosts, and another 38 species in 19 families have been experimentally infected (12). In contrast, *S. kunkelii*, the causal agent of corn stunt, has a narrow host range and only corn (*Zea mays*) and teosinte (*Z. mays mexicana* and *Z. mays parviglumis*) are known as natural hosts (111). The only known natural host of *S. phoeniceum* is periwinkle (92). The symptoms induced by spiroplasmas include stunting (shortened internodes), reduction in leaf size and yellowing. These symptoms are similar to those induced by several phytoplasmas.

However, symptoms such as shoot proliferation, virescence and phyllody, which are typical for many phytoplasmoses, have never been observed.

### 3.1.2    Insect vector relationships

Phytoplasmas and spiroplasmas are spread to their plant hosts by phloem-feeding homopterous insects, primarily of the family *Cicadellidae* (leafhoppers), and less common by planthoppers (*Fulgoroidea*) of the family *Cixiidae* and psyllids (*Psylloidea*) (33, 105). It is hypothesised that plant-pathogenic mollicutes evolved from insect-inhabiting organisms. Through feeding of infested insects, mollicutes might have been transmitted to plants and might have developed parasitism of internal plant tissue (30). Obviously nonphytopathogenic and nonculturable mycoplasma-like organisms, that might be related to plant-pathogenic mollicutes, have been observed in several insect orders that feed on vascular fluids (48).

The transmission modus by phytoplasma and spiroplasma vectors is circulative, propagative, and persistent. This means that mollicutes are ingested during phloem feeding, must traverse the wall of the mid-gut portion of the intestinal tract, multiply in the hemolymph, and pass through the salivary glands, in which they multiply further. Then, the pathogens are introduced, along with saliva, into sieve tubes of a new host plant. The insects remain inoculative for life (85, 87).

In the transmission process, highly specific mollicute-insect interactions seem to be involved. Although their molecular basis is poorly understood, there is indication that specific attachment reactions between mycoplasmas and receptors in insect tissue are required for penetration of the gut and salivary gland barriers of the vector. Most advanced evidence comes from the system *S. citri/Circulifer tenellus*, where the spiroplasmal surface protein P89 seems to play a role in spiroplasma adhesion. The failure to penetrate either one of the barriers results in the failure of the entire transmission process (23, 24, 61, 120). From recent PCR and electron microscopical studies there is indication for transovarial transmission of AY-type phytoplasmas by the leafhoppers *Scaphoideus titanus* and *Hishimonoides sellatiformis* (1, 46). However, most types of phytoplasmas were not believed to infect genital organs; therefore, phytoplasmas have not been believed to transmit vertically to the next generation of vector insects (46).

The level of specificity of the mycoplasma-vector interaction affects the number of vectors that are capable of transmitting a given pathogen, and the number of vectors and their feeding behavior greatly affects plant host range of phytoplasmas and spiroplasmas. For example, AY-group phytoplasmas, a large and pathologically diverse group, can be transmitted by >30 presumably polyphagous vectors to >200 plant species belonging to 45 families (for review see 58). California AY phytoplasmas, which belong to RFLP subgroup 16SrI-B of the AY subclade (see below), are transmitted by 24 leafhopper species (105). Thus, the AY phytoplasmas are considered to have a low vector specificity. Also, *S. citri* is transmitted by several leafhopper vectors (12). The X-disease agent which predominantly infects *Prunus* species is transmitted by at least 13 leafhopper species (87). It seems that other plant-pathogenic mycoplasmas have a higher vector specificity. Phytoplasmas for which only one or a few vectors are known include the rubus stunt, elm yellows, ash yellows, and peach yellows agents (85, 105). The flavescence dorée phytoplasma is in nature only known in grapevine and is transmitted by the monophagous vector *Scaphoideus titanus* (15), and for *S. kunkelii*, *Dalbulus maidis* is the principal vector (111). In contrast, some leafhoppers such as *Fieberiella florii* and *Macrosteles fascifrons* can efficiently transmit several distinctly different phytoplasmas, while others including *C. tenellus* and *D. maidis*, are able to transmit both phytoplasmas and spiroplasmas (18, 48, 105).

While most phytoplasmas are transmitted by leafhoppers, each of the fruit tree phytoplasmas of the apple proliferation group (apple proliferation, pear decline, European stone fruit yellows, and peach yellow leaf roll phytoplasmas; see Fig. 2) appear to be transmitted by one or two psyllids (13, 14, 27, 43, 87). Leafhoppers are not known to vector phytoplasmas of this group. Planthoppers also seem to have a relatively narrow range of transmission. For example, *Hyalesthes obsoletus*, *Myndus crudus* and *Oliarus atkinsoni* are known to only transmit stolbur-related, coconut lethal yellowing, and phormium yellow leaf phytoplasmas, respectively (17, 63, 68).

Phytoplasma and spiroplasma vectors are differently affected by the pathogens they transmit. For example, *Colladonus montanus* leafhoppers infected with the X-disease phytoplasma lived approximately half as long as uninfected leafhoppers (41). The same leafhopper infected with the same phytoplasma and six infected leafhopper species which transmit the maize bushy stunt phytoplasma produced fewer offspring than did healthy leafhoppers (42, 67, 81). Also, several *Dalbulus* spp. which poorly transmit the corn stunt agent were adversely affected by *S. kunkelii*, whereas the spiroplasma was not pathogenic to the efficient principal vector *D. maidis* (111). A beneficial effect was observed when the feral aster leafhopper, *Macrosteles quadrilineatus*, fed on AY phytoplasma infected plants. The exposed leafhoppers lived longer and laid more eggs than nonexposed leafhoppers. This reaction is explained by a long evolutionary association of the two organisms (4). Another beneficial effect was observed on aster, a non-host of *D. maidis*. The survival of this leafhopper was considerably increased by AY phytoplasma infection (86).

## 3.2    Molecular diversity of phytoplasmas

### 3.2.1    Comprehensive differentiation and classification by RFLP and sequence analysis of ribosomal DNA

DNA-based methods were introduced into phytoplasmology only about a decade ago when it became possible to enrich phytoplasmal DNA from infected plants and insects (50, 53, 99). Of the DNA sequences used to characterise, differentiate and classify microorganisms, the conserved 16S rRNA gene is of paramount importance and is widely used in phytoplasma research. RFLP analysis of 16S rDNA was employed in the first comprehensive phytoplasma classification in which 12 major groups and several subgroups could be differentiated   (60, 94). In the most recent attempts, 14 groups, 40 subgroups and several unclassified and or undesignated phytoplasmas or groups of phytoplasmas could be distinguished by RFLP analysis of 16S rDNA (58, 59).

Major efforts to comprehensively classify phytoplasmas on the basis of sequence analysis of 16S rDNA, the standard method for phylogenetic analysis, were undertaken by several groups (29, 80, 101, 102). These and other studies in which, over time, more than 100 phytoplasmas were included, showed that these plant pathogens represent a coherent, monophyletic group in which the evolutionary distance as expressed in the sequence similarity of the 16S rRNA genes is not greater than about 14%. These differences are smaller, for example, than in groups of the genus *Mycoplasma* where in both the hominis and pneumoniae group 16S rDNA sequence differences of up to 18% were observed (84).

Within the phytoplasma clade, subclades have been defined which largely agree with the groups established by RFLP analysis. As, at the beginning of phylogenetic phytoplasma classification, no threshold of sequence similarity was defined for assigning a species rank, the idea was to consider each subclade as the equivalent of distinct species under the provisional status '*Candidatus*' (36, 37). With the increasing number of isolates examined, the subclades became so complex that further differentiation was required. In a recent comprehensive classification, in which 48 isolates were included, it was proposed to distinguish 20 subgroups and provisional species in the phytoplasma clade (101). On the basis of the phylogram included in this chapter (Fig. 2), which comprises 82 isolates, it is proposed to distinguish, in addition to 6 described putative species, 26 other subgroups and strains at the putative species level. This proposition is based on the recommendation that threshold levels of about 2.5% difference in 16S rDNA sequence similarity may be sufficient to define provisional phytoplasma species. In case that other criteria such as additional molecular data, serological results, plant host range and insect vector specificity support separation, species may be defined even if they differ by less than 2.5% in their 16S rDNA sequence (2).

This basis for defining provisional phytoplasma species seems to be close to the existing classification of culturable bacteria. An important criterion for defining bacterial species is that strains sharing at least 70% homology of their total DNA with 5°C or less ΔTm should be included in the same species whilst strains showing a higher divergence should be assigned to different species (37, 110). Although there is no close correlation between 16S rDNA similarity and total DNA homology, it is unlikely that at 16S rDNA sequence similarity values below 97.5%, the total DNA homology will be higher than 60 to 70% (104). In a similar study it was shown that bacteria which share about 99% 16S rDNA sequence homology may vary in their total DNA sequence similarity as determined by DNA-DNA hybridisation between 23.5 and 93% (26). In practise it is not uncommon that bacterial species differ in their 16S rDNA sequence by only about 1% (26, 84) (see also chapter on Taxonomy of *Mollicutes*, this volume).

### 3.2.2    Diversity within subclades

There is no doubt that phytoplasmas consist of more taxonomic entities than those proposed in the grouping of Fig. 2. This has not only been shown by RFLP analysis of PCR-amplified 16S rDNA (see above) but also by RFLP analysis of the 16S/23S rDNA spacer region, genes encoding ribosomal proteins and elongation factor Tu (*tuf* gene), and randomly cloned DNA fragments. Further evidence was obtained by sequence analysis of the *tuf* gene and ribosomal protein L22 gene, dot and Southern blot analysis using randomly cloned DNA fragments as probes, and serological studies. Sequence analyses revealed that the 16S/23S rDNA spacer, *tuf* gene, *rpl22*, and all randomly cloned DNA fragments used to date as probes and in RFLP analysis are less conserved than 16S rDNA and allow a more detailed differentiation than this gene (for reviews see 58, 101). Even higher conserved than 16S rDNA seems to be the phytoplasmal 23S rRNA gene. RFLP analysis allowed differentiation of primary groups but not of members of the X-disease phytoplasma group which could be distinguished according to their 16S rDNA restriction profiles (29a).

Due to the great amount of data on molecular diversity of phytoplasmas, only examples can be discussed in some detail in this chapter. Within the AY group, one of the largest and most intensively studied phytoplasma

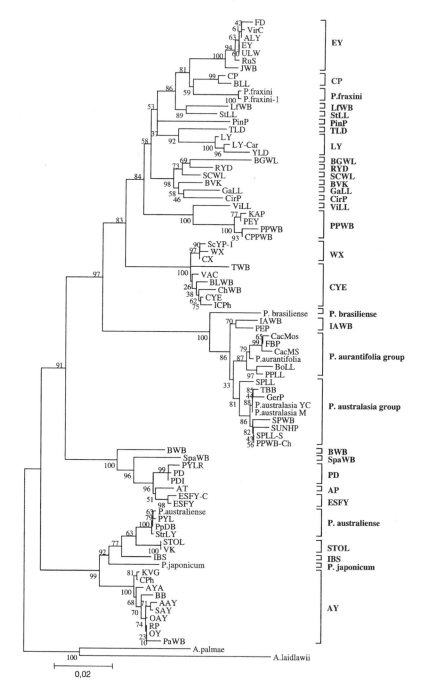

*Figure 2.* Comprehensive phylogenetic analysis of phytoplasmas. Proposed grouping is shown in the right hand column. Provisional taxa, strain acronyms and sequence accession numbers (from bottom): PaWB, paulownia witches' broom (AF279271); OY, onion yellows (D12569); RP, rape phyllody (U89378); OAY, oenothera aster yellows (M30790); SAY, western severe aster yellows (M86340); AAY, American aster yellows (X68373); BB, tomato big bud/Arkansas (L33760); AYA, aster yellows from apricot (X68338); CPh, clover phyllody/Canada (AF222066); KVG, clover phyllody/Germany (X83870); *Candidatus (Ca.)* Phytoplasma japonicum (AB010425); IBS, Italian bindweed stolbur (Y16391); VK, grapevine yellows/Germany (X76428); STOL, stolbur of pepper, (X76427); StrLY, strawberry lethal yellows (AJ243045); PpDB, papaya dieback (Y10095); PYL, phormium yellow leaf (U43570); *Ca. P.* australiense (X95706); ESFY, European stone fruit yellows/Germany (X68374); ESFY-C, European stone fruit yellows/Czech Republic (Y11933); AT, apple proliferation (X68375); PD, pear decline/Germany (X76425); PDI, pear decline/Italy (Y16392); PYLR, peach yellow leaf roll (Y16394); SpaWB, spartium WB (X92869); BWB, buckthorn WB (X76431); SPLL-S, sweet potato little leaf (from *Stylosanthes* sp.) (AJ289193); PPWB-Ch, pigeon pea WB/China (AF028813); SUNHP, sunn hemp WB (X76433); SPWB, sweet potato WB (L33770); *Ca. P.* australasia M (papaya mosaic) (Y10096); *Ca. P.* australasia YC, (papaya yellow crinkle) (Y10097); TBB, tomato big bud/Australia (Y08173); GerP, gerbera phyllody (AB026155); SPLL, sweet potato little leaf (X90591); PPLL, pigeon pea little leaf (AJ289191); BoLL, bonamia little leaf (Y15863); *Ca. P.* aurantifolia (U15442); CacMS, cactus phytoplasma 'Martinez-Soriano' 1999 (AF200718); CacMos, cactus mosaic (AF320575); FBP, faba bean phyllody (X83432); PEP, picris echiodes phyllody (Y16393); IAWB, Italian alfalfa WB (Y16390); *Ca. P.* brasiliense (AF147708); ICPh, Italian clover phyllody (X77482); CYE; clover yellow edge (AF173558); ChWB, chayote WB (AF147706); BLWB, black locust WB (AF244363); VAC, vaccinium WB (X76430); TWB; tsuwabuki WB (D12580); CX, Canadian X-disease (L33733); WX, western X-disease (L04682); ScYP-1, sugarcane yellows type I (AF056094); CPPWB, Caribbean pigeon pea WB (U18763); PPWB, pigeon pea WB/Florida (L33735); PEY, picris echioides yellows (Y16389); KAP, knautia arvensis phyllody (Y18052); ViLL, vigna lanceolata little leaf (Y15866); GaLL, galactia tenuiflora little leaf (Y15865); CirP, cirsium phyllody (X83438); BVK, from leafhopper *Psammotettix cephalotes* (X76429); SCWL, sugarcane white leaf (X76432); RYD, rice yellow dwarf (D12581); BGWL, Bermuda grass white leaf (Y16388); PinP, from *Pinus sylvestris* (AJ310849); TLD, Tanzanian lethal decline (X80117); YLD, Yucatan lethal decline (U18753); LY-Car, carludovica palmata lethal yellowing (AF237615); LY, coconut lethal yellowing (U18747); StLL, stylosanthes little leaf (AJ289192); LfWB, loofah WB (L33764); *Ca. P.* fraxini (X68339); *Ca. P.* fraxini-1 (AF189215); BLL, brinjal (eggplant) little leaf (X83431); CP, clover proliferation (L33761); JWB, jujube WB (AF305240); RuS, rubus stunt (Y16395); ULW, elm yellows/France (X68376); EY, elm yellows/USA (AF189214); VirC, from Virginia creeper (*Parthenocissus quinquefolia*) (AF305198); ALY, alder yellows (Y16387); FD, flavescence dorée (X76560). *Acholeplasma palmae* (L33734) and *A. laidlawii* (M23932) were used as outgroup. Bar represents a phylogenetic distance of 2%.

subclade, only relatively small differences in the 16S rDNA sequence exist which make differentiation on this basis difficult (Fig. 2). However, using RFLP analysis of 16S rDNA, 7 subgroups (16SrI-A, B, C, D, E, F, K) can be distinguished (58, 59). The 16S rDNA results were largely confirmed and consolidated by sequence and RFLP analysis of the *tuf* gene (71). A more detailed differentiation, resulting in 9 subgroups, was possible by combined

RFLP analysis of the 16S rDNA and a fragment of a ribosomal protein gene operon that includes *rpl22* (58, 59). Some of the subgroups identified, including those represented by the maize bushy stunt, paulownia witches' broom, and blueberry stunt agents, respectively, seem to affect only a single plant species (59, 71). Thus, their differentiation from the other AY agents appears to be appropriate from a phytopathological point of view. However, other subgroups, in particular 16SrI-A and 16SrI-B, affect a wide and often overlapping range of plant species and induce a wide variety of symptoms. The pathological and biological relevance of their separation is therefore questionable. It is possible that for these phytoplasma groups plant host and/or insect vector specificity are low. However, additional resolution of subgroup 16SrI-B was obtained by serological studies, using 4 monoclonal antibodies, which resulted in subdivision of this subgroup into 4 clusters (47). Also, Southern blot analysis showed a strong variation of the RFLP patterns within subgroup-A and subgroup-B members (57, 98). However, the relationship of these results to pathological and epidemiological traits is not clear.

The apple proliferation subclade which is composed of the temperate fruit tree phytoplasmas and strains SpaWB and BWB is, in contrast to the AY phytoplasma group, proposed to be taxonomically resolved in Fig. 2. In this subclade, 16S rDNA sequence similarity between strains SpaWB, BWB and the fruit tree phytoplasmas is below the defined threshold of 2.5%. In contrast, the apple proliferation, European stone fruit yellows and pear decline/peach yellow leaf roll agents represented by strains AT, ESFY/ESFY-C, PD/PDI/PYLR (see Fig. 2), respectively, differ from each other in less than 1.5% of their 16S rDNA nucleotide positions. However, in accordance with the guidelines for defining provisional phytoplasma species (2), plant host and vector specificity (13, 14, 27, 43, 87), primer specificity and RFLP analyses of ribosomal and nonribosomal DNA (6, 39, 40, 65, 66, 100) and serological comparisons (6) indicate that the three temperate fruit tree phytoplasmas are distinct taxonomic entities. Within the apple proliferation agent genetic polymorphisms have been observed (39, 51) the taxonomic and pathological relevance of which is unclear.

Besides the AY phytoplasmas, the *Candidatus* (*Ca.*) P. australasia, *Ca*. P. aurantifolia, clover yellow edge (CYE) and elm yellows (EY) groups are poorly resolved in Fig 2. However, from these clusters, too, there are results from RFLP analyses of ribosomal and nonribosomal DNA, DNA-DNA hybridisation and serological studies which indicate that these groups contain candidates for separation at provisional species level (20, 58, 75, 91, 95, 96, 101). Thus the number of putative *Candidatus* species is certainly higher than that proposed in Fig. 2. In a previous attempt to classify phytoplasmas, 264 isolates were included, for which molecular data was available. Without considering AY serotypes and AY strains differing in Southern blot hybridisation patterns, 75 types of phytoplasmas could be distinguished (101). All this data indicate that within the coherent phytoplasma clade a large number of taxonomic entities exist.

### 3.2.3 Diversity based on other markers

The genome of the yet uncultured phytoplasmas is poorly characterised. Data on the guanine and cytosine (G+C) composition of phytoplasma DNA were obtained by HPLC and buoyant density centrifugation analysis of DNA enriched by CsCl-bisbenzimide gradient centrifugation. Values between 23.0 and 29.5 mol% for AY phytoplasma strains and an apple proliferation isolate have been determined (52, 99) which are in the range of that of the culturable mycoplasmas that usually extend from 24 to 33 mol% (90).

Pulsed-field gel electrophoresis was used to determine the genome size of many isolates from most phylogenetic groups. The sizes identified are in the range of those of the genus *Mycoplasma* and extend from 530 to 1350 kb (72, 82, 83, 122). The largest chromosomes were identified within the stolbur-type agents and members of the AY group which appear, according to the branching pattern shown in Fig. 2 (see also under 2.), to be the phylogenetic earliest phytoplasmas. Relatively small chromosomes with sizes below 700 kb were observed in all phylogenetic groups except for stolbur. Within the AY phytoplasmas sizes from 660 to 1130 kb were estimated (72). The smallest genome (530 kb) was shown by the Bermuda grass white leaf phytoplasma. It seems to represents the smallest chromosome of any known living cell (72).

Mapping of phytoplasma chromosomes was done with isolates from three phylogenetic groups. Maps constructed from the western X-disease (22), apple proliferation (56), European stone fruit yellows (73) and sweet potato little leaf (83) phytoplasma chromosomes revealed that the restriction sites are not conserved and that different enzymes were required to obtain suitable macrorestriction fragments for mapping. The distribution of the G+C-rich restriction sites recognised by the rare cutting enzymes used was relatively uniform in the chromosomes of the X-disease and the sweet potato little leaf agents whereas a pronounced clustering was observed in the apple proliferation and European stone fruit yellows phytoplasma chromosomes. This may indicate differences in the overall base pair composition of the genome. Within two strains of the apple proliferation phytoplasma, which differed in genome size, a polymorphism of the restriction fragments was observed. Also, the closely related pear decline and European stone fruit yellows agents showed restriction profiles which differed from each other and the apple proliferation strains (56). Similarly, the closely related sweet potato little leaf and tomato big bud phytoplasmas did not show common restriction fragments following digestion with 4 enzymes (83). These results indicate that macrorestriction mapping might be a useful tool for strain characterisation, a method increasingly used for typing bacterial pathogens (34).

Like their close phylogenetic relatives, the acholeplasmas, the phytoplasmas have two sets of rRNA genes (49, 55, 97). In the only completely sequenced phytoplasma rRNA operon, that of the loofah witches' broom agent, the three rRNA genes are organised in the typical prokaryotic order: 5'-16S RNA-23S RNA-5S rRNA-3'. The size of these genes is 1538, 2846, and 113 bp, respectively, and the 5S rDNA is followed by a tRNA$^{Val}$ and a tRNA$^{Asn}$ gene (35). In all partially sequenced rRNA operons, the 16S rRNA gene was followed by the 23S rDNA gene (55, 101). In the 16S/23S rDNA spacer, a single tRNA$^{Ile}$ gene was supposed to be present in all phytoplasmas (55, 103). However, like in most other mollicutes, this gene could not be detected in the spacer of the stylosanthes little leaf phytoplasma (StLL in Fig. 2) (96). Also, the 16S/23S rDNA spacer of one of the loofah witches' broom phytoplasma rRNA operons does not include a tRNA$^{Ile}$ gene (35). By examining 16S rDNA or a fragment including this gene and the 16S/23S rDNA spacer region, sequence heterogeneity of the two operons has been observed (7, 62, 71). This is relatively common among the AY phytoplasmas and is responsible for the characteristic RFLP patterns of several AY subgroups and is thus a useful taxonomic tool (71).

Polyclonal and monoclonal antibodies prepared to several phytoplasmas recognise in Western blot analysis one or two immunodominant proteins (IMPs) with a molecular mass between 15 and 32 kDa. IMP genes from seven phytoplasmas of four distinct phylogenetic phytoplasma subclades have been isolated, cloned and sequenced. Nucleotide and deduced amino acid sequence analysis revealed no significant similarities between phytoplasmas from the AY, apple proliferation and X-disease groups. However, the IMP gene and its product from the apple proliferation agent showed significant overall homology to gene and gene product of the phylogenetic distant sweet potato witches' broom phytoplasma. By Southern blot and DNA sequence analysis of IMP genes, homologies were also identified between closely related organisms such as the apple proliferation group members and between X-disease and peach rosette phytoplasmas (3, 6, 9, 121). The predicted properties of the putative translation products suggested three different kinds of IMPs, represented by the AY-group, apple proliferation group/sweet potato witches'-broom, and X-disease phytoplasmas. In all these groups a single major hydrophilic domain was identified, exposed to the outside of the phytoplasma cell. This domain, which is supposed to be the most important part of the protein in terms of external interactions, did not show significant amino acid sequence similarity between the closely related but pathologically distinct AY and clover phyllody phytoplasma. Also, similarity between apple proliferation group members and sweet potato witches'-broom phytoplasmas was low. Within the apple proliferation group, sequence similarity between the two pome fruit pathogens causing apple proliferation and pear decline was much higher than between these phytoplasmas and the European stone fruit yellows agent that infects stone fruits (3). This data indicate that the IMPs may not only be a valuable tool to study phytoplasma interactions with their plant and insect hosts but also used to characterise and distinguish pathologically different phytoplasmas. It appears that analysis of IMP genes and their products enables phytoplasma differentiation at the same sensitive taxonomic level as serological tests.

## 3.3    Taxonomic and molecular diversity of spiroplasmas

In the most recent classification of the spiroplasmas, 34 groups and several subgroups were distinguished (115). Within these groups and subgroups, 34 species have been defined at present (J. G. Tully, pers. commun.). All 3 plant pathogenic spiroplasmas belong to group I, which is sub-divided into 8 subgroups or serovars. *S. citri*, *S. kunkelii* and *S. phoeniceum* belong to subgroup I-1, I-3 and I-8, respectively. Interestingly, *S. melliferum*, a honeybee pathogen but not pathogenic to plants, is also a member of group I (subgroup I-2) and is closely related to the three plant pathogenic spiroplasmas. The other four spiroplasmas in this group are either plant surface- or arthropod-associated organisms. Plant surface spiroplasmas are thought to be deposited by insects, which are the major hosts of members of groups II to XXXIV (115).

Most insect spiroplasmas have been identified in species of six evolutionary advanced orders including *Hymenoptera, Coleoptera, Diptera, Lepidoptera, Homoptera,* and *Hemiptera* (108), and one (*S. platyhelix*) in the ancient order *Odonata* (113). Spiroplasmas are most commonly found in the gut, less frequently in the hemolymph, and occasionally in various organs. In some insects, spiroplasmas that are ingested can prove highly invasive and pathogenic or lethal to the host. In addition to *S. melliferum*, lethal infections in honeybees were also observed following exposure to *S. apis* (108). *S. poulsonii* is associated with sex ratio abnormalities in the progeny of *Drosophila* spp. (114) and *S. floricola* is associated with a lethargy disease of the scarabaeid beetle *Melolontha melolontha* (10). Spiroplasmas were also isolated from blood sucking insects such as mosquitoes and tabanid flies (108, 115).

The genus *Spiroplasma* comprises organisms with the largest genomes within the class *Mollicutes*. Genome sizes of at least one strain of each of the 34 groups have been determined by pulsed-field gel electrophoresis (for review see (10). Spiroplasma genome sizes are generally above 1000 kb with only 4 exceptions: spiroplasma strain W115 (980 kb), *S. monobiae* (940 kb), spiroplasma strain TUIS-1 (840 kp), and *S. platyhelix* (780 kp). The largest genome (2220 kb) is that of *S. ixodetis*, a spiroplasma isolated from ticks. Group I organisms, which include the plant-pathogenic spiroplasmas, have genome sizes ranging from 1460 (*S. melliferum*) to 2020 kp (strain LB-12, isolated from a plant bug). Intra-species genome size variations from 1650 to 1910 kp were evidenced among 12 *S. citri* strains (118). This polymorphism may have resulted from genomic re-arrangements. RFLP analysis of total genomic DNA showed variations in the digestion profiles, and extrachromosomal DNA (either virus replicative forms or plasmids) was evidenced in several strains (5). The G+C content of spiroplasma DNA varies within the range of other mollicutes from 24 to 31% (90).

Physical and genomic maps were constructed for the two closely related spiroplasmas *S. citri* and *S. melliferum* whose genomes differ by about 360 kp and shares 68% DNA/DNA homology (117, 119). The size reduction affected mainly one *Apa*I restriction fragment (ApA). A conserved distribution of the gene markers on the chromosomes of the two spiroplasmas was found, especially for the genes close to the origin of replication (*ori*C). However, the distance between the tRNA$^{Trp}$ and the spiralin genes was much shorter in *S. melliferum*, and a large chromosome fragment had an inverted orientation in the two chromosomes. Both genomes are characterised by the presence of multiple regions homologous to replicative forms of two single stranded DNA nonlytic SpV1 type viruses (SpV1-R8A2B and SpV1-S102).

Variability was also studied in 12 *S. citri* strains after *Apa*I and *Bss*HII digestions of the chromosomes followed by hybridisation with 14 probes (118). Each of these enzymes cut Morocco *S. citri* R8A2 strain chromosome in 9 fragments, while strains from Syria, Spain, Israel and other locations lead to different restriction maps with one or both enzymes. One strain from California had no *Bss*HII site. However, despite this obvious heterogeneity, it seemed that the gene order was conserved in all strains and certain restriction sites were also conserved. It is possible that mutations and DNA methylation accounted for restriction differences among the strains. Variability in the 16S and 23S rDNA restriction profiles could also be demonstrated within the 12 *S. citri* strains.

Strains of *S. citri* were shown to have a different electrophoretic mobility of spiralin, the most abundant and immunodominant membrane lipoprotein of this species (116). Sequencing of the spiralin genes of 8 *S. citri* strains revealed that all spiralins except one were 241 amino acids in length. Point mutations occurring in the various spiralin genes allowed separation of the strains into 5 groups. The molecular masses calculated from the sequences could not explain the differences in electrophoretic mobility, nor was there an indication that the variations affected the posttranslational palmitoylation of this acylated protein. This suggested that the polymorphism was either due to a structural property of the whole protein or to an unidentified posttranslational modification (11, 25).

Using various primer pairs defined on spiralin sequences of *S. citri*, similar genes could be amplified by PCR from spiroplasmas other than those of group I. Thus, this protein is probably characteristic for all members of the genus *Spiroplasma* (11). Comparison of *S. melliferum, S. phoeniceum* and *S. kunkelii* spiralin gene sequences showed 88.6%, 80% and 78% homology, respectively, with *S. citri*. Comparison of the spiralin amino acid sequence showed that there are two conserved regions in the proteins. Most of the N-terminal amino acids are conserved and represent a protein region very similar to a typical eubacterial lipoprotein signal peptide. The central regions between positions 107 and 132 are also conserved. A sequence repetition (close to positions 50 and 150) is present in all spiralin genes but different from one species to another except for a consensus sequence VTKXE present in the repetition. Similar repeats occur in cell wall associated proteins of Gram positive bacteria and are involved in eukaryotic cell recognition. The spiralin repeats might have a similar function (11).

## 4. CONCLUDING REMARKS

As phytoplasmas are the most important group among plant and insect mollicutes, they are treated most extensively in this chapter. Molecular methods allow us to differentiate and classify this large and pathologically diverse group of plant pathogens. Although the groupings proposed are not always satisfactory from the phytopathological point of view, the tremendous progress made in differentiation and classification, and also in detection and identification, is currently the major achievement of molecular phytoplasmology. This progress is mainly based on examination of conserved genes whereas there is still very little or no information on the genetic background of phytoplasmas involved in host-pathogen interactions and metabolism. Even such an elementary question as whether or not plant host specificity of phytoplasmas exists cannot be answered with certainty. It appears that at least in some cases the insect vector host range is responsible for the pathological diversity of phytoplasmas. Most of these and other problems are difficult to study as long these pathogens cannot be cultured under axenic conditions. However, after it became possible to isolate and purify full-length phytoplasma chromosomes at a preparative scale, it may be possible in the near future to clone and sequence complete phytoplasma genomes. This would provide a wealth of valuable information on the genetic background of phytoplasmas including their metabolic capacity and factors involved in pathogenicity. Information on phytoplasma metabolism would greatly improve the chances to cultivate phytoplasmas.

Due to space limitation, some aspects of the diversity of plant-pathogenic mollicutes could not be treated. These subjects include geographic diversity, diversity in extrachromosomal DNA, differences in virulence within given pathogens, and possible differences in pathogen-host interactions as evidenced by symptomatology, histopathology and physiological, biochemical and genetic investigations.

## ACKNOWLEDGMENTS

We thank J. G. Tully for information on mycoplasma taxonomy, and M. Maixner and C. Marcone for critical reading of the manuscript.

## REFERENCES

1.  **Alma, A., D. Bosco, A. Danielli, A. Bertaccini, M. Vibio, and A. Arzone**. 1997. Identification of phytoplasmas in eggs, nymphs and adults of *Scaphoideus titanus* Ball reared on healthy plants. Insect Mol. Biol. **6**:115-121.
2.  **Anonymous**. 2000. Phytoplasma, spiroplasma, mesoplasma, and entomoplasma working team of International Research Programme on Comparative Mycoplasmology (IRPCM) of the International Organization for Mycoplasmology (IOM). Report of consultations, Fukuoka, Japan, July 2000.
3.  **Barbara, D. J., A. Morton, M. F. Clark, and D. L. Davies**. 2001. Molecular variation in immunodominant membrane proteins from phytoplasmas. Acta Hortic. **550**:405-408.
4.  **Beanland, L., C. W. Hoy, S. A. Miller, and L. R. Nault**. 2000. Influence of aster yellows phytoplasma on the fitness of aster leafhopper (Homoptera: Cicadellidae). Ann. Entomol. Soc. Am. **93**:271-276.
5.  **Bebear, C. M., P. Aullo, J. M. Bové, and J. Renaudin**. 1996. *Spiroplasma citri* virus SpV1: Characterization of viral sequences present in the spiroplasma host chromosome. Curr. Microbiol. **32**:134-140.
6.  **Berg, M., D. L. Davies, M. F. Clark, H. J. Vetten, G. Maier, C. Marcone, and E. Seemüller**. 1999. Isolation of the gene encoding an immunodominant membrane protein of the apple proliferation phytoplasma, and expression and characterization of the gene product. Microbiology **145**:1937-1943.
7.  **Berges, R., M. T. Cousin, J. Roux, R. Mäurer, and E. Seemüller**. 1997. Detection of phytoplasma infections in declining *Populus nigra* 'Italica' trees and molecular differentiation of the aster yellows phytoplasmas identified in various *Populus* species. Eur. J. For. Pathol. **27**:33-43.

8.    **Berges, R., M. Rott, and E. Seemüller**. 2000. Range of phytoplasma concentrations in various plant hosts as determined by competitive polymerase chain reaction. Phytopathology **90**:1145-1152.

9.    **Blomquist, C. L., D. J. Barbara, D. L. Davies, M. F. Clark, and B. C. Kirkpatrick**. 2001. An immunodominant membrane protein gene from the Western X-disease phytoplasma is distinct from those of other phytoplasmas. Microbiology **147**:571-580.

10.   **Bové, J. M.** 1997. Spiroplasmas: Infectious agents of plants, arthropods and vertebrates. Wien. Klin. Wochenschr. **109**:604-612.

11.   **Bové, J. M., X. Foissac, and C. Saillard**. 1993. Spiralins, p. 203-223. *In* S. Rottem and I. Kahane (ed.), Subcellular Biochemistry, Vol. 20: Mycoplasma Cell Membranes. Plenum Press, New York, N.Y.

12.   **Calavan, E. C. and J. M. Bové**. 1989. Ecology of *Spiroplasma citri*, p. 425-485. *In* R. F. Whitcomb and J. G. Tully (ed.), The Mycoplasmas, Vol. V, Spiroplasmas, Acholeplasmas, and Mycoplasmas of Plants and Arthropods. Academic Press, San Diego, CA.

13.   **Carraro, L., N. Loi, P. Ermacora, A. Gregoris, and R. Osler**. 1998. Transmission of pear decline by using naturally infected *Cacopsylla pyri*. Acta Hortic. **472**:665-668.

14.   **Carraro, L., R. Osler, N. Loi, P. Ermacora, and E. Refatti**. 1998. Transmission of European stone fruit yellows phytoplasma by *Cacopsylla pruni*. J. Plant Pathol. **80**:233-239.

15.   **Caudwell, A.** 1988. Flavescence dorée, p. 45-46. *In* R. C. Pearson and A. C. Goheen (ed.), Compendium of Grape Diseases. APS Press, St. Paul, MN.

16.   **Chang, C. J.** 1989. Nutrition and cultivation of spiroplasmas, p. 201-239. *In* R. F. Whitcomb and J. G. Tully (ed.), The Mycoplasmas, Vol. V, Spiroplasmas, Acholeplasmas, and Mycoplasmas of Plants and Arthropods. Academic Press, San Diego, CA.

17.   **Cherry, R. H. and F. W. Howard** . 1984. Sampling for adults of the planthopper *Myndus crudus*, a vector of lethal yellowing of palms. Trop. Pest Manag. **30**:22-25.

18.   **Chiykowski, L. N. and R. C. Sinha**. 1990. Differentiation of MLO diseases by means of symptomatology and vector transmission. Zbl. Bakt. **Suppl. 20**:280-287.

19.   **Davis, R. E., E. L. Dally, D. E. Gundersen, I.-M. Lee, and N. Habili**. 1997. "*Candidatus* Phytoplasma australiense," a new phytoplasma taxon associated with Australian grapevine yellows. Int. J. Syst. Bacteriol. **47**:262-269.

20.   **Davis, R. I., B. Schneider, and K. S. Gibb**. 1997. Detection and differentiation of phytoplasmas in Australia. Austral. J. Agr. Res. **48**:535-544.

21.   **Doi, Y., M. Teranaka, K. Yora, and H. Asuyama**. 1967. Mycoplasma- or PLT group-like microorganisms found in the phloem elements of plants infected with mulberry dwarf, potato witches' broom, aster yellows, or paulownia witches' broom. Ann. Phytopathol. Soc. Jpn. **33**:259-266.

21a.  **Esau, K.** 1969. The phloem. *In*: W. Zimmermann, P. Ozenda, and H. P. Wulff, Encyclopedia of Plant Anatomy, Vol. V/2. Gebrüder Bornträger, Berlin and Stuttgart.

22.   **Firrao, G., C. D. Smart, and B. C. Kirkpatrick**. 1996. Physical map of the western X-disease phytoplasma chromosome. J. Bacteriol. **178**:3985-3988.

23.   **Fletcher, J., A. C. Wayadande, U. Melcher, and F. Ye**. 1998. The phytopathogenic mollicute-insect vector interface: A closer look. Phytopathology **88**:1351-1358.

24.   **Foissac, X., J. L. Danet, C. Saillard, P. Gaurivaud, F. Laigret, C. Pare, and J. M. Bové**. 1997. Mutagenesis by insertion of Tn4001 into the genome of *Spiroplasma citri*: Characterization of mutants affected in plant pathogenicity and transmission to the plant by the leafhopper vector *Circulifer haematoceps*. Mol. Plant-Microbe Interact. **10**:454-461.

25. **Foissac, X., C. Saillard, J. Gandar, L. Zreik, and J. M. Bové.** 1996. Spiralin polymorphism in strains of *Spiroplasma citri* is not due to differences in posttranslational palmitoylation. J. Bacteriol. **178**:2934-2940.

26. **Fox, G. E., J. D. Wisotzkey, and P. J. Jurtshuk.** 1992. How close is close: 16S rRNA sequence identity may not be sufficient to guarantee species identity. Int. J. Syst. Bacteriol. **42**:166-170.

27. **Frisinghelli, C., L. Delatti, M. S. Grando, D. Forti, and M. E. Vindimian.** 2000. *Cacopsylla costalis* (Flor 1861), as a vector of apple proliferation in Trentino. J. Phytopathol. **148**:425-431.

28. **Griffiths, H. M., W. A. Sinclair, C. D. Smart, and R. E. Davis.** 1999. The phytoplasma associated with ash yellows and lilac witches'-broom: '*Candidatus* Phytoplasma fraxini'. Int. J. Syst. Bacteriol. **49**:1605-1614.

29. **Gundersen, D. E., I.-M. Lee, S. A. Rehner, R. E. Davis, and D. T. Kingsbury.** 1994. Phylogeny of mycoplasmalike organisms (phytoplasmas): a basis for their classification. J. Bacteriol. **176**:5244-5254.

29a. **Guo, Y. H., Z.-M. Cheng, and J. A. Walla.** 2000. Amplification of the 23S rRNA gene and its application in differentiation and detection of phytoplasmas. Can. J. Plant Pathol. **22**:380-386.

30. **Hackett, K. J. and T. B. Clark.** 1989. Ecology of spiroplasmas, p. 113-200. *In* R. F. Whitcomb and J. G. Tully (ed.), The Mycoplasmas, Vol V, Spiroplasmas, Acholeplasmas, and Mycoplasmas of Plants and Arthropods. Academic Press, San Diego, CA.

31. **Hackett, K. J. and D. E. Lynn.** 1985. Cell-assisted growth of a fastidious spiroplasma. Science **230**:825-827.

32. **Hackett, K. J., D. E. Lynn, D. L. Williamson, A. S. Ginsberg, and R. F. Whitcomb.** 1986. Cultivation of the *Drosophila* sex-ratio spiroplasma. Science **232**:1253-1255.

33. **Harris, K. F.** 1979. Leafhoppers and aphids as biological vectors: Vector-virus relationships, p. 217-308. *In* K. Maramorosch and K. F. Harris (ed.), Leafhopper Vectors and Plant Disease Agents. Academic Press, New York, N.Y.

34. **Hielm, S., J. Björkroth, E. Hyytia, and H. Korkeala.** 1998. Genomic analysis of *Clostridium bolutinum* group II by pulsed-field gel electrophoresis. Appl. Environ. Microbiol. **64**:703-708.

35. **Ho, K. C., C. C. Tsai, and T. L. Chung.** 2001. Organization of ribosomal RNA genes from a loofah witches' broom phytoplasma. DNA and Cell Biology **20**:115-122.

36. **International Committee on Systematic Bacteriology - Subcommittee on the Taxonomy of Mollicutes.** 1993. Minutes of the interim meeting, August 1992, Ames, Iowa. Int. J. Syst. Bacteriol. **43**:394-397.

37. **International Committee on Systematic Bacteriology - Subcommittee on the Taxonomy of Mollicutes.** 1995. Revised minimum standards for description of new species of the class *Mollicutes* (Division *Tenericutes*). Int. J. Syst. Bacteriol. **45**:605-612.

38. **International Committee on Systematic Bacteriology - Subcommittee on the Taxonomy of Mollicutes.** 1997. Minutes of the interim meeting, July 1996, Orlando, FL. Int. J. Syst. Bacteriol. **47**:911-914.

39. **Jarausch, W., C. Saillard, J. M. Broquaire, M. Garnier, and F. Dosba.** 2000. PCR-RFLP and sequence analysis of a non-ribosomal fragment for genetic characterization of European stone fruit yellows phytoplasmas infecting various *Prunus* species. Mol. Cell. Probes **13**.

40. **Jarausch, W., C. Saillard, F. Dosba, and J. M. Bové.** 1994. Differentiation of mycoplasmalike organisms (MLOs) in European fruit trees by PCR using specific primers derived from the sequence of a chromosomal fragment of the apple proliferation MLO. Appl. Environ. Microbiol. **60**:2916-2923.

41. **Jensen, D. D.** 1959. A plant virus lethal to its vector. Virology **8**:164-175.
42. **Jensen, D. D.** 1971. Vector fecundity reduced by western X-disease. J. Invert. Pathol. **17**:389-394.
43. **Jensen, D. D., W. H. Griggs, C. Q. Gonzales, and H. Schneider.** 1964. Pear decline virus transmission by pear psylla. Phytopathology **54**:1346-1351.
44. **Kartte, S. and E. Seemüller**. 1988. Variable response within the genus *Malus* to the apple proliferation disease. Z. PflKrankh. PflSchutz **95**:25-34.
45. **Kartte, S. and E. Seemüller**. 1991. Susceptibility of grafted *Malus* taxa and hybrids to apple proliferation disease. J. Phytopathol. **131**:137-148.
46. **Kawakita, H., T. Saiki, W. Wei, W. Mitsuhashi, K. Watanabe, and M. Sato.** 2000. Identification of mulberry dwarf phytoplasmas in the genital organs and eggs of leafhopper *Hishimonoides sellatiformis* . Phytopathology **90**:909-914.
47. **Keane, G., A. Edwards, and M. F. Clark.** 1996. Differentiation of group 16Sr-IB aster yellows phytoplasmas with monoclonal antibodies. Proc. BCPC Symp.: Diagnostic in Crop Production, 1.- 3.4.1996, Warwick, UK, **65**:263-268.
48. **Kirkpatrick, B. C.** 1991. Mycoplasma-like organisms - plant and invertebrate pathogens, p. 4050-4067. *In* A. Balows, H. G. Trüper, M. Dworkin, W. Harder, and K. H. Schleifer (ed.), The Procaryotes, Vol 3. Springer-Verlag, New York, N.Y.
49. **Kirkpatrick, B. C., G. A. Fisher, J. D. Fraser, and A. H. Purcell.** 1990. Epidemiological and phylogenetic studies on western X-disease mycoplasma-like organism. Zbl. Bakt. **Suppl. 20**:287-297.
50. **Kirkpatrick, B. C., D. C. Stenger, T. J. Morris, and A. H. Purcell.** 1987. Cloning and detection of DNA from a nonculturable plant pathogenic mycoplasma-like organism. Science **238**:197-200.
51. **Kison, H., B. Schneider, and E. Seemüller.** 1994. Restriction fragment length polymorphism within the apple proliferation mycoplasmalike organism. J. Phytopathol. **141**:395-401.
52. **Kollar, A. and E. Seemüller** . 1989. Base composition of the DNA of mycoplasmalike organisms associated with various plant diseases. J. Phytopathol. **127**:177-186.
53. **Kollar, A., E. Seemüller, F. Bonnet, C. Saillard, and J. M. Bové.** 1990. Isolation of the DNA of various plant pathogenic mycoplasmalike organisms from infected plants. Phytopathology **80**:233-237.
54. **Konai, M., K. J. Hackett, D. L. Williamson, J. J. Lipa, J. D. Pollack, G. E. Gasparich, E. A. Clark, D. C. Vacek, and R. F. Whitcomb.** 1996. Improved cultivation systems for isolation of the Colorado potato beetle spiroplasma. Appl. Environ. Microbiol. **62**:3453-3458.
55. **Kuske, C. R. and B. C. Kirkpatrick.** 1992. Phylogenetic relationships between the western aster yellows mycoplasmalike organism and other prokaryotes established by 16S rRNA gene sequence. Int. J. Syst. Bacteriol. **42**:226-233.
56. **Lauer, U. and E. Seemüller.** 2000. Physical map of the chromosome of the apple proliferation phytoplasma. J. Bacteriol. **182**:1415-1418.
57. **Lee, I.-M., R. E. Davis, T. A. Chen, L. N. Chiykowski, J. Fletcher, C. Hiruki, and D. A. Schaff.** 1992. A genotype-based system for identification and classification of mycoplasmalike organisms (MLOs) in the aster yellow MLO strain cluster. Phytopathology **82**:977-986.
58. **Lee, I.-M., Davis, R. E., and Gundersen-Rindal, D. E.** 2000. Phytoplasma: Phytopathogenic mollicutes. Annu. Rev. Microbiol. **54**:221-255.
59. **Lee, I.-M., D. E. Gundersen-Rindal, R. E. Davis, and I. M. Bartoszyk.** 1998. Revised classification scheme of phytoplasmas based on RFLP analysis of 16S rRNA and ribosomal protein gene sequences. Int. J. Syst. Bacteriol. **48**:1153-1169.

60.    **Lee, I.-M., R. W. Hammond, R. E. Davis, and D. E. Gundersen.** 1993. Universal amplification and analysis of pathogen 16S rDNA for classification and identification of mycoplasmalike organisms. Phytopathology **83**:834-842.

61.    **Lefol, C., A. Caudwell, J. Lherminier, and J. Larrue.** 1993. Attachment of the flavescence dorée pathogen (MLO) to leafhopper vectors and other insects. Ann. Appl. Biol. **123**:611-622.

62.    **Liefting, L. W., M. T. Andersen, R. E. Beever, R. C. Gardener, and R. L. S. Forster.** 1996. Sequence heterogeneity in the two 16S rRNA genes of phormium little leaf phytoplasma. Appl. Environ. Microbiol. **62**:3133-3139.

63.    **Liefting, L. W., R. E. Beever, C. J. Winks, M. N. Pearson, and R. L. S. Forster.** 1997. Planthopper transmission of phormium yellow leaf phytoplasma. Australas. Plant Patholog. **26**:148-154.

64.    **Lim, P.-O. and B. B. Sears.** 1989. 16S rRNA sequence indicates that plant-pathogenic mycoplasmalike organisms are evolutionarily distinct from animal mollicutes. J.Bacteriol. **171**:5901-5906.

65.    **Lorenz, K.-H., F. Dosba, C. Poggi Pollini, G. Llacer, and E. Seemüller.** 1994. Phytoplasma diseases of *Prunus* species in Europe are caused by genetically similar organisms. Z. PflKrankh. PflSchutz **101**:567-575.

66.    **Lorenz, K.-H., B. Schneider, U. Ahrens, and E. Seemüller.** 1995. Detection of the apple proliferation and pear decline phytoplasmas by PCR amplification of ribosomal and nonribosomal DNA. Phytopathology **85**:771-776.

67.    **Madden, L. V. and L. R. Nault.** 1983. Differential pathogenicity of corn stunting mollicutes to leafhopper vectors in *Dalbulus* and *Baldulus* species. Phytopathology **73**:1608-1614.

68.    **Maixner, M.** 1994. Transmission of German grapevine yellows (Vergilbungskrankheit) by the planthopper *Hyalesthes obsoletus* (Auchenorrhyncha: Cixiidae). Vitis **33**:103-104.

69.    **Maixner, M., M. Rüdel, X. Daire, and E. Boudon-Padieu.** 1995. Diversity of grapevine yellows in Germany. Vitis **34**:235-236.

70.    **Maniloff, J.** 1992. Phylogeny of mycoplasmas, p. 549-560. *In* J. Maniloff, R. N. McElhaney, L. R. Finch, and J. B. Baseman (ed.), Mycoplasmas. Molecular Biology and Pathogenesis. American Society for Microbiology, Washington, D.C.

71.    **Marcone, C., I.-M. Lee, R. E. Davis, A. Ragozzino, and E. Seemüller.** 2000. Classification of aster yellows-group phytoplasmas based on combined analysis of rRNA and *tuf* gene sequences. Int. J. Syst. Evol. Microbiol. **50**:1703-1713.

72.    **Marcone, C., H. Neimark, A. Ragozzino, U. Lauer, and E. Seemüller.** 1999. Chromosome sizes of phytoplasmas composing major phylogenetic groups and subgroups. Phytopathology **89**:805-810.

73.    **Marcone, C. and E. Seemüller.** 2001. A chromosomal map of the European stone fruit yellows phytoplasma. Microbiology **147**:1213-1221.

74.    **Marwitz, R.** 1990. Diversity of yellows disease agents in plant infections. Zbl. Bakt. Suppl. **20**:431-434.

75.    **Mäurer, R., E. Seemüller, and W. A. Sinclair.** 1993. Genetic relatedness of mycoplasmalike organisms affecting elm, alder, and ash in Europe and North America. Phytopathology **83**:971-976.

76.    **McCoy, R. E.** 1979. Mycoplasmas and yellows diseases, p. 229-264. *In* R. F. Whitcomb and J. G. Tully (ed.), The Mycoplasmas, Vol. III, Plant and Insect Mycoplasmas. New York, N.Y.

77.    **McCoy, R. E., A. Caudwell, C. J. Chang, T. A. Chen, L. N. Chiykowski, M. T. Cousin, J. L. Dale, G. T. N. De Leeuw, D. A. Golino, K. J. Hackett, B. C. Kirkpatrick, R. Marwitz, H. Petzold, R. C. Sinha, M. Sugiura, R. F. Whitcomb, I. L. Yang, B. M. Zhu, and E. Seemüller.** 1989. Plant diseases associated with mycoplasma-like organisms, p. 545-640. *In* R. F. Whitcomb and J. G. Tully (ed.), The

Mycoplasmas, Vol. V, Spiroplasmas, Acholeplasmas, and Mycoplasmas of Plants and Arthropods. Academic Press, San Diego, CA.

78.  **Montano, H. G., R. E. Davis, E. L. Dally, S. Hogenhout, J. P. Pimentel, and P. S. T. Brioso**. 2001. '*Candidatus* Phytoplasma brasiliense', a new phytoplasma taxon associated with hibiscus witches' broom disease. Int. J. Syst. Evol. Microbiol. **51**:1109-1118.

79.  **Murray, R. G. E. and K. H. Schleifer**. 1994. Taxonomic notes: A proposal for recording the properties of putative taxa of procaryotes. Int. J. Syst. Bacteriol. **44**:174-176.

80.  **Namba, S., S. Oyaizu, S. Kato, S. Iwanami, and T. Tsuchizaki**. 1993. Phylogenetic diversity of phytopathogenic mycoplasmalike organisms. Int. J. Syst. Bacteriol. **43**:461-467.

81.  **Nault, L. R., L. V. Madden, W. E. Styer, B. W. Triplehorn, G. F. Shambaugh, and S. E. Heady**. 1984. Pathogenicity of corn stunt spiroplasma and maize bushy stunt mycoplasma to their vector, *Dalbulus longulus*. Phytopathology **74**:977-979.

82.  **Neimark, H. and B. C. Kirkpatrick**. 1993. Isolation and characterization of full-length chromosomes from non-culturable plant-pathogenic mycoplasma-like organisms. Mol. Microbiol. **7**:21-28.

83.  **Padovan, A. C., G. Firrao, B. Schneider, and K. S. Gibb**. 2000. Chromosome mapping of the sweet potato little leaf phytoplasma reveals genome heterogeneity within the phytoplasmas. Microbiology **146**:893-902.

84.  **Pettersson, B., M. Uhlén, and K.-E. Johansson**. 1996. Phylogeny of some mycoplasmas from ruminants based on 16S rRNA sequences and definition of a new cluster within the hominis group. Int. J. Syst. Bacteriol. **46**:1093-1098.

85.  **Purcell, A. H.** 1982. Insect vector relationship with procaryotic plant pathogens. Annu. Rev. Phytopathol. **20**:397-417.

86.  **Purcell, A. H.** 1988. Increased survival of *Dalbulus maidis* Delong & Wolcott, a specialist on maize, on non-host plants infected with mollicute plant pathogens. Entomol. Exp. Appl. **46**:187-196.

87.  **Purcell, A. H.** 1996. Vector transmission and epidemiology of indigenous and exotic prokaryotic pathogens of plants. Atti Con. Ann. Soc. Italiana di Patologia Vegetale (SIPaV), Udine, 26-27 Sept. 1996, p.5-11.

88.  **Razin, S.** 1992. Mycoplasma taxonomy and ecology, p. 3-22. *In* J. Maniloff, R. N. McElhaney, L. R. Finch, J. B. Baseman (ed.), Mycoplasmas. Molecular Biology and Pathogenesis. American Society for Microbiology, Washington, D.C.

89.  **Razin, S.** 2000. The genus *Mycoplasma* and related generea (class *Mollicutes*). *In* M. Dworkin, S. Falkow, E. Rosenberg, K. H. Schleifer, and E. Stackebrandt (ed.), The Prokaryotes. An evolving electronic source for the microbial community, 3rd edition. Springer-Verlag, New York, N.Y. (http//:www.procarytes.com).

90.  **Razin, S., D. Yogev, and Y. Naot**. 1998. Molecular biology and pathogenicity of mycoplasmas. Microbiol. Mol. Biol. Rev. **62**:1094-1156.

91.  **Saeed, E. M., N. Sarindu, D. L. Davies, M. F. Clark, J. Roux, and M. T. Cousin**. 1994. Use of monoclonal antibodies to identify mycoplasmalike organisms (MLOs) from the Sudan and from Thailand. J. Phytopathol. **142**:345-349.

92.  **Saillard, C., J. C. Vignault, J. M. Bové, J. G. Tully, D. L. Williamson, and A. Fos**. 1986. *Spiroplasma phoeniceum* sp. nov., a new plant pathogenic species from Syria. Int. J. Syst. Bacteriol. **37**:106-115.

93.  **Sawayanagi, T., N. Horikoshi, T. Kanehira, M. Shinohara, A. Bertaccini, M. T. Cousin, C. Hiruki, and S. Namba**. 1999. '*Candidatus* Phytoplasma japonicum', a new phytoplasma taxon associated with Japanese hydrangea phyllody. Int. J. Syst. Bacteriol. **49**:1275-1285.

94. **Schneider, B., U. Ahrens, B. C. Kirkpatrick, and E. Seemüller**. 1993. Classification of plant-pathogenic mycoplasma-like organisms using restriction-site analysis of PCR-amplified 16S rDNA. J. Gen. Microbiol. **139**:519-527.
95. **Schneider, B., M. T. Cousin, S. Klinkong, and E. Seemüller**. 1995. Taxonomic relatedness and phylogenetic positions of phytoplasmas associated with disease of faba bean, sunnhemp, sesame, soybean, and eggplant. Z. PflKrankh. PflSchutz **102**:225-232.
96. **Schneider, B., A. C. Padovan, S. De La Rue, R. Eichner, R. I. Davis, A. Bernuetz, and K. S. Gibb**. 1999. Detection and differentiation of phytoplasmas in Australia: an update. Austral. J. Agr. Res. **50**:333-342.
97. **Schneider, B. and E. Seemüller**. 1994. Presence of two sets of ribosomal genes in phytopathogenic mollicutes. Appl. Environ. Microbiol. **60**:3409-3412.
98. **Schneider, B. and E. Seemüller**. 1994. Studies on the taxonomic relationships of mycoplasmalike organisms by Southern blot analysis. J. Phytopathol. **141**:173-185.
99. **Sears, B. B., P.-O. Lim, N. Holland, B. C. Kirkpatrick, and K. L. Klomparens**. 1989. Isolation and characterization of DNA from a mycoplasmalike organism. Mol. Plant-Microbe Interact. **2**:175-180.
100. **Seemüller, E., H. Kison, K.-H. Lorenz, B. Schneider, C. Marcone, C. D. Smart, and B. C. Kirkpatrick**. 1998. Detection and identification of fruit tree phytoplasmas by PCR amplification of ribosomal and nonribosomal DNA, p. 56-66. *In* C. Manceau (ed.), COST 823: New technologies to improve phytodiagnosis. European Community, Luxembourg.
101. **Seemüller, E., C. Marcone, U. Lauer, A. Ragozzino, and M. Göschl**. 1998. Current status of molecular classification of the phytoplasmas. J. Plant Pathol. **80**:3-26.
102. **Seemüller, E., B. Schneider, R. Mäurer, U. Ahrens, X. Daire, H. Kison, K.-H. Lorenz, G. Firrao, L. Avinent, B. B. Sears, and E. Stackebrandt**. 1994. Phylogenetic classification of phytopathogenic mollicutes by sequence analysis of 16S ribosomal DNA. Int. J. Syst. Bacteriol. **44**:440-446.
103. **Smart, C. D., B. Schneider, C. L. Blomquist, L. J. Guerra, N. A. Harrison, U. Ahrens, K.-H. Lorenz, E. Seemüller, and B. C. Kirkpatrick**. 1996. Phytoplasma-specific PCR primers based on sequences of the 16S-23S rRNA spacer region. Appl. Environ. Microbiol. **62**:2988-2993.
104. **Stackebrandt, E. and B. M. Goebel**. 1994. Taxonmomic note: A place for DNA-DNA reassociation and the 16S rRNA sequence analysis in the present species definition in bacteriology. Int. J. Syst. Bacteriol. **44**:846-849.
105. **Tsai, J. H.** 1979. Vector transmission of mycoplasmal agents of plant diseases, p. 265-307. *In* R. F. Whitcomb and J. G. Tully (ed.), The Mycoplasmas, Vol. III, Plant and Insect Mycoplasmas. Academic Press, New York, N.Y.
106. **Tully, J. G.** 1996. Mollicute-host interrelationships: Current concepts and diagnostic implications, p. 1-21. *In* J. G. Tully and S. Razin (ed.), Molecular and Diagnostic Procedures in Mycoplasmology, Vol. II, Diagnostic Procedures. Academic Press, San Diego, CA.
107. **Tully, J. G., J. M. Bové, F. Laigret, and R. F. Whitcomb**. 1993. Revised taxonomy of the class *Mollicutes*: proposed elevation of a monophyletic cluster of arthropod-associated mollicutes to ordinal rank (*Entomoplasmatales* ord. nov.), with provision for familial rank to separate species with nonhelical morphology (*Entomoplasmataceae* fam. nov.) from helical species (*Spiroplasmataceae*), and emended descriptions of the order *Mycoplasmatales*, family *Mycoplasmataceae*. Int. J. Syst. Bacteriol. **43**:378-385.
108. **Tully, J. G. and R. F. Whitcomb**. 1991. The genus *Spiroplasma*, p. 1961-1980. *In* A. Balows, H. G. Trüper, M. Dworkin, W. Harder, and K. H. Schleifer (ed.), The Prokaryotes, Vol. 2. Springer-Verlag, New York, N.Y.

109. **Tully, J. G., R. F. Whitcomb, K. J. Hackett, D. L. Rose, R. B. Henegar, J. M. Bové, P. Carle, D. L. Williamson, and T. B. Clark.** 1994. Taxonomic description of eight new non-sterol-requiring mollicutes assigned to the genus *Mesoplasma*. Int. J. Syst. Bacteriol. **44**:685-693.

110. **Wayne, L. G., D. J. Brenner, R. R. Colwell, P. A. D. Grimont, O. Kandler, M. I. Krichevsky, L. H. Moore, W. E. C. Moore, R. G. E. Murray, E. Stackebrandt, M. P. Starr, and H. G. Trüper.** 1987. Report of the ad hoc committee on reconciliation of approaches to bacterial systematics. Int. J. Syst. Bacteriol. **37**:463-464.

111. **Whitcomb, R. F.** 1989. *Spiroplasma kunkelii*: Biology and ecology, p. 487-544. *In* R. F. Whitcomb and J. G. Tully (ed.), The Mycoplasmas, Vol. V, Spiroplasmas, Acholeplasmas, and Mycoplasmas of Plants and Arthropods. Academic Press, San Diego, CA.

112. **White, D. T., L. L. Blackall, P. T. Scott, and K. B. Walsh.** 1998. Phylogenetic positions of phytoplasmas associated with dieback, yellow crinkle and mosaic diseases of papaya, and their proposed inclusion in 'Candidatus Phytoplasma australiense' and a new taxon, 'Candidatus Phytoplasma australasia'. Int. J. Syst. Bacteriol. **48**:941-951.

113. **Williamson, D. L., J. R. Adams, R. F. Whitcomb, J. G. Tully, P. Carle, M. Konai, J. M. Bové, and R. B. Henegar.** 1997. *Spiroplasma platyhelix* sp. nov, a new mollicute with unusual morphology and genome size from the dragonfly *Pachydiplax longipennis*. Int. J. Syst. Bacteriol. **47**:763-766.

114. **Williamson, D. L., B. Sakaguchi, K. J. Hackett, R. F. Whitcomb, J. G. Tully, P. Carle, J. M. Bové, J. R. Adams, M. Konai, and R. B. Henegar.** 1999. *Spiroplasma poulsonii* sp. nov., a new species associated with male-lethality in *Drosophila willistoni*, a neotropical species of fruit fly. Int. J. Syst. Bacteriol. **49**:611-618.

115. **Williamson, D. L., R. F. Whitcomb, J. G. Tully, G. E. Gasparich, D. L. Rose, P. Carle, J. M. Bové, K. J. Hackett, J. R. Adams, R. B. Henegar, M. Konai, C. Chastel, and F. E. French.** 1998. Revised group classification of the genus *Spiroplasma*. Int. J. Syst. Bacteriol. **48**:1-12.

116. **Wroblewski, H., K.-E. Johansson, and S. Hjerten.** 1977. Purification and characterization of spiralin, the main protein of the *Spiroplasma citri* membrane. Biochim. Biophys. Acta **465**:275-289.

117. **Ye, F., F. Laigret, and J. M. Bové.** 1994. A physical and genomic map of the prokaryote *Spiroplasma melliferum* and its comparison with the *Spiroplasma citri* map. C. R. Acad. Sci. Paris, Life Sci. **317**:392-398.

118. **Ye, F., F. Laigret, P. Carle, and J. M. Bové.** 1995. Chromosomal heterogeneity among various strains of *Spiroplasma citri* . Int. J. Syst. Bacteriol. **45**:729-734.

119. **Ye, F., F. Laigret, J. C. Whitley, C. Citti, L. R. Finch, P. Carle, J. Renaudin, and J. M. Bové.** 1992. A physical and genetic map of the *Spiroplasma citri* genome. Nucleic Acids Res. **20**:1559-1565.

120. **Yu, J., A. C. Wayadande, and J. Fletcher.** 2000. *Spiroplasma citri* surface protein P89 implicated in adhesion to cells of the vector *Circulifer tenellus*. Phytopathology **90**:716-722.

121. **Yu, Y.-L., K. W. Yeh, and C. P. Lin.** 1998. An antigenic protein gene of a phytoplasma associated with sweet potato witches' broom. Microbiology **144**:1257-1262.

122. **Zreik, L., P. Carle, J. M. Bové, and M. Garnier.** 1995. Characterization of the mycoplasmalike organism associated with witches'-broom disease of lime and proposition of a *Candidatus* taxon for the organism, "Candidatus Phytoplasma aurantifolia". Int. J. Syst. Bacteriol. **45**:449-453.

# Chapter 6

# Cell Division

MAKOTO MIYATA
*Department of Biology, Graduate School of Science, Osaka City University, Sumiyoshi-ku, Osaka 558-8585 JAPAN*

## 1. INTRODUCTION

Cell division is the final step of cell reproduction as a well-grown cell divides into daughter cells. However, this phenomenon cannot occur by itself and requires preceding cellular events. A cell must essentially duplicate its mass, namely the amount of small and large biomolecules, if the cell is going to divide by binary fission. The replication of chromosomal DNA is done with the highest priority; however, it should occur in synchronicity with other reproduction events. The replicated chromosomes must also be partitioned for delivery to the daughter cells properly. Cytokinesis, the final step of cell division, may depend on the formation of a special apparatus. Mycoplasmas have the distinct ability to adhere to cell surfaces of their hosts. For some mycoplasma species, this ability depends on a terminal structure located at a cell pole. Duplication of this structure before cell division is for them an indispensable event. In this chapter, individual steps of cell division will be discussed focusing on recent studies.

## 2. DIVISION MODE

The small and fragile cell structures of mycoplasmas hamper the analysis of division schemes. Several modes of reproduction have been proposed, including binary fission, fragmentation of an elongated cell, and budding. Binary fission has been demonstrated in *M. pneumoniae*, *M. gallisepticum*, and *M. mobile*. These mycoplasmas have a polarized cell morphology

*Molecular Biology and Pathogenicity of Mycoplasmas*, Edited by Razin and Herrmann, Kluwer Academic/Plenum Publishers, New York, 2002

characterized by a protruding membrane extension responsible for adhesion (5, 38, 42, 49, 50) (see "Duplication of the attachment organelle" section in this chapter). Binary fission has also been reported for mycoplasmas without cell polarity. For instance, *M. capricolum* has a rod cell morphology without obvious polarity (51). In larger *M. capricolum* cells observed in a fast growing culture, constriction was seen at the center of the cell, suggesting reproduction by binary fission. This assumption was supported by the observations that both DNA mass and cell length rarely exceeded twice their minimum values. *Spiroplasma citri*, which features filamentous helical cell morphology, can be monitored for cell elongation by the number of helical turns (12, 13). Cells with two helical turns were the major fraction of a cell population in a fast growing culture, and only occasionally were cells seen with more than four turns. Constriction could be seen with a periodicity of two helical turns. Pulse-labeling of the membrane protein revealed that elongation occurs at a cell pole and at the center where the constriction site would be formed. These observations lead to the conclusion that reproduction of *S. citri* occurs by binary fission after cell elongation at one cell pole.

These recent analyses have shown binary fission as the division mode for mycoplasmas. However, a variety of division modes of mycoplasmas in nature cannot be conclusively ruled out because growth conditions in their natural hosts may be different from those in the laboratory. Moreover, division modes of mollicutes may be diverse among the almost 200 species.

## 3. DNA REPLICATION

Generally, the rate of DNA replication should coincide with the duplication of other biomolecules, otherwise the daughter cells will not be similar to the parent cell, and the number of cell divisions would be limited. Actually, the relative DNA-to-protein content in a cell population is quite stable at all stages of a batch culture of *Mycoplasma capricolum* (51). In the case of this species, the replication reaction proceeds along the parental DNA strands at a rate of 6 kb/min, ten times slower than that of *Escherichia coli*. The reaction is bidirectional from the initiation site (36, 37, 39, 40) as with other eubacteria. The time required for one round of DNA replication is calculated to be 95 min, which corresponds to the time measured to double the number of CFU. Additionally, it was found in batch culture that the amount of DNA located near the *ori* region was twice that present at the terminus region. The value of 2.0 can be explained given the assumption that the replication procedure takes most of the time of one cell division interval, and that most DNA molecules in the culture are replicating intermediates.

Taken together, this suggests that DNA replication occurs in an interval which precisely corresponds to the time between cell divisions (51).

It is known that the chromosome content in *Escherichia coli* and *Bacillus subtilis* is adjusted by modifying the initiation frequency of DNA replication. When biomolecular synthesis is proceeding slowly, the initiation frequency is reduced; when it is going fast, the frequency is increased (16). The change of the initiation frequency in mycoplasma has not been examined. However, considering the good agreement between DNA and protein content of *M. capricolum*, it is likely that DNA replication can be adjusted also in mycoplasmas. This agreement is unlikely to be explained by regulation of protein synthesis because depletion of nucleotides inhibited DNA replication but did not affect protein synthesis (51).

How is the frequency of initiation regulated? Even in the well studied bacteria, the mechanism is not yet clear (16). The DnaA protein is known as an essential protein for initiation in walled bacteria (4, 54). The protein binds to repeat sequences (DnaA-box) around the initiation site, unwinds the DNA duplex, and recruits DNA gyrase and other factors to form replication complexes. The ATP-bound DnaA protein will initiate replication, but in the ADP-bound form it will not. The ratio of ATP- to ADP-bound forms is believed to play a key role in controlling initiation. The *dnaA* gene is conserved in all mycoplasma genomes examined so far (6, 10, 11, 14, 18) and is confirmed to be expressed in *M. capricolum* (53). Detailed analyses of the progression of the replication reaction revealed that the initiation site is adjacent to the *dnaA* gene in the *M. capricolum* chromosome (36, 37). These observations suggest the participation of DnaA in initiation in mycoplasma. Initiation occurs in the region flanking the *dnaA* gene and is dependent on the function of the *dnaA* gene product in most walled bacteria. The participation of the DnaA protein in the initiation of chromosome replication is also suggested by the analysis of initiation activity of a DNA fragment in *S. citri* (47, 61), a spiroplasma phylogenetically closely related to *M. capricolum* (58). A 1.3 kb DNA fragment containing the 3' terminal region of the *dnaA* gene and its flanking region had the ability to initiate DNA replication when the fragment was inserted into a plasmid harboring an antibiotic resistance gene. Three repeats of the DnaA-box consensus sequence found in walled bacteria are identified in the 3' non-coding region flanking the *dnaA* gene of *S. citri*. Five repeats of the consensus sequence can be found also in non-coding regions flanking the *dnaA* gene in *M. pulmonis* (6), but they are absent in *M. pneumoniae* (17), and *M. genitalium* (10), where the DnaA-box sequences might be diverse. Considering that the ATP/GTP binding motif (P-loop) is conserved in the sequences of mycoplasmal *dnaA* genes (55), it is possible that DnaA protein also plays a role in controlling the initiation frequency in mycoplasmas.

4.      SEGREGATION OF THE REPLICATED
         CHROMOSOME

There are two possibilities regarding segregation of the mycoplasma chromosome. The first is that the replicated chromosomes are randomly distributed to the daughter cells. The second possibility is that the replicated chromosomes are faithfully segregated by a special mechanism. Visualization of the chromosomal DNA by staining with DAPI (4',6-diamidino-2-phenylindole), a DNA specific fluorescent dye, has helped to address this question. DAPI staining of *M. pneumoniae* revealed that no cells lacked a chromosome in the cell population of a normally growing culture. Also, a small constriction was observed in elongated cells, suggesting that the chromosome is segregated into daughter cells faithfully (50). The segregation of the nucleoid could be followed in *M. capricolum* because the position of the nucleoid could be localized, using phase-combined fluorescense microscopy, the visible light reducing the weak fluorescence of DAPI staining of cytoplasmic RNA, unmasking the strongly fluorescent nucleoides (52). In a fast growing culture of *M. capricolum*, most cells had one or two nucleoids in a cell, and no cells without nucleoids were found. The nucleoids were positioned in the center in mononucleoid cells or at one-quarter and three-quarters of the cell length in binucleoid cells. These observations suggest the *M. capricolum* has a method of ensuring delivery of replicated DNA to daughter cells.

What mechanism is responsible for the nucleoid segregation? Nucleoid segregation is generally divided into three steps: the first is the partitioning of replicated chromosomes, the second is the movement of the two nucleoids, and the last is their positioning in the cell body ready to divide (Fig. 1). The movement of replicated nucleoids may be coupled with a mechanism responsible for cell elongation in *M. capricolum* because the distance between the replicated nucleoids is proportional to the total cell length (52). Generally, the elongation process of the wall-less cells (such as cells in higher eukaryotes) is composed of two events. One is membrane expansion via the insertion of new membrane components, and the other is extension of cytoskeletal structures. The partitioning of the nucleoid seems to be linked to the extension of the cytoskeleton rather than expansion of the membrane, because nucleoid partitioning was still observed after cell elongation was totally stopped by inhibition of lipid synthesis. The nucleoid positions were significantly biased towards the cell poles when protein synthesis was inhibited by chloramphenicol, showing that proper positioning depends on *de novo* protein synthesis. Proteins marking the precise position for the partitioned nucleoids may be synthesized in a manner coupled to the progression of the cell division cycle.

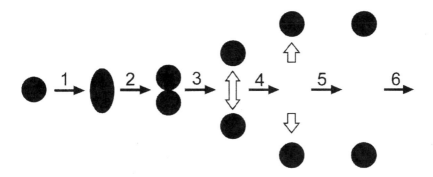

*Figure 1.* Definition of each step of chromosome segregation. 1: DNA replication, 2: partitioning, 3, 4: movement, 5: positioning, 6: cell division.

## 5. DUPLICATION OF THE ATTACHMENT ORGANELLE

### 5.1 Scheme for migration of the attachment organelle

Mycoplasmas with a polarized morphology have a membrane protruding extension at one pole. The extension usually works as an attachment organelle, enriched for adhesion proteins (2, 8, 20) (see chapter Cytadherence and the cytoskeleton). In a population of cells in culture, no cells can be found without the terminal structure, showing that the formation of the organelle is tightly coupled to the cell reproduction cycle (22, 42, 49, 50). Images of elongated cells with a constriction at the center showed two such organelles, distributed bipolarly. This suggests that duplication of the organelle is a prerequisite for cell division in mycoplasma species possessing this structure.

Where is the nascent organelle formed? Bredt recognized the attachment organelle as a small knob on the cell surface under phase contrast microscopy (5). Observations of the cell division process of *M. pneumoniae* suggest that two attachment organelles are actually adjacent to each other at an early stage of cell division. This hypothesis was substantiated by microscopic analysis of cells fluorescently labeled for the P1 adhesin. Seto *et al.* (50) classified *M. pneumoniae* cells into three types based on the position of the attachment organelle and concomitantly measured the DNA content for each cell by the signal intensity of DAPI. The DNA content was significantly different among the cell types. Cells with a single organelle at one cell pole had a lower DNA content than cells with two foci. Those with one focus at each cell pole had the highest DNA content. This observation suggests that the nascent organelle is formed next to the old one and migrates to the opposite cell pole before binary fission (Fig. 2). This scheme may be similar for all flask-shaped mycoplasmas but possibilities of other scenarios cannot be ruled out because this behavior has not yet been examined in other species.

While there is agreement that the terminal membrane extension is formed coinciding with the cell division cycle, there is a controversy about the role of the organelle in chromosome segregation. The behavior of the organelle in *M. pneumoniae* resembles the behavior of the replication origin in walled bacteria (19, 59). In *E. coli*, the duplicated copies of the *oriC* locus (responsible for initiation of DNA replication) begin at the center of the cell during early stages of cell division, and then migrate to opposite cell poles before cytokinesis. The *oriC* may be oriented by an apparatus reminiscent of the eukaryotic mitotic machinery (7). Quinlan and Maniloff found that newly replicated DNA could be enriched in a cell fraction containing the terminal membrane extension of *M. gallisepticum*, suggesting association of chromosomal DNA with the membrane (44). The attachment organelle might work as a mitotic-like apparatus in mycoplasmas.

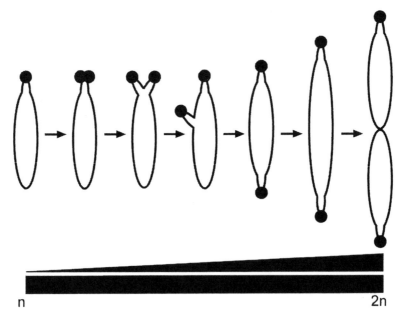

*Figure 2.* Duplication of the terminal structure in *M. pneumoniae*. Lower solid lines indicate DNA content of each cell type (50).

## 5.2    Formation of nascent organelle

A model can be proposed for the duplication of the attachment organelle. Formation of the new organelle seems to occur near the old one in *M. pneumoniae*, and is coupled to the progression of cell division. Seven proteins have been reported to be essential for cytadherence in *M. pneumoniae* from mutant studies: HMW1, HMW2, HMW3, P1 adhesin, P40, P90, and P30 (23, 25, 27, 29, 50) (see "Cytadherence and cytoskeletal elements" chapter). P1 adhesin is known to be a membrane spanning protein with at least three putative transmembrane stretches and plays the key role in adherence as a direct receptor of animal cell surfaces (45). These proteins are all coded in three unlinked loci in the genome with other 11 proteins whose functions are not yet known. Cytological studies have demonstrated their localization at the attachment organelle (24, 50). P65, which is co-transcribed with HMW2, is also localized at the attachment organelle. Chemical crosslinking studies suggest that HMW1, HMW3, P40, P90, P30, and P65 may have physical interactions with the P1 adhesin (28, 30). These observations suggest that the proteins are working as a large complex or are in very close proximity in the cell.

Is there any sequential scheme for assembly of these proteins? Immunofluorescence experiments incorporating double staining for the P1 adhesin in combination with other cytadherence proteins lacked sufficient temporal resolution to discern any particular order of assembly of the proteins (50). However, examination of mutants missing a cytadherence protein showed a hierarchy for the localization of cytadherence proteins. For example, a mutant missing HMW2 does not show any focused fluorescent antibody staining for other cytadherence proteins but another mutant defective for P30 shows foci for other cytadherence proteins, i.e. HMW3, P1, P40, and P90. Assuming that the hierarchy is based on the order of incorporation of each protein into the tip complex, we have suggested a scheme for assembly as shown in Fig. 3 (50) (unpublished data). We cannot know the assembly timing of HMW2, because its antibody is not available yet due to its low immunogenicity. We assigned this protein at the start of the assembly line based on the observation that HMW2 is required for stability of HMW1, HMW3, and P65 (9, 24, 43). A slightly different assembly scheme has been proposed by Krause and Balish (24), where P40 and P90 have an independent pathway for incorporation into the complex because HMW2 is not required for the stability of these proteins. However, these proteins can be placed in the same pathway with other proteins because HMW2 and HMW1 have clear effects on the localization of these proteins (50).

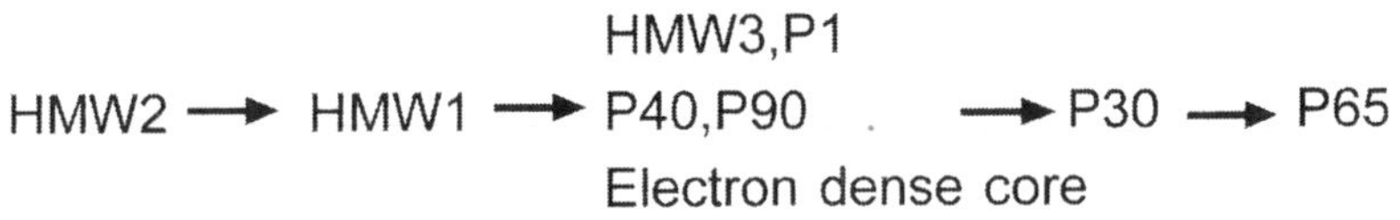

                                    HMW3,P1
HMW2 ⟶ HMW1 ⟶ P40,P90      ⟶ P30 ⟶ P65
                              Electron dense core

*Figure 3.* Assembly scheme of cytadherence proteins and the electron-dense core of the terminal structure of *M. pneumoniae.*

## 6.    CYTOSKELETAL STRUCTURES

Mycoplasmas have a variety of non-spherical cell morphologies, suggesting the existence of a cytoskeletal structure supporting the membrane. Filamentous protein structures have been reported for some species. For a summary see the review by S. Trachtenberg (56). These filamentous protein structures should be duplicated before cell division. Tight coupling of cell division and protein filament formation is suggested in *M. pneumoniae.* The attachment organelle of *M. pneumoniae* is believed to be supported by a rod-like structure. This structure can be seen under electron microscopy by thin sectioning as an electron-dense core (60). TritonX-100 extraction revealed that this structure may be associated with a bundle of protein filaments which branch out into finer filaments in the cell body, supporting the membrane in places other than the terminal structure (15, 35). Attempts to identify the proteins contained in this structure have not been successful yet, although candidates have been identified by performing mass spectrometry on proteins isolated from the Triton X-100-insoluble cytoskeletal fraction of *M. pneumoniae* (46). As this structure is missing in some cytadherence mutants, we can place formation of the electron-dense core in the assembly scheme of the attachment organelle (Fig. 3). Thin-section TEM of several mutants revealed that the electron-dense core depends on HMW2 and HMW1 but not on other known cytadherence proteins, suggesting that its formation is coupled to duplication of the attachment organelle and thus to the cell division cycle (unpublished data).

## 7.    CYTOKINESIS

### 7.1    Constriction

The last stage of the reproduction cycle is the formation of a constriction followed by the final division of the cell. In many mycoplasma species, a constriction can be observed in larger cells. As the surface area of the membrane increases dramatically during formation of a constriction, it is reasonable to imagine that membrane components are supplied to this site. Pulse-chase experiments in *S. citri* showed that newly synthesized proteins were inserted around the division site (12). Inhibition of lipid synthesis by depleting fatty acids from the medium stopped cell division of *M. capricolum* immediately, suggesting that a supply of lipid components is also required for constriction of mycoplasmas (52). These observations also suggest the existence of a special mechanism which inserts new components into the cell membrane.

The FtsZ protein is considered as the homolog of the eukaryotic tubulin and the corresponding gene is well conserved in bacterial genomes (31). Extensive analysis of some walled-bacteria revealed that this protein forms a ring structure in a complex with some other proteins at the division plane (32, 34). All mollicute genomes examined so far except *Ureaplasma urealtyticum* (14) have a homolog of this gene (10, 18, 26, 57). Although mollicute genomes have homologs of the *ftsZ* gene, they do not have any other of the known genes required for cytokinesis of walled bacteria. Since the peripheral structures of mycoplasmas are totally different from those of walled bacteria, the mechanism of constriction in mycoplasmas may be different from that of walled bacteria while the core functionality is dependent on FtsZ.

## 7.2    Gliding motility

In the case of certain flask-shaped mycoplasma species, gliding motility should be considered as a force utilized to complete cell division (see below). Such species can bind to and glide on glass and other solid surfaces. They always move with the attachment organelle at the leading end. The attachment organelle thus seems to be associated also with gliding. Gliding velocity averages 0.4, 0.1, 0.6, 0.04, and 3.0 µm/s for *M. pneumoniae*, *M. genitalium*, *M. pulmonis*, *M. gallisepticum*, and *M. mobile*, respectively (21). The gliding force of *M. mobile* reaches to 26-28 pN, which is 5-7 times that of motor proteins such as myosin and kinesin (36a). Analysis of the genomic sequences of some gliding mycoplasmas did not show any homologs of motor proteins such as myosin and kinesin, bacterial flagella, or bacterial pili (6, 10, 18, 41). Moreover, no extracellular structures can be found on the cell surface by scanning electron microscopy (21). These facts suggest that mycoplasmas glide by a totally unknown mechanism.

Mutants of *M. mobile* with altered gliding motility can be isolated by using an ultra-soft agar assay developed for this purpose (41). In these mutants, it was found that the ability to cytadhere closely mirrored the ability of this mycoplasma to bind to glass. Because cytadherence is linked to the tip structure, it is likely that the attachment organelle plays a major role in the gliding mechanism in addition to static adhesion to animal cells. This assumption is supported by the observations that the cytadherence-deficient mutants of *M. pneumoniae* cannot bind to a glass surface, and consequently cannot glide. When a cell of a gliding mycoplasma is ready to divide, it has attachment organelles at both cell poles and the two organelles seem to move in opposite directions during cytokinesis (49, 50). Therefore, it is likely that some tension is applied to the cell body and has an effect on cell division. Observations of a living cell of *M. pneumoniae* suggest participation of gliding in cell division, namely through both organelle attachment to the glass surface and in pulling of the cell body in opposite directions (5). Although the cytadherence mutants of *M. pneumoniae* can grow normally, they often show branched structures (48, 50). Abnormal morphology can also be seen in the gliding mutants of *M. mobile*. Four non-motile mutant strains of *M. mobile* which cannot bind to glass tend to show abnormal cell morphology, although branch formation is not observed in this species (41). The abnormal morphology may suggest that normal cell division requires the tension caused by gliding of daughter cells in opposite directions.

## 7.3 Branch formation

What induces branching when normal cell division is inhibited? Abnormal branch formation of *M. capricolum* may suggest one possible mechanism (51, 52). When DNA replication of the wild-type strain of *M. capricolum* is inhibited by depletion of nucleotides, branches are formed with high frequency. In the branched cells, the replication intermediate of the chromosome is always found around the cell center. Presumably, the abnormal positioning of the chromosome interrupts cell division, and the excess membrane components supplied at the potential position of constriction form a branch.

## 8. PERSPECTIVES

Recent evidence has shown that the cell division processes of mycoplasmas are more organized than expected. It is an especially misleading concept that mycoplasmas have vague cellular structure and function (1). We should not ignore the advances in the field of bacterial cell biology in general, where the study of bacterial cell events has been moved forward by examination of subcellular architecture (i.e., chromosome segregation, cytokinesis, organelle formation, etc.). The concepts and methods from the field as a whole often have impacts on mycoplasmology. However, this does not mean that the cell biology of mycoplasmas is analogous to that of walled bacteria. Mycoplasmas have distinct features, i.e. lack of a cell wall, limited genetic information, and a parasitic lifestyle. These features make mycoplasma cells different from those of walled bacteria. On the contrary, a concept obtained in the mycoplasma field might have an impact on bacterial cell biology in general. The concept of subcellular protein localization and the use of immunofluorescence microscopy have been hallmarks of the mycoplasma field since the early 80s, when P1 protein localization was established for *M. pneumoniae* (2, 8, 20). This precedes their acceptance in the study of walled bacteria by about 10 years (3, 33). The simplicity of mycoplasmas may have worked well in this case. We can expect that another important discovery in mycoplasmology will foreshadow an influential contribution to the field of bacterial cell biology as a whole.

## ACKNOWLEDGMENTS

I am very grateful to Dr. Shintaro Seto (with whom I have enjoyed cell cycle studies of *M. capricolum* and *M. pneumoniae*) for invaluable comments on this manuscript, and also to Mr. Jake Jaffe for reading this manuscript.

## REFERENCES

1. **Alberts, B., D. Bray, J. Lewis, M. Raff, K. Roberts, and J. D. Watson.** 1994. Molecular biology of the cell., Third ed. Garland Publishing, Inc., New York & London.
2. **Baseman, J. B., R. M. Cole, D. C. Krause, and D. K. Leith.** 1982. Molecular basis for cytadsorption of *Mycoplasma pneumoniae*. J. Bacteriol. **151**:1514-1522.
3. **Bi, E. F., and J. Lutkenhaus.** 1991. FtsZ ring structure associated with division in *Escherichia coli*. Nature. **354**:161-4.
4. **Bramhill, D., and A. Kornberg.** 1988. A model for initiation at origins of DNA replication. Cell. **54**:915-8.
5. **Bredt, W.** 1968. Motility and multiplication of *Mycoplasma pneumoniae*. A phase contrast study. Pathol. Microbiol. **32**:321-6.
6. **Chambaud, I., R. Heilig, S. Ferris, V. Barbe, D. Samson, F. Galisson, I. Moszer, K. Dybvig, H. Wroblewski, A. Viari, E. P. Rocha, and A. Blanchard.** 2001. The complete genome sequence of the murine respiratory pathogen *Mycoplasma pulmonis*. Nucleic Acids Res. **29**:2145-53.
7. **Errington, J.** 1998. Dramatic new view of bacterial chromosome segregation. ASM News. **64**:210-217.
8. **Feldner, J., U. Göbel, and W. Bredt.** 1982. *Mycoplasma pneumoniae* adhesin localized to tip structure by monoclonal antibody. Nature. **298**:765-7.

9.   Fisseha, M., H. W. Gohlmann, R. Herrmann, and D. C. Krause. 1999. Identification and complementation of frameshift mutations associated with loss of cytadherence in *Mycoplasma pneumoniae*. J. Bacteriol. **181**:4404-10.

10.  Fraser, C. M., J. D. Gocayne, O. White, M. D. Adams, R. A. Clayton, R. D. Fleischmann, C. J. Bult, A. R. Kerlavage, G. Sutton, J. M. Kelley, R. D. Fritchman, J. F. Weidman, K. V. Small, M. Sandusky, J. Fuhrmann, D. Nguyen, R. Utterback, D. M. Saudek, C. A. Phillips, J. M. Merrick, J.-F. Tomb, B. A. Dougherty, K. F. Bott, P.-C. Hu, T. S. Lucier, S. N. Peterson, H. O. Smith, C. A. Hutchison III, and J. C. Venter. 1995. The minimal gene complement of *Mycoplasma genitalium*. Science. **270**:397-403.

11.  Fujita, M. Q., H. Yoshikawa, and N. Ogasawara. 1992. Structure of the *dnaA* and DnaA-box region in the *Mycoplasma capricolum* chromosome: conservation and variations in the course of evolution. Gene. **110**:17-23.

12.  Garnier, M., M. Clerc, and J. M. Bove. 1984. Growth and division of *Spiroplasma citri*: elongation of elementary helices. J. Bacteriol. **158**:23-8.

13.  Garnier, M., M. Clerc, and J. M. Bove. 1981. Growth and division of spiroplasmas: morphology of *Spiroplasma citri* during growth in liquid medium. J. Bacteriol. **147**:642-52.

14.  Glass, J. I., E. J. Lefkowitz, J. S. Glass, C. R. Heiner, E. Y. Chen, and G. H. Cassell. 2000. The complete sequence of the mucosal pathogen *Ureaplasma urealyticum*. Nature. **407**:757-62.

15.  Göbel, U., V. Speth, and W. Bredt. 1981. Filamentous structures in adherent *Mycoplasma pneumoniae* cells treated with nonionic detergents. J. Cell Biol. **91**:537-543.

16.  Helmstetter, F. C. 1996. Timing of synthetic activities in the cell cycle, p. 1627-1639. *In* F. C. Neidhardt (ed.), *Escherichia coli* and *Salmonella*. ASM press, Washington D. C.

17.  Hilbert, H., R. Himmelreich, H. Plagens, and R. Herrmann. 1996. Sequence analysis of 56 kb from the genome of the bacterium *Mycoplasma pneumoniae* comprising the dnaA region, the atp operon and a cluster of ribosomal protein genes. Nucleic Acids Res. **24**(4):628-39.

18.  Himmelreich, R., H. Hilbert, H. Plagens, E. Pirkl, B.-C. Li, and R. Herrmann. 1996. Complete sequence analysis of the genome of the bacterium *Mycoplasma pneumoniae*. Nucleic Acids Res. **24**:4420-4449.

19.  Hiraga, S. 2000. Dynamic localization of bacterial and plasmid chromosomes. Annu. Rev. Genet. **34**:21-59.

20.  Hu, P. C., R. M. Cole, Y. S. Huang, J. A. Graham, D. E. Gardner, A. M. Collier, and W. A. Clyde, Jr. 1982. *Mycoplasma pneumoniae* infection: role of a surface protein in the attachment organelle. Science. **216**:313-5.

21.  Kirchhoff, H. 1992. Motility, p. 289-306. *In* J. Maniloff, R. N. McElhaney, L. R. Finch, and J. B. Baseman (ed.), Mycoplasmas- Molecular Biology and Pathogenesis. American Society for Microbiology, Washington, D.C.

22.  Krause, D. C. 1998. *Mycoplasma pneumoniae* cytadherence: organization and assembly of the attachment organelle. Trends in Microbiol. **6**:15-18.

23.  Krause, D. C. 1996. *Mycoplasma pneumoniae* cytadherence: unravelling the tie that binds. Mol. Microbiol. **20**:247-53.

24.  Krause, D. C., and M. F. Balish. 2001. Structure, function, and assembly of the terminal organelle of *Mycoplasma pneumoniae*. FEMS Microbiol. Lett. **198**:1-7.

25.  Krause, D. C., D. K. Leith, R. M. Wilson, and J. B. Baseman. 1982. Identification of *Mycoplasma pneumoniae* proteins associated with hemadsorption and virulence. Infect. Immun. **35**:809-17.

26. **Kukekova, A. V., A. Y. Malinin, J. A. Ayala, and S. N. Borchsenius.** 1999. Characterization of *Acholeplasma laidlawii ftsZ* gene and its gene product. Biochem Biophys Res. Commun. **262**:44-9.

27. **Layh-Schmitt, G., and M. Harkenthal.** 1999. The 40- and 90-kDa membrane proteins (ORF6 gene product) of *Mycoplasma pneumoniae* are responsible for the tip structure formation and P1 (adhesin) association with the Triton shell. FEMS Microbiol. Lett. **174**:143-9.

28. **Layh-Schmitt, G., and R. Herrmann.** 1994. Spatial arrangement of gene products of the P1 operon in the membrane of *Mycoplasma pneumoniae*. Infect. Immun. **62**:974-9.

29. **Layh-Schmitt, G., R. Himmelreich, and U. Leibfried.** 1997. The adhesin related 30-kDa protein of *Mycoplasma pneumoniae* exhibits size and antigen variability. FEMS Microbiol. Lett. **152**:101-8.

30. **Layh-Schmitt, G., A. Podtelejnikov, and M. Mann.** 2000. Proteins complexed to the P1 adhesin of *Mycoplasma pneumoniae*. Microbiology. **146**:741-747.

31. **Lowe, J., and L. A. Amos.** 1998. Crystal structure of the bacterial cell-division protein FtsZ. Nature. **391**:203-6.

32. **Lutkenhaus, J., and S. G. Addinall.** 1997. Bacterial cell division and the Z ring. Annu. Rev. Biochem. **66**:93-116.

33. **Maddock, J. R., and L. Shapiro.** 1993. Polar location of the chemoreceptor complex in the *Escherichia coli* cell. Science. **259**:1717-23.

34. **Margolin, W.** 1999. The bacterial cell divisin machine. ASM News. **65**:137-143.

35. **Meng, K. E., and R. M. Pfister.** 1980. Intracellular structures of *Mycoplasma pneumoniae* revealed after membrane removal. J. Bacteriol. **144**:390-399.

36. **Miyata, M., and T. Fukumura.** 1997. Asymmetrical progression of replication forks just after initiation on *Mycoplasma capricolum* chromosome revealed by two-dimensional gel electrophoresis. Gene. **193**:39-47.

36a. **Miyata, M., W. S. Ryu and H.C. Berg.** 2002. Force and velocity of *Mycoplasma mobile* gliding. J. Bacteriol. **184:** in press

37. **Miyata, M., K. Sano, R. Okada, and T. Fukumura.** 1993. Mapping of replication initiation site in *Mycoplasma capricolum* genome by two-dimensional gel-electrophoretic analysis. Nucleic Acids Res. **21**:4816-23.

38. **Miyata, M., and S. Seto.** 1999. Cell reproduction cycle of mycoplasma. Biochimie. **81**:873-8.

39. **Miyata, M., L. Wang, and T. Fukumura.** 1993. Localizing the replication origin region on the physical map of the *Mycoplasma capricolum* genome. J. Bacteriol. **175**:655-660.

40. **Miyata, M., L. Wang, and T. Fukumura.** 1991. Physical mapping of the *Mycoplasma capricolum* genome. FEMS Microbiol. Lett. **63**:329-334.

41. **Miyata, M., H. Yamamoto, T. Shimizu, A. Uenoyama, C. Citti, and R. Rosengarten.** 2000. Gliding mutants of *Mycoplasma mobile*: relationships between motility and cell morphology, cell adhesion and microcolony formation. Microbiology. **146**:1311-1320.

42. **Morowitz, H. J., and J. Maniloff.** 1966. Analysis of the life cycle of *Mycoplasma gallisepticum*. J. Bacteriol. **91**:1638-44.

43. **Popham, P. L., T. W. Hahn, K. A. Krebes, and D. C. Krause.** 1997. Loss of HMW1 and HMW3 in noncytadhering mutants of *Mycoplasma pneumoniae* occurs post-translationally. Proc. Natl. Acad. Sci. U S A. **94**:13979-84.

44. **Quinlan, D. C., and J. Maniloff.** 1972. Membrane association of the deoxyribonucleic acid growing-point region in *Mycoplasma gallisepticum*. J. Bacteriol. **112**:1375-9.

45. **Razin, S., and E. Jacobs.** 1992. Mycoplasma adhesion. J. Gen. Microbiol. **138**:407-22.

46.  **Regula, J. T., G. Boguth, A. Görg, J. Hegermann, F. Mayer, R. Frank, and R. Herrmann.** 2001. Defining the mycoplasma 'cytoskeleton': the protein composition of the Triton X-100 insoluble fraction of the bacterium *Mycoplasma pneumoniae* determined by 2-D gel electrophoresis and mass spectrometry. Microbiology. **147**:1045-1057.

47.  **Renaudin, J., A. Marais, E. Verdin, S. Duret, X. Foissac, F. Laigret, and J. M. Bove.** 1995. Integrative and free *Spiroplasma citri oriC* plasmids: expression of the *Spiroplasma phoeniceum* spiralin in *Spiroplasma citri*. J. Bacteriol. **177**:2870-7.

48.  **Romero-Arroyo, C. E., J. Jordan, S. J. Peacock, M. J. Willby, M. A. Farmer, and D. C. Krause.** 1999. *Mycoplasma pneumoniae* protein P30 is required for cytadherence and associated with proper cell development. J. Bacteriol. **181**:1079-87.

49.  **Rosengarten, R., and H. Kirchhoff.** 1989. Growth morphology of *Mycoplasma mobile* 163K on solid surfaces: Reproduction, aggregation, and microcolony formation. Curr. Microbiol. **18**:15-22.

50.  **Seto, S., G. Layh-Schmitt, T. Kenri, and M. Miyata.** 2001. Visualization of the attachment organelle and cytadherence proteins of *Mycoplasma pneumoniae* by immunofluorescence microscopy. J. Bacteriol. **183**:1621-30.

51.  **Seto, S., and M. Miyata.** 1998. Cell reproduction and morphological changes in *Mycoplasma capricolum*. J. Bacteriol. **180**:256-64.

52.  **Seto, S., and M. Miyata.** 1999. Partitioning, movement, and positioning of nucleoids in *Mycoplasma capricolum*. J. Bacteriol. **181**:6073-80.

53.  **Seto, S., S. Murata, and M. Miyata.** 1997. Characterization of *dnaA* gene expression in *Mycoplasma capricolum*. FEMS Microbiol. Lett. **150**:239.

54.  **Skarstad, K., and E. Boye.** 1994. The initiator protein DnaA: evolution, properties and function. Biochim. Biophys. Acta. **1217**:111-130.

55.  **Suzuki, K., M. Miyata, and T. Fukumura.** 1993. Comparison of the conserved region in the *dnaA* gene from three mollicute species. FEMS Microbiol. Lett. **114**:229-233.

56.  **Trachtenberg, S.** 1998. Mollicutes-wall-less bacteria with internal cytoskeletons. J. Struct. Biol. **124**:244-56.

57.  **Wang, X., and J. Lutkenhaus.** 1996. Characterization of the *ftsZ* gene from *Mycoplasma pulmonis*, an organism lacking a cell wall. J. Bacteriol. **178**:2314-9.

58.  **Weisburg, W. G., J. G. Tully, D. L. Rose, J. P. Petzel, H. Oyaizu, D. Yang, L. Mandelco, J. Sechrest, T. G. Lawrence, J. Van Etten, J. Maniloff, and C. R. Woese.** 1989. A phylogenetic analysis of the Mycoplasmas: Basis for their classification. J. Bacteriol. **171**:6455-6467.

59.  **Wheeler, R. T., and L. Shapiro.** 1997. Bacterial chromosome segregation: is there a mitotic apparatus? Cell. **88**:577-9.

60.  **Wilson, M. H., and A. M. Collier.** 1976. Ultrastructural study of *Mycoplasma pneumoniae* in organ culture. J. Bacteriol. **125**:332-9.

61.  **Ye, F., J. Renaudin, J. M. Bove, and F. Laigret.** 1994. Cloning and sequencing of the replication origin (*oriC*) of the *Spiroplasma citri* chromosome and construction of autonomously replicating artificial plasmids. Curr. Microbiol. **29**:23-9.

# Chapter 7

# The Cell Membrane and Transport

ÅKE WIESLANDER and MARIA ROSÉN
*Department of Biochemistry and Biophysics, Arrhenius Laboratories, Stockholm University, 116 91 Stockholm, Sweden*

## 1. INTRODUCTION

*Membranes define cellular units.* Membranes are integral parts in the basic concept of the cell as the smallest unit of life. This is structurally evident for the smallest and simplest of cells, i.e. *Mollicutes* and certain *Buchnera* symbionts, as well as for the largest and most elaborate eukaryotic cells with many membrane-surrounded compartments and organelles housing various specialized functions and systems. Analyses of the potential numbers of encoded membrane proteins in a number of various, sequenced genomes substantiates the importance of membrane proteins, indicating that proteins with one, or more, transmembrane segments constitute 20 to 30% of all cellular protein species[97, 108]. Furthermore, certain membrane transporters and receptors constitute the major protein classes in prokaryotes and eukaryotes.

*Membrane functions.* Membrane proteins in prokaryotes function as transporters, enzymes, energy metabolism, sensors, and motility devices, according to the many genomes now sequenced, see e.g. *Bacillus subtilis*[50]. The membrane lipid bilayers provide the ultimate barrier and permeability functions, supported by various outside lipopolysaccharide and polysaccharide-capsule layers. These latter molecules are only indirectly encoded in the genomes as enzyme genes, often organized in operons, and hence their composition and structure cannot easily be deduced from translated sequences. Historically, studies of the mycoplasma membrane functions have been focused on rather few areas, e.g. lipid bilayer composition related to permeability, and transport of certain compounds.

*Molecular Biology and Pathogenicity of Mycoplasmas*, Edited by Razin and
Herrmann, Kluwer Academic/Plenum Publishers, New York, 2002

Much of this work was done with species of less medical importance, and not on the four mollicutes presently (2001) sequenced. Unfortunately, these early membrane studies have not been followed up, or confirmed in the sequenced species. *Mycoplasma genitalium*, and to a certain extent *Mycoplasma pneumoniae*, due to their minimal genome size have been the focus of a number of sequence prediction efforts to establish the types and numbers of protein folds and structural groups (see summary in Teichmann *et al.*[98]). However, since rather few membrane protein structures have been determined, these structural genomics effort contribute substantially less to the understanding of membrane protein structure and functions. Hence, we are left with the trials to associate "old" biochemical data for certain species and functions, with many "deduced" functions from the sequenced genomes of other *Mollicutes*. In addition to the problems of correctly annotating functions to genes on the basis of sequence similarity to established genes in other bacteria, at least 30% of the genes in the four sequenced Mollicute genomes have only hypothetical/unknown functions. Here, we will try to summarize important new, basic concepts that substantially add to the data published in previous books and reviews about mycoplasma cell and molecular biology[4, 57, 78]. The reader is also referred to a special volume on mycoplasma membranes[85], as well as to an update on Mollicutes metabolism[75].

## 2.    STRUCTURE AND FUNCTION OF MEMBRANE PROTEINS

### 2.1    General properties

*Protein types.* Membrane proteins are usually divided into two groups on the basis of the type of their interaction with the lipid bilayer, which in turn relates to the methods used for their release or solubilization. The peptide chains of *integral proteins* cross the hydrophobic lipid core one, several, or many times. Most often the transmembrane (TM) segments are composed of approximately 20 fairly hydrophobic amino acid residues in an α-helix configuration. Such segments are easily recognized at the sequence level by computer programs. The so called "positive-inside-rule", i.e. a certain enrichment of positively charged residues close to the TM segments on the cytoplasmic side (*vs* the exterior side)[107], also often makes prediction of the orientation of even a multispanning protein possible. However, the distinction of less hydrophobic, or more amphiphilic, TM segments (α-helix or β-strand) are more difficult. Likewise, a clear separation between the N-terminal, TM anchor segments and N-terminal signal peptides, i.e. the cleavable address tags for protein secretion or export, is still not properly achieved by the computer algorithms.

*Peripheral proteins* lack TM segments, and hence are soluble, but are associated with the membrane surfaces (protein and/or lipid) by mainly electrostatic (charge/charge) interactions. A certain fraction of these proteins can also have their anchoring enhanced by one, two, or three hydrophobic (usually acyl) chains, where more chains yield a stronger association. Such lipoproteins are more common in Gram-positive bacteria, due to the lack of an entrapping periplasmic space, and are especially prevalent in Mollicutes (see below).

*Protein charge.* With respect to their charge properties prokaryotic and eukaryotic proteins from organisms with sequenced genomes seem to form two or three clusters, respectively, as seen on both 2D-gels and their computed isoelectric points. A recent analysis clearly revealed that the clusters with high pI's in both groups contain mainly membrane-bound or – associated proteins[89]. A high pI, e.g. 9 to 11-12 or so, implies that the protein will have a net positive charge at physiological pH ($\approx$7) and hence be prone to bind to a negatively charged surface, like a phospholipid bilayer surface. The fractions of such encoded proteins are large in all the genomes analyzed, but in Mollicutes they are even larger and seem to constitute the majority[40].

## 2.2    Types and numbers of Mollicute proteins

*Membrane protein numbers.*    All genomes analyzed so far seem to encode a substantial fraction of intergral membrane proteins, 20-30% of all ORFs, depending on the algorithms used[108]. Large eukaryotic cells with more extensive and elaborate membrane systems have larger percentages than small ones and prokaryotic cells. Table 1 presents numbers compiled from the published Mollicute genomes. For a comparision to be valid (similar algorithms etc.), the figures were taken from sources where many genomes are compared. Furthermore, comparisons with other Gram-positive organisms are more informative because of their close genetic relatedness to Mollicutes; *Escherichia coli* (and other Gram-negative bacteria) also have a substantial number of amphiphilic β-barrel proteins (e.g. porins) in their outer membranes absent from the typical Gram-positive bacteria. It is evident that the Mollicutes have a substantial fraction of integral membrane proteins, perhaps slightly more than the common, larger bacteria. The larger fraction of integral proteins in *M. pulmonis*, one of the four Mollicutes in Table 1, was attributed to a higher number of membrane transport proteins[13]. In most genomes, including those of mycoplasmas, encoded proteins with 2 to 4 TM segments seem to be the dominant membrane protein classes. For

*M. genitalium* and *M. pnenumoniae* an analysis revealed an exception, i.e. the 6 TM segments class is the second largest[108]. This appears to represent the so called ABC-transporters (see below).

*Table 1.* Proteins Containing Transmembrane Segments (TM) and Signal Peptides (SP) According to Computer Analyses of Complete Genomes

| Organism | ORFs with / without TM[a] | SP[b] | Lipoproteins[b] |
|---|---|---|---|
| *Bacillus subtilis* | 1331 / 2769 | 294 | 114 |
| *M. pneumoniae* | 188 / 489 | 116 | 49 |
| *M. genitalium* | 75 / 405 | 113 | 21 |
| *M. pulmonis* | 274 / 508 | --c | 56 |
| *U. urealyticum* | 193 / 418 | 119 | 74 |

[a]From the PEDANT homepage of genomes at the Munich Information Center for Protein Sequences (http://pedant.gsf.de).
[b]According to algorithms by Saleh *et al.*[87]. For *B. subtilis* the figures derive from detailed analyses of type I and II signal peptides[102], yielding lower numbers than the algorithm. The *M. pulmonis* and *U. urealyticum* lipoproteins are derived from the published genomes. Lipoproteins are included in the SP numbers.
[c]Data not available.

*Lipoproteins.* High numbers of lipoproteins (Table 1) in the Mollicutes, detected by the presence of type II signal peptides with a cysteine residue at the potential cleavage site, confirm earlier observations based on fatty acid labelling, and on visualization of such proteins on 2D-gels[67]. Genome analyses have indicated even higher numbers of lipoproteins for larger bacteria, like *B. subtilis* (Table 1) and more than 90 in *E. coli*[99]. Evidence is accumulating that several mycoplasma lipoproteins lack the third acyl chain, attached by an amide bond to the free amino group of the N-terminal Cys residue in other bacteria[14]. The gene encoding the N-acyltransferase responsible for the amide linkage seems to be absent from the four sequenced mollicute genomes (Table 1), whereas the genes for the two enzymes responsible for adding the diacylglycerol moiety were identified. This may be a characteristic feature typical for certain bacterial groups; the N-acyltransferase has not been identified in several larger genomes, including *B. subtilis*[102]. However, the typical glyceryl-cysteine moiety carrying the acyl chains is present in a number of Mollicutes[51], and many of their lipoproteins seem to have two ester-linked acyl chains[51, 67]. Two acyl chains must yield weaker anchoring than three chains, as was indicated by studies of certain eukaryotic proteins, facilitating perhaps the release of these proteins from the outer membrane surface of the mollicutes. In *B.subtilis* several lipoproteins were found in the extracellular surroundings, potentially released also by proteolytic processing[3]. Numbers and extent of released proteins could be coupled to the physiological state, and fragments of some transmembrane proteins were also found[3]. *E. coli* has a special ABC transporter system for releasing intact lipoproteins into the periplasmic space for further transfer to the outer membrane[113]. As could be expected no genes for the transmembrane or periplasmic components of this system seem to be present in *M. pneumoniae*, but the intracellular ATP-binding protein (LolD) is very similar to the mycoplasma orf MPN081 according to the COG database at NCBI (unpublished observation), annotated as a glutamine transport ATP-binding protein[21].

*Valid numbers.* Unfortunately, membrane proteins are strongly under-represented (for technical reasons) in the "proteome" sets visualized by 2D gel electrophoresis, as was clearly illustrated for *M. pneumoniae*[79]. Hence, the numbers in Table 1 should be interpreted as rough guides, awaiting confirmation by biochemical and genetic analyses. It is also anticipated that improved prediction algorithms[44] and the impact of the pI values (see above), may slightly decrease the predicted numbers for TM proteins with one or two helices, and increase the peripheral protein category, respectively.

## 2.3    Major membrane protein functions

*Established functions.* A number of membrane proteins from Mollicutes have been purified a long time ago, many lipoproteins were detected and visualized by radioactive labelling, and a substantial number of membrane proteins have been specifically cloned (summarized in Wieslander *et al.*[109]). However, for essentially none of these proteins have their structure and function been established. The few exceptions are the lipoprotein spiralin from *Spiroplasma melliferum*, for which a membrane topology and possible function have been suggested (Wroblewski and coworkers), as well as for some proteins of the cytoskeleton and those functioning in *M. pneumoniae* adherence (e.g.[49]; see also "Cytadherence and cytoskeleton" chapter). We will therefore discuss some general functional features and principles that can be deduced from analyses and comparisons of membrane protein sequences.

### 2.3.1    Protein secretion and the Sec system

*Secretion signals.* A general, key process in cells is the way protein components are targeted to, inserted into, translocated (secreted) through one or more membranes, and eventually released into (exported) the outside compartment. It has been established that a number of components in the "general secretion pathway" (Sec) are evolutionary conserved in prokaryotes, and in the endoplasmic reticulum of eukaryotes. Essential factors of the system are the N-terminal signal peptides (type I), certain chaperons and receptors, and a translocation pore assembly. Although the bacterial components of this system are best known for *E. coli*, certain differences are recorded for Gram-positive organisms[103] and therefore a comparison with the machinery in *B. subtilis* is more relevant for Mollicutes. In *B.subtilis* four types of N-terminal secretion/export signals are known[102]. (i) The archaetypical, prokaryotic and eukaryotic signal peptide, cleaved by a type I signal peptidase; six of the latter have been identified at the genome level. A specific subgroup is formed by the so-called twin-arginine (RR) motif ones, directing certain partially folded proteins into the distinct Tat (twin-arginine) pathway. (ii) Lipoprotein signal peptides, cleaved by a type II signal peptidase at a conserved Cys-containing motif (one present). (iii) Special signal peptides of prepilins, cleaved by the ComC peptidase (type IV secretion). (iv) The special signal peptides found on certain bacteriocins and pheromones exported by certain ABC transporters, and cleaved by a subunit of the latter. The type I signal peptides of several *Bacillus* species are substantially longer (average 28 aa) than their *E. coli* counterparts (average 22.6 aa), and also longer than the mollicutes ones (average 25 aa)[25, 102]. However, the mollicutes lipoprotein peptides are of equal length as the non-lipoprotein ones, whereas the *Bacillus* and *E. coli* lipoprotein peptides are 19-22 aa (average), i.e. shorter than the non-lipoprotein ones. In addition, patterns of hydrophobicity were clearly different between both the peptides from various cellular compartment proteins, and between different bacterial species, respectively[25]. In *E. coli*, these patterns were different for signal peptides on proteins headed for different cellular compartments, and certain Bacillus enzymes had peptide property patterns conserved over the species borders. Hence, the molecular interactions between various components in this protein traffic route may not be identical in various bacteria. This illustrates the presence of important differences, for a very conserved signal in Nature, between the Mollicutes and both Gram-positive and Gram-negative bacteria.

*Secretory machinery.* Many of the components in the general Sec-dependent protein secretion pathway of *B. subtilis* can be identified in Mollicutes, but the similarities are weak for certain proteins. Table 2 illustrates the classes and types of components. Several of the proteins are conserved in all bacteria, whereas a substantial fraction is typical for the Gram-positive ones. Generally, there are substantially fewer "accessory" factors (e.g. chaperons) present in Mollicutes. Note that the *E. coli* chaperon SecB, potentially interacting with outer membrane preproteins, is absent in *B.subtilis*[103] and was not identified in the Mollicutes. Protein Ffh ("fiftyfour homolog") together with a typical, small RNA forms the signal recognition particle (SRP), which in *E. coli* seems preferentially engaged in recognizing integral membrane proteins as they protrude from the ribosome. SRP then interacts with the FtsY receptor, and eventually secretion proceeds through the SecEYG channel. Only in *M. pulmonis* a homologous type I signal peptidase was clearly identified, raising doubts that standard signal peptides, observed in the sequences of many proteins in Mollicutes, are processed as in other bacteria. We must await properly executed biochemical analyses to solve this issue. The presence of a gene for a lipoprotein signal peptidase (type II) in all the mollicute genomes analysed so far, is supported by several functional analyses (reviewed by Chambaud *et al*[14].). An integral translocation channel factor ("YidC"), conserved in most bacteria, mitochondria and chloroplasts, needed for the insertion of integral membrane proteins[23, 88], is also encoded in the Mollicutes genomes according to a directed BLAST search (unpublished observation), cf. Table 2. In *E. coli* this protein is much more abundant than the SecY channel component. *B. subtilis* "YidC" (i.e. SpoIIIJ) is a lipoprotein, and also an integral protein spanning the membrane, which is secreted by the special Tat (twin-arginine) pathway[45]. The latter can handle the transfer of partially folded proteins and is conserved in many bacteria and in chloroplasts. A twin-arginine motif can be found in a few mollicutes signal peptides (unpublished observation). Furthermore, lipid-modified membrane-spanning proteins are indicated in the mollicutes genome sequences, and biochemically verified for the b-subunit of the *M.pneumoniae* $F_OF_1$-ATPase[77] and in *A. laidlawii* membranes[67]. However, only very weak sequence similarities to the *B. subtilis* TatCd protein (cf.[45]) were recorded for two ORFs; MG384.1 and MPN565 in *M. genitalium* and *M. pneumoniae*, respectively. The latter was annotated as a conserved hypothetical protein[21]. A number of pathogenic Gram-negative bacteria have a special export system for transferring proteins directly into their eukaryotic host cells (e.g.[73]), called type III secretion. Weak similarities were recorded for two *Yersinia pestis* type III components and the UU400 and UU384 ORFs in *U. urealyticum*. Likewise, type IV secretion components (cf. (iii) above) could not be identified in mollicutes by comparison with *B. subtilis* Com components.

*Post-translational processing.* In Gram-positive bacteria proteins can also be anchored to various polymeric components of the cell envelope, outside of the cytoplasmic membrane[66]. A conserved mechanism involves a transpeptidase ("sortase") for cleaving membrane protein chains and attaching the soluble part of the protein to an amino acid residue of the peptide repeat in the peptidoglycan mesh[60, 66]. The *Staphylococcus aureus* and *B. subtilis* (YwpE) sortases have some, but not strong sequence similarities to ORF MPN369, which is related to a number of *M. pneumoniae* lipoproteins. All of these lipoproteins are encoded close to or within REP segments in the chromosome. This is also the case for several variants/fragments of the major P1 adhesin, a protein exposed to a post-translational proteolytic processing[49]. The MPN369 protein is different from

MPN386, which was recently proposed as an alternative to the missing type I signal peptidase in *M. pneumoniae*[21].

*Table 2.* Components of Sec-dependent Protein Export

| Component[a] | *B. subtilis*[a] | Mollicutes[b] |
|---|---|---|
| Secretion-dependent chaperons | Ffh / FtsY / scRNA<br>CsaA / MrgA | MPN061 / 425 / +<br>- / - |
| General chaperons | GroEL / GroES<br>DnaK / DnaJ / GrpE<br>Trigger factor | MPN573 / 574<br>MPN434 / 021 / 120<br>MPN331 |
| Translocation motor | SecA | MPN210 |
| Translocation channel | SecY / G / E<br>SecDF / YrbF ("YajC")<br>"YidC" (SpoIIIJ) | MPN184 / 242 / 068<br>MPN396 (SecD) / -<br>MPN680 |
| Signal peptidases | SipS / T / U / V / P / W<br>LspA (type II) | MPU6300 / (MPN627/684)<br>MPN293 |
| SP-degrading peptidases | SppA / TepA | - / - |
| Foldases | PrsA<br>BdbA / B / C | (UU558)<br>MPN263 / - / - |

[a]As outlined in Tjalsma *et al.*[102], and van Wely *et al.*[103].
[b]*M. pneumoniae* homologs (analogs) were identified by BLAST searches in the COG database at NCBI, using the *B. subtilis* proteins as search probes, and presented as the *M. pneumoniae* MPN gene numbers[21]. Most often homologs were also identified in *M. genitalium* and *U. urealyticum* (*M. pulmonis* is not in the database; MPU6300 hit from Chamboud *et al.*[13]). +, the scRNA gene is present in all four sequenced genomes. Numbers within parantheses indicate weak similarity; MPN627 have another assigned function[21], and MPN263 encodes thioredoxin similar to the BdbA gene of *B. subtilis*. "YajC" and "YidC" are the corresponding *E. coli* names for YrbF and SpoIIIJ, respectively.

*In conclusion*, the Mollicutes seem to have the core components of the conserved Sec secretion machinery of Gram-positive bacteria, but lack many of the accessory proteins (chaperons). So far, no other of the "specialized" secretion systems in other bacteria have been clearly observed in Mollicutes; however, mollicute genomes reveal weak sequence similarities to some components of these systems.

### 2.3.2    Membrane transport systems

*Transporter profiles.* Membrane transporter proteins constitute a major class of proteins in prokaryotes. The numbers, types and substrate specificities of the transporters in a bacterium, as deduced from genome sequence analyses, seem to yield a corresponding image to the metabolic traits determined by conventional biochemical means[69]. That is, for compounds used or metabolized a transport system is often found or indicated at the genome level. However, sequence analyses have revealed far more transport systems than was known (or expected) from biochemical analyses, especially for bacteria with larger genomes, e.g *E. coli, B. subtilis* and others. It was also anticipated that Mollicutes, with their small genomes and complex or extensive nutritional demands, should have a larger fraction of membrane transporters (on a relative basis) to be able to take up all essential nutrients needed. However, the number of transporters in the Mollicutes seem to be average, on basis of chromosome length, when compared to the numbers for other genomes[69]. For several important metabolites needed by cells, the numbers of transport systems can vary from one up to several (or even many). In order to compare transporter composition ("profiles") in different groups of prokaryotes, it is more fruitful to use an approach based on the structural organisation and energy demand of the transport system. Furthermore, for Mollicutes a comparison with fermentative Gram-positive bacteria is more relevant, but too few complete sequences of the latter have been analyzed yet. In Table 3 transport systems, indicated at the sequence level in *M. pneumoniae* and *B. subtilis*, are grouped according to the recent nomenclature for transport systems[86]. Figures and distribution are essentially identical between *M. genitalium* and *M. pneumoniae*, containing 20 and 22 recognized systems, respectively[69].

*Channel proteins* (group A in Table 3) comprise of α-helical channels, β-barrel proteins, pore-forming toxins, and non-ribosomally synthesized channels. These usually catalyze the movement of solutes by an energy-independent process by passage through a transmembrane aqueous pore. This group is a minor one in both the organisms shown in the table as well as in many other prokaryotes[69], with one member of the MIP family (glycerol facilitators and aquaporins) present in *M. pneumoniae* (Table 3).

*Secondary transporters* (group B in Table 3) are driven by a transmembrane electrochemical potential derived from protons, sodium or other ions[86]; this potential is hence the secondary energy source, derived from the primary metabolism. It is the largest group of transporters in conventional bacteria, and mechanistically these systems can function as uniporters, symporters or antiporters. A correlation between the numbers of these systems and the ability of the organism to generate an efficiently high electrochemical potential from an electron transport chain seems to be valid[69], as shown here for *B. subtilis*. Consequently, this type of transporters are less abundant in bacteria with less efficient energy-generating mechanisms like the Mollicutes (Table 3). This situation is also reflected in the numbers of transmembrane (TM) segments observed in the membrane proteins. Proteins with many TM segments (10 or more), typical for secondary transporters, are substantially less abundant among the Mollicutes membrane proteins than in e.g. *B. subtilis* and *E. coli* [69, 108].

*Table 3.* Distribution of Membrane Transport Systems

| Protein families[a] | *M. pneumoniae* | *B.subtilis* |
|---|---|---|
| A. *Channel proteins* (4 families)[b] | 1  (5%)[c] | 3 (1%)[c] |
|     MIP (major intrinsic protein fam.) | 1 | 1 |
| | | |
| B. *Secondary transporters* (54 families) | 5  (23%) | 168 (63%) |
|     MFS (major facilitator sup.fam.) | 1 | 65 |
|     APC (aa-polyamine-org. cation fam.) | 3 | 18 |
|     Trk ($K^+$ transporter fam.) | 1 | 2 |
| | | |
| C. *Primary active transporters* (4 families) | 13  (59%) | 71  (27%) |
|     ABC (ATP-binding-casette superfam.) | 11 | 64 |
|     F-ATPase ($H^+/Na^+$ transloc., F-type) | 1 | 1 |
|     P-ATPase (P-type ATPase superfam.) | 1 | 4 |
| | | |
| D. *Group translocators* (6 families) | 3  (14%) | 17  (6%) |
|     Glc (PTS glucose/glucoside fam.) | 1 | 10 |
|     Fru (PTS fructose-mannitol fam.) | 1 | 3 |
|     Man (PTS man/fru/sor fam.) | 1 | 1 |
| | | |
| E. *Unclassified* (8 families) | 0  (0%) | 6  (2%) |
|              Total | 22 | 265 |

[a]Transporter families outlined from 18 prokaryotic genomes, of various bacterial systematic groups, by especially BLAST analysis of translated sequences, and by comparison with established and/or cloned systems from *E. coli* and *B. subtilis*[69]. The systems are grouped according to the recent, official IUBMB transport nomenclature; see[86] for an extensive description.
[b]Number of families found for each of the major groups A to E in all the 18 genomes. Names of *individual families* are only given for the ones present in *M. pneumoniae*. Note that *B. subtilis* contains substantially more families (not shown), see[69].
[c]Figures within parentheses denote the approximate percentages for each of the transporter groups in relation to the total number of transport systems suggested in the two species (i.e. 22 and 265), respectively.

*Primary active transporters* use a chemical, electrical, or solar source, to directly drive solute transport[86]. In Mollicutes, as well as in other "energy-poor" bacteria, the energy source is ATP derived mainly from substrate-level phosphorylation (e.g. glycolysis). The importance of these systems in Mollicutes is illustrated by the comparison in Table 3. Although the ABC transporters are the most numerous groups of all proteins in both *B. subtilis* and *E. coli*[10, 50], their percentage in Mollicutes is much higher than in *B. subtilis* (Table 3). This is also reflected in the number of TM segments in

Mollicutes membrane proteins in general; proteins with six segments typical for the ABC proteins constitute a major class in Mollicutes[108]. An ABC import transporter complex often consists of two transmembrane ("channel") proteins, two ATP-hydrolyzing proteins driving the transport process, and two external substrate-binding proteins, the latter being often lipoproteins in Gram-positive bacteria. Frequently, the ABC transporter genes are organized in operons but other chromosomal arrangements also seem common, especially for the binding proteins. For ABC exporters, the ATP-binding subunits are often fused to the transmembrane part, and the substrate-binding partners are absent. They are involved in transporting a large variety of small and large molecules, and are the molecular sites for a number of inherited human diseases. It should be pointed out that the F-ATPase (group C in Table 3) of most Mollicutes acts as an outward proton pump, creating a fairly small proton gradient by the expenditure of ATP[75]; the magnitude of the electrical potential can be larger[16, 17, 91]. A $Na^+$-dependent electrochemical potential, found in a large number of bacterial pathogens, is probably not present in Mollicutes as evidenced by the absence of a gene coding for a key primary $Na^+$ pump in *M. genitalium* and *M. pneumoniae* genome sequences[37]. The transmembrane b-subunit of the ATPase, acting as a stator between the $F_0$-$F_1$ units in the rotatory processes, is lipid-modified in *M. pneumoniae* (see above) and most likely also in *M. genitalium, M. gallisepticum* and *U. urealyticum* as indicated by the presence of a lipoprotein signal peptide cleavage site (with a Cys residue), but not in *M. pulmonis*, according to the genome sequences (data not shown). This may be an improved anchor property related to the mechanism of "reverse" action for the ATPases, e.g. extent or absense of proton pumping. In *M. pulmonis* there are also more genes for the ATPase subunits, which are differently organized on the chromosome[13].

*Group translocator systems.* The presence and function of the phosphoenol-pyruvate sugar phosphotransferase systems (PTS), belonging to the group translocator transport systems (group D in Table 3), have been functionally studied and verified in a number of *Spiroplasma* and *Mycoplasma* species, whereas they seem to be absent in *Acholeplasma* species (reviewed in Cirillo[16]). These systems are also absent from archea and eukaryotes, and contrary to expectations seem not to be widespread in eubacteria; only six of 14 bacterial genomes contained a PTS system as identified at the genomic level[69]. Their relative high fraction in the *M. pneumoniae* genome, i.e. 14 % (Table 3), adds to the importance of the primary transporters (59 %) and shows that transport systems driven by glycolytic intermediates in *M. pneumoniae* constitute almost three quarters of all systems.

*Types and specificities.* The fraction of unclassified systems (group E in Table 3) is 0 % in *M. pneumoniae*. However, this relates to transporter types and structural features; with respect to specificities for substrates the uncertainty is substantially higher. For *M. pneumoniae* and the other sequenced Mollicutes, transporter substrate specificities can only be proposed on the basis of amino acid sequence similarities to other, established systems; for none of the compounds listed in Table 4 has the corresponding transporter been biochemically (functionally) verified in *M. pneumoniae*. Still, some characteristic features have been evident: (i) Systems for sugars, amino acids and their derivatives are identified, but not for organic cations. (ii) The number of systems for inorganic ions is low. This may be typical for some pathogens; archaea and free-living eubacteria have a much higher fraction. (iii) No systems were yet identified in *M. pneumoniae* for nucleic acid precursors and vitamin transport. These are indeed essential nutrients, and the failure to demonstrate them illustrates the problem of correlating biochemical data with sequence annotations[74]. (iv) No system for drug efflux has been properly recognized, as well as no system for macromolecule (e.g. lipopolysaccharide and/or capsule precursor) efflux. The latter is a unique deficiency noted for *M. pneumoniae* and *M. genitalium* of the 18 genomes analyzed[69]. However, capsules have been clearly identified in many species, including *M. pneumoniae*, *M. pulmonis* and *U. urealyticum*[64]. Drug efflux, not recognized at the gene level in *M. pneumoniae* (Table 3 and 4), may be more common for soil bacteria.

*Ecological niche and genome size.* The comparisons in Table 3 clearly show the profound differences in transporter profiles, and hence metabolic capacities and ecological niches, between organisms with a fermentative and an oxidative energy metabolism, respectively. Table 3 and 4 together also illustrate the differences between small and large prokaryotic genomes; for the latter cells there are more transporter families and more paralogs within the families, yielding more transporters per DNA unit length. This leads to a broader range of substrates that can be transported but also to a redundancy, e.g. 7 systems for glucose transport in *E. coli*[69]. Correspondence between ecological niches and transporter profiles is also obvious when comparing *M. pneumoniae* with *M. pulmonis* and *U. urealyticum*, which are all mucosal pathogens in mammals but located in different tissues. The two latter species, one having a larger and the other a smaller genome than *M. pneumoniae*, respectively, have more transport systems[13, 32]. The increase in number concerns especially the ABC primary transporters category (see above), reflected also in an increase in the number of membrane lipoproteins, i.e. the ABC substrate-binding proteins in both species (Table 1). For *U. urealyticum* several systems not present in *M. pneumoniae* were tentatively identified at the sequence level, most prominent are six variants of a $Fe^{3+}$/haemin ABC transporter system[32]. However, no biochemical verification or function of these are presently known, but they may represent a "fitness island" typical for certain pathogens and symbionts[36]. In both *U. urealyticum* and *M. pulmonis* the apparent lack of certain expected transport systems (such as for nickel, an essential urease cofactor in *U. urealyticum*), or difficulties in annotating specific functions to identified ABC systems, were pointed out by the authors[13, 32]. It has been argued that the *M. genitalium* transporters, and hence the *M. pneumoniae* ones, differ in specificities from the more common *E. coli* transporters, being able to transport many related, essential compounds[70]. This would hamper recognition of substrate specificities based on sequence similarities. For *M. pneumoniae*, a new approach based on sequence "property-pattern" analyses and structural features[25, 94], could suggest functions on the basis of structural similarities for most of the ABC transporters (Table 3), including three identified export systems[26].

*Table 4.* Substrate Specificities of the Transport Systems

| Compounds | Transporters[a] in | *M. pneumoniae* | *B.subtilis* |
|---|---|---|---|
| Inorganic molecules | | | |
|     Cations | | 4 | 37 |
|     Anions | | 1 | 19 |
|     Water | | 0 | 0 |
| Carbon sources | | | |
|     Sugars & derivatives | | 6 | 40 |
|     Mono/di/tri-carboxylates | | 0 | 13 |
| Amino acids & derivatives | | | |
|     Amino acids (aa) | | 3 | 45 |
|     Aa/amides/polyamines | | 1 | 3 |
|     Peptides | | 1 | 5 |
| Bases & derivatives | | | |
|     Nucleobases | | 0 | 5 |
|     Nucleosides | | 0 | 3 |
|     Nucleotides | | 0 | 0 |
| Vitamins & cofactors | | 0 | 0 |
| Drugs & toxic compounds | | 3 | 45 |
| Macromolecules | | 0 | 5 |
| Unknown | | 3 | 42 |

[a]Numbers of transport systems calculated from[69]. For *M. pneumoniae* no system has been functionally verified, and substrate specificities are based on transporter sequence similarities.

## 3.      MEMBRANE LIPID BILAYERS

*Early findings.*   A lipid bilayer is the structural base of biological membranes. Cells and membranes of Mollicutes appeared early as useful models for the study of membrane lipid properties. Several major findings of general biological importance were first revealed using *A. laidlawii* (then *M. laidlawii*) as the biological model system; these are reviewed in detail by McElhaney[61].   Important findings (not in rank or order) were: (i) The establishment of the lipid bilayer as the structural base of a biological membrane (by differential scanning calorimetry and x-ray diffraction). (ii) Solubilization of membranes to micelles by common detergents, and the reappearance of a bilayer upon dialysis. (iii) The presence of a gel to liquid crystalline phase transition for the lipids, the temperature region of which could be dramatically modified by the type of fatty acids incorporated *in vivo* as lipid acyl chains, and by incorporation of cholesterol. (iv) The strong influence of the type of lipid acyl chains, and cholesterol, on the passive transmembrane diffusion ("permeability") of several small molecules, with a nice correlation between *in vitro* liposome models and cells. (v) The magnitude and appearance of certain "mobility" properties of the bilayer acyl chains (in liposomes and membranes), most detailed by a so-called order parameter profile established by $^2$H-NMR. (vi) The identification, analysis and regulation of nonbilayer-prone membrane lipids, maintaining certain important bilayer packing properties. Several of the early findings were incorporated in the "fluid mosaic" model for biological membranes by Singer and Nicolson[93]. We will review here new findings of the last decade, adding on the new findings based on genome sequences and extended analyses of lipid-synthesizing enzymes as well as discussing the physico-chemical packing properties of membrane lipids, brought down to the enzyme and gene level.

## 3.1      Biosynthetic pathways: lipids, enzymes and genes

*Membrane lipid and capsule composition.*   The extensive compilation of membrane lipid composition in Mollicutes by Smith[95] shows that all species analyzed (21 in number) contained the phospholipid phosphatidylglycerol (PG) and most also diphosphatidylglycerol (DPG; cardiolipin), in addition to other less common lipids. Furthermore, all the 8 of *Acholeplasma* species tested contain glycolipids, as do many *Mycoplasma* species (4 out of 9),

*S. citri* and *U. urealyticum*. Only in *Acholeplasma* are the glycolipids major constituents in the membranes. A polysaccharide capsule is present in many Mollicutes[64], and *Acholeplasma* have specific lipoglycans[95]. In related Gram-positive bacteria, a capsule and teichoic acid polymers are frequently covalently anchored to the membrane by glycolipids[30]. Genes encoding the enzymes for the lipid and polysaccharide pathways have not been consistently identified in the Mollicutes genomes, and are not accounted for in the analysis and discussions of the essential contents in the so called "minimal cell" cf.[42, 48, 71]. In the small intracellular symbiont *Buchnera* essentially no genes for lipid synthesis are encoded[90], suggesting that the bacterium aquires all of its lipids by uptake from the eukaryotic host. A similar strategy is partially used by Mollicutes for certain lipids and precursors (see McElhaney[62], and below).

*Metabolic pathways.* Figure 1 shows the tentative, simplified pathways for membrane lipid biosynthesis expected for *M. pneumoniae* and similar species and includes also some comments on *A. laidlawii* lipid biosynthesis. Table 5 lists the enzymes and genes potentially involved. In concordance with biochemical data, no enzymes for fatty acid synthesis were encoded in the four *Mycoplasma* and *Ureaplasma* genomes, whereas the pathway is present in *A. laidlawii*. However, all four *Mycoplasma* and *Ureaplasma* genomes contain a gene for acyl carrier protein (ACP) which is likely to act as an acyl chain carrier/donor in lipid biosynthesis (step 1 + 2 in Figure 1). Fatty acids may be released from extracellular host lipids by lipases, several of which are encoded in *M. pneumoniae*[41], and taken up by the Mollicutes by an active process[62]. The presence of only one initial acyltransferase (Table 5) may be related to the observed changes in position specificities of saturated and unsaturated chains in the lipids found by Rottem (summarized in McElhaney[62]). In the PG pathway, a PGP-phosphatase gene (step 5 in Figure 1; Table 5) has not been found in any of the four genomes using two related *E. coli* genes as search probes. However, all Mollicutes definitely synthesize the PG lipid, and hence must use a hitherto undetected enzyme for this step. Cardiolipin was not detected in *M. pneumoniae* by Plackett *et al*[72], which corroborates the absence of the synthase gene (step 7 in Figure 1; Table 5). However, a cardiolipin synthase gene is present in *M. pulmonis* and *U. urealyticum*. Acyl-PG (step 6 in Figure 1; Table 5) is present in several Mollicutes[95]. Due to its three acyl chains this lipid is bound to be nonbilayer-prone as is cardiolipin, especially in the presence of divalent cations (see below).

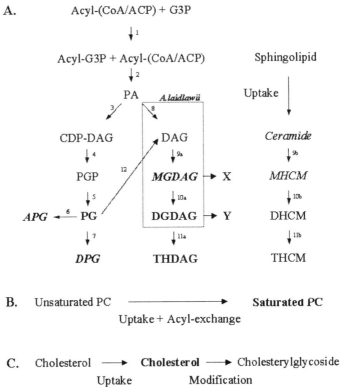

**A.**        Acyl-(CoA/ACP) + G3P

                  ↓ 1

     Acyl-G3P + Acyl-(CoA/ACP)          Sphingolipid

                  ↓ 2

             PA      *A.laidlawii*        Uptake │

        3 ╱      ╲ 8

   CDP-DAG      DAG              *Ceramide*

      ↓ 4        ↓ 9a               ↓ 9b

     PGP    12   *MGDAG* → X        *MHCM*

      ↓ 5        ↓ 10a               ↓ 10b

  *APG* ← 6 PG   DGDAG → Y         DHCM

      ↓ 7        ↓ 11a               ↓ 11b

     *DPG*       THDAG             THCM


**B.**   Unsaturated PC ─────────────→ **Saturated PC**

                    Uptake + Acyl-exchange


**C.**   Cholesterol ──→ **Cholesterol** ──→ Cholesterylglycoside

                 Uptake        Modification

*Figure 1.* Metabolic pathways for membrane lipid biosynthesis in Mollicutes. Simplified routes as they apply specifically to *M. pneumoniae* and *A. laidlawii*. (**A**) Synthesis of phospho- and glycolipids. Glycolipids differ in structures between the two species, and may initiate from either DAG or PG in *M. pneumoniae in vitro*. Names in bold refer to major species, italics to nonbilayer-prone lipids, and numbers to individual enzymes indicated at gene level or functionally verified/cloned, see Table 5. X and Y, modification of the glycolipids with a third acyl chain (*A. laidlawii* MAMGlcDAG and MADGlcDAG), or for Y alternatively also to glycerophosphorylated variants (*A. laidlawii* GPDGlcDAG and MABGPDGlcDAG). (**B**) Potential synthesis of sphingolipids in *M. pneumoniae*. (**C**) and (**D**) Uptake and modification of phosphatidylcholine (*M. pneumoniae*) and cholesterol, respectively. ACP, acyl carrier protein; G3P, *sn*-glycerol-3-phosphate; PA, phosphatidic acid; DAG, 1,2-diacylglycerol; MG, monoglycosyl; DG, diglycosyl; MH, monohexose; DH, dihexose; TH, trihexose; PGP, phosphatidylglycerolphosphate; PG, phosphatidylglycerol; APG, acylphosphatidylglycerol; DPG, diphosphatidylglycerol; PC, phosphatidylcholine; and CM, ceramide. Based on the presence of membrane lipids in Mollicutes (reviewed by Smith[95]), their potential biosyntheses[62, 75], structure determinations by Smith and by Lindblom and coworkers, and on our enzyme studies, cloning work and sequence analyses.

*Glycolipid-synthesizing enzymes.* One may anticipate that the Mollicutes, as a clearly defined group of very small and related bacteria, should be equally homogenous with respect to glycolipid composition and structure as is the case for the phospholipids (see above). However, early chemical characterizations indicated that this is not the case[95], and it is strongly corroborated by our recent analyses of the enzymes involved. In the right half of Figure 1 some general features of glycolipid synthesis are outlined, as they may appear in *M. pneumoniae* and *A. laidlawii*. Both species contain DAG-based glycolipids with one and two glycoside-linked hexoses, but the source of the DAG and the sugar anomeric configuration ($\alpha$ or $\beta$) plus carbon atoms involved are different[39, 95]. The structures of the two major *A. laidlawii* glycolipids (see box in Figure 1) are 1,2-diacyl-3-O-($\alpha$-D-glucopyranosyl)-*sn*-glycerol (MGlcDAG), and 1,2-diacyl-3-O-[$\alpha$-D-glucopyranosyl-(1$\rightarrow$2)-O-$\alpha$-D-glucopyranosyl]-*sn*-glycerol (DGlcDAG) as determined by NMR techniques[39]; see also the structures in Smith[95]. The corresponding *M. pneumoniae* lipids contain either galactose or glucose, or both, tentatively linked in $\beta$ configuration where the carbons involved are not known[83]. The *A. laidlawii* enzymes involved (8, 9a and 10a in Figure 1 / Table 5) have been purified[8, 46, 105], and the two glucosyltransferases 9a and 10a were also cloned, sequenced and overproduced with functions maintained[7, 24]. Both enzymes belong to glycosyltransferase family 4 (out of 55 families in the CAZy systematics) of glycosyltransferases[11], and each recognized a number of partially similar sequences particularily in the genomes of Gram-positive pathogens, but also in thermophiles, archaea, and even in eukaryotes (cf.[7]); they must hence be evolutionary very conserved. Many of these organisms contain glycolipids as major membrane constituents, and three identified, related sequences from *Borrelia burgdorferi* and *Streptococcus pneumoniae* were PCR-cloned into *E. coli* and were shown to encode functionally active lipid glycosyltranferases. Under certain growth conditions (see below) *A. laidlawii* MGlcDAG and DGlcDAG may partially be esterified with a third acyl chain at carbon 6 of the (inner) glucose moiety[38, 39], yielding MAMGlcDAG and MADGlcDAG (X and Y in Figure 1A).

*M. pneumoniae* lipid labelling *in vivo* correlated with enzyme assays *in vitro*, showing three different glycolipids containing one, two and three hexoses[83] (see Figure 1A). Several phosphoglycolipids were also detected. Extensive searches in the *M. pneumoniae* and *M. genitalium* genomes, using many lipid and capsule enzyme genes as search probes, identified three intracellular glycosyltransferases as candidates for the synthesis of the membrane glycolipids, and potentially also of a lipid capsule. These genes i.e. MPN028, 075, 483 and their *M. genitalium* counterparts, were all similar to capsule enzymes from Gram-positive bacteria but not to the two *A.*

*laidlawii* ones mentioned above, nor to genes of *M. pulmonis* and *U. urealyticum*[83]. A similarity to enzymes able to use ceramide (instead of DAG) as the lipid acceptor was also noted; this probably explains the ability of *M. pneumoniae* to synthesize ceramide-based glycolipids recorded *in vitro* as well as *in vivo*[83] (see Figure 1B). In the latter case a release of the ceramide glycolipid into the growth medium was observed. All three enzymes were related sequence-wise and belong to CAZy systematics family 2 of glycosyltransferases, adding sugars in β-configuration[11]. A processive character, i.e. an ability to add several sugars consecutively to the same acceptor, a property which occurs in other members of CAZy family 2[12]. The presence of one or more of these three enzymes may also result in the synthesis of a capsule component. The latter would also have to be complemented by an export system, perhaps one of the three ABC export transporters (see above).

*Table 5.* Lipid-Synthesizing Enzymes and Genes

| Enzyme[a]/Figure 1 | Function[b]/Gene name | Comment[c] |
|---|---|---|
| 1 | G3P acyltransferase (plsA) | Not in *M. pneumoniae* |
| 2 | 1-acyl-G3P acyltransferase (plsB) | Homology found |
| 3 | PA cytidyltransferase (cdsA) | Homology found |
| 4 | GP phosphatidyltransferase (pgsA) | Homology found |
| 5 | PGP phosphatase | Homologs not found |
| 6 | PG acyltransferase | Lipid labelled |
| 7 | Cardiolipin synthase (cls) | *M.pulmonis* (not in Mpn) |
| 8 | PA phosphatase (pap) | *A. laidlawii* (not in Mpn) |
| 9a | DAG UDP-Glc-transferase (mgs) | Cloned from *A. laidlawii* |
| 9b | CM UDP-Gal-transferase | Activity in *M. pneumoniae* |
| 10a | MGlcDAG UDP-Glc-transferase (dgs) | Cloned from *A. laidlawii* |
| 10b | MHCM UDP-Gal-transferase | Activity in *M. pneumoniae* |
| 11a | DHDAG UDP-Gal-transferase | Activity in *M. pneumoniae* |
| 11b | DHCM UDP-Gal-transferase | Activity in *M. pneumoniae* |
| 12 | PG phosphatase/UDP-Gal-transferase | *M. pneumoniae in vitro* |

[a]Numbers from schematic metabolic pathways in Figure 1.
[b]Short description of function. Gene names as in *E. coli* and *A. laidlawii*.
[c]Similar genes (sequence similarity), enzyme activity, or metabolic labelling detected, respectively.

*Lipid enzymes and evolution.* The lack of sequence similarities between the two *A. laidlawii* and the three *M. pneumoniae* glycosyltransferases is also evident at the structural level. The enzymes could be modelled to be similar to two completely different, solved structures formed by the *E. coli* MurG and *Bacillus* SpsA glycosyltransferases, respectively[7, 83]. Since sequences closely related to both the *A. laidlawii* and *M. pneumoniae* glycosyltransferases occur also in *Streptococcus pneumoniae*, it suggests the way how the ancestors of *Acholeplasma* and *Mycoplasma* may have received (or kept) different sets of glycolipid-encoding genes during their evolution. The number of capsule glycosyltransferases in different *S. pneumoniae* strains is very large, making the potential variation in such a selection procedure rather excessive. This may also be related to the ecological niche occupied. *M. pneumoniae* and *M. pulmonis*, both lung parasites, have a gene for a potential carnitine palmitoyltransferase II (MPN114) of eukaryotic signature, whereas this gene is lacking from *M. genitalium* and *U. urealyticum*. Resembling many other Mollicutes, the former two have the ability to take up and incorporate exogenous phosphatidylcholine and the ceramide lipid sphingomyelin, but in contrast to other species they also exchange the unsaturated acyl chains on the PC with saturated acyl chains[84], as illustrated in Figure 1C. The lung surfactant phospholipids have unique "mechanical" (surface tension) properties partly due to their saturated chains[33, 104], and perhaps the Mollicutes inhabiting this niche have learnt to mimic this by selectively aquiring proper eukaryotic lipids and enzymes, including the àcyl transferase MPN114. Cholesterol may also provide *Mycoplasma* membranes with certain mechanical properties, and it seems to be incorporated by an active process (reviewed by Dahl[19]), indicated in Figure 1D. In *A. laidlawii* and *M. pneumoniae* no radioactive label was transferred from cholesterol to other membrane constituents. The lower amounts of cholesterol incorporated into *A. laidlawii* membranes may partially depend on its lower solubility in the dominant glucolipids[27, 63], and on the ability to regulate the bilayer-nonbilayer propensities of the membrane (see below).

## 3.2 Regulation of lipid bilayer packing properties

*Lipid chain "melting"*. Only the sterol-nonrequiring *Acholeplasma* of the Mollicutes group, but none of the others, has the ability to synthesize saturated fatty acids *de novo*, but not essential unsaturated ones[62, 75]. All Mollicutes species must therefore, more or less, rely on the host supply of fatty acids and cannot, or only marginally (*A. laidlawii*), affect or regulate essential acyl chain properties in the membrane. Hence, the biologically common response of adjusting the lipid "melting" (gel to liquid-crystalline phase transition) temperature[92] by changing the metabolic balance between saturated and unsaturated acyl chains[59] is absent in Mollicutes and must be executed by their hosts. This ability is thus obviously dispensable for these parasites in their ecological niches.

*Lipid "molecular shape"*. By the pioneering work of especially Luzzati and coworkers it has long been known that biological membranes contain substantial amounts of polar lipids that by themselves cannot form bilayers (or lamellar liquid-crystalline phases). Based on experimental analyses of lipid packing preferences and observed metabolic variations among the polar lipid species it was proposed that *A. laidlawii* can regulate the balance between bilayer and nonbilayer-prone lipids[110]. This hypothesis was then supported by analyses of total membrane lipid extracts[55] and by the impact of a number of foreign guest molecules in the membrane with established abilities to affect phase equilibria[82, 112]. Additionally, a "constant" lipid surface charge density (and surface potential) was maintained for the bilayer mixture[15]. In simple terms, the bacterium strives at maintaining certain membrane packing properties, visualized as a more or less constant bilayer/nonbilayer phase transition temperature at a certain distance above or close to the growth temperature, and metabolically counteracting any lipid "packing" changes brought upon the bacteria. This was visualized as a relationship between the volume of the chains ($v$), and the headgroup interfacial area ($a$) times chain length ($l$) of the lipids (i.e. $v / a \times l$), when attractive and repulsive forces are at equilibrium[43]. Lipids with small or uncharged headgroups (i.e. small $a$) will have an "inverted cone" packing shape, and prefer to assemble into (reversed) nonbilayer-type structures with curved surfaces, like reversed cubic or hexagonal phases[81]. Larger headgroups (large $a$) yield cylindrical-like packing geometries, forming bilayers or even micelles for larger $a$. For a given headgroup ($a$), substantially shorter chains ($l$) yield micelles and longer ones reversed-type (nonbilayer) aggregates, relative to the chain length in the bilayer state[43]. Most importantly, other molecules present (e.g. sterols, solvents, or even proteins) will affect (or add to) the $v$, $a$, and $l$ parameters, and consequently modify or change the packing preferences in the membrane bilayer.

*Spontaneous curvature and chain order.* During the last decade the above described model have been extensively analyzed and extended in a substantial number of publications, more than space permits to be discussed here. Important features of the present principles are outlined in Table 6. The effect of various proportions of saturated and unsaturated acyl chains, and the resulting metabolic lipid adjustments on "mechanical" (spontaneous curvature) properties of the *A. laidlawii* lipid bilayer have been established. Each monolayer of the bilayer has a certain strive to bend outwards (relax concavely) due to the presence of substantial amounts of nonbilayer-prone lipids (with small heads; see ($a$) above). Compensatory metabolic changes in bilayer/nonbilayer-prone lipids aim at maintaining the same spontaneous curvature for the in vivo mixture, while the values for individual MGlcDAG (nonbilayer) and DGlcDAG (bilayer) are dramatically different from each other, but of similar magnitude as the common membrane lipids phosphatidylethanolamine (nonbilayer-prone) and phosphatidylcholine (bilayer), respectively[68]. Nonbilayer-prone lipids, due to their smaller headgroup areas, force the acyl chains to become more extended (more ordered) which is clearly seen in *A. laidlawii* bilayer lipid mixtures and membranes[29, 100]. However, this "order" (order parameter) is not kept constant *in vivo*, in contrast to the curvature (see above). An association with cell size in *A. laidlawii* may however apply to chain order[111]. The lipids in these bilayers are thus maintained in a state of controlled elastic stress (or chain "frustration"), and the potential energy in a monolayer can probably be sensed by proteins and also drive conformational changes of the latter[28, 35].

*More nonbilayer lipids.* The chain length is important for the packing properties of lipids and amphiphiles (see above). *A. laidlawii* can tolerate (grow with) substantial variations in acyl chain length, from ≈14.5 carbons to almost 20 carbons (cf. Table 6)[111]. This corresponds to a variation in bilayer hydrophobic thickness from ≈23 to ≈31 Å[100]. However, only in the 16 to 18 carbon window were 100% mono-unsaturated chains allowed; at shorter or longer lengths increased chain saturation was essential. The major nonbilayer lipid MGlcDAG turns bilayer-prone with shorter and/or saturated chains[56, 58]. Still, fairly similar phase equilibria are formed by the membrane lipids over the entire thickness range[2]. This is achieved by the sequential synthesis of other (new), more nonbilayer-prone lipids at conditions when the properties of MGlcDAG are not sufficient[1, 56]; see details for DAG, MAMGlcDAG and MADGlcDAG in Table 6. Even a lipid that forms a micellar aggregate, i.e. the phosphoglucolipid GPDGlcDAG, can be synthesized[22,52]. *A. laidlawii* strain B can also synthesize a unique, nonbilayer-prone polyprenyl-glucoside[31, 53].

Hence, *A. laidlawii* is using elaborate and extensive mechanisms with several, sequentially synthesized, nonbilayer-prone lipids to maintain certain lipid bilayer packing properties. The numbers of components involved allows the adaptation to many different "lipid supply" conditions, and also a careful fine-tuning of the bilayer packing properties. Obviously, these principles are integrated into the metabolic pathways, and a number of enzymatic steps must be involved (Figure 1).

*Simulation at enzyme level in vitro.* The consecutive synthesis steps to MGlcDAG (nonbilayer) and DGlcDAG (bilayer) in *A. laidlawii* (step 9a/10a in Figure 1) are crucial because these lipids are the major ones under most conditions (Table 6). Analysis of kinetic and lipid-packing parameters in liposomes and mixed-micelles systems showed that some of the regulatory features recorded *in vivo* were confined to the purified enzymes *in vitro*. Most important, both the MGlcDAG and DGlcDAG synthases (glucosyltransferases; mgs and dgs, respectively) demanded the presence of substantial amounts of an activator lipid for activity[46, 105]. For mgs most anionic lipids were sufficient whereas dgs were best stimulated by phosphatidylglycerol or cardiolipin. In this manner the PG pathway "controls" the flow into the glucolipid pathway (see Figure 1A), and this is part of the surface charge density regulation (see above). Similar features may operate in *M. pneumoniae*[83]. Both surface binding and activity of these

*Åke Wieslander and Maria Rosén*

peripheral *A. laidlawii* enzymes seem to be governed by the anionic lipids[54]. Dgs and the mixed enzymes could mimic the *in vivo* output to nonbilayer-prone additives[20, 47, 105]. Furthermore, activity of dgs, but not mgs, was substantially stimulated by certain important phosphorylated glycolytic metabolites (like fructose-1,6-bisphosphate and ATP), and even by dsDNA, indicating a coupling to the metabolic status of the cell and potentially also to cell division[106]. The predicted 3D-structures of mgs and dgs are very similar to each other (above) and also to the glycosyltransferase making the major lipid on this planet, i.e. monogalactosyl-DAG of all photosynthetic membranes[24].

*Table 6.* Regulation of Membrane Lipid Packing Properties in *Acholeplasma laidlawii.*

The amounts and types of membrane polar lipids are extensively regulated. Similar *spontaneous curvature*, bilayer to nonbilayer *phase equilibria*, and a certain *surface charge density*, are maintained for the lipid membrane. Bilayer and nonbilayer-prone glycolipids and anionic phospholipids (Figure 1), are synthesized ("adjusted") as a function of: (i) Type of acyl chains incorporated (length, unsaturation, branching); (ii) Foreign "guest" molecules in the membrane (sterols, solvents, detergents, pigments); (iii) Growth temperature; (iv) Cellular ionic environment; (v) Membrane electrochemical potential / energy metabolism.

| Acyl chain length $(C_n)$ | PG pathway (mol%) | Glycolipid pathway [a] (mol%) | Lipid phosphates (mol/100 mol lipid) |
|---|---|---|---|
| $\approx$14.5 to $\approx$20 | 9 to 22 | 91 to 78 | 34 to 34 |

| Unsaturated chains (%UFA) | | | |
|---|---|---|---|
| <20 to >80 | 10 to 16 | 90 to 84 | 25 to 43 |

([a]including phosphoglucolipids)

| Acyl chain length / unsaturated chains $(C_n)$ | Nonbilayer-prone lipids[b] (mol%) | DGlcDAG (mol%) |
|---|---|---|
| 15 | 20 | $\approx$80 | $\approx$5 |
| 18 | 98 | $\approx$12 | $\approx$55 |

([b]sum of DAG, **MGlcDAG**, MAMGlcDAG and MADGlcDAG)

1. **MGlcDAG** is the major nonbilayer-prone species, and **DGlcDAG** the major bilayer-forming species under most conditions.

2. Sterols, solvents and detergents in the membrane (in $C_n$ 16 to 18 range) induce changes in the metabolic MGlcDAG / DGlcDAG balance, in relation to the ability of the additives to shift phase equilibria in the *nonbilayer* or *bilayer* direction. This *compensatory* (potentially *"homeostatic"*) mechanism is supported by extensive physico-chemical analyses of lipid and bilayer mixtures.

3. MGlcDAG turns bilayer-prone with shorter and/or saturated chains, and is gradually (partially) replaced by the more nonbilayer-prone MADGlcDAG *in vivo*. With longer and saturated chains MGlcDAG is partially replaced by the more nonbilayer-prone DAG (precursor) and MAMGlcDAG. This indicates a *fine-tuning* of the *packing properties.*

4. The *radius of spontaneous curvature* for the lipid bilayer is fairly constant ($C_n$ 16 to 18 range) but with very short and long radii for MGlcDAG and DGlcDAG, and of similar magnitude as phosphatidylethanolamine and phosphatidylcholine, respectively.

*Other systems.* Various kinds of antibacterial peptides are produced by many organisms. Many seem to act by electrostatic attraction to the lipid bilayer surface, followed by more or less deep penetration into the membrane, and eventually thinning of the bilayer or forming pores resulting in collapse of the electrochemical gradient. Wrobléwski and coworkers have analysed the effects of a large number of natural and synthetic peptides against a number of Mollicutes[9]. Gramicidin S was by far the most active peptide of the natural products, efficiently lowering the membrane potential. The mechanism, as analyzed with *A. laidlawii* and *E. coli* lipids, seems to be a thinning of the lipid bilayer causing an increase in the curvature stress, and eventually membrane rupture by nonbilayer-aggregate formation[76, 96].

A choline-containing phosphoglycolipid (MfGL-II) is the major polar lipid of *M. fermentans* membranes (see chapter "Invasion of mycoplasmas into and fusion with host cells"). Its backbone is an α-MGlcDAG, similar to the *A. laidlawii* one, and its synthesis probably proceeds along the X-route in Figure 1A. Specific antibodies indicate that the complete lipid is not synthesized by *M. penetrans*, *M. capricolum, M. gallisepticum*, or *A. laidlawii*[6], and neither by *M. pneumoniae* according to labelling[83]. This lipid provides the cell with unique permeability and resistance to osmotic swelling, due to its substantially higher "melting" temperature[5].

*General importance and validity of findings.* Similar bilayer packing principles as discussed here for *A. laidlawii* can be inferred from old data for *Pseudomonas fluorescens*[18], and definitely operate in certain *Clostridium* species[34] and *B. megaterium*[81]. They are also valid for the *E. coli* wild type and lipid mutants[65, 80]. For all these other species much of the regulation of packing properties, i.e. bilayer/nonbilayer-balance, curvature stress, takes place at the acyl chain level with more or less constant polar heads (usually phospholipids). However, the general importance of the curvature properties is illustrated by the recent findings that the glucolipid MGlcDAG (nonbilayer-prone) from *A. laidlawii*, introduced by cloning of mgs, can restore a cell-division deficiency and transporter malfunction of a phosphatidylethanolamine-minus (i.e. nonbilayer-deficient) *E. coli* mutant (Wikström, Dowhan and Wieslander, unpublished observation). It may be proposed here that cholesterol and cholesterol esters, both prone to induce nonbilayer structures with certain other lipids[81, 101], fulfill analogous functions in all sterol-requiring Mollicutes. This would add to the nonbilayer properties provided by cardiolipin, acylphosphatidylglycerol, and monoglycolipids (Figure 1) present in small amounts in many species.

## 4.    CONCLUSIONS

Analysis of of the occurrence of encoded membrane protein sequences in the genomes of *M. genitalium, M. pneumoniae, U. urealyticum* and *M. pulmonis* have revealed that they have a fairly large number of integral membrane proteins. However, few functional analyses of these proteins have been made hitherto. Membrane lipoprotein numbers are large in mollicutes, but less than in the larger conventional bacteria; many of these appear to be transport substrate-binding proteins. Prominent among the integral proteins are membrane transporters, especially of the ABC type. The latter group, a major one among the primary transporters, is common in cells with a restricted ability to generate energy from metabolism, and clearly is a typical trait for the Mollicutes. Numbers of transporters in relation to chromosome length are average. The majority of the basal components of the Sec machinery for protein secretion are present in mollicutes, whereas several of the accessory, chaperon-like and post-translational ones seem to be more or less absent but their numbers may increase upon further analysis.

Lipid bilayer packing properties in *A. laidlawii* membranes have been analysed in more detail with respect to physico-chemical principles, than for any other cell. Important findings regarding the bilayer/nonbilayer lipid balance and molecular-mechanical properties (curvature stress) provided by the polar lipids have later been found valid in several other (larger) bacteria including *E. coli*. However, Mollicutes lack the ability to efficiently regulate properties of the lipid acyl chains. Purification, cloning, sequencing and functional expression of the *A. laidlawii* glycolipid enzymes have revealed the connection between lipid biosynthesis and these biophysical packing properties. Comparisons show that the variation of Mollicutes glycolipids and enzyme sequences seem to be large, and that the *M. pneumoniae* enzymes are related to lipid polysaccharide capsule ezymes in Gram-positive bacteria. The *A. laidlawii* glycolipid enzymes are related to other, conserved enzymes in especially Gram-positive pathogens, but also in thermophiles and archaea. Lipid composition and sequence data have been integrated into more comprehensive metabolic pathways, also revealing dependency on host lipid delivery.

## ACKNOWLEDGEMENTS

Work in our laboratory has been supported by the Swedish Natural Science Research Council, Umeå University Biotechnology Foundation, and by the K & A Wallenberg Foundation. We thank past and present coworkers for their contributions.

# REFERENCES

1.  **Andersson, A. S., L. Rilfors, R. N. Lewis, R. N. McElhaney, and G. Lindblom.**
    1998. Occurrence of monoacyl-diglucosyl-diacyl-glycerol and monoacyl-bis-
    glycerophosphoryl-diglucosyl-diacyl-glycerol in membranes of *Acholeplasma laidlawii*
    strain B-PG9. Biochim. Biophys. Acta **1389**:43-49.
2.  **Andersson, A. S., L. Rilfors, G. Orädd, and G. Lindblom.** 1998. Total lipids with
    short and long acyl chains from *Acholeplasma laidlawii* form nonlamellar phases.
    Biophys. J. **75**:2877-2887.
3.  **Antelmann, H., H. Tjalsma, B. Voigt, S. Ohlmeier, S. Bron, J. M. van Dijl, and M.
    Hecker.** 2001. A proteomic view on genome-based signal peptide predictions. Genome
    Research **11**:1484-502.
4.  **Barile, M. F., and S. Razin.** 1979. The Mycoplasmas, vol. 1. Academic Press, New
    York.
5.  **Ben-Menachem, G., T. Byström, H. Rechnitzer, S. Rottem, L. Rilfors, and G.
    Lindblom.** 2001. The physico-chemical characteristics of the phosphocholine-
    containing glycoglycerolipid MfGL-II govern the permeability properties of
    *Mycoplasma fermentans.* Eur. J. Biochem. **268**:3694-3701.
6.  **Ben-Menachem, G., F. Wagner, U. Zähringer, E. T. Rietschel, and S. Rottem.**
    1997. Antibody response to MfGL-II, a phosphocholine-containing major lipid of
    *Mycoplasma fermentans* membranes. FEMS Microbiol. Lett.. **154**:363-369.
7.  **Berg, S., M. Edman, L. Li, M. Vikström, and A. Wieslander.** 2001. Sequence
    properties of the 1,2-diacylglycerol 3-glucosyltransferase from *Acholeplasma laidlawii*
    membranes. Recognition of a large group of lipid glycosyltransferases in eubacteria and
    archaea. J. Biol. Chem. **276**:22056-22063.
8.  **Berg, S., and Å. Wieslander.** 1997. Purification of a phosphatase which hydrolyzes
    phosphatidic acid, a key intermediate in glucolipid synthesis in *Acholeplasma laidlawii*
    A membranes. Biochim. Biophys. Acta **1330**:225-232.
9.  **Beven, L., and H. Wrobléwski.** 1997. Effect of natural amphipathic peptides on
    viability, membrane potential, cell shape and motility of mollicutes. Res. Microbiol.
    **148**:163-175.
10. **Blattner, F. R., G. Plunkett, C. A. Bloch, N. T. Perna, V. Burland, M. Riley, J.
    Collado-Vides, J. D. Glasner, C. K. Rode, G. F. Mayhew, J. Gregor, N. W. Davis,
    H. A. Kirkpatrick, M. A. Goeden, D. J. Rose, B. Mau, and Y. Shao.** 1997. The
    complete genome sequence of *Escherichia coli* K-12. Science **277**:1453-1474.
11. **Campbell, J. A., G. J. Davies, V. Bulone, and B. Henrissat.** 1997. A classification of
    nucleotide-diphospho-sugar glycosyltransferases based on amino acid sequence
    similarities. Biochem. J. **326**:929-939.
12. **Cartee, R. T., W. T. Forsee, J. W. Jensen, and J. Yother.** 2001. Expression of the
    *Streptococcus pneumoniae* type 3 synthase in *Escherichia coli.* Assembly of type 3
    polysaccharide on a lipid primer. J. Biol. Chem. **276**:48831-48839.
13. **Chambaud, I., R. Heilig, S. Ferris, V. Barbe, D. Samson, F. Galisson, I. Moszer, K.
    Dybvig, H. Wrobléwski, A. Viari, E. P. Rocha, and A. Blanchard.** 2001. The
    complete genome sequence of the murine respiratory pathogen *Mycoplasma pulmonis.*
    Nucleic Acids Res. **29**:2145-2153.
14. **Chambaud, I., H. Wrobléwski, and A. Blanchard.** 1999. Interactions between
    mycoplasma lipoproteins and the host immune system. Trends Microbiol. **7**:493-499.
15. **Christiansson, A., L. E. Eriksson, J. Westman, R. Demel, and Å. Wieslander.** 1985.
    Involvement of surface potential in regulation of polar membrane lipids in
    *Acholeplasma laidlawii.* J. Biol. Chem. **260**:3984-3990.
16. **Cirillo, V. P.** 1993. Transport systems in mycoplasmas. Subcellular Biochemistry
    **20**:293-310.

17. **Clementz, T., A. Christiansson, and Å. Wieslander.** 1986. Transmembrane electrical potential affects the lipid composition of *Acholeplasma laidlawii*. Biochemistry. **25**:823-830.
18. **Cullen, J., M. C. Phillips, and G. G. Shipley.** 1971. The effects of temperature on the composition and physical properties of the lipids of *Pseudomonas fluorescens*. Biochemical J. **125**:733-742.
19. **Dahl, J.** 1993. The role of cholesterol in mycoplasma membranes. Subcellular Biochemistry **20**:167-188.
20. **Dahlqvist, A., S. Nordström, O. P. Karlsson, D. A. Mannock, R. N. McElhaney, and Å. Wieslander.** 1995. Efficient modulation of glucolipid enzyme activities in membranes of *Acholeplasma laidlawii* by the type of lipids in the bilayer matrix. Biochemistry. **34**:13381-13389.
21. **Dandekar, T., M. Huynen, J. T. Regula, B. Ueberle, C. U. Zimmermann, M. A. Andrade, T. Doerks, L. Sanchez-Pulido, B. Snel, M. Suyama, Y. P. Yuan, R. Herrmann, and P. Bork.** 2000. Re-annotating the *Mycoplasma pneumoniae* genome sequence: adding value, function and reading frames. Nucleic. Acids. Res. **28**:3278-3288.
22. **Danino, D., A. Kaplun, G. Lindblom, L. Rilfors, G. Orädd, J. B. Hauksson, and Y. Talmon.** 1997. Cryo-TEM and NMR studies of a micelle-forming phosphoglucolipid from membranes of *Acholeplasma laidlawii* A and B. Chem. Phys. Lipids. **85**:75-89.
23. **de Gier, J. W., and J. Luirink.** 2001. Biogenesis of inner membrane proteins in *Escherichia coli*. Mol. Microbiol. **40**:314-322.
24. **Edman, M., S. Berg, P. Storm, M. Wikström, S. Vikström, A. Öhman, and Å. Wieslander.** 2002. Similar structure features of two glycosyltransferases synthesizing the major non-bilayer and bilayer glycolipids lipids in *Acholeplasma laidlawii*. Submitted for publication. .
25. **Edman, M., T. Jarhede, M. Sjöström, and Å. Wieslander.** 1999. Different sequence patterns in signal peptides from mycoplasmas, other gram-positive bacteria, and *Escherichia coli:* a multivariate data analysis. Proteins **35**:195-205.
26. **Edman, M., M. Sjöström, and Å. Wieslander.** 2002. Multivariate analysis of ABC-dependent membrane transport proteins from five genomes reveals group-specific features in sequence properties. Under revision.
27. **Efrati, H., Y. Wax, and S. Rottem.** 1986. Cholesterol uptake capacity of *Acholeplasma laidlawii* is affected by the composition and content of membrane glycolipids. Arch. Biochem. Biophys. **248**:282-288.
28. **Epand, R. M., and R. F. Epand.** 1994. Calorimetric detection of curvature strain in phospholipid bilayers. Biophys. J. **66**:1450-1456.
29. **Eriksson, P. O., L. Rilfors, Å. Wieslander, A. Lundberg, and G. Lindblom.** 1991. Order and dynamics in mixtures of membrane glucolipids from *Acholeplasma laidlawii* studied by $^2$H NMR. Biochemistry **30**:4916-4924.
30. **Fischetti, V. A., R. P. Novick, J. J. Ferretti, D. A. Portnoy, and J. I. Rood.** 2000. Gram-Positive Pathogens. ASM Press, Washington D.C.
31. **Foht, P. J., Q. M. Tran, R. N. Lewis, and R. N. McElhaney.** 1995. Quantitation of the phase preferences of the major lipids of the *Acholeplasma laidlawii* B membrane. Biochemistry **34**:13811-13817.
32. **Glass, J. I., E. J. Lefkowitz, J. S. Glass, C. R. Heiner, E. Y. Chen, and G. H. Cassell.** 2000. The complete sequence of the mucosal pathogen *Ureaplasma urealyticum*. Nature **407**:757-762.
33. **Goerke, J.** 1998. Pulmonary surfactant: functions and molecular composition. Biochim. Biophys. Acta **1408**:79-89.
34. **Goldfine, H.** 1993. Phospholipid biosynthetic enzymes of butyric acid-producing clostridia. In Sebald, M. (ed.), Genetics and Molecular Biology of Anaerobes, Springer Verlag, New York.

35. **Gruner, S. M.** 1989. Stability of lyotropic phase with curved interfaces. J. Phys. Chem. **93**:7562-7570.

36. **Hacker, J., and E. Carniel.** 2001. Ecological fitness, genomic islands and bacterial pathogenicity. A Darwinian view of the evolution of microbes. EMBO reports **2**:376-381.

37. **Hase, C. C., N. D. Fedorova, M. Y. Galperin, and P. A. Dibrov.** 2001. Sodium ion cycle in bacterial pathogens: evidence from cross-genome comparisons. Microbiol. Mol. Biol. Rev. **65**:353-370.

38. **Hauksson, J. B., G. Lindblom, and L. Rilfors.** 1994. Structures of glucolipids from the membrane of *Acholeplasma laidlawii* strain A-EF22. II. Monoacylmonoglucosyldiacylglycerol. Biochim. Biophys. Acta. **1215**:341-345.

39. **Hauksson, J. B., L. Rilfors, G. Lindblom, and G. Arvidson.** 1995. Structures of glucolipids from the membrane of *Acholeplasma laidlawii* strain A-EF22. III. Monoglucosyldiacylglycerol, diglucosyldiacylglycerol, and monoacyldiglucosyldiacylglycerol. Biochim. Biophys. Acta **1258**:1-9.

40. **Herrmann, R., and B. Reiner.** 1998. *Mycoplasma pneumoniae* and *Mycoplasma genitalium*: a comparison of two closely related bacterial species. Curr. Opin. Microbiol. **1**:572-579.

41. **Himmelreich, R., H. Hilbert, H. Plagens, E. Pirkl, B. C. Li, and R. Herrmann.** 1996. Complete sequence analysis of the genome of the bacterium *Mycoplasma pneumoniae*. Nucleic. Acids Res. **24**:4420-4449.

42. **Hutchison, C. A., S. N. Peterson, S. R. Gill, R. T. Cline, O. White, C. M. Fraser, H. O. Smith, and J. C. Venter.** 1999. Global transposon mutagenesis and a minimal *Mycoplasma* genome. Science **286**:2165-2169.

43. **Israelachvili, J. N., D. J. Mitchell, and B. W. Ninham.** 1976. Theory of self-assembly of hydrocarbon amphiphiles into micelles and bilayers. J. Chem. Soc. Faraday Trans. II. **72**:1525-1568.

44. **Jayasinghe, S., K. Hristova, and S. H. White.** 2001. Energetics, stability, and prediction of transmembrane helices. J. Mol. Biol. **312**:927-934.

45. **Jongbloed, J. D., U. Martin, H. Antelmann, M. Hecker, H. Tjalsma, G. Venema, S. Bron, J. M. van Dijl, and J. Muller.** 2000. TatC is a specificity determinant for protein secretion via the twin-arginine translocation pathway. J. Biol. Chem. **275**:41350-41357.

46. **Karlsson, O. P., A. Dahlqvist, S. Vikström, and Å. Wieslander.** 1997. Lipid dependence and basic kinetics of the purified 1,2-diacylglycerol 3-glucosyltransferase from membranes of *Acholeplasma laidlawii*. J. Biol. Chem. **272**:929-936.

47. **Karlsson, O. P., M. Rytömaa, A. Dahlqvist, P. K. Kinnunen, and Å. Wieslander.** 1996. Correlation between bilayer lipid dynamics and activity of the diglucosyldiacylglycerol synthase from *Acholeplasma laidlawii* membranes. Biochemistry **35**:10094-10102.

48. **Koonin, E. V.** 2000. Bridging the gap between sequence and function. Trends Genetics **16**:16.

49. **Krause, D. C., and M. F. Balish.** 2001. Structure, function, and assembly of the terminal organelle of *Mycoplasma pneumoniae*. FEMS Microbiol. Lett. **198**:1-7.

50. **Kunst, F., N. Ogasawara, I. Moszer, A. M. Albertini, G. Alloni, V. Azevedo, M. G. Bertero, P. Bessieres, A. Bolotin, S. Borchert, R. Borriss, L. Boursier, A. Brans, M. Braun, S. C. Brignell, S. Bron, S. Brouillet, C. V. Bruschi, B. Caldwell, V. Capuano, N. M. Carter, S. K. Choi, J. J. Codani, I. F. Connerton, and A. Danchin.** 1997. The complete genome sequence of the gram-positive bacterium *Bacillus subtilis*. Nature **390**:249-256.

51. **Le Henaff, M., A. Chollet, and C. Fontenelle.** 2001. Chemical analysis of lipid-modified membrane proteins in *Acholeplasma laidlawii*. Curr. Microbiol. **43**:424-428.

52.   **Lewis, R. N., and R. N. McElhaney.** 1995. *Acholeplasma laidlawii* B membranes contain a lipid (glycerylphosphoryldiglucosyldiacylglycerol) which forms micelles rather than lamellar or reversed phases when dispersed in water. Biochemistry **34:**13818-13824.

53.   **Lewis, R. N., A. W. Yue, R. N. McElhaney, D. C. Turner, and S. M. Gruner.** 1990. Thermotropic characterization of the 2-O-acyl,polyprenyl alpha-D-glucopyranoside isolated from palmitate-enriched *Acholeplasma laidlawii* B membranes. Biochim. Biophys. Acta **1026:**21-28.

54.   **Li, L., O. P. Karlsson, P. Storm, S. Berg, and Å. Wieslander.** 2002. Binding governs activity of the 1,2-diacylglycerol 3-glucosyltransferase from *Acholeplasma laidlawii* at the lipid bilayer surface. To be submitted.

55.   **Lindblom, G., I. Brentel, M. Sjölund, G. Wikander, and Å. Wieslander.** 1986. Phase equilibria of membrane lipids from *Acholeplasma laidlawii:* importance of a single lipid forming nonlamellar phases. Biochemistry **25:**7502-7510.

56.   **Lindblom, G., J. B. Hauksson, L. Rilfors, B. Bergenståhl, Å. Wieslander, and P. O. Eriksson.** 1993. Membrane lipid regulation in *Acholeplasma laidlawii* grown with saturated fatty acids. Biosynthesis of a triacylglucolipid forming reversed micelles. J. Biol. Chem. **268:**16198-16207.

57.   **Maniloff, J., R. N. McElhaney, L. R. Finch, and J. B. Baseman (eds.).** 1992. Mycoplasmas, molecular biology and pathogenesis. ASM Press, Washington D.C.

58.   **Mannock, D. A., P. E. Harper, S. M. Gruner, and R. N. McElhaney.** 2001. The physical properties of glycosyl diacylglycerols. Calorimetric, X-ray diffraction and Fourier transform spectroscopic studies of a homologous series of 1,2-di-O-acyl-3-O-(beta-D-galactopyranosyl)-sn-glycerols. Chem. Phys. Lipids **111:**139-161.

59.   **Marr, A. G., and J. L. Ingraham.** 1962. Effect of temperature on the composition of fatty acids in *Escherichia coli.* J. Bacteriol. **84:**1260-1267.

60.   **Mazmanian, S. K., G. Liu, H. Ton-That, and O. Schneewind.** 1999. *Staphylococcus aureus* sortase, an enzyme that anchors surface proteins to the cell wall. Science **285:**760-763.

61.   **McElhaney.** 1992. Membrane structure, p. 113-157. *In* J. Maniloff, R. N. McElhaney, L. R. Finch, and J. B. Baseman (ed.), Mycoplasmas, Molecular biology and pathogenesis. ASM Press, Washington D.C.

62.   **McElhaney.** 1992. Lipid incorporation, biosynthesis, and metabolism, p. 231-259. *In* J. Maniloff, R. N. McElhaney, L. R. Finch, and J. B. Baseman (ed.), Mycoplasmas, Molecular biology and pathogenesis. ASM Press, Washington D.C.

63.   **McMullen, T. P., B. C. Wong, E. L. Tham, R. N. Lewis, and R. N. McElhaney.** 1996. Differential scanning calorimetric study of the interaction of cholesterol with the major lipids of the *Acholeplasma laidlawii* B membrane. Biochemistry **35:**16789-16798.

64.   **Minion, F. C., and R. F. Rosenbusch.** 1993. Extramembranous structure in mycoplasmas. Subcellular Biochemistry **20:**189-201.

65.   **Morein, S., A. Andersson, L. Rilfors, and G. Lindblom.** 1996. Wild-type *Escherichia coli* cells regulate the membrane lipid composition in a "window" between gel and non-lamellar structures. J. Biol. Chem. **271:**6801-6809.

66.   **Navarre, W. W., and O. Schneewind.** 1999. Surface proteins of gram-positive bacteria and mechanisms of their targeting to the cell wall envelope. Microbiol. Mol. Biol. Rev. **63:**174-229.

67.   **Nyström, S., P. Wallbrandt, and Å. Wieslander.** 1992. Membrane protein acylation. Preference for exogenous myristic acid or endogenous saturated chains in *Acholeplasma laidlawii.* Eur. J. Biochem. **204:**231-240.

68.   **Österberg, F., L. Rilfors, Å. Wieslander, G. Lindblom, and S. M. Gruner.** 1995. Lipid extracts from membranes of *Acholeplasma laidlawii* A grown with different fatty

acids have a nearly constant spontaneous curvature. Biochim. Biophys. Acta **1257**:18-24.

69. **Paulsen, I. T., L. Nguyen, M. K. Sliwinski, R. Rabus, and M. H. Saier.** 2000. Microbial genome analyses: comparative transport capabilities in eighteen prokaryotes. J. Mol. Biol. **301**:75-100.

70. **Paulsen, I. T., M. K. Sliwinski, and M. H. Saier.** 1998. Microbial genome analyses: global comparisons of transport capabilities based on phylogenies, bioenergetics and substrate specificities J. Mol. Biol. **277**:573-592.

71. **Peterson, S. N., and C. M. Fraser.** 2001. The complexity of simplicity. Genome Biology **2**: COMMENT2002

72. **Plackett, P., B. P. Marmion, E. J. Shaw, and R. M. Lemcke.** 1969. Immunochemical analysis of *Mycoplasma pneumoniae*. 3. Separation and chemical identification of serologically active lipids. Aust. J .Exp. Biol. Med. Sci. **47**:171-195.

73. **Plano, G. V., J. B. Day, and F. Ferracci.** 2001. Type III export: new uses for an old pathway. Mol. Microbiol. **40**:284-293.

74. **Pollack, J. D.** 1997. Mycoplasma genes: a case for reflective annotation. Trends Microbiol. **5**:413-419.

75. **Pollack, J. D., M. V. Williams, and R. N. McElhaney.** 1997. The comparative metabolism of the mollicutes (Mycoplasmas): the utility for taxonomic classification and the relationship of putative gene annotation and phylogeny to enzymatic function in the smallest free-living cells. Crit. Rev. Microbiol. **23**:269-354.

76. **Prenner, E. J., R. N. Lewis, K. C. Neuman, S. M. Gruner, L. H. Kondejewski, R. S. Hodges, and R. N. McElhaney.** 1997. Nonlamellar phases induced by the interaction of gramicidin S with lipid bilayers. A possible relationship to membrane-disrupting activity. Biochemistry **36**:7906-7916.

77. **Pyrowolakis, G., D. Hofmann, and R. Herrmann.** 1998. The subunit b of the F0F1-type ATPase of the bacterium *Mycoplasma pneumoniae* is a lipoprotein. J. Biol. Chem. **273**:24792-24796.

78. **Razin, S., D. Yogev, and Y. Naot.** 1998. Molecular biology and pathogenicity of mycoplasmas. Microbiol. Mol. Biol. Rev. **62**:1094-1156.

79. **Regula, J. T., B. Ueberle, G. Boguth, A. Gorg, M. Schnolzer, R. Herrmann, and R. Frank.** 2000. Towards a two-dimensional proteome map of *Mycoplasma pneumoniae*. Electrophoresis **21**:3765-3780.

80. **Rietveld, A. G., J. A. Killian, W. Dowhan, and B. de Kruijff.** 1993. Polymorphic regulation of membrane phospholipid composition in *Escherichia coli*. J. Biol Chem. **268**:12427-12433.

81. **Rilfors, L., G. Lindblom, Å. Wieslander, and A. Christiansson.** 1984. Lipid bilayer stability in biological membranes, p. 205-245. *In* M. Kates, and M. L.A. (ed.), Membrane fluidity. Plenum Press, New York.

82. **Rilfors, L., G. Wikander, and Å. Wieslander.** 1987. Lipid acyl chain-dependent effects of sterols in *Acholeplasma laidlawii* membranes. J. Bacteriol. **169**:830-838.

83. **Rosén, M., and Å. Wieslander.** 2002. Extensive membrane glycolipid synthesis in the small *Mycoplasma pneumoniae*: Recognition of three glycosyltransferases. Under revision.

84. **Rottem, S., L. Adar, Z. Gross, Z. Ne'eman, and P. J. Davis.** 1986. Incorporation and modification of exogenous phosphatidylcholines by mycoplasmas. J. Bacteriol. **167**:299-304.

85. **Rottem, S., and I. Kahane (eds.).** 1993. Mycoplasma Cell Membranes. Subcellular Biochemistry vol. 20.

86. **Saier, M. H.** 2000. A functional-phylogenetic classification system for transmembrane solute transporters. Microbiol. Mol. Biol. Rev. **64**:354-411.

87. **Saleh, M. T., M. Fillon, P. J. Brennan, and J. T. Belisle.** 2001. Identification of putative exported/secreted proteins in prokaryotic proteomes. Gene **269**:195-204.

88.  **Samuelson, J. C., M. Chen, F. Jiang, I. Moller, M. Wiedmann, A. Kuhn, G. J. Phillips, and R. E. Dalbey.** 2000. YidC mediates membrane protein insertion in bacteria. Nature **406**:637-641.

89.  **Schwartz, R., C. S. Ting, and J. King.** 2001. Whole proteome pI values correlate with subcellular localizations of proteins for organisms within the three domains of life. Genome Research **11**:703-709.

90.  **Shigenobu, S., H. Watanabe, M. Hattori, Y. Sakaki, and H. Ishikawa.** 2000. Genome sequence of the endocellular bacterial symbiont of aphids *Buchnera* sp. APS. Nature **407**:81-86.

91.  **Shirvan, M. H., and S. Rottem.** 1993. Ion pumps and volume regulation in mycoplasma. Subcellular Biochemistry **20**:261-292.

92.  **Sinensky, M.** 1974. Homeoviscous adaptation--A homeostatic process that regulates the viscosity of membrane lipids in *Escherichia coli.* Proc. Nat. Acad. Sci. USA **71**:522-525.

93.  **Singer, S. J., and G. L. Nicolson.** 1972. The fluid mosaic model of the structure of cell membranes. Science **175**:720-731.

94.  **Sjöström, M., S. Rännar, and Å. Wieslander.** 1995. Polpeptide sequence property relationships in *Escherichia coli* based on auto cross covariances. Chemom. Intell. Lab. Syst. **29**:295-305.

95.  **Smith, P. F.** 1992. Membrane lipid and lipopolysaccharide structures, p. 79-93. *In* J. Maniloff, R. N. McElhaney, L. R. Finch, and J. B. Baseman (ed.), Mycoplasmas, Molecular biology and pathogenesis. ASM Press, Washington D.C.

96.  **Staudegger, E., E. J. Prenner, M. Kriechbaum, G. Degovics, R. N. Lewis, R. N. McElhaney, and K. Lohner.** 2000. X-ray studies on the interaction of the antimicrobial peptide gramicidin S with microbial lipid extracts: evidence for cubic phase formation. Biochim. Biophys. Acta **1468**:213-230.

97.  **Stevens, T. J., and I. T. Arkin.** 2000. Do more complex organisms have a greater proportion of membrane proteins in their genomes? Proteins **39**:417-420.

98.  **Teichmann, S. A., C. Chothia, and M. Gerstein.** 1999. Advances in structural genomics. Curr. Opin. Struct. Biol. **9**:390-399.

99.  **Terada, M., T. M. Kuroda, Si. S., and H. Tokuda.** 2001. Lipoprotein sorting signals evaluated as the LolA-dependent release of lipoproteins from the cytoplasmic membrane of *Escherichia coli.* J. Biol. Chem. **276**:47690-47694.

100.  **Thurmond, R. L., A. R. Niemi, G. Lindblom, Å. Wieslander, and L. Rilfors.** 1994. Membrane thickness and molecular ordering in *Acholeplasma laidlawii* strain A studied by $^2$H NMR spectroscopy. Biochemistry **33**:13178-13188.

101.  **Tilcock, C. P., M. J. Hope and P. R. Cullis.** 1984. Influence of cholesterol esters of varying unsaturation on the polymorphic phase preference of egg phosphatidylethanolamine. Chem. Phys. Lipids **35**: 363-370.

102.  **Tjalsma, H., A. Bolhuis, J. D. Jongbloed, S. Bron, and J. M. van Dijl.** 2000. Signal peptide-dependent protein transport in *Bacillus subtilis:* a genome-based survey of the secretome. Microbiol. Mol. Biol. Rev. **64**:515-547.

103.  **van Wely, K. H., J. Swaving, R. Freudl, and A. J. Driessen.** 2001. Translocation of proteins across the cell envelope of Gram-positive bacteria. FEMS Microbiol. Rev. **25**:437-454.

104.  **Veldhuizen, R., K. Nag, S. Orgeig, and F. Possmayer.** 1998. The role of lipids in pulmonary surfactant. Biochim. Biophys. Acta **1408**:90-108.

105.  **Vikström, S., L. Li, O. P. Karlsson, and Å. Wieslander.** 1999. Key role of the diglucosyldiacylglycerol synthase for the nonbilayer-bilayer lipid balance of *Acholeplasma laidlawii* membranes. Biochemistry **38**:5511-5520.

106.  **Vikström, S., L. Li, and A. Wieslander.** 2000. The nonbilayer/bilayer lipid balance in membranes. Regulatory enzyme in *Acholeplasma laidlawii* is stimulated by metabolic

phosphates, activator phospholipids, and double-stranded DNA. J Biol Chem. **275:** 9296-9302.

107. **von Heijne, G.** 1992. Membrane protein structure prediction. Hydrophobicity analysis and the positive-inside rule. J. Mol. Biol. **225:**487-494.

108. **Wallin, E., and G. von Heijne.** 1998. Genome-wide analysis of integral membrane proteins from eubacterial, archaean, and eukaryotic organisms. Protein Sci. **7:**1029-1038.

109. **Wieslander, Å., M. J. Boyer, and Wróblewski.** 1992. Membrane protein structure, p. 93-113. *In* J. Maniloff, R. N. McElhaney, L. R. Finch, and J. B. Baseman (ed.), Mycoplasmas, Molecular biology and pathogenesis. ASM Press, Washington D.C.

110. **Wieslander, Å., A. Christiansson, L. Rilfors, and G. Lindblom.** 1980. Lipid bilayer stability in membranes. Regulation of lipid composition in *Acholeplasma laidlawii* as governed by molecular shape. Biochemistry **19:**3650-3655.

111. **Wieslander, Å., S. Nordström, A. Dahlqvist, L. Rilfors, and G. Lindblom.** 1995. Membrane lipid composition and cell size of *Acholeplasma laidlawii* strain A are strongly influenced by lipid acyl chain length. Eur. J Biochem. **227:**734-744.

112. **Wieslander, Å., L. Rilfors, and G. Lindblom.** 1986. Metabolic changes of membrane lipid composition in *Acholeplasma laidlawii* by hydrocarbons, alcohols, and detergents: arguments for effects on lipid packing. Biochemistry **25:**7511-7517.

113. **Yakushi, T., K. Masuda, S. Narita, S. Matsuyama, and H. Tokuda.** 2000. A new ABC transporter mediating the detachment of lipid-modified proteins from membranes. Nature Cell Biol. **2:**212-218.

# Chapter 8

## Central Carbohydrate Pathways: Metabolic Flexibility and the Extra Role of Some "Housekeeping" Enzymes

J. DENNIS POLLACK
*Department of Molecular Virology, Immunology and Medical Genetics, The Ohio State University, Columbus, Ohio 43210, USA*

## 1. INTRODUCTION

The biologic simplicity personified by Mollicutes cells was exposed by several revolutionary studies that established the utility of examining simple prokaryotes in studying life [13,17,46,53,59]. Others have continued to analyze these data and further reveal and illuminate their biochemical wealth [5,30,57]. Proteomic studies extended the findings beyond the immediate genome to the expressed proteins [5,123,168]. Mollicutes metabolism and molecular biology have been reviewed [36,85,116,122].

Cell-free independent life is unarguably complex. The best opportunity to understand the complexities of interacting components of biochemistry-metabolism, genomic sequence, expression and structural data in such cells is clearly offered by analysis of the simple free-living life-system personified by Mollicutes. We keep in mind the caution of G. Schoolnik that "housekeeping genes may confer virulence" and believe that scrutiny of central carbohydrate metabolism will reveal an unappreciated and significant role in pathogenesis, as intimated by cell culture studies, especially those of Gabridge *et al.* [48]

## 2. THE MOST ANCIENT WAY

The earliest envisioned prokaryotes, the simplest, most primitive, most ancestral cellular forms are arguably comparable to the extant low %G+C

*Molecular Biology and Pathogenicity of Mycoplasmas*, Edited by Razin and Herrmann, Kluwer Academic/Plenum Publishers, New York, 2002

*Clostridium* spp., the putative progenitor of the Mollicutes [133,169,175]. This hypothetical prokaryote ancestor was anaerobic and fermentative (heterotrophic). Fermentative in this sense meaning the process where the electrons of catabolism are "trapped" (accepted) by organic acceptors rather than by oxygen. Subsequent evolution and the appearance of oxygen beginning near the Early Proterozoic era led to the appearance of other forms of anaerobic and aerotolerant bacteria.

The initiating steps in biochemical and then cellular evolution were arguably the appearance of sugars: the rationale is based upon the ubiquity of the carbon units ("C-C") of sugars, as hexoses (glucose) and pentoses (ribose), in, *e.g.*, cell walls, polymers, nucleic acids, flavins, and ATP [52]. Glucose is often described as the universal – the preferential fuel. The anaerobic fermentative process using glucose is the most ubiquitous of the biological energy-conversion processes [52].

## 3. BUT WHY GLUCOSE?

Glucose is the most stable of the hexose isomers [10]. The anaerobic processing (fermentation) of glucose to lactic acid with the concomitant production of ATP is mechanistically the simplest type of energy conservation [52]. This form of conservation is called substrate phosphorylation. Further, support for the antiquity of the role of glucose and the fermentative process is suggested by the fact that ATP synthesis occurs in a soluble system without membranous involvement and that the relatively low level of energy conserved as ATP is taken as compatible with the supposed inefficiency of primitive mechanisms [52]. Further, the primitiveness of anaerobic fermentation is also compatible with observations that its components (ATP, NAD) are present in almost every extant cellular process that supplies or depends upon utilizable energy [71].

## 4. CARBOHYDRATE CATABOLISM

Carbohydrate metabolism in the Mollicutes was specifically reviewed in 1992 by Miles [85] and by others [106,116]. The present exposition is an extension of the above emphasizing major and new perspectives.

Glycolysis, *i.e.*, the catabolism of glucose by the Embden-Meyerhof-Parnas (EMP), or part of the EMP, is perhaps the most ancient path of energy metabolism as well as the most completely understood [42]. Contrary views propose that the Entner-Doudoroff (ED) pathway is more ancient [134] or that the citric acid (TCA) cycle is equally old or of more ancient lineage [165]. Incidentally, there is no evidence for a functional ED pathway or TCA cycle in Mollicutes [116]. Although the absence of both the TCA and cytochromes is not unique: spirochaetes, *Thermotoga* and *Pyrococcus* lack coding sequences for these systems as revealed by the COG (Clusters of Orthologous Groups) database (http://www.ncbi.nlm.nih.gov/cgi-bin/COG/palox?sys=all), their absence emphasizes the metabolic flexibility of the complete dissimilation of carbohydrate [31]. Most intriguing is the suggestion that glycine fermentation was the most ancient catabolic route [26].

The glycolytic path is often viewed as ubiquitous and when present is complete – this is not entirely correct. It is completely absent in *Rickettsia* [4]. In other cells, parts of the upper and lower portions of the path are absent, for example, pyruvate kinase (PK) [31]. Differences exist not only in content but also in enzyme mechanism and glycolytic regulation [162]. In Mollicutes, *in aqua* analyses revealed that the upper parts (hexose portion) of glycolysis (from hexose to G3P), *e.g.,* activities of 6-phosphofructokinase (6PFK) and fructose-1-6-*bis*phosphate (F-1,6-P) aldolase are undetectable in *Mycoplasma hominis* and *Mycoplasma bovigenitalium* [34]. *Asteroleplasma anaerobium* lacks PK, whose function is presumably replaced by a pyrophosphate (PPi)-dependent pyruvate:orthophosphate dikinase (PPDK) [101]. In the absence of PK its substrate phospho*enol*pyruvate (PEP) would presumably accumulate and levels of its product pyruvate would fall. PEP might be converted to oxaloacetate (OAA) by the action of PEP-carboxykinase or PEP-carboxylase. The former conserving energy as GTP, the latter, reported in *A. laidlawii* B, releases Pi and $CO_2$ [79]. Significantly, in these Mollicutes, pyruvate may be produced during glucose transport as a "by-product" of PEP-phosphotransferase system (PTS) activity. The production of pyruvate during transport may obviate some of the presumed metabolic responsibilities of PK. Another Mollicutes exception is *M. bovigenitalium* that lacks detectable phosphoglycerate kinase activity [34]. Otherwise, all tested fermentative Mollicutes have the complete triose portion of the glycolytic path to pyruvate [116].

Further, as there are numerous examples of single "promiscuous" proteins with multiple enzyme activities suggest the possibility that the ability to catabolize one sugar (glucose or mannose) may intimate the ability to catabolize the other or similar hexose isomers, although unlikely, such as: allose, altrose, gulose, indose, galactose or talose.

# 5.    FUELS, THEIR PRIMING AND THEIR ENTRY INTO THE CELL

Predictions made solely from *in silico* analyses are always putative and may be incorrect [107,108]. However, *in silico* findings have led to consequential *in aqua* findings not apparently envisioned prior to genome annotation. For example, the 1-phosphofructokinase (1PFK) activity was first identified only <u>after</u> prompting by genomic annotation in *Mycoplasma genitalium* [46], *Mycoplasma pneumoniae* [59] and *Spiroplasma citri* [51]. 1PFK enzyme activity in cytoplasmic extracts was then demonstrated in *M. genitalium* G37 and *M. pneumoniae* FH (unpublished data, J. D. Pollack). 1PFK activity obviates the role of glucose-6-phosphate isomerase and 6PFK activity by phosphorylating fructose-1-phosphate (F1P) to a component of glycolysis F-1,6-P. The sequence that is apparently essential for fructose assimilation follows a pattern as for glucose (Figure 1, modified after G. Gottschalk) [55]. The figure does not completely describe or distinguish between membranous and cytoplasmic components of the PEP-phosphotransferase system for carbohydrate. Fructose catabolism was reported to be useful in distinguishing a number of *M. fermentans* strains [96].

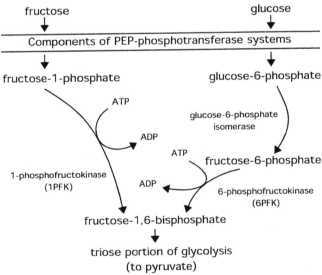

*Figure 1.* Scheme for fructose assimilation, modifed after Gottschalk [55]

In the case of *Spiroplasma citri* the role of fructose metabolism assumes special importance. Gaurivaud *et al.* [50] for the first time functionally characterized an operon for sugar utilization in Mollicutes. They demonstrated that the ability of their *S. citri* strain to use fructose was related to its phytopathogenicity. They indicated that fructose utilization by infecting phytopathogenic *S. citri* might impair the ability of the plant to load sucrose resulting in deleterious accumulation of leaf carbohydrates.

Besides of glucose and fructose, other examples of sources for carbon fuel and differentiating metabolism in Mollicutes have been reported, *e.g.*, glycerol, galactose, mannose, organic acids, sucrose, N-acetylglucosamine and maltose (see Section 20) [85,96,128,164]. Organic acids replace arginine as energy sources in some non-fermentative mycoplasmas [150]. Correlating the ability of 14 avian Mollicutes to oxidize carbohydrates or organic acids, consume oxygen and produce $H_2O_2$, a potential pathogenicity factor, led to the observation that oxygen consumers oxidizing organic acids or catabolizing glucose had higher levels of NADH oxidase activity and $H_2O_2$ production than those isolates that did not [149]. Enhanced glycerol uptake and its oxidation were reported in virulent African strains but not in less virulent European strains of *Mycoplasma mycoides* subsp. *mycoides* SC [127,164]. Utilization of maltose is an important means of rapidly differentiating *Mycoplasma mycoides* subsp. *mycoides* SC, associated with high virulence (African strains), from other members of the *M. mycoides* gene cluster [127,128]. Some bovine isolates of *Mycoplasma canis* metabolize sucrose, an infrequently reported occurrence [82].

Genomic annotation has implicated xylulose in *M. pneumoniae* as carbon sources; its metabolic role is uncertain because of a lack of biochemical details [126]. An intriguing observation but without any immediately recognized metabolic role is the ability of some *Acholeplasma* and *Mycoplasma* spp. to oxidize alcohols to acetones [1]. The observation may be associated with the detection of alcohol dehydrogenase activity in *A. laidlawii* [54].

## 6. PRIMING ENTRY OF CARBOHYDRATE (GLUCOSE)

In most anaerobic or facultative anaerobes, but in neither aerobic bacteria nor eukaryotes, the transfer of carbohydrates into the cell involves the accumulation of phosphorylated sugars. The phosphoryl donor is PEP; the process is characterized as a "group translocation" [55,117,171]. Although glucose is generally preferred, the phosphorylation is not specific for glucose. In *E. coli,* phosphorylation (group transfer) occurs in the transport of a variety of sugars: fructose, mannose and other hexoses. The overall reaction is mediated by the cell's phospho*enol*pyruvate:sugar phosphotransferase system (PTS) and is known in Mollicutes [23,60,125,177]; it is most simply:

$$PEP + carbohydrate_{outside} \rightarrow pyruvate + P\text{-}carbohydrate_{inside} \,^{[171]} \quad (Eq. 1)$$

In a significant analysis of the PTS of Mollicutes, Reizer *et al.*[125] described the process as involving five components. There are two energy-coupling proteins: Enzyme I and the three-domain sugar specific carrier complex HPr. The HPr protein of *M. capricolum* subsp. *capricolum* is structurally similar to those in Gram-positive bacteria and was taken as supportive to the view that Mollicutes evolved from the Gram-positive bacteria [103].

There is a perceived reduction in the number of Mollicutes transporters thought to reflect their non-specific requirements for carbohydrates [45,138]. Paulsen *et al.* reported that the PTS is present in a minority of 18 organisms they studied, however, they identified transporters of the glucose and fructose PTS families in *M. genitalium* and *M. pneumoniae*, the latter also had representative(s) of the mannose family [97].

Not all carbohydrates are phosphorylated during transport into their cells by the PTS system. In the absence of the PTS, the neutral sugar after transport is apparently phosphorylated by ATP. Hexokinase has been reported in *Acholeplasma* spp. but not in *Mycoplasma* spp. [34]. Conversely, the PTS has been reported in fermentative *Mycoplasma* spp. but not in *Acholeplasma* spp. [24]. *Acholeplasma florum* (now *Mesoplasma florum*) has a non-glucose fructose-PTS [91]. The PTS process and metabolite transport in Mollicutes has been analyzed in numerous reports and is a topic discussed in another chapter of this volume. The role of transport cannot be overestimated, as it appears to regulate the metabolic potential of an organism and dictates in what tissues or environment an organism can survive [45].

## 7.  GLUCOSE INSIDE THE CELL: ATP FORMATION AND THE INVOLVED PATHS TO OXYGEN AND PYRUVATE

Glucose or fructose after transport into the cell appears in the form of glucose-6-phosphate or F1P, respectively. Eventually each of these hexoses is converted to two units of pyruvate. Depending on the mollicute a concomitant net synthesis of one or two units of ATP occurs. Pyruvate is considered the terminus of the glycolytic sequence (glycolysis). Glycolysis is also called the Embden-Meyerhof-Parnas (EMP) pathway. Conservation of free-energy during glycolysis in the form of ATP, by direct transfer of a phosphoryl group to ADP without the direct intervention of oxygen, is called substrate phosphorylation.

## 8. THE INVOLVEMENT OF OXYGEN

An alternate manner of energy conservation common to eukaryotes and most prokaryote and archaean cells involves $O_2$ and is known as oxidative phosphorylation (oxphos). Oxphos involves cytochrome pigments and the transfer of electrons to oxygen in the production of ATP. Mollicutes lack cytochrome pigments and hence apparently only conserve energy by some form of substrate phosphorylation [116]. Mollicutes contain components generally associated with oxphos, iron-labile sulfur, FAD and FMN containing NADH oxidases and dehydrogenases [69,116,124]. Genes for NADH oxidases have been annotated in the genomes of *M. genitalium* (*nox*, MG275) [46], *M. pneumoniae* (*nox*, MPN394/444) [59] and *M. pulmonis* (*nox*, MYPU0230) [17] but not in *U. urealyticum* [53]. NADH oxidase activity is found in all Mollicutes except *Ureaplasma* [116].

Mollicutes then, except perhaps ureaplasmas, are all considered to have a flavin terminated respiratory system with an "obligatory", as presently but not completely understood, requirement for oxygen as the terminal electron acceptor [116]. There is no report concerning the presence or absence of other acceptors of electrons (perhaps inorganic). In the anaerobic genera (*i.e.*, *Anaeroplasma* and *Asteroleplasma*) the passage of electrons to some unknown terminal sink is obviously or apparently not to atmospheric oxygen.

NADH oxidase activity (NOA) has another particularly important metabolic role in the cytochrome-less Mollicutes. That is, the regeneration of $NAD^+$. $NAD^+$ is the recognized electron acceptor in many biochemical reactions. By acting as an electron acceptor $NAD^+$ is reduced to NADH. There are at least two mollicute classes of NOA. The membrane-localized oxygen-dependent subunit FMN:Cu:Fe:labile S = 1:1:6:6 NOA of *Acholeplasma* spp. [124] and the cytoplasmic-localized oxygen-dependent FAD:protein = 1:1 NOA of *M. capricolum* subsp. *capricolum* that lacks iron, copper, manganese, and molybdenum and is not composed of subunits [69].

In growing cells of *Acholeplasma* spp. the NOA-like activity, with one exception, mediates the dramatic transfer of electrons from NADH to the colorless free radical generator benzyl viologen ($E'_0$ -0.359 V) to produce a soluble deep purple color in its culture fluids. Species of *Spiroplasma*,

*Mycoplasma, Ureaplasma, Entomoplasma* cannot do this, nor can four out of nine *Mesoplasma* (Table 1) [109].

Mollicutes, excluding ureaplasmas and the anaerobes, are capable of transfer of either one or two electrons from reduced flavin-NADH to oxygen. In turn, producing (in theory) water or more certainly a variety of toxic oxygen enriched components, as for example superoxide or $H_2O_2$. Generally, these oxygenated compounds are toxic and contribute to the death or pathology of the Mollicutes cell and any associated host tissues [83,135,157]. Both the Mollicutes and any host cell can probably limit the deleterious action of reactive oxygen species by elaborating the enzymes superoxide dismutase (SOD) and/or catalase (CAT). A crucial clarification of the status of these activities in Mollicutes was made by Meier and Habermehl [83]. They determined that the inability of some earlier investigators to find, for example, SOD and possibly CAT in Mollicutes was due to the lack of dialysis of the cell fraction being examined. They found SOD and CAT in a number of Mollicutes in which the activities were reported absent as well as in new strains. Recently, the presence of a Cu/Zn SOD in *Mycoplasma hyopneumoniae* was reported [19]. The pathogenic role of oxygenated species and other components or products of Mollicutes have been reviewed [157]. Subsequently, Kannan and Baseman provided the first evidence that an invasive mollicute, *Mycoplasma penetrans* had both hemolytic and hemoxidative activities [67].

In the absence or paucity of SOD and CAT activities, some Mollicutes may have alternate mechanisms for counteracting the toxic effects of reactive oxygen species. For example, Ben-Menachen *et al.* [9] reported the presence of the thioredoxin reductase system (NTS) in *Mycoplasma* spp.

## 9.   THE CONTRIBUTORY ROLE OF THE THIOREDOXIN REDUCTASE SYSTEM

Although the magnitude of the role of thioredoxin-reductase system (NTS) in Mollicutes metabolism is presently not certain, we imagine that it is significant. Ben-Menachem *et al.* [9] were the first to describe, clone, sequence and isolate the thioredoxin and the thioredoxin reductase components in *M. pneumoniae* and *M. capricolum* subsp. *capricolum*. The relevant genes were putatively annotated in *M. genitalium* (MG124/102) [46], *M. pulmonis* (MYPU1840/7070/7130) [17] and in *U. urealyticum (parvum)* (UU074 and UU589) [53]. The NTS system as envisioned by the authors transfers electrons from NADPH to protein disulfides oxidized by toxic oxygenated compounds [9]. This detoxifying function may serve Mollicutes lacking or deficient in protective catalase or peroxidase activity. The NTS system may also act to replenish the cells reductive capacity for other reactions, such as the reduction of diphosphate ribonucleotides to diphosphate deoxyribonucleotides.

However, the source of NADPH is not clear for in the Mollicutes with the NTS system no G6P dehydrogenase or 6-phosphogluconate dehydrogenase activities nor the corresponding genes have been detected. These two enzymes of the oxidative portion of the pentose phosphate pathway are the presumed major source of NADPH. Further, there is no information as to the relevant presence or involvement of NAD kinase or NADH-NADPH transhydrogenase activity in Mollicutes that might act in supplying NADPH. Although, the activity of glutamate dehydrogenase in supplying reducing equivalents has been suspected [136,146]. The NTS system has been discussed more completely elsewhere [116].

## 10. THE DIRECT ROUTE TO PYRUVATE

The ten reactions of glycolysis have all been detected in fermentative-Mollicutes, but not all ten in other Mollicutes [116]. The initial reactions are: glucokinase (hexokinase), glucose-6-phosphate isomerase and 6-phosphofructokinase (6PFK), also include is the aldolase reaction that cleaves the F-1,6-P into the trioses (G3P and DHAP). The presence of some of these activities in a single comparative study of the EMP pathway in Mollicutes was reported [34].

## 11. THE REVERSIBLE PPI-6PFK

A reversible pyrophosphate (PPi)-dependent 6PFK has been isolated from *A. laidlawii* and detected in other acholeplasmas [34,114]. The presence of a reversible PFK activity in acholeplasmas remotely suggests that these Mollicutes are capable of synthesizing glucose, a process known in other cells as gluconeogenesis but requiring additional enzymatic activities. The above possibility may be a good example of flexible metabolism, a point made by Dandekar *et al.* [31] that reflects the "high plasticity and versatility" of glycolysis. We emphasize that such metabolic flexibility makes prediction of pathways and their inhibition, based only on genomic construction or, for that matter, enzymatic data, problematic.

## 12.    THE ROLES OF PYROPHOSPHATE IN MOLLICUTES

Pyrophosphate has a dual role in Mollicutes metabolism: a biosynthetic one leading toward substrate phosphorylation, and an energetic one resulting from the thermodynamically favorable hydrolysis of PPi when it is a reaction product in the presence of inorganic pyrophosphatase (PPase), in the presence of water and pyrophosphatase it is hydrolyzed to 2 Pi. A third role, a nutritional one, a requirement for the addition of PPi to the growth medium, is not known in Mollicutes. In some *Clostridium* and *Desulfotomaculum* spp., PPi serves as a major source of energy when added to the growth medium[72,176]. PPase activity has been reported in Mollicutes[33], and annotated in the genomes of *M. genitalium* (*ppa*, MG351)[46], *M. pneumoniae* (*ppa*, MPN528/314)[59], *M. pulmonis* (*ppa*, MYPU4700)[17] and *U. urealyticum* (*ppa*, UU315)[53]. PPase is essential for the growth of *E. coli*[18].

Pyrophosphate (PPi) as a biosynthetic reactant is required by *Acholeplasma* spp. and *Anaeroplasma intermedium*, since their 6PFKs are entirely dependent of PPi and cannot use ATP[101,114]; for them glycolysis and survival are not possible without PPi. PPi has another role. It serves *Asteroleplasma anaerobium* that lacks PK activity by utilizing the PPi-dependent pyruvate, orthophosphate dikinase in order to convert PEP to pyruvate[101]. *Acholeplasma* spp., a few strains of *M. hominis* and *Anaerolplasma intermedium,* use PPi but not ATP to phosphorylate nucleoside in the synthesis of mononucleotide[81,100,101,156]. *Spiroplasma* spp. (8/9) can phosphorlate deoxygunosine with either PPi or ATP[111,116]. This PPi-dependent nucleoside kinase reaction is unknown in any other cell. Petzel *et al.* has suggested that PPi-dependent enzymes occur more frequently among the Mollicutes than other prokaryotes[100]. There are other PPi dependent enzymes, such as, PEP-carboxykinase activity reported in *A. intermedium*[101]. The biosynthetic roles of PPi and PPi-dependent kinases have been reviewed[72,176].

A second role of PPi, involves the action of PPase. The hydrolysis of PPi by this enzyme is frequently regarded as one that "draws" the PPi generating "half-reaction" forward to 2 Pi. Many biosynthetic reactions produce PPi, *e.g.*, the synthesis of aminoacyl-tRNAs, the action of polymerases, formation of acyl-CoA-esters, and the uridylate activation of carbohydrates. In Mollicutes also, the action of PPases may be significantly involved in the salvage of purine bases and other reactions involving phosphoribosyltransferase in the presence of PPase activity[116].

The role of PPi as an energy source (phosphorylating donor) is taken to represent a remnant of our ancient metabolism[16,73]. It is the simplest compound containing the -P-O-P- linkage and is thought to be present on earth in its early evolution toward biogenesis, being found in extant life-forms [88]. However, PPi-metabolism, such as, PPi-6PFK activity in some Mollicutes, may not be ancient and is apparently the result of advantageous convergent evolution [84]. The introduction of a single mutation converting a PPi-dependent to an ATP-dependent PFK has led workers to propose that the ancestral PFK was an ATP-dependent PFK and that PPi-PFK evolved later[22].

Less appreciated is the modulating role that PPi or ATP-dependent 6PGK activities have on the overall glycolytic process. Although the concept of the single "rate limiting step" in a pathway is now less regarded, it is clear that in *A. laidlawii* the product of PPi-6PFK activity, *i.e.*, F-1,6-P is essential for the activation of its LDH [93]. Such activation effectively controls pyruvate metabolism by favoring lactate production and the necessary oxidation of NADH to NAD$^+$. When LDH is not activated because of low levels of F-1,6-P, it may effect glycolysis, the formation of acetyl-CoA and/or oxaloacetate from pyruvate is then presumably favored and the oxidation of NADH to NAD$^+$ diminished. This "control" may exist in other fermentative Mollicutes, almost all of which have 6PFK and lactate dehydrogenase activities [116].

## 13. NON-FERMENTATIVE MOLLICUTES: THEIR GLYCOLYTIC DEFICITS AND ROLE OF THE PENTOSE PHOSPHATE PATHWAY

The non-fermentative Mollicutes that were studied *in aqua* (*M. hominis* and *M. bovigenitalium* PG11) lack hexokinase, 6PFK and the aldolase that cleaves the F-1,6-P to two trioses and separates the "upper" (hexose) from the "lower" (triose) portion of the EMP [34,116]. The non-fermentatives also lack glucose-6-phosphate dehydrogenase (G6PD) that links the pentose phosphate pathway (PPP) and the EMP. The PPP rejoins the EMP at G3P and F6P after bypassing the missing EMP activities [34]. Therefore, our earlier speculation that non-fermentatives bypass these deficiencies via any G6PD activity has been withdrawn [116]. Schuster *et al.* [140] have suggested, based on an analysis of "elementary flux modes" (minimal sets of enzymes that can generate valid steady states), that there is a "missing link" in the metabolism of *M. hominis*. Dandekar *et al.* [31] suggested that an alternate mode of the PPP circumventing the "missing" aldolase and 6PFK might involve transaldolase, transketolase and a F6P phosphoketolase. However, the absence of transaldolase and transketolase activities reported for non-fermentative Mollicutes (*M. bovigenitalium* and in five strains of *M. hominis*) suggests otherwise [34].

The *M. bovigenitalium* strain mentioned above, later identified as *M. agalactiae* PG45, was grown in the Edward's-type medium. It did not consume the added glucose nor did pyruvate or lactate accumulate in the medium. The pH of the medium was not altered during growth, while a constant $EC_A$ (adenylate energy charge) was maintained and a 5 x $10^9$ cfu/ml at log phase [8]. This suggests that the EMP path (or part) was not operative and an unknown source of carbon, presumably proteinaceous, or purine/pyrimidine, was catabolized for growth. This strain of *M. bovigenitalium* lacks PGK activity, and this step may be the crucial missing component [34]. Perhaps, glycerol, pyruvate, or α-ketobutyrate are involved as they were shown to act as energy precursors in non-fermentative non-arginine hydrolyzing Mollicutes including *M. bovigenitalium* [166].

## 14.    BEYOND PHOSPHOFRUCTOKINASE TO PYRUVATE AND THE PYRUVATE "ROUNDHOUSE"

Following 6PFK, there are seven enzymatic steps to pyruvate, in sequence: aldolase, triose phosphate isomerase, G3P dehydrogenase, phosphoglycerate kinase (PGK), phosphoglycerate mutase, enolase and PK. The first four enzyme activities have been detected in many Mollicutes [116] and are annotated in the genomes of *M. pneumoniae* [59], *M. genitalium* [46], *M. pulmonis* [17], and *U. urealyticum* (*parvum*) [53]. PK (and PFK) was annotated in *S. citri* [21]. Recent research pertaining to these activities has centered around the activities of PGK and PK that produce ATP without involving oxygen consumption, *i.e.,* by substrate phosphorylation.

## 15.    PHOSPHOGLYCERATE KINASE (PGK) AND PYRUVATE KINASE (PK)

In Mollicutes the enzymatic action of PGK and PK may be the primary mechanisms by which ATP is generated [116]. We studied the biochemical properties of partially purified PGK and PK, as well as 6PFK and acetate kinase (AK) from *M. genitalium*, *M. pneumoniae*, *M. capricolum* subsp. *capricolum* and *M. fermentans* [112]. Additionally, in this study [112], the nucleic acid sequence of the PGK of *M. capricolum* subsp. *capricolum* was derived from the N-terminal amino acid sequence of the purified enzyme. Appropriate PCR-primers were constructed and the gene was sequenced by standard techniques. The protein and nucleic acid sequences were compared to other PGKs by phylogenetic analysis. The most significant observation was biochemical: the promiscuous nature of the kinase activity. It was reported that the partially purified 6PFK (Reaction 3 of glycolysis, Eq. 2) and the reversible PGK (Reaction 7 of glycolysis, Eq.3) and PK (Reaction 10, the final reaction of glycolysis, Eq. 4) from four *Mycoplasma* spp. can all mediate the production of a variety of ribo- and deoxyribo- nucleotides. These syntheses are most frequently attributed to the action of nucleoside diphosphate kinase (NDK). NDK activity was not found in *M. pneumoniae* nor has it been annotated in *M. pneumoniae, M. genitalium* and *U. urealyticum (parvum)*. In particular, the PK activity of each of the four *Mycoplasma* spp. was described as capable of synthesizing, when offered the respective nucleoside diphosphate and PEP: ATP, ITP, GTP, dATP, dGTP, CTP, TTP, UTP, dCTP and dUTP. These so-called non-specific activities of glycolytic kinases may obviate or reduce the necessity for NDK activity.

**6PFK**

Fructose-6-phosphate (F6P) + nucleoside triphosphate (NTP) $\rightarrow$

$$Mg^{2+}$$

Fructose-1,6-*bis*phosphate (F-1,6-P) + nucleoside diphosphate (NDP) + H$^+$   (Eq. 2)

**PGK**

1,3-Bisphosphoglycerate + NDP $\rightarrow$ 3-Phosphoglycerate (G3P) + NTP     (Eq. 3)

$$Mg^{2+}$$

**PK**

Phospho*enol*pyruvate (PEP) + NDP + H$^+$ $\rightarrow$ Pyruvate + NTP         (Eq. 4)

$$Mg^{2+}, K^+$$

A product of Eq.3, *i.e.*, G3P deserves an additional comment. In other cells G3P is the reactant that may lead to the synthesis of phosphoserine and serine, and in reaction with $N^5,N^{10}$-methylene-tetrahydrofolate to glycine. Serine itself may lead directly to pyruvate. In Mollicutes, the metabolic linkages to amino acids and their interconversions are mostly unstudied; a notable exception is the arginine deiminase pathway [116].

# 16.     THE PYRUVATE "ROUNDHOUSE"

Pyruvate and phospho*enol*pyruvate metabolism is complex [170]. Many of the known reactions involving these intermediates have not been reported in the Mollicutes literature. On the other hand, no sequenced Mollicutes genome has been completely annotated so that unidentified or unsuspected activities and pathways may exist. Further, evidence suggesting the existence of single proteins with multiple enzymic activities infers that it is difficult to declare with confidence that a biochemical conversion is absent from any microbe or cell because an unsuspected enzyme may be mediating two reactions [107,129]. Additionally, non-orthologous gene displacement (similar enzymatic function by unrelated proteins) compounds the problems inherent in predicting function and the existence or description of metabolic pathways [49]. Direct metabolic evidence is the best evidence.

Pyruvate and associated "roundhouse" functionalities have additional metabolic roles in Mollicutes. One example is the five sequential reactions of the pyruvate dehydrogenase complex; the overall stoichiometry is seen in Eq. 5.

$$\text{Pyruvate} + \text{CoA} + \text{NAD}^+ \rightarrow \text{Acetyl-CoA} + CO_2 + \text{NADH} \qquad \text{(Eq. 5)}$$

The synthesis of acetyl-CoA by the pyruvate dehydrogenase complex has been frequently and extensively studied by Zhu and Peterkofsky [177]. Their studies revealed three operons: one encoding NADH oxidase and lipoate protein ligase, another encoding pyruvate dehydrogenase EIα and EIβ, and a third encoding pyruvate dehydrogenase EII and dihydrolipoamide dehydrogenase. Phosphotransacetylase and acetate kinase constituted an additional transcription unit of the region. Evolutionarily the genes were generally similar to the corresponding genes in *M. genitalium* and *A. laidlawii*, except for an unusual gene coding for the dihydrolipoamide dehydrogenase [46,167]. Enzyme activity of the pyruvate dehydrogenase complex has been frequently reported in Mollicutes: *Mycoplasma*, *Acholeplasma*, *Spiroplasma* and *Ureaplasma* [25,32,79,111], but was absent in *Anaeroplasma* and *Asteroleplasma* [101].

In Mollicutes, other relatively unrecognized roles for pyruvate "roundhouse" components may include the degradation of glucogenic (alanine, cysteine, glycine, serine, threonine, tryptophan) and ketogenic (isoleucine, leucine, lysine, threonine) amino acids to pyruvate and acetyl-CoA, respectively. Glasfeld *et al.* [54] reported the activity of alanine dehydrogenase (deaminating) in *A. laidlawii*, the product of the reaction being pyruvate + $NH_3$ and NADH. Similar reactions may act as glycolytic entry points, routes by which apparently non-glucose utilizing Mollicutes gain both carbon and ATP. The same authors also detected alcohol dehydrogenase in *A. laidlawii* [54].

The report of aspartate transaminase activity in *Acholeplasma* and *Mycoplasma* spp [79,146] suggests that there may be a general relationship to α-ketoglutarate after degradation (if there is degradation) of the glucogenic amino acids arginine, glutamate, glutamine, histidine and proline. These amino acids, in turn, may be involved in biosynthetic and interconversion reactions. However, strikingly, no genomic annotation or proteomic evidence for transaminase activity has been reported in the Mollicutes analyzed [17,46,59,116,123,168].

Lactate is the major product of pyruvate metabolism and fermentation in most Mollicutes. It is formed by the action of cytoplasmic LDH. Lactate is released or secreted by the Mollicutes into the growth medium and the concomitant decrease in pH is associated with cell death. However, a most important product, as noted above, of this reaction is the reoxidation of NADH to $NAD^+$. $NAD^+$ may be considered the primary electron acceptor of the flavin-terminated respiratory system of Mollicutes The interplay of factors involving PEP, pyruvate, ATP, OAA. NADH and CoASH in the metabolism of Mollicutes have been discussed more thoroughly [116].

LDH has been reported in all tested Mollicutes tested but probably not in ureaplasmas. These include six *Acholeplasma* spp., two *Entomoplasma* spp., three *Mesoplasma* spp., ten *Mycoplasma* spp. (including two strains of the non-fermentative *M. hominis*), nine *Spiroplasma* spp. and a single *Anaeroplasma* and *Asteroleplasma* sp. [79,101,111,115]. One study finds LDH in *U. urealyticum* [94], but this was not confirmed by others [105]. Further, there is no annotation for LDH in the *U. ureaplasma* (*parvum*) genome [53]. The role of LDH in Mollicutes metabolism has been specifically discussed elsewhere [116] and below in relation to malate dehydrogenase.

In summary, pyruvate is immediately relatable to numerous metabolic activities in different mollicutes: the expulsion of metabolic "waste" (lactate) and perhaps electrons, the generation and consumption of ATP and NADH and the latter's essential (electron accepting) oxidized state ($NAD^+$), glycolysis, the partial reversal of glycolysis, the transport of carbohydrates (PTS), the synthesis and degradation of proteins via oxaloacetate and amino acids, through acetyl-CoA (discussed in another chapter) to fatty acid synthesis and β-oxidation, and possibly the synthesis of purines via aspartate and IMP [79]. These roles have all been demonstrated or are suspected in different Mollicutes [86,116].

## 17.  THE TCA CYCLE IS ABSENT IN MOLLICUTES BUT FOR MALATE DEHYDROGENASE

Mollicutes like few other organisms lack both the tricarboxylic acid cycle (TCA or citric acid or Krebs cycle) and cytochrome pigments. These deficiencies reflect an inefficient use of carbohydrate, since in those Mollicutes capable of fermentation, the major end-product, *i.e.*, lactate, is still relatively reduced. That is, lactate has electrons whose energy could be trapped if its precursor pyruvate was instead diverted to the synthesis of acetyl-CoA. Acetyl-CoA could then be processed either as in other cells to lipid or by the TCA cycle to $CO_2$. Although Mollicutes lack the TCA cycle they do possess one of its recognized components: malate dehydrogenase (MDH). What is the role of MDH?

Oxaloacetate and $NADH + H^+$ are convertible by MDH to malate and $NAD^+$. In effect, MDH activity acts as an important source of the essential intermediate and electron acceptor $NAD^+$. Although, all Mollicutes except for ureaplasmas have NADH oxidase activity whose action, like that of MDH, oxidizes NADH to $NAD^+$, but such activity is linked to oxygen and to the production of toxic oxygenated products. MDH modulates $NAD^+$ production without the direct involvement of $O_2$. MDH activity may also reduce lactate production and in this way decreases the amount of this deleterious acid in the growth medium, but the process expends ATP in producing OAA and the role of the end-product malate is not clearly established. The process in the ureaplasmas is unclear.

We have demonstrated that MDH and LDH activities are present in all the Mollicutes examined except for the ureaplasmas [79,108,116]. Only *ldh* has been annotated in *M. genitalium* (MG460) [46], *M. pneumoniae* (MPN674) [59] and *M. pulmonis* (MYPU7590) [17], *mdh* has not. The *ldh* or *mdh* genes were not annotated in *U. urealyticum* (*parvum*) [53]. Cordwell *et al.* suggested that in Mollicutes the LDH protein would also function as an MDH [29]. Such activities would support the view that the small-genome Mollicutes possess multienzyme proteins. In other cells the two activities are distinguishable proteins and are coded for by different genes [29]. Boerncke *et al.* [11] and Wilks *et al.* [172] showed that the LDH or the MDH activities of *E. coli* or *B. stearothermophilus* were interchangeable by a single amino acid substitution. Madern [77] found that the putative LDH gene coded for a product that was not LDH but enzymaticaly identified as an NADPH-dependent MDH; he emphasized the importance of biochemical studies in the assignment of function to gene sequence. Our efforts (unpublished) to clone the *M. genitalium* gene for *ldh* (MG460) into an *mdh⁻ E. coli* have been unsuccessful.

## 18.    LINKAGE OF THE PPP AND EMP

The role of the pentose phosphate pathway (PPP) (or the hexose monophosphate shunt) in Mollicutes has been reviewed both in relation to the presence or absence of genomic annotation of its components and its enzymatic activity [100,107,108,116]

The PPP and the EMP can connect at G3P and F6P. This permits considerable metabolic flexibility in those organisms that possess the PPP because the reversibility of their reactions permit the synthesis of ribose as R5P from G3P and hexoses as F6P from pentoses as R5P [116].

## 19. THE ROLE OF THE PPP AND DEOYXRIBOSE-5-PHOSPHATE ALDOLASE IN MOLLICUTES

Previous reviews of Mollicutes' carbohydrate metabolism did not emphasize the role of the PPP. A connection between glycolysis and nucleic acid metabolism in *M. mycoides*, subsp. *mycoides* and in *U. urealyticum* is known to involve the PPP [27,89,89,92,131]. DeSantis *et al.* [34] reported in a comparative study of the PPP and EMP in 18 Mollicutes that there were two general paths. *A. laidlawii, Acholeplasma equifetale* and *Acholeplasma axanthum* had activities of the entire PPP. *Mesoplasma florum, M. capricolum* subsp. *capricolum, M. gallisepticum, M. hyopneumoniae, M. pneumoniae,* and *M. genitalium* lacked the early PPP steps where $NADP^+$ + $CO_2$ + R5P are produced, but were capable of interconverting PPP pentoses and EMP hexoses.

The non-fermentative Mollicutes, *M. bovigenitalium* and five strains of *M. hominis* possessed only the ability to convert ribulose-5-phosphate to R5P and X5P. The role of this activity is unclear, since there is no apparent role for the product – R5P. The same authors also showed that all of the Mollicutes had deoxyribose-5-phosphate aldolase activity that was also reported by Cocks *et al.* in *U. urealyticum* [27,34]. This latter activity is not part of the PPP, but is important as it links DNA catabolism to the EMP by way of G3P.

In *M. genitalium* the ribose-5-phosphate isomerase (MG396), D-ribulose-5-phosphate 3-epimerase (MG112) and the transketolase (*tktA*, MG066) genes have been annotated [46]. In *U. urealyticum (parvum)* the epimerase (*rpe*, UU456) and transketolase (*tktA*, UU586) were annotated [53], while in *M. pneumoniae* the epimerase (*araD*, MPN498) and the transketolase 1 (*tktB*, MPN082) [59] were annotated. In *M. pulmonis*, TK (*tkt*, MYPU5110), the epimerase (*rpe*, MYPU6830) and the isomerase (MYPU6120) were annotated [17].

The absence of annotation for transaldolase (TA) in Mollicutes is surprising. In the traditional understanding of the PPP, TA activity converts the sedoheptulose-7-phosphate (S7P) and G3P arising from transketolase (TK) activity to erythrose-4-phosphate (E4P) and F6P. In Mollicutes it is not clear how E4P is generated without TA; E4P besides its role in the PPP is necessary for the synthesis of aromatic amino acids. DeSantis *et al.* [34] found TA activity only in the direction toward both S7P + G3P. They detected TA in: *M. pneumoniae, M. genitalium, A. laidlawii, A equifetale,* but not in *A.*

*axanthum, M. capricolum* subsp. *capricolum, M. gallisepticum, M. hyopneumoniae, M. bovigenitalium* and 5 strains of *M. hominis*. These Mollicutes presumably have an unrecognized gene sequence coding for TA-like activity. TK activity was detected in all Mollicutes except for the *M. hominis* strains and *M. bovigenitalium*. Hence, the non-fermentative Mollicutes, *M. hominis* and *M. bovigenitalium* have no reported TK or TA activity.

## 20.    OTHER CARBOHYDRATE PATHS IN MOLLICUTES

The complete routes by which Mollicutes metabolize carbohydrates have not all been confidently established. Efforts to describe entire routes and their ancillary passages by *in aqua* analyses are hampered by the necessity of employing numerous, different and complex chemistries. In many cases it is not always clear what enzymes to assay for; there is a certain amount of insightful presumption in designing metabolic protocols. The availability of annotation of sequenced genomes considerably reduces such guesswork. Gaps in enzymaticaly derived or deduced circuits are frequently filled by knowledge that the presumed missing enzyme or alternate activity is annotated [28]. Such combined analyses may lead to a more confidently complete picture of metabolic status, may indicate the best choices of confirmatory experiments and most agreeably substantiate putative annotation and enzymatic identity. These amalgamated relationships were categorized under the term "combinatorial genomics" [108].

As an example, in *M. mycoides* subsp. *mycoides* glycerol metabolism is essential for aerobic growth in a defined medium [132]. Enzymatic studies involving glycerol oxidase and the concomitant production of $H_2O_2$ were reviewed [86]. The glycerol kinase gene was annotated in *M. genitalium* (*glpK*, MG038) [46] and in *M. pneumoniae* (*glpK*, MPN050) [59] and *M. pulmonis* (*glpK*, MYPU2210) [17]. The $NAD^+$ dependent glycerol-3-phosphate dehydrogenase gene was also annotated in *U. urealyticum* (*parvum*) (*gpdA*, UU382) [53] and *M. pneumoniae* (*glpD*, MPN051) [59]. The combined enzymatic and genomic data sets present a more complete metabolic picture of glycerol metabolism as recognized in other cells; it also has some predictive value.

Galactose utilization by some *Acholeplasma* spp. and *Anaeroplasma* sp. is known [130,158]. The putative UDP-glucose 4-epimerase gene (*galE*, MG118) was reported in *M. genitalium* [46] the gene product acts to interconvert UDP-glucose and UDP-galactose. The ribose/galactose ABC transporters (*rbsC-2*, UU013, *rbsC-1*, UU014) and the presumably galactose interacting phosphoriboseisomerase gene (*rpiB*, UU006) were annotated in *U.*

*urealyticum* (*parvum*) [53]. The sequences in *U. urealyticum* (*parvum*) (*rpiB*, UU006) [53] and *M. pneumoniae* (*lacA*, MPN595) [59] are similar to defined domains reported in a galactose-6-phosphate isomerase subunit sequence of *M. genitalium* (*rpi*, MG396) [46]. The combined genomic and enzymatic data set is more revealing of the probable metabolic pathway than either set alone.

Mannose utilization is less frequently observed; it has been reported in *M. mycoides* subsp. *mycoides* [86]. The putative phosphomannomutase gene was reported in *U. urealyticum* (*parvum*) (*manB*, UU530) [53], *M. genitalium* (*manB*, MG053) [46], *M. pneumoniae* (*cpsG*, MPN066) [59,123] and *Mycoplasma pirum* (*cpsG*) [154]. Mannitol-1-phosphate dehydrogenase has been putatively annotated in *M. pneumoniae* [59] and *M. pulmonis* (*mtlD*, MYPU7500) [17]. Mannose-6-phosphate isomerase is also annotated in *M. pulmonis* (*pmi*, MYPU7250) [17].

In the ExPasy database the P72 lipoprotein of *M. mycoides* subsp. *mycoides* is assigned as mannitol-1-phosphate dehydrogenase and is 91% identical to the P67 protein of *Mycoplasma* sp. bovine group 7 and has 53% identity to an unannotated gene in *M. capricolum* subsp. *capricolum* [20,47]. It is somewhat uncertain if these genes are reported to code for mannitol-1-phosphate dehydrogenase. Their role is also uncertain: mannitol-1-phosphate is known to accumulate in *Staphylococcus aureus* and other bacteria and the accumulation probably involves mannitol-1-phosphate dehydrogenase [37]. Speculatively, mannose metabolism may be involved in the supply of fructose to glycolysis or in glycoprotein secretion.

Genomic data considerably extends our understanding of metabolic activity but I believe that the view drawn from only this parameter can be problematic because of insufficient inclusion of growth and enzymatic data. The prediction of metabolic activity should be drawn from a combination of a variety of data as, for example, enzyme expression, enzyme activity (elucidated by *in aqua* analyses), the presence and breadth of function of multienzyme proteins, and enzyme structure. Such a combinatorial analysis has considerably stronger predictive value than the one deduced by fewer parameters.

## 21. METABOLIC TAXONOMY

The taxonomy of Mollicutes emphasizes genomic analyses based on 16S rRNA relationships, an emphasis that diminishes older schemes relying on *in vitro* chemical responses of whole cells or their extracts [3,14,64,120] (see Taxonomy of *Mollicutes* chapter). 16S rRNA analyses may not, however, nor has it been claimed, to distinguish species within genera with any reliability [43]. There are very few purely biochemical tests that distinguish the majority of the burgeoning number of Mollicutes species within any genera from each other. Distinguishing Mollicutes species from each other is still mostly accomplished by immuno-chemical procedures. However, our present taxonomic understanding still relies on some classical biochemical assays even when deficiencies of the *in aqua* approach are appreciated, *i. e.,* we still insist for the classification of Mollicutes at the generic level knowledge of the isolates ability to grow in air, catabolize glucose, arginine and urea [7,118,121]. Glucose catabolism is least consequential, taxonomically. This may be because comprehensive studies of all strains have not been performed. Presently, then except perhaps at the generic level, metabolism is correctly viewed as insufficient to accurately characterize or differentiate Mollicutes from each other. Metabolism, as for other research modalities, serves as a component in either a combinatorial or polyphasic hierarchy [108,163].

Mollicutes taxonomy as derived from phylogenetic analyses, primarily ssu rRNA analyses, is discussed elsewhere in this volume (see Taxonomy of *Mollicutes* chapter) and I will only focus on some reports describing Mollicutes taxonomies based on aspects of metabolism, enzymes or molecules other than RNA's. Although *end-product* analyses ("acid and gas from glucose") is quite helpful, my emphasis is that knowledge of the presence of enzyme activity (*effector* analyses) is often more desirous, as the effector-proteins, *e.g.*, enzymes, are closely related to DNA sequence, transcription and translation.

A particularly valuable example of end-product-analyses were the highly sensitive assays of Sjöström and Kenny [143]. They examined whole washed cells of 22 species of *Mycoplasmatales* (14 were glycolytic) and found by 2D-polyamide TLC chromatography the end-products of the arginine dihydrolase (deiminase) pathway. They used radioactive arginine and ornithine in their assays and detected the enzymatic products by dansylation and radioautography. They found that the classification of fermentative *M. capricolum*, *M. putrefaciens*, and Bovine group VII of the glycolytic serogroup I as arginine non-utilizing mycoplasmas, was probably incorrect. Prior opinions concerning the ability of these strains to utilize arginine or not were based on apparently insensitive assays. However, although, it was clear that these mycoplasmas had active arginine deiminase and ornithine transcarbamylase, the magnitude of the enzyme activities was not assessed. A glycolytic strain of *M. gallisepticum* showed evidence of only arginine deiminase activity.

Examples of effector-analyses are limited. In Mollicutes the differential presence of adenosine triphosphatases and NAD dehydrogenases detected by quantitative immunoelectrophoresis were used to categorize 14 strains of

Mollicutes, representing three genera [144]. A study of Ball *et al.* [6] is noteworthy. These authors used 20 aminoacyl β-naphthylamide substrates in fluorometric assays of arginine utilizing and non-utilizing *Mycoplasma* spp. (16 species) and *Acholeplasma* spp. (6 species). They were able in some cases to rapidly characterize the strains by their aminopeptidase activity.

I believe that certain enzymatic assays as shown in Table 1 may eventually permit assignment of all Mollicutes isolates to genera [109,116]. At present the limited data only permit differentiation of *Acholeplasma*, *Spiroplasma*, *Anaeroplasma*, *Asteroleplasma* and *Ureaplasma* from each other and other Mollicutes. The data do not distinguish *Mycoplasma*, *Entomoplasma* and some *Mesoplasma* spp. from each other. Urease and anaerobiosis are established differential criteria [130,142]; the diagnostic utility of NADH oxidase localization (Table 1) has received some acceptance [158,159].

Table 1 includes *Acholeplasma multilocale* PN525 [58]. It was claimed on metabolic grounds that it was not an *Acholeplasma* sp. because unlike 12 other *Acholeplasma* spp. it lacked hexokinase activity, glucose-6-phosphate dehydrogenase activity, PPi-dependent 6PFK activity, and dUTPase and uracil-DNA glycosylase activities [116]. It has no membrane localized NADH oxidase activity and could not reduce benzyl viologen like all other acholeplasmas. Further, unlike all *Acholeplasma* spp., *A. multilocale* PN525, like *Spiroplasma* spp., can use both PPi and ATP in the deoxyguanosine kinase (dGUOK) assay [116]. Based on these data it was considered that *A. multilocale* is not a *bona fide Acholeplasma* sp.[116].

## 22.    PHYLOGENY, BUT NOT WITH rRNA

The phylogeny of Mollicutes has been extensively studied, analysed and reviewed [78,119] based on 16S rRNA genes (see Taxonomy of Mollicutes chapter). The controversies that surround the use of ssu rRNA to discern phylogenetic relationships and the view of C. Fraser that "the story is by far more complicated than is suggested by the ribosomal DNA tree" are challenging [41,74,75,90,98,102]. Disparities between phylogeny and physiology have been reported [2]. Regardless of the "controversies" involving the role of 16S (ssu) rRNA's they are probably the best presently available single marker for inferring phylogenetic relationships among bacteria (and certainly Mollicutes) [35,119].

Our observations may serve to widen the field of discussion. In this section, I am primarily concerned with the role of Mollicutes enzymes as phylogenetic markers. Other marker molecules (non-rRNA, non-enzyme) have been considered for phylogenetic analyses of Mollicutes [38,66,68,145,174]. A phylogenetic tree using the ribosomal protein *rps3* was shown to be consistent with 5S and 16S rRNA trees and positioned *Acholeplasma* (now *Mesoplasma*) *florum* with the 16S rRNA-Mycoplasma grouping [155]. The study supported the view that the phytoplasma are more closely related to *Acholeplasma* than to other Mollicutes. The study also reported that the *Anaeroplasmataceae* are closely related to the *Acholeplasmataceae* and suggested that the sterol requirement of *Anaeroplasma abactoclasticum* was a derived trait resulting from a loss of functions that permitted growth without cholesterol.

*Table 1.* Comparison of Metabolic Activities of 76 Species Representing the Eight Genera of Mollicutes: the Presence of dUTPase Activity, the Requirement of Pyrophosphate or ATP in the 6PFK Reaction, the Ability of Growing Cultures to Reduce Benzyl Viologen (BV), the Ratio of NADH Oxidase Activity in Membrane (M) and Soluble (S = cytoplasmic) Fractions and the Ability of Growing Cells to Synthesize Lipids From [1-C[14]]Sodium Acetate[1,2]

|  | dUTPase[3] | ATP or PPi dependent 6PFK[4] | BV[5] | M/S[6] (ratios) | Synthesis of lipids from acetate |
|---|---|---|---|---|---|
| *Acholeplasma* (12 species)[7] | + (9/9) | PPi+ (9/9) | +(12/12) | 8.7-763.2 (9/9) | + (9/9) |
| *Acholeplasma multilocale*[7] | - | ATP+ | - | <.001 | - |
| *Anaeroplasma intermedium*[8] | + | PPi+ | + | ? | ? |
| *Asteroleplasma anaerobium*[8] | + | ? | ? | ? | ? |
| *Entomoplasma* (5 species) | + (2/2) | ATP+ (2/2) | - (4/5), ± (1/5) | <.001 (2/2) | - (1/1) |
| *Mesoplasma* (9 species) | - (6/8), + (2/8) | ATP+ (3/3) | - (4/9), + (4/9), ± (1/9) | <.001 (3/3) | - (3/3) |
| *Mycoplasma* (27 species) | - (12/13), + (1/13)[9] | ATP+ (13/13) | - (21/21) | .002-.23 (14/14) | - (9/9) |
| *Spiroplasma* (17 species) | + (8/9), ±(1/9)[10] | ATP+ (8/9), NA[11] (1/9) | - (13/13) | .02-.05 (3/3) | ? |
| *Ureaplasma* (4 species)[12] | + (1/1) | ATP+ (1/1) | - (4/4) | NA (1/1)[13] | - (1/1) |

[1]General reference for the data in this table is reference 116.

[2]The value(s) in each cell is/are a composite of all the published data [116]. For example, in the first row, of the total 12 species of *Acholeplasma* that have been studied for any of the five properties in this Table only 9 have been tested for dUTPase activity and all were positive. In another example, row six, 9 *Mesoplasma* species were tested for BV activity: 4 were negative, 4 were positive, and 1 was variable. But only 3 of these same 9 *Mesoplasma* species were studied for M/S and ATP/PPi dependent PFK activities, in the former instance all 3 M/S ratios were <.001 and in the latter case all 3 PFK activities were ATP dependent.

[3]dUTPase activity mediates the reaction: dUTP = dUMP + PPi.

[4]ATP or PPi dependent 6PFK activity mediates the reaction: F6P + ATP or
PPi = F-1,6-P + ADP or Pi, respectively.

[5]BV = Benzyl viologen. Colorless benzyl viologen is reduced to the highly colored free
radical form by the action of NADH oxidase involving $O_2$ [109].

[6]A ratio of specific enzyme activities indicating the cellular localization of NADH oxidase
activity: $\mu$mol NADH oxidized min[-1] mg[-1] protein in the membrane fraction protein /$\mu$mol
NADH oxidized min[-1] mg[-1] protein in the soluble (cytoplasmic) fraction. NADH is
presumably oxidized by $O_2$ to $NAD^+$. Ratios <0.3 indicate that the NADH oxidase activity
is localized in the soluble (cytoplasmic) fraction; ratios >0.3 indicate a membrane
localization.

[7]Excluding *Acholeplasma multilocale* PN525 [58] (See text).

[8]Species of *Anaeroplasma* and *Asteroleplasma* are the only anaerobic Mollicutes.

[9]*Mycoplasma mycoides* subsp. *mycoides* is the only (1/13) *Mycoplasma* spp. reported to be
dUTPase positive [40,173], in this and other metabolic respects it resembles *Spiroplasma* spp.

[10] *Spiroplasma floricola* 23-6 has a variable response.

[11]NA = No 6PFK activity was detected in either *Spiroplasma floricola* OBMG or
*Spiroplasma floricola* 23-6 [111].

[12]The only Mycoplasmatales with urease activity are *Ureaplasma* spp. [142,160].

[13]No NADH oxidase activity was detected in either fraction.

Mollicutes' ability to synthesize lipids from radioactive acetate (Table 4)
is also considered to be a discriminatory property [56,104,110,113]. An exception to
the rule that all and only *Acholeplasma* spp. are capable of this synthesis is
*Acholeplasma palmae* J233 that did not convert acetate to lipid [110,161].

We studied phylogenies derived from comparisons of single genes.
These "gene (phylogeny) trees" are criticized as rarely consistent with each
other due to horizontal gene transfer, unrecognized paralogy and highly
variable rates of evolution [147]. Our present opinion is that such acceptable
objections may be relatable to the metabolic essentiality of the gene being
studied and the closeness of its relationship to the universal ancestor. That is,
newer and more rapidly evolving genes are less congruent to each other than
those genes that are older and universal and metabolically essential
(housekeeping genes), with highly conserved catalytic and binding centers.
This hypothesis remains to be proved.

The Ouzounis and Krypides list, based on EcoCyc and PROSITE
analysis published in 1996, included thirty-seven universal metabolic
enzyme families and patterns [95]. Our own studies suggest that the number
may be 32-35, since, *e.g.*, not all Mollicutes have detectable GMP synthase
or adenylosuccinate synthase and no Mollicutes has detectable citrate
synthase activities. Furthermore, genes for any of these activities have not
been reported in any Mollicutes. Interestingly, MDH is listed but LDH is
not. The LDH-MDH multifunctional protein anomaly described in
Mollicutes is discussed above.

We studied PGK activity in Mollicutes. From *M. capricolum* subsp.
*capricolum* we isolated and sequenced the active PGK protein; then using

deduced primers and purified DNA by PCR determined its nucleotide sequence [112]. Phylogenetic analyses of the PGK gene (*pgk*) were performed by a computational neighbor-joining (NJ) method. One hundred complete *pgk* gene sequences were examined from different life forms: Bacteria, Archaea and Eukarya. The analyses included *pgk* sequences of *M. pneumoniae*, *M. genitalium*, *U. urealyticum*, and the aforementioned *M. capricolum* subsp. *capricolum* and unpublished and donated *pgk* sequences of *M. pulmonis*, *M. hyopneumoniae*, *Spiroplasma citri*, and *M. mycoides* subsp. *mycoides* SC carried out in other laboratories [113]. The associative power of the analyses was striking. As there are about 410 amino acids coded by the *pgk* gene it was surprising that there was enough evolutionary signal in that number of residues to separate the examples clearly into Bacteria, Archaea and Eukarya. Examinations of these three major clusters show even higher biological affinity, subgroups were formed that reflected relationships already associated by non-molecular traditionally Linnaean (phenotypic) criteria. That is, the derived *pgk* sequence phylogeny mimiced in many aspects established biologic groupings, not only within the major clades of Eukarya, Bacteria, Archaea but within taxonomically homogeneous subclades [112]. Of the 100 different sequences included in the phylogenetic study five ciliates were selected and grouped together in the neighbor-joining (N-J) tree. Immediately following them was the only flagellate included in the analysis and then a group of all the 14 fungi included. The Archaea were entirely and separately grouped at the bottom of the tree. Within the Archaea the Crenarchaeota were entirely separable from the Euarchaeota. Other individual and entire groupings were the *Trypanosomatideae*, mammals, *Spirochaetaceae*, plants, *Actinomycetales*, and the alpha, beta and most (4/5) of the gamma divisions of the prokaryotes (eubacteria) [112].

The branching order of the eight different Mollicutes in the *pgk*-tree followed exactly the same 16S rRNA groupings described by Johansson *et al.* [65]. The *pgk*-tree associates *M. mycoides* subsp. *mycoides*, *M. capricolum* subsp. *capricolum* and *S. citri* to Johansson *et al.* Group II-Spiroplasma, *M. pulmonis* and *M. hyopneumoniae* to Group IV-Hominis, and *M. pneumoniae*, *M. genitalium* and *U. urealyticum* (*parvum*) to Group III-Pneumoniae [112].

Of additional interest to mycoplasmologists is the closer relationship of the Mollicutes cluster to the low G+C non-spore formers *Staphylococcus aureus* and *Lactobacillus delbruekii*, *bulgaricus* and its greater distance from the low G+C spore formers *Bacillus* and *Clostridium* spp. The *pgk*-tree suggests that *Mycoplasma* spp. are more closely related to the *Streptococcus-Lactobacillus* subline than to the *Bacillus* and *Clostridium* spp. The relationship of Mollicutes to the *Streptococcus-Lactobacillus* was first made by Neimark [93] who concluded that the Mollicutes (acholeplasmas) descended from this group. His view was based on the similarity of their F-1,6-P activated LDHs and the immunologic homology of their aldolases.

Unfortunately, the *pgk* sequence from *Clostridium innocuum*, a bacterium relatable to Mollicutes, is not known [169,175]. The exact depth of the *pgk*-tree branches should be more convincingly established by sensitive and elaborate computational means other than NJ techniques. However, even with large amounts of data it may not be possible to reconstruct prokaryote phylogeny using standard methods [152].

## 23.    EPILOGUE: *QUO VADIS*, MYCOPLASMATALES?

Speaking only from the point of view of metabolism, even the study of "housekeeping" genes and their activities may hold surprises, they "may confer virulence" [44]. Although there is some emphasis on identifying the minimal number of genes essential for independent cellular life, a task now viewed by this reviewer as not foreseeable, it may be of some, if not substantial value, considering the caveats described in this review, to shift, a bit, the goal's focus or attention towards first recognizing what enzymatic activities are essential, and then attempt to identify the genes that code for such activities. The general emphasis and perhaps a more revealing approach might be to define first which genes, proteins, and enzyme activities are relatable to viability in a particular environment [148]. G. Schoolnik is quoted as saying that *Mycobacterium tuberculosis* upon macrophage entry dramatically shifts its metabolism towards fatty acids for carbon and energy, a significant departure from its metabolism in culture [44]. So what genes are essential?

The genome of *E. coli* has been characterized as a model of efficiency; its study is claimed to place science within "sight of the holy grail" of an understanding of the integrated genetic and metabolic patterns of an organism that "will probably be *E. coli*" [139]. *Escherichia coli* has this position because of the wealth of information already available and "knowledge builds on knowledge" [139]. However, it may be that "the massive biochemical literature on *Escherichia coli* has prejudiced the general biochemical perspective of bacteria for some time" [15]. *Escherichia coli* has nearly 4400 genes, ten times more than *M. genitalium*; the metabolic complexity in such a cell, considering the presence of multifunctional components and the imponderables already mentioned, may be undecipherable. A "simple" system is better, even studying *M. genitalium* there is complexity enough [99]. However, *E. coli* can be grown in a completely defined medium and generally Mollicutes disadvantageously cannot. Is it certain that a completely defined environment, *i.e.*, growth medium, is necessary for a complete or reliable understanding of "function" *in vivo*? The simplest cell metabolism seems to be the most reasonable analytical object.

In fact, as "the goal [is to] bridge the gap between phenotype and genotype" [12] it may be more easily accomplished using a mollicute such as *M. genitalium*. *M. genitalium* is the simplest organism with about 480 annotated coding regions, a number approximately one-tenth that of *E. coli*'s number of encoding genes. Studies to characterize the complete and essential protein-enzymatic content of Mollicutes cells and their interactions have been a constant area of enquiry with a significant volume of literature for decades [61,70,116,122]. The minimal-gene-set concept is a paradigm built by Mollicutes [116] (see Mycoplasmas and the Minimal Cell Concept chapter). Recognition of the role that Mollicutes already play in reaching a complete biochemical understanding of a free-living cell is acknowledged: *M. genitalium* is characterized as "the standard benchmarking genome in (both) computational and experimental genomics" as well as in structural predictions [62,151].

Complete understanding of the content, regulation and evolution of cell metabolism will require an amalgamation of physiologic, genomic, microarray analyses, enzymatic function, computational technology, the services of different disciplines and certainly structural assignments of genome sequences [153]. Especially when understanding is incomplete, predictive opinions require the incorporation of all available data. A simple model system will be valuable, the simpler the better, especially if a polyphasic approach is chosen [163].

There are new untraditional approaches to the study of cellular metabolism. The development of theoretical and experimental techniques for phenotype characterization of metabolic routes claimed to be stoichiometrically and thermodynamically feasible may prove to be particularly illuminating and definitive. This new concept ("fluxes" or "flux modes") has recently been tested in a number of applications, *e.g.,* it has been shown to be able to reveal which metabolic pathways are active [39,137,140,141]. Another alternative leading to a complete metabolic understanding and gene assignment is that proposed by Martzen *et al.* [80]. Their technical *tour de force* bypasses the need to purify polypeptides as a "backward" route to the identification of the coding DNA sequence, their procedure is called "expression cloning": the introduction of cDNA pools into various host cells, followed by screening for activity and identifying the responsible DNA. MacBeath and Schreiber [76] describe a miniaturized technique using small sample volumes to examine thousands of proteins simultaneously in order to screen for protein-protein interactions, identify enzyme substrates and protein targets of small molecules. Idekar *et al.* [63]

used a comprehensive integrated analysis involving mRNA expression profiles to study yeast metabolism after genetic or environmental changes, a methodology that may become generally useful.

To reprise the philosophy set forth over forty-years ago by Professors Harold Morowitz, then of Yale University, the late Robert C. Cleverdon at the University of Connecticut and their students who began the search for a model of independent cellular life: the simplest organisms seem to be the most logically profitable examples to study – and they are the Mollicutes. Our present understanding of Mollicutes metabolism is still a prologue to future studies.

## ACKNOWLEDGMENTS

I thank Casey O'Stroske for drawing the figure. I apologize to all those authors whose many contributions were not cited. I would thank and acknowledge receipt of Public Health Service Grant RO1-A133193 from the National Institutes of Health for partial support of the research content of this Chapter.

## REFERENCES

1. **Abu-Amero, K. K., E. A. Abu-Groun, M. A. Halablab, and R. J. Miles**. 2000. Kinetics and distribution of alcohol oxidizing activity in *Acholeplasma* and *Mycoplasma* species. FEMS Microbiol. Lett. 183:147-151.
2. **Achenbach, L. A., and J. D. Coates**. 2000. Disparity between bacterial phylogeny and physiology. ASM News 65:752-757.
3. **Aluotto, B. B., R. G. Wittler, C. O. Williams, and J. E. Faber**. 1970. Standardized bacteriologic techniques for the characterization of *Mycoplasma* species. Int. J. Syst. Bacteriol. 20:35-58.
4. **Andersson, S. G., A. Zomorodipour, J. O. Andersson, T. Sicheritz-Ponten, U. C. Alsmark, R. M. Podowski, A. K. Näslund, A. S. Eriksson, H. H. Winkler and C. G. Kurland**. 1998. The genome sequence of *Rickettsia prowazekii* and the origin of mitochondria. Nature 396:133-140.
5. **Balasubramanian, S., T. Schneider, M. Gerstein, and L. Regan**. 2000. Proteomics of *Mycoplasma genitalium*: identification and characterization of unannotated and atypical proteins in a small model genome. Nucl. Acids Res. 28:3075-3082.
6. **Ball, H. J., S. D. Neill, and L. R. Reid**. 1982. Use of arginine aminopeptidase activity in characterization of arginine-utilizing mycoplasmas. J. Clin. Microbiol. 15:28-34.
7. **Barile, M**. 1983. Arginine hydrolysis, p. 345-349. *In* S. Razin and J. G. Tully (ed.), *Methods in Mycoplasmology*, vol. 1. Academic Press, Inc., NY.
8. **Beaman, K. D. and J. D. Pollack**. 1983. Synthesis of adenylate nucleotides by Mollicutes (mycoplasmas). J. Gen. Microbiol. 129:3103-3110.

9. **Ben-Menachem, G., R. Himmelreich, R. Herrmann, Y. Abramowitz, and S. Rottem**. 1997. The thioredoxin reductase system of mycoplasmas. Microbiology. 143:1933-1940.

10. **Bloch, K. E.** 1979. Speculations on the evolution of sterol structure and function. CRC Crit. Rev. Biochem. 7:1-5.

11. **Boerncke, W. E., C. S. Millard, P. W. Stevens, S. N. Kakar, F. J. Stevens, and M. I. Donnelly**. 1995. Stringency substrate specificity of *Escherichia coli* malate dehydrogenase. Arch. Biochem. Biophys. 322:43-52.

12. **Bork, P., T. Dandekar, Y. Diaz-Lazcoz, F. Eisenhaber, M. Huynen, and Y. Yuan**. 1998. Predicting function: from genes to genomes and back. J. Mol. Biol. 283:707-725.

13. **Bork, P., C. Ouzounis, G. Casari, R.Schneider, C. Sander, M. Dolan, W. Gilbert, and P. M. Gillevet**. 1995. Exploring the *Mycoplasma capricolum* genome: a minimal cell reveals its physiology. Mol. Microbiol. 16:955-967.

14. **Bradbury, J.** 1977. Rapid biochemical test for characterization of the Mycoplasmatales. J. Clin. Microbiol. 5:531-534.

15. **Byng, G. S., J. F. Kane, and R. A. Jensen**. 1982. Diversity in the routing and regulation of complex biochemical pathways as indicators of microbial relatedness. CRC Crit. Rev. Microbiol. 9:227-252.

16. **Calvin, M.** 1956. Chemical evolution and the origin of life. Am. Scient. 44:248-263.

17. **Chambaud, I., R. Heilig, S. Ferris, V. Barbe, D. Samson, F. Galisson, I. Mozed, K. Dybvig, H. Wróblewski, A. Viari, E. P. C. Rocha, and A. Blanchard**. 2001. The complete genome sequence of the murine respiratory pathogen *Mycoplasma pulmonis*. Nucl. Acids Res. 29:2145-2153.

18. **Chen, J., A. Brevet, M. Fromant, F. Lévêque, J.-M. Schmitter, S. Blanquet, and P. Plateau**. 1990. Pyrophosphatase is essential for the growth of *Escherichia coli*. J. Bacteriol. 172:5686-5689.

19. **Chen, J. R., C. N. Weng, T. Y. Ho, I. C. Cheng, and S. S. Lai**. 2000. Identification of the copper-zinc superoxide dismutase activity in *Mycoplasma hyopneumoniae*. Vet. Microbiol. 73:301-310.

20. **Cheng, X., J. Nicolet, R. Miserez, P. Kuhnert, M. Krampe, T. Pilloud, E. M. Abdo, C. Griot, and J. Frey**. 1996. Characterization of the gene for an immunodominant 72 kDa lipoprotein of *Mycoplasma mycoides* subsp. *mycoides* small colony type. Microbiology 142:3515-3524.

21. **Chevalier, C., C. Saillard, and J. M. Bové**. 1990. Organization and nucleotide sequences of the *Spiroplasma citri* genes for ribosomal protein S2, elongation factor Ts, spiralin, phosphofructokinase, pyruvate kinase, and an unidentified protein. J. Bacteriol. 172:2693-2703.

22. **Chi, A., and R. G. Kemp**. 2000. The primordial high energy compound: ATP or inorganic pyrophosphate? J. Biol. Chem. 275:35677-35679.

23. **Chung, T. L., L. Farh, Y. L. Chen, and D. Shiuan**. 2000. Molecular cloning and characterization of a unique 60 kDa/72 kDa antigen gene encoding enzyme I of the phosphoenolpyruvate:sugar phosphotransferase system (PTS) of *Mycoplasma*

24. **Cirillo, V., and S. Razin**. 1973. Distrbution of a phosphoenolpyruvate dependent sugar phosphotransferase system in mycoplasmas. J. Bacteriol. 113:212-217.

25. **Clark, A. F., D. F. Farrell, W. Burke, and C. R. Scott**. 1978. The effect of mycoplasma contamination on the in vitro assay of pyruvate dehydrogenase in cultured fibroblasts. Clin. Chim. Acta 82:119-124.

26. **Clarke, P. H., and S. R. Elsden**. 1980. The earliest catabolic pathways. J. Mol. Evol. 15:333-338.

27.    **Cocks, B. G., F. A. Brake, A. Mitchell, and L. R. Finch**. 1985. Enzymes of intermediary carbohydrate metabolism in *Ureaplasma urealyticum* and *Mycoplasma mycoides* subsp. *mycoides*. J. Gen. Microbiol. 131:2129-2135.
28.    **Cordwell, S. J.** 1999. Microbial genomes and "missing" enzymes: redefining biochemical pathways. Arch. Microbiol. 172:269-279.
29.    **Cordwell, S., D. J. Basseal, J. D. Pollack and I. Humphery-Smith**. 1997. Malate/lactate dehydrogenase in mollicutes: evidence for a multienzyme protein. Gene 195:113-120.
30.    **Dandekar, T., M. Huynen, J. T. Regula, B. Ueberle, C. U. Zimmermann, M. A. Andrade, T. Doerks, L. Sánchez-Pulido, B. Snel, M. Suyama, Y. P. Yuan, R. Herrmann, and P. Bork**. 2000. Re-annotating the *Mycoplasma pneumoniae* genome sequence: adding value, function and reading frames. Nucl. Acids Res. 28:3278-3288.
31.    **Dandekar, T., S. Schuster, B. Snell, M. Huynen and P. Bork**. 1999. Pathway alignment: application to the comparative analysis of glycolytic enzymes. Biochem. J. 343:115-124.
32.    **Davis, J. W. Jr., J. T. Manolukas, B. E. Capo, and J. D. Pollack**. 1990. Pyruvate metabolism and the absence of a tricarboxylic cycle in *Ureaplasma urealyticum*, p. 666-669. *In* G. Stanek, G. H. Cassell, J. G. Tully, and R. Whitcomb (ed.), *Recent Advances in Mycoplsmology*, Zbt. Bakt. Suppl. 20, pp. 962. Gustav Fischer Verlag, Stuttgart, Germany.
33.    **Davis, J. W., Jr., I. S. Moses, C. Ndubuka, and R. Ortiz**. 1987. Inorganic pyrophosphatase activity in cell-free extracts of *Ureaplasma urealyticum*. J. Gen. Microbiol. 133:1453-1459.
34.    **DeSantis, D., V.V. Tryon and J. D. Pollack**. 1989. Metabolism of Mollicutes: the Embden-Meyerhof-Parnas pathway and the hexose monophosphate shunt. J. Gen. Microbiol. 135:683-691.
35.    **Doolittle, W. F.** 1999. Phylogenetic classification and the universal tree. Science 284:2124-2128.
36.    **Dybvig, K., and L. L. Voelker**. 1996. Molecular biology of mycoplasmas. Annu. Rev. Microbiol. 50:25-57.
37.    **Edwards, K. G., H. J. Blumenthal, M. Khan, and M. E. Slodkl**. 1981. Intracellular mannitol, a product of glucose metabolism in staphylococci. J. Bacteriol. 146:1020-1029.
38.    **Falah, M., and R. S. Gupta**. 1997. Phylogenetic analysis of mycoplasmas based on Hsp70 sequences: cloning of the *dnaK* (*hsp70*) gene region of *Mycoplasma capricolum*. Int. J. Syst. Bacteriol. 47:38-45.
39.    **Fell, D**. 1997. *Understanding the Control of Metabolism*, pp. 301. *Frontiers in Metabolism*, 2. Portland Press, London.
40.    **Finch, L. R., and A. Mitchell**. 1992. Sources of nucleotides, p. 211-230. *In* J. Maniloff, R. N. McElhaney, L. R. Finch, and J. B. Baseman (ed.), *Mycoplasmas: Molecular Biology and Pathogenesis*. ASM Press, Washington, DC
41.    **Forterre, P**. 1996. Protein versus rRNA: problems in rooting the universal tree of life. ASM News 63:89-95.
42.    **Fothergill-Gilmore, L. A., and P. A. Michels**. 1993. Evolution of glycolysis. Prog. Biophys. Mol. Biol. 59:105-235.
43.    **Fox, G. E., J. D. Wisotzkey, and P. Jurtshak**. 1992. How close is close: 16S rRNA sequence identity may not be sufficient to guarantee species identity. Int. J. Syst. Bacteriol. 42:166-170.
44.    **Fox, J. L**. 2001. Microbial genomics reuniting with microbiology. ASM News 67:247-252.

45. **Fraser, C., J. Eisen, R. D. Fleischmann, K. A. Ketchum, and S. Peterson**. 2000. Comparative genomics and understanding of microbial biology. Emerg. Inf. Dis. 6:505-512.

46. **Fraser, C., J. D. Gocayne, O. White, M. D. Adams, R. A. Clayton, R. D. Fleischmann, C. J. Bult, A. R. Kerlavage, G. Sutton, J. M. Kelley, J. L. Fritchman, J. F. Wiedman, K. V. Small, M. Sandusky, J. Fuhrmann, D. Nguyen, T. R. Utterback, D. M. Saudek, C. A. Phillips, J. M. Merrick, J.-F. Tomb, B. A. Dougherty, K. F. Bott, P.-C. Hu, T. S. Lucier, S. N. Peterson, H. O. Smith, C. A. Hutchison III and C. Venter**. 1995. The minimal gene complement of *Mycoplasma genitalium*. Science 270:397-403.

47. **Frey, J., X. Cheng, M. P. Monnerat, E. M. Abdo, M. Krawinkler, G. Bolske, and J. Nicolet**. 1998.Genetic and serologic analysis of the immunogenic 67 67-kDa lipoprotein of *Mycoplasma* sp. bovine group 7. Res. Microbiol. 149:55-64.

48. **Gabridge, M, D. K. F. Chandler, and M. Barile**.1985. Pathogenic factors in mycoplasmas and spiroplasmas, p. 313-351. *In* S. Razin and M. F. Barile (ed.), *The Mycoplasmas,* vol. IV. *Mycoplasma Pathogenicity*. Academic Press, NY.

49. **Galperin, M. Y., and E. V. Koonin**. 1999. Functional genomics and enzyme evolution, p. 265-284. *In* E. M. Bradbury, and S. Pongor (ed.), *Structural Biology and Functional Genomics*. NATO Science Series 3. High Technology, vol. 71. Kluwer Academic Publishers, Boston, MA.

50. **Gaurivaud, P., J. L. Danet, F. Laigret, M. Garnier, and J. M. Bové**. 2000. Fructose utilization and phytopathogenicity of *Spiroplasma citri*. Mol. Plant Microbe Interact. 13:1145-1155.

51. **Gaurivaud, P., F. Laigret, M. Garnier, and J. Bové**. 2000. Fructose utilization and pathogenicity of *Spiroplasma citri*: characterization of the fructose operon. Gene 252:61-69.

52. **Gest, H., and J. W. Schopf**. 1983. Biochemical evolution of anaerobic energy conversion: the transition from fermentation to anoxygenic photosynthesis, p. 135-148. . *In* J. W. Schopf (ed.), *Earth's Earliest Biosphere. Its Origin and Evolution*. Princeton University Press, NJ.

53. **Glass, J. I., E. J. Lefkowitz, J. S. Glass, C. R. Heiner, E. Y. Chen and G. H. Cassell**. 2000. The complete sequence of the mucosal pathogen *Ureaplasma urealyticum*. Nature 407:757-762.

54. **Glasfeld, A., G. F. Leanz, and S. A. Benner**. 1990. The stereospecificities of seven dehydrogenases from *Acholeplasma laidlawii*. J. Biol. Chem. 265:11692-11699.

55. **Gottschalk, G**. 1986. Aerobic growth of *Escherichia coli* on substrates other than glucose, p. 96-103. *In* G. Gottschalk, *Bacterial Metabolism*. Springer-Verlag, NY.

56. **Herring, P. K., and J. D. Pollack**. 1974. Utilization of [1-C$^{14}$]acetate in the synthesis of lipids by acholeplasmas. Int. J. Syst. Bacteriol. 24:73-78.

57. **Herrmann, R., and B. Reiner**. 1998. *Mycoplasma pneumoniae* and *Mycoplasma genitalium*: a comparison of two closely related bacterial species. Curr. Opin. Microbiol. 1:572-579.

58. **Hill, A. C., A. A. Polak-Vogelzang, and A. F. Angulo**. 1992. *Acholeplasma multilocale* sp. nov., isolated from a horse and a rabbit. Int. J. Syst. Bacteriol. 42:513-517.

59. **Himmelreich, R., H. Hilbert, H. Plagens, E. Pirkl, B. C. Li and R. Herrmann**. 1996. Complete sequence analysis of the genome of the bacterium *Mycoplasma pneumoniae*. Nucl. Acids Res. 24:4420-4449.

60.     **Huang, K., G. Kapadia, P.-P. Zhu, A. Peterkofsky, and O. Herzberg**. A promiscuous binding surface: crystal structure of the IIA domain of the glucose-specific permease from *Mycoplasma capricolum*. Structure 6:697-710.

61.     **Hutchison III, C. A., S. N. Peterson, S. R. Gill, R. T. Cline, O. White, C. M. Fraser, H. O. Smith, and J. C. Venter**. 1999. Global transposon mutagenesis and a minimal mycoplasma genome. Science 286:2165-2169.

62.     **Huynen, M., B. Snel, W. Lathe III, and P. Bork**. 2000. Predicting protein function by genomic context: quantitative evaluation and qualitative inferences. Cancer Res. 10:1204-1210.

63.     **Idekar, T., V. Thorsson, J. A. Ranish, R. Christmas, J. Buhler, J. K. Eng, R. Bumgarner, D. R. Goodlott, R. Aebersold, and L. Hood**. 2001. Integrated genomic and proteomic analyses of a systematically perturbed metabolic network. Science 292:929-934.

64.     **James, A. L**. 1994. Enzymes in taxonomy and diagnostic bacteriology, p.471-492. *In*, M. Goodfellow and A. G. O'Donnell (ed.), *Chemical Methods in Prokaryotic Systematics*. John Wiley & Sons, NY.

65.     **Johansson, K.-E., M. U. K. Heldtander, and B. Pettersson**. 1998. Characterization of mycoplasmas by PCR and sequence analysis with universal 16S rDNA primers, p. 145-165. *In* R. J. Miles and R. A. J. Nicholas (ed.), *Methods in Molecular Biology*, vol. 104, *Mycoplasma Protocols*. Humana Press, Inc., Totowa, NJ.

66.     **Kamla, V., B. Heinrich, and U. Hadding**. 1996. Phylogeny based on elongation factor Tu reflects the phenotypic features of mycoplasmas better than that based on 16S rRNA. Gene 171:83-87.

67.     **Kannan, T. R., and J. B. Baseman**. 2000. Hemolytic and hemoxidative activities in *Mycoplasma penetrans*. Infect. Immun. 68:6419-6422.

68.     **Karlin, S., G. M. Weinstock, and V. Brendel**. 1995. Bacterial classification derived from RecA protein sequence comparisons. J. Bacteriol. 177:6881-6893.

69.     **Klomkes, M., R. Altdorf, and H.-D. Ohlenbusch**. 1985. Purification and properties of an FAD-containing NADH oxidase from *Mycoplasma capricolum*. Biol. Chem. (Hoppe-Seyler's) 366:963-969.

70.     **Koonin, E. V**. 2000. How many genes can make a cell: the minimal-gene-set concept. Annu. Rev. Genomics Hum. Genet. 1:99-116.

71.     **Krebs, H. A., and H. L. Kornberg**. 1957. Energy transformation in living matter. Ergeb. Physiol. Biol. Chem. Exp. Pharmakol. 49:212-298.

72.     **Kulaev, I. S., and V. M. Vagabov**. 1983. Polyphosphate metabolism in micro-organisms. Adv. Microb. Physiol. 24:83-171.

73.     **Lipman, F**. 1984. Pyrophosphate as a possible precursor of ATP, p.133-135. *In* K. Matsuno, K. Dose, K. Harada, and D. L. Rohlfling (ed.), *Molecular Evolution and Protobiology*. Plenum Press, NY.

74.     **Ludwig, W., and K.-H. Schleifer**. 1999. Phylogeny of *Bacteria* beyond the 16S rRNA standard. ASM News 65:752-757.

75.     **Ludwig, W., O. Strunk, S. Klugbauer, N. Klugbauer, M. Weizenegger, J. Neumaier, M. Bachleitner, and K.-H. Schleifer**. 1998. Bacterial phylogeny based on comparative sequence analysis. Electrophoresis 19:554-568.

76.     **MacBeath, G., and S. L. Schreiber**. 2000. Printing proteins as microarrays for high-throughput function determination. Science 289:1760-1763.

77.     **Madern, D**. 2000. The putative L-lactate dehydrogenase from *Methanococcus jannaschii* is an NADPH-dependent L-malate dehydrogenase. Mol. Microbiol. 37:1515-1520.

78.   **Maniloff, J**. 1992. Mycoplasma phylogeny, p.549-559. *In* J. Maniloff, R. N. McElhaney, L. R. Finch, and J. B. Baseman (ed.), *Mycoplasmas. Molecular Biology and Pathogenesis*. ASM Press, Washington, DC.

79.   **Manolukas, J. M. F. Barile, D. K. F. Chandler, and J. D. Pollack**. 1988. Presence of anaplerotic reactions and transamination, and the absence of the tricarboxylic acid cycle in Mollicutes. J. Gen. Microbiol. 134:791-800.

80.   **Martzen, M. R., S. M. McCraith, S. L. Spinelli, F. M. Torres, S. Fields, E. J. Grayhack, and E. M. Phizicky**. 1999. A biochemical genomics approach for identifying genes by the activity of their products. Science 286:1153-1155.

81.   **McElwain, M. C., D. F. K. Chandler, M. F. Barile, T. F. Young, V. V. Tryon, J. W. Davis, Jr., J. P. Petzel, C. J. Chang, M. V. Williams, and J. D. Pollack**. 1988. Purine and pyrimidine metabolism in Mollicutes species. Int. J. Syst. Bacteriol. 38:417-423.

82.   **Megid, R., R. A. Nicholas, and R. J. Miles**. 2001. Biochemical characterization of *Mycoplasma bovirhinis*, *Mycoplasma dispar* and recent bovine isolates of *Mycoplasma canis*. Vet. Res. Commun. 25:1-12.

83.   **Meier, B., and G. G. Habermehl**. 1990. Evidence for superoxide dismutase and catalase in mollicutes and release of reactive oxygen. Arch. Biochem. Biophys. 277:74-79.

84.   **Mertens, E**. 1991. Pyrophosphate-dependent phosphofructokinase, an anaerobic glycolytic enzyme? FEBS Lett. 285:1-5.

85.   **Miles, R. J**. 1992. Catabolism in Mollicutes. J. Gen. Microbiol. 138:1773-1783.

86.   **Miles, R. J., A. E. Beezer, and D. H. Lee**. 1985. Kinetics of utilization of organic substrates by *Mycoplasma mycoides* subsp. *mycoides* in a salts solution: a flow calorimetric study. J. Gen. Microbiol. 131:1845-1852.

87.   **Miller, S. L., and M. Parris**. 1964. Synthesis of pyrophosphate under primitive earth conditions. Nature 204:1248-1250.

88.   **Mitchell, A., and L. R. Finch**. 1977. Pathways of nucleotide biosynthesis in *Mycoplasma mycoides*, subsp. *mycoides*. J. Bacteriol. 130:1047-1054.

89.   **Mitchell, A., and L. R. Finch**. 1979. Enzymes of pyrimidine metabolism in *Mycoplasma mycoides* subsp. *mycoides*. J. Bacteriol. 137:1073-1080.

90.   **Moreira, D., and H. Philippe**. 2000. Molecular phylogeny: pitfalls and progress. Int. Microbiol. 3:9-16.

91.   **Navas-Castillo, J., F. Laigret, A. Hocquellet, C. J. Chang, and J.-M. Bové**. 1993. Evidence for a phosphoenolpyruvate dependent sugar-phosphotransferase system in the mollicute *Acholeplasma florum*. Biochimie 75:675-679.

92.   **Neale, G. A. M., A. Mitchell, and L. R. Finch**. 1983. Pathways of pyrimidine deoxyribonucleotide biosynthesis in *Mycoplasma mycoides* subsp. *mycoides*. J. Bacteriol. 154:17-22.

93.   **Neimark, H**. 1979. Phylogenetic relationships between mycoplasmas and other prokaryotes, p. 43-62. *In* M. F. Barile and S. Razin (ed.), *The Mycoplasmas*, vol. I, *Cell Biology*. Academic Press, NY.

94.   **O'Brien, S., J. M. Simonson, S. Razin, and M. F. Barile**. 1983. On the distribution and characteristics of isozyme expression in *Mycoplasma*, *Acholeplasma*, and *Ureaplasma* species. Yale J. Biol. Med. 56:701-708.

95.   **Ouzounis, C., and N. Kyrpides**. 1996. The emergence of major cellular processes in evolution. FEBS Lett. 390:119-123.

96.   **Ozcan, R., and R. Miles**. 1999. Biochemical diversity of *Mycoplasma fermentans* strains. FEMS Microbiol. Lett. 176:177-181.

97. **Paulsen, I., L. Nguyen, M. K. Sliwinski, R. Rabus, and M. Saier, Jr**. 2000. Microbial genome analyses: comparative transport capabilities in eighteen prokaryotes. J. Mol. Biol. 301:75-100.

98. **Pennisi, E.** 1999. Is it time to uproot the tree of life? Science 284:1305-1307.

99. **Peterson, S. N., and C. M. Fraser**. 2001. The complexity of simplicity. Genome Biol. 2:2002.1-2002.7.

100. **Petzel, J. P., P. A. Hartman, and M. J. Allison**. 1989. Pyrophosphate-dependent enzymes in walled bacteria phylogenetically related to the wall-less bacteria of the class *Mollicutes*. Int. J. Syst. Bacteriol. 39:413-419.

101. **Petzel, J. P., M. C. McElwain, D. DeSantis, J. Manolukas, M. V. Williams, P. A. Hartman, M. J. Allison, and J. D. Pollack**. 1989. Enzymic activities of carbohydrate, purine, and pyrimidine metabolism in the *Anaeroplasmataceae* (class *Mollicutes*). Arch. Microbiol. 152:309-316.

102. **Phillipe, H., and P. Forterre**. 1999. The rooting of the universal tree of life is not reliable. J. Mol. Evol. 49:509-523.

103. **Pieper, U., G. Kapadia, P.-P. Zhu, A. Peterkofsky, and O. Herzberg**. 1995. Structural evidence for the evolutionary divergence of mycoplasma from gram-positive bacteria: the histidine-containing phosphocarrier protein. Structure 3:781-790.

104. **Pollack, J. D.** 1978. Differentiation of *Mycoplasma* and *Acholeplasma*. Int. J. Syst. Bacteriol. 28:425-426.

105. **Pollack, J. D.** 1986. Metabolic distinctiveness of ureaplasmas. Pediatr. Infect. Dis. 5:S305-S307.

106. **Pollack, J. D.** 1992. Carbohydrate metabolism and energy conservation, p.181-200. *In* J. Maniloff, R. N. McElhaney, L. R. Finch, and J. B. Baseman (ed.), *Mycoplasmas: Molecular Biology and Pathogenesis*. ASM Press, Washington, DC.

107. **Pollack, J. D.** 1997. *Mycoplasma* genes: a case for reflective annotation. Trends Microbiol. 5:413-419.

108. **Pollack, J. D.** 2001. *Ureaplasma urealyticum*: an opportunity for combinatorial genomics. Trends Microbiol. 9:169-175.

109. **Pollack, J. D., J. Banzon, K. Donelson, J. G. Tully, J. W. Davis, Jr., K. J. Hackett, C. Agbanyim, and R. J. Miles**. 1996. Reduction of benzyl viologen distinguishes genera of the class Mollicutes. Int. J. Syst. Bacteriol. 46:881-884.

110. **Pollack, J. D., K. D. Beaman, and J. A. Robertson**. 1984. Synthesis of lipid from acetate is not characteristic of *Acholeplasma* or *Ureaplasma*. Int. J. Syst. Bacteriol. 34:124-126.

111. **Pollack, J. D., M. C. McElwain, D. DeSantis, J. T. Manolukas, J. G. Tully, C. J. Chang, R. F. Whitcomb, K. J. Hackett, and M. V. Williams**. 1989. Metabolism of members of the *Spiroplasmataceae*. Int. J. Syst. Bacteriol. 39:406-412.

112. **Pollack, J. D., M. A. Meyers, T. Dandekar, and R. Herrmann**. Submitted. The utility of glycolytic kinases with multifunctional enzyme activity in small genome *Mycoplasma* species: the replacement of nucleoside diphosphate kinase activity and the phylogenetic relevance of phosphoglycerate kinase.

113. **Pollack, J. D., and M. E. Tourtellotte**. 1967. Synthesis of saturated long chain fatty acids from sodium acetate-1-$C^{14}$ by *Mycoplasma*. J. Bacteriol. 93:636-641.

114. **Pollack, J. D. and M. V. Williams**. 1986. PPi-dependent phosphofructotransferase (phosphofructokinase) activity in the mollicutes (mycoplasma) *Acholeplasma laidlawii*. J. Bacteriol. 165:53-60.

115.  **Pollack, J. D., M. V. Williams, J. Banzon, M. A. Jones, L. Harvey, and J. G. Tully.** 1996. Comparative metabolism of *Mesoplasma, Entomoplasma, Mycoplasma,* and *Acholeplasma.* Int. J. Syst. Bacteriol. 46:885-890.

116.  **Pollack, J. D., M. V. Williams, and R. N. McElhaney.** 1997. The comparative metabolism of the mollicutes (mycoplasmas): the utility for taxonomic classification and the relationship of putative gene annotation and phylogeny to enzymatic function in the smallest free-living cells. Crit. Rev. Microbiol. 23:269-354.

117.  **Postma, P. W., J. W. Lengeler, and G. R. Jacobson.** 1993. Phosphoenolpyruvate:carbohydrate phosphotransferase systems of bacteria. Microbiol. Rev. 57:543-594.

118.  **Razin, S.** 1983. Urea hydrolysis, p. 351-353. *In* S. Razin and J. G. Tully (ed.), *Methods in Mycoplasmology,* vol. 1. Academic Press, Inc., NY.

119.  **Razin, S.** 1989. Molecular approach to mycoplasma phylogeny, p. 33-69. *In* R. F. Whitcomb and J. G. Tully (ed.), *The Mycoplasmas,* vol. V, *Spiroplasmas, Acholeplasmas, and Mycoplasmas of Plants and Arthropods.* Academic Press, Inc., NY.

120.  **Razin, S.** 1992. Mycoplasma taxonomy and ecology, p. 3-22. *In* J. Maniloff, R. N. McElhaney L. R. Finch, and J. B. Baseman (ed.), *Mycoplasmas: Molecular Biology and Pathogenesis.* ASM Press, Washington, DC.

121.  **Razin, S., and V. P. Cirillo.** 1983. Sugar fermentation, p. 337-343. *In* S. Razin and J. G. Tully (ed.), *Methods in Mycoplasmology,* vol. 1. Academic Press, Inc., NY.

122.  **Razin, S., D. Yogev, and Y. Naot.** 1998. Molecular biology and pathogenicity of mycoplasmas. Microbiol. Mol. Biol. Rev. 62:1094-1156.

123.  **Regula, J. T., B. Ueberle, G. Boguth, A. Görg, M. Schnölzer, R. Herrmann, and R. Frank.** 2000. Towards a two-dimensional proteome map of *Mycoplasma pneumoniae.* Electrophoresis 21:3765-3780.

124.  **Reinards, R., J. Kubicki, and H.-D. Ohlenbusch.** 1981. Purification and characterization of NADH oxidase from membranes of *Acholeplasma laidlawii,* a copper-containing iron-sulfur flavoprotein. Eur. J. Biochem. 120:329-337.

125.  **Reizer, J., I. T. Paulsen, A. Reizer, F. Titgemeyer, and M. H. Saier, Jr.** 1996. Novel phosphotransferase system genes revealed by bacterial genome analysis: the complete complement of pts genes in *Mycoplasma genitalium.* Microbiol. Comp. Genomics 1:151-164.

126.  **Reizer, J., A. Reizer, and M. H. Saier, Jr.** 1997. Is the ribulose monophosphate pathway widely distributed in bacteria? Microbiology 143:2519-2520.

127.  **Rice, P., B. M. Houshaymi, E. A. Abu-Groun, R. A. Nicholas, and R. J. Miles.** 2001. Rapid screening of $H_2O_2$ production by *Mycoplasma mycoides* and differentiation of European subsp. *mycoides* SC (small colony) isolates. Vet. Microbiol. 78:343-351.

128.  **Rice, P., B. M. Houshaymi, R. A. Nicholas, and R. J. Miles.** 2000. A rapid biochemical test to aid identification of *Mycoplasma mycoides* subsp. *mycoides* small colony (SC) strains. Lett. Appl. Microbiol. 30:70-74.

129.  **Riley, M., and B. Labedan.** 1996. Gene products: physiological functions and common ancestries, p. 2118-2202. *In* F. C. Neidhardt, R. Curtiss III, J. L. Ingraham, E. C. C. Lin, B. Magasink, W. S. Reznikoff, M. Riley, M. Scheachter, and E. Umbarger (ed.), *Escherichia coli* and *Salmonella: Cellular and Molecular Biology,* 2nd ed. ASM Press, Washington, D.C.

130.  **Robinson, I.** 1979. Special features of anaeroplasmas, p. 515-528. *In* M. F. Barile and S. Razin (ed.), *The Mycoplasmas,* vol. 1, *Cell Biology.* Academic Press, NY.

131. **Rodwell, A. W.** 1960. Nutrition and metabolism of *Mycoplasma mycoides* var. *mycoides.* Ann. N. Y. Acad. Sci. 79:499-507.
132. **Rodwell, A. W., and A. Mitchell.** 1979. Nutrition, growth and reproduction, p. 103-139. *In* M. F. Barile and S. Razin (ed.), *The Mycoplasmas,* vol. I, *Cell Biology.* Academic Press, NY.
133. **Rogers, M. J., J. Simons, R. T. Walker, W. G. Weisburg, C. R. Woese, R. S. Tanner, I. M. Robinson, D. A. Stahl, G. Olsen, R. H. Leach, and J. Maniloff.** 1985. Construction of the mycoplasma evolutionary tree from 5S rRNA sequence data. Proc. Nat. Acad. Sci. USA 82:1160-1164.
134. **Romano, A. H. and T. Conway.** 1996. Evolution of carbohydrate metabolic pathways. Res. Microbiol. 147:448-455.
135. **Rottem, S., and M. F. Barile.** 1993. Beware of mycoplasmas. Trends Biotechnol. 11:143-151.
136. **Salih, M. M., H. Ernø, and V. Simonsen.** 1983. Electrophoretic analysis of isoenzymes of mycoplasma species. Acta Vet. Scand. 24:14-33.
137. **Sauer, U., D. R. Lasko, J. Fiaux, M. Hochuli, R. Glaser, T. Szyperski, K. Wuthrich, and J. E. Bailey.** 1999. Metabolic flux ratio analysis of genetic and environmental modulations of *Escherichia coli* central carbon metabolism. J. Bacteriol. 181:6679-6688.
138. **Saurin, W., and E. Dassa.** 1996. In search of *Mycoplasma genitalium* lost substrate-binding proteins: sequence divergence could be the result of a broader substrate specificity. Mol. Microbiol. 22:389-390.
139. **Schaechter, M.,** and the View From Here Group. 2001. *Escherichia coli* and *Salmonella* 2000: the view from here. Microbiol. Mol. Biol. Rev. 65:119-130.
140. **Schuster, S., T. Dandekar, and D. A. Fell.** 1999. Detection of elementary flux modes in biochemical networks: a promising tool for pathway analysis and metabolic engineering. TIB Tech. 17:53-60.
141. **Schuster, S., and D. A. Fell, and T. Dandekar.** 2000. A general definition of metabolic pathways useful for systematic organization and analysis of complex metabolic networks. Nat. Biotechnol. 18:326-332.
142. **Shepard, M. C., C. D. Lunceford, D. K. Ford, R. H. Purcell, D. Taylor-Robinson, S. Razin, and F. T. Black.** 1979. *Ureaplasma urealyticum* gen. nov., sp. nov.: proposed nomenclature for the human T (T-strain) mycoplasma. Int. J. Syst. Bacteriol. 24:160-171.
143. **Sjöström, K., K. C. S. Chen, and G. E. Kenny.** 1986. Detection of end products of the arginine dihydrolase pathway in both fermentative and non-fermentative *Mycoplasma* species by thin-layer chromatography. Int. J. Syst. Bacteriol. 36:60-65.
144. **Sjöström, K., and G. E. Kenny.** 1983. Distinctive antigenic specificities of adenosine triphosphatases and reduced nicotinamide dehydrogenases as means for classification of the order Mycoplasmtales. Int. J. Syst. Bacteriol. 33:218-228.
145. **Skrypal, I. H., A. M. Onyshchenko, L. P. Malynovskya, I. P. Tokovenko, and L. O. Havrylko.** 1995. The phenotypic traits of Mollicutes as their possible phylogenetic markers. Mikrobiol. Z. 57:3-8. (Read in abstract translation only).
146. **Smith, P.** 1971. Relationship of structure to function, Table IV.3, p. 172. *In* P. Smith, *The Biology of the Mycoplasmas.* Academic Press, NY.
147. **Snel, B., P. Bork, and M. A. Huynen.** 1999. Genome phylogeny based on gene content. Nat. Genet. 21:108-110.
148. **Tao, H., C. Bausch, C. Richmond, F. R. Blattner, and T. Conway.** 1999. Functional genomics: expression analysis of *Escherichia coli* growing on a minimal and rich media. J. Bacteriol. 181:6425-6440.

149.   **Taylor, R. R., K. Mohan, and R. J. Miles**. 1996. Diversity of energy-yielding substrates and metabolism in avian mycoplasmas. Vet. Microbiol. 51:291-304.

150.   **Taylor, R. R., H. Varsani, and R. J. Miles**. 1994. Alternatives to arginine as energy sources for the non-fermentative *Mycoplasma gallinarum*. FEMS Microbiol. Lett. 115:163-167.

151.   **Teichmann, S., C. Chothia, and M. Gerstein**. 1999. Advances in structural genomics. Curr. Opin. Struct. Biol. 9:3990-399.

152.   **Teichmann, S. A., and G. Mitchison**. 1999. Is there a phylogenetic signal in prokaryote proteins? J. Mol. Evol. 49:98-107.

153.   **Teichmann, S. A., J. Park, and C. Clothia**. 1998. Structural assignments to the *Mycoplasma genitalium* proteins show extensive gene duplication and domain rearrangements. Proc. Natl. Acad. Sci. USA 95:14658-14663.

154.   **Tham, T. N., S. Ferris, R. Kovacic, L. Montagnier, and A. Blanchard**. 1993. Identification of *Mycoplasma pirum* genes involved in the salvage pathways for nucleosides. J. Bacteriol. 175:5281-5285.

155.   **Toth, K. F., N. Harrison, and B. B. Sears**. 1994. Phylogenetic relationship among members of the class Mollicutes deduced from the *rps*3 gene sequences. Int. J. Syst. Bacteriol. 44:119-124.

156.   **Tryon, V. V. and J. D. Pollack**. 1984. Purine metabolism in *Acholeplasma laidlawii* B: a novel PPi-dependent nucleoside kinase activity. J. Bacteriol.159:265-270.

157.   **Tryon, V. V., and J. Baseman**. 1992. Pathogenic determinants and mechanisms, p. 457-471. *In* J. Maniloff, R. N. McElhaney L. R. Finch, and J. B. Baseman (ed.), *Mycoplasmas: Molecular Biology and Pathogenesis*. ASM Press, Washington, DC.

158.   **Tully, J. G**. 1984. Family II. Acholeplasmataceae, Edward and Freundt, p. 775-781. *In* N. R. Krieg and J. G. Holt (ed.), *Bergey's Manual of Systematic Bacteriology*, vol. 1. Williams and Wilkins, Baltimore.

159.   **Tully, J. G**. 1989. Class Mollicutes: new perspectives from plant and arthropod studies, p. 1-31. *In* R. F. Whitcomb and J. G. Tully (ed.), *The Mycoplasmas*, vol. V, *Spiroplasmas, Acholeplasmas, and Mycoplasmas of Plants and Arthropods*. Academic Press, Inc., NY.

160.   **Tully, J. G., and S. Razin**. 1977. The Mollicutes (Mycoplasmas and Ureaplasmas), p. 417-443. *In* A. I. Laskin, and H. A. Lechavalier (ed.), *CRC Handbook of Microbiology*, 2$^{nd}$ ed., vol. 1. CRC Press, Cleveland.

161.   **Tully, J. G., R. F. Whitcomb, D. L. Rose, J. M. Bove, P. Carle, N. L. Somerson, D. L. Williamson, and S. Eden-Green**. 1994. *Acholeplasma brassicae* sp. nov. and *Acholeplasma palmae* sp. nov., two non-sterol-requiring Mollicutes from plant surfaces. Int. J. Syst. Bacteriol. 44:680-684.

162.   **van der Oost, J., G. Schut, S. W. M. Kengen, W. R. Hagen, M. Thomm, and W. M. de Vos**. 1998. The ferredoxin-dependent conversion of glyceraldehyde-3-phosphate in the hyperthermophilic archaeon *Pyrococcus furiosus* represents a novel site of glycolytic regulation. J. Biol. Chem. 273:28149-28154.

163.   **Vandamme, P., B. Pot, M. Gillis, P. de Vos, K. Kersters, and J. Swings**. 1996. Polyphasic taxonomy, a consensus approach to bacterial systematics. Microbiol. Rev. 60:407-438.

164.   **Vilei, E. M., and J. Frey**. 2001. Genetic and biochemical characterization of glycerol uptake in *Mycoplasma mycoides* subsp. *mycoides* SC: its impact on $H_2O_2$ production and virulence. Clin. Diagn. Lab. Immunol. 8:85-92.

165.   **Wächterhauser, G**. 1990. Evolution of the first metabolic cycles. Proc. Natl. Acad. Sci. USA 87:200-204.

166. **Wadher, B. J., H. Varsani, E. A. Abu-Groun, and R. J. Miles.** 1990. Biochemical characterization of bovine, caprine and ovine mycoplasmas. IOM Lett. 1:517-518.

167. **Wallbrandt, P. H. M., V. Tegman, B. H. Jonsson, and Å. Wieslander.** 1992. Identification and analysis of the genes coding for the putative pyruvate dehydrogenase enzyme complex in *Acholeplasma laidlawii.* J. Bacteriol. 174:1388-1396.

168. **Wasinger, V., J. D. Pollack and I. Humphery-Smith.** 2000. The proteome of *Mycoplasma genitalium*; CHAPS soluble component. Eur. J. Biochem. 267:1571-1582.

169. **Weisburg, W. G., J. G. Tully, D. L. Rose, J. P. Petzel, H. Oyaizu D. Yang, L. Mandelco, J. Sechrest, T.G. Lawrence, J. Van Etten, J. Maniloff, and C. R. Woese.** 1989. A phylogenetic analysis of the mycoplasmas: basis for their classification. J. Bacteriol. 171:6455-6467.

170. **White, D.** 2000. Central metabolic pathways, p.180-211. *In* D. White, *The Physiology and Biochemistry of Prokaryotes,* 2nd ed. Oxford University Press, NY.

171. **White, D.** 2000. Solute transport, p. 398-416. *In* D. White, *The Physiology and Biochemistry of Prokaryotes,* 2nd ed. Oxford University Press, NY.

172. **Wilks, H. M., K. W. Hart, R. Feeney, C. R. Dunn, H. Muirhead, W. N. Chia, D. A. Barstow, T. Atkinson, A. R. Clarke, and J. J. Holbrook.** 1988. A specific, highly active malate dehydrogenase by redesign of a lactate dehydrogenase framework. Science 242:1541-1544.

173. **Williams, M. V., and J. D. Pollack.** 1990. The importance of differences in pyrimidine metabolism of the mollicutes. Zentralbl. Bakteriol. Suppl. 20:163-171.

174. **Woese, C., G. J. Olsen, M. Ibba, and D. Soll.** 2000. Aminoacyl-tRNA synthetases, the genetic code, and the evolutionary process. Microbiol. Mol. Biol. Rev. 64:202-236.

175. **Woese, C. R., E. Stackebrandt, and W. Ludwig.** 1985. What are mycoplasmas: the relationship of tempo and mode in bacterial evolution. J. Mol. Evol. 21:305-316.

176. **Wood, H. G.** 1985. Inorganic pyrophosphate and polyphosphates as sources of energy. Curr. Topics Cell. Regul. 26:355-369.

177. **Zhu, P.-P., N. Nosworthy, A. Ginsburg, M. Miyata, Y. J. Seok, and A. Peterkofsky.** 1997. Expression, purification, and characterization of enzyme IIA(*glc*) of the phosphoenolpyruvate:sugar phosphotransferase system of *Mycoplasma capricolum.* Biochemistry 36:6947-6953.

Chapter 9

# Database Systems for the Analysis of Biochemical Pathways

ISABEL ROJAS-MUJICA* and ERICH BORNBERG-BAUER[#]
*European Media Laboratory, Heidelberg, Germany; [#]UMBER - Bioinformatics Group, School of Biological Sciences, University of Manchester, UK

## 1.     INTRODUCTION

The mycoplasmas were amongst the first organisms for which the complete genome sequence was obtained and made available to the public domain (13; 10; 6). Complementary information, such as the proteome of *mycoplasma pneumoniae* (26) and reconstructed pathways, has also been made available and can be queried or downloaded from various locations on the web. However, database systems that allow scientists to work with these data in an integrated manner, together with other relevant information, are still not common. This is a wide-spread problem that not only applies to mycoplasma. Current biological research uses a wide range of interacting software which in turn uses a large number of disparate data source. These problems have arisen for several reasons which are, amongst others, the specialisation of biological disciplines, the lack of unified interfaces and the variations in the interpretation of the data. Bioinformatic tools need to overcome these difficulties. For biochemical pathways this implies the development of a system which is based on the functional roles of molecular objects such as reactions and pathways. From a data modelling point of view the modelling of biochemical processes is a complex problem. Fuzziness of the definitions, exceptions and complex relations, are some of the common characteristics that are found when attempting to model these processes. Models for storage and representation of pathways and data on expression and the genome, must reflect the logics of the cellular machinery. At the

*Molecular Biology and Pathogenicity of Mycoplasmas*, Edited by Razin and Herrmann, Kluwer Academic/Plenum Publishers, New York, 2002

same time they must be designed to offer a platform for optimal storage, retrieval and analysis of such data for the biological researcher.

In this chapter, we provide an overview of available data resources and relevant tools for extracting pathway-related information from the most widely used databases and briefly explain their usage. We will also discuss some of the systems and methods, both established and new, to support the process of examination and understanding of the data. We will revise several of the methods and problems related to the integration, handling and interpretation of biochemical data. As an introduction to data modelling, we present a small example related to the modelling of enzymatic reactions. This is meant to raise awareness of the importance of structuring data. It should give researchers, who have no experience in creating databases, a rough idea of what they could gain from putting their data into an integrated system.

## 2.        BIOCHEMICAL DATABASES

The genomes of *Mycoplasma pneumoniae* (*M.pn.*) and *Mycoplasma genitalium* (*M.gen.*) are widely viewed as a blueprint for a cell with a minimal metabolism. The availability of data generated from the sequencing of their genomes and results from both comparative genome analysis and experiments, such as global transposon mutagenesis (14), make them a sensible choice to study methods for pathway analysis and data integration.

There are several databases – accessible through the World Wide Web (WWW) – that contain information about biochemical pathways, to different extends and with different focuses (see Appendix). With the growing amount of information related to biochemical reactions and thus with the growing interest towards the study of biochemical networks, the number of projects and databases (or data collections)[i] dealing with biochemical pathways is rapidly expanding. This means that, like any other study, this short description of some of the existing biochemical pathway databases can only be a snapshot of the current stand. We will mention a few of these databases, not all of which contain explicit information about mycoplasmas.

---

[i] We will use the term database to refer to data collections and their presentation. From a strictly computational view, most biological "data-bases" are actually data sources or repositories as they lack a formal definition of their structure. The difference will be made clear further on in this chapter.

## 2.1 General resources

Existing approaches to provide integrated access to biological data are widely based on a common principle of database technology, known as *data-warehouse*. Basically, they do not integrate (i.e. reformat) the underlying primary resources, such as GenBank or SWISS-PROT. Data-warehouses define interfaces to these databases and incorporate links between their data at a higher–integration–layer. For example, one can access information from GenBank and link this to information in SWISS-PROT, through the use of pre-defined links. Generally, it is also possible to "join" the information from two or more databases by using these links. Such systems are said to create "wraps" around the primary sources, as they leave them in their original structure. Best known examples are SRS (9), Entrez (27), and DBGET (15). SRS focuses on the integration of as many databases as possible and has a relatively open architecture. For SRS alone, there exist worldwide at least 40 different web available versions, each with a different collection of integrated databases. Recent versions can also be customised through the internal scripting language Ikarus. Entrez' major strength is probably its integrated access to PubMed. DBGET in turn is tightly linked to KEGG. Most of these approaches pre-index the data sources. This means that all entries are sorted according to some predefined criteria. This allows the search of information in huge amounts of data to be carried out in seconds. Integrating another database into your own SRS version requires the pre-indexing of the new resource. To query the data the user can choose the search space (i.e. the data bases to be scanned) and may enter one or more expressions or key-phrases. Search expressions can be combined with the use of Boolean operators (AND, OR) to form more sophisticated search criteria. The indexing approach is in general easier to use for the non-computer expert or occasional users since it is based on flat files.[ii] However, such an a-posteriori integration of data is inefficient and tedious and the formulation of complex queries is very complicated and in some cases close to impossible. The query formulation depends on skills and knowledge of the user. While data retrieval is certainly possible and efficient, knowledge deduction is virtually impossible.

Another system that offers an integrated access to biochemical data is MARGBench (11). MARGBench offers mechanisms for the integration of data from multiple sources in such a way that the user can create a local database with the data extracted from the specified sources.

---

[ii] Flat files are widely used by most of the bio-community for the storage of data. Flat files are files with text, one unit of text in each file, where the text is only structured by separating blocks for example using white-spaces (blanks). They have no predefined and documented semantics.

A different approach is TAMBIS (2), which is based on an *ontology*. An *ontology* can be defined as a formal description of a set of concepts and relationships in a domain of interest, in such a way that it is rigorous enough to support reasoning. All objects such as domains, proteins and compounds, are defined and associated mainly in terms of IsA or HasA relationships. With the use of logical statements it is possible to define complex relationships, constraints or even deduction rules. The main goal of the TAMBIS project is to provide transparent access for the biological user to existing biological data sources. This means that the user formulates a query to the system and the system will transparently access and retrieve the data from the relevant data sources. The user does not have to specify the source of the data nor the terms used to retrieve the data from each source. The query mechanism is guided by the concepts (the objects and their relationships) defined in the ontology. For example, to retrieve all proteins of *M.pn.*, we could create a new concept "Protein which has Organism Classification Species", which in turn used the concepts 'Protein' and 'Species'. By setting the value of species to be *Mycoplasma pneumoniae* and submitting the query, TAMBIS uses the knowledge it contains to determine from which database and which field it can retrieve the data. For example, it will use SWISS-PROT's organism species field to retrieve the appropriate proteins. One does not need to stick to the keywords or identifiers contained in the data sources.

In the area of bioinformatics, ontologies have become very popular. This has been mainly motivated by the need to integrate and reason about the vast amount of available biological data (17; 28). Ontologies can be used to describe controlled vocabularies, and thus to assist the exchange of knowledge at a terminological level. Sharing of knowledge requires a clear semantics (3). The Gene Ontology (GO) (31) is an example of an ontology being produced to share knowledge of gene and protein roles (especially for eukaryotes). They are generating three independent ontologies covering: biological processes, molecular functions and cellular components. This classification is now being used by biologists, for example to assign functions to genes.

Another well known ontology in the biological domain is the EcoCyc Ontology (17), which is used to describe the genes and metabolism of *E.coli*. SABioDB (29) also has an associated ontology, that focuses on the representation of metabolic pathways in general, covering concepts ranging from the genetics up to the general descriptions of pathways. With the objective to exchange experiences and definitions of ontologies in the area of molecular biology a group of researchers in this area has set up the BioOntologies work group (4).

## 2.2     Existing biochemical pathway specific resources

Biochemical pathway databases mainly provide information about reactions and pathways in general (organism independent information) and organism-specific data (such as genes, proteins, data about enzymatic activities, kinetic data, etc.). The graphical presentation of the biochemical pathways is often used as a platform to display information related to the reactions such as enzymes (for metabolic pathways), compound names or information about expression data of an enzyme in a given organism.

The most commonly used pathway databases that contain organism specific information include:

- WIT (24),
- KEGG (16),
- Metacyc (18),
- ENZYME (30),
- PathDB (23),
- EMP (1)

Many more databases are linked to these biochemical pathway databases, containing more detailed information about proteins (e.g. SCOP, PDB or PIR), genomes (e.g. GenBank, EMBL Database, TIGR or MGDB), chemical prop-erties of compounds (e.g. Beilstein, ChemFinder) or literature data (e.g., Med-line/PubMed) (for a summary see (34) and (12)).

Data about the biochemical pathways specific to mycoplasmas is present in KEGG and partly in WIT. SABioDB, a database system which focuses on data related to biochemical pathways in *Mycoplasma pneumoniae* and related organisms, is being developed at the EML (European Media Laboratory, Heidelberg, Germany) as part of the ELSA (8) project. All of these databases contain data about genes and genomes, enzymes, reactions and pathways. They mainly differ in the level of detail of the information, query mechanisms and presentation of information.

WIT extracts information about enzymes from the ENZYME database and from the Metabolic Pathways (EMP) database, which is embedded within the system. Pathways are classified based on their metabolic function and according to their starting and/or terminating compounds. For example, querying for `glycolysis in mycoplasma` will return all metabolic pathways that contain glycolysis, even those that include the uptake of glucose (or fructose) from extracellular medium. Information about the compartments where the reactions occur is included in the graphical representation of the pathway, where the user can obtain more information about compounds and/or enzymes by mouse clicks. The link to the EMP database allows the retrieval of literature-related and organism specific enzyme information.

The KEGG system contains all known metabolic pathways and a limited number of regulatory and transport mechanisms (16). It consists of three strongly connected databases, namely:

- LIGAND, with information about compounds, enzymes and reactions,
- PATHWAY, which contains graphical representations of the pathways and lists of enzymes and reactions within the pathways; and
- GENES, which contains organism related genome and gene information.

Using the DBGET integrated database retrieval system, KEGG offers many links to other databases. Pathways are classified according to the chemical structures of the main compounds, e.g. carbohydrates, lipids, amino acids, etc.

SABioDB has being created to support the analysis and modelling of biochemical pathways, taking as case study *Mycoplasma pneumoniae*. As in KEGG, pathways can have multiple classifications, but the main classification used is based on the structures of the main compounds, e.g., Carbohydrate metabolism.

Although the remaining databases mentioned do not contain information related to mycoplasmas, all of them are capable of storing such information. Probably it is just a matter of time until this information will be integrated. Therefore, it is interesting to briefly describe the data they contain and the properties of the related system.

MetaCyc describes metabolic pathways, reactions, enzymes, and substrate compounds from a variety of different organisms (the majority from micro-organisms), but does not provide genomic data. Its "complement" system EcoCyc (18) describes the genome and the biochemical machinery of *E. coli*. The data contained in MetaCyc were gathered from a variety of literature and on-line sources, and contain citations to the source of each pathway.

PathDB is a general metabolic pathway database, representing compounds, metabolic proteins (enzymes or transport proteins), reactions and pathways. It supports the inclusion of information on the location within the cell or organism and displays it whenever available. In addition, it contains descriptions of the kinetic, thermodynamic and physicochemical properties of pathway components based on the scientific literature.

Expasy-biochemical pathways consists of two databases, namely ENZYME (30) and Biochemical Pathways. The ENZYME database is a repository of information relative to the nomenclature of enzymes with references to the SWISS-PROT protein sequence entries that correspond to the enzyme. The biochemical pathways database is an electronic version of the Boehringer Mannheim Biochemical Pathways Chart, indexed and with links to the ENZYME database.

## 3. QUERYING AND EXTRACTING THE INFORMATION

All the above mentioned databases are accessible through the WWW. Some, such as PathDB, MetaCyc and SABioDB, can also be used as stand-alone systems, with remote access to the database. The user can navigate through the information by the use of links, or jump to a specific subset of the data via the submission of a query or using search facilities. The level of specification of the search criteria differs from one database to the other.

### 3.1 Querying the data

The most common ways to query data are by
- providing a word or part of one (probably plus some wild character like '*'). The database is then searched for a word that matches this pattern. In most cases the range of the search should be specific to a particular type of information, e.g. protein name.
- selecting the type of information that one would like to see from a given menu of alternatives.
- specifying some of the characteristics of a certain type of objects given a general description of the object type (query by example). The system will then search for the subset of objects that satisfy these specifications.
- using a domain[iii] specific query language to formulate queries.
- clicking on (hyper-)links to navigate through the data.

Most database systems provide a combination of the above mentioned mechanisms. One interesting feature that is offered in nearly all of them is the possibility to find a pathway between two given compounds. The classical description of a pathway as a pre-defined sequence of reactions is too limited when one wants to query the existence of alternative pathways for the production of a certain substance or execution of a certain biochemical task.

Another interesting feature is offered by PathVis (29), a pathway visualisation tool complementing SABioDB. It enables the user to compare pathways in terms of the reactions that they contain and the enzymes (in the

---

[iii] Where a domain denotes a specialised field of research.

case of comparison of pathways from different organisms) that catalyse the reactions.

KEGG, MetaCyc and WIT also display the interconnectivity between path-ways (predefined pathways). This is done using predefined metabolisms as an unerlying grid. This gives a clearer view of the task or influence of the studied pathway to/and from other pathways. An overview of all pathways as in Meta-path is very difficult to understand but the information about the immediate neighbours is a very useful piece of information.

## 3.2    Extracting the information for its use in your own project

Most of the WWW accessible databases present the data in HTML (HyperText Markup Language) (36) formated pages. Extracting information from HTML pages is a very tedious and cumbersome process. HTML focuses of on the display of data and not on its description. The eXtendible Markup Language (XML) (35) was designed to describe data and to focus on what the data is (33). It is emerging as a standard for the exchange and description of documents. One can also store document is XML format. An XML file is nothing but a plain text file where the information in the file is tagged. Using meaningful tags we can identify each element (information token), in this way adding "intelligence" to the documents.

A small, but constantly growing, number of biochemical databases offers the possibility to view the data in XML format. Others, like KEGG and EcoCyc, allow the user to download the data in a flat file format, in the form of multiple files. In all these cases, to extract the information from the documents, it is necessary to write a program (parser) that reads (parses) the pages and extracts the data. For HTML pages and flat file formats the parser is tightly coupled with the format of the page. This means that changes in the format of the file will most probably lead to changes in the program that extracts the data. In the case of XML formatted pages there are already standard parsers provided (for example SAX (21) or DOM (37)).

We would like to take this opportunity to encourage biologists to publish their data in a structured way, in order to facilitate the development of parsing programs for their data. As an introduction to the formatting of data in XML in Fig. 1 we present a simple XML page, containing the data for the first reaction of glycolysis in mycoplasma (we have commented each line with '##' to facilitate the understanding). As mentioned above, every piece of information is identified by a tag. XML tags are not predefined. One must define one's own tags. So, for example, under the tag "[reaction]" we include all attributes or information about a given reaction. The tag reaction. We can include more than one reaction, each starting with a "`[reaction]`" tag, or more substrates or products for a reaction. The tag is used to identify/describe the information included.

There are a series of projects for the establishment of XML specifications for the biochemical and bioinformatics domains have risen, amongst them: Bioinformatics Sequence Markup Language -BSML (19), BIOpolymer Markup language -BIOML (25), GAME (20), CML (22), DAS (7) and BioDOM (32). A good idea is to have a look at some of these to use some of the specifications already defined. An introduction to creating your XMLdocument can be found at http://www.w3schools.com/xml/xml whatis.asp.

```
## Identification of an XML document
<?XML VERSION="1.0">
<organism> ## We describe the pathway of a given organism
   <organism_name> mycoplasma pneumoniae </organism_name>
   <pathway> ## Pathway within the organism
      <pathway_name> Glycolysis </pathway_name>
      <reaction> ## Reaction in the pathway
         <substrate> ## Substrate of the reaction
            <compound_name> alpha D-Glucose-6-phosphate </compound_name>
            <stoichiometry> 1 </stoichiometry>
         </substrate> ## end of substrate description
         <product> ## Product of the reaction
            <compound_name> beta D-Fructose-6-phosphate </compound_name>
            <stoichiometry> 1 </stoichiometry>
         </product>
         <enzyme> ## enzyme information
            <ECClassification> 5.3.1.9 </ECClassification>
            <enzyme_Name> Glucose-6-phosphase isomerase </enzyme_Name>
         </enzyme> ## end of enzyme information
      </reaction> ## end of reaction information
   </pathway> ## end of pathway information
</organism> ## end of organism information
</XML> ## end of XML document
```

*Figure 1.* Example of XML entry for information related to a *Mycoplasma pneumoniae* pathway

# 4.   PRINCIPLES FOR BUILDING YOUR OWN PATHWAY DATABASE

In this section we want to provide the reader with a brief introduction to the modelling of a database and its implementation. This will be done by means of a short example related to enzymatic reactions.

Building databases is a non-trivial task and starting in the wrong direction can mean the loss of much time. The first step in building a database for a given domain is to create a model that describes the characteristics of the elements of interest and the relationships amongst these elements. For example in the case of biochemical pathways, we are interested in including information related to reactions, the elements that participate in these reactions, the catalyst, etc. This model should be a formalisation by which data and their relationships can be unambiguously described and discussed among the experts. The model should be independent of the structure of the data that we want to store in the database, and must be implementation independent. This means that to begin with the modelling process is a matter of pencil and paper and then we address implementation issues, i.e. the translation of the model into an specific database system. The data that will be stored in the database can be retrieved from a variety of sources (such as the ones mentioned in the previous chapter). A database as such performs no calculations or analysis of the data.

Let us briefly describe how a data model is built along a case study for a subset of the concepts related to enzymatic reactions.

First of all it is necessary to have a clear definition of the concept that we want to model. In general, an *enzymatic reaction* can be defined as *a biochemical reaction that is catalyzed by an enzyme*, where:

- A *biochemical reaction* is an event in which one or more compounds (substrates) are chemically processed to "produce" one or more compounds (products). Substrates and products can be referred to as reactants.
- A *catalyst* is a substance (compound) that increases the velocity of a chemical reaction, without undergoing any permanent chemical change. Therefore it is not to be considered as a reactant.
- An *enzyme* is a protein (or group of proteins) that acts as a catalyst for a biochemical reaction (leaving aside RNA enzymes for brevity).

If we consider that the same biochemical reaction can be catalyzed by more than one enzyme, then each combination reaction-enzyme will be considered as an *enzymatic reaction*.

These concepts and their relationships can now be cast into a conceptual data model. We will use UML (Universal Modelling Language) (5), to represent our model (see Fig. 2).

Pointed arrows indicate that a type of object is a subtype of another type of object, e.g. a `protein/peptide` IsA (type of) `compound`. The lines indicate an *association relationship*. The numbers at each side of the line indicate the cardinality of the relation. Therefore, to indicate that a biochemical reaction can have one or more reactants, we place a `1..*` on the side of the `reactant` entity.

Let us define as catalytic reaction the combination of a biochemical reaction with a catalyst. Thus, a catalytic reaction can only exist if the corresponding biochemical reaction and catalyst exist. Therefore, in the line between `biochemicalReaction` and `catalyticReaction` we place a `1..1` next to the rectangle corresponding to a biochemical reaction, indicating that a catalytic reaction is associated to one and only one biochemical reaction. On the side of the catalytic reaction we place a `0..*` indicating that a biochemical reaction can be associated with zero or many catalytic reactions, or in other words that a biochemical reaction can be catalyzed by none or many catalysts. Similar is the relation between the catalyst and a catalytic reaction. This indicates that a catalyst catalyses at least one reaction, but may also catalyse other catalytic reactions. To be able to say that a compound is a catalyst, it must be known that it catalyses at least one reaction. An example would be any peroxidase. Peroxidases can act as catalases as well (and vice versa), although with a lower efficiency.

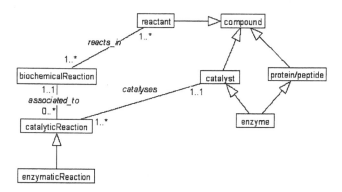

*Figure 2.* Initial UML diagram describing the objects ("entity types") related to metabolic reactions. Entity types are drawn as rectangles and lines between the enities represent relationships (see text for further explanation).

Compounds can be classified in different manners. `Protein`, for example, would be an entity type for which there exist instances in the database that correspond to "real" (physical) objects, such as crambin, hemoglobin or hexokinase. Under this same criteria we can add other compound types such as `nucleotideSequences`, `ionicAtoms` and `simpleMolecules` or even complex compounds formed by combinations of compounds. On the other hand, `reactant` and `catalyst` are so called *roles* that a compound may play and they do not correspond to actual instances in the database. However, they can be viewed as relations between `compound` and `biochemicalReaction`. These conceptual differences are important to keep in mind at the time of implementation, as we will see further on.

Now we can extend the model to include information on the EC classification of enzymatic reactions. In reality, this classification is used to classify both enzymatic reactions (with respect to the type of enzyme that can catalyse it), as well as enzymes (with respect to the type of reactions that they can catalyse). We need to make this distinction because in the existing data sources we will find that in some cases a reaction is attributed an EC number, although no protein/peptide has been identified, or vice-versa a protein/peptide is classified under a given EC number, without having identified the reaction that it catalyses in the given organism. To include information about the ECclassification we then have to slightly modify our model (see Fig. 3). The object `ecNumber` is placed between `enzymaticReaction` and `enzyme`. The association `enzymaticReaction` and `ecNumber` is `0..*` - `0..1` because an enzymatic reaction can have at most one `ecNumber` associated with it. Conversely, every EC number may refer to many reactions. The zero next to `enzymaticReaction` appears for practical reasons. Although, in principle, there is at least one enzymatic reaction associated to every EC classification, it can be the case that this reaction is not included in the database (we assume that we want to allow this, specially if we are working on a particular organism). This now has an impact on the cardinality between `catalyticReaction` and `catalyst`. Since `enzymaticReaction` is a subtype of `catalyticReaction`, `enzyme` a subtype of `catalyst` and we have just stated that for both types there may be associations without instances, the association changes from `1..*` `--` `1..1` (see Fig. 2) to `0..*` `--` `0..1`(see Fig. 3).

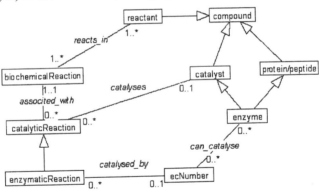

*Figure 3.* UML diagram describing the objects (or entities) related to metabolic reactions including EC numbers

In the following we want to refine the model (see Fig. 5). `reactant`, `catalyst` and `enzyme` (derived from `catalyst`) are roles that can also be represented as relationships. For example, for a compound to play the role of a reactant in a reaction, the reaction object must exist. The `reactant` object can then be substituted by a many to many (`0..*` `--` `1..*`) association between `biochemicalReaction` and `compound`. This association can be represented as an entity (`reactant`) with two associations:

- one to `biochemicalReaction`, indicating that a biochemical reaction can have one or many compounds as reactants, and
- another one to `compound`, indicating that a compound can be a reactant in zero or many reactions.

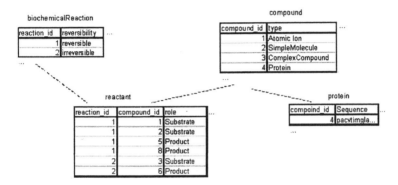

*Figure 4*. Table representation of the relationship biochemicalReaction-compound, and of the protein entity. Notice that the table reactant has the keys of both the biochemical reaction table and the compound table. In this way each row (combination of keys) refers to a reactant of a biochemicalReaction. The same compound cannot participate twice in the same biochemicalReaction. Compounds that are proteins are included in the protein table, using the same compound id entry as key.

Fig. 4 shows an example of the implementation of the entities protein and reactant in a relational database, i.e. as tables. Notice that the protein table has the same key (`compound id`) as the compound table, this is because every protein is a compound. The reactant table contains the keys of the biochemical reaction table and of the compound table, for each compound in a reaction it is then possible to identify if this compound acts as a substrate or as a product in the reaction.

This splitting of the association is necessary when we want to add additional information to the reactant relation, for example to specify the type of reactant "substrate" or "product"[iv]. Splitting of many to many relations is also necessary when implementing the model in relational databases.

---

[iv] This is of course an artificial division if we consider that all biochemical reactions should be treated as reversible. However, to be able to identify which compounds react to produce another compound or set of compounds, it is necessary to do this distinction.

In the same way that we represented the many to many association between `compound` and `biochemicalReaction` with the entity reactant, we can create a new entity (`potentialEnzReaction`) to represent the many to many relation between `enzyme` and `ecNumber`.

We will assume that we do not want to say anything in particular about a catalyst, so we will not represent this concept by a table when implementing. All relevant information about a compound as a catalyst should be related to the reactions that it is acting upon, thus we can eliminate this object from the model represented in Fig. 3 and directly associate `catalyticReaction` with `compound` (see Fig. 5).

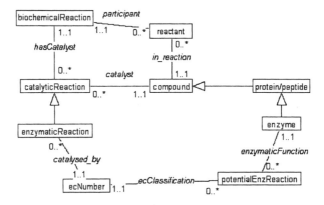

*Figure 5.* UML diagram describing the objects (or entities) related to metabolic reactions

The implementation of this model in a relational database requires the definition of *key attributes (identifiers)* for each of the main entity types (`compound, biochemicalReaction, ecNumber`), such that we can uniquely identify each entry. Although a sequence could be used as a unique identifier for a protein, this is not recommendable given the size of the sequences. In many cases consecutive integer numbers are recommended. One to many relationships are implemented by including the key of the "one"-side as a key in the table of the "many"-side. For example, the reactant table will have as key values the keys of the `biochemicalReaction` table and of the `compound` table, both also defined as being foreign keys (not proper of the table). Inheritance relationships (the arrows) are resolved by giving the subtype object the same key as the supertype. So, for example the compound id that defines a protein will be the same id used in the protein table. In Figure 6 the final implementation (only with a limited number of attributes per table) of the database model in Fig. 5 is shown.

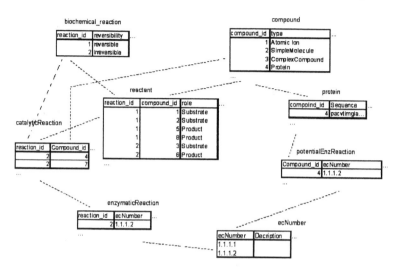

*Figure 6.* Relational database implementation

# 5. CONCLUSIONS

In this chapter we have provided a brief introduction to the current state of biochemical pathway databases: existing resources, how to use them, and introduction of some of the terms and technologies most commonly used. We hope to have provided enough information to create the basis with which a researcher, wanting to start a project on the analysis of biochemical pathways, can begin. First of all, by knowing where to find the data, how to work with this data, either by using one of the existing systems or by building his or her own database system, to which new data can be added. It was not our intention to provide a deep insight into the contents of the available databases or of database technology. This chapter should only be taken as a starting point to obtain a general overview of the field and as pointer to sources with more detailed information. The area of databases in bioinformatics has still many problems to address, mainly related to data integration, data exchange and even data modelling. Therefore, to keep up to date with the development of this technology we recommend the regular review of database systems available on the internet.

Once again, we would like to take the opportunity to encourage researchers in the biomolecular field, to try to store their data in a structured format, where the semantics of the data can be better worked out.

## ACKNOWLEDGMENTS

The authors thank Esther Ratsch for her comments and review, Ulrike Wittig for her contribution to the work and Victoria Kelly for her web-review. The authors gratefully acknowledge that this work, as part of the ELSA project, was supported by the Klaus Tschira Stiftung (KTS) and the BMBF (Project Bioregio 12212).

## REFERENCES

1.  **Argonne National Laboratory**, USA. The Enzymology Database. http://wit.mcs.anl.gov/WIT2/EMP/.
2.  **Baker, P., Brass, A., Goble, C., Paton, N. and R. Stevens** (1998). TAMBIS-Transparent Access to Multiple Biological Information Sources. In *Int. Conf. on Inteligent Systems for Molecular Biology*, pages 25–34, Montreal, Canada. The American Association for Artificial Intelligence (AAAI) Press.
3.  **Baker, P., Goble, C., Bechhofer, S., Paton, N., Stevens, R. and A. Brass** (1999). An ontology for bioinformatics applications. *Bioinformatics*, 15(6):510–520.
4.  **BioOntology Workgroup**. The BioOntology Workgroup. http://smi-web. stanford.edu/projects/bio-ontology/.
5.  **Booch, G., Rumbaugh, J. and I. Jacobson** editors (1999). *The Unified Modelling Language User Guide*. Addison-Wesley.
6.  **Dandekar, T., Huynen, M., Regula, J.T., Ueberle, B., Zimmermann, C., Andradae, M.A., Doerks, T., Sanches-Pulido, L., Snel, B., Suyama, M., Yuan, Y.P., Herrmann, R. and P. Bork** (2000). Re-annotating the *My-coplasma Pneumoniae* genome sequence: adding value, function and reading frames. *Nucleic Acids Res.*, 28:3278–3288.
7.  **Dowell, R.D., Jokerst, R.M., Day, A., Eddy, S.R. and L. Stein** (2001). The Distributed Annotation System (DAS). *BMC Bioinformatics*, in press.
8.  **ELSA**, ELektronischer Stoffwechsel-Atlas. http://projects.eml.org/elsa.
9.  **Etzold, T., Ulyanov, A. and P. Argos** (1996). Srs: information retrieval system for biology data banks. *Methods Enzymology*, 266:114–128.
10. **Fraser, C.M., Gocayne, J.D.** et al. (1995) The minimal gene compliment of *Mycoplasma genitalium*. *Science*, 270:397–403.
11. **Freier, A., Hofestaedt, R., Lange, M. and U. Scholz** (1998). An integrated architecture for modelling and simulation of metabolic networks. In *BIOTECHNOLOGY 2000*, pages 210–211.
12. **Frishman, D., Heumann, K., Lesk, A. and H.W. Mewes** (1998). Comprehensive, comprehensible, distributed and intelligent databases: Current status. *Bioinformatics*, 14:551–561.
13. **Himmelreich, R., Hilbert, H., Plagens, H., Pirkl, E., Li, B.C. and R. Herrmann** (1996). Complete sequence analysis of the genome of of the bacterium mycoplasma pneumoniae. *Nucl. Acids. Res.*, 24:4420–4449.
14. **Hutchison, C.A., Peterson, S.N., Gill, S.R., Cline, R.T., White, O., Fraser, C.M., Smith, H.O. and J.C. Venter** (1999). Global transposon mutagenesis and a minimal mycoplasma genome. *Science*, 286:2165–9.
15. **Kanehisha, M.** (1997). *Trends Biochem Sci.*, 22:442–444. http://www.genome.ad.jp/dbget.
16. **Kanehisha, M. and S. Goto** (2000). KEGG: Kyoto encyclopedia of genes and genomes. *Nucleic Acids Research*, 28:27–30.

17. **Karp, P**. (2000). An Ontology for Biological Function Based on Molecular Interactions. *Bioinformatics*, 16(3):269–285.

18. **Karp, P., Riley, M., Saier, M., Paulsen, I.T., Paley, S.M. and A. Pellegrini-Toole** (2000). The EcoCyc and MetaCyc databases. *Nucleic Acids Research*, 28:56–59.

19. **Inc. Labbook**. Bioinformatic Sequence Markup Language (BSML). http://www.labbook.com/.

20. **Lewis, S.E. and E. Frise**. Genome Annotation Markup Elements (GAME). http://www.bioxml.org/Projects/game/game0.1.html.

21. **Megginson Technologies** (1999). Simple Api for XML. http://www.megginson.com.

22. **Murray-Rust, P**.. XML and the Launch if Chemical Markup Language. http://www.xml-cml.org/.

23. **National Center for Genome Resources**. PathDB. http://www.ncgr.org-/pathdb.

24. **Overbeek, R., Larsen, N., Pusch, G.D., D'Souza, M., Selkov Jr, E., Kyrpi-des, N., Fonstein, M., Maltsev, M. and E. Selkov** (2000). Wit: integrated system for high-throughput genome sequence analysis and metabolic re-construction. *Nucl. Acids Res.*, 28:123 – 125.

25. **Protometrics**. The BIOpolymer Markup Language (BIOML). http://www.bioml.com/.

26. **Regula, J., Ueberle, B., Boguth, G., Georg, A., Schnoelzer, M., Herrmann, R. and R. Frank** (2000). Towards a two dimensional proteome map of *Mycoplasma pneumonia*e. In *Electrophoresis*.

27. **Schuler, G.D., Epstein, J.A., Ohkawa, H. and J.A. Kans** (1996). Entrez: Molecualr biology database and retrieval systems. *Methods Enzymology*, 266:141–162.

28. **Schulze-Kremer, S**. (1998). Ontologies for Molecular Biology. In *Proceedings of the third Pacific Symposium on Biocomputing*, Hawaii, pages 693–704. AAAI Press.

29. **Scientific Databases and Visualisation Group**, EML. System for the Analysis of Biochemical Data. http://projects.eml.org/SDBV/projects.

30. **Swiss Institute of Bioinformatics**. Enzyme nomenclature database. http://www.expasy.ch/enzyme/.

31. **The Gene Ontology Consortium** (2000). Gene Ontology: tool for the unification of biology. *Nature Genetics*, 25:25–29.

32. **Virtual School of Molecular Sciences**. Biological Document Object Models. http://ala.vsms.nottingham.ac.uk/biodom/.

33. **W3Schools**. Introduction to XML. http://www.w3schools.com/xml/.

34. **Wittig, W. and A. De Beuckelaer** (2001). Analysis and comparison of metabolic pathway databases. *Briefings in Bioinformatics*, 2:126 – 142.

35. **World Wide Web Consortium** (W3C). Extensible Markup Language (XML). http://www.w3.org/XML/.

36. **World Wide Web Consortium** (W3C). HyperText Markup Language (HTML). http://www.w3.org/MarkUp/.

37. **World Wide Web Consortium** (W3C) (1998). Document Object Model (DOM). http://www.w3c.org.

# APPENDIX

*Table 1.* Some of the most commonly used biochemical databases

| Database/ System Type | Description | Web Address |
| --- | --- | --- |
| **Protein Databases** | | |
| Swiss-Prot | A curated protein sequence database providing a high level of annotations, minimal redundancy and high level of integration with other databases. | http://www.expasy.ch/sprot/sprot-top.html |
| Tr-EMBL | A computer-annotated supplement of SWISS-PROT that contains all the translations of EMBL nucleotide sequence entries not yet integrated in SWISS-PROT. | http://www.expasy.ch/sprot/sprot-top.html |
| SCOP | Structural Classication of Proteins for the investigation of sequences and structures. | http://scop.mrc-lmb.cam.ac.uk/scop/ |
| PDB | (Brookhaven Protein Databank) a repository for the processing and distribution of 3-D biological macromolecular structure data. | http://molbio.info.nih.gov/doc/mrus/pdb.html |
| PIR | (Protein Information Resource) a comprehensive, non-redundant, expertly annotated, fully classifed and extensively cross-referenced protein sequence database in the public domain. The PIR-PSD, iProClass and other PIR auxiliary databases provide an integration of sequences, functional, and structural information to support genomics and proteomics research. | http://pir.georgetown.edu/ |
| **Genome Databases** | | |
| GenBank | The NIH genetic sequence database; an annotated collection of all publicly available DNA sequences. | http://www.ncbi.nlm.nih.gov/Genbank/ |
| **Database Retrieval Systems** | | |
| SRS | A data integration platform, providing access to the large volumes of diverse and heterogeneous internal and public domain databases. | available at: http://srs.ebi.ac.uk/ |

| Database/ System Type | Description | Web Address |
|---|---|---|
| DBGET/ LinkDB | An integrated bioinfomatics data-base retrieval system. | www.genome.ad.jp/dbget/ |
| Entrez | A retrieval system for searching several linked databases. | www.ncbi.nlm.nih.gov/Entrez/ |
| EMBL Database | The EMBL Nucleotide Sequence Database constitutes one of Europe's primary nucleotide sequence resource. | http://www.ebi.ac.uk/embl/index.html |
| **Metabolic Pathway DBs** | | |
| KEGG | (Kyoto Encyclopedia of Genes and Genomes) provides information in the field of molecular and cellular biology. Contains three strongly connected databases: PATHWAY, LIGANDS and GENES. | http://www.genome.ad.jp/kegg/ |
| PATHWAY | Contains graphical representations of the pathways and lists of enzymes and reactions within a pathway. | http://www.genome.ad.jp/kegg/kegg2.html #pathway |
| LIGAND | Contains information about com-pounds, enzymes and reactions. | http://www.genome.ad.jp/dbget/ligand.html |
| GENES | Contains organism related genome and gene information. | http://www.genome.ad.jp/dbget-bin/www bfind?genes |
| ENZYME | (Enzyme nomenclature database) a repository of information relative to the nomenclature of enzymes. It describes each type of characterized enzyme for which an EC (Enzyme Commission) number has been provided. | http://www.expasy.ch/enzyme |
| WIT | (What is there?) a www-based system to support the curation of function assignments made to genes and the development of metabolic models. | http://wit.mcs.anl.gov/WIT2/ |
| MetaCyc | A metabolic-pathway database, describing pathways, reactions, and enzymes of a variety of organisms, with a microbial focus. | http://ecocyc.pangeasystems.com/ecocyc/met acyc.html |
| PathDB | A functional prototype research tool for biochemistry and functional genomics. | http://www.ncgr.org/pathdb/index.html |

| Database/ System Type | Description | Web Address |
|---|---|---|
| EMP | (Database of Enzymes and Metabolic pathways) a comprehensive electronic source of biochemical data, covering all aspects of enzymology and metabolism. | http://emp.mcs.anl.gov/ |
| Brenda | A comprehensive enzyme information system containing enzyme functional data. | http://www.brenda.uni-koeln.de/ |
| EXPASY- Biochemical Pathway | Digitized version of the Boehringer Mannheim \Biochemical Pathways" wall chart. | http://www.expasy.ch |
| **Literature Databases** PubMed | A service of the National Library of Medicine provides access to over 11 million MEDLINE citations and additional life science journals. Includes links to many sites providing full text articles. | http://www4.ncbi.nlm.nih.gov/PubMed/ |

# Chapter 10

# Mycoplasmas and the Minimal Genome Concept

CLYDE A. HUTCHISON III and MICHAEL G. MONTAGUE
*Department of Microbiology and Immunology, The University of North Carolina, C.B. 7290, Mary Ellen Jones Building, Chapel Hill NC, 27599, USA.*

## 1.     MYCOPLASMAS AS SIMPLE MODEL CELLS

The ultimate goal of biochemistry and molecular biology is the complete description of biological systems in terms of the laws of chemistry and physics. Consequently the mycoplasmas attracted the attention of biologists because of their small size and apparent simplicity, just as the hydrogen atom provided the simplest model for the development of physicists' theories of the atom. But, while hydrogen atoms are primitive, fusing within the interiors of stars to form the heavier elements, the mycoplasmas are simple for quite the opposite reason. They evolved from complex bacteria through the loss of functions that are unnecessary in their habitats as parasites on multicellular organisms. This evolution has proceeded far down the path toward distilling a minimal set of essential cellular genes.

Interest in minimal life forms long predates recent developments in genome sequencing. Harold Morowitz and colleagues started a search for the simplest cells in the 1950s (reviewed in Morowitz, 1984)[17]. These studies soon focused on mycoplasmas because they possess both the smallest physical dimensions and the smallest genome sizes of any known cells. The smallest known genome for a cell capable of autonomous replication is that of *M. genitalium*[2].

The small size of the *M. genitalium* genome (580 kb) made it an attractive candidate when cellular genome sequencing became a reality, and it was the second cellular genome to be completely sequenced[4]. Analysis of the sequence revealed only 480 protein coding genes plus 37 genes for RNA species. Soon the genome sequence of *M. pneumoniae* was completed[7,8], and two additional mollicute genome sequences *Ureaplasma urealyticum*[5] and *M. pulmonis*[1] have since become available. Additional projects are underway, making the mollicutes one of the most extensively studied groups at the level of complete genome sequencing. Already the comparison of these sequences yields many insights into what it means to be a mollicute (see the chapter by Dandekar et al., Comparative genome analysis of Mollicutes).

The availability of complete genome sequences has put the quest for a minimal organism in a new light. One important question posed by the availability of complete genomic sequence is how many genes are essential for cellular life. We are now in a position to approach this problem by reframing the age-old question "What is life?" in genomic terms as: "What is a minimal set of essential cellular genes?"

## 2.    THE RANGE OF GENOME COMPLEXITY

The DNA content of cellular genomes ranges from 580 kb for *M. genitalium* to about 100 billion bp for the lily, a range of about $10^5$-fold. However, the actual number of genes encoded apparently covers a considerably smaller range, from about 500 for *M. genitalium* to only 30-40,000 for human[14,24], the largest genome for which we have a reasonable estimate. By comparison the large bacterial genome of *Pseudomonas aeruginosa* contains some 5,570 protein coding genes. Within the bacteria gene density is quite constant, at close to one gene per kilobase on the average.

## 3.    WHAT IS A MINIMAL GENOME AND CAN IT BE DEFINED?

A minimal genome is one that contains the minimum number of genes necessary for cell growth and replication. Although this appears to be a simple concept, it is not one that is easily approached in a rigorous manner, either experimentally or in theory. One difficulty is that the genetic functions required by a cell depend on the environment in which the cell must survive. Another is that there can be different solutions to particular biological requirements such as energy production, transport, and replication mechanisms. Neither experiments nor evolution can test all possible gene combinations for viability because of the very large numbers involved. It will therefore be difficult to discover simple solutions to biological requirements that have never been discovered by evolution, if such solutions exist. Similarly, it will be difficult to unearth primitive genetic solutions that may have existed on the early Earth but are noncompetitive today. For these reasons a *global* solution to the problem of minimizing the cellular genome seems impractical, if not impossible, with present methods.

THE FAR SIDE® By GARY LARSON

"Hey! I got news for you, sweetheart! ... I *am*
the lowest form of life on earth!"

*Figure 1.* The minimal cell concept in popular culture. Is this Gary Larson's vision of mycoplasmas?

In spite of these difficulties there has been considerable recent interest in the minimal genome concept (see reviews[12,15,18,20] and Fig.1). Both theoretical and experimental approaches have focused on finding a *local* solution to the problem, starting from the reduced gene set found in parasites. Consideration of the assumptions implicit in the design of these studies makes comparison of their results more informative, and aids interpretation of the biological implications of specific proposals for minimal gene sets.

## 4.    COMPARATIVE GENOMICS AND THE MINIMAL GENOME

The determination of the second complete genome sequence, that of *M. genitalium*, allowed a comparative genomic approach to the minimal genome problem[19]. By analyzing the set of gene functions common to *H. influenzae* and *M. genitalium,* and assuming that genes conserved across large phylogenetic distances are likely to be essential, Mushegian and Koonin[19] made a specific proposal of a minimal gene set consisting of 256 genes (see Appendix). Because these organisms are mammalian parasites, this approach represents an attempt to find the local solution to the minimal genome problem for parasites growing in the environment provided by their hosts.

Now that complete genome sequences are available for some 50 microbes, including more than a dozen parasites, it is interesting to look more broadly at the process of gene loss associated with adaptation to a parasitic lifestyle. One approach is the construction of whole genome trees based upon gene content[3,16,21,22,25].

In this method the genes are classified into orthologous groups, and the presence or absence of a gene in each of the orthologous groups is treated as a genetic trait for each of the genomes under analysis. Any of the conventional tree making procedures used for producing phylogenetic trees from molecular sequence data can be used in the analysis. This method clusters genomes with similar gene content profiles, and results in a tree which gives a combination of phylogenetic and lifestyle information. Such a tree is shown (Fig.2) for 19 species including 9 parasitic ones. The tree clearly separates the Eukaryotes, the Eubacteria and the Archaea. The parasitic Eubacteria with reduced genome size form a distinct branch within the Eubacteria. The mycoplasmas cluster with the other parasites, rather than the Gram-positive bacteria to which they are phylogenetically related (represented by *B. subtilis* in this tree). This is consistent with the notion of a common pattern of gene loss during evolution toward parasitism.

The comparative genomic approach to defining the minimal gene[18] set has been refined in light of the availability of more complete genome sequences[12]. The set of orthologous gene families shared by all genomes gives a first approximation to the minimal gene set defined by this comparative approach. But this is complicated by the phenomenon of non-orthologous displacement (NOD). NOD is a term used to describe situations in which the same function is performed in different organisms by genes that are not orthologous by descent. There are many readily identifiable examples of NOD in essential functions where the biochemistry is well understood[12].

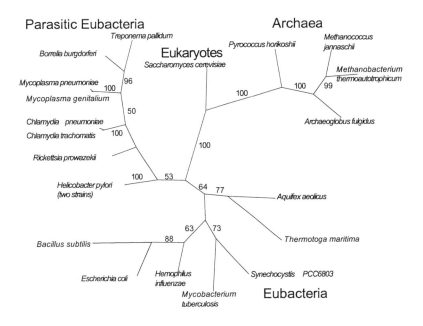

*Figure 2*. Whole genome gene content tree using COG analysis[16] to identify orthologous genes. The tree was produced from the COGs posted on the NCBI web site on Jan. 1, 2000, by using the program PAUP Version 4.0b by D. L. Swofford. The Neighbor-joining method was used, and bootstrap numbers were obtained from 100 replicates using the full heuristic search option.

Some NOD events involve displacement of an individual protein function; for example, the lysyl-tRNA synthetases of Archaea are unrelated to the type found in most Bacteria. Other cases of NOD involve the use of alternative genetic solutions to performing essential functions in different organisms; for example, *Mycoplasma* species generate energy by glycolysis whereas *Ureaplasma* species use urea hydrolysis to generate ATP. If all NOD events can be identified, then a minimal gene set can be constructed as outlined in Figure 3. The genes from the smallest genome that have orthologs in all other genomes (the intersection I in Fig. 3) are combined with the genes that have non-orthologous functional equivalents in the other genomes (NOD cases) to form a minimal gene set. Analysis of the many genomes available today shows that NOD is a much more prevalent than originally supposed. Only 80 of the 256 *M. genitalium* genes in the minimal gene set originally proposed had orthologs in all of 25 genomes analyzed[12] (see Appendix). The great majority of the other 176 genes appear to be essential in mycoplasma, but are involved in NOD events in one or more of the analyzed genomes.

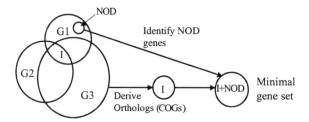

*Figure 3.* Identification of a minimal gene set by comparing genomes according to the procedure described by Koonin[12]. G1, G2, G3; three genomes. COGs; clusters of orthologous groups. NOD; non-orthologous displacement.

It should be noted that the minimal gene set proposal based on comparative genomics does not include any structural genes involved in the mycoplasma cytoskeleton or the tip structure that mediates cellular adhesion. Adhesion is an essential function in the natural habitat of mycoplasmas. It is also possible that some of these genes may perform mycoplasma-specific functions that are essential even for laboratory growth.

## 5.    EXPERIMENTAL APPROACHES TO DEFINING A MINIMAL GENOME

Experimental work related to the minimal genome concept has so far focused on defining the subset of essential genes in naturally occurring genomes. Before the era of genome sequencing, estimates of the fraction of non-minimal genomes that is essential for cell growth and replication have been made in yeast[6] (12%), as well as the bacterium *Bacillus subtilis*[10] (9%). The indispensable portion of the *B. subtilis* genome was estimated to be 562 kb, remarkably close to the size of the *M. genitalium* genome. It should be noted that these studies estimate the size of the essential gene set required for growth in the laboratory, in a rich medium.

The most extensive study aimed at cataloging non-essential genes, for the purpose of defining a minimal genome, has been conducted using *M. genitalium* and *M. pneumoniae*[9]. In this study a derivative of the transposon Tn4001[13] was used to disrupt genes randomly. Populations of cells that were viable while harboring a transposon insertion were analyzed by sequencing the junction between the transposon and the chromosome. Insertions that appeared capable of disrupting gene function were observed in 129 different genes in *M. genitalium* or their *M. pneumoniae* orthologs. Statistical analysis of the experiment, which was not continued to saturation, leads to an estimate of 265 – 350 for the number of essential genes in *M. genitalium*.

Again, these results bear on a special case of the minimal genome concept. The experiments were conducted with small parasitic bacteria, but grown in an extremely rich medium in the laboratory (SP-4 medium). So we expect the minimal genome under these conditions to contain the genes essential for parasites, except for a) those needed only for association with the host and, b) those needed to deal with stresses found in the natural habitat but absent in the laboratory. One difficulty with this experimental system is the lack of a chemically defined medium for growth of these organisms, making it difficult to assess the potential importance of genes coding for metabolic enzymes.

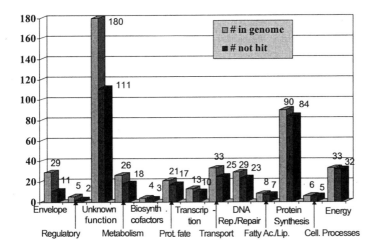

*Figure 4.* Distribution of non-essential genes among functional categories in *M. genitalium*. For each role category, the total number of genes in *M. genitalium* is shown, along with the number in which a disruptive transposon insertion has not been observed. Categories are arranged in order of decreasing fraction of genes that are dispensable, from left to right. This figure is based upon annotation of the genome at TIGR that has been updated since the original publication (see the information on "TIGR role category" in the spreadsheet posted at http://www.unc.edu/~clyde/MycGene.xls). The category "Unknown function" in this figure includes genes annotated as both "unknown function" and "conserved hypothetical".

The distribution of essential genes among functional categories is of interest (Fig. 4). The genes in which no transposon disruption has been observed provide a first approximation to the essential gene set for mycoplasma. Although bioinformatic predictions of gene function differ somewhat in analyses performed by different approaches (see Appendix and http://www.unc.edu/~clyde/mycoplasmadbs.html), it is clear that the genome of *M. genitalium* is closer to a minimal genome for some functions than for others. Energy metabolism appears to be very close to a local minimum in the genome of *M. genitalium*. In contrast, genes involved with the cell envelope are mostly non-essential under the conditions of the experiment. Categories related to gene expression, and replication of genetic information, are among those with high fractions of essential genes. The largest category is genes of unknown function. This category includes genes that are mycoplasma-specific, as well as genes of undefined, or poorly defined function which are also found in other bacteria. Although many of these genes are clearly non-essential, over 100 were not disrupted by transposon mutagenesis and it appears that most of these are essential[9].

## 6.    IMPLICATIONS OF THE PROPOSED MINIMAL GENE SETS

It is interesting to compare experimental and theoretical approaches to defining the minimal gene set for mycoplasmas. An Appendix to this chapter lists the annotated genes of *M. genitalium* sorted into two groups. Group 1 consists of 338 genes that deserve consideration as potential candidates for inclusion in a minimal mycoplasma gene set, on the basis of both experimental and bioinformatic approaches. The group includes the minimal gene set originally proposed by Mushegian and Koonin[19], derived by comparing the *H. influenzae* and *M. genitalium* gene complements. Also included are all *M. genitalium* genes that are members of gene families (COGs) extending beyond the mycoplasmas, and for which no viable transposon disruption events were observed[9]. Putatively disruptive transposon insertions were observed[9] in 38 of the 256 genes (15%) of the minimal gene set originally proposed by Mushegian and Koonin (see Appendix). Reanalysis in light of subsequent genome sequences shows that only a few of these 38 genes are conserved universally, or among all Eubacteria. Most have quite scattered phylogenetic profiles, and are presumably not really essential genes[12]. On the other hand, a few genes with transposon insertions appear to be universally conserved (see Appendix), and additional experimental evidence will be required to determine whether these genes are actually dispensable or not. It should be kept in mind that the gene disruption approach is aimed at defining the essential gene set for growth under laboratory conditions. The minimal gene set defined by the comparative genomic approach should also include genes required for survival over evolutionary time periods in the natural habitat of the parasite-host relationship. Also, the set of all essential genes, as defined indirectly by disruption of non-essential genes, is not necessarily sufficient for a minimal genome. Genes that are individually dispensable might perform the same essential function, and therefore not be simultaneously dispensable. The set of essential genes therefore sets a lower bound on the minimal gene set. Considering these differences in the underlying assumptions, there is quite good agreement between the results of theoretical and experimental approaches to defining the minimal gene concept.

Group 2 in the Appendix consists of 146 genes that are unlikely to be required in a minimal genome for growth in the laboratory. Transposon mutagenesis indicates that most of these are non-essential. The rest are mycoplasma-specific genes. We are unaware of evidence that any of these genes is essential for growth in the laboratory.

A frustrating aspect of microbial genomics is the very large number of genes for which reliable functional assignments cannot be made, based upon sequence similarity to genes of known function in other organisms. These include both genes that are apparently species-specific and, more interestingly, genes found in many species but without a functional assignment in any of them. Both the comparative genomics approach[12,18] and the experimental approach using transposon mutagenesis[9] indicate that there are a significant number of such genes in the minimal gene set. This raises the question of whether there are significant essential cellular functions yet to be discovered. This question can be approached experimentally by focusing attention on genes of unknown cellular function which are members of the minimal gene set deduced by comparative genomics, and also appear to be essential in the transposon gene disruption experiments.

## 7. POTENTIAL APPLICATIONS OF MINIMAL GENOMES

Although a minimal gene set for a mycoplasma has not been completely defined, much progress has been made on the problem both theoretically and experimentally. It is clear that even the smallest naturally occurring genome contains many genes that are not essential under laboratory growth conditions. Work on the minimal genome concept is a good example of the way that theoretical work, using bioinformatic techniques, can interact with experimental work, in the rapidly developing science of genomics.

In the future it may be possible to develop "genome engineering" techniques that will allow the convenient construction of chromosomes to test the sufficiency of specific proposals for minimal gene sets. Such substantially reduced genomes may be useful in a number of applications, whether or not they are truly minimal. One such area is the computational modeling of cellular processes[23], where the computational difficulty may be a steep function of the number of interacting genes involved.

Another area in which proposals for the minimal genome are already useful is the new field of structural genomics. The Berkeley Structural Genomics Center is based on the concept of determining a complete set of structures for the gene products of a minimal cell[11], using *M. genitalium* and *M. pneumoniae* as model organisms. Work on the minimal gene set allows structure determinations to focus on genes that are essential but have no known homologs. This approach aims to bring the power of structural biology to bear upon a global understanding of the molecular basis of life.

## ACKNOWLEDGMENTS

We thank the members of the Hutchison laboratory, especially Gwynedd Smith and Bradford Powell for helpful discussions. We thank Scott Peterson and other colleagues at TIGR with whom we have collaborated on experimental approaches to the minimal genome problem. We also thank Kenneth Bott and P.-c. "Ed" Hu for introducing us to *M. genitalium*. This work is funded, in part, by support from NIH to the Berkeley Structural Genomics Center.

## REFERENCES

1.  **Chambaud, I., R. Heilig, S. Ferris, V. Barbe, D. Samson, F. Galisson, I. Moszer, K. Dybvig, H. Wroblewski, A. Viari, E. P. Rocha, and A. Blanchard.** 2001. The complete genome sequence of the murine respiratory pathogen *Mycoplasma pulmonis*. Nucleic Acids Res. 29:2145-2153.
2.  **Colman, S. D., P.-c. Hu, W. Litaker, and K. F. Bott.** 1990. A physical map of the *Mycoplasma genitalium* genome. Mol. Microbiol. 4:683-687.
3.  **Fitz-Gibbon, S. T., and C. H. House.** 1999. Whole genome-based phylogenetic analysis of free-living microorganisms. Nucl. Acids Res. 27:4218-4222.
4.  **Fraser, C. M., J. D. Gocayne, O. White, M. D. Adams, R. A. Clayton, R. D. Fleischmann, C. J. Bult, A. R. Kerlavage, G. Sutton, J. M. Kelley, J. L. Fritchman, J. F. Weidman, K. V. Small, M. Sandusky, J. Fuhrman, D. Nguyen, T. R. Utterback, D. M. Saudek, C. A. Phillips, J. M. Merrick, J.-F. Tomb, B. A. Dougherty, K. F. Bott, P.-c. Hu, T. S. Lucier, S. N. Peterson, H. O. Smith, C. A. III Hutchison, and J. C. Venter.** 1995. The minimal gene complement of *Mycoplasma genitalium*. Science 270:397-403; Updated annotation of the genome is available from the TIGR Comprehensive Microbial Resource at http://www.tigr.org/tigr-scripts/CMR2/GenomePage3.spl?database=gmg
5.  **Glass, J. I., E. J. Lefkowitz, J. S. Glass, C. R. Heiner, E. Y. Chen, and G. H. Cassell.** 2000. The complete sequence of the mucosal pathogen *Ureaplasma urealyticum*. Nature 407:757-762.
6.  **Goebl, M. D. and T. D. Petes.** 1986. Most of the yeast genomic sequences are not essential for cell growth and division. Cell 46:983-992.

7.  **Himmelreich, R., H. Hilbert, H. Plagens, E. Pirkl, B. C. Li, and R. Herrmann.** 1996. Complete sequence analysis of the genome of the bacterium *Mycoplasma pneumoniae*. Nucl. Acids Res. **24**:4420-4449.
8.  **Himmelreich, R., H. Plagens, H. Hilbert, B. Reiner, and R. Herrmann.** 1997. Comparative analysis of the genomes of the bacteria *Mycoplasma pneumoniae* and *Mycoplasma genitalium*. Nucl. Acids Res. **25**:701-712.
9.  **Hutchison, C. A. III, S. N. Peterson, S. R. Gill, R. T. Cline, O. White, C. M. Fraser, H. O. Smith, and J. C. Venter.** 1999. Global transposon mutagenesis and a minimal mycoplasma genome. Science **286**:2165-2169.
10. **Itaya, M.** 1995. An estimation of minimal genome size required for life. FEBS Lett. **362**:257-260.
11. **Kim, S.-H.** 2000. Structural genomics of microbes: an objective. Current Opinion in Structural Biology **10**:380-383.
12. **Koonin, E. V.** 2000. How many genes can make a cell: the minimal-gene-set concept. Annu. Rev. Genomics Hum. Genet. **1**:99-116.
13. **Knudtson, K. L., and F. C. Minion.** 1993. Construction of Tn4001lac derivatives to be used as promoter probe vectors in mycoplasmas. Gene **137**:217-222.
14. **Lander, E. S., L. M. Linton, B. Birren, et al.** 2001. Initial sequencing and analysis of the human genome. Nature. **409**:860-921.
15. **Maniloff, J.** 1996. The minimal cell genome: "On being the right size". Proc. Natl. Acad. Sci. U.S.A. **93**:10004-10006.
16. **Montague, M. G., and C. A. Hutchison III.** 2000. Gene content phylogeny of herpesviruses. Proc. Natl. Acad. Sci. U.S.A. **97**:5334-5339.
17. **Morowitz, H. J.** 1984. The completeness of molecular biology. Isr. J. Med. Sci. **20**:750-753.
18. **Mushegian, A.** 1999. The minimal genome concept. Curr. Opin. Genet. Dev. **9**:709-714.
19. **Mushegian, A. R., and E. V. Koonin.** 1996. A minimal gene set for cellular life derived by comparison of complete bacterial genomes. Proc. Natl. Acad. Sci. U.S.A. **93**:10268-10273.
20. **Peterson, S. N., and C. M. Fraser.** 2001. The complexity of simplicity. Genome Biology **2**: comment 2002.1-2002.8.
21. **Snel, B., P. Bork, and M. A. Huynen.** 1999. Genome phylogeny based on gene content. Nature Genetics **21**:108-110.
22. **Tekaia, F., A. Lazcano, and B. Dujon.** 1999. The genomic tree as revealed from whole proteome comparisons. Genome Research **9**:550-557.
23. **Tomita, M., K. Hashimoto, K. Takahashi, T. S. Shimizu, Y. Matsuzaki, F. Miyoshi, K. Saito, S. Tanida, K. Yugi, J. C, Venter, and C. A. Hutchison III.** 1999. E-CELL: software environment for whole-cell simulation. Bioinformatics **15**:72-84.
24. **Venter, J. C., M. D, Adams, E. W. Myers, et al.** 2001. The sequence of the human genome. Science **291**:1304-1351.
25. **Wolf, Y. I., I. B. Rogozin, N. V. Grishin, R. L. Tatusov, and E. V. Koonin.** 2001. Genome trees constructed using five different approaches suggest new major bacterial clades. BMC Evol. Biol. **1**:8.

# APPENDIX

### Candidates for the minimal mycoplasma gene set

The genes of *M. genitalium* are listed in two groups separated by a horizontal line. The first group (338 genes) includes the most likely candidates for a minimal mycoplasma gene set. This group consists of all *M. genitalium* genes that are either 1) included in the minimal gene set originally proposed by Mushegian and Koonin[19], 2) assigned to a COG that was not disrupted by a transposon insertion[9], or 3) members of COGs that are universally represented in sequenced genomes[12]. The second group consists of all other annotated genes. This group contains most of the genes that are non-essential in the laboratory. Within each group the rows were sorted by *M. genitalium* gene number and then by the NCBI function category. This information, plus much more, is available as an Excel spreadsheet at http://www.unc.edu/~clyde/Xref.html.

| *M. gen.* | *M. pn.* | Role category (TIGR) | Function Category (NCBI) | MGS | Tn hits | Univ. | Predicted gene function |
|---|---|---|---|---|---|---|---|
| (a) | (b) | (c) | (d) | (e) | (f) | (g) | (h) |
| | | | | | | | GROUP 1 GENES |
| MG038 | MPN050 | energy metabolism | C | C | | | Glycerol kinase |
| MG271 | MPN390 | energy metabolism | C | C | | | Dihydrolipoamide dehydrogenase/glutathione oxidoreductase and related enzymes |
| MG272 | MPN391 | energy metabolism | C | C | | | Dihydrolipoamide acyltransferases |
| MG273 | MPN392 | energy metabolism | C | C | | | Thiamine pyrophosphate-dependent dehydrogenases, E1 component beta subunit |
| MG274 | MPN393 | energy metabolism | C | C | | | Thiamine pyrophosphate-dependent dehydrogenases, E1 component alpha subunit |
| MG293 | MPN420 | central intermediary metabolism | C | I | g6 | | Glycerophosphoryl diester phosphodiesterase |
| MG299 | MPN428 | central intermediary metabolism | C | C | p1 | | Phosphotransacetylase |
| MG337 | MPN488 | conserved hypothetical | C | | | | NifU homologs involved in Fe-S cluster formation |
| MG351 | MPN528 | central intermediary metabolism | C | L | | | Inorganic pyrophosphatase |
| MG357 | MPN533 | central intermediary metabolism | C | C | | | Acetate kinase |
| MG398 | MPN597 | energy metabolism | C | C | | | F0F1-type ATP synthase epsilon subunit (mitochondrial delta subunit) |
| MG399 | MPN598 | energy metabolism | C | C | | | F0F1-type ATP synthase beta subunit |

| (a) | (b) | (c) | (d) | (e) | (f) | (g) | (h) |
|---|---|---|---|---|---|---|---|
| MG400 | MPN599 | energy metabolism | C | C | | | F0F1-type ATP synthase gamma subunit |
| MG401 | MPN600 | energy metabolism | C | C | | | F0F1-type ATP synthase alpha subunit |
| MG402 | MPN601 | energy metabolism | C | C | | | F0F1-type ATP synthase delta subunit (mitochondrial oligomycin sensitivity protein) |
| MG403 | MPN602 | energy metabolism | C | C | | | F0F1-type ATP synthase b subunit |
| MG404 | MPN603 | energy metabolism | C | C | | 1 | F0F1-type ATP synthase c subunit/Archaeal/vacuolar-type H+-ATPase subunit K |
| MG405 | MPN604 | energy metabolism | C | C | | | F0F1-type ATP synthase a subunit |
| MG460 | MPN674 | energy metabolism | C | | | | Malate/lactate dehydrogenases |
| MG084 | MPN222 | conserved hypothetical | D | | | 1 | Predicted ATPase of the PP-loop superfamily implicated in cell cycle control |
| MG224 | MPN317 | cellular processes | D | M | | | Cell division GTPase |
| MG298 | MPN426 | unknown function | D | | | | Chromosome segregation ATPases |
| MG379 | MPN557 | dna metabolism | D | RS | | | NAD/FAD-utilizing enzyme apparently involved in cell division |
| MG043 | MPN056 | transport and binding proteins | E | F | | | ABC-type spermidine/putrescine transport system, permease component I |
| MG044 | MPN057 | transport and binding proteins | E | F | | | ABC-type spermidine/putrescine transport system, permease component II |
| MG077 | MPN215 | transport and binding proteins | E | E | | | ABC-type dipeptide/oligopeptide/nickel transport systems, permease components |
| MG078 | MPN216 | transport and binding proteins | E | E | | | ABC-type dipeptide/oligopeptide/nickel transport systems, permease components |
| MG079 | MPN217 | transport and binding proteins | E | E | | | ABC-type dipeptide/oligopeptide/nickel transport system, ATPase component |
| MG080 | MPN218 | transport and binding proteins | E | E | | | ABC-type dipeptide/oligopeptide/nickel transport system, ATPase component |
| MG225 | MPN318 | conserved hypothetical | E | | | | Amino acid transporters |
| MG324 | MPN470 | protein fate | E | | | 1 | Xaa-Pro aminopeptidase |
| MG336 | MPN487 | biosynthesis of cofactors, etc. | E | J, E | | | Selenocysteine lyase |

| (a) | (b) | (c) | (d) | (e) | (f) | (g) | (h) |
|---|---|---|---|---|---|---|---|
| MG391 | MPN572 | protein fate | E | E | | | Leucyl aminopeptidase |
| MG394 | MPN576 | central intermediary metabolism | E | H | g1 | 1 | Glycine hydroxymethyltransferase |
| MG006 | MPN006 | purines, pyrimidines, etc. | F | F | | 1 | Thymidylate kinase |
| MG030 | MPN033 | purines, pyrimidines, etc. | F | F | | | Uracil phosphoribosyltransferase |
| MG034 | MPN044 | purines, pyrimidines, etc. | F | | | | Thymidine kinase |
| MG049 | MPN062 | purines, pyrimidines, etc. | F | F | g1 | | Purine-nucleoside phosphorylase |
| MG050 | MPN063 | energy metabolism | F | C | | | Deoxyribose-phosphate aldolase |
| MG052 | MPN065 | purines, pyrimidines, etc. | F | F | g2 | | Cytidine deaminase |
| MG058 | MPN073 | purines, pyrimidines, etc. | F | F | | | Phosphoribosylpyrophosphate synthetase |
| MG107 | MPN246 | purines, pyrimidines, etc. | F | F | | | Guanylate kinase |
| MG171 | MPN185 | purines, pyrimidines, etc. | F | F | | | Adenylate kinase and related kinases |
| MG227 | MPN320 | purines, pyrimidines, etc. | F | F | p1 | | Thymidylate synthase |
| MG229 | MPN322 | purines, pyrimidines, etc. | F | F | | | Ribonucleotide reductase beta subunit |
| MG230 | MPN323 | unkown function | F | | | | Protein involved in ribonucleotide reduction |
| MG231 | MPN324 | purines, pyrimidines, etc. | F | F | p1 | 1 | Ribonucleotide reductase alpha subunit |
| MG264 | MPN382 | conserved hypothetical | F | | g3 | 1 | Predicted nucleotide kinase |
| MG268 | MPN386 | conserved hypothetical | F | F | g1 | | Deoxyguanosine/deoxyadenosine kinase |
| MG276 | MPN395 | purines, pyrimidines, etc. | F | F | | | Adenine/guanine phosphoribosyltransferases and related PRPP-binding proteins |

| (a) | (b) | (c) | (d) | (e) | (f) | (g) | (h) |
|---|---|---|---|---|---|---|---|
| MG330 | MPN476 | purines, pyrimidines, etc. | F | F | | | Cytidylate kinase |
| MG382 | MPN561 | purines, pyrimidines, etc. | F | F | | | Uridine kinase |
| MG434 | MPN632 | purines, pyrimidines, etc. | F | F | | | Uridylate kinase |
| MG458 | MPN672 | purines, pyrimidines, etc. | F | F | | | Hypoxanthine-guanine phosphoribosyltransferase |
| MG023 | MPN025 | energy metabolism | G | C | | | Fructose/tagatose biphosphate aldolases |
| MG033 | MPN043 | transport and binding proteins | G | C | g2 | | Glycerol uptake facilitator and related permeases |
| MG041 | MPN053 | transport and binding proteins | G | | | | Phosphotransferase system, HPr-related proteins |
| MG042 | MPN055 | transport and binding proteins | G | F | | | ABC-type sugar/spermidine/putrescine transport systems, ATPase component |
| MG053 | MPN066 | energy metabolism | G | C | | 1 | Phosphomannomutase |
| MG061 | MPN076 | conserved hypothetical | G | | | 2 | Permeases of the major facilitator superfamily |
| MG063 | MPN079 | energy metabolism | G | C | | | Fructose-1-phosphate kinase and related fructose-6-phosphate kinase (PfkB) |
| MG066 | MPN082 | energy metabolism | G | M | | | Transketolase |
| MG069 | MPN207 | transport and binding proteins | G | | | | Phosphotransferase system IIA components |
| MG111 | MPN250 | energy metabolism | G | C | | | Glucose-6-phosphate isomerase |
| MG112 | MPN251 | energy metabolism | G | C | | | Pentose-5-phosphate-3-epimerase |
| MG119 | MPN258 | transport and binding proteins | G | C | | | ABC-type sugar (aldose) transport system, ATPase component |
| MG129 | MPN268 | conserved hypothetical | G | | | | Phosphotransferase system IIB components |
| MG187 | MPN134 | transport and binding proteins | G | C | | | ABC-type multidrug transport system, ATPase component |
| MG188 | MPN135 | transport and binding proteins | G | | | | ABC-type sugar transport systems, permease components |

| (a) | (b) | (c) | (d) | (e) | (f) | (g) | (h) |
|-----|-----|-----|-----|-----|-----|-----|-----|
| MG189 | MPN136 | transport and binding proteins | G | | | | Sugar permeases |
| MG215 | MPN302 | energy metabolism | G | C | | | 6-phosphofructokinase |
| MG216 | MPN303 | energy metabolism | G | C | | | Pyruvate kinase |
| MG294 | MPN421 | conserved hypothetical | G | | g5/p1 | 2 | Permeases of the major facilitator superfamily |
| MG300 | MPN429 | energy metabolism | G | C | | | 3-phosphoglycerate kinase |
| MG301 | MPN430 | energy metabolism | G | C | | | Glyceraldehyde-3-phosphate dehydrogenase/erythrose-4-phosphate dehydrogenase |
| MG396 | MPN595 | energy metabolism | G | | | | Ribose 5-phosphate isomerase RpiB |
| MG407 | MPN606 | energy metabolism | G | C | | | Enolase |
| MG429 | MPN627 | transport and binding proteins | G | | | | Phosphoenolpyruvate-protein kinase (PTS system EI component in bacteria) |
| MG430 | MPN628 | energy metabolism | G | C | | | Phosphoglyceromutase |
| MG431 | MPN629 | energy metabolism | G | C | | | Triosephosphate isomerase |
| MG012 | MPN016 | conserved hypothetical | H | J | | | Glutathione synthase/Ribosomal protein S6 modification enzyme |
| MG013 | MPN017 | central intermediary metabolism | H | H | | | 5,10-methylene-tetrahydrofolate dehydrogenase |
| MG037 | MPN047 | conserved hypothetical | H | | | | Nicotinic acid phosphoribosyltransferase |
| MG047 | MPN060 | central intermediary metabolism | H | H | | | S-adenosylmethionine synthetase |
| MG145 | MPN158 | biosynthesis of cofactors, etc. | H | H | | | FAD synthase |
| MG228 | MPN321 | purines, pyrimidines, etc. | H | H | | | Dihydrofolate reductase |
| MG240 | MPN336 | conserved hypothetical | H | | | | Predicted HD superfamily hydrolase involved in NAD metabolism |
| MG245 | MPN348 | conserved hypothetical | H | H | | | 5-formyltetrahydrofolate cyclo-ligase |
| MG270 | MPN389 | biosynthesis of cofactors, etc. | H | H | p1 | | Lipoate-protein ligase A |

| (a) | (b) | (c) | (d) | (e) | (f) | (g) | (h) |
|-----|-----|-----|-----|-----|-----|-----|-----|
| MG383 | MPN562 | biosynthesis of cofactors, etc. | H | H | | | NAD synthase |
| MG114 | MPN253 | fatty acid and phospholipid | I | I | | | Phosphatidylglycerophosphate synthase |
| MG211.1 | MPN298 | conserved hypothetical | I | | | | Phosphopantetheinyl transferase (holo-ACP synthase) |
| MG212 | MPN299 | fatty acid and phospholipid | I | I | | | 1-acyl-sn-glycerol-3-phosphate acyltransferase |
| MG287 | MPN406 | fatty acid and phospholipid | I | I | | | Acyl carrier protein |
| MG333m | MPN479 | fatty acid and phospholipid | I | I | | | Acyl carrier protein phosphodiesterase |
| MG368 | MPN546 | fatty acid and phospholipid | I | | | | Fatty acid/phospholipid biosynthesis enzyme |
| MG437 | MPN637 | hypothetical - not conserved | I | I | | 1 | CDP-diglyceride synthetase |
| MG459 | MPN673 | conserved hypothetical | I | | | | 2C-methyl-D-erythritol 2,4-cyclodiphosphate synthase |
| MG005 | MPN005 | protein synthesis | J | J | | 1 | Seryl-tRNA synthetase |
| MG021 | MPN023 | protein synthesis | J | J | | 1 | Methionyl-tRNA synthetase |
| MG026 | MPN029 | protein synthesis | J | J | | 1 | Translation elongation factor P/translation initiation factor eIF-5A |
| MG035 | MPN045 | protein synthesis | J | J | | 1 | Histidyl-tRNA synthetase |
| MG036 | MPN046 | protein synthesis | J | J | | | Aspartyl-tRNA synthetase |
| MG055.1 | | protein synthesis | J | | | | Ribosomal protein L33 |
| MG070 | MPN208 | protein synthesis | J | J | | 1 | Ribosomal protein S2 |
| MG081 | MPN219 | protein synthesis | J | J | | 1 | Ribosomal protein L11 |
| MG082 | MPN220 | protein synthesis | J | J | | 1 | Ribosomal protein L1 |
| MG083 | MPN221 | protein synthesis | J | J | | | Peptidyl-tRNA hydrolase |
| MG087 | MPN225 | protein synthesis | J | J | | 1 | Ribosomal protein S12 |
| MG088 | MPN226 | protein synthesis | J | J | | 1 | Ribosomal protein S7 |
| MG089 | MPN227 | protein synthesis | J | J | | 1 | Translation elongation and release factors (GTPases) |
| MG090 | MPN228 | protein synthesis | J | J | | | Ribosomal protein S6 |
| MG092 | MPN230 | protein synthesis | J | J | | | Ribosomal protein S18 |

| (a) | (b) | (c) | (d) | (e) | (f) | (g) | (h) |
|-----|-----|-----|-----|-----|-----|-----|-----|
| MG093 | MPN231 | protein synthesis | J | J | | | Ribosomal protein L9 |
| MG098 | MPN236 | protein synthesis | J | | | | Asp-tRNAAsn/Glu-tRNAGln amidotransferase C subunit |
| MG099 | MPN237 | protein synthesis | J | | | | Asp-tRNAAsn/Glu-tRNAGln amidotransferase A subunit and related amidases |
| MG100 | MPN238 | protein synthesis | J | | | | Asp-tRNAAsn/Glu-tRNAGln amidotransferase B subunit (PET112 homolog) |
| MG106 | MPN245 | protein fate | J | J | g1 | | N-formylmethionyl-tRNA deformylase |
| MG113 | MPN252 | protein synthesis | J | J | | | Aspartyl/asparaginyl-tRNA synthetases |
| MG126 | MPN265 | protein synthesis | J | J | | 1 | Tryptophanyl-tRNA synthetase |
| MG136 | MPN277 | protein synthesis | J | J | | | Lysyl-tRNA synthetase class II |
| MG142 | MPN155 | protein synthesis | J | J | | | Translation initiation factor 2 (GTPase) |
| MG143 | MPN156 | transcription | J | J | | | Ribosome-binding factor A |
| MG150 | MPN164 | protein synthesis | J | J | | 1 | Ribosomal protein S10 |
| MG151 | MPN165 | protein synthesis | J | J | | 1 | Ribosomal protein L3 |
| MG152 | MPN166 | protein synthesis | J | J | | 1 | Ribosomal protein L4 |
| MG153 | MPN167 | protein synthesis | J | J | | 1 | Ribosomal protein L23 |
| MG154 | MPN168 | protein synthesis | J | J | | 1 | Ribosomal protein L2 |
| MG155 | MPN169 | protein synthesis | J | J | | 1 | Ribosomal protein S19 |
| MG156 | MPN170 | protein synthesis | J | J | | 1 | Ribosomal protein L22 |
| MG157 | MPN171 | protein synthesis | J | J | | 1 | Ribosomal protein S3 |
| MG158 | MPN172 | protein synthesis | J | J | | 1 | Ribosomal protein L16/L10E |
| MG159 | MPN173 | protein synthesis | J | J | | 1 | Ribosomal protein L29 |
| MG160 | MPN174 | protein synthesis | J | J | | 1 | Ribosomal protein S17 |
| MG161 | MPN175 | protein synthesis | J | J | | 1 | Ribosomal protein L14 |
| MG162 | MPN176 | protein synthesis | J | J | | 1 | Ribosomal protein L24 |
| MG163 | MPN177 | protein synthesis | J | J | | 1 | Ribosomal protein L5 |

| (a) | (b) | (c) | (d) | (e) | (f) | (g) | (h) |
|---|---|---|---|---|---|---|---|
| MG164 | MPN178 | protein synthesis | J | J | | 1 | Ribosomal protein S14 |
| MG165 | MPN179 | protein synthesis | J | J | | 1 | Ribosomal protein S8 |
| MG166 | MPN180 | protein synthesis | J | J | | 1 | Ribosomal protein L6 |
| MG167 | MPN181 | protein synthesis | J | J | | 1 | Ribosomal protein L18 |
| MG168 | MPN182 | protein synthesis | J | J | | 1 | Ribosomal protein S5 |
| MG169 | MPN183 | protein synthesis | J | J | | 1 | Ribosomal protein L15 |
| MG172 | MPN186 | protein fate | J | J | | 1 | Methionine aminopeptidase |
| MG173 | MPN187 | protein synthesis | J | J | | 1 | Translation initiation factor IF-1 |
| MG174 | MPN188 | protein synthesis | J | J | | | Ribosomal protein L36 |
| MG175 | MPN189 | protein synthesis | J | J | | 1 | Ribosomal protein S13 |
| MG176 | MPN190 | protein synthesis | J | J | | 1 | Ribosomal protein S11 |
| MG178 | MPN192 | protein synthesis | J | J | | | Ribosomal protein L17 |
| MG182 | MPN196 | protein synthesis | J | J | g5 | | Pseudouridylate synthase (tRNA psi55) |
| MG194 | MPN105 | protein synthesis | J | J | | 1 | Phenylalanyl-tRNA synthetase alpha subunit |
| MG196 | MPN115 | protein synthesis | J | J | | | Translation initiation factor IF3 |
| MG197 | MPN116 | protein synthesis | J | J | | | Ribosomal protein L35 |
| MG198 | MPN117 | protein synthesis | J | J | | | Ribosomal protein L20 |
| MG209 | MPN292 | unknown function | J | J | g1/p1 | | Pseudouridylate synthases, 23S RNA-specific |
| MG210a | | | J | | | | Ribosomal protein S21 |
| MG232 | MPN325 | protein synthesis | J | J | | | Ribosomal protein L21 |
| MG233 | MPN326 | conserved hypothetical | J | | | | Predicted ribosomal protein |
| MG234 | MPN327 | protein synthesis | J | J | | | Ribosomal protein L27 |
| MG251 | MPN354 | protein synthesis | J | J | | | Glycyl-tRNA synthetase, class II |
| MG252 | MPN355 | protein synthesis | J | J | p1 | | rRNA methylases |
| MG253 | MPN356 | protein synthesis | J | J | | | Cysteinyl-tRNA synthetase |

| (a) | (b) | (c) | (d) | (e) | (f) | (g) | (h) |
|---|---|---|---|---|---|---|---|
| MG257 | MPN360 | protein synthesis | J | J | | | Ribosomal protein L31 |
| MG258 | MPN361 | protein synthesis | J | J | | | Protein chain release factor A |
| MG259 | MPN362 | conserved hypothetical | J | L | | | Putative translation factor (SUA5) |
| MG266 | MPN384 | protein synthesis | J | J | | 1 | Leucyl-tRNA synthetase |
| MG283 | MPN402 | protein synthesis | J | J | | 1 | Prolyl-tRNA synthetase |
| MG292 | MPN419 | protein synthesis | J | J | | 1 | Alanyl-tRNA synthetase |
| MG295 | MPN422 | conserved hypothetical | J | RS | g1 | | Predicted tRNA methyltransferase, contains the PP-loop ATPase domain |
| MG311 | MPN446 | protein synthesis | J | | | 1 | Ribosomal protein S4 and related proteins |
| MG325 | MPN471 | protein synthesis | J | J | | | Ribosomal protein L33 |
| MG334 | MPN480 | protein synthesis | J | J | | 1 | Valyl-tRNA synthetase |
| MG345 | MPN520 | protein synthesis | J | J | g1 | 1 | Isoleucyl-tRNA synthetase |
| MG346 | MPN521 | protein synthesis | J | J | g1/p2 | | Predicted rRNA methylase (SpoU class) |
| MG361 | MPN538 | protein synthesis | J | J | | 1 | Ribosomal protein L10 |
| MG362 | MPN539 | protein synthesis | J | J | | | Ribosomal protein L7/L12 |
| MG363 | MPN540 | protein synthesis | J | J | | | Ribosomal protein L32 |
| MG363.1 | MPN541 | protein synthesis | J | J | | | Ribosomal protein S20 |
| MG365 | MPN543 | protein synthesis | J | J | | | Methionyl-tRNA formyltransferase |
| MG375 | MPN553 | protein synthesis | J | J | | 1 | Threonyl-tRNA synthetase |
| MG378 | MPN556 | protein synthesis | J | J | | 1 | Arginyl-tRNA synthetase |
| MG417 | MPN616 | protein synthesis | J | J | | 1 | Ribosomal protein S9 |
| MG418 | MPN617 | protein synthesis | J | J | | 1 | Ribosomal protein L13 |
| MG424 | MPN622 | protein synthesis | J | J | | 1 | Ribosomal protein S15P/S13E |
| MG426 | MPN624 | protein synthesis | J | J | g1 | | Ribosomal protein L28 |
| MG433 | MPN631 | protein synthesis | J | J | | | Translation elongation factor Ts |

| (a) | (b) | (c) | (d) | (e) | (f) | (g) | (h) |
|---|---|---|---|---|---|---|---|
| MG435 | MPN636 | protein synthesis | J | J | | | Ribosome recycling factor |
| MG444 | MPN658 | protein synthesis | J | J | | | Ribosomal protein L19 |
| MG445 | MPN659 | protein synthesis | J | J | | | tRNA-(guanine-N1)-methyltransferase |
| MG446 | MPN660 | protein synthesis | J | J | | | Ribosomal protein S16 |
| MG451 | MPN665 | protein synthesis | J | J | | 1 | GTPases - translation elongation factors |
| MG455 | MPN669 | protein synthesis | J | J | p1 | 1 | Tyrosyl-tRNA synthetase |
| MG462 | MPN678 | protein synthesis | J | J | | 1 | Glutamyl- and glutaminyl-tRNA synthetases |
| MG463 | MPN679 | protein synthesis | J | J | | 1 | Dimethyladenosine transferase (rRNA methylation) |
| MG465 | MPN681 | transcription | J | J | | | RNase P protein component |
| MG466 | MPN682 | protein synthesis | J | J | | | Ribosomal protein L34 |
| MG018 | MPN020 | transcription | K | | | | Superfamily II DNA/RNA helicases, SNF2 family |
| MG022 | MPN024 | transcription | K | | | | DNA-directed RNA polymerase delta subunit |
| MG027 | MPN030 | conserved hypothetical | K | | | | Transcription termination factor |
| MG054 | MPN067 | conserved hypothetical | K | K | | 1 | Transcription antiterminator |
| MG104 | MPN243 | cellular processes | K | J | | | Exoribonucleases |
| MG141 | MPN154 | transcription | K | K | | | Transcription terminator |
| MG177 | MPN191 | transcription | K | K | | 1 | DNA-directed RNA polymerase alpha subunit/40 kD subunit |
| MG205 | MPN124 | regulatory functions | K | | | | Transcriptional regulator of heat shock gene |
| MG214 | MPN301 | conserved hypothetical | K | | | | Predicted transcriptional regulator containing the HTH domain |
| MG249 | MPN352 | transcription | K | K | p1 | | DNA-directed RNA polymerase sigma subunits (sigma70/sigma32) |
| MG282 | MPN401 | transcription | K | K | | | Transcription elongation factor |
| MG340 | MPN515 | transcription | K | K | | 1 | DNA-directed RNA polymerase beta' subunit/160 kD subunit |
| MG341 | MPN516 | transcription | K | K | | 1 | DNA-directed RNA polymerase beta subunit/140 kD subunit |
| MG367 | MPN545 | transcription | K | K | | | dsRNA-specific ribonuclease |
| MG001 | MPN001 | dna metabolism | L | L | | 1 | DNA polymerase III beta subunit |
| MG003 | MPN003 | dna metabolism | L | L | | | DNA gyrase (topoisomerase II) B subunit |

| (a) | (b) | (c) | (d) | (e) | (f) | (g) | (h) |
|---|---|---|---|---|---|---|---|
| MG004 | MPN004 | dna metabolism | L | L | | | DNA gyrase (topoisomerase II) A subunit |
| MG007 | MPN007 | conserved hypothetical | L | | | | ATPase involved in DNA replication |
| MG009 | MPN009 | conserved hypothetical | L | RS | g9 | | Mg-dependent DNase |
| MG031 | MPN034 | dna metabolism | L | | | | DNA polymerase III alpha subunit, the Gram-positive type |
| MG057 | MPN072 | unkown function | L | | | | Small primase-like proteins (Toprim domain) |
| MG073 | MPN211 | dna metabolism | L | L2 | | | Helicase subunit of the DNA excision repair complex |
| MG091 | MPN229 | dna metabolism | L | L | | | Single-stranded DNA-binding protein |
| MG094 | MPN232 | dna metabolism | L | L | | | Replicative DNA helicase |
| MG097 | MPN235 | dna metabolism | L | L2 | | | Uracil DNA glycosylase |
| MG122 | MPN261 | dna metabolism | L | L | | 1 | Topoisomerase IA |
| MG199 | MPN118 | dna metabolism | L | | | | Ribonuclease HII Family 2 |
| MG203 | MPN122 | dna metabolism | L | L | | | DNA gyrase (topoisomerase II) B subunit |
| MG204 | MPN123 | dna metabolism | L | L | | | DNA gyrase (topoisomerase II) A subunit |
| MG206 | MPN125 | dna metabolism | L | L2 | | | Nuclease subunit of the exconuclease complex |
| MG235 | MPN328 | dna metabolism | L | | | | Endonuclease IV |
| MG244 | MPN341 | dna metabolism | L | L | p2 | | Superfamily I DNA and RNA helicases |
| MG250 | MPN353 | dna metabolism | L | L | | | DNA primase (bacterial type) |
| MG254 | MPN357 | dna metabolism | L | L | | | NAD-dependent DNA ligase (contains BRCT domain type II) |
| MG261 | MPN378 | dna metabolism | L | L | p1 | | DNA polymerase III alpha subunit |
| MG262 | MPN379 | dna metabolism | L | L | | 1 | 5'-3' exonuclease (including N-terminal domain of PolI) |
| MG262.1 | | dna metabolism | L | L2 | | | Formamidopyrimidine-DNA glycosylase |
| MG291.1 | MPN418 | conserved hypothetical | L | | | | Predicted endonuclease involved in recombination |
| MG315 | MPN450 | conserved hypothetical | L | | | | DNA polymerase III delta subunit |
| MG339 | MPN490 | dna metabolism | L | L2 | g3 | 1 | RecA/RadA recombinase |

| (a) | (b) | (c) | (d) | (e) | (f) | (g) | (h) |
|---|---|---|---|---|---|---|---|
| MG353 | MPN529 | conserved hypothetical | L | L | | | Bacterial nucleoid DNA-binding protein |
| MG358 | MPN535 | dna metabolism | L | L2 | | | Holliday junction resolvasome DNA-binding subunit |
| MG359 | MPN536 | dna metabolism | L | L2 | | | Holliday junction resolvasome helicase subunit |
| MG360 | MPN537 | conserved hypothetical | L | | | | Nucleotidyltransferase/DNA polymerase involved in DNA repair |
| MG419 | MPN618 | dna metabolism | L | | | | DNA polymerase III, gamma/tau subunits |
| MG420 | | dna metabolism | L | L | | | DNA polymerase III, gamma/tau subunits |
| MG421 | MPN619 | dna metabolism | L | L2 | g1/p1 | | Excinuclease ATPase subunit |
| MG425 | MPN623 | transcription | L | J | | | Superfamily II DNA and RNA helicases |
| MG469 | MPN686 | dna metabolism | L | L | p1 | | ATPase involved in DNA replication initiation |
| MG060 | MPN075 | conserved hypothetical | M | M | | | Glycosyltransferases involved in cell wall biogenesis |
| MG086 | MPN224 | cell envelope | M | M | | | Prolipoprotein diacylglyceryltransferase |
| MG118 | MPN257 | energy metabolism | M | M | p1 | | UDP-glucose 4-epimerase |
| MG222 | MPN315 | conserved hypothetical | M | RS | | | Predicted S-adenosylmethionine-dependent methyltransferase involved in cell envelope biogenesis |
| MG335.2 | | conserved hypothetical | M | | | | Glycosyltransferases involved in cell wall biogenesis |
| MG356 | MPN532 | conserved hypothetical | M | | | | Predicted choline kinase involved in LPS biosynthesis |
| MG380 | MPN558 | dna metabolism | M | RS | g1 | | Predicted S-adenosylmethionine-dependent methyltransferase involved in bacterial cell division |
| MG453 | MPN667 | purines, pyrimidines, etc. | M | M | | | UDP-glucose pyrophosphorylase |
| MG048 | MPN061 | protein fate | N | N | | 1 | Signal recognition particle GTPase |
| MG072 | MPN210 | protein fate | N | N | | | Preprotein translocase subunit SecA (ATPase, RNA helicase) |
| MG138 | MPN279 | unknown function | N | RS | | | Membrane GTPase LepA |
| MG170 | MPN184 | protein fate | N | N | | 1 | Preprotein translocase subunit SecY |
| MG210 | MPN293 | protein fate | N | M | | | Lipoprotein signal peptidase |
| MG297 | MPN425 | protein fate | N | O, N | | 1 | Signal recognition particle GTPase |
| MG464 | MPN680 | conserved hypothetical | N | | | | Preprotein translocase subunit YidC |
| MG019 | MPN021 | protein fate | O | O | | | Molecular chaperones (contain C-terminal Zn finger domain) |
| MG046 | MPN059 | protein fate | O | | | 1 | Metal-dependent proteases with possible chaperone activity |

| (a) | (b) | (c) | (d) | (e) | (f) | (g) | (h) |
|-----|-----|-----|-----|-----|-----|-----|-----|
| MG059 | MPN074 | unknown function | O | M | p1 | | tmRNA-binding protein |
| MG102 | MPN240 | energy metabolism | O | F | | 1 | Thioredoxin reductase/alkyl hydroperoxide reductase |
| MG124 | MPN263 | energy metabolism | O | F | | 1 | Thiol-disulfide isomerase and thioredoxins |
| MG200 | MPN119 | unknown function | O | | | | Molecular chaperones (contain C-terminal Zn finger domain) |
| MG201 | MPN120 | protein fate | O | O | | | Molecular chaperone GrpE (heat shock protein) |
| MG238 | MPN331 | cellular processes | O | O | | | FKBP-type peptidyl-prolyl cis-trans isomerase (trigger factor) |
| MG239 | MPN332 | protein fate | O | O | | | ATP-dependent Lon protease, bacterial type |
| MG305 | MPN434 | protein fate | O | O | | | Molecular chaperone |
| MG355 | MPN531 | protein fate | O | O | p3 | | ATPases with chaperone activity, ATP-binding domain |
| MG392 | MPN573 | protein fate | O | O | p2 | 1 | Chaperonin GroEL (HSP60 family) |
| MG393 | MPN574 | protein fate | O | O | | | Co-chaperonin GroES (HSP10) |
| MG408 | MPN607 | regulatory functions | O | O | g2 | | Peptide methionine sulfoxide reductase |
| MG448 | MPN662 | conserved hypothetical | O | O | | | Conserved domain frequently associated with peptide methionine sulfoxide reductase |
| MG454 | MPN668 | conserved hypothetical | O | | | | Stress-induced protein |
| MG457 | MPN671 | cellular processes | O | O | | | ATP-dependent Zn proteases |
| MG071 | MPN209 | transport and binding proteins | P | P | | | Cation transport ATPases |
| MG127 | MPN266 | conserved hypothetical | P | F | | | Arsenate reductase and related proteins, glutaredoxin family |
| MG179 | MPN193 | transport and binding proteins | P | | | | ABC-type cobalt transport system, ATPase component |
| MG180 | MPN194 | transport and binding proteins | P | E | | | ABC-type cobalt transport system, ATPase component |
| MG181 | MPN195 | conserved hypothetical | P | | | | ABC-type cobalt transport system, permease component CbiQ and related transporters |
| MG290 | MPN416 | transport and binding proteins | P | | | | ABC-type cobalamin/Fe3+-siderophores transport systems, ATPase components |
| MG302 | MPN431 | conserved hypothetical | P | | | | ABC-type cobalt transport system, permease component CbiQ and related transporters |
| MG303 | MPN432 | transport and binding proteins | P | | | | ABC-type cobalt transport system, ATPase component |

| (a) | (b) | (c) | (d) | (e) | (f) | (g) | (h) |
|---|---|---|---|---|---|---|---|
| MG304 | MPN433 | transport and binding proteins | P | | | | ABC-type cobalt transport system, ATPase component |
| MG322 | MPN460 | transport and binding proteins | P | P | | | Trk-type K+ transport systems, membrane components |
| MG323 | MPN461 | conserved hypothetical | P | | | | K+ transport systems, NAD-binding component |
| MG409 | MPN608 | regulatory functions | P | | | | Phosphate uptake regulator |
| MG410 | MPN609 | transport and binding proteins | P | P | g1/p2 | | ABC-type phosphate transport system, ATPase component |
| MG411 | MPN610 | transport and binding proteins | P | P | g4 | | ABC-type phosphate transport system, permease component |
| MG015 | MPN019 | transport and binding proteins | Q | N | | | ABC-type multidrug/protein/lipid transport system, ATPase component |
| MG447 | MPN661 | conserved hypothetical | Q | | | | Na+-driven multidrug efflux pump |
| MG008 | MPN008 | cellular processes | R | RS | | | Predicted GTPase |
| MG020 | MPN022 | protein fate | R | | | | Predicted hydrolases or acyltransferases (alpha/beta hydrolase superfamily) |
| MG024 | MPN026 | unknown function | R | RS | p2 | 1 | Predicted GTPase |
| MG056 | MPN071 | conserved hypothetical | R | RS | | | Predicted methyltransferases |
| MG065 | MPN081 | transport and binding proteins | R | P | | | ABC-type (unclassified) transport system, ATPase component |
| MG115 | | conserved hypothetical | R | | | | Uncharacterized protein (competence- and mitomycin-induced) |
| MG120 | MPN259 | conserved hypothetical | R | C | | | Uncharacterized ABC-type transport system, permease component |
| MG121 | MPN260 | conserved hypothetical | R | | | | Uncharacterized ABC-type transport system, permease component |
| MG125 | MPN264 | conserved hypothetical | R | RS | | | Predicted hydrolases of the HAD superfamily |
| MG128 | MPN267 | conserved hypothetical | R | | | | Predicted kinase |
| MG139 | MPN280 | conserved hypothetical | R | | | | Predicted hydrolase of the metallo-beta-lactamase superfamily |
| MG146 | MPN159 | conserved hypothetical | R | | | | Uncharacterized CBS domain-containing proteins |
| MG190 | MPN140 | cell envelope | R | | | | Exopolyphosphatase-related proteins [Adhesin mgpA] |

| (a) | (b) | (c) | (d) | (e) | (f) | (g) | (h) |
|---|---|---|---|---|---|---|---|
| MG195 | MPN106 | protein synthesis | R | J | | 1,2 | Phenylalanyl-tRNA synthetase beta subunit |
| MG248 | MPN351 | conserved hypothetical | R | | | | Predicted SAM-dependent methyltransferase |
| MG265 | MPN383 | conserved hypothetical | R | RS | | | Predicted hydrolases of the HAD superfamily |
| MG275 | MPN394 | energy metabolism | R | C | | | Uncharacterized NAD(FAD)-dependent dehydrogenases |
| MG327 | MPN473 | fatty acid and phospholipid | R | | | | Predicted acetyltransferases and hydrolases with the alpha/beta hydrolase fold |
| MG329 | MPN475 | conserved hypothetical | R | | | | Predicted GTPases |
| MG335 | MPN481 | conserved hypothetical | R | RS | g1 | | Predicted GTPases |
| MG342 | MPN517 | conserved hypothetical | R | | | | Predicted flavoprotein |
| MG344 | MPN519 | fatty acid and phospholipid | R | | | | Predicted acetyltransferases and hydrolases with the alpha/beta hydrolase fold |
| MG347 | MPN522 | conserved hypothetical | R | RS | | | Predicted S-adenosylmethionine-dependent methyltransferase |
| MG369 | MPN547 | conserved hypothetical | R | | | | Predicted kinase related to dihydroxyacetone kinase |
| MG371 | MPN549 | conserved hypothetical | R | | | | Exopolyphosphatase-related proteins |
| MG384 | MPN563 | unkown function | R | RS | | | Predicted GTPase |
| MG387 | MPN568 | unkown function | R | RS | | | GTPases |
| MG388 | MPN569 | conserved hypothetical | R | | | | Predicted metal-dependent hydrolase |
| MG449 | MPN663 | conserved hypothetical | R | O | | 2 | EMAP domain |
| MG461 | MPN677 | conserved hypothetical | R | | | | HD superfamily phosphohydrolases |
| MG468.1 | | transport and binding proteins | R | | | | ABC-type (unclassified) transport system, ATPase component |
| MG105 | MPN244 | conserved hypothetical | S | | | | Uncharacterized ACR |
| MG134 | MPN275 | conserved hypothetical | S | | | | Uncharacterized BCR |
| MG221 | MPN314 | conserved hypothetical | S | RS | | | Uncharacterized BCR |
| MG246 | MPN349 | conserved hypothetical | S | | | | Uncharacterized BCR |
| MG247 | MPN350 | conserved hypothetical | S | RS | | | Predicted membrane protein |
| MG326 | MPN472 | conserved hypothetical | S | | | | Uncharacterized BCR |

| (a) | (b) | (c) | (d) | (e) | (f) | (g) | (h) |
|---|---|---|---|---|---|---|---|
| MG332 | MPN478 | conserved hypothetical | S | RS | | | Uncharacterized ACR |
| MG432 | MPN630 | conserved hypothetical | S | | | | Uncharacterized BCR |
| MG450 | MPN664 | conserved hypothetical | S | | | | Uncharacterized BCR |
| MG108 | MPN247 | protein fate | T | | | | Protein serine/threonine phosphatases |
| MG109 | MPN248 | protein fate | T | | | | Serine/threonine protein kinases |
| MG278 | MPN397 | regulatory functions | T | K | p7 | | Guanosine polyphosphate pyrophosphohydrolases/synthetases |
| MG045 | MPN058 | cell envelope | - | F | p1 | | |
| MG406 | MPN605 | conserved hypothetical | - | C | | | |
| | | | | | | | GROUP 2 GENES |
| MG385 | MPN566 | conserved hypothetical | C | | | g4/p3 | Glycerophosphoryl diester phosphodiesterase |
| MG470 | MPN688 | cellular processes | D | | | g2 | ATPases involved in chromosome partitioning |
| MG183 | MPN197 | protein fate | E | | | g1 | Oligoendopeptidase F |
| MG226 | MPN319 | conserved hypothetical | E | | | g2/p2 | Amino acid transporters |
| MG051 | MPN064 | purines, pyrimidines, etc. | F | | | g1 | Thymidine phosphorylase |
| MG132 | MPN273 | unkown function | F | | | g6 | Diadenosine tetraphosphate (Ap4A) hydrolase and other HIT family hydrolases |
| MG062 | MPN078 | transport and binding proteins | G | | | g4/p1 | Phosphotransferase system fructose-specific component IIB |
| MG011 | MPN015 | conserved hypothetical | H | | | g8 | Glutathione synthase/Ribosomal protein S6 modification enzyme |
| MG213 | MPN300 | conserved hypothetical | H | | | g1 | Dihydrofolate reductase |
| MG372 | MPN550 | conserved hypothetical | H | | | g1/p1 | Thiamine biosynthesis ATP pyrophosphatase |
| MG116 | MPN255 | conserved hypothetical | I | | | p4 | 4-diphosphocytidyl-2-methyl-D-erithritol synthase |
| MG370 | MPN548 | conserved hypothetical | J | | | g5 | Pseudouridylate synthases, 23S RNA-specific |
| MG010 | MPN014 | conserved hypothetical | L | | | g1 | DNA primase (bacterial type) |
| MG140 | MPN153 | conserved hypothetical | L | | | g4/p1 | Superfamily I DNA and RNA helicases and helicase subunits |
| MG186 | MPN133 | cell envelope | L | | | g2 | Micrococcal nuclease (thermonuclease) homologs |
| MG308 | MPN443 | transcription | L | | | g1 | Superfamily II DNA and RNA helicases |
| MG438 | MPN638 | conserved hypothetical | L | | | g2/p1 | Restriction endonuclease S subunits |

| (a) | (b) | (c) | (d) | (e) | (f) | (g) | (h) |
|---|---|---|---|---|---|---|---|
| MG025 | MPN028 | conserved hypothetical | M | | g1 | | Glycosyltransferases involved in cell wall biogenesis |
| MG137 | MPN278 | cell envelope | M | | p2 | | UDP-galactopyranose mutase |
| MG040 | MPN052 | cell envelope | N | | p1 | | Surface lipoprotein |
| MG055 | MPN068 | conserved hypothetical | N | | p1 | | Preprotein translocase subunit SecE |
| MG002 | MPN002 | unknown function | O | | g1 | | Molecular chaperones (contain C-terminal Zn finger domain) |
| MG208 | MPN291 | conserved hypothetical | O | | p2 | | Inactive homologs of metal-dependent proteases, putative molecular chaperones |
| MG236 | MPN329 | conserved hypothetical | P | | p4 | | Fe2+/Zn2+ uptake regulation proteins |
| MG412 | MPN611 | cell envelope | P | | g3 | | ABC-type phosphate transport system, periplasmic component |
| MG014 | MPN018 | transport and binding proteins | Q | | g1 | | ABC-type multidrug/protein/lipid transport system, ATPase component |
| MG390 | MPN571 | transport and binding proteins | Q | | g1/p2 | | ABC-type bacteriocin/lantibiotic exporters, contain an N-terminal double-glycine peptidase domain |
| MG029 | MPN032 | conserved hypothetical | R | | g5 | | Putative intracellular protease/amidase |
| MG039 | MPN051 | conserved hypothetical | R | | p2 | | Predicted dehydrogenase |
| MG110 | MPN249 | conserved hypothetical | R | | g1 | | Predicted GTPases |
| MG130 | MPN269 | conserved hypothetical | R | | g1 | | Predicted HD superfamily hydrolase |
| MG207 | MPN126 | conserved hypothetical | R | | g4/p1 | | Predicted phosphoesterase |
| MG263 | MPN381 | conserved hypothetical | R | | p2 | | Predicted hydrolases of the HAD superfamily |
| MG306 | MPN435 | conserved hypothetical | R | | p1 | | Predicted integral membrane protein |
| MG310 | MPN445 | fatty acid and phospholipid | R | | g5 | | Predicted acetyltransferases and hydrolases with the alpha/beta hydrolase fold |
| MG316 | MPN451 | conserved hypothetical | R | | g3 | | Predicted multitransmembrane, metal-binding protein |
| MG352 | MPN528a | conserved hypothetical | R | | g1 | | Penicillin-binding protein-related factor A, putative recombinase |
| MG423 | MPN621 | conserved hypothetical | R | | g1 | | Predicted hydrolase of the metallo-beta-lactamase superfamily |
| MG442 | MPN656 | conserved hypothetical | R | | g1 | | Predicted GTPases |
| MG467 | MPN683 | transport and binding proteins | R | | g1 | | ABC-type (unclassified) transport system, ATPase component |
| MG103 | MPN241 | conserved hypothetical | S | | g1 | | Uncharacterized BCR |

| (a) | (b) | (c) | (d) | (e) | (f) | (g) | (h) |
|-----|-----|-----|-----|-----|-----|-----|-----|
| MG443 | MPN657 | conserved hypothetical | S | | p1 | | Uncharacterized BCR |
| MG085 | MPN223 | regulatory functions | T | | g1 | | Serine kinase of the HPr protein, regulates carbohydrate metabolism |
| MG016 | | conserved hypothetical | - | | | | |
| MG017 | | conserved hypothetical | - | | g2 | | |
| MG028 | MPN031 | conserved hypothetical | - | | | | |
| MG032 | MPN042 | conserved hypothetical | - | | g5 | | |
| MG055.2 | MPN070 | conserved hypothetical | - | | | | |
| MG064 | MPN080 | conserved hypothetical | - | | | | |
| MG067 | MPN083 | cell envelope | - | | g3 | | |
| MG068 | MPN084 | cell envelope | - | | | | |
| MG074 | MPN212 | conserved hypothetical | - | | p1 | | |
| MG075 | MPN213 | | - | | | | |
| MG076 | MPN214 | conserved hypothetical | - | | | | |
| MG095 | MPN233 | cell envelope | - | | | | |
| MG096 | MPN234 | conserved hypothetical | - | | g4 | | |
| MG101 | MPN239 | conserved hypothetical | - | | | | |
| MG117 | MPN256 | conserved hypothetical | - | | | | |
| MG123 | MPN262 | conserved hypothetical | - | | | | |
| MG131 | | hypothetical - not conserved | - | | g1 | | |
| MG133 | MPN274 | conserved hypothetical | - | | | | |
| MG135 | MPN276 | conserved hypothetical | - | | | | |
| MG144 | MPN157 | conserved hypothetical | - | | | | |
| MG147 | MPN160 | conserved hypothetical | - | | | | |
| MG148 | MPN161 | conserved hypothetical | - | | | | |
| MG149 | MPN162 | cell envelope | - | | g1 | | |
| MG149.1 | | conserved hypothetical | - | | | | |
| MG184 | MPN198 | dna metabolism | - | | | | |

| (a) | (b) | (c) | (d) | (e) | (f) | (g) | (h) |
|-----|-----|-----|-----|-----|-----|-----|-----|
| MG185 | MPN199 | cell envelope | - | | g1 | | |
| MG191 | MPN141 | cell envelope | - | | g1/p1 | | [Adhesin P1 (mgpB)] |
| MG192 | MPN142 | cell envelope | - | | g1/p2 | | [mgpC protein – adhesin operon] |
| MG202 | MPN121 | conserved hypothetical | - | | g1/p1 | | |
| MG210.1 | MPN295 | conserved hypothetical | - | | | | |
| MG211 | MPN297 | conserved hypothetical | - | | | | |
| MG217 | MPN309 | cell envelope | - | | | | |
| MG218 | MPN310 | cell envelope | - | | p4 | | [cytadherence accessory protein (hmw2)] |
| MG218.1 | MPN311 | conserved hypothetical | - | | | | |
| MG219 | | hypothetical - not conserved | - | | | | |
| MG220 | MPN313 | conserved hypothetical | - | | | | |
| MG223 | MPN316 | conserved hypothetical | - | | | | |
| MG237 | MPN330 | conserved hypothetical | - | | g1/p1 | | |
| MG241 | MPN337 | conserved hypothetical | - | | p1 | | |
| MG242 | MPN338 | conserved hypothetical | - | | | | |
| MG243m | MPN339 | conserved hypothetical | - | | | | |
| MG255 | MPN358 | conserved hypothetical | - | | g4 | | |
| MG255.1 | | conserved hypothetical | - | | | | |
| MG256 | MPN359 | conserved hypothetical | - | | g2 | | |
| MG260 | MPN364 | cell envelope | - | | | | |
| MG267 | MPN385 | conserved hypothetical | - | | | | |
| MG269 | MPN387 | conserved hypothetical | - | | g40 | | |
| MG277 | MPN396 | conserved hypothetical | - | | | | |
| MG279 | MPN398 | conserved hypothetical | - | | g8/p3 | | |
| MG280 | MPN399 | conserved hypothetical | - | | g10/p13 | | |
| MG281 | MPN400 | conserved hypothetical | - | | g28/p5 | | |
| MG284 | MPN403 | conserved hypothetical | - | | p1 | | |

| (a) | (b) | (c) | (d) | (e) | (f) | (g) | (h) |
|---|---|---|---|---|---|---|---|
| MG285 | MPN404 | conserved hypothetical | - | | g3 | | |
| MG286 | MPN405 | conserved hypothetical | - | | g1 | | |
| MG288m | | conserved hypothetical | - | | g4 | | |
| MG289 | MPN415 | transport and binding proteins | - | | | | |
| MG291 | MPN417 | transport and binding proteins | - | | g1 | | |
| MG296 | MPN423 | conserved hypothetical | - | | g2 | | |
| MG307 | MPN436 | cell envelope | - | | p1 | | |
| MG309 | MPN444 | cell envelope | - | | | | |
| MG312 | MPN447 | cell envelope | - | | | | [cytadherence accessory protein (hmw1)] |
| MG313 | MPN448 | conserved hypothetical | - | | | | |
| MG314 | MPN449 | conserved hypothetical | - | | | | |
| MG317 | MPN452 | cell envelope | - | | g1 | | [cytadherence accessory protein (hmw3)] |
| MG318 | MPN453 | cell envelope | - | | | | |
| MG319 | MPN454 | conserved hypothetical | - | | | | |
| MG320 | MPN455 | conserved hypothetical | - | | p1 | | |
| MG320.1 | | hypothetical - not conserved | - | | g1 | | |
| MG321 | MPN456 | cell envelope | - | | | | |
| MG323.1 | MPN469 | conserved hypothetical | - | | g1 | | |
| MG328 | MPN474 | conserved hypothetical | - | | g9 | | |
| MG331 | MPN477 | conserved hypothetical | - | | | | |
| MG335.1 | | hypothetical - not conserved | - | | | | |
| MG338 | MPN489 | cell envelope | - | | g1 | | |
| MG343 | MPN518 | conserved hypothetical | - | | g1 | | |
| MG348 | MPN523 | cell envelope | - | | | | |
| MG349 | MPN525 | conserved hypothetical | - | | | | |
| MG350 | MPN526 | conserved hypothetical | - | | | | |
| MG350.1 | | conserved hypothetical | - | | | | |

| (a) | (b) | (c) | (d) | (e) | (f) | (g) | (h) |
|---|---|---|---|---|---|---|---|
| MG354 | MPN530 | conserved hypothetical | - | | | | |
| MG364 | MPN542 | conserved hypothetical | - | | | | |
| MG366 | MPN544 | conserved hypothetical | - | | g1 | | |
| MG373 | MPN551 | conserved hypothetical | - | | | | |
| MG374 | MPN552 | conserved hypothetical | - | | | | |
| MG376 | MPN554 | conserved hypothetical | - | | | | |
| MG377 | MPN555 | conserved hypothetical | - | | | | |
| MG381 | MPN559 | conserved hypothetical | - | | | | |
| MG384.1 | | conserved hypothetical | - | | | | |
| MG386 | MPN567 | cell envelope | - | | | | |
| MG389 | MPN570 | conserved hypothetical | - | | p1 | | |
| MG395 | MPN592 | cell envelope | - | | g40 | | |
| MG397 | MPN596 | conserved hypothetical | - | | | | |
| MG414 | MPN612 | conserved hypothetical | - | | g2/p9 | | |
| MG415 | MPN613 | conserved hypothetical | - | | g1/p2 | | |
| MG422 | MPN620 | conserved hypothetical | - | | | | |
| MG427 | MPN625 | conserved hypothetical | - | | | | |
| MG428 | MPN626 | conserved hypothetical | - | | g1 | | |
| MG439 | MPN639 | cell envelope | - | | | | |
| MG440 | MPN643 | cell envelope | - | | g2 | | |
| MG441 | MPN648 | conserved hypothetical | - | | | | |
| MG452 | MPN666 | conserved hypothetical | - | | g2 | | |
| MG456 | MPN670 | conserved hypothetical | - | | | | |
| MG468 | MPN684 | conserved hypothetical | - | | g1 | | |

(a) the *M. genitalium* gene designation.

(b) the orthologous gene in *M. pneumoniae*.

(c) role category for the gene according to the TIGR/DOE Comprehensive Microbial Resource, http://www.tigr.org/tigr-scripts/CMR2/GenomePage3.spl?database=gmg.

(d) functional category according to NCBI COG system, see http://www.ncbi.nlm.nih.gov/cgi-bin/COG/palox?fun=all. Translation, ribosomal structure and biogenesis=J; Transcription=K; DNA replication, recombination and repair=L; Cell division and chromosome partitioning=D; Posttranslational modification, protein turnover, chaperones=O; Cell envelope biogenesis, outer membrane=M; Cell motility and secretion=N; Inorganic ion transport and metabolism=P; Signal transduction mechanisms=T; Energy production and conversion=C; Carbohydrate transport and metabolism=G; Amino acid transport and metabolism=E; Nucleotide transport and metabolism=F; Coenzyme metabolism=H; Lipid metabolism=I; Secondary metabolites biosynthesis, transport and catabolism=Q; General function prediction only=R; Function unknown=S

(e) genes of the minimal gene set of Mushegian and Koonin[19] are indicated by a code for the functional group in which they were placed. Translation = J; Replication = L; Recombination and repair = L2; Transcription = K; Chaperone functions = O; Nucleotide metabolism = F; Amino acid metabolism = E; Lipid metabolism = I; Energy = C; Coenzyme metabolism and utilization = H; Exopolysaccharides = M; Uptake of inorganic ions = P; Secretion, receptors = N; Other conserved proteins = RS.

(f) an entry indicates observation of putatively disruptive Tn4001 insertion(s) in the gene[9]. The number of distinct sites in *M. genitalium* and *M. pneumoniae* (X and Y) at which such insertions have been observed are indicated by gX and/or pY.

(g) an entry indicates that a gene is a member of a universally conserved family[12]. A "1" indicates families with a single representative in *M. genitalium*, and "2" indicates genes for which *M. genitalium* contains two members of a conserved family.

(h) predicted functions are taken from the NCBI COG system, see http://www.ncbi.nlm.nih.gov/cgi-bin/Entrez/coxik?gi=26 for *M. genitalium* COGs. Functions for adhesins and structural proteins studied directly in mycoplasmas are given [in brackets].

# Chapter 11

# Comparative Genome Analysis of the Mollicutes

THOMAS DANDEKAR, BEREND SNEL, STEFFEN SCHMIDT, WARREN LATHE, MIKITA SUYAMA, MARTIJN HUYNEN and PEER BORK
*EMBL, Postfach 102209, D-69012 Heidelberg, Germany*

## 1.    OVERVIEW

The Mollicutes are Eubacteria that have probably been derived from Lactobacilli, Bacilli, and Streptococci by regressive evolution and genome reduction to produce the smallest and simplest free-living and self-replicating cells. The life style is in general parasitic. Structurally, the Mollicutes are characterized by the complete lack of a cell wall, and the presence of an internal cytoskeleton[27, 46].

As in other comparative studies, comparative genomics of Mollicutes requires the availability of a sufficient number of species and genomes to draw solid conclusions[42]. Several Mollicute genomes have been partially sequenced and many will be completely determined in the near future. This includes *Mycoplasma capricolum, Mycoplasma mycoides subsp. mycoides SC* (The Royal institute of Technology, Stockholm; National Veterinary Institute, Uppsala; still in progress) and *Mycoplasma hyopneumoniae* (in progress, University of Washington). However, for comparative genome analysis it is desirable to have these genome sequences complete, well annotated, and available in public databases. Until recently, this did only apply to *Mycoplasma genitalium*[12], *Mycoplasma pneumoniae*[19], and *Ureaplasma urealyticum*[15]. At the time of writing, the genomic analysis of *Mycoplasma pulmonis* became available too[7]. Moreover, the number of total genomes analyzed, and in particular those of prokaryotes has rapidly increased in the last years (see e.g. www.tigr.org/tdb/mdb/mdbinprogress.html for an overview). Table 1a

*Molecular Biology and Pathogenicity of Mycoplasmas*, Edited by Razin and Herrmann, Kluwer Academic/Plenum Publishers, New York, 2002

summarizes some useful comparative genomics WEB pointers for Mollicutes.

*Table 1a.* Some useful Mollicute genome WEB pointers

| | |
|---|---|
| Microbial genomes in progress | www.tigr/tdb/mdb/mdbinprogress.html |
| Annotation for *Escherichia coli* | www.genome.wisc.edu |
| Annotation for *Mycoplasma genitalium* | www.bork.embl-heidelberg.de/Annot/MG |
| Annotation for *Mycoplasma pneumoniae* | www.bork.embl-heidelberg.de/Annot/MP |
| Annotation for *Mycoplasma pulmonis* | Genolist.pasteur.fr/MypuList |
| Annotation for *Ureaplasma urealyticum* | Genome.microbio.uab.edu/uu/uugen.html |
| *M. pneumoniae* Genome and proteome project | www.zmbh.uni-heidelberg.de/M_pneumoniae/MP_HOME.html mail.zmbh.uni-heidelberg.de/M_pneumoniae/genome/Results.html |
| *Mycoplasma genitalium* genome page | www.tigr.org/tigr-scripts/CMR2/GenomePage3.spl?database=gmg |
| Mycoplasma metabolism | www.med.ohio-state.edu/medmicro/pages/jdpollack.htm |
| *Mycoplasma gallisepticum* (veterinary site) | members.aol.com/FinchMG/MGLinks.htm #Consolidated |
| Clusters of orthologous genes | www.ncbi.nlm.nih.gov/COG |
| gene context revealed by STRING | www.bork.embl-heidelberg.de/STRING |
| Global transposon mutagenesis of *M.pneumoniae* and *M.genitalium*[21] | www.sciencemag.org/feature/data/1042937.shtml (supplementary material) |

*Table 1b.* Basic genome parameters

| | size[1] | genes[2] | density[3] | families[4] |
|---|---|---|---|---|
| *Mycoplasma genitalium* | 580,074 | 480 | 0.83 | 393 |
| *Mycoplasma pneumoniae* | 816,394 | 687 | 0.84 | 462 |
| *Ureaplasma urealyticum* | 751,719 | 611 | 0.81 | 429 |
| *Mycoplasma pulmonis* | 963,879 | 782 | 0.81 | 519 |

[1] Given is the size in basepairs
[2] Given is the number of identified protein reading frames
[3] Given is the ratio of genes per kilobase
[4] Given is the number of different protein families. Note that many of these families contain only one protein because no further related protein was found in the Mollicute genome. We use an E-value cut-off of 0.01. Our results are comparable to those of Teichmann *et al.* (1999) who refer to a similar procedure to assist structure predictions.

The position of the Mollicute genomes from *M. pneumoniae, M. genitalium, U. urealyticum,* and *M. pulmonis* in the phylogenetic tree shows the Mollicutes to be a well defined, monophyletic group with the *M. genitalium* and *M. pneumoniae* genomes being closely related sister species, the *U. urealyticum* genome a little bit more distant and *M. pulmonis* having diverged the earliest from the four (Figure 1a-c). Figure 1a shows a genome phylogeny based on shared gene content between completely sequenced genomes. This measure is an alternative to measures that are based on levels of sequence identity and has the advantage that it better reflects the evolution of the whole genome[40]. An alternative to reconstructing genome phylogenies based on gene content, is to reconstruct them on the basis of gene order. As conservation of gene order decreases almost linearly with time over short evolutionary distances[42], this might be appropriate for comparisons within the Mollicutes. Figure 1b shows the genome phylogeny based on shared gene order. Genome phylogenies as the ones above can easily be derived by calculating pairwise distances between all pairs of genomes based on the number of shared genes (or based on the number of conserved gene pairs in the case of gene order). Subsequently, a simple, distance-based phylogenetic inference method is used (such as neighbor-joining) to construct the phylogeny.

Figure 1a shows that for the phylogeny based on gene content, the Mollicute genomes examined are monophyletic and positioned on a branch of the tree within the Gram-positive bacteria. The Gram-positive bacteria themselves are however non-monophyletic, falling apart in the low- and high-GC content branch. However, when using gene order as measure (Figure 1b) all gram-positive bacteria do appear to be monophyletic. In Figure 1c we compare genome phylogenies with the results from a standard ribosomal RNA tree (based on 23 S rRNA). Despite the differences regarding early evolutionary events, all three independent measures confirm that *M.pulmonis* has diverged the earliest of the four Mollicute genomes compared. An overview of basic genome parameters such as size and number of genes is given in Table 1b. For example, the smallest genome, *M.genitalium*, contains 480 genes that fall into 393 gene families. Although there is substantial variation in the number of genes and gene families per genome, gene density is similar, varying between 81 to 84%.

There is a considerable variation among the Mollicute genome sizes. Already early studies[37] with contour clamped homogeneous field (CHEF) agarose gel electrophoresis of DNA showed genome size variation for different serotype standard strains of *Ureaplasma urealyticum* isolates. Genome sizes (in kbp) were 760 for four biotype 1 strains (characterized by temporary inhibition of growth in broth by manganese) and 840-1140 for eleven biotype 2 (permanent growth inhibition by manganese) strains. Other estimates were: 720 for *Mycoplasma hominis*, 1070 for *Mycoplasma hyopneumoniae*, 890 for *Mycoplasma flocculare*, 1180 and 1350 for *Mycoplasma mycoides subsp. mycoides* Y and GC1176-2, respectively, and 1650 and 1580 for *Acholeplasma laidlawii* B and PG 8, respectively[32, 36].

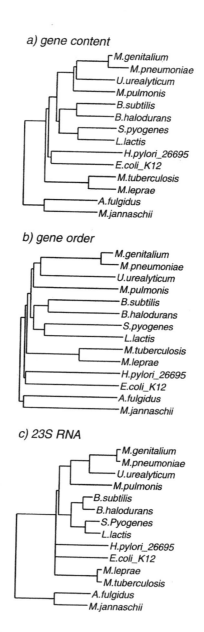

*Figure 1.* Phylogenetic trees for the four well annotated Mollicute genomes (a) drawn according to gene content. (b) drawn according to gene order. (c) based on ribosomal RNA content.

These data illustrate that Mollicute genomes are diverse in size with some larger than estimated from DNA renaturation kinetics. The genome inventory may vary quite significantly between different strains of the same species, as seems to be the case for many prokaryotic genomes in general. This clear genetic diversity in individual Mollicute species has also been confirmed in current studies. Thus *Mycoplasma capricolum subsp. capripneumoniae* strains were analyzed for amplified fragment length polymorphisms (AFLP[25]). AFLP fingerprints of 38 strains derived from different countries in Africa and the Middle East consisted of over 100 bands in the size range of 40-500bp. The similarity between individual AFLP profiles, calculated by Jaccard's coefficient, ranged from 0.92 to 1.0. On the basis of the polymorphisms detected, the analysed strains can explicitly be grouped into two major clusters, equivalent to two evolutionary lines of the species found by 16S rDNA analysis.

Horizontal gene transfer has also to be considered in comparative genomics of prokaryotes[13, 43, 40]. Archaea and nonpathogenic bacteria seem to have the highest percentages of such genes, pathogenic bacteria, except for *Mycoplasma genitalium*, have the lowest. Furthermore, genes involved in transcription and translation are less likely to be transferred than metabolic household genes.

As illustrated in the next section, one has to bear in mind that any sequencing project and its genome annotation provide only a momentary picture of the genome. First, every species is evolving in time, caused by adaptation and other selective processes (e.g. rapid co-evolution of host and parasite) or by neutral evolution. Even if genome sequences would be static, our knowledge on those is evolving, as are the methods to analyse them.

## 2. EVOLVING MOLLICUTE ANNOTATION: THE *M. PNEUMONIAE* EXAMPLE

The Mollicute genomes contain all basic features of a living cell and many specific modifications, though most of these genomes are quite compact and small. The genome of *M.genitalium*[12] was the first one to be sequenced and published, followed a few months later by *M.pneumoniae*[19]. The second genome was also extremely valuable for the annotation of the first one[18] as the tools for identifying functional features are not perfect. Even when reassessing the functional information available for the very same genome, numerous differences become apparent. For example, the *Mycoplasma pneumoniae* genome was meticulously re-annotated four years after the original publication to incorporate novel data applying latest software[8]. The total number of identified ORFs increased from 677 to 688: Ten new

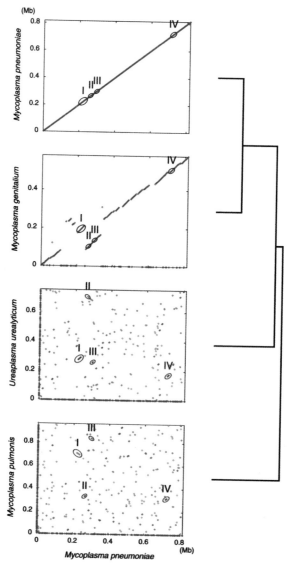

*Figure 2.* Dotplots of orthologous genes in four genome pairs. *M.pneumoniae* is always in the horizontal axis in the four panels. The top panel is the comparison with itself. Directional similarity is indicated by colors: green, pairs of genes with the same direction; red, those with opposite directions. The open reading frames (ORFs) without significant similarity to the other compared genome, even in local DNA sequence level, are defined as the species specific ORFs and indicated by blue dots on each axis. Four syntenic regions are marked with ellipses with Roman numerals; I, II, III, and IV (see Table 3).

proteins were predicted in intergenic regions, and two were newly identified by mass-spectroscopy. One protein ORF was discarded. The predicted number of RNAs was increased from 39 to 42 genes. For 19 of the now 35 tRNAs and for six other functional RNAs the exact genome positions were re-annotated, two new Leu tRNAs and a small 200nt RNA were identified, 16 protein reading frames were extended and 8 shortened. A consistent annotation vocabulary was introduced and annotation reasoning, categories and comparisons to other published data on protein function assignments were given. Experimental evidence included 2D gel chromatography in combination with mass-spectroscopy and new gene expression data. Compared to the original annotation, this study increased the number of proteins with predicted functional features from 349 to 458. This involved 36 new predictions and 73 protein assignments confirmed by published literature. 23 previous annotations were to broad in their definitions and had to be reduced to better describe the function of the encoded protein. For 30 protein genes we found additions to their predicted function compared to the previous annotation. mRNA expression data support transcription of 184 of the functionally unassigned reading frames. Proteins missing in the first annotation were identified such as the subunit A of glutamine-tRNA amidotransferase.

**Reannotation examples** include:

♦ **Molecular functions** for several proteins were clarified (in the following MPN denotes the new numbering, MP the original numbering of the *M.pneumoniae* genome[19]. An example are MPN558(MP284) and MPN557(MP285), previously annotated as glucose-inhibited cell-division proteins B and A, which are a methyltransferase (MPN558(MP284)) and an NADH-oxidoreductase (MPN557(MP285)), and that have homologs with known structure (1BHJ and 1FEA, respectively).

♦ In 36 cases the functional assignment was completely new. An example is the **protein secretion system** in *M.pneumoniae* (Figure 3). The system has been well characterized in *Escherichia coli*. Cytosolic chaperons or regulators (trigger factor, SecB, DnaK, bacterial signal recognition particle and FtsY) deliver the protein to a membrane transporter (SecA). The receptor should also function as a motor to push the protein across the membrane via specific protein channels (SecY, SecG, SecE, SecD and SecF). Himmelreich *et al.* (1996) noted that they had identified trigger factor, DnaK, SRP and FtsY as well as SecA, whereas from the channel-forming proteins only SecY could be assigned, leaving the secretion pathway incomplete. Protein reading frames similar to SecD, SecE and SecG have now been identified, yielding a new and more complete picture of this secretory pathway in

*M.pneumoniae.* Since several pathogenicity factors are secreted, the respective protein channels are potential drug targets. No homologous sequence has been found for SPase I in the secretory pathway in *M.pneumoniae.* SPase I would cleave the signal peptide before secretion. Suitable cleavage sites have been identified for several *M.pneumoniae* proteins. One of the proteases identified may contain this function, e.g. the new annotated intracellular protease MPN386(MP542). The dihydroxyacetone kinase domain from MPN547(MP295) could yield ATP for *M.pneumoniae* by transforming dihydroxyacetone phosphate and ADP into dihydroxyacetone and ATP. The predicted activity can be metabolically connected to the phospholipid metabolism in *Mycoplasma pneumoniae* and the necessary supply of dihydroxyacetone phosphate via MPN051(MP103) (glycerol 3-phosphate dehydrogenase reading frame, confirmed in re-annotation).

♦ **Carbohydrate metabolizing operons** were known previously for fructose (MPN078(MP077); MPN079(MP076)) and mannitol (MPN651(MP191) to MPN653(MP189)). It is now apparent that Ribulose is transported (MPN496(MP346), MPN494(MP347)) and channeled via D-arabinose 6-hexulose 3-phosphate synthase (MPN493(MP348)) and D-arabinose 6-hexulose 3-phosphate isomerase MPN492(MP349) into fructose 6-phosphate and glycolysis.

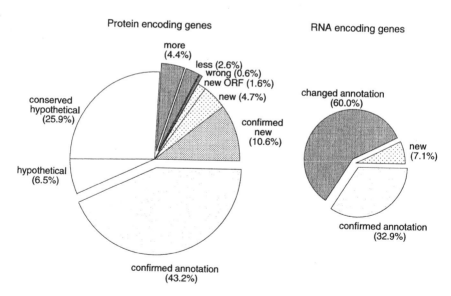

*Figure 3.* Annotation gain after re-annotation in *M.pneumoniae.* Different re-annotation categories are given and represented according to size. Both the re-annotation of proteins and of RNAs are compared.

# 3. EVOLVING METHODS FOR MOLLICUTE GENOME ANALYSIS: THE *M.GENITALIUM* EXAMPLE

Mollicute genome annotation heavily relies on comparative sequence analysis methods. Due to their compactness Mollicute genomes have been frequently used as a benchmark for method development. This, in turn, adds further to our knowledge of the Mollicute genomes.

For example, the *M.genitalium* genes have been used as a benchmark for three-dimensional proteins structure prediction[11]. A dramatic sensitivity and selectivity increase could be shown when iterative homology search techniques have been applied to the same genome[23, 38]. Other groups developed methods for this purpose further and also used *M.genitalium* as benchmark. This led to the estimate that currently for more than 50% of all the proteins encoded in *M.genitalium*, reliable assignments to domains with known three dimensional structure can be made (cf. reviews of Teichmann[45, 44]).

Genome analysis of *M.pneumoniae* and *M.genitalium* is also a good test case for examining the predictive power of the so-called genome context methods to predict functional interactions of encoded proteins (Table 2). Various qualitatively new methods (independent of homology based assignments) have been recently proposed to predict functional interactions between proteins based on the genomic context of their genes. Important types of genomic context that one can examine are (i) the fusion of genes; (ii) the conservation of gene-order or co-occurrence of genes in potential operons and (iii) the co-occurrence of genes across genomes (phylogenetic profiles). These types were compared by Huynen *et al.* (2000). Their coverage, their correlations with various types of functional interaction, and their overlap with homology-based function assignment was analysed in *M.genitalium.* Quantitatively, conservation of gene order is the technique with the highest coverage, applying to 37% of the genes. By combining gene order conservation with gene fusion (6%), the co-occurrence of genes in operons in the absence of gene order conservation (8%), and the co-occurrence of genes across genomes (11%), significant context information can be obtained for 50% of the genes (the categories overlap). Qualitatively, it was observed in this computational analysis of *M.genitalium* that the functional interactions between genes became stronger as the requirements for physical neighbourhood were set more stringent, while the fraction of potential false positives decreased. In cases in which gene order was conserved in a substantial fraction of the genomes (equal or more than in six out of twenty-five genomes) a single type of functional interaction clearly dominated, namely physical interaction (>80%). In the other cases, complementary function information from homology searches, available for most of the genes with significant genomic context, remained essential for prediction of the type of interaction. Using a combination of genome context and homology searches, new functional features could be predicted for 10% of the *M. genitalium* genes.

Table 2. *M. genitalium* genes for which genomic context adds functional information[1]

| Protein | conserved with | proposed role |
|---------|---------------|---------------|
| MG008 | MG466 (rib. Protein L34) | translation |
| MG009 | MG006 (thymidilate kinase) | nucleotide. metabolism (dCTPase?) |
| MG053 | MG052 (cytidine deaminase) | phosphoribomutase |
| MG134 | MG240 (dnaX), recR | physical interaction with DNA pol. gamma subunits |
| MG233 | MG232, MG234 (rib. proteins) | ribosomal protein |
| MG464 | MG465 (ribonul. P component) | ribonuclease |

[1] In some cases some information about the molecular function can already be retrieved by homology searches: MG008 encodes the so-called thiophene and furan oxidation protein. Its genomic association with ribosomal protein L34 indicates a role in translation. The inability of a species to oxidize thiophene or furan in the absence of MG008 might be a secondary effect, caused by the inability to translate an mRNA into the protein required for the oxidation. MG009 is homologous to deaminases, dehydratases and phosphohydrolases that generally have pyrimidines as substrate. Its genomic link with thymidilate kinase indicates a role in nucleotide metabolism, possibly in the creation of a precursor of thymidilate. MG052 is homologous to phosphohexomutases, and is generally annotated as a phosphomannomutase. The location of MG053 in a nucleoside salvage pathway operon from which deoB, a phosphoribomutase, is missing, indicates that MG053 might have acquired the phosphoribose as substrate. Note that in the genome-based metabolic map of *M. pneumoniae*[19] a phosphohexomutase is however still required to fill the gap between glucose-1-phosphate and glucose-6-phosphate. There is no other candidate for this function than MG053, which leads us to propose that MG053 has two substrate specificities: glucose-1,6-phosphate and deoxy-ribose-1,5-phosphate. Also for proteins for which no information can be derived by homology searches, information can be gained by context searches. MG134 is a hypothetical protein that tends to occur with MG240 and recR, coding for DNA polymerase subunit gamma/tau and a protein that is involved in recombinational repair respectively. The high frequency of occurrence with dnaX and recR indicate a physical interaction between the proteins. MG464 encodes an inner membrane protein that has a strong genomic link with the ribonuclease P protein component, indicating a role as a ribonuclease, or possibly in translation.

**As an example** may serve the two *Mycoplasma genitalium* proteins MG246 and MG130 that do not have orthologs with experimentally determined functions. Information about their role in the cell can be gained by a combination of homology searches and genome context searches. Using sensitive, profile-based homology searches (PSI-BLAST[1]), MG246 can be shown to be homologous to the catalytic domain of 5-prime nucleotidase from *Escherichia coli* (UshA)[22]. It is a good candidate for the nucleotidase activity that has been measured in *M. genitalium*[17] but for which no gene has yet been identified. MG130 contains a KH single-stranded ribonucleotide-binding domain (protein domain which is homologous to hnRNP-K)[30] and an phosphohydrolase domain (HD-domain) that hydrolyses phosphates from nucleotides[2]. The functions of these proteins can be linked to each other and to other proteins in the cell using the conservation of genomic context (Figure 5). MG246 and MG130 tend to occur in potential operons with each other, and with 1) MG244, a type II DNA helicase orthologous to the PcrA helicase of *Bacillus subtilis* that is involved in DNA repair and in rolling circle replication. 2) 5-formyl tetrahydrofolate cyclo-ligase, involved in the synthesis of tetrahydrofolate (a co-factor in nucleotide metabolism 3) recA, a single stranded DNA binding protein involved in DNA repair, 4) cinA, a competence-induced protein and 5) phosphatidylglycero-phosphate synthase. The homology information indicates that MG246 and MG130 play a role in nucleotide metabolism. The unifying theme in the context information is that of DNA repair. One possibility would be that MG246 and MG130 are involved in the degradation of nucleotides that are removed in DNA repair.

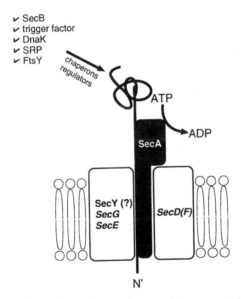

*Figure 4.* Identification of complete pathways after re-annotation. Several missing components of the secretion system in *M. pneumoniae* could be identified after re-annotation of the *M. pneumoniae* genome[8]. The original genome annotation[19] identified the components shown in black. The new genome analysis added the components shown in grey.

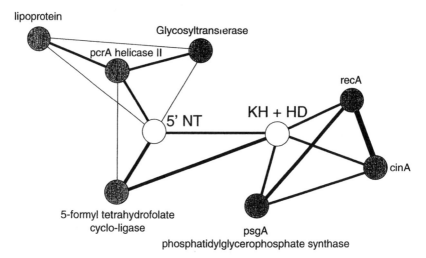

*Figure 5.* The functions of two hypothetical *Mycoplasma genitalium* proteins MG130 (containing a KH and a HD domain) and MG246 (homologous to 5' nucleotidase) can be related to each other and to a number of other Mycoplasma and Ureaplasma proteins[22]. Specifically to MG244 (a helicase II), MG245 (5-formyl tetrahydrofolate cyclo-ligase, MG339 (recA), MG115 (cinA) and MG114 (pgsA phosphatidylglycerophosphate synthase). The thickness of the lines indicates the number of times the genes neighbour each other on a genome. See the text for further information.

## 4.     COMPARATIVE GENOME ANALYSIS TO DETECT FEATURES OF SPECIFIC MOLLICUTES

The previous section showed that comparative genome analysis such as using the conservation of genome context is a powerful tool in the analysis of Mollicute reading frames and in predicting previously unknown structures or functions. This section will now focus on genome specific features apparent from comparative analysis of Mollicutes.

An initial genome analysis[6] of 214kb from *Mycoplasma capricolum* detected 287 putative proteins representing about a third of the estimated total. A large fraction of these (75%) could be assigned a likely function as a result of homology searches. There is a relatively large number of enzymes involved in metabolic transport and activation suitable for efficient use of host cell nutrients. Bork *et al.* (1995) found in addition an unexpected diversity of enzymes involved in DNA replication and repair and several anabolic enzymes. A sizeable number of orthologs in *E. coli* was identified (82, corresponding to 28.6% of the proteins analysed).

The subsequent sequencing and analysis of the genomes of *Mycoplasma genitalium* and *Mycoplasma pneumoniae* made it possible to define the essential functions of a self-replicating minimal cell and the specific features of a Mollicute. Glass *et al.* (2000) subsequently reported the complete sequence (751,719bp) of *Ureaplasma urealyticum* (parvum biovar), another mucosal human pathogen. It is a common commensal of the urogenital tract but it can cause opportunistic infections, e.g. during pregnancy. Differential genome analysis becomes more powerful considering these three well-annotated public genomes together. Several features make *U. urealyticum* unique among Mollicutes and all bacteria. Almost all ATP synthesis in this organism is the result of urea hydrolysis, which generates an energy-producing electrochemical gradient. Some highly conserved eubacterial enzymes appear not to be encoded by *U. urealyticum*, including the cell-division protein FtsZ, chaperonins GroES and GroEL, and ribonucleoside-diphosphate reductase. *U. urealyticum* has six closely related iron transporters, which apparently arose through gene duplication, suggesting that it has a kind of respiratory system not present in other small bacteria. The genome is only 25.5% G+C in nucleotide content, and the G+C content of individual genes may be used to predict how essential those genes are to Ureaplasma survival. There are 613 *U. urealyticum* protein-coding genes[15], only 324 are homologous to *M. genitalium* genes or *M. pneumoniae* genes. No function could be predicted for 77 of the genes shared by all three Mollicutes; for a list see genome.microbio.uab.edu/uu.

Using rapid automated procedures for genome comparisons unique and specific genes for this Mollicute genome triplet or quartet (considering then in addition *M .pulmonis*, see below) but not present in 49 other genomes (Eucaryota, Archaea, Eubacteria with known genome sequence) are calculated and shown in Table 3. Furthermore, some Mollicute specific conserved small gene clusters become apparent (Table 4). This interesting set of genes should be further characterized experimentally, their unique nature makes it difficult (legend to Figure 3) to deduce their function by sequence comparisons alone.

Figure 2 shows dotplot comparisons of the orthologous genes in all four Mollicute genomes using always *M. pneumoniae* as the horizontal axis. The tree shown, based on 263 orthologues shared between all four genomes confirms the Mollicute phylogeny found in Figure 1a-c by different methods.

*Table 3.* Synteny regions in four compared Mollicute genomes[1]

| Group I. Ribosomal proteins | | | | |
|---|---|---|---|---|
| MPN164 | MG150 | UU232 | MYPU_5900 | S10 |
| MPN165 | MG151 | UU233 | MYPU_5890 | L3 |
| | | | MYPU_5880 | (unknown ORF) |
| MPN166 | MG152 | UU234 | MYPU_5870 | L4 |
| MPN167 | MG153 | UU235 | MYPU_5860 | L23 |
| MPN168 | MG154 | UU236 | MYPU_5850 | L2 |
| MPN169 | MG155 | UU237 | MYPU_5840 | S19 |
| MPN170 | MG156 | UU238 | MYPU_5830 | L22 |
| | | | MYPU_5820 | (unknown ORF) |
| MPN171 | MG157 | UU239 | MYPU_5810 | S3 |
| MPN172 | MG158 | UU240 | MYPU_5800 | L16 |
| MPN173 | MG159 | UU241 | MYPU_5790 | L29 |
| MPN174 | MG160 | UU242 | MYPU_5780 | S17 |
| MPN175 | MG161 | UU243 | MYPU_5770 | L14 |
| MPN176 | MG162 | UU244 | MYPU_5760 | L24 |
| MPN177 | MG163 | UU245 | MYPU_5750 | L5 |
| MPN178 | MG164 | UU246 | MYPU_5740 | S14 |
| MPN179 | MG165 | UU247 | MYPU_5730 | S8 |
| MPN180 | MG166 | UU248 | MYPU_5720 | L6 |
| MPN181 | MG167 | UU249 | MYPU_5710 | L18 |
| MPN182 | MG168 | UU250 | MYPU_5700 | S5 |
| MPN183 | MG169 | UU251 | MYPU_5690 | L15 |
| MPN184 | MG170 | UU252 | MYPU_5680 | SecY |
| MPN185 | MG171 | UU253 | MYPU_5670 | adenylate kinase |
| MPN186 | MG172 | UU254 | MYPU_5660 | methionine amino peptidase |
| MPN187 | MG173 | UU255 | MYPU_5650 | initiation factor 1 |
| MPN188 | MG174 | UU256 | MYPU_5640 | L36 |
| MPN189 | MG175 | UU257 | MYPU_5630 | S13 |
| MPN190 | MG176 | UU258 | MYPU_5620 | S11 |
| MPN191 | MG177 | UU259 | MYPU_5610 | RNA polymerase alpha core subunit |
| MPN192 | MG178 | UU260 | MYPU_5600 | L17 |

| Group II (oligopeptide transporter) | | | | |
|---|---|---|---|---|
| MPN215 | MG077 | UU568 | MYPU_2830 | oligopeptide transport system permease |
| MPN216 | MG078 | UU567 | MYPU_2840 | oligopeptide transport system permease |
| MPN217 | MG079 | UU566 | MYPU_2850 | Oligopeptide transport ATP-binding protein |
| MPN218 | MG080 | UU565 | MYPU_2860 | oligopeptide transport ATP-binding |

| Group III (kinase region) | | | | |
|---|---|---|---|---|
| MPN246 | MG107 | UU214 | MYPU_6870 | guanylate kinase |
| | | UU215 | | (conserved hypothetical) |
| MPN247 | MG108 | UU216 | MYPU_6860 | protein phoshatase |
| MPN248 | MG109 | UU217 | MYPU_6850 | Ser/Thr protein kinase |
| MPN249 | MG110 | UU218 | MYPU_6840 | conserved hypothetical |

| Group III (kinase region) | | | | |
|---|---|---|---|---|
| Group IV (ATP synthase) | | | | |
| MPN597 | MG398 | UU127 | MYPU_2650 | ATP synthase ε-chain |
| MPN598 | MG399 | UU128 | MYPU_2660 | ATP synthase β-chain |
| MPN599 | MG400 | UU129 | MYPU_2670 | ATP synthase γ-chain |
| | | UU130 | | (unique hypothetical) |
| MPN600 | MG401 | UU131 | MYPU_2680 | ATP synthase α-chain |
| | | UU132 | | ATP synthase δ-chain (C-term) |
| | | | MYPU_2690 | ATP synthase δ-chain |
| MPN601 | MG402 | UU133 | | ATP synthase δ-chain |
| MPN602 | MG403 | UU134 | MYPU_2700 | ATP synthase B chain |
| MPN603 | MG404 | UU135 | MYPU_2710 | ATP synthase C chain |
| MPN604 | MG405 | UU136 | MYPU_2720 | ATP synthase A chain |

[1]MPN – *M. pneumoniae*; MG – *M. genitalium*; UU – *U. urealyticum*;
MYPU – *M. pulmonis*

Furthermore, readily apparent from this plot for all four genomes are the conserved regions for ribosomal proteins and three other synteny regions (*Table 3*): Oligopeptide transporter synteny region, ATP synthase coding region and an interesting regulatory locus with kinases, a phosphatase and a conserved hypothetical protein. *U. urealyticum* is less conserved in genome order than are *M. pneumoniae* and *M. genitalium* amongst each other, *M. pulmonis* is even more distant (see Figure 2). Thus no other big synteny regions stand out. Genome specific genes are shown as blue circles on the axis, but are not shown in the *M. pneumoniae* self comparison. These are also not shown for *M. genitalium* as it contains no open reading frames without similarity to *M. pneumoniae*[20]. *M. genitalium* has been described as the smallest living organism. However, there are 74 genes in *M. genitalium* that do not have orthologues in *U. urealyticum*. Ten of these genes concern energy metabolism[15], which is probably a result of the adaptation of *U. urealyticum* to produce ATP by urea hydrolysis. This is simpler than glycolysis in *M. genitalium*. Global transposon mutagenesis in *M. genitalium*[21] showed that 129 of the 480 protein encoding genes from *M. genitalium* were not essential: Any of them could be deleted without causing death of the cell. However, one has to stress that gene deletion combinations were not tested and may be lethal in many cases. From the 351 genes remaining, 265 were found to be with high probability essential. In *U. urealyticum* there are 69 proteins homologous to the unessential genes from *M. genitalium* and there are 255 of the possibly essential genes also contained in *U. urealyticum*. Moreover, 289 genes from *U. urealyticum* have no homologues in *M. genitalium*. *Mycoplasma pulmonis* causes respiratory disease. It is now[7] the fourth pathogenic Mollicute whose sequence is available in a well annotated form. The strain sequenced, UAB CTIP is

*Thomas Dandekar et al.*

composed of a single circular chromosome (length: 963 879bp) with a very low GC content (26.6 mol%), in the same range as *Ureaplasma urealyticum*.

*Table 4.* Unique Mollicute genes

(a)Unique genes in the Mollicute Triplet *U. urealyticum*, *M. genitalium* and *M. pneumoniae*.

| | | |
|---|---|---|
| RMG055 | MPN068 | UU580 |
| MG068 | MPN084 | UU045 |
| MG144 | MPN157 | UU322 |
| MG149_1 | MPN163 | UU092 |
| MG255 | MPN358 | UU572 |
| MG284 | MPN403 | UU352 |
| MG286 | MPN405 | UU505 |
| MG306 | MPN435 | UU450 |
| MG350_1 | MPN527 | UU122 |
| MG377 | MPN555 | UU068 |
| MG384_1 | MPN565 | UU462 |
| MG397 | MPN596 | UU292 |
| MG423 | MPN621 | UU509 |

(b) mollicutes specific clusters[1]

Extra results from 2 iteration(s); converged after 1 iterations

| | | | |
|---|---|---|---|
| 1 | MG046 | MPN059 | UU411 |
| 2 | MG047 | MPN060 | UU412 |

Extra results from 2 iteration(s); converged after 1 iterations

| | | | |
|---|---|---|---|
| 1 | MG127 | MPN266 | UU176 |
| 2 | MG128 | MPN267 | UU177 |
| 3 | MG129 | MPN268 | UU178 |

Extra results from 2 iteration(s); converged after 1 iterations

| | | | |
|---|---|---|---|
| 1 | MG313 | MPN448 | UU330 |
| 2 | MG315 | MPN450 | UU328 |

Extra results from 2 iteration(s); converged after 1 iterations

| | | | |
|---|---|---|---|
| 1 | MG324 | MPN470 | UU532 |
| 2 | MG325 | MPN471 | UU533 |

Extra results from 2 iteration(s); converged after 1 iterations

| | | | |
|---|---|---|---|
| 1 | MG371 | MPN549 | UU417 |
| 2 | MG372 | MPN550 | UU418 |
| 3 | MG373 | MPN551 | UU419 |

[1] These conserved clusters were found by carrying out a large scale STRING analysis[39] on the whole *M. genitalium* genome where we required that the whole detected cluster only occurred in the Mollicutes and in none of the other 40 complete genomes (in contrast to the larger synteny regions shown in Figure 2 which are found also in other genomes).

(c) Unique genes in the Mollicute quartet: three genomes above and in addition that of *M. pulmonis*)

| | | | |
|---|---|---|---|
| MG068 | MPN581 | MYPU_5040 | UU045 |
| MG350_1 | MPN527 | MYPU_5910 | UU122 |
| MG423 | MPN621 | MYPU_7050 | UU509 |
| MG045 | MPN058 | MYPU_4220 | UU110 |
| MG373 | MPN551 | MYPU_4770 | UU419 |

The above mentioned genes (gene identifier given) are all commonly present in the compared mollicute genomes but **not** in the other 49 known genomes compared to (including eucaryotes). This type of specific unique conservation makes it difficult to identify the function by genome analysis and only suggestions for function and reason for conservation can be given for the specific clusters shown in b: MG234 and MG235 are an aminopeptidase and rpL33, respectively; both are involved in translation. In MG046 and MG047, one is a sialoglycoprotease and the other is the S-adenosylmethione synthesis protein metK. MG371 is homologous to phosphoesterases. MG372 is homologous to thiI.

There are 782 protein reading frames covering 91.4% of the genome. A function could be assigned to 486 ORFs (62.2% of all). Of the remaining 37.8%, 92 ORFs (11.8%) are conserved hypothetical proteins. 204 open reading frames remained without any significant database match (26%). Only 29 tRNAs genes and one ribosomal operon are present. Chambaud *et al.* (2001) identified several pathogenicity factors by their genome analysis: sequence polymorphisms within stretches of repeated nucleotides generate phase-variable protein antigens. A recombinase gene is likely to catalyse the site-specific DNA inversions in major *M. pulmonis* surface antigens (see chapter on Genetic Mechanisms of Antigenic Variation by Yogev *et al.*). Predicted virulence factors include a hemolysin, secreted nucleases and a glycoprotease. Eight genes previously reported to be essential for a self replicating cell in *M. genitalium* are missing from the larger *M. pulmonis* genome (spoT, cpsG, gtaB, fruK, groEL, groES, udk, galE); the lack of the stringent response gene in *M. pulmonis* (spoT) perhaps being the most striking. GroEL and GroES are also missing in *Ureaplasma urealyticum*. The set of unique genes shared by all Mollicutes in the quartet is small (Table 3c).

## 5. COMPARATIVE ANALYSIS TO REVEAL COMMON FEATURES OF MOLLICUTES

After characterizing genome specific features apparent in different Mollicutes it is appropriate to discuss several general features present in most of the sequenced and annotated Mollicute genomes. More general statements have to be considered with care, as this presents only a minority of all mollicutes. The wall-less Mollicutes as suggested by their phylogenetic position clearly descended from low GC content gram-positive bacteria. This is further supported by the phylogenetic analysis as discussed above (Figure 1a-c; Figure 2). Some Mollicutes are exceedingly small (0.3 micron in diameter). Their genomes are equally small. However, their gene density is roughly similar (Table 1b): The smallest, *Mycoplasma genitalium* contains 580,070 base pairs and currently 480 ORFs, corresponding to 0.83 protein genes per kilobase (according to our updated data, www.bork.embl-heidelberg.de/Annot/MG/). The *Escherichia coli* genome has 4,639,000bp and contains 4290 identified protein ORFs and 115 RNA genes[4]; latest genome annotation at www.genome.wisc.edu). This corresponds to 0.92 protein ORFs/kb and 0.94 genes/kb including RNA genes. Mollicutes are considered models for describing the minimal metabolism necessary to sustain independent life[31]. Nevertheless, *M. pneumoniae* contains approximately 25% of the total number of known protein folds with single domains[47]. The Mollicutes sequenced to date have no cytochromes or the TCA cycle except for malate dehydrogenase activity. Some uniquely require cholesterol for growth, some require urea and some are anaerobic. They fix $CO_2$ in anaplerotic or replenishing reactions. Some require pyrophosphate instead of ATP as an energy source for reactions, including the rate-limiting step of glycolysis: 6-phosphofructokinase. Mollicutes scavenge for nucleic acid precursors and apparently do not synthesize pyrimidines or purines *de novo*. Some species uniquely lack dUTPase activity such as *Entomoplasma ellychniae* ELCN-1T, *Entomoplasma melaleucae* M-1T, *Mesoplasma seiffertii* F7T, *Mesoplasma entomophilum* TACT, *Mesoplasma florum* L1T, *Mycoplasma fermentans* PG18T, *M.pneumoniae*, *M.genitalium* and *Acholeplasma multilocale* PN525T, but still have uracil-DNA glycosylase. Some other species also lack DNA uracil-DNA glycosylase such as *Mesoplasma entomophilum*. The absence of the latter two reactions that limit the incorporation of uracil or remove it from DNA may be related to the marked mutability of the Mollicutes and their tachytelic or rapid evolution (e.g. these enzyme activities are lacking in *M. genitalium* and *M. pneumoniae*). Approximately 150 cytoplasmic activities have been identified in these organisms, 225 to 250 are presumed to be present (according to Pollack *et al.*, 1997, see also his chapter Central Carbohydrate Pathways: Metabolic Flexibility and the Extra Role of Some "Housekeeping" Enzymes). About 100 of the core reactions can be graphically linked to a metabolic map. This includes glycolysis, pentose phosphate pathway, arginine dihydrolase pathway, transamination, and purine, pyrimidine, and lipid metabolism. Reaction sequences or loci of particular importance are: phosphofructokinases, NADH oxidase, thioredoxin complex, deoxyribose-5-phosphate aldolase, and lactate, malate, and glutamate dehydrogenases. Enzymatic activities of the Mollicutes can be grouped according to metabolic similarities that are taxonomically discriminating. A pathway listing of all relevant enzymes encoded in the *M. pneumoniae* genome including available *M. genitalium* orthologues is found at www.bork.embl-heidelberg.de/Annot/MP/.

Mollicute genomes have developed specialized cell reproduction cycles adapted to the limited genome information and a parasitic life style[29]. Thus

DNA replication in *Mycoplasma capricolum* starts at a fixed site neighboring the dnaA gene and proceeds to both directions after a short arrest in one direction. The initiation frequency fits the slow speed of the replication fork, setting the DNA content constant. The replicated chromosomes migrate to one and three quarters of cell length before cell division to ensure delivery of the replicated DNA to daughter cells. The cell reproduction is based on binary fission but a branch is formed when DNA replication is inhibited. *Mycoplasma pneumoniae* has a terminal structure, designated as an attachment organelle, responsible for both host cell adhesion and gliding motility[27]. Behaviour of the organelle in a cell implies coupling of organelle formation to the cell reproduction cycle. Several proteins coded in three operons are delivered sequentially to a position neighboring the previous organelle and a nascent one is formed. One of the duplicated attachment organelles migrates to the opposite pole of the cell before cell division (see chapters Cell Division and Cytoskeleton, and Cell Division).

Karlin and Campbell (1994) collected data on oligonucleotide abundance (see also the chapters on cell division and cytosceleton). Though we do not support their hypothesis that a Mollicute- or *Sulfolobus*-like endosymbiont rather than an alpha-proteobacterium is the ancestor of animal mitochondrial genomes, there are pronounced similarities in extremes of oligonucleotide abundance common to animal mtDNA, *Sulfolobus*, and *Mycoplasma capricolum*. Furthermore, there are pronounced discrepancies of these relative abundance values with respect to alpha-proteobacteria. In addition, genomic dinucleotide relative abundance measures place *Sulfolobus* and *M. capricolum* among the closest to animal mitochondrial genomes, whereas the classical eubacteria, especially the alpha-proteobacteria, are at excessive distances. There are also considerable molecular and cellular phenotypic analogies among mtDNA, *Sulfolobus*, and *M. capricolum*.

Another observation related to nucleotide composition is the moderate avoidance of palindromes in Mollicutes, examined in detail for *Mycoplasma genitalium*[14]. Short palindromic sequences (4, 5 and 6bp palindromes) are avoided at a statistically significant level in the genomes of several bacteria, including the completely sequenced *Haemophilus influenzae* and *Synechocystis* sp. genomes and in the complete genome of the archaeon *Methanococcus jannaschii*. In contrast, there is no detectable avoidance in the genomes of chloroplasts and mitochondria. The sites for type II restriction-modification enzymes detected in the above given species tend to be among the most avoided palindromes in a particular genome, indicating a direct connection between the avoidance of short oligonucleotide words and restriction-modification systems with the respective specificity. Palindromes corresponding to sites for restriction enzymes from other species are also avoided, albeit less significantly, suggesting that in the course of evolution bacterial DNA has been exposed to a wide spectrum of restriction enzymes. This is probably the result of lateral transfer mediated by mobile genetic elements, such as plasmids and prophages. Palindromic words appear to accumulate in DNA once it becomes isolated from restriction-modification systems, as demonstrated by the case of organellar genomes. We note furthermore that there is good genome evidence for an elaborate type I restriction enzyme system both in *M. pneumoniae*[19, 8] and *M. pulmonis*[16].

## 6.    COMBINING GENOMIC FEATURES WITH BIOCHEMICAL KNOWLEDGE

Several of the conclusions apparent from Mollicute genome analysis can be enhanced or critically re-examined when combined with biochemical knowledge and functional analysis. Thus *M. genitalium* is often pictured as a minimal genome[31, 21]. However, this is a relative definition as this depends on the biochemical context of the environment. A good case can only be made for genes essential for a self-replicating cell under any circumstances. For instance, primordial tRNA modifications are required to prevent frame-shifting just besides the anti-codon. A suitable G37-tRNA-methyltransferase gene is also found in Mollicute genomes including *M. genitalium*[3].

The minimalistic setup of the Mollicutes is interesting from a biotechnological point of view. Potentially they could be engineered just to produce more and more of the desired protein with only a small burden of their own standard architecture.

The small Mollicute genomes should use their genomic information efficiently[35]. Evidence for this are the broad aminotransferase acitivities in *M. pneumoniae* and the multifunctional enzyme encoded by MPN158[MP674], (a riboflavin kinase, an FMN adenylyltransferase; a predicted nicotinate-nucleotide adenylyltransferase would furthermore complete the synthesis from imported nicotinic acid to NAD, a pathway otherwise incomplete). *M. pneumoniae* pseudogenes for arginine deiminase, MPN305[MP531] and MPN304[MP532] are an example for remnant enzymes[35] from genome reduction due to the parasitic life style.

For more detailed genome studies, homologous recombination should be developed further. For this, a plasmid that replicates in *Escherichia coli* but not in *M. genitalium* was constructed and used to disrupt the cytadherence-related gene MG218 of *M. genitalium*[9].

There are discrepancies between genes annotated in the genome and biochemical activities measured. This is partly caused by strain variation (see above) and partly because it is difficult to assign and detect function just by inference from sequence similarities for those open reading frames only occurring in Mollicutes. This includes the biochemically measured enzymes such as aspartate aminotransferase activity (E.C. 2.6.1.1.) well measured biochemically previously in *M. pneumoniae*[28] or 5 prime nucleotidase activity in *M. genitalium*. Both were not detected in the respective original genome annotation, however candidate genes are now available[8, 23]. There are also examples for the opposite case, thus there is genome evidence for ammonia production (arginine deaminase MPN560$_{MP282}$) in *M. pneumoniae* but enzyme activities of this sort have not yet been unequivocally determined yet[33]. Examination of genomic or enzymatic activity data alone neither provides a complete picture of metabolic function or potential, nor confidently reveals sites amenable to inhibition. *Ureaplasma urealyticum* (parvum) provides several examples[34]. Combining evidence from genomic sequence, transcription, translational phenomena, structure and enzymatic activity gives the best picture of the organism's metabolic capabilities.

Non-orthologous gene displacement complicates function assignment by sequence analysis. For example polymerase type I has been duplicated in *M. pneumoniae*[26] and seems to replace the missing function for DNA repair. Biochemical experiments are necessary to confirm this hypothesis. However, complete genome sequences from close by non-mollicute genomes will enhance bioinformatical analyses such as those of *Streptococcus pyogenes*[10] and *Lactococcus lactis*[5]. The combination of analysis tools should in particular give more complete information about pathogenicity factors, the cytoskeleton related proteins and proteins comprising species-specific adaptations such as the attachment organelle in *M. pneumoniae* (besides the well known cytadherence operon a total of 27 proteins in the genome are predicted to be involved in cytadherence in *M. pneumoniae*; see chapter on Cytadherence and Cytoskeleton).

Comparative genome analysis of Mollicutes is an ongoing exercise. In the light of new genomes, more experimental data and better software, more genome information can be deciphered[15, 22, 8]. The exercise of re-annotation and re-evaluation of gene information will become more important. The genome variety of Mollicutes is, despite their small genomes rather considerable and includes many specific adaptations. Thus, comparative genome analysis will remain an important tool to broaden our knowledge on these organisms.

# REFERENCES

1.  Altschul, S. F., T. L. Madden, A. A. Schaffer, J. Zhang, Z. Zhang, W. Miller, and D. J. Lipman. 1997. Gapped BLAST and PSI-BLAST: a new generation of protein database search programs. Nucleic Acids Res. **25**:3389–3402.

*Thomas Dandekar et al.*

2.  Aravind, L. and E. V. Koonin. 1998. The HD domain defines a new superfamily of metal-dependent phosphohydrolases. Trends Biochem. Sci. **23**:469–472.

3.  Bjork, G. R., K. Jacobsson, K. Nilsson, M. J. Johansson, A. S. Bystrom, and O. P. Persson. 2001. A primordial tRNA modification required for the evolution of life? EMBO J. **20**:231–239.

4.  Blattner, F. R., , C. A. Bloch, N. T. Perna, V. Burland, M. Riley, J. Collado-Vides, J. D. Glasner, C. K. Rode, G. F. Mayhew, J. Gregor, N. W. Davis, H. A. Kirkpatrick, M. A. Goeden, D. J. Rose, B. Mau, and Y. Shao. 1997. The complete genome sequence of *Escherichia coli* K-12. Science **277**:1453–1474.

5.  Bolotin, A., S. Mauger, K. Malarme, S. D. Ehrlich, and A. Sorokin. 1999. Low-redundancy sequencing of the entire *Lactococcus lactis* IL1403 genome. Antonie Van Leeuwenhoek **76**:27–76.

6.  Bork, P., C. Ouzounis, G. Casari, R. Schneider, C. Sander, M. Dolan, W. Gilbert, and P. M. Gillevet. 1995. Exploring the *Mycoplasma capricolum* genome: a minimal cell reveals its physiology. Mol. Microbiol. **16**:955–967.

7.  Chambaud, I., R. Heilig, S. Ferris, V. Barbe, D. Samson, F. Galisson, I. Moszer, K. Dybvig, H. Wroblewski, A. Viari, E. P. Rocha, and A. Blanchard. 2001. The complete genome sequence of the murine respiratory pathogen *Mycoplasma pulmonis*. Nucleic Acids Res. **29**:2145–2153.

8.  Dandekar, T., M. Huynen, J. T. Regula, B. Ueberle, C. U. Zimmermann, M. A. Andrade, T. Doerks, L. Sanchez-Pulido, B. Snel, M. Suyama, Y. P. Yuan, R. Herrmann, and P. Bork. 2000. Re-annotating the *Mycoplasma pneumoniae* genome sequence: adding value, function and reading frames. Nucleic Acids Res. **28**:3278–3288.

9.  Dhandayuthapani, S., W. G. Rasmussen, and J. B. Baseman. 1999. Disruption of gene mg218 of *Mycoplasma genitalium* through homologous recombination leads to an adherence-deficient phenotype. Proc. Natl. Acad. Sci. USA **96**:5227–5232.

10. Ferretti, J. J., W. M. McShan, D. Ajdic, D. J. Savic, G. Savic, K. Lyon, C. Primeaux, S. Sezate, A. N. Suvorov, S. Kenton, H. S. Lai, S. P. Lin, Y. Qian, H. G. Jia, F. Z. Najar, Q. Ren, H. Zhu, L. Song, J. White, X. Yuan, S. W. Clifton, B. A. Roe, and R. McLaughlin. 2001. Complete genome sequence of an M1 strain of *Streptococcus pyogenes*. Proc. Natl. Acad. Sci. USA **98**:4658–4663.

11. Fischer, D. and D. Eisenberg. 1997. Assigning folds to the proteins encoded by the genome of *Mycoplasma genitalium*. Proc. Natl. Acad. Sci. USA **94**:11929–11934.

12. Fraser, C. M., J. D. Gocayne, O. White, M. D. Adams, R. A. Clayton, R. D. Fleischmann, C. J. Bult, A. R. Kerlavage, G. Sutton, and J. M. K. and. 1995. The minimal gene complement of *Mycoplasma genitalium*. Science **270**:397–403.

13. Garcia-Vallve, S., A. Romeu, and J. Palau. 2000. Horizontal gene transfer in bacterial and archaeal complete genomes. Genome Res. **10**:1719–1725.

14. Gelfand, M. S. and E. V. Koonin. 1997. Avoidance of palindromic words in bacterial and archaeal genomes: a close connection with restriction enzymes. Nucleic Acids Res. **25**:2430–2439.

15. Glass, J. I., E. J. Lefkowitz, J. S. Glass, C. R. Heiner, E. Y. Chen, and G. H. Cassell. 2000. The complete sequence of the mucosal pathogen *Ureaplasma urealyticum*. Nature **407**:757–762.

16. Gumulak-Smith, J., A. Teachman, A. H. Tu, J. W. Simecka, J. R. Lindsey, and K. Dybvig. 2001. Variations in the surface proteins and restriction enzyme systems of *Mycoplasma pulmonis* in the respiratory tract of infected rats. Mol. Microbiol. **40**:1037–1044.

17. **Hamet, M., C. Bonissol, and P. Cartier.** 1979. Activities of enzymes of purine and pyrimidine metabolism in nine Mycoplasma species. Adv. Exp. Med. Biol. **122B**:231–235.

18. **Herrmann, R. and B. Reiner.** 1998. *Mycoplasma pneumoniae* and *Mycoplasma genitalium*: a comparison of two closely related bacterial species. Curr. Opin. Microbiol. **1**:572–579.

19. **Himmelreich, R., H. Hilbert, H. Plagens, E. Pirkl, B. C. Li, and R. Herrmann.** 1996. Complete sequence analysis of the genome of the bacterium *Mycoplasma pneumoniae*. Nucleic Acids Res. **24**:4420–4449.

20. **Himmelreich, R., H. Plagens, H. Hilbert, B. Reiner, and R. Herrmann.** 1997. Comparative analysis of the genomes of the bacteria *Mycoplasma pneumoniae* and *Mycoplasma genitalium*. Nucleic Acids Res. **25**:701–712.

21. **Hutchison III, C. A., S. N. Peterson, S. R. Gill, R. T. Cline, O. White, C. M. Fraser, H. O. Smith, and J. C. Venter.** 1999. Global transposon mutagenesis and a minimal Mycoplasma genome. Science **286**:2165–2169.

22. **Huynen, M., B. Snel, , and P. Bork.** 2000. Predicting protein function by genomic context: quantitative evaluation and qualitative inferences. Genome Res. **10**:1204–1210.

23. **Huynen, M., T. Doerks, F. Eisenhaber, C. Orengo, S. Sunyaev, Y. Yuan, and P. Bork.** 1998. Homology-based fold predictions for *Mycoplasma genitalium* proteins. J. Mol. Biol. **280**:323–326.

24. **Karlin, S. and A. M. Campbell.** 1994. Which bacterium is the ancestor of the animal mitochondrial genome? Proc. Natl. Acad. Sci. USA **91**:12842–12846.

25. **Kokotovic, B., G. Bolske, P. Ahrens, and K. Johansson.** 2000. Genomic variations of *Mycoplasma capricolum* subsp. capripneumoniae detected by amplified fragment length polymorphism (AFLP) analysis. FEMS Microbiol. Lett. **184**:63–68.

26. **Koonin, E. V. and P. Bork.** 1996. Ancient duplication of DNA polymerase inferred from analysis of complete bacterial genomes. Trends Biochem. Sci. **21**:128–129.

27. **Krause, D. C.** 1996. *Mycoplasma pneumoniae* cytadherence: unravelling the tie that binds. Mol. Microbiol. **20**:247–253.

28. **Manolukas, J. T., M. F. Barile, D. K. Chandler, and J. D. Pollack.** 1988. Presence of anaplerotic reactions and transamination, and the absence of the tricarboxylic acid cycle in mollicutes. J. Gen. Microbiol. **134 ( Pt 3)**:791–800.

29. **Miyata, M. and S. Seto.** 1999. Cell reproduction cycle of mycoplasma. Biochimie **81**:873–878.

30. **Musco, G., G. Stier, C. Joseph, M. A. Castiglione Morelli, M. Nilges, T. J. Gibson, and A. Pastore.** 1996. Three-dimensional structure and stability of the KH domain: molecular insights into the fragile X syndrome. Cell **85**:237–245.

31. **Mushegian, A. R. and E. V. Koonin.** 1996. A minimal gene set for cellular life derived by comparison of complete bacterial genomes. Proc. Natl. Acad. Sci. USA **93**:10268–10273.

32. **Neimark, H. C. and C. S. Lange.** 1990. Pulse-field electrophoresis indicates full-length Mycoplasma chromosomes range widely in size. Nucleic Acids Res. **18**:5443–5448.

33. **Pollack, J. D.** 1997. Mycoplasma genes: a case for reflective annotation. Trends Microbiol. **5**:413–419.

34. **Pollack, J. D.** 2001. *Ureaplasma urealyticum*: an opportunity for combinatorial genomics. Trends Microbiol. **9**:169–175.

35. **Pollack, J. D., M. V. Williams, and R. N. McElhaney.** 1997. The comparative metabolism of the mollicutes (Mycoplasmas): the utility for taxonomic classification and the relationship of putative gene annotation and phylogeny to enzymatic function in the smallest free-living cells. Crit. Rev. Microbiol. **23**:269–354.

36. **Pyle, L. E., L. N. Corcoran, B. G. Cocks, A. D. Bergemann, J. C. Whitley, and L. R. Finch**. 1988. Pulsed-field electrophoresis indicates larger-than-expected sizes for mycoplasma genomes. Nucleic Acids Res. **16**:6015–6025.

37. **Robertson, J. A., L. E. Pyle, G. W. Stemke, and L. R. Finch**. 1990. Human ureaplasmas show diverse genome sizes by pulsed-field electrophoresis. Nucleic Acids Res. **18**:1451–1455.

38. **Rychlewski, L., B. Zhang, and A. Godzik**. 1998. Fold and function predictions for *Mycoplasma genitalium* proteins. Fold Des. **3**:229–238.

39. **Snel, B., G. Lehmann, P. Bork, and M. A. Huynen**. 2000. STRING: a web-server to retrieve and display the repeatedly occurring neighbourhood of a gene. Nucleic Acids Res. **28**:3442–3444.

40. **Snel, B., P. Bork, and M. A. Huynen**. 1999. Genome phylogeny based on gene content. Nat. Genet. **21**:108–110.

41. **Sokal, R. R. and C. D. Michener**. 1958. A statistical method of evaluating systematic relationships. Univ. Kansas Sci. bull. **28**:1409–1438.

42. **Suyama, M. and P. Bork**. 2001. Evolution of prokaryotic gene order: genome rearrangements in closely related species. Trends Genet. **17**:10–13.

43. **Teichmann, S. A. and G. Mitchison**. 1999. Is there a phylogenetic signal in prokaryote proteins? J. Mol. Evol. **49**:98–107.

44. **Teichmann, S. A., A. G. Murzin, and C. Chothia**. 2001. Determination of protein function, evolution and interactions by structural genomics. Curr. Opin. Struct. Biol. **11**:354–363.

45. **Teichmann, S. A., C. Chothia, and M. Gerstein**. 1999. Advances in structural genomics. Curr. Opin. Struct. Biol. **9**:390–399.

46. **Trachtenberg, S**. 1998. Mollicutes-wall-less bacteria with internal cytoskeletons. J. Struct. Biol. **124**:244–256.

47. **Wolf, Y. I., N. V. Grishin, and E. V. Koonin**. 2000. Estimating the number of protein folds and families from complete genome data. J. Mol. Biol. **299**:897–905.

# Chapter 12

# Transcriptome and Proteome Analyses of Mollicutes

JANUARY WEINER 3RD, CARL-ULRICH ZIMMERMANN, BARBARA
UEBERLE, and RICHARD HERRMANN
*Zentrum für Molekulare Biologie Heidelberg, ZMBH, Universität Heidelberg, Im
NeuenheimerFeld 282, 69120 Heidelberg, Germany*

## 1. INTRODUCTION

The publication of the first complete genome sequence of a bacterium
*(Haemophilus influenzae)* in 1995 (33) marked a turning point in the genetic
analysis of cells and organisms. Before that, at the most, only subsets of the
cellular gene repertoire were known and could be analyzed. Nowadays,
having access to sequence data of entire genomes, much more
comprehensive studies became possible aiming at the understanding of the
two key processes of all living systems, reproduction and metabolism,
defined as the sum of all chemical reactions taking place in a cell.

A genome sequence and its annotation initially provide a list of proposed
open reading frames (ORFs). About 50-70% of these ORFs can be assigned
to function through bioinformatics, that is mainly by similarity searches in
databases for proteins with known function from other organisms. However,
the similarity based function prediction is being complemented by other
methods (8) based on different criteria like protein structure prediction (85)
or co-occurrence of genes in the same genome (52). Depending on the size
of a genome and the number of proposed ORFs one is left with functionally
unassigned ORFs in the range of 200 ORFs for a bacterium with a small
genome like *Mycoplasma pneumoniae* (genome size: 816,394 base pairs
(bp), 688 ORFs, (22, 50)) or in the range of about 1500 ORFs for
conventional bacteria like *Escherichia coli* (genome size 4,639,221 bp, 4288
ORFs, (7)) or *Bacillus subtilis* (genome size 4,214,810 bp, 4100 ORFs,
(61)). But to really understand the biology of an entire cell, at least for a few

*Molecular Biology and Pathogenicity of Mycoplasmas*, Edited by Razin and
Herrmann, Kluwer Academic/Plenum Publishers, New York, 2002

model organisms, the functions of all genes should be deciphered (58, 85, 86). Since functional analysis, based on experimental evidence is laborious and often time consuming, it is reasonable to apply procedures which help to select a smaller subset of ORFs to be studied. This could be done either in association with a specific problem or, in a more general approach, correlated to gene expression. For instance, one could determine which of the functionally unassigned ORFs are expressed or repressed under certain environmental conditions. An example for a specific problem could be the identification of the proteins forming a cytoskeleton-like structure in a mollicute species (87).

There are two ways to study gene expression, either at the level of transcription or at the level of translation. If these analyses are being done with intact cells the new terms transcriptome (90) and proteome (17) are used. The transcriptome is defined as the sum of all transcripts derived from a genome (90) and the proteome refers to the total protein complement of a cell (17). The established genome sequence of the organism under investigation is essential for both analyses.

## 1.1    Transcriptome analysis

The most widely used method for transcriptome analysis is the microarray technology (13, 23, 72, 91). The experimental concept is rather simple, it is a kind of a dot blot procedure. The main problem is the large number of different genes which have to be handled in parallel (4, 21). Gene specific probes have first to be generated and immobilized on a matrix (Fig. 1). These probes are then hybridized with cellular labeled total RNA or, as in most cases, with total RNA derived complementary DNA (cDNA) synthesized by reverse transcriptase. The DNA is usually labeled by incorporation of radioactive or fluorescent labeled precursers during cDNA synthesis. To detect hybridization signals, the matrix is exposed either to a sensitive film or to a PhosphorImager screen or, in case of fluorescent labels, scanned by a confocal laser scanner (Fig. 2).

The generation of gene specific probes can be done in different ways. The simplest way is by amplification of complete genes or parts of them by PCR and immobilization of the resulting ds DNA on nylon membranes or glass slides. Instead of PCR, chemically synthesized 17-mer to 20-mer oligonucleotides are also frequently used for the same purpose (16). In a technically more advanced procedure, combining photolithography with oligonucleotide chemistry, oligonucleotides are directly synthesized in a highly combinatorial fashion on a solid substrate providing high-density oligonucleotide arrays (14, 34, 35, 63, 69). Oligonucleotide densities of $10^4$ -

a.)

**ORF-probes in microtiter-plates**

b.) **Cell-culture: standard & experimental condition**

**Robotic transfer of probes**

**RNA isolation**

**Radiolabeled cDNA**

**Hybridization-profile**

c.)

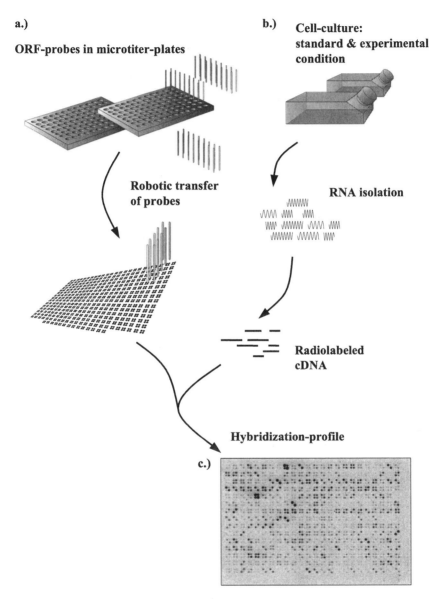

*Figure 1.* Scheme of transcriptome analysis: a.) Construction of microarray-filter. ORF-probes are collected in microtiter-plates and are robotically transferred onto nylon membranes with a pin-tool. b.) Probe synthesis: cells are grown at different conditions, their RNA is isolated and reverse transcribed. c.) Microarray-profile: radioactive labeled cDNA is hybridized to a microarray-filter.

$10^6$ sequences/cm$^2$ (even more) can be generated by this method. It is particularly advantageous if the number of gene probes exceeds the gene pool of eubacteria.

The signal strength in a microarray-based system depends strongly on hybridization parameters, and therefore - among other factors - on the sequence of the DNA used as hybridization probe. Therefore, expression signals of different genes within one experimental condition cannot be directly compared. It is not legitimate to assume that the signal strengths of two different genes reflect their absolute expression or minuscule differences in expression. This limitation, however, does not apply to the comparison of the expression of a single gene tested under various experimental conditions. In any case, extensive and careful statistical analysis is necessary to extract biologically meaningful data, estimate the influence of different error sources, and reduce random noise (21).

Fluorescent labels allow direct comparisons of two profiles. If two sets of cDNAs, each derived from total RNA of cells grown under different conditions, are labeled with differently colored dyes - a red and a green one – both cDNAs can be mixed and hybridized to the same microarray. In such case the genes, which do not change expression between the two conditions will give a yellow signal, and genes up- or down-regulated will give a red or a green signal.

It is beyond the scope of this chapter to give a complete overview on the available methods for detecting and quantifying mRNAs. The interested reader can find the details in references 1, 67 and 91.

## 1.2    Proteome analysis

The present standard procedure for proteome analysis combines two-dimensional (2-D) gel electrophoresis (41, 57) for separation of complex protein mixtures with mass spectrometry (82) as a means to characterize and identify individual proteins (Fig. 3). The 2-D gel system separates proteins in the first dimension according to charge and in the second dimension according to molecular mass, and provides a highly reproducible protein pattern of an organism or a cell (Fig. 4). The bottleneck of this method is the separation in the first dimension. Since the separation has to be done under non-denaturing conditions, many membrane proteins are excluded from the protein separation pattern (78). Membrane proteins are defined in this context as proteins with one or more predicted transmembrane segments (15, 68), due to stretches of hydrophobic amino acids. To separate proteins according to charge a pH gradient has to be established across the gel. This could consist of either a stable preformed immobilized pH gradient (IPG; 41) or a gradient developing during the electrophoresis (57). In either type of

## Profile evaluation

**32°C**

**HS**

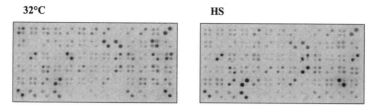

## Array data

| MPN# | ORFMP# | 32°C | | | | HS | | | |
|------|--------|------|------|---------|------------|------|------|---------|------------|
| | | Mean1 | Mean2 | Average | % of total | Mean1 | Mean2 | Average | % of total |
| MPN532 | ORFMP310 | 42,53 | 45,21 | 43,87 | 0,14% | 32,69 | 32,64 | 32,66 | 0,17% |
| MPN531 | ORFMP311 | 27,06 | 27,03 | 27,04 | 0,22% | 79,05 | 73,22 | 76,13 | 1,70% |
| MPN530 | ORFMP312 | 47,25 | 45,94 | 46,60 | 0,10% | 34,98 | 35,17 | 35,08 | 0,12% |
| MPN529 | ORFMP313 | 29,21 | 30,08 | 29,65 | 0,04% | 25,53 | 26,82 | 26,17 | 0,10% |
| MPN528 | ORFMP314 | 28,92 | 28,87 | 28,89 | 0,06% | 21,28 | 20,50 | 20,89 | 0,07% |
| MPN527 | ORFMP315 | 25,82 | 25,42 | 25,62 | 0,03% | 18,73 | 17,41 | 18,07 | 0,04% |
| MPN526 | ORFMP316 | 37,64 | 36,83 | 37,24 | 0,03% | 25,84 | 25,13 | 25,49 | 0,06% |
| MPN525 | ORFMP317 | 29,02 | 30,97 | 29,99 | 0,02% | 20,43 | 20,95 | 20,69 | 0,05% |
| MPN524 | ORFMP318 | 29,12 | 29,14 | 29,13 | 0,16% | 21,41 | 21,81 | 21,61 | 0,18% |
| MPN523 | ORFMP319 | 32,29 | 30,01 | 31,15 | 0,18% | 47,63 | 43,58 | 45,60 | 0,32% |
| MPN522 | ORFMP320 | 28,02 | 27,10 | 27,56 | 0,03% | 17,65 | 17,62 | 17,63 | 0,02% |
| MPN521 | ORFMP321 | 27,45 | 28,04 | 27,74 | 0,06% | 20,05 | 19,58 | 19,81 | 0,05% |
| MPN520 | ORFMP322 | 27,51 | 27,26 | 27,39 | 0,04% | 19,16 | 20,21 | 19,68 | 0,08% |
| MPN519 | ORFMP323 | 23,34 | 22,81 | 23,08 | 0,02% | 17,97 | 16,32 | 17,15 | 0,02% |
| MPN518 | ORFMP324 | 24,85 | 25,45 | 25,15 | 0,42% | 17,52 | 16,76 | 17,14 | 0,42% |
| MPN517 | ORFMP325 | 23,43 | 24,22 | 23,83 | 0,14% | 19,19 | 18,14 | 18,66 | 0,18% |

## Data evaluation

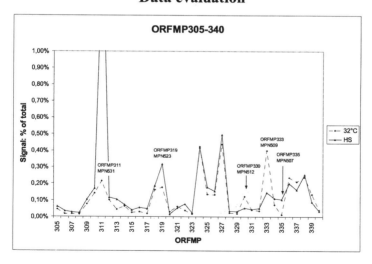

*Figure 2.* Transcriptome analysis. Comparison of two experimental conditions (continuous growth at 32°C vs heat shock (HS) for 15' at 43°C). Here are shown parts from two different micoarray-profiles, a section from the acquired array-data and a histogram comparing the evaluated data. The example is taken from an analysis of *M. pneumoniae*.

gradient, proteins with isoelectric points (pI) above 10.5 are not well separated, and preparation of pH gradients in the alkaline region (pH 10.5) is still a technical challenge. After separation, the proteins are stained and the visible spots or bands are excised from the acrylamide gel. The proteins in the gel slices are digested by protease, e.g. trypsin. The generated peptides were eluted and characterized by using a mass spectrometer (65, 81, 97, 98). Two kinds of data are generated, the molecular weight of the tryptic peptides and, depending on the method, partial amino acid sequences from individual tryptic peptides (28). These data are then compared either directly with protein databases or with databases constructed by translating the genomic DNA sequences in all six possible reading frames into protein sequences independent of the original annotation. In the latter case, any open reading frame located between two stop codons will be included in the similarity search (56, 81). This approach guarantees that even proteins which were overlooked in the original annotation can be identified in a proteome analysis.

## 1.3    Transcriptome versus proteome

While the genome is well defined by the number of basepairs, a transcriptome or proteome is not, because they are by definition dependent on individual gene expression. It is indeed very difficult to prove that a proposed gene will be silent – not expressed – under any growth conditions. The reasons for failures to detect expression are manifold, e.g. low copy number of a mRNA or protein species, unfavorable features of a protein making its isolation or separation impossible, or the strict dependence of expression on very specific environmental conditions.

In general, transcription is much easier to monitor, since, based on a complete genome sequence, very specific probes for any gene and also for any intergenic region can be synthesized. The technology is advanced to such an extent that the number of genes is not any more a limiting factor in a transcriptome analysis. The situation is different for proteome analysis, which is hampered by the deficiencies of existing technology, despite remarkable improvements in the separation of complex protein mixtures (27, 41, 57). It is still not feasible to routinely separate several thousand proteins in quantities, large enough to allow their identification by mass spectrometry. In addition, very basic or very hydrophobic proteins are excluded from separation by 2-D gel electrophoresis (36, 78), the most powerful method presently applied for the separation of protein mixtures.

Another serious problem concerns the analysis of a mixture of different cells. Investigating the interaction of mycoplasmal cells with a host cell, e.g. epithelium cells in culture, any isolation of macromolecules (DNA, RNA

## 2-D electrophoresis

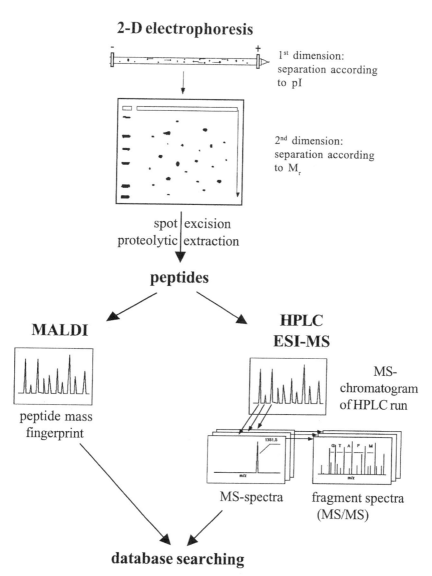

1st dimension:
separation according
to pI

2nd dimension:
separation according
to $M_r$

spot | excision
proteolytic | extraction

**peptides**

**MALDI**

**HPLC
ESI-MS**

MS-
chromatogram
of HPLC run

peptide mass
fingerprint

1351.5

m/z

G T A F M

m/z

MS-spectra          fragment spectra
(MS/MS)

**database searching**

*Figure 3.* Scheme of proteome analysis. Identification of the proteins separated on 1-D or 2-D gels by mass spectrometry. After cutting the spots or bands out of the gel, proteins are in-gel digested with trypsin. Two different methods of mass spectrometry can be applied. In MALDI mass spectrometry, the masss of the tryptic peptides are determined and used for database searching (peptide mass fingerprinting). In LC-MS/MS, additional sequence information of the tryptic peptides is obtained, improving the reliability of the results (65, 82, 97). MALDI, matrix-assisted laser desorption/ionization. ESI-MS, electrospray ionization mass spectrometry. LC, liquid chromatography.

proteins) will yield a mixture of both mycoplasmal and eukaryotic cell products. In a transcriptome analysis, the high specificity of gene probes allows to identify mRNA species from mycoplasmas even if there is a large excess of host RNA and vice versa, but in proteome analysis there is no simple way to differentiate bacterial and host proteins in a 2-D gel pattern. An assignment of the protein to its organism can only be made after the mass spectrometral analysis.

It has been frequently shown that the expression of bacterial genes is turned on and off under different growth conditions. Since regulation can take place at the level of transcription, translation or co- and post-translational modification, one has to study the products of these processes and quantify their concentration in cells to understand gene expression and regulation at the cellular level. Presently the only possible way to measure expression of all genes of an organism is by transcriptome analysis. Although the absolute quantification is not always possible, the method is good enough to show that the concentration of individual mRNA species does not always reflect the concentration of the corresponding proteins (44). However, because regulation of gene expression is also frequently mediated at the transcriptional level, e.g. initiation of transcription, mRNA modification and degradation, changes in mRNA concentration are often causing changes in protein concentration. Normally, proteins and often their modified versions are the functional end products in the cell, therefore the proteins have to be analyzed to understand the cellular processes, even if the methods are still not advanced enough to include all proteins of a cell in such an analysis.

Expression analyses are very valuable tools for many purposes. First of all, the identification of gene products provides an overview of how many genes are expressed and confirms that proposed ORFs reflect indeed genes. This seems rather trivial, but in the course of genome annotations one is rather frequently confronted with the choice between several possible ORFs coded for on both strands. Presently, the highest benefit will be gained from comparative expression analyses of cells which have been grown under different environmental conditions, because it allows the study of the response of the organism to external stimuli. Depending on the applied stimuli it is also possible to make preliminary functional assignments of ORFs functionally unassigned if they are co-regulated together with genes with known function. Further improvement of the existing methodology will eventually permit the quantification of transcripts from individual gene.

Considering the experimental operations required for transcriptome and proteome analyses, it is obvious that simple cells, like members of the Mollicutes offer considerable advantages. First of all, the number of genes of many Mollicute species is well below 1000 (73), and second, among the

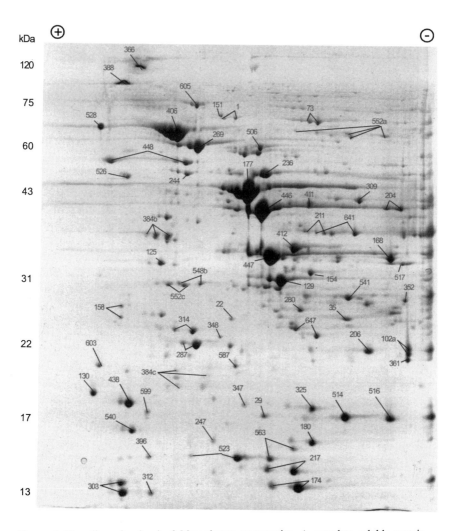

*Figure 4.* Two-dimensional gel of *Mycoplasma pneumoniae*. A complete soluble protein extract was separated in an immobilized pH-gradient from 3-10 in the first dimension and in a 12.5% acrylamide SDS-gel in the second dimension. Protein spots were visualized by staining with colloidal Coomassie blue. For a gel stained with the more sensitive silver method, see (75) or www.mpiib-berlin.mpg.de/2D-PAGE/microorganisms/index.html.

Mollicutes four species: *(M. genitalium* (480 genes, 37), *M. pneumoniae* (688 genes, 22, 50), *Ureaplasma urealyticum* (611 genes, 42), *M. pulmonis* (782 genes, 12)) have already been sequenced and several more are in the process of being sequenced, providing the largest pool of complete genome information for a single bacterial class.

So far, transcriptome analysis has been done only for *M. pneumoniae* (39) and proteome analyses for *M. genitalium* (17, 94, 95), *Spiroplasma melliferum* (17, 18) and *M. pneumoniae* (74, 75, 88). But since the technology for the analyses of cells with small genomes is less demanding, one can expect more of these global expression analyses from Mollicutes in the near future.

# 2.    THE TRANSCRIPTOME OF *M. PNEUMONIAE*

The experimental procedure of generating microarray-derived data is outlined in Figures 1 and 2. Specific gene probes for all proposed 688 ORFs (22, 50) were synthesized by PCR and aliquots of each were immobilized in doublets on a nylon membrane. After hybridization with $^{33}$P labeled complementary DNA (cDNA) derived form total RNA extracted from *M. pneumoniae* cells, a pattern of signals with different intensities appeared (Fig. 1). The intensity of individual signals correlates with the number of corresponding cDNA molecules in the hybridization mixture, although we can not yet correlate signals to an exact number of mRNA molecules in the original total RNA preparation.

Signal strengths of individual spots depend on various factors, e.g. exposure time, amount and quality of the RNA used, activity of the $[\gamma - ^{33}P]$-ATP etc. There are also significant differences between the membranes used. Therefore, to compare individual profiles, the raw data from hybridization experiments first has to be normalized. Unloaded spots scattered over the membranes are used for estimating the average amount of non-specific binding of cDNA molecules to a naked membrane, and in the first step of data normalization this value is subtracted from all collected data points. Individual signals are normalized either by representing them as a percentage of the sum of all signals, or by subtracting the mean and dividing by standard deviation; both methods give similar results. For each gene, the comparisons of its expression in different experimental conditions were done using a non-parametric statistical test. It has to be kept in mind that when doing so many (~1000) comparisons the confidence level must be kept on the conservative side. We used $p = 0.01$ in all conducted analyses.

In this way a standard transcription profile was derived from *M. pneumoniae* cells grown under standard laboratory conditions at 37°C

(Fig. 2). These standard profiles were compared to profiles derived from *M. pneumoniae* cells grown either at 32°C or cells exposed to a heat shock (HS) of 15 min. at 43°C prior to the isolation of total RNA (39). A growth temperature of 32°C was included in this study, because temperature dependent expression profiles by jumping from 32°C to 43°C instead of from 37°C to 43°C facilitated the identification of heat induced gene expression. In addition, it was observed that a temperature of 37°C already exerts heat shock stress on a significant number of genes. The results showed that about 600 of the 688 proposed genes are transcribed and that transcription of many genes is temperature dependent and therefore regulated. There are also numerous genes, which are constitutively expressed at all three temperatures. A very important result is the finding that a substantial proportion of genes with high transcription signals have no known function. Their high expression level suggests that they code for important products and that they serve as attractive candidates for future functional analyses. Since *M. pneumoniae* codes for one sigma factor only, the obvious question concerns the mechanisms regulating the rate of expression. Four of the conserved heat shock genes (lonA, clpB, dnaK, dnaJ) carry a CIRCE-element (46) in front, which strongly suggests, that those are regulated by the heatshock repressor HrcA (MPN124). However, there are several more genes with increased transcription at 43°C, which have no CIRCE element and must therefore be regulated in a different way. The same also applies to genes which are either up or down regulated at 32°C. Since the annotation of the genome sequence of *M. pneumoniae* and further searches for DNA binding proteins with regulatory function indicated that the number of candidate regulatory proteins is rather small, a study was initiated to define experimentally promoters of *M. pneumoniae* and examine whether certain promoter sequences/structures could be correlated with signal strength of individual genes within a transcriptome analysis. Twenty-two promoters of genes with different transcription levels (as judged by the microarray analysis, ref. 39) were experimentally defined by primer extension (96). Together with older promoter studies (26, 40, 53, 55, 92), a set of 35 known transcription sites in *M. pneumoniae* could be included in the analysis. The consensus sequence turned out to be highly conserved only in the -10 region, but very weakly in the -35 region, and altogether proved to be very dissimilar to the known promoter sequences of *E. coli* and *B. subtilis* (see also chapter "Transcription and translation"). Due to the low conservation of mycoplasmal promoter sequences it was necessary to apply a numerical algorithm allowing an *in silicio* identification of new promoters.

The information contained in the collected promoter sequences can be represented as a so-called frequency matrix. First, the three regions (+1, i.e. transcription start region and the -35 and -10 regions) of the known promoter sequences are aligned, and the first consensus sequences are constructed. For a given position in such an abstract consensus sequence, the frequencies of each of the four nucleotides in the aligned promoters are calculated and divided by the expected frequency of this nucleotide, given by the GC contents of the genome. The logarithmized values become the matrix scores. The three constructed matrices can be used to search for promoters in arbitrary sequences. Gaps between the -35, -10 and +1 may vary, and the frequencies of different gap sizes are also recorded (gap penalties).

Given an arbitrary sequence, it is scanned with the matrices, and all possible alignments are tried out. For a given alignment, each nucleotide in the sequence is assigned the given score for that nucleotide and that position in the promoter matrix; also, the gap penalties are substracted from the score. The alignment with the highest total score wins: this is the *in silico* calculated best promoter found in the given sequence. This procedure has been described by Hertz and Stormo (48) and was successfully applied to a large set of known *E. coli* promoters. It works also with the sequences from *M. pneumoniae* and *M. genitalium*, and is a useful additional tool for the identification of genes and operons (see the web page http://www.zmbh.uni-heidelberg.de/Matrix).

Although such algorithm proposes a promoter in virtually any sequence tried, the sequences which do not contain promoters get statistically lower scores; therefore, the scores only represent the similarity of the *in silico* calculated promoter sequences with the consensus sequence derived from experimental data. The sequences used to construct the matrix, i.e. the experimentally defined promoter sequences, get especially high scores. It could thus be proposed that the promoter scores which reflect the similarity of a sequence to an abstract consensus promoter sequence, correlate with gene expression. This, however, is not the case. A high score denotes conserved promoter sequences, but this does not necessarily indicate a high level of transcription. This leads to an important conclusion, namely that concentration of mRNA molecules of individual genes in *M. pneumoniae* cannot be correlated directly with the promoter sequence.

## 3.    THE PROTEOME OF *SPIROPLASMA MELLIFERUM*

The pioneering work on proteome analysis of Mollicutes was carried out by Humphery-Smith and his colleagues on *Spiroplasma melliferum* (17, 18) and *Mycoplasma genitalium* (17, 94, 95). At the time of their first publications complete genome sequences of the two microorganisms were not available and also the sensitivity and accuracy of mass spectrometral methods were not at the high level as they are today. Since a genome sequence which could be translated in all six possible reading frames was absent, one had to depend on cross-species identification on the basis of percentage amino acid composition and of molecular mass of peptide fragments generated by tryptic digestion and searching for similarities to defined proteins in databases like PIR/SWISS-PROT and MOWS. In addition, N-terminal protein sequencing was applied for further confirmation. A computer-generated ranking according to given score values was used to identify the protein in search. By this approach, nine of twelve proteins of *S. melliferum* and eight of eleven proteins of *M. genitalium* were identified. Striking was the identification of a tubulin ß chain in *M. genitalium*, a result, which was not confirmed by the annotation of the genome sequence (37). In retrospect, this protein could have been FtsZ, which is considered as a bacterial tubulin-like protein (29). Two years later a much more extended study on *S. melliferum* was published. Still without a genome sequence, the authors had to depend again on cross-species identification as described above. A reference gel composed from several 2-D gels, showed 456 silver-stained protein spots, of which about 1/3 were further analyzed and 65 proteins were identified (18).

At the time of publication (1997) this kind of study was laborious and based on limited methods. However, the fast increasing number of complete bacterial genome sequences had a marked impact on the further development of proteome analyses, since they provide a much more precise basis for the interpretation of protein data generated by mass-spectrometry. Nevertheless, for all cells or organisms without known genome sequence, cross-species identification is still the only choice, but one has to keep in mind that the confidence level of these data is lower.

## 4.   THE PROTEOME OF *M. GENITALIUM*

The recent publication of the proteome of *Mycoplasma genitalium* (95) signifies a major progress. Based on the complete genome sequence, proteins of individual spots could be identified by comparing the experimentally determined molecular mass of tryptic peptides with the theoretically predicted, DNA sequence derived, molecular masses of peptides in eubacterial and species-specific protein databases.

Using four overlapping pH windows for the separation in the first dimension (pH 2-5, pH 4-7, pH 6-11, pH 8.5-12) and SDS-PAGE with two different size ranges for separation in the second dimension, all together 427 different distinct protein spots could be visualized by silver-staining. The cells showing these 427 spots were collected during logarithmic growth.

From these 427 protein spots, a total of 201 were analyzed and 158 were identified. These 158 proteins were encoded by only 112 different genes. As in all other proteome analyses, the number of different protein spots exceeds the number of genes they are derived from, because many proteins are partly cotranslationally or posttranslationally modified in a way which changes the mobility in a 2-D gel. Among others, well known examples of modification are phosphorylation, acylation or processing of a large precursor protein into smaller subfragments. In addition to cells collected in the logarithmic phase, a proteome analysis was also done with *M. genitalium* collected in the stationary phase. The results indicated that the number of proteins in cells from the stationary phase is significantly lower and that the intensity of several protein spots relative to each other also has changed.

Out of the 112 genes with identified gene products, 17 coded for hypothetical proteins supporting the original annotation of the genome sequence (37) indicating that they were most probably of functional relevance. Otherwise they would not have been synthesized in a cell with such a small genome. The residual 95 other genes are members of almost all functional classes, formulated by M. Riley for the better understanding of functions in bacteria (76).

## 5.    THE PROTEOME OF *M. PNEUMONIAE*

The proteome of *M. pneumoniae* was analyzed by methods similar to those applied to proteome analysis of *M. genitalium*. The only difference was that the data generated by mass spectrometry included partial amino acid sequences of the tryptic peptides for many of the identified proteins in addition to the molecular mass information. This contributed considerably to the reliability of the data (75).

The proteome analysis was done by applying two different experimental approaches. In the first approach, complete soluble protein extracts were used as starting material for 2-D gel electrophoresis (Fig. 4). About 450 different protein spots could be assigned to 224 genes. Since this was the maximum of spots we could separate, different enrichment procedures were applied. These included differential centrifugation, anion and cation exchange chromatography, solubility in Triton X-100, affinity chromatography with heparin as a ligand and isolation of biotinylated surface exposed proteins by binding to immobilized streptavidin (88). The subcellular fractions were not as complex as the complete soluble cellular protein fraction and therefore, they could be separated as individual bands in 1-D SDS acrylamide gels. The advantage was that membrane proteins, proteins with a *pI* above 10.5 or proteins with a molecular mass below 10000

Da or above 100000 Da, could be much better solubilized and separated under the denaturing conditions of the SDS-PAGE compared to the non-denaturing conditions of the first dimension or the 2-D gel. The enrichment procedures made it possible to extend the number of genes with proven gene products from 224 to 305 (Table 1). This corresponds to about 44 % of the proposed set of 688 genes coding for proteins in *M. pneumoniae*. From these 305 genes, 63 were functionally not assigned, recommending them as preferred candidates for further functional studies. One of the subcellular fractions was of particular interest. This was the fraction which contained the Triton-X 100 insoluble proteins (74). It is believed, in analogy to many eukaryotic cells, to be enriched in proteins participating in the formation of a mycoplasmal cytoskeleton-like structure (38, 59, 60, 66, 87). Fifty to one hundred protein spots could be visualized in this fraction on 2-D gels after staining with colloidal Coomassie blue or with the more sensitive silver method (74). Of 50 Coomassie blue stained spots, 41 were identified. Among these were seven proteins which have been already proposed to function either as structural proteins in a cytoskeleton-like structure or cytadherence associated proteins (HMW1-3 (25), P65, P200 (70), P1 (84), the gene products of the ORF6 gene of the P1 operon, P40 and P90 (55, 83)) and P30 (20); see also the chapter "Cytadherence and cytoskeleton".

*Table 1.* Classification of proposed and verified genes (88)

| Predicted isoelectric point | No. of transmembrane segments | | | | Number of proteins |
|---|---|---|---|---|---|
| | 0 | 1-3 | 4-6 | 7-14 | |
| <9.50 | 140 a /15 b (96 c ) | 54/7 (44) | 2 (3) | 1 (2) | 197/24 (45) |
| 9.51 – 10.50 | 29/11 (73) | 26/6 (63) | 2/1 /13) | 2/2 (18) | 59/20 (167) |
| 10.51 – 11.50 | 31/21 (24) | 4/3 (14) | -/- (16) | 1/1 (8) | 36/25 (62) |
| 11.51 – 12.31 | 13/13 (7) | -/- (2) | -/- (-) | -/- (-) | 13/13 (9) |
| | | | | | Σ = 305/80 (383) |

a complete number of protein coding genes identified in both 1-D- and 2-D-gels
b number of protein coding genes identified in 1D gels only
c proposed genes not yet verified experimentally.

In addition, 11 proteins were functionally not assigned but some of them appeared to be good candidates to be considered as structural proteins. Structural prediction programs indicated coiled-coil (64) structures or transmembrane segments (15, 68), both are useful features for building extended protein structures in cells. Unexpected was the high concentration of the heat shock protein DnaK and the elongation factor Tu (EF-Tu) and the presence of several enzymes or enzyme subunits participating normally in different metabolic pathways. Although it seems unlikely that these enzymes are also involved in the formation of a structural network, the possible role of DnaK and EF-Tu has to be further analyzed in this respect.

The benefits of proteome analysis for defining genes were demonstrated by the finding that five proteins which were identified by mass spectrometry (22, 88) were not included in the original annotation of the *M. pneumoniae* genome sequence. Three of these proteins were smaller than the 100 amino acids cut-off point in the first annotation (50) for ORFs without functional assignment. The other two genes code for 142 and 157 amino acid long proteins without known functions. They were located in former "intergenic regions", and have probably been overlooked during the original annotation.

*Table 2.* Comparison of gene expression on protein and mRNA levels at 37°C

| MPN[1] | ORF[2] | Annotation[3] | Protein[4] | RNA[5] |
|---|---|---|---|---|
| 434 | 406 | heat shock protein DnaK | 18.28 | 1.12 |
| 665 | 177 | elongation factor TU (tuf) | 17.26 | 2.29 |
| 393 | 446 | pyruvate dehydrogenase E1-alpha subunit (pdhA) | 7.32 | 1.75 |
| 573 | 269 | heat shock protein GroEL | 6.41 | 1.56 |
| 392 | 447 | pyruvate dehydrogenase E1-beta subunit (pdhB) | 6.01 | 2.08 |
| 606 | 236 | enolase (eno) (EC 4.2.1.11) | 3.76 | 1.91 |
| 323 | 514 | ribonucleotide reductase type I (nrdI) | 2.40 | 1.19 |
| 428 | 412 | phosphotransacetylase (pta) | 2.33 | 0.66 |
| 314 | 523 | conserved hypothetical | 2.23 | 1.36 |
| 668 | 174 | conserved hypothetical | 1.79 | 1.34 |
| 389 | 450 | lipoate protein ligase (lplA) | 1.50 | -0.29 |
| 25 | 129 | fructose-bisphosphate aldolase (tsr) | 1.46 | 0.40 |
| 555 | 287 | conserved hypothetical | 1.28 | 0.85 |
| 598 | 244 | ATP synthase beta chain (atpD) | 0.92 | 0.86 |
| 191 | 641 | RNA polymerase alpha core subunit (rpoA) | 0.88 | 0.40 |
| 539 | 303 | ribosomal protein L7/L12 ('A' type) (rpL7/L12) | 0.88 | -0.48 |
| 331 | 506 | trigger factor (tig) | 0.85 | 2.02 |
| 429 | 411 | phosphoglycerate kinase (pgk) | 0.80 | 0.82 |
| 401 | 438 | transcription elongation factor (greA) | 0.79 | 1.38 |
| 600 | 242 | ATP synthase alpha chain (atpA) | 0.72 | 1.67 |
| 533 | 309 | acetate kinase (ackA) | 0.71 | 0.21 |
| 674 | 168 | L-lactate dehydrogenase (ldh) | 0.63 | 1.40 |
| 295 | 541 | conserved hypothetical | 0.62 | 0.73 |
| 227 | 605 | elongation factor G (fus) | 0.62 | -0.66 |
| 120 | 35 | heat shock protein GrpE | 0.58 | 0.40 |
| 528 | 314 | inorganic pyrophosphatase (ppa) | 0.57 | -0.18 |

[1]Gene numbering according to Dandekar et al. (2000), ref. 22
[2]Old gene numbering according to Himmelreich et al. (1996), ref. 50
[3]Annotated function
[4]Measured spot volumes of proteins from a 2D-gel (75,88)
[5]Relative expression strength of mRNA (39)

Although the proteome analysis of *M. pneumoniae* is far from being completed, it already permits us to draw some general conclusions concerning the methodology based on statistical evaluation of proteins which could be identified and those which were missed (Table 1). When the proteins were classified according to *pI* and the number of proposed transmembrane segments as a single criterion for membrane proteins, pattern emerged strongly suggesting that proteins with more than three transmembrane segments and $pI$'s above 10.5 were significantly under-represented in a proteome analysis based on 2-D gel electrophoresis. But the results also hinted that at least 276 proposed *M. pneumniae* proteins, not yet identified, with a $pI$ below 10.5 and less than four transmembrane segments (Table 1), have a realistic chance to be defined, provided that they are synthesized in sufficient amounts (50-100 femtomoles per spot in a 2-D gel) and are not masked by other overlapping proteins (88).

In general, the positions of individual proteins in the 2-D gels corresponded to the DNA sequence-based predicted $pI$'s and molecular masses. Exceptions were the abnormally migrating proline-rich acidic proteins (HMW1, HMW3, P65, P200, see ref. 70) and the lipoproteins of the murein lipoprotein type of *Escherichia coli* (9) which, in most cases, were identified as subfragments. Of 20 lipoproteins, 15 appeared to be processed. One of the not processed lipoproteins was subunit b of the $F_0F_1$-type ATPase (71). It is not yet decided whether the fragmentation of lipoproteins is a *M. pneumoniae* specific processing step or a result of proteolytic activities in the growth medium.

Finally, comparing the data from the transcriptome and proteome analysis of *M. pneumoniae* shows clearly that the most abundant proteins in total protein extracts from *M. pneumoniae* are not translated from mRNA species with the highest signal strength. This stresses again the importance of the proteome analysis (Table 2).

For more detailed information on transcriptome and proteome analyses of *M. pneumoniae* see http://www.zmbh.uni-heidelberg.de/M_pneumoniae.

## 6. COMPARISON OF THE PROTEOMES OF *M. GENITALIUM* AND *M. PNEUMONIAE*

Comparing the data of the proteome analysis of *M. genitalium* and *M. pneumoniae* reveals some interesting differences, although both analyses are still incomplete. Almost all of the gene products identified in protein extracts at *M. genitalium* were also found in *M. pneumoniae*, except for the two proteins RecA and DnaA. These proteins were not only expressed in *M. genitalium*, but also in other mycoplasmas. Thus, DnaA is a highly expressed protein in *Mycoplasma capricolum* (80) and RecA was identified in several mollicute species (see also the chapter "DNA replication, repair and stress response"). Based on the predicted $pI$ of 9.77 for RecA and 9.57 for DnaA and the absence of transmembrane segments, both proteins should have been identified in *M. pneumoniae* assuming a "normal" state of expression. The absence of RecA in *M. pneumoniae* would explain the failure to transform *M. pneumoniae* by inserting additional DNA via homologous recombination into the genome, a procedure, which worked for *M. genitalium* (24). In support of these protein data, the transcriptional analyses of *M. pneumoniae* (39) showed also only weak signals for the recA and dnaA genes indicating a low number of mRNA copies.

*Mycoplasma genitalium* and *M. pneumoniae* are phylogenetically very closely related species and all the proposed ORFs from *M. genitalium* are contained within the larger genome of *M. pneumoniae* (47, 49). It seems justified, therefore, to extend results on orthologous genes between both species. For instance, Hutchison and colleagues (51) in their interpretation of the results of their global transposon mutagenesis of *M. genitalium* concluded that orthologous genes which were inactivated in one species were also dispensable in the other one and vice versa. Following this reasoning, we argue that orthologous genes which are expressed in *M. pneumoniae* are also expressed in *M. genitalium*. This would mean that 284

of the 305 protein-coding genes of *M. pneumoniae* (21 genes are *M. pneumoniae* – specific) are also translated in *M. genitalium*. Adding the RecA and DnaA proteins, the expression of 286 of the 480 proposed ORFs(~ 60%) from *M. genitalium* would be indirectly confirmed.

## 7.    CONCLUSION

So far, transcriptome and proteome analyses of Mollicutes are mainly concerned with proving the expression of the proposed genes, which is an important question to answer for such small cells. However, there are many more applications of the transcriptome and proteome methodology not discussed in this chapter. To provide some ideas about the potential of these methods the reader is referred to several selected references concerning either transcriptome (2, 4, 19, 31, 45, 63, 77, 79, 99) or proteome (3, 5, 6, 10, 11, 32, 43, 54, 62, 89, 93) analyses or studies combining both approaches (30).

## ACKNOWLEDGEMENTS

Work in our laboratory has been supported by the Graduiertenkolleg "Pathogene Mikroorganismen: Molekulare Mechanismen und Genome", grants from the Deutsche Forschungsgemeinschaft and by the Fonds der Chemischen Industrie.

# REFERENCES

1.  **Adams, M.D., J.M. Kelley, J.D. Gocayne, M. Dubnick, M.H. Polymeropoulos, H. Xiao, C.R. Merril, A. Wu, B. Olde, R.F. Moreno and et al.** 1991. Complementary DNA sequencing: expressed sequence tags and human genome project. *Science*, **252**:1651-1656.

2.  **Ang, S., C.Z. Lee, K. Peck, M. Sindici, U. Matrubutham, M.A. Gleeson and J.T. Wang.** 2001. Acid-induced gene expression in *Helicobacter pylori*: study in genomic scale by microarray. *Infect. Immun.*, **69**:1679-1686.

3.  **Antelmann, H., C. Scharf and M. Hecker.** 2000. Phosphate starvation-inducible proteins of *Bacillus subtilis*: proteomics and transcriptional analysis. *J. Bacteriol.*, **182**: 4478-4490.

4.  **Arfin, S.M., A.D. Long, E.T. Ito, L. Tolleri, M.M. Riehle, E.S. Paegle and G.W. Hatfield.** 2000. Global gene expression profiling in *Escherichia coli* K12. The effects of integration host factor. *J. Biol. Chem.*, **275**:29672-29684.

5.  **Backert, S., E.-C. Müller, P. R. Jungblut and T. F. Meyer.** 2001. Tyrosine phosphorylation patterns and size modification of the *Helicobacter pylori* CagA protein after translocation into gastric epithelial cells. *Proteomics*, **1**:608-617.

6.  **Bernhardt, J., K. Buttner, C. Scharf and M. Hecker.** 1999. Dual channel imaging of two-dimensional electropherograms in *Bacillus subtilis*. *Electrophoresis*, **20**:2225-2240.

7.  **Blattner, F.R., G. Plunkett 3rd, C.A. Bloch, N.T. Perna, V. Burland, M. Riley, J. Collado-Vides, J.D. Glasner, C.K. Rode, G.F. Mayhew, J. Gregor, N.W. Davis, H.A. Kirkpatrick, M.A. Goeden, D.J. Rose, B. Mau and Y. Shao.** 1997. The complete genome sequence of *Escherichia coli* K-12. *Science*, **277**:1453-1474.

8.  **Bork, P., T. Dandekar, Y. Diaz-Lazcoz, F. Eisenhaber, M. Huynen and Y. Yuan.** 1998. Predicting function: from genes to genomes and back. *J. Mol. Biol.*, **283**:707-25.20

9.  **Braun, V. and K. Rehn.** 1969. Chemical characterization, spacial distribution and function of a lipoprotein of the *E. coli* cell wall. *Eur. J. Biochem.*, **10**:426-438.

10. **Bumann, D., T.F. Meyer and P.R. Jungblut.** 2001. Proteome analysis of the common human pathogen *Helicobacter pylori*. *Proteomics*, **1**:473-479.

11. **Cash, P.** 2000. Proteomics in medical microbiology. *Electrophoresis*, **21**:1187-1201.

12. **Chambaud, I., R. Heilig, S. Ferris, V. Barbe, D. Samson, F. Galisson, I. Moszer, K. Dybvig, H. Wroblewski, A. Viari, E.P. Rocha and A. Blanchard.** 2001. The complete genome sequence of the murine respiratory pathogen *Mycoplasma pulmonis*. *Nucleic Acids Res.*, **29**:2145-2153.

13. **Chee, M., R. Yang, E. Hubbell, A. Berno, X.C. Huang, D. Stern, J. Winkler, D.J. Lockhart, M.S. Morris and S.P. Fodor.** 1996. Accessing genetic information with high-density DNA arrays. *Science*, **274**:610-614.

14. **Cho, R.J., M. Fromont Racine, L. Wodicka, B. Feierbach, T. Stearns, P. Legrain, D.J. Lockhart and R.W. Davis.** 1998. Parallel analysis of genetic selections using whole genome oligonucleotide arrays. *Proc. Natl. Acad. Sci. USA*, **95**:3752-3757.

15. **Claros, M.G. and G. von Heijne.** 1994. TopPred II: an improved software for membrane protein structure predictions. *Comput. Appl. Biosci.*, **10**:685-686.

16. **Cohen, G., J. Deutsch, J. Fineberg and A. Levine.** 1997. Covalent attachment of DNA oligonucleotides to glass. *Nucleic Acids Res.*, **25**:911-912.

17.  **Cordwell, S.J., M.R. Wilkins, A. Cerpa-Poljak, A.A. Gooley, M. Duncan, K.L. Williams and I. Humphery-Smith.** 1995. Cross-species identification of proteins separated by two-dimensional gel electrophoresis using matrix-assisted laser desorption ionisation/time-of-flight mass spectrometry and amino acid composition. *Electrophoresis*, **16**:438-443.

18.  **Cordwell, S.J., D.J. Basseal and I. Humphery-Smith.** 1997. Proteome analysis of *Spiroplasma melliferum* (A56) and protein characterisation across species boundaries. *Electrophoresis*, **18**:1335-1346.

19.  **Cox, J.M., C.L. Clayton, T. Tomita, D.M. Wallace, P.A. Robinson and J.E. Crabtree.** 2001. cDNA Array analysis of cag pathogenicity island-Associated *Helicobacter pylori* epithelial cell response genes. *Infect. Immun.*, **69**:6970-6980.

20.  **Dallo, S.F. and J. Morrow.** 1990. Characterization of the gene for a 30-kilodalton adhesin-related protein of *Mycoplasma pneumoniae*. *Infect. Immun.*, **58**:4163-4165.

21.  **Danchin, A. and A. Sekowska.** 2000. Expression profiling in reference bacteria: dreams and reality. *Genome Biol.*, **1**:1024.1-1024.5.

22.  **Dandekar, T., M. Huynen, J.T. Regula, B. Ueberle, C.U. Zimmermann, M. Andrade, T. Doerks, L. Sanchez-Pulido, B. Snel, M. Suyama, Y.P. Yuan, R. Herrmann and P. Bork.** 2000. Re-annotating the *Mycoplasma pneumoniae* genome sequence: adding value, function and reading frames. *Nucleic Acids Res.*, **28**:3278-3288.

23.  **Devaux, F., P. Marc and C. Jacq.** 2001. Transcriptomes, transcription activators and microarrays. *FEBS Lett.*, **498**:140-144.

24.  **Dhandayuthapani, S., W. G. Rasmussen and J. B. Baseman.** 1999. Disruption of gene *218* of *Mycoplasma genitalium* through homologous recombination leads to an adherence-deficient phenotype. *Proc. Natl. Acad. Sci. USA*, **96**:5227-5232.

25.  **Dirksen, L.B., T. Proft, H. Hilbert, H. Plagens, R. Herrmann and D.C. Krause.** 1996. Sequence analysis and characterization of the hmw gene cluster of *Mycoplasma pneumoniae*. *Gene*, **171**:19-25.

26.  **Duffy, M.F., I.D. Walker and G.F. Browning.** 1997. The immunoreactive 116 kDa surface protein of *Mycoplasma pneumoniae* is encoded in an operon. *Microbiology*, **143**: 3391-3402.

27.  **Dunn, M.J. (ed.).** 2001. From genome to proteome. Proceedings of the fourth Siena 2D electrophoresis meeting. *Electrophoresis*, **22**:1593-1877.

28.  **Eng, J.K., A.L. McCormack and J.R.I. Yates.** 1994. An approach to correlate tandem mass spectral data of peptides with amino acid sequences in a protein database. *J. Amer. Soc. Mass Spectrometry*, **5**:976-989.

29.  **Erickson, H.P.** 1995. FtsZ, a prokaryotic homolog of tubulin. *Cell*, **80**:367-370.

30.  **Evers, S., K. Di Padova, M. Meyer, H. Langen, M. Fountoulakis, W. Keck and C. P. Gray.** 2001. Mechanism-related changes in the gene transcription and protein synthesis patterns of *Haemophilus influenzae* after treatment with transcriptional and translational inhibitors. *Proteomics*, **1**:522-544.

31.  **Fawcett, P., P. Eichenberger, R. Losick and P. Youngman.** 2000. The transcriptional profile of early to middle sporulation in *Bacillus subtilis*. *Proc. Natl. Acad. Sci. USA*, **97**:8063-8068.

32.  **Fey, S.J., A. Nawrocki, M.R. Larsen, A. Gorg, P. Roepstorff, G.N. Skews, R. Williams and P.M. Larsen.** 1997. Proteome analysis of *Saccharomyces cerevisiae*: a methodological outline. *Electrophoresis*, **18**:1361-1372.

33.  **Fleischmann, R.D., M.D. Adams, O. White, R.A. Clayton, E.F. Kirkness, A.R. Kerlavage, C.J. Bult, J.F. Tomb, B.A. Dougherty, J.M. Merrick and et al.** 1995.

Whole-genome random sequencing and assembly of *Haemophilus influenzae* Rd. *Science*, **269**:496-512.

34. **Fodor, S.P.** 1997. Massively parallel genomics. *Science*, **277**:393-395.

35. **Fodor, S.P.A., J.L. Read, M.C. Pirrung, L. Stryer, A.T. Lu and D. Solas.** 1991. Light-directed, spatially addressable parallel chemical synthesis. *Science*, **251**:767-773.

36. **Fountoulakis, M., B. Takacs and H. Langen.** 1998. Two-dimensional map of basic proteins of *Haemophilus influenzae*. *Electrophoresis*, **19**:761-766.

37. **Fraser, C.M., J.D. Gocayne, O. White, M.D. Adams, R.A. Clayton, R.D. Fleischmann, C.J. Bult, A.R. Kerlavage, G. Sutton, J.M. Kelley and et al.** 1995. The minimal gene complement of *Mycoplasma genitalium*. *Science*, **270**:397-403.

38. **Göbel, U., V. Speth and W. Bredt.** 1981. Filamentous structures in adherent *Mycoplasma pneumoniae* cells treated with nonionic detergents. *J. Cell. Biol.*, **91**:537-543.

39. **Göhlmann, H.W., C.U. Zimmermann, J. Weiner 3rd and R. Herrmann.** 2002. Transcription profiles of *Mycoplasma pneumoniae* grown at different temperatures. Submitted.

40. **Göhlmann, H. W. H., J. Weiner 3rd, A. Schön and R. Herrmann.** 2000. Identification of a Small RNA within the pdh Gene Cluster of *Mycoplasma pneumoniae* and *Mycoplasma genitalium*. *J. Bacteriol.*, **182**:3281-3284.

41. **Görg, A., C. Obermaier, G. Boguth, A. Harder, B. Scheibe, R. Wildgruber and W. Weiss.** 2000. The current state of two-dimensional electrophoresis with immobilized pH gradients. *Electrophoresis*, **21**:1037-1053.

42. **Glass, J.I., E.J. Lefkowitz, J.S. Glass, C.R. Heiner, E.Y. Chen and G.H. Cassel.** 1998. The complete sequence of the mucosal pathogen *Ureaplasma urealyticum*. *Nature*, **407**:757-62.

43. **Grunenfelder, B., G. Rummel, J. Vohradsky, D. Roder, H. Langen and U. Jenal.** 2001. Proteomic analysis of the bacterial cell cycle. *Proc. Natl. Acad. Sci. USA*, **98**:4681-4686.

44. **Gygi, S.P., Y. Rochon, B.R. Franza and R. Aebersold.** 1999. Correlation between protein and mRNA abundance in yeast. *Mol. Cell Biol.*, **19**:1720-1730.

45. **Hauser, N.C., M. Vingron, M. Scheideler, B. Krems, K. Hellmuth, K.D. Entian and J.D. Hoheisel.** 1998. Transcriptional profiling on all open reading frames of *Saccharomyces cerevisiae*. *Yeast*, **14**:1209-1221.

46. **Hecker, M., W. Schumann and U. Volker.** 1996. Heat-shock and general stress response in *Bacillus subtilis*. *Mol. Microbiol.*, **19**:417-428.

47. **Herrmann, R. and B. Reiner.** 1998. *Mycoplasma pneumoniae* and *Mycoplasma genitalium*: a comparison of two closely related bacterial species. *Curr. Opin. Microbiol.*, 1:572-579.

48. **Hertz, G.Z. and G.D. Stormo.** 1999. Identifying DNA and protein patterns with statistically significant alignments of multiple sequences. *Bioinformatics*, **15**:563-577.

49. **Himmelreich, R., H. Plagens, H. Hilbert, B. Reiner and R. Herrmann.** 1997. Comparative analysis of the genomes of the bacteria *Mycoplasma pneumoniae* and *Mycoplasma genitalium*. *Nucleic Acids Res.*, **25**:701-712.

50. **Himmelreich, R., H. Hilbert, H. Plagens, E. Pirkl, B.C. Li and R. Herrmann.** 1996. Complete sequence analysis of the genome of the bacterium *Mycoplasma pneumoniae*. *Nucleic Acids Res.*, **24**:4420-4449.

51.  **Hutchison, C., S. Peterson, S. Gill, R. Cline, O. White, C. Fraser, H. Smith and J. Venter.** 1999. Global transposon mutagenesis and a minimal Mycoplasma genome. *Science*, **286**:2165-2169.

52.  **Huynen, M., B. Snel, W. Lathe 3rd and P. Bork.** 2000. Predicting protein function by genomic context: quantitative evaluation and qualitative inferences. *Genome Res.*, **10**: 1204-10.

53.  **Hyman, H.C., R. Gafny, G. Glaser and S. Razin.** 1988. Promoter of the Mycoplasma pneumoniae rRNA operon. *J. Bacteriol.*, **170**:3262-3268.

54.  **Ideker, T., V. Thorsson, J.A. Ranish, R. Christmas, J. Buhler, J.K. Eng, R. Bumgarner, D.R. Goodlett, R. Aebersold and L. Hood.** 2001. Integrated genomic and proteomic analyses of a systematically perturbed metabolic network. *Science*, **292**:929-934.

55.  **Inamine, J.M., S. Loechel and P.C. Hu.** 1988. Analysis of the nucleotide sequence of the P1 operon of *Mycoplasma pneumoniae*. *Gene*, **73**:175-183.

56.  **James, P., M. Quadroni, E. Carafoli and G. Gonnet.** 1994. Protein identification in DNA databases by peptide mass fingerprinting. *Protein Sci.*, **3**:1347-1350.

57.  **Jungblut, P., B. Thiede, U. Zimny-Arndt, E.-C. Müller, C. Scheler, B. Wittmann-Liebold and A. Otto.** 1996. Resolution power of two-dimensional electrophoresis and identification of proteins from gels. *Electrophoresis*, **17**:839-847.

58.  **Koonin, E.V., R.L. Tatusov and M.Y. Galperin.** 1998. Beyond complete genomes: from sequence to structure and function. *Curr. Opin. Struct. Biol.*, **8**:355-363.

59.  **Krause, D.C.** 1996. *Mycoplasma pneumoniae* cytadherence: unraveling the tie that binds. *Mol. Microbiol.*, **20**:247-253.

60.  **Krause, D. and M. Balish.** 2001. Structure, function, and assembly of the terminal organelle of *Mycoplasma pneumoniae*. *FEMS Microbiol. Lett.*, **198**:1-7.

61.  **Kunst, F., N. Ogasawara, I. Moszer, A.M. Albertini, G. Alloni, V. Azevedo, M.G. Bertero, P. Bessieres, A. Bolotin, S. Borchert, R. Borriss, L. Boursier, A. Brans, M. Braun, S.C. Brignell, S. Bron, S. Brouillet, C.V. Bruschi, B. Caldwell, V. Capuano,N.M. Carter, S.K. Choi, J.J. Codani, I.F. Connerton, A. Danchin and et al.** 1997.The complete genome sequence of the gram-positive bacterium *Bacillus subtilis*. *Nature*,**390**:249-256.

62.  **Langen, H., B. Takacs, S. Evers, P. Berndt, H.W. Lahm, B. Wipf, C. Gray and M. Fountoulakis.** 2000. Two-dimensional map of the proteome of *Haemophilus influenzae*. *Electrophoresis*, **21**:411-429.

63.  **Lockhart, D.J., H. Dong, M.C. Byrne, M.T. Follettie, M.V. Gallo, M.S. Chee, M. Mittmann, C. Wang, M. Kobayashi, H. Horton and E.L. Brown.** 1996. Expression monitoring by hybridization to high-density oligonucleotide arrays. *Nat Biotechnol.*, **14**: 1675-1680.

64.  **Lupas, A.** 1996. Coiled coils: new structures and new functions. *Trends Biochem. Sci.*, **21**:375-382.

65.  **Mann, M. and M. Wilm.** 1995. Electrospray mass spectrometry for protein characterization. *Trends Biochem. Sci.*, **20**:219-224.

66.  **Meng, K.E. and R.M. Pfister.** 1980. Intracellular structures of *Mycoplasma pneumoniae* revealed after membrane removal. *J. Bacteriol.*, **144**:390-399.

67.  **Okubo, K., N. Hori, R. Matoba, T. Niiyama, A. Fukushima, Y. Kojima and K. Matsubara.** 1992. Large scale cDNA sequencing for analysis of quantitative and qualitative aspects of gene expression. *Nat Genet*, **2**:173-179.

68.  **Pasquier, C., V.J. Promponas, G.A. Palaios, J.S. Hamodrakas and S.J.**

Hamodrakas. 1999. A novel method for predicting transmembrane segments in proteins based on a statistical analysis of the SwissProt database: the PRED-TMR algorithm. *Protein Eng.*, **12**:381-385.

69. **Pease, A.C., D. Solas, E.J. Sullivan, M.T. Cronin, C.P. Holmes and S.P.A. Fodor.** 1994. Light-generated oligonucleotide arrays for rapid DNA sequence analysis. *Proc. Natl. Acad. Sci. USA*, **91**:5022-5026.

70. **Proft, T., H. Hilbert, H. Plagens and R. Herrmann.** 1996. The P200 protein of *Mycoplasma pneumoniae* shows common features with the cytadherence-associated proteins HMW1 and HMW3. *Gene*, **171**:79-82.

71. **Pyrowolakis, G., D. Hofmann and R. Herrmann.** 1998. The subunit b of the F0F1-type ATPase of the bacterium *Mycoplasma pneumoniae* is a lipoprotein. *J. Biol. Chem.*, **273**:24792-24796.

72. **Ramsay, G.** 1998. DNA chips: state-of-the art. *Nat Biotechnol.*, **16**:40-44.

73. **Razin, S., D. Yogev and Y. Naot.** 1998. Molecular Biology and Pathogenicity of Mycoplasmas. *Microbiol. Mol. Biol. Rev.*, **62**:1094-1156.

74. **Regula, J.T., G. Boguth, A. Gorg, J. Hegermann, F. Mayer, R. Frank and R. Herrmann.** 2001. Defining the mycoplasma "cytoskeleton": the protein composition of the Triton X-100 insoluble fraction of the bacterium *Mycoplasma pneumoniae* determined by 2-D gel electrophoresis and mass spectrometry. *Microbiology*, **147**:1045-1057.

75. **Regula, J.T., B. Ueberle, G. Boguth, A. Görg, M. Schnölzer, R. Herrmann and R. Frank.** 2000. Towards a Proteome Map of *Mycoplasma pneumoniae*. *Electrophoresis*, **21**:3765-3780.

76. **Riley, M.** 1993. Functions of the gene products of *Escherichia coli*. *Microbiol. Rev.*, **57**:862-952.

77. **Rimini, R., B. Jansson, G. Feger, T.C. Roberts, M. de Francesco, A. Gozzi, F. Faggioni, E. Domenici, D.M. Wallace, N. Frandsen and A. Polissi.** 2000. Global analysis of transcription kinetics during competence development in *Streptococcus pneumoniae* using high density DNA arrays. *Mol. Microbiol.*, **36**:1279-1292.

78. **Santoni, V., M. Molloy and T. Rabilloud.** 2000. Membrane proteins and proteomics: un amour impossible. *Electrophoresis*, **21**:1054-1070.

79. **Sekowska, A., S. Robin, J-J. Daudin, A Hénaut and A. Danchin.** 2001. Extracting biological informaiton from DNA arrays: an unexpected link betwen arginine and methionine metabolism in *Bacillus subtilis*. *Genome Biol.*, **2(6)**:0019.1-0019.12.

80. **Seto, S., S. Murata and M. Miyata.** 1997. Characterization of dnaA gene expression in *Mycoplasma capricolum*. *FEMS Microbiol. Lett.*, **150**:239-247.

81. **Shevchenko, A., O.N. Jensen, A.V. Podtelejnikov, F. Sagliocco, M. Wilm, O. Vorm, P. Mortensen, H. Boucherie and M. Mann.** 1996. Linking genome and proteome by mass spectrometry: large-scale identification of yeast proteins from two dimensional gels. *Proc. Natl. Acad. Sci. USA*, **93**:14440-14445.

82. **Siuzdak, G.** 1994. The emergence of mass spectrometry in biochemical research. *Proc. Natl. Acad. Sci. USA*, **91**:11290-11297.

83. **Sperker, B., P. Hu and R. Herrmann.** 1991. Identification of gene products of the P1 operon of *Mycoplasma pneumoniae*. *Mol. Microbiol.*, **5**:299-306.

84. **Su, C.J., V.V. Tryon and J.B. Baseman.** 1987. Cloning and sequence analysis of cytadhesin P1 gene from *Mycoplasma pneumoniae*. *Infect. Immun.*, **55**:3023-3029.

85. **Teichmann, S.A., A.G. Murzin and C. Chothia.** 2001. Determination of protein function, evolution and interactions by structural genomics. *Curr. Opin. Struct. Biol.*, **11**:354-363.

86. **Thornton, J.M.** 2001. From genome to function. *Science*, **292**:2095-2097.

87. **Trachtenberg, S.** 1998. Mollicutes-wall-less bacteria with internal cytoskeletons. *J. Struct. Biol.*, **124**:244-256.

88. **Ueberle, B., R. Frank and R. Herrmann.** 2002. The Proteome of the bacterium *Mycoplasma pneumoniae*: comparing predicted open reading frames to identify gene products. *Proteomics*. Accepted for publication.

89. **VanBogelen, R.A., E.E. Schiller, J.D. Thomas and F.C. Neidhardt.** 1999. Diagnosis of cellular states of microbial organisms using proteomics. *Electrophoresis*, **20**:2149-2159.

90. **Velculescu, V.E., L. Zhang, W. Zhou, J. Vogelstein, M.A. Basrai, D.E. Bassett, Hieter Jr., B. Vogelstein and K.W. Kinzler.** 1997. Characterization of the yeast transcriptome. *Cell*, **88**:243-251.

91. **Velculescu, V.E., L. Zhang, B. Vogelstein and K.W. Kinzler.** 1995. Serial analysis of gene expression. *Science*, **270**:484-487.

92. **Waldo, R.H. 3rd, P.L. Popham, C.E. Romero-Arroyo, E.A. Mothershed, K.K. Lee and D.C. Krause.** 1999. Transcriptional analysis of the hmw gene cluster of *Mycoplasma pneumoniae*. *J. Bacteriol.*, **181**:4978-4985.

93. **Washburn, M.P. and J.R. Yates 3rd.** 2000. Analysis of the microbial proteome. *Curr. Opin. Microbiol.*, **3**:292-297.

94. **Wasinger, V.C., S.J. Cordwell, A. Cerpa-Poljak, J.X. Yan, A.A. Gooley, M.R. Wilkins, M.W. Duncan, R. Harris, K.L. Williams and I. Humphery-Smith.** 1995. Progress with gene-product mapping of the Mollicutes: *Mycoplasma genitalium*. *Electrophoresis*, **16**:1090-1094.

95. **Wasinger, V.C., J.D. Pollack and I. Humphery-Smith.** 2000. The proteome of *Mycoplasma genitalium*. Chaps-soluble component. *Eur. J. Biochem.*, **267**:1571-1582.

96. **Weiner, J. 3rd, R. Herrmann and G. F. Browning.** 2000. Transcription in *Mycoplasma pneumoniae*. *Nucleic Acids Res.*, **28**:4488-4496.

97. **Wilm, M.** 2000. Mass spectrometric analysis of proteins. *Adv. Protein. Chem.*, **54**:1-30.

98. **Wilm, M., A. Shevchenko, T. Houthaeve, S. Breit, L. Schweigerer, T. Fotsis and M. Mann.** 1996. Femtomole sequencing of proteins from polyacrylamide gels by nano-electrospray mass spectrometry. *Nature*, **379**:466-469.

99. **Ye, R.W., W. Tao, L. Bedzyk, T. Young, M. Chen and L. Li.** 2000. Global gene expression profiles of *Bacillus subtilis* grown under anaerobic conditions. *J. Bacteriol.*, **182**: 4458-4465.

# Chapter 13

## DNA Replication, Repair and Stress Response

NIANXIANG ZOU and KEVIN DYBVIG
*Department of Genomics and Pathobiology, University of Alabama at Birmingham,
Birmingham, Alabama, USA*

## 1.     INTRODUCTION

Chromosomal replication is a key event in the life cycle of all living organisms including mycoplasmas, a group closely related to Gram-positive bacteria. The small genomes of this group have attracted attention for the study of a minimum gene set required for a living organism, and the study of mycoplasma DNA replication is important for this purpose (12, 47). The most thoroughly studied replication system is that of *Escherichia coli*. It has served as a model to understand the DNA replication machine (28). The era of complete genome sequencing has opened a new stage for research on the molecular biology of mycoplasmas by identification of mycoplasmal gene homologs that are well-studied in other bacteria. The goal of this chapter is to consolidate findings on mycoplasma DNA replication and repair and put them in the overall picture delineated by studies with other microorganisms, especially *E. coli*.

## 2.     REPLICATION INITIATION

The first short chromosomal DNA sequence that can lead to autonomous DNA replication was identified for *E. coli* by cloning the origin of chromosomal DNA replication, *oriC*. The *oriC* regions have been identified for a broad array of bacteria (64). Although many protein factors have been shown to be involved in the initiation of *E. coli* chromosome replication, the

*Molecular Biology and Pathogenicity of Mycoplasmas*, Edited by Razin and
Herrmann, Kluwer Academic/Plenum Publishers, New York, 2002

DnaA protein is the rate-limiting and most important *trans*-acting factor in this process (40).

## 2.1    DnaA protein

*E. coli* DnaA protein binds to relaxed sequences with a consensus TTATCCACA (DnaA box) present in four repeated copies in *oriC*. DnaA binding to the DnaA boxes leads to opening of the DNA double helix and formation of the "initiation complex". This allows for entry of the DnaB and DnaC proteins to form the "pre-priming complex" that leads to further strand separation. The entry of the replication machinery is then followed to form replication forks (40).

The activity of *E. coli* DnaA is regulated by binding to adenine nucleotides. *E. coli* DnaA has a high affinity for ATP and ADP and also interacts with membrane bilayers (16, 45). The replicatively inactive ADP-bound DnaA can be converted to the active ATP-bound form in the presence of acidic phospholipids, *oriC* and ATP. The opening of the DNA double helix by DnaA is dependent on the energy provided by hydrolyzing the bound ATP to ADP by the intrinsic ATPase activity. Besides initiation functions, DnaA is also a transcription factor regulating the expression of genes downstream of the DnaA boxes (39). The DnaA protein of *M. capricolum* was mainly detected in the membrane preparation (53) suggesting that initiation of chromosomal replication in mycoplasma is also a membrane-affiliated event.

Suzuki *et al.* used degenerate oligonucleotide primers to amplify the conserved region of *dnaA* genes in 15 different mollicute species (56). PCR fragments were amplified only from *Mycoplasma mycoides* subsp. *mycoides*, *Spiroplasma apis* and *Spiroplasma citri*. A comparison of the predicted DnaA proteins from these species and the DnaA proteins of mycoplasmas for which the complete genome sequence is known indicates that the mycoplasmal DnaA proteins are highly diverse. DnaA protein of *M. capricolum* was produced at about 10-fold higher level than DnaA in *E. coli* as analyzed by immunoblotting assay.

## 2.2    Replication origin (*oriC*)

The *oriC* and adjacent DNA sequences normally contain clusters of DnaA boxes, a few A+T-rich sequences, and the *dnaA* and *dnaN* genes (40). Organization of the mycoplasmal *oriC* regions is similar to that of *B. subtilis*. For several mycoplasma species examined, there are multiple copies of DnaA boxes flanking the *dnaA* gene, and *dnaN* is localized close to *dnaA* (Figure 1). *M. pulmonis*, *M. capricolum*, and *S. citri* have a consensus DnaA box sequence similar to that of *E. coli*, TTATCCACA (Table 1). Miyata *et al.* mapped the initiation site of *M. capricolum* by two-dimensional gel-electrophoretic analysis (42-44). Replication was shown to initiate in an 875-bp region downstream of the *dnaA* gene, while candidate DnaA boxes are present in the non-coding region upstream of *dnaA*, at least 655 bp from the initiation site. This large distance between the DnaA binding sites and initiation site may be possible because of the high A+T content of the *oriC* region. The region around the *dnaA* gene of *M. pneumoniae* is marked by an uneven distribution of G and C. The region between nucleotides 4000 and 4850 is non-coding for peptides larger than 40 amino acids and has a G+C content of 30 mol%, well below the average G+C for *M. pneumoniae*, 41 mol% (23). In *M. genitalium*, the untranscribed A+T-rich region between *dnaA* and *dnaN* was deduced by a mathematical analysis to be the initiation region of chromosome replication (32). However, for *M. genitalium, M. pneumoniae,* and *Ureaplasma urealyticum,* consensus DnaA-box sequences have not been identified. Presumably, DnaA proteins of these species bind to sequences that have significantly diverged from the consensus. The identification of DnaA boxes and the analysis of *oriC* of these species would be of interest.

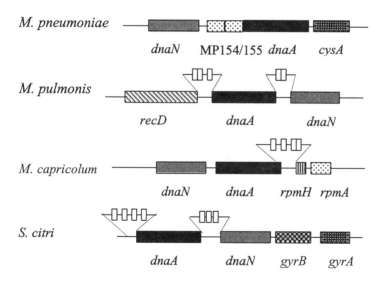

*Figure 1.* Organization of the *oriC* region of select mycoplasma species designating the location of DnaA boxes (small rectangles).

*Table 1.* Nucleotide sequences of putative DnaA boxes

| M. pulmonis | M. capricolum | S. citri |
|---|---|---|
| TTATCCAAA | TTATACACA | CTTTCCACA |
| TTATCAACA | TTTTCCACA | TTTTCCACA |
| TTATCCACA | TTATCAACA | TTTTCCACA |
| TTATCCAAA | TTATCTACA | TTTTCCACA |
| TTATCCAAA |  | TTTTCCACG |
|  |  | TTTTCCACA |
|  |  | TTTTCTACA |

Although some effort has been made to construct autonomously replicating plasmids using the *oriC* (minichromosomes) for mycoplasmas, the system in *S. citri* is the only reported case that functions at present. Ye *et al.* cloned a *S. citri* DNA fragment containing the entire *dnaA* gene and the seven 5'- and 3'-flanking DnaA boxes in a vector containing the tetracycline resistant gene, *tetM* (63). The plasmid was maintained in *S. citri* for at least 15 passages. Renaudin *et al.* constructed a plasmid comprising the *dnaA* region of *S. citri* and *colE1* of *E. coli* (48). They showed that this plasmid can be shuttled between *S. citri* and *E. coli* and that only the DnaA boxes downstream of *dnaA* are essential for plasmid replication in *S. citri*. However, this shuttle plasmid is not stable in *S. citri*. Upon passage, it integrated into the *S. citri* chromosome resulting in duplication of the *oriC* sequence. (see Extrachromosomal Elements and Gene Transfer chapter)

## 3.    REPLICATION FORKS

At the replication origin site, two replication forks are formed by the cooperation of a series of DNA replication proteins. The replication forks progress bidirectionally. At 37°C, the processivity of the DNA replication fork is about $10^2$-$10^3$ nucleotides (nt)/sec. Seto and Miyata (1998) determined that replication fork progression in *M. capricolum* was about $10^2$ nt/sec, which is about 10 times slower than that of *E. coli* (52). The DNA polymerase can only synthesize nascent chains in the 5' to 3' direction. Due to the double helix nature of DNA strand, the elongation process is asymmetrical. A single priming event is required for the leading strand synthesis. Repeated synthesis of RNA primers is needed for the lagging strand to initiate synthesis of short DNA chains (Okazaki fragments). Completion of the lagging strand requires a set of enzymes to remove the primers, fill in the resulting gaps, and join together the short nascent DNA strands (37).

## 3.1 DNA polymerase III holoenzyme

The main DNA polymerase involved in prokaryotic chromosomal DNA replication is the polymerase III holoenzyme (PolIII HE). In *E. coli*, PolIII HE is composed of two identical functional units, each containing 10 different subunits, $\alpha$ (DnaE, polymerase subunit), $\varepsilon$ (DnaQ, 3'-5' editing-exonuclease), $\theta$ (HolE, core component), $\tau$ (DnaX, dimerizes core), $\gamma$ (DnaX, clamp loader), $\delta$ (HolA, clamp loader), $\delta$' (HolB, clamp loader), $\chi$ (HolC, clamp loader), $\psi$ (HolD, clamp loader), $\beta$ (DnaN, sliding clamp). Different subassemblies can be isolated in the purification of the *E. coli* PolIII HE. In ascending order of processivities, they are: the core ($\alpha\varepsilon\theta$), PolIII' ($\alpha\varepsilon\theta$)$_2\tau_2$, and PolIII* ($\alpha\varepsilon\theta$)$_2\tau_2\gamma_2\delta\delta$'$\chi\psi$. The HE is completed by the addition of ($\beta_2$)$_2$ to PolIII* (37).

In Gram-positive bacteria, DNA polymerase and 3'-5' editing-exonuclease activities are in the same PolIII subunit, PolC. In contrast, the polymerase (DnaE) and the 3'-5' exonulcease (DnaQ) are located on separate PolIII subunits in *E. coli*. PolC has been purified from several Gram-positive bacteria (2, 34), and *M. pulmonis* (3). In each case, the final preparation contains only a single subunit possessing polymerase and proofreading exonuclease activity. Table 2 compares the subunit compositions of PolIII HE of *E. coli* and *B. subtilis*. Peptides homologous to *E. coli* $\theta$, $\delta$, $\chi$, and $\psi$ (core component and three clamp loaders) are not encoded by the genomes of *B. subtilis* and mycoplasmas. It remains unclear whether Gram-positive bacteria express the $\gamma$ subunit from the *dnaX*, which, in *E. coli*, encodes both $\tau$ subunit and $\gamma$ subunit through a translational frameshift (59). As deduced from genome sequence analysis, *M. genitalium* (14), *M. pulmonis* (7), *M. pneumoniae* (24), and *U. urealyticum* (18) are each predicted to have the same repertoire of PolIII HE subunits as does *B. subtilis*.

*Table 2.* Subunit composition of PolIII HE.

| E. coli | B. subtilis | Functions |
|---|---|---|
| DnaE ($\alpha$) | DnaE | DNA polymerase |
| | PolC | DNA polymerase and 3' to 5' exonuclease |
| DnaQ ($\varepsilon$) | | 3' to 5' exonuclease |
| HolE ($\theta$) | | Core component, stimulates $\varepsilon$ |
| DnaX ($\tau$) | DnaX | Dimerizes core, binds $\gamma$ complex |
| DnaX ($\gamma$) | | Clamp loader, binds ATP |
| HolA ($\delta$) | | Clamp loader, binds to $\beta$ |
| HolB ($\delta$') | HolB | Clamp loader, binds to $\gamma$ and $\delta$' |
| HolC ($\chi$) | | Clamp loader, binds to SSB |
| HolD ($\psi$) | | Clamp loader, binds to $\chi$ and $\gamma$ |
| DnaN ($\beta$) | DnaN | Sliding clamp |

Experimentally, five proteins of the Gram-positive bacterium *Streptococcus pyogenes* PolIII, PolC and the τ, δ, δ', and β subunits, are sufficient to achieve rapid DNA synthesis, reaching a high speed of 700 nt/sec (5). *S. pyogenes* only produces a full-length τ by *dnaX*. The *S. pyogenes* τ binds PolC weakly as compared to that of *E. coli* (5). This may account for the inability to purify the PolIII HE from Gram-positive bacteria.

PolC is an exclusive target for a family of nucleotide analogs including 6-(*p*-Hydroxyphenylazo) uracil (HPUra) (17). Barnes *et al.* used *M. pulmonis* as a model system to study the relationship of chromosomal DNA replication in mycoplasmas to Gram-positive bacteria, especially *B. subtilis* (3). HPUra and three other drugs that belong to the same family inhibited polymerase activity and the growth of mycoplasma cells. The deduced amino acid sequence of the *M. pulmonis* PolC protein revealed three exonuclease domains and a well-separated polymerase domain in the same ORF, similar to PolC of *B. subtilis*.

Koonin and Bork did a comparative analysis between the sequences of the Gram-negative bacterium *Haemophilus influenzae* and *M. genitalium*. In addition to PolC, *M. genitalium* encodes a second protein (DnaE) homologous to the DNA PolIII α subunit (27). The mycoplasmal DnaE protein is highly similar to Gram-negative bacterial DnaE, which lacks the editing-exonuclease domain. In contrast, *H. influenzae* encodes only a single PolIII α subunit (DnaE). They suggested that a horizontal transfer of a *dnaE* gene from a Gram-negative bacterium to Gram-positive bacteria could account for this gene duplication. Indeed, the complete genome sequences of a variety of Gram-positive bacterial and mycoplasmal species has confirmed the presence of two different PolIII α subunits, PolC and DnaE. Hence, the putative gene transfer occurred near the diverging point of the Gram-positive and Gram-negative bacteria.

Assuming PolC is the main replicase for Gram-positive bacteria and mycoplasmas, what is the role of DnaE? Hutchison III *et al.* screened a transposon library of *M. genitalium* for mutations in various genes (26). They reported the identification of a library member in which the transposon had inserted into *dnaE* at a site that should effectively knockout gene activity, suggesting that *dnaE* is not essential for cell viability. DnaE in mycoplasmas may have a role in nonessential DNA replication processes such as DNA repair. This hypothesis provides an explanation for mycoplasmas to have dispensed with the synthetic activity of DNA polymerase I, as discussed below. However, it has recently been shown that both DnaE and PolC are essential DNA polymerases required for DNA replication in *B. subtilis* (10a). It is likely that DnaE along with PolC are similarly required for DNA replication in mycoplasmas, which is contrary to the report in *M. genitalium* (26).

## 3.2    Primosomal proteins

Primosome is a multienzyme replication machine that locates at the tip of a replication fork. It translocates along DNA synthesizing short RNA primers to be used to initiate synthesis of the nascent DNA strand. The basic primosome proteins in *E. coli* include DnaB, DnaC, DnaG, PriA, PriB, PriC, and DnaT. DnaB is the replication fork helicase, forming a hexamer as the active form in solution and at the replication fork, and is a DNA-dependent NTPase. DnaB interacts with the DnaC protein in a stoichiometric manner. DnaB also interacts with DnaG and the polymerase subunit (DnaE) in the holoenzyme. By forming the DnaC-DnaB complex, DnaC loads the DnaB helicase to the initiation site in *oriC*, opened by DnaA protein, and has been referred to as "molecular matchmaker". The DnaG primase is an essential enzyme that synthesizes small primer RNAs for the initiation of nascent DNA chain. Other protein factors, DnaT, PriA, PriB and PriC, are involved in *E. coli* φX-type primosome formation (37).

The *M. pneumoniae* (24) and *M. genitalium* (14) genomes both encode a 473-amino-acid peptide that is homologous to DnaB, a 620-amino-acid peptide for primase (DnaG), and a 212-amino-acid truncated primase peptide. The fact that both *M. pneumoniae* and *M. genitalium* encode the truncated primase makes it unlikely that this peptide is a result of functionless gene duplication. Similarly, *M. pulmonis* (7) and *U. urealyticum* (18) encode a DnaB-type helicase and a DnaG-type primase. All mycoplasmas sequenced lack the DnaC-type "matchmaker" protein as well as other primosome protein factors.

## 3.3    Proteins involved in sealing Okazaki fragments

Multiple enzymes are involved in the process of sealing Okazaki fragments. The DNA polymerase I (PolA) protein of *E. coli* possesses three enzymatic activities, a 5'-3' exonuclease activity located in the N-terminal part of the protein (the small domain) and a DNA polymerase activity which, together with a 3'-5' exonuclease activity, is located in the C-terminal part of the protein (the Klenow fragment). The combination of the 5'-3' exonuclease and the polymerase activities results in the so-called nick-translation activity, which is responsible for the removal of the RNA primers and filling the primer gaps in the lagging strand (28). In *E. coli*, the 5'-3' exonuclease activity is essential, while the polymerase domain can be mutated (9). Removing the RNA primers also involves RNase H. RNase H is a family of divalent cation-dependent nucleases specific for the RNA strand of RNA DNA hybrids. DNA ligase is the enzyme to finally join the Okazaki fragments together.

In *M. genitalium*, *M. pneumoniae*, *M. pulmonis*, and *U. urealyticum*, the *polA* gene only encodes the N-terminal 5'-3' exonuclease domain and lacks the synthetic polymerase activity. In a way, this is consistent with the observations that in *E. coli* PolA polymerase activity is not required for viability and other DNA polymerases can compensate for the PolA mutation. The only synthetic DNA polymerase enzymes encoded by the mycoplasmas species are DnaE and PolC. It is almost certain that either or both of these enzymes are responsible for filling the primer gaps between Okazaki fragments in these mycoplasma species. The genomes of *U. ureaplasma* and *M. pulmonis* encode RNase HII (7, 18). The original annotations of the *M. pneumoniae* and *M. genitalium* genomes failed to identify any RNase H gene (14, 24), but these genomes were subsequently shown to encode RNase H (4). Homologs to the *E. coli* DNA ligase gene (*lig*) are identified in all sequenced mycoplasma species.

## 3.4      Replication fork arrest and termination of DNA replication

In circular chromosomes that replicate bidirectionally, two replication forks must eventually meet when DNA replication is about to finish. *E. coli* and *B. subtilis* contain similar sequence-specific replication termination systems (6). The terminator sequences are located about $180^{\circ}$ from the *oriC*. The termination region contains two clusters of *ter* sites of opposed polarity. The terminator protein binds to *ter* sites with high affinity, and the protein-DNA complex arrests replication forks in an orientation-dependent manner. Replication forks approaching from one direction pass through the *ter* cluster that is in the non-arresting direction but cannot pass the second cluster of inverted orientation. The replication fork arrest is achieved by antagonizing helicase-catalyzed DNA unwinding in a polar mode (22).

Homologs of terminator proteins as well as candidates of *ter* sites, have not been found in mycoplasmas whose genomes are completely sequenced. This is not surprising, since in *E. coli* the entire terminus region including *ter* sites and terminator protein gene (*tus*) can be deleted without significantly affecting cell growth, an indication that the sequence-specific termination is not essential (21).

# 4. OTHER PROTEINS INVOLVED IN DNA REPLICATION

## 4.1 Single-stranded DNA binding protein (SSB)

SSB performs a vital role in replication, recombination and repair by binding to intermediate single-stranded DNA and protecting the strands from refolding or nucleolytic attack. *E. coli* SSB is a homotetramer consisting of monomers of 177 amino acids. SSB binds single-stranded DNA (ssDNA) tightly and cooperatively and can lower the melting temperature of double-stranded DNA (dsDNA) significantly. SSB also functions through direct interactions with other proteins including helicase and DNA polymerases (33). All sequenced mycoplasma species encode SSB protein.

## 4.2 Topoisomerases

Topoisomerases catalyze transient strand breaks to release the positive superhelical tensions caused by chromosomal DNA replication and other cellular processes such as transcription, recombination, and DNA repair (31, 49). Two types of topoisomerases are present in prokaryotic organisms. The type I topoisomerases act by making a single-stranded nick, passing another strand through the nick, and changing the linking number by one. Type II topoisomerases act by transiently nicking both strands, passing another double-stranded DNA through the gap and changing the linking number by two. In *E. coli*, type I topoisomerases include topoisomerase I (topo I) and topoisomerase III (topo III). The type II topoisomerases include gyrase (topo II) and topoisomerase IV (topo IV). Topoisomerases have been intensely studied because of their potential as drug targets for antibiotic treatment and in antitumor treatment with inhibitors of eukaryotic topoisomerases (19).

### 4.2.1 Type I topoisomerases

From genome sequencing data, only one type I topoisomerase homolog (topo I) is identified in *M. genitalium*, *M. pneumoniae*, *M. pulmonis*, and *U. urealyticum*. Homologs to the other *E. coli* type I topoisomerase, topo III, have not been identified in mycoplasmas.

### 4.2.2 Type II topoisomerases

Homologs to the *E. coli* DNA gyrase subunit A (GyrA), subunit B (GyrB), topo IV subunit A (ParE), and topo IV subunit B (ParC), have been identified in all species of mycoplasma examined and are most likely indispensable. Both gyrase and topo IV are composed of A and B subunits in the form of $A_2B_2$. Horowitz *et al.* studied the topoisomerase activities in extracts of *Mycoplasma fermentans* K7, incognitus, and *Mycoplasma pirum* (25). They showed that two peptides of 66 and 180 kDa were associated with the topoisomerase activity. The topoisomerase activity of these microorganisms included ATP-dependent supercoil relaxation that could be inhibited by topoisomerase inhibitors.

## 5.      DNA REPAIR AND STRESS RESPONSE

Mycoplasmas encode the smallest gene sets for most biological functions, including DNA repair and stress responses. Table 3 lists genes found in the sequenced mycoplasmal genomes that might be involved in DNA repair, stress response systems, and homologous recombination. Discussed below are the DNA repair and stress response pathways that are possibly utilized by mycoplasmas.

## 5.1      SOS stress response

The SOS stress response system induces basic repair pathways and error-prone translesion DNA synthesis, making it possible for organisms to complete replication in the presence of a certain amount of DNA damage. The SOS response of *E. coli* involves the action of at least 30 genes regulated at the transcriptional level by the LexA repressor protein. LexA inhibits transcription by binding to operator sequences, called SOS boxes, located upstream of the SOS genes. Following damage to DNA, the LexA repressor undergoes proteolysis, in which RecA protein acts as a coprotease, inducing expression of SOS genes (61). Two LexA-regulated genes, *umuD* and *umuC*, are required for damage-induced translesion DNA synthesis. DNA damage induces formation of the dimer $UmuD_2$, which undergoes a RecA-mediated cleavage reaction generating an active $UmuD'_2$ protein. The complex $UmuD'_2C$ functions as an error-prone DNA polymerase (Pol V) (55, 57). A UmuC family protein in *E. coli*, DinB, was also found to function as an SOS translesion DNA polymerase (Pol IV) (60).

*Table 3.* Mycoplasma DNA repair genes

| Repair process and genes | Mycoplasma species with gene homolog[a] |
|---|---|
| SOS response and recombinational repair | |
| *recA* (Recombinase) | Mg, Mpn, Mpu, Uu |
| *mucB* | Mpn, Mpu |
| *recR* | Mpu, Uu |
| *ruvA* | Mg, Mpn, Mpu, Uu |
| *ruvB* | Mg, Mpn, Mpu, Uu |
| *ruvC* | Absent |
| *recB,C* (Exo V subunits) | Absent |
| *recD* (ExoV, subunit) | Mpu |
| Base excision repair | |
| *ung* (Uracil-DNA glycosylase) | Mg, Mpn, Mpu, Uu |
| *fpg* (*mutM*) (Formamidopyrimidine-DNA glycosylase) | Mg, Mpn, Mpu, Uu |
| *nfo* (Endonuclease IV) | Mg, Mpn, Mpu, Uu |
| Nucleotide excision repair | |
| *uvrA,B,C* (Excinuclease ABC) | Mg, Mpn, Mpu, Uu |
| *uvrD* (DNA helicase II) | Mg, Mpn, Mpu, Uu |
| *dnaE* (DNA polymerase with repair role?) | Mg, Mpn, Mpu, Uu |
| *lig* (DNA ligase) | Mg, Mpn, Mpu, Uu |

[a]Mycoplasma species are abbreviated as Mg (*M. genitalium*), Mpn (*M. pneumoniae*), Mpu (*M. pulmonis*) and Uu (*U. urealyticum*).

In all the mycoplasma species examined, the common gene that is involved in the SOS system is *recA* (47). Homologs of the regulatory gene *lexA* as well as most other important SOS genes are not present. Therefore, the presence of a SOS stress response system similar to that of *E. coli* is unlikely. There is little experimental evidence of inducible processes for DNA repair pathways in mycoplasmas. Both *M. pneumoniae* and *M. pulmonis* encode a gene homologous to *mucB* belonging to the *umuC* gene family. However, whether the mycoplasmal MucB protein has translesion DNA polymerase activity is unknown.

## 5.2    Base excision repair

Base excision repair (BER) is an organism's primary pathway to repair DNA lesions induced by a wide variety of genotoxic agents including active oxygen species, alkylating agents, and UV irradiation. BER is a multiprotein process. The major players are DNA glycosylases. DNA glycosylases catalyze the breakage of the glycosyl bond between the damaged base and the DNA sugar-phosphate backbone producing apurinic/apyrimidinic (AP) sites. These sites are processed by AP endonuclease/AP lyase that cleaves the phosphodiester bond either 5' or 3' to the AP site (38). In *E.coli*, both exonuclease III, encoded by the *xthA* gene, and endonuclease IV, encoded by the *nfo* gene, account for the AP endonuclease activity (10).

The enzyme activity of uracil-DNA glycosylase of mycoplasmas is usually low when assayed experimentally, and some mycoplasma species reportedly lack any detectable activity (46). Uracil-DNA glycosylase was purified from *M. lactucae* and DNA containing uracil residues was shown to be the only substrate for the enzyme (62). The product of the reaction was uracil, which was also a noncompetitive inhibitor. The enzyme has a higher Km for uracil-containing DNA than those of the glycosylases of other prokaryotic organisms. The decreased capacity to remove uracil residues from DNA may contribute to the low G+C content of mycoplasmal genomes because incorporated uracil residues, if not repaired, will mispair with A and lead to A+T in daughter chromosomes.

From the complete genome sequences, homologs for the uracil-DNA glycosylase gene (*ung*), the *fpg* gene, and the *nfo* gene have been identified in *M. genitalium*, *M. pneumoniae*, *U. urealyticum*, and *M. pulmonis*. Fpg (formamidopyrimidine-DNA glycosylase) repairs oxidized G residues (7, 8-dihydro-8-oxoguanine), which if not removed would mispair with A leading to G to T mutations.

## 5.3      Nucleotide excision repair

Nucleotide excision repair (NER) has a wide variety of substrates including lesions repaired by BER, but primarily operates on large lesions and bulky adducts that cause significant distortions in DNA conformation (50). Thus, NER is a more elementary repair system for an organism including mycoplasmas. Given the coding capacity of mycoplasmas, NER is also the most efficient pathway for repairing DNA lesions.

In *E. coli*, NER requires UvrA, UvrB, UvrC, DNA helicase II, DNA polymerase I, and DNA ligase. UvrA is a DNA binding protein, while UvrB only associates with DNA in the presence of UvrA. The dimer form, $UvrA_2$, binds to UvrB to form a heterotrimer, $UvrA_2B$. The $UvrA_2B$ complex has DNA helicase activity and induces unwinding of DNA double helix in an ATP-dependent manner. This helicase activity of the $UvrA_2B$ complex enables it to actively scan DNA in search of distortions induced by DNA damage. Upon encountering a damaged site, $UvrA_2$ dissociates from the $UvrA_2B$-DNA complex resulting in a stable UvrB-DNA complex that serves as a binding site for UvrC. The excinuclease activity of UvrB and UvrC makes incisions on the 3' side and 5' side of the lesion, respectively. DNA helicase II and DNA polymerase I function together to remove the short DNA strand between the incision sites. The resulting gap is filled in by DNA polymerase I and ligated by DNA ligase.

Homologs to *uvrA, B,* and *C* are found in mycoplasmas that have sequenced genomes. Mycoplasmal genomes also encode DNA helicase II (UvrD) and DNA ligase. However, PolA in mycoplasmas lacks the synthetic DNA polymerase domain. Therefore, this enzyme is not responsible for filling the gap during NER. For mycoplasmas to have an intact NER pathway, another DNA polymerase must have this role. DnaE is a likely candidate.

## 5.4 Recombinational repair system

It has been argued that the primary role of homologous recombination is DNA repair that is vital for short time survival, while its role in horizontal transfer of genetic material has only later evolved for long-term advantage of the organism (8). Recombination is now known to initiate at a double-strand break (DSB) or single-strand gap in one of the pairing DNA molecules. A recombination-dependent replication process has been proposed to describe the repair of DSBs in *E. coli* (29). This process initiates at D-loops and requires RecA, RecBCD, the recombination hotspot *chi* and the PriA protein. The intrusion by the 3' single-stranded tail of a DSB (produced by RecBCD) serves as the origin-independent replication initiation for repairing of the DSB.

### 5.4.1 RecA protein

RecA is a key protein involved in recombination and repair (41). Two major properties of RecA are promoting DNA strand-exchange and stimulating DNA repair and stress responses. These are two competitive, highly-related processes (20). RecA is capable of forming a helical, multimeric nucleoprotein filament around single-stranded DNA. RecA is activated following DNA damage. The activated form of RecA functions as a coprotease to stimulate autodigestion of some SOS proteins (15).

The *recA* genes are found in all mycoplasma species analyzed thus far. Despite the important functions of RecA, the gene is mutable in most bacteria including mycoplasma. With a DNA fragment homologous to *recA*, Dybvig and Woodard disrupted the *recA* gene of *A. laidlawii* (13). A naturally occurred *recA* mutant due to ochre mutation in the gene was also identified in one strain of *A. laidlawii*. Mutants of *recA* in *A. laidlawii* were shown to be more susceptible to UV irradiation and treatment with the alkylating agent methylmethanesulfonate. Marais *et al.* reported that RecA protein of *S. citri* R8A2 contains only the N-terminal 130 amino acid residues (36). The truncated C-terminal region contains the putative DNA-binding domain. Surprisingly, all strains of *S. citri* and *S. melliferum* analyzed have truncations in the *recA* gene (35). These strains of spiroplasma have high sensitivity to UV irradiation, in agreement with the involvement of RecA in DNA repair (30, 35).

### 5.4.2    RecBCD

RecBCD is an important enzyme for homologous recombination in *E. coli*. It is a heterotrimer composed of one copy of each of the products of the *recB*, *recC*, and *recD* genes (58). The enzyme is an ATP-dependent exonuclease and a DNA helicase. RecBCD enzyme binds to the end of a dsDNA substrate and initiates unwinding and at the same time degrades the 3' terminated strand (11), and the new ssDNA end is loaded with RecA (1). A gene homologous to *recD* has been identified in *M. pulmonis*, while homologs to *recB* and *C* are absent. In *M. genitalium*, *M. pneumoniae*, and *U. urealyticum*, homologs to *recB*, *C* and *D* have not been found.

### 5.4.3    RuvABC

RuvA of *E. coli* specifically binds Holliday junctions in tetrameric form. RuvB is a hexameric ring helicase that, in complex with RuvA, catalyzes branch migration of Holliday junctions. RuvC is an endonuclease specific for Holliday junctions and introduces symmetrical strand cleavage across the point of strand exchange. The RuvAB complex acts at arrested replication forks (51). Holliday junctions are formed at the stalled replication forks in presence of RecBCD and RuvABC. In cells lacking RecBC, RuvABC causes DSBs. In cells proficient for RecBC, the RuvAB complex is uncoupled from RuvC and is involved in restoring stalled replication forks independently of the RuvC endonuclease (51).

Genes encoding homologs of RuvA and RuvB are widespread in bacteria including the mycoplama species that have sequenced genomes, whereas homologs of RuvC are less common. *ruvC* is absent from the complete genome sequences of mycoplasmas, *B. subtilis*, and *Borrelia burgdorferi* (54). The fact that mycoplasmas lack RuvC and RecBC is consistent with the observation in *E. coli* that in absence of RecBC, RuvC causes fatal DSBs at stalled replication forks. It is not known how mycoplasmas deal with stalled replication forks. However, DSB formation may be reduced due to the lack of RuvC, and RuvAB may play an important role in restoring stalled replication forks.

## 6.     CONCLUSIONS

The genome size of most mycoplasma species is roughly 1 megabase, about a quarter the genome size of *E. coli* or *B. subtilis*. The small coding capacity of mycoplasmal genomes allows little redundancy in most cellular functions, even vital ones such as DNA replication. The subunit composition of mycoplasmal DNA polymerase III HE is identical to that of *B. subtilis* and other Gram-positive bacteria, indicating this composition is essential and is highly conserved during divergent evolution. Mycoplasmas have a strong tendency to lose genes that are nonessential for their walled ancestors. This tendency is well represented by the absence of such functions as the DnaC "match-maker" protein and many other primosome factors, DNA polymerase I synthetic activity, sequence-specific termination system for DNA replication, most SOS response proteins, and proteins involved in basic DNA repair processes. The remaining genes provide an important clue to the basic mechanisms of DNA replication and repair processes. An intact NER and a functional BER is retained in mycoplasmas, indicating that these two repair systems are essential for the existence and stability of a bacterial species. The SOS system and the recombinational repair system are truncated. Except for *recA*, other SOS genes exist only sporadically and apparently cannot perform significant stress response functions because other subunits needed to form functional complexes are usually not present. It is possible that some SOS and DNA repair genes are still in the process of being evolutionarily lost through the process of genome reduction. The reduced DNA repair capability of mycoplasmas may cause an increase in genetic instability and contribute to the high species diversification of mycoplasmas.

## ACKNOWLEDGMENTS

We would like to thank Dr. Makoto Miyata for helpful suggestions and Dr. Alain Blanchard for providing information on the *M. pulmonis* genome annotations and *oriC* analysis.

## REFERENCES

1.     **Anderson, D. G., and S. C. Kowalczykowski.** 1997. The translocating RecBCD enzyme stimulates recombination by directing RecA protein onto ssDNA in a *chi*-regulated manner. Cell **90**:77-86.

2.    **Barnes, M. H., and N. C. Brown.** 1979. Antibody to *B. subtilis* DNA polymerase III: use in enzyme purification and examination of homology among replication-specific DNA polymerases. Nucleic Acids Res, **6:**1203-1219.

3.    **Barnes, M. H., P. M. Tarantino, Jr, P. Spacclapoli, N. C. Brown, H. Yu, and K. Dybvig.** 1994. DNA polymerase III of *Mycoplasma pulmonis*: isolation and characterization of the enzyme and its structural gene, *polC.* Mol. Microbiol. **13:**843-854.

4.    **Bellgard, M. I., and T. Gojobori.** 1999. Identification of a ribonuclease H gene in both *Mycoplasma genitalium* and *Mycoplasma pneumoniae* by a new method for exhaustive identification of ORFs in the complete genome sequences. FEBS Lett. **445:**6-8.

5.    **Bruck, I., and M. O'Donnell.** 2000. The DNA replication machine of a gram-positive organism. J. Biol. Chem **275:**28971-28983.

6.    **Bussiere, D. E., and D. Bastia.** 1999. Termination of DNA replication of bacterial and plasmid chromosomes. Mol. Microbiol. **31:**1611-1618.

7.    **Chambaud, I., R. Heilig, S. Ferris, V. Barbe, D. Samson, F. Galisson, I. Moszer, K. Dybvig, H. Wroblewski, A. Viari, E. Rocha, and A. Blanchard.** 2001. The complete genome of the murine respiratory pathogen *Mycoplasma pulmonis*. Nucleic Acids Res. **29:**In press.

8.    **Cox, M. M.** 1993. Relating biochemistry to biology: how the recombinational repair function of RecA protein is manifested in its molecular properties. Bioessays **15:**617-623.

9.    **DeLucia, P., and J. Cairns.** 1969. Isolation of an *E. coli* strain with a mutation affecting DNA polymerase. Nature **224:**164-1166.

10.   **Demple, B., A. Johnson, and D. Fung.** 1986. Exonuclease III and endonuclease IV remove 3' blocks from DNA synthesis primers in $H_2O_2$-damaged *Escherichia coli.* Proc. Natl. Acad. Sci. USA **83:**7731-7735.

10a.  **Dervyn, E., C. Suski, R. Daniel, C. Bruand, J. Chapuis, J. Errington, L. Janniere, and S. D. Ehrlich.** 2001. Two essential DNA polymerases at the baterial replication fork. *Science* **294:**1716-1719.

11.   **Dixon, D. A., and S. C. Kowalczykowski.** 1993. The recombination hotspot *chi* is a regulatory sequence that acts by attenuating the nuclease activity of the *E. coli* RecBCD enzyme. Cell **73:**87-96.

12.   **Dybvig, K., and L. L. Voelker.** 1996. Molecular biology of mycoplasmas. Annu. Rev. Microbiol. **50:**25-57.

13.   **Dybvig, K., and A. Woodard.** 1992. Construction of *recA* mutants of *Acholeplasma laidlawii* by insertional inactivation with a homologous DNA fragment. Plasmid **28:**262-266.

14.   **Fraser, C. F., J. D. Gocayne, O. White, M. D. Adams, R. A. Clayton, R. D. Fleischmann, C. J. Bult, A. R. Kerlavage, G. Sutton, J. M. Kelley, J. L. Fritchman, J. F. Weidman, K. V. Small, M. Sandusky, J. Fuhrmann, D. Nguyen, T. R. Utterback, D. M. Saudek, C. A. Phillips, J. M. Merrick, J.-F. Tomb, B. A. Dougherty, K. F. Bott, P.-C. Hu, T. S. Lucier, S. N. Peterson, H. O. Smith, C. A. Hutchison, III, and J. C. Venter.** 1995. The minimal gene complement of *Mycoplasma genitalium*. Science **270:**397-403.

15.   **Friedberg, E. C., G. C. Walker, and W. Siede.** 1995. DNA Repair and Mutagenesis. ASM Press, Washington, D.C.

16.   **Garner, J., and E. Crooke.** 1996. Membrane regulation of the chromosomal replication activity of *E. coli* DnaA requires a discrete site on the protein. EMBO J. **15:**3477-3485.

17.     **Gass, K. B., R. L. Low, and N. R. Cozzarelli.** 1973. Inhibition of a DNA polymerase from *Bacillus subtilis* by hydroxyphenylazopyrimidines. Proc. Natl. Acad. Sci. USA **70:**103-107.

18.     **Glass, J. I., E. J. Lefkowitz, J. S. Glass, C. R. Heiner, E. Y. Chen, and G. H. Cassell.** 2000. The complete sequence of the mucosal pathogen *Ureaplasma urealyticum.* Nature **407:**757-762.

19.     **Guichard, S. M., and M. K. Danks.** 1999. Topoisomerase enzymes as drug targets. Curr. Opin. Oncol. **11:**482-489.

20.     **Harmon, F. G., W. M. Rehraue, and S. C. Kowalczykowski.** 1996. Interaction of *Escherichia coli* RecA protein with LexA repressor. II. Inhibition of DNA strand exchange by the uncleavable LexA S119A repressor argues that recombination and SOS induction are competitive processes. J. Biol. Chem. **27:**23874-23883.

21.     **Henson, J. M., and P. L. Kuempel.** 1985. Deletion of the terminus region (340 kilobase pairs) from the chromosome of *Escherichia coli.* Proc. Natl. Acad. Sci. USA **82:**3766-3770.

22.     **Hiasa, H., and K. J. Marians.** 1992. Differential inhibition of the DNA translocation and DNA unwinding activities of DNA helicases by the *Escherichia coli* Tus protein. J. Biol. Chem. **267:**11379-11385.

23.     **Hilbert, H., R. Himmelreich, H. Plagens, and R. Herrmann.** 1996. Sequence analysis of 56 kb from the genome of the bacterium *Mycoplasma pneumoniae* comprising the *dnaA* region, the *atp* operon and a cluster of ribosomal protein genes. Nucleic Acids Res, **24:**628-639.

24.     **Himmelreich, R., H. Hilbert, H. Plagens, E. Pirkl, B.-C. Li, and R. Herrmann.** 1996. Complete sequence analysis of the genome of the bacterium *Mycoplasma pneumoniae.* Nucl. Acids Res. **24:**4420-4449.

25.     **Horowitz, S., R. Maor, and E. Priel.** 1997. Characteriztion of DNA topoisomerase activity in two strains of *Mycoplasma fermentans* and in *Mycoplasma pirum.* J. Bacteriol. **179:**6626-6632.

26.     **Hutchinson, C. A., III, S. N. Peterson, S. R. Gill, R. T. Cline, O. White, C. M. Fraser, H. O. Smith, and J. C. Venter.** 1999. Global transposon mutagenesis and a minimal mycoplasma genome. Science **286:**2165-2169.

27.     **Koonin, E. V., and P. Bork.** 1996. Ancient duplication of DNA polymerase inferred from analysis of complete bacterial genomes. TIBS **21:**128-129.

28.     **Kornberg, A., and T. A. Baker.** 1992. DNA Replication. W. H. Freeman and Co., New York.

29.     **Kuzminov, A.** 1999. Recombinational repair of DNA damage in *Escherichia coli* and bacteriophage. Microbiol. Mol. Biol. Rev. **63:**751-813.

30.     **Labarere, J., and G. Barroso.** 1989. Lethal and mutation frequency responses of *Spiroplasma citri* cells to UV irradiation. Mutat. Res. **210:**135-141.

31.     **Linn, S.** 1996. The DNases, Topoisomerases, and Helicases of *Escherichia coli*, p. 764-772. *In* F. C. Neidhardt, R. Curtiss, III, E. C. C. Lin, K. B. Low, B. Magasanik, W. S. Reznikoff, M. Riley, M. Schaechter, and H. E. Umbarger (ed.), *Escherichia coli* and *Salmonella*: Cellular and Molecular Biology. ASM Press, Washington, D.C.

32.     **Lobry, J. R.** 1996. Origin of replication of *Mycoplasma genitalium*. Science **272:**745-746.

33.     **Lohman, T. M., and M. E. Ferrari.** 1994. *Escherichia coli* single-stranded DNA-binding protein: multiple DNA-binding modes and cooperativities. Annu. Rev. Biochem. **63:**527-570.

34.  **Low, R. L., S. A. Rashbaum, and N. R. Cozzarelli.** 1976. Purification and characterization of DNA polymerase III from *Bacillus subtilis*. J. Biol. Chem. **251:**1311-1325.

35.  **Marais, A., J. M. Bove, and J. Renaudin.** 1996. Characterization of the *recA* gene regions of *Spiroplasma citri* and *Spiroplasma melliferum*. J. Bacteriol. **178:**7003-7009.

36.  **Marais, A., J. M. Bove, and J. Renaudin.** 1996. *Spiroplasma citri* virus SpV1-derived cloning vector: deletion formation by illegitimate and homologous recombination in a spiroplasmal host strain which probably lacks a functional *recA* gene. J. Bacteriol. **178:**862-870.

37.  **Marians, K. J.** 1996. Replication fork propagation, p. 749-763. *In* F. C. Neidhardt, R. Curtiss, III, E. C. C. Lin, K. B. Low, B. Magasanik, W. S. Reznikoff, M. Riley, M. Schaechter, and H. E. Umbarger (ed.), *Escherichia coli* and *Salmonella*: Cellular and Molecular Biology. ASM Press, Washington, D.C.

38.  **McCullough, A. K., M. L. Dodson, and R. S. Lloyd.** 1999. Initiation of base excision repair: glycosylase mechanisms and structures. Annu. Rev. Biochem. **68:**255-285.

39.  **Messer, W., and C. Weigel.** 1997. DnaA initiator—also a transcription factor. Mol. Microbiol. **24:**1-6.

40.  **Messer, W., and C. Weigel.** 1996. Initiation of chromosome replication, p. 1579-1601. *In* F. C. Neidhardt, R. Curtiss, III, E. C. C. Lin, K. B. Low, B. Magasanik, W. S. Reznikoff, M. Riley, M. Schaechter, and H. E. Umbarger (ed.), *Escherichia coli* and *Salmonella*: Cellular and Molecular Biology. ASM Press, Washington, D.C.

41.  **Miller, R. V., and T. A. Kokjohn.** 1990. General microbiology of *recA*: environmental and evolutionary significance. Annu. Rev. Microbiol. **44:**365-394.

42.  **Miyata, M., and T. Fukumura.** 1997. Asymmetrical progression of replication forks just after initiation on *Mycoplasma capricolum* chromosome revealed by two-dimensional gel electrophoresis. Gene **193:**39-47.

43.  **Miyata, M., K.-I. Sano, R. Okada, and T. Fukumura.** 1993. Mapping of replication initiation site in *Mycoplasma capricolum* genome by two-dimensional gel-electrophoretic analysis. Nucl. Acids Res. **21:**4816-4823.

44.  **Miyata, M., L. Wang, and T. Fukumura.** 1993. Localizing the replication origin on the physical map of the *Mycoplasma capricolum* genome. J. Bacteriol. **175:**655-660.

45.  **Newman, G., and E. Crooke.** 2000. DnaA, the initiator of *Escherichia coli* chromosomal replication, is located at the cell membrane. J. Bacteriol. **182:**2604-2610.

46.  **Pollack, J. D., M. V. Williams, and R. N. McElhaney.** 1997. The comparative metabolism of the mollicutes (Mycoplasmas): the utility for taxonomic classification and the relationship of putative gene annotation and phylogeny to enzymatic function in the smallest free-living cells. Crit. Rev. Microbiol. **23:**269-354.

47.  **Razin, S., D. Yogev, and Y. Naot.** 1998. Molecular biology and pathogenicity of mycoplasmas. Microbiol. Mol. Biol. Rev. **62:**1094-1156.

48.  **Renaudin, J., A. Marais, E. Verdin, S. Duret, X. Foissac, F. Laigret, and J. M. Bove.** 1995. Integrative and free *Spiroplasma citri oriC* plasmids: expression of the *Spiroplasma phoeniceum* spiralin in *Spiroplasma citri*. J. Bacteriol. **177:**2870-2877.

49.  **Roca, J.** 1995. The mechanisms of DNA topoisomerases. Trends Biochem. Sci. **20:**156-160.

50.  **Sancar, A.** 1996. DNA excision repair. Annu. Rev. Biochem. **65:**43-81.

51.  **Seigneu, M., V. Bidnenko, S. D. Ehrlich, and B. Michel.** 1998. RuvAB acts at arrested replication forks. Cell **95:**419-430.

52.   **Seto, S., and M. Miyata.** 1998. Cell reproduction and morphological changes in *Mycoplasma capricolum*. J. Bacteriol. **182:**256-24.

53.   **Seto, S., S. Mutata, and M. Miyata.** 1997. Characterization of *dnaA* gene expression in *Mycoplasma capricolum*. FEMS microbiol. Lett. **150:**239-247.

54.   **Sharples, G. J., S. M. Ingleston, and R. G. Lloyd.** 1999. Holliday junction processing in bacteria: insights from the evolutionary conservation of RuvABC, RecG, and RusA. J. Bacteriol. **181:**5543-5550.

55.   **Smith, B. T., and G. C. Walker.** 1998. Mutagenesis and more: *umuDC* and the *Escherichia coli* SOS response. Genetics **148:**1599-1610.

56.   **Suzuki, K., M. Miyata, and T. Fukumura.** 1993. Comparison of the conserved region in the *dnaA* gene from three mollicute species. FEMS Microbiol. Lett. **114:**229-233.

57.   **Tang, M., X. Shen, E. G. Frank, M. O'Donnell, R. Woodgate, and M. F. Goodman.** 1999. UmuD'2C is an error-prone DNA polymerase, *Escherichia coli* pol V. Proc. Natl. Acad. Sci. USA **96:**8919-8924.

58.   **Taylor, A. F., and G. R. Smith.** 1995. Monomeric RecBCD enzyme binds and unwinds DNA. J. Biol. Chem. **270:**24451-24458.

59.   **Tsuchihashi, Z., and A. Kornberg.** 1990. Translational frameshifting generates the gamma subunit of DNA polymerase III holoenzyme. Proc. Natl. Acad. Sci. USA **87:**2516-2520.

60.   **Wagne, J., P. Gruz, S. R. Kim, M. Yamada, K. Matsui, R. P. Fuchs, and T. Nohmi.** 1999. The *dinB* gene encodes a novel *E. coli* DNA polymerase, DNA pol IV, involved in mutagenesis. Mol. Cell. Biol. **4:**281-286.

61.   **Walker, G. C.** 1996. The SOS response of *Escherichia coli*, p. 1400-1416. *In* F. C. Neidhardt, R. Curtiss, III, E. C. C. Lin, K. B. Low, B. Magasanik, W. S. Reznikoff, M. Riley, M. Schaechter, and H. E. Umbarger (ed.), *Escherichia coli* and *Salmonella*: Cellular and Molecular Biology. ASM Press, Washington, D.C.

62.   **Williams, M. V., and J. D. Pollack.** 1990. A mollicute (mycoplasma) DNA repair enzyme: purification and characterization of uracil-DNA glycosylase. J. Bacteriol. **172:**2979-2985.

63.   **Ye, F., J. Renaudin, J.-M. Bove, and F. Laigret.** 1994. Cloning and sequencing of the replication origin (*oriC*) of the *Spiroplasma citri* chromosome and construction of autonomously replicating artificial plasmids. Current Microbiol. **29:**23-29.

64.   **Yoshikawa, H., and N. Ogasawara.** 1991. Structure and function of DnaA and the DnaA-box in eubacteria: evolutionary relationships of bacterial replication origins. Mol. Microbiol. **5:**2589-2597.

# Chapter 14

# Transcription and Translation

AKIRA MUTO and CHISATO USHIDA
*Department of Biochemistry and Biotechnology, Faculty of Agriculture and Life Science, Hirosaki University, Bunkyo-cho 3, Hirosaki 036-8561, Japan*

## 1.      INTRODUCTION

Most of our current knowledge of transcription and translation mechanisms in prokaryotes comes from the studies of *Escherichia coli*. These studies have provided detailed insights into many aspects of RNA and protein synthesis, including the components involved in these processes, individual reactions, regulation, kinetics *etc*. It is generally accepted that the mechanisms elucidated in the *E. coli* system can be extrapolated to those in other eubacterial species. The genetic and biochemical studies of mollicutes including the recent completion of the total genome sequences of *Mycoplasma pneumoniae*[21], *M. genitalium*[13], *M. pulmonis*[9a] and *Ureaplasma urealyticum*[16], together with 214Kbp (about one/fifth of the total genome) sequence of *M. capricolum*[7], have revealed that the basic machineries and fundamental processes in transcription and translation are essentially the same as those of *E. coli* and other eubacteria. Since mycoplasmas belong to the gram-positive bacterial group having DNAs with a low G+C content, they share a common ancestor with bacilli, clostridia, enterococci, lactobacilli, staphylococci and streptococci[73]. Therefore, many aspects of transcription and translation mechanisms of mycoplasmas are similar to those of the gram-positive rather than the gram-negative bacteria. However, several mycoplasma specific features, some of which are very important, have also been found. For example, most genetic systems of the species in the class *Mollicutes* use UGA, one of the stop codons in the universal code, as a tryptophan (Trp) codon; the anticodon compositions of tRNAs and codon recognition patterns of mycoplasmas are also distinguished from those of other bacteria. This chapter describes the several features of mycoplasmal

*Molecular Biology and Pathogenicity of Mycoplasmas*, Edited by Razin and Herrmann, Kluwer Academic/Plenum Publishers, New York, 2002

transcription and translation systems. Reviews covering different aspects of the molecular biology and genetics of mycoplasmas are also available[8, 11, 12, 38, 43, 52, 53].

## 2.  GENERAL FEATURES OF MYCOPLASMA GENOMES AND GENES

The genomes of mycoplasmas are distinguished by their small size and low G+C content[52, 53]. The genome size among members of the class *Mollicutes* varies from about 580 to about 2,200 kbp (kilobase-pairs); the genome of *M. genitalium* is 580,070 bp[13], which is the smallest of all living organisms so far known. With a few exceptions, the G+C content of mycoplasmal genomes is within the range of 24 to 33%. Many mollicutes, including *M. capricolum*, *M. pulmonis*, *Spiroplasma citri*, and *U. urealyticum*, have a G+C content as low as 25%, the lowest in eubacteria. The G+C content of *M. pneumoniae* is 40%, which is the highest among mycoplasmas. Mycoplasmas are regarded as degenerated forms of gram-positive bacteria[43, 73]. They seem to have evolved under two evolutionary constraints, minimizing genome size and reducing genomic G+C content[43].

The small size of the genome reflects the number of the genes encoded. The genome of *M. genitalium* conains only 483 ORFs[13], that of *M. pneumoniae* contains 688 ORFs[10, 21], 782 ORFs in *M. pulmonis*[9a] and 613 ORFs in *U. urealyticum*[16], compared to 4,288 in *E. coli* K12[6] and 4,100 in *Bacillus subtilis* 168[34]. The genomes of *M. genitalium*, *M. pneumoniae*, and *U. urealyticum* contain no genes for amino acid metabolism, and a much smaller number of genes for lipid and nucleotide metabolism, for energy metabolism and for various cellular processes compared to those of *E. coli* and *B. subtilis*. On the other hand, the number of genes for the transcription and the translation apparatus of mycoplasmas    is comparable to those of other bacteria, and comprises more than 20% of the total ORFs[13, 16, 21, 22]. All mollicute genomes so far analyzed have one or two sets of rRNA genes, with one gene each for 23S, 16S and 5S rRNAs, while the *E. coli* and the *B. subtilis* genomes contain 7 and 10 rRNA gene sets, respectively. The organization of rRNA genes of most mycoplasmas is the same as the one for the corresponding eubacterial genes, 16S rRNA-23S rRNA-5S rRNA, forming an operon. However, several exceptions have been reported: the 5S rRNA genes is located apart from 16S rRNA-23S rRNA   genes   in *M. hyopneumoniae* and *M. flocculare*[63, 67]; one of the 16S rRNA genes   is separated from the 23S rRNA-5S rRNA genes in *M. gallisepticum* S6[59]; *M. gallisepticum* A5969 contains a third truncated 16S rRNA gene[59], which is probably a pseudogene; in *M. fermentans*, two 16S rRNA-23S rRNA gene sets are arranged in the unusual tail-to-tail orientation, and the 5S rRNA genes are located separately[23]. The *M. capricolum* genome contains 30 genes encoding 29 tRNA species[2, 38], while *M. pneumoniae*,*M. genitalium* and *U. urealyticum* genomes contain 35, 33 and 30 tRNA genes[13, 16, 21], respectively, the numbers are much smaller than those of *E.coli* with 86 and *B. subtilis* with 89 genes[6, 34]. The set of 29 tRNA genes identified recently in the *M. pulmonis* genome represents the smallest set of tRNA genes among the sequenced genomes[9a]. Like other eubacteria, many tRNA genes of mycoplasmas are clustered on the chromosome forming large operons. Mycoplasmal genomes contain the genes for other ubiquitous small RNA species, RNase P RNA, tmRNA and 4.5S RNA. In addition, *M. capricolum* contains at least three other small RNA species, MCS2, MCS3 and MCS4 RNAs[70], *M. pneumoniae* contains a 205- and 210-base long RNA[17], and in *M. genitalium* a short version of this RNA has been found, the functions of which are unknown.

The mycoplasmal genome has been considered to contain a small set of genes mainly indispensable for autonomous growth. Comparison of ORFs from *M. genitalium* and *M. pneumoniae* showed that almost all the proposed ORFs in the smaller genome of *M. genitalium* are present in the larger genome of *M. pneumoniae*, and that more than 100 ORFs present in, *M. pneumoniae* have no counterpart in *M. genitalium*[22]. Furthermore, random gene-disruption analysis by transposon insertion in the *M. genitalium* genome has suggested that 265 to 350 genes among the total 483 genes coding for proteins are essential for growth, including about 100 genes of unknown function[24] (see the Mycoplasmas and the minimal cell concept chapter). These results indicate the presence of many non-essential genes even in the small genome of *M. genitalium*.

The low G+C content of the genome may be due to the strong AT-biased mutation pressure (AT pressure) that has been operating during the evolution of mycoplasmas[42]: the mutations from GC pairs to AT pairs have preferentially occurred probably due to the tendency of the DNA replication and/or repair system. High A+T content can be seen in various parts of the genomes, especially in the selectively neutral parts, such as spacers between genes. The codon third positions, are strongly biased to A or T. A comparison of G+C contents between various genes of *M. genitalium* and *U. urealyticum* has shown that genes with lower G+C contents are likely to be expendable, while essential genes maintain higher G+C contents[16]. The change of the genetic code UGA from stop to Trp can be explained as the result of the strong AT pressure (see below).

## 3. TRANSCRIPTION

### 3.1 RNA polymerase

The subunit composition of the mycoplasmal RNA polymerase (DNA-dependent RNA polymerase) resembles those of other eubacteria consisting of $\alpha_2\beta\beta$' core subunits and one $\sigma$ factor, as deduced from their gene sequences[13, 16, 21]. Despite structural similarities among the RNA polymerases from mycoplasmas and other eubacteria, the mycoplasmal RNA polymerase is relatively insensitive to rifampicin and streptolydigin, antibiotics that block transcription initiation by binding to the $\beta$ subunit[14, 51]. They share this property with the RNA polymerases of clostridia species, which are phylogenetically related closely to mycoplasmas[73]. Some differences in amino acid sequences in the Rif region of the $\beta$ subunit of mycoplasmal polymerases have been suggested to be responsible for rifampicin resistance[14].

Only one type of $\sigma$ factor, homologous to the $\sigma^{70}$ type facor of *E. coli* or the $\sigma^A$ vegetative type of gram-positive bacteria, has been found in mycoplasmas, while *E. coli* and *B. subtilis* contain at least 6 and 18 $\sigma$ factors, respectively[6, 34]. This suggests that global gene expression changes in mycoplasmas responding to different environmental conditions are not controlled by the promoter selectivity of polymerases with alternative $\sigma$ factors, that is common in *E. coli*, *B. subtilis* and many other eubacteria. However, the genome sequences of *M. pneumoniae*, *M. genitalium*, and *U. urealyticum* , but not that of *M. pulmonis*, have revealed the presence of the gene homologous to the $\delta$ subunit or $\sigma^E$ (*rpoE*)[9a, 13, 16, 21], known to be required for the expression of several stress-response genes in *E. coli* and some other bacteria, suggesting that a part of the genes in mollicutes are also controlled by the promoter selectivity with an alternative $\sigma$ factor.

### 3.2 Promoters and transcription initiation

A DNA-dependent RNA polymerase recognizes a promoter sequence located upstream of a gene or an operon and initiates transcription. Although very few promoters have been experimentally identified, mycoplasmal RNA polymerases seem to recognize a promoter similar to those of other eubacteria. Promoters of mycoplasmal genes have the bacterial consensus sequence for promoters recognized by $\sigma^{70}$ type factor of *E. coli* or $\sigma^A$ vegetative type of gram-positive bacteria: a –35 region (TTGACA) and a –10 region (Pribnow box: TATAAT) with about 17 bp-spacer separating the two regions[43, 53]. The transcription initiation sites of the rRNA operons and some of the tRNA operons of *M. capricolum* have been analyzed by S1 mapping and/or primer extension methods: all the initiation sites are 6 to 8 bp downstream of the putative −10 boxes[66, 75], as in the case of *E. coli* or *B. subtilis*. In fact, *M. capricolum rrnA* and a tRNA gene cluster were recognized by *E. coli* polymerase both in vivo and in vitro[66]. Weiner *et al.* analyzed transcriptional start points of 22 genes of *M. pneumoniae*, and found that their promoters have a strongly conserved −10 region, but only a weak consensus −35 region[72]. They also showed that the transcripts of genes have heterogeneous 5'-ends, indicating that the initiations occur at several different sites at the 3'-downstream region of the promoter[72]. It is not known whether the heterogeneous initiation correlates with the regulation of gene expression. Waldo *et al.* also analyzed transcription initiation sites of four genes in the *hmw* gene cluster of *M. pneumoniae*, and found that there are conserved −10-like sequences TAAAAT upstream of the start sites, but −35 regions are not conserved[71].

## 3.3    Termination and attenuation

Termination of transcription seems to be independent of the termination factor Rho, since the corresponding gene or protein has not been found in any mycoplasma species. Instead, Rho-independent type of terminators, having sequences with a short stem-loop followed by a run of U(T) residues, have been found in the 3'-downstream regions of many mycoplasmal genes and operons[13, 16, 21, 43], showing that this type of transcription termination is common in mycoplasmas. Analysis of *in vitro* transcription of the *M. capricolum* tRNA$^{Trp}$ gene cluster indicated that termination takes place at the T-stretch region soon after the dyad-symmetrical structure[75]. The genes for the homologs of termination modulating proteins NusA and NusG were found in the genomes of *M. genitalium*, *M. pneumoniae*, and *U. urealyticum*[13, 16, 21].

The genes for tRNA$^{Trp}$(UCA) and tRNA$^{Trp}$(CCA) in *M. capricolum* are arranged tandemly on the chromosome forming a single operon[74, 75]. Promoter and terminator sequences are located upstream of tRNA$^{Trp}$(UCA) and downstream of tRNA$^{Trp}$(CCA) gene, respectively. In addition, there is a terminator-like sequence, a dyad-symmetrical sequence followed by a T-stretch in the 40-bp spacer between the two tRNA genes. *In vitro* transcription of the DNA fragment containing this operon with partially purified *M. capricolum* RNA polymerase produced two transcripts of 240- and 120-bases. S1 mapping analysis showed that initiation of these two transcripts occurred at the same position, 10-bp upstream of the tRNA$^{Trp}$(UCA) gene[75]. This finding indicates that the 240-base transcript is a dimeric precursor containing two tRNA sequences, while the 120-base transcript is a precursor to tRNA$^{Trp}$(UCA). The amount of the dimeric precursor is about one-fifth of that of the monomeric precursor. In fact, the intracellular level of tRNA$^{Trp}$(UCA) is 5 to 10 times as high as that of tRNA$^{Trp}$(CCA). The terminator-like sequence between the two genes thus operates as an attenuator of transcription, and more than 80% of the transcripts that start from the initiation site are terminated by the structure. This is the only case of attenuation so far analyzed in a mycoplasmal operon.

## 3.4     Processing of RNA gene transcripts

Many rRNA and tRNA genes in bacteria are clustered forming operons. Like other eubacteria, the rRNA genes of most mycoplasmas are organized in the order of 16S rRNA-23SrRNA-5SrRNA, functioning as an operon. The rRNA operons of phytoplasmas and one of the operons of *A. laidlawii* carry one or two tRNA genes in the intergenic region between 16S and 23S rRNA genes[44, 53, 60, 65], while no tRNA genes are found in this region in other mycoplasmas. Many mycoplasmal tRNA genes are also clustered forming operons. In *M. capricolum*, 22 out of 30 tRNA genes are organized in five tRNA gene clusters consisting of 9, 5, and 4 genes and two sets of 2 genes[38]. The clustering of tRNA genes was also found in the genomes of of *M. mycoides*[56], *M. pneumoniae*[21], *M. genitalium*[13], and *U. urealyticum*[16], *S. meliferum*[55] and *A. laidlawii*[65]. These genes are transcribed as a single transcript followed by processing for the production of mature rRNAs or tRNAs. Detailed processing pathways of mycoplasmal rRNAs and tRNAs are unknown. Several ribonucleases (RNases) are known to be involved in rRNA and tRNA processing in *E. coli*. The homologs of two RNases are so far identified in the genomes of *M. pneumoniae*, *M. genitalium*, and *U. urealyticum*[13, 16, 21]: RNase III which cuts double stranded stems formed between the 5' and 3'-flanking sequences of 16S and 23S rRNA precursor, and RNase P that removes the leader sequences from precursor tRNAs generating mature 5' ends of tRNAs. In *M. capricolum*, only 13 types of modified nucleosides were found among the total tRNAs[2], in contrast to 23 so far known in *E. coli*[62]. The 5' end nucleoside of the TφC-loop of all the *M. capricolum* tRNAs is an unmodified U[2], whereas the corresponding nucleoside of most of the known tRNAs in other eubacteria is ribothymidine (T). Thus, the modification level of mycoplasmal RNAs is lower than that of other bacteria[5], due to the lack of the genes coding for many modification enzymes in the genome. So far, the genes for the homologs of tRNA modification enzymes, pseudouridylate synthase I, tRNA (guanine N-1)-methyltransferase and tRNA-(5-methylaminomethyl-2-thiouridylate) methyltransferase, and that for rRNA methylases, were found in the genomes of *M. genitalium*[13], *M. pneumoniae*[21], and *U. urealyticum*[16].

## 3.5    Regulation

The transcription initiation is one of the major steps to control gene expression in bacteria. The genomes of mycoplasmas almost completely lack the genes for enzymes involved in amino acid biosynthesis, *de novo* nucleotide biosynthesis and fatty acid biosynthesis, while the genes required for self-replication, such as for DNA replication, transcription and translation, are well conserved. Thus, many genes in mycoplasmas may be expressed constitutively during growth, and the refined switching mechanisms for transcriptional adaptation to environments may be limited. In fact, a few genes for repressor-like proteins have been found in mycoplasmal genomes[7, 13, 16, 21]. The major control in mycoplasmas might be quantitative regulation rather than on/off control switch for gene expression. The presence of only one type of σ- subunit ($\sigma^{70}$ type) in mycoplasmas supports this view. In *E. coli*, intracellular amounts of many constitutively expressed gene products, such as rRNAs, tRNAs, ribosomal proteins, RNA polymerase, aminoacyl-tRNA synthetases *etc.*, are coordinately controled depending on the growth rate of the cell[18]. One of the major control elements for this regulation is guanosine tetraphosphate (ppGpp) or guanosine pentaphosphate (pppGpp), which operate as negative factors for transcription initiation for several genes. In amino acid auxoptrophic mutants of *E. coli* and other eubacteria, the synthesis of rRNA, tRNA and many other constitutive proteins is inhibited by deprivation of the required amino acid, accompanied by simultaneous accumulation of ppGpp and pppGpp (stringent control[9]). The rRNA synthesis in *M. capricolum* is reported to be inhibited by the omission of amino acid supplement from the partially defined medium with simultaneous accumulation of ppGpp and pppGpp[15]. The genes for a stringent control response protein, SpoT/RelA (ppGpp 3'-pyrophosphohydrolase) were detected in *M. pneumoniae*[21] and *U. urealyticum*[16] genomes, but not in the *M. genitalium* genome. These findings suggest that stringent control operates at least in some species. Since mycoplasmas, as parasites, may not usually face such a sudden change in environmental conditions, the physiological meaning of the stringent control in mycoplasmas is unclear. It is possible, however, that the intracellular level of ppGpp or pppGpp plays some role in the coordinate regulation of constitutively expressed genes in mycoplasmas[18].

Several genes for the heat shock proteins, DnaK, GroEL, GroES, GrpE and DnaJ, which are known to function as molecular chaperons, have been found in the genomes of *M. pneumoniae*[21], *M. genitalium*[13], *M. capricolum*[7] and *U. urealyticum*[16]. The cis-acting conserved palindromic repeated sequences, similar to the 'CIRCE' element found in *B. subtilis* and other gram-positive bacteria, are detected in front of four heat shock genes in *M. pneumoniae*[21]. The gene (*hrcA*) for a homolog of the repressor protein that is proposed to bind on the CIRCE element was also found in this genome.

These findings suggest that these heat-shock protein synthesis is negatively regulated, although not all heat-shock protein genes in this species have a CIRCE element.

## 4.      TRANSLATION

### 4.1      Ribosomes, translation factors and aminoacyl tRNA synthetases

About 15% of the total genes predicted from the genome sequences of *M. genitalium*, *M. pneumoniae*, and *U. urealyticum* have been assigned as components involved in translation, including ribosomal components, tRNAs, aminoacyl tRNA synthetases, and translation factors[13, 16, 21].

Mycoplasmas resemble typical eubacteria in having 70S ribosomes consisting of 50S and 30S subunits. The 50S subunit contains two rRNA species, 5S and 23S, and about 31-35 ribosomal proteins, and the 30S subunit consists of 16S rRNA and about 20 proteins, which are comparable in numbers to other eubacterial ribosomes. The genome sequences of *M. genitalium*, *M. pneumoniae*, and *U. urealyticum* confirmed the existence of 19-20 proteins for the 30S subunit (lacking S1 resembling ribosomes of gram-positive bacteria) and 31-34 proteins for the 50S subunit[13, 16, 21]. Many of the mollicute ribosomal protein genes are organized in operons, retaining the gene orders found in *E. coli* and other eubacteria[13, 16, 21, 48]. The amino acid sequences of the proteins are also well conserved (40-60% identity to those of *B. subtilis* proteins) among eubacteria[48]. The sequences of rRNAs and tRNAs are also conserved in all eubacteria.

The genes for three initiation factors, IF1, IF2, and IF3, and for four elongation factors, EF-Tu, EF-Ts, EF-G and EF-P, were identified in *M. genitalium, M. pneumoniae, M. capricolum* and *U. urealyticum* genomes[7, 13, 16, 21]. Most eubacteria contain two release factors, RF-1 for UAA and UAG and RF-2 for UAA and UGA. Since most mycoplasmas use UGA as Trp codon instead of stop codon, it has been expected that these species do not carry RF-2. In fact, the *M. genitalium, M. pneumoniae* and *U. urealyticum* genomes contain only one release factor homologous to RF-1[13, 16, 21]. *M. capricolum* is also reported to have only RF-1[25, 26]. The gene for RF-3 found in *E. coli* has not been detected in mycoplasmal genomes; the genes for ribosome recycling factor (RRF) were found in *M. genitalium, M. pneumoniae* and *U. urealyticum* genomes[13, 16, 21]. The *M. pneumoniae, M. genitalium* and *U. urealyticum* genomes contain the gene for the homolog of elongation factor P (EFP)[13, 16, 21], which has been found in *E. coli*. It stimulates the peptidyltransferase activity of fully assembled 70S prokaryotic ribosomes and enhances the synthesis of certain dipeptides initiated by N-formylmethionine[4]. The gene methionyl-tRNA formyltransferase (*fmt*) for formylation of initiator Met-tRNA has been found in the genome sequences of *M. genitalium*[13], *M. pneumoniae*[21], *M. capricolum*[7] and *U. urealyticum*[16]. Overall similarities of mycoplasmal ribosomal components and translation factors to those of other bacteria described above suggest that general translational machineries and processes of mollicutes are essentially the same as those of other eubacteria

The genome sequences of *M. genitalium*, *M. pneumoniae*, and *U. urealyticum* contain 19 aminoacyl tRNA synthetase genes, one for each of 19 amino acids; glutaminyl-tRNA synthetase is the only one not found in them[13, 16, 21]. This situation is shared with *B. subtilis* and other gram-positive bacteria, in which tRNA$^{Glu}$ is first charged with Gln by glutamyl-tRNA synthetase and the charged Gln is subsequently converted to Glu by glutamyl-tRNA amidotransferase. The genes for two subunits of the glutamyl-tRNA amidotransferase have been found in their genomes[13, 16, 21]. Recent global transposon mutagenesis experiments with *M. genitalium* revealed that aminoacyl tRNA synthetases for Ile and Tyr might be dispensable for growth of this species[24].

## 4.2 Shine-Dalgarno sequence and translational initiation

Nucleotide sequence data indicate that the translation of most mycoplasmal genes begins with a AUG initiation codon, while the alternative initiation codons (GUG and UUG) are also used in several genes. At least three intitation factors, IF1, IF2 and IF3, are required for initiating translation in all eubacteria. The specific nucleotide sequence, known as Shine-Dalgarno sequence (SD-sequence) upstream of the initiater AUG codon, pairing with the 3'-terminal sequence of 16S rRNA (anti-SD sequence), is generally required for correct initiation. Initiation occurs with 30S subunits combining first with mRNA through SD- and anti-SD-sequence interaction, and then with f-met-tRNA and IF2-GTP (30S initiation complex). The SD-sequences have been found in the 5'-upstream regions of most mycoplasmal genes, from *M. capricolum*[43] and *S. citri*[8]. However, the absence of SD-sequence has been observed for many genes in *M. pneumoniae* and *M. genitalium*. The 5'-ends of transcripts of some *M. pneumoniae* genes are heterogeneous, and a high proportion of transcripts lack a 5' untranslated leader region[71, 72]. Thus, the translation initiation mechanism of these species remains elusive. Recently, an alternative process initiating translation without a 5'-leader sequence was found in *E. coli*, in which IF2 and f-met-tRNA first make a complex with the 30S subunit and then AUG at the 5'-extreme end of a mRNA binds to the complex[19]. This pathway is analogous to those of archaeal and eukaryotic translation initiation systems. It is conceivable that *M. pneumoniae* and *M. genitalium* use this pathway for initiation in the translation process. The other possibility is that some other sequences downstream the initiation codon, just like the downstream-box found in *E. coli*[61a], is involved in the translation initiation of mollicute mRNA lacking the 5'-leader region.

## 4.3    Codon usage

The genomic G+C content of bacteria is reflected in its codon usage[41]. Most mollicultes have genomes with a very low G+C content, and thus the codon usage of them is strongly biased to use A and U(T). Among mollicutes, the genomic G+C content varies from about 25% to 40%. The average G+C contents of protein coding sequences of *M. capricolum, M. genitalium* and *M. pneumoniae* are 26%, 33% and 41%[7, 13, 21], respectively, reflecting their genomic G+C contents, 25%, 32% and 40%, respectively. Table 1 shows the codon usage of the three species. The outstanding feature is the extremely high usage of the codons having A or U(T) at the third codon position in *M. capricolum*. More than 90% of codons end in A or U in *M. capricolum*[41, 43]. The average G+C contents at the third codon position in *M. genitalium* and *M. pneumoniae* are 28% and 42%, respectively[13, 21, 22], which are much lower than that in *E. coli* (56%), showing that mycoplasmas preferentially use A- or U-rich codons among synonymous codons. The preferential use of A- or U-rich codons can also be seen at the codon first position. For example, among six synonymous codons for Leu, UUR (R=A or G) codons dominate over CUN (N=U, C, A or G) codons in mycoplasmas. The average G+C contents of the codon first and second position are also biased to A and U (39%, 42%, and 47% at the first, and 31%, 30% and 33% at the second position in *M. capricolum, M. genitalium* and *M. pneumoniae*, respectively[7, 13, 21]), which are significantly lower than those in *E. coli* (62% at the first and 42% at the second position).

Such a biased use of A- and U-rich codons may be largely brought about by AT pressure[28, 41, 49, 50]. Directional mutations from GC to AT pairs in the DNA must have occurred equally at all three codon positions. However, the silent codon positions, mainly the third and partly the first positions, evolved fastest, because many mutations at these positions are non-deleterious. Thus, the bias to A and U richness in mycoplasmal genes is the strongest in the third position and then in the first position[28, 41, 50]. The second codon position is the most invariable, because mutations at this position always cause amino acid substitutions in proteins.

The most prominent feature in codon usage in most mollicutes is the use of the codon UGA, one of the stop codons in the universal code, as a Trp codon. All mollicutes so far analyzed, except for *A. laidlawii* [64, 65] and phytoplasmas[35, 68, 76], use UGA, together with the universal UGG, as a Trp

Table 1. Codon usage in Mycoplasma capricolum, Mycoplasma pneumoniae and Mycoplasma genitalium

| Codon | M.c | M.p | M.g | Codon | M.c | M.p | M.g | Codon | M.c | M.p | M.g | Codon | M.c | M.p | M.g |
|---|---|---|---|---|---|---|---|---|---|---|---|---|---|---|---|
| Phe (UUU) | 39.5 | 43.0 | 52.6 | Ser (UCU) | 15.7 | 8.2 | 12.4 | Tyr (UAU) | 25.1 | 14.3 | 24.0 | Cys (UGU) | 6.0 | 5.4 | 6.6 |
| Phe (UUC) | 4.1 | 12.8 | 8.2 | Ser (UCC) | 0.1 | 9.6 | 4.0 | Tyr (UAC) | 3.5 | 17.9 | 8.3 | Cys (UGC) | 1.5 | 2.1 | 1.6 |
| Leu (UUA) | 63.5 | 39.2 | 50.1 | Ser (UCA) | 24.8 | 8.7 | 16.4 | Stop (UAA) | – | – | – | Trp (UGA) | 6.2 | 6.1 | 6.3 |
| Leu (UUG) | 3.2 | 21.5 | 14.2 | Ser (UCG) | 0.7 | 6.4 | 1.1 | Stop (UAG) | – | – | – | Trp (UGG) | 1.0 | 5.8 | 3.4 |
| Leu (CUU) | 5.3 | 10.1 | 19.9 | Pro (CCU) | 9.7 | 8.3 | 14.5 | His (CAU) | 10.6 | 6.2 | 5.5 | Arg (CGU) | 7.9 | 9.7 | 6.9 |
| Leu (CUC) | 0.0 | 12.2 | 5.0 | Pro (CCC) | 0.4 | 9.1 | 3.6 | His (CAC) | 3.2 | 11.9 | 10.3 | Arg (CGC) | 0.6 | 10.7 | 3.0 |
| Leu (CUA) | 9.1 | 10.6 | 12.7 | Pro (CCA) | 17.6 | 10.9 | 10.8 | Gln (CAA) | 37.6 | 37.9 | 38.3 | Arg (CGA) | 0.3 | 2.5 | 1.3 |
| Leu (CUG) | 0.1 | 9.5 | 4.4 | Pro (CCG) | 0.4 | 6.7 | 0.9 | Gln (CAG) | 1.5 | 15.7 | 8.9 | Arg (CGG) | 0.0 | 5.0 | 1.0 |
| Ile (AUU) | 70.4 | 46.0 | 51.6 | Thr (ACU) | 30.2 | 19.3 | 25.4 | Asn (AAU) | 64.6 | 25.1 | 46.0 | Ser (AGU) | 17.2 | 21.0 | 25.7 |
| Ile (AUC) | 8.2 | 14.4 | 17.9 | Thr (ACC) | 1.2 | 21.9 | 10.3 | Asn (AAC) | 11.4 | 37.0 | 28.9 | Ser (AGC) | 2.9 | 10.6 | 6.7 |
| Ile (AUA) | 16.3 | 5.5 | 12.7 | Thr (ACA) | 22.2 | 10.4 | 16.6 | Lys (AAA) | 107.1 | 46.3 | 70.4 | Arg (AGA) | 27.7 | 4.0 | 14.0 |
| Met (AUG) | 22.5 | 15.6 | 15.2 | Thr (ACG) | 0.1 | 7.9 | 1.6 | Lys (AAG) | 10.4 | 39.1 | 24.4 | Arg (AGG) | 0.4 | 2.8 | 4.6 |
| Val (GUU) | 41.7 | 21.2 | 37.7 | Ala (GCU) | 33.5 | 25.2 | 27.4 | Asp (GAU) | 43.1 | 30.4 | 42.3 | Gly (GGU) | 26.4 | 27.9 | 22.9 |
| Val (GUC) | 1.2 | 11.0 | 3.4 | Ala (GCC) | 0.4 | 16.5 | 4.1 | Asp (GAC) | 3.5 | 19.2 | 6.8 | Gly (GGC) | 0.7 | 11.8 | 5.0 |
| Val (GUA) | 23.0 | 13.7 | 13.1 | Ala (GCA) | 20.5 | 13.8 | 21.4 | Glu (GAA) | 54.7 | 42.0 | 45.5 | Gly (GGA) | 29.2 | 6.4 | 11.5 |
| Val (GUG) | 2.3 | 18.7 | 7.1 | Ala (GCG) | 0.6 | 11.1 | 2.7 | Glu (GAG) | 4.3 | 14.7 | 11.2 | Gly (GGG) | 2.3 | 9.0 | 6.8 |

Codon usage of *M. capricolum* (M.c) is mainly from ribosomal protein genes[43,48], those of *M. pneumoniae* (M.p) and *M. genitalium* (M.g) are from the complete set of predicted open reading frames[13,21,22]. All values are calculated in thousands.

codon. As discussed below, this change of codon assignment may be caused by AT pressure. Another change of a codon assignment has been found for the CGG codon in *M. capricolum*: This GC-rich codon seems to be a nonsense (or unassigned) codon in this species[47] (see below). This change may also be a consequence of AT pressure exerted during the evolution of *M. capricolum*. Since CGG codons are reported to be used in *M. genitalium, M. pneumoniae, U. urealyticum, M. pulmonis* and several other mycoplasmas, this change seems to be so far specific for *M. capricolum*.

## 4.4     tRNA anticodons and codon-recognition

The sequences of all tRNA species (including nucleoside modifications) and their genes in *M. capricolum* have been determined[2, 39]. In fact, *M. capricolum* is the only genetic system among all organisms and organelles for which the complete sequences of all tRNAs have been determined at both RNA and DNA levels. The genome sequences of *M. genitalium, M. pneumoniae* and *U. urealyticum* also provide their total tRNA composition in the cells. *M. capricolum* contains 30 genes for 29 tRNA species[2, 39]; *M. pneumoniae, M. genitalium,* and *M. pulmonis* have 35, 33 and 29 tRNA genes[13, 21, 9a], respectively , and *U. urealyticum* has 30 tRNA genes[16]. The numbers of the genes are much smaller than those in *E. coli* and *B. subtilis*, which have 86 and 89 tRNA genes, respectively[6, 34].

Table 2 shows the anticodons of total tRNAs from *M. capricolum*[43], *M. genitalium* (and *M. pneumoniae*)[22, 58] and *U. urealyticum*[16], and their possible correspondence to codons. The numbers of anticodons for *M. capricolum, M. genitalium* (and *M. pneumoniae*) and *U. urealyticum* are 28, 32 and 29, respectively[2, 13, 16, 21]. These numbers are only slightly higher than the 24 anticodons found in yeast mitochondria and much fewer than the numbers in *E. coli* or *B. subtilis* for both of which at least 39 anticodons have been deduced from the tRNA gene sequences[6, 34]. These facts suggest that many tRNA genes have been discarded in the evolution of mycoplasmas, allowing codons to be read by as small a number of anticodons as possible. Only five types of modified nucleosides were found in the first position of anticodons in *M. capricolum*[2], while at least eight different modifications have been found in *E. coli* tRNAs[62]. The anticodon compositions of mollicutes are strikingly different from those of *E. coli* and other eubacteria. The most prominent difference is seen in the anticodon compositions of family boxes (or four-codon boxes), in which there are four codons for a single amino acid. In *E. coli*, four synonymous codons in family boxes are read by at least two anticodons, GNN and *UNN (* represents a derivative of 5-hydroxyuridine or its equivalent), except for the Arg family box (see below). Anticodon GNN reads codons NNC and NNU, and anticodon *UNN reads codons NNU, NNA and NNG by wobbling. Anticodon CNN is present in

Table 2. Anticodons of tRNAs in *Mycoplasma capricolum*, *Mycoplasma pneumoniae* and *Ureaplasma urealyticum* [a]

| Codon | M.c | M.p[b] | U.u |
|---|---|---|---|
| Phe (UUU) | GAA | GAA | GAA |
| Phe (UUC) | | | |
| Leu (UUA) | UAA[c] | TAA | TAA |
| Leu (UUG) | CAA[d] | | CAA |
| Leu (CUU) | | | |
| Leu (CUC) | | | |
| Leu (CUA) | UAG | TAG | TAG |
| Leu (CUG) | | | |
| Ile (AUU) | GAU | GAT | GAT |
| Ile (AUC) | | | |
| Ile (AUA) | CAU[e] | CAT | CAT |
| Met (AUG) | CAU[f] | CAT | CAT |
| Val (GUU) | | | |
| Val (GUC) | | | |
| Val (GUA) | UAC | TAC | TAC |
| Val (GUG) | | | |

| Codon | M.c | M.p[b] | U.u |
|---|---|---|---|
| Ser (UCU) | | | |
| Ser (UCC) | | | |
| Ser (UCA) | UGA | TGA | TGA |
| Ser (UCG) | CGA | CGA | |
| Pro (CCU) | | | |
| Pro (CCC) | | | |
| Pro (CCA) | UGG | TGG | TGG |
| Pro (CCG) | | | |
| Thr (ACU) | AGU | GGT | GGT |
| Thr (ACC) | | | |
| Thr (ACA) | UGU | TGT | TGT |
| Thr (ACG) | | | |
| Ala (GCU) | | | |
| Ala (GCC) | | | |
| Ala (GCA) | UGC | TGC | TGC |
| Ala (GCG) | | | |

| Codon | M.c | M.p | U.u |
|---|---|---|---|
| Tyr (UAU) | GUA | GTA | GTA |
| Tyr (UAC) | | | |
| Stop (UAA) | | – | – |
| Stop (UAG) | | | |
| His (CAU) | GUG | GTG | GTG |
| His (CAC) | | | |
| Gln (CAA) | UUG[g] | TTG | TTG |
| Gln (CAG) | | | |
| Asn (AAU) | GUU | GTT | GTT |
| Asn (AAC) | | | |
| Lys (AAA) | UUU[g] | TTT | TTT |
| Lys (AAG) | CUU | CTT | CTT |
| Asp (GAU) | GUC | GTC | GTC |
| Asp (GAC) | | | |
| Glu (GAA) | UUC[g] | TTC | TTC |
| Glu (GAG) | | | |

| Codon | M.c | M.p | U.u |
|---|---|---|---|
| Cys (UGU) | GCA | GCA | GCA |
| Cys (UGC) | | | |
| Trp (UGA)[h] | UCA[c] | – | – |
| Trp (UGG) | CCA[d] | TCA / CCA | TCA / CCA |
| Arg (CGU) | | | |
| Arg (CGC) | ICG[j] | GCG | ACG |
| Arg (CGA) | | | |
| Arg (CGG)[i] | | TCG | TCG |
| Ser (AGU) | GCU | GCT | GCT |
| Ser (AGC) | | | |
| Arg (AGA) | UCU[g] | TCT | TCT |
| Arg (AGG) | | | |
| Gly (GGU) | GCC | GCC | GCC |
| Gly (GGC) | | | |
| Gly (GGA) | UCC | TCC | TCC |
| Gly (GGG) | | | |

[a] *Mycoplasma capricolum* (M.c) anticodons are from RNA sequencings (including modified nucleotides) by Andachi *et al.*[2], *M. pneumoniae* (M.p) and *Ureaplasma urealyticum* (U.u) anticodons are from genomic DNA sequences[6,21], therefore codons from M.c are written with U and those from M.p and M.g with T.
[b] The anticodon composition of *M. genitalium* tRNAs[13] is identical to that of *M. pneumoniae*.
[c] Modification of anticodon first nucleotide: cmnm$^5$Um
[d] Modification of anticodon first nucleotide: partially 2'-O-methylcytidine
[e] Modification of anticodon first nucleotide: lysidine
[f] Initiator and elongator methionine tRNAs have the same anticodon.
[g] Modification of anticodon first nucleotide: cmnm$^5$U
[h] UGA is tryptophan codon.
[i] CGG is probably an unassigned codon in *M. capricolum*.
[j] I (inosine: modification from A)

some boxes and reads only codon NNG. Strikingly, anticodons GNN and CNN do not exist in any of the eight family boxes of *M. capricolum*[2, 39]; GNN anticodons have been so far found in Ser, Thr, Arg, and Gly family boxes in *M. genitalium*[13] and *M. pneumoniae*[21], and only in the Gly family box in *U. urealyticum*[16]; CNN anticodons have been found only in Ser and Thr family boxes in *M. genitalium*[13] and *M. pneumoniae*[21]. Anticodon UNN, with unmodified U, is a sole anticodon for each of six family boxes, excluding the Thr and Arg family boxes in *M. capricolum*[2]. The same situations are predicted in many family boxes in *M. genitalium*, *M. pneumoniae* and *U. urealyticum*, since only a tRNA gene with anticodon TNN has been found for each box[13, 21] (modifications in tRNA anticodons are not known in these species). Thus, the four synonymous codons in each family box codons must be read by a single anticodon UNN by four-way wobbling. This situation is analogous to that in most family boxes of various mitochondria (except for plant mitochondria). Recent complete genome sequencing of *Rickettsia prowazekii* strain Madrid E has revealed that there exists only one tRNA species with a TNN anticodon for Val and Pro family boxes[3], suggesting that these codons are read by a single anticodon UNN with unmodified U at the first position. It should be noted that *Rickettsia* is a parasitic intracellular bacterium having a small genome with a low G+C content like the mycoplasmas. In the Thr family box of *M. capricolum*, in addition to anticodon UGU, there is a tRNA with anticodon AGU, with the first position A unmodified[1]. The anticodon AGU seems to read all the four codons in the CAN family box[27]. An unmodified A residue at the anticodon first position occurs for yeast mitochondrial tRNA$^{Arg}$(ACG)[57] and *Aspergillus* mitochondrial tRNA$^{Gly}$(ACC)[32]. There are three anticodons (GNN, TNN and CNN) in the Ser and Thr family boxes of *M. genitalium*[13] and *M. pneumoniae*[21], while only one tRNA with anticodon TNN exists in these boxes in *U. urealyticum*[16]. In the Arg family box, three codons, CGU, CGC and CGA, are read by anticodon ICG (I= inosine: modification of A) by three way wobbling, and CGG is read by anticodon GCG in *E. coli* and other eubacteria. Interestingly, in *M. capricolum*, only one tRNA$^{Arg}$(ICG) has been found in the Arg family box[2, 39]. In accordance with this finding, codon CGG has not been detected in the reading frames so far sequenced. Furthermore, CGG codon is read neither as Arg nor as any other amino acids in the *M. capricolum in vitro* translation systems[47]. These data suggest that CGG is an unassigned (nonsense) codon in *M. capricolum*[47]. Perhaps strong AT-biased mutation pressure has converted all CGG codons to the synonymous Arg codons by silent mutations; as a result, anticodon CCG may have become unnecessary and disappeared. However, other mollicutes including *M. genitalium*, *M. pneumoniae* and *U. urealyticum*, use CGG as an Arg codon with low frequency, and contain the genes for tRNA$^{Arg}$ (TCG), which could read CGG, in their genomes[13, 16, 21]. In *M. pneumoniae*, the CGN-box codons are read by two tRNAs with anticodons GCG and TCG[21], while in *U. urealyticum* they are read by anticodons ACG (probably ICG) and TCG[16]. The tRNA with the anticodon CCG, which is common in many eubacteria, has not been found in mycoplasmas.

Two synonymous codons in all NNY-type (Y=U or C) two-codon sets, for Asn, Asp, Cys, His, Phe, Ser, and Tyr, are translated by a single anticodon, GNN. The G residue of these anticodons is sometimes modified to queuosine (Q) or 2'-O-methylguanosine in *E. coli* and other eubacteria. But in *M. capricolum*, all these Gs are unmodified[2]. Two synonymous codons in all NNR-type (R=A or G) two-codon sets are translated in eubacteria by anticodon UNN, with U modified. The modified U is 5-methyl-2-thiouridine derivative ($xm^5s^2U$) or its equivalent, which pairs mainly with NNA and weakly with NNG, in *E. coli*. The second anticodon CNN pairing only with the NNG codon is present in some boxes. The U residue at the first anticodon position for NNR-type two-codon sets in *M. capricolum* is always modified: it is 5-carboxymethylaminomethyluridine ($cmnm^5U$) for Gln, Glu, Lys and Arg[2], in contrast to $xm^5s^2U$ (for Gln, Glu and Lys) and 5-methoxycarbonylmethyluridine (for Arg) in *E. coli*, and 5-carboxymethylaminomethyl-2'-O-methyluridine ($cmnm^5Um$) for Leu, as in the case of *E. coli* tRNA$^{Leu}$(UAA)[62]. Yeast mitochondria use $cmnm^5U$ at the first position of the anticodon, for at least three NNR-type two-codon sets[57]. There are two Trp tRNAs with anticodon CCA and UCA in most mollicutes, while all other eubacteria have only one Trp tRNA with anticodon CCA (see below). The first nucleoside of the anticodon UCA is modified to $cmnm^5Um$ in *M. capricolum*[2]. The occurrence of of tRNA$^{Typ}$(UCA), which reads UGA and UGG codons by wobbling, has been found only in mitochondria. In addition to these UNN anticodons, CNN anticodons, which read only NNG codons, are found in some NNR-type two-codon sets, Leu(CAA), Lys(CUU) and Trp(CCA). The first nucleosides of the anticodons of tRNA$^{Leu}$ (CAA) and tRNA$^{Typ}$(CCA) are partially 2'-O-methylated in *M. capricolum*[2].

Like in other eubacteria, there are two tRNAs for Met in mollicutes, initiator and elongator. Both initiator and elongator tRNA$^{Met}$ of *M. capricolum* have an unmodified C residue at the anticodon first position[2], in contrast to *E. coli* elongator tRNA$^{Met}$, in which the anticodon first nucleoside C is N-4 acetylated[62]. There are two tRNA$^{Ile}$ species, tRNA$^{Ile}$(GAU) for codons AUU and AUC and tRNA$^{Ile}$(*CAU) for codon AUA, where *C is 4-amino-2-($N^6$-lysino)-1-($\beta$-D-ribofuranosyl)pyrimidinium, or lysidine, in *M. capricolum*[2] as in *E. coli*. The anticodon compositions for two-codon sets of *M. genitalium*, *M. pneumoniae* and *U. urealyticum* are essentially the same as that of *M. capricolum* (see Table 2).

In *M. capricolum*, of 62 amino acid codons, 57 are translated by a single anticodon. Only four codons UUG (Leu), AAG (Lys), UGG (Trp) and ACU (Thr), may be read by two anticodons; nonobligate (redundant) anticodons for these codons are CAA (Leu), CUU (Lys), CCA (Trp) and AGU (Thr), respectively[2,39].

Anticodon composition and codon recognition patterns of mycoplasmas described above resemble those of mitochondria rather than those of other eubacteria for the following reasons: (i) the use of UNN, with U unmodified, anticodons in family boxes; (ii) the disappearance of many nonobligate GNN and CNN anticodons; (iii) the presence of anticodon UCA that can read universal stop codon UGA as Trp; (iv) the use of cmnm$^5$U at the anticodon first position of tRNAs for NNR-type two-codon sets; and (v) the low content of modified nucleotides. The mitchondrial genome is small, implying that it has discarded many genes during evolution, like the mycoplasmal genomes. The G+C content of mitochondrial DNA of simpler eukaryotes is very low, suggesting that strong AT pressure has been exerted at an early stage of mitochondrial evolution. Thus, the genomes of mitochondria and mycoplasmas seem to have developed under similar evolutionary constraints, gene economization and AT pressure, resulting in similarities in their tRNAs[43].

## 4.5    UGA as a tryptophan codon

Most mollicutes use UGA, an universal stop codon, as a Trp codon. All species belonging to the genera *Mycoplasma, Spiroplasma, Ureaplasma*, and *Mesoplasma* so far analyzed use UGA as a Trp codon; *Acholeplasma laidlawii*[64, 65] and phytoplasmas[35, 68, 76] are the only mollicutes known to use the universal genetic code, in which UGA is a stop codon. Sequence comparison analysis showed that the genera *Acholeplasma* and *Phytoplasma* are phylogentically closely related to each other, and they are branched from the other mycoplasmas during the early stage of evolution of mollicutes[35, 68, 73, 76]. Thus, the change of the UGA codon assignment must have occurred in the major mycoplasmal lineage after the separation of the acholeplasma-phytoplasma branch.

In accordance with the appearance of UGA Trp codon, a new tRNA$^{Trp}$ with anticodon UCA occurred. *A. laidlawii* contains a single tRNA$^{Trp}$ with anticodon CCA for codon UGG, the gene for which is present as a single copy[64, 65]. *M. genitalium, M. pneumoniae, M. capricolum* and *U. urealyticum* genomes contain the genes for Trp tRNA with anticodon UCA in addition to tRNA$^{Trp}$(CCA)[2, 13, 16, 21]. In *M. capricolum*, the genes of two Trp tRNAs are arranged tandemly in a single operon in the order tRNA$^{Trp}$(UCA)-tRNA$^{Trp}$(CCA), separated by a short (40-bp) spacer[74, 75]. The tRNA$^{Trp}$(UCA) gene could have emerged by duplication of the tRNA$^{Trp}$(CCA) gene, since the two genes are closely related, both in their linkage on the chromosome and in their high sequence similarity (78% identity). During the evolutionary change of the UGA codon assignment, a simple emergence of anticodon UCA is unlikely to have happened, because UGA codon must have been used as a stop codon in ancestral bacteria. It is likely that a series of changes led to the establishment of UGA as a regular Trp codon under AT pressure during mycoplasma evolution[43, 49, 50]: First, AT pressure would lead to the mutational conversion of the stop codon UGA to synonymous UAA. At this stage, the gene for RF-2, which recognizes UGA as stop codon, would have been deleted from the genome, so that UGA would have become unassigned. Then, duplication of tRNA$^{Trp}$(CCA) must have occurred, followed by the mutational change of one of the duplicates to tRNA$^{Trp}$(UCA). The appearance of tRNA$^{Trp}$(UCA) enable the UGG codon, under AT pressure, to mutate to UGA Trp codons. These evolutionary steps could have taken place by a series of silent (neutral) changes from GC to AT pairs on DNA.

## 4.6    Translation in vivo and in vitro

Since the transcriptional and translational machineries and signals in mycoplasmas are only similar but not identical to those of *E. coli* and other eubacteria, not all mycoplasmal protein genes cloned into vector DNAs can be expressed in *E. coli* cells. In addition, most genes are not fully expressed because UGA is a Trp codon in many mycoplasmas, wheras it is a stop codon in *E. coli*. Only a few mycoplasmal genes having no Trp residue can be fully expressed in *E. coli*. One way to overcome this difficulty is to use *E. coli* with opal suppressor gene for expression, although the efficiency of expression considerably decreases with increasing numbers of UGAs in the mycoplasmal mRNA[37, 45, 61]. Alternatively, the cloned DNA, in which all TGAs are converted to TGGs by in vitro mutagenesis, will be translatable in *E. coli* cells[54]. In *B. subtilis*, the efficiency of UGA codon readthrough is relatively high due to the presence of a tRNA that reads it as Trp[36]. Furthermore, mutations in the *B. subtilis* RF-2 gene (*prfB*) enhance the level of UGA readthrough[29]. Taking advantage of these facts, the genes of P30 adhesin (containing one UGA) from *M. pneumoniae* and of methionine sulfoxide reductase (two UGAs) from *M. genitalium* integrated in the *B. subtilis* genome were succesfully expressed[29].

The in vitro expression of mycoplasmal genes using *E. coli* extracts is also difficult due to the UGA codon. However, it was demonstrated that a S-30 fraction of *M. capricolum* stimulates RNA-dependent peptide synthesis as efficiently as that of *E. coli*, showing that this system is useful for expression of mycoplasmal genes in vitro[46, 47].

## 4.7    tmRNA and trans-translation

A unique function of bacterial tmRNA (transfer-messenger RNA, also known as 10Sa RNA or SsrA RNA) has recently been elucidated (see for reviews refs.30, 42). The RNA contains both a tRNA-like structure in the 5'- and 3'-end sequences and an internal reading frame encoding a 'tag' peptide, functioning both as a tRNA and an mRNA[20, 69]. It has been shown that tmRNA is employed in a trans-translation reaction process to add a C-terminal peptide tag to the incomplete nascent protein product translated from a broken mRNA lacking a stop codon[31]. tmRNA can be aminoacylated with alanine by the alanyl-tRNA synthetase[69]. When a ribosome is stalled at the 3'-end of a truncated mRNA with peptidyl-tRNA on the P-site, alanyl-tmRNA enters the A-site, and the ribosome switches the translation from this truncated mRNA to the tag-coding sequence on tmRNA. The translation proceeds up to the stop codon in the RNA producing chimaeric peptides and releasing them from the ribosome. The tag is the target for specific proteases[31]. Therefore, tmRNA plays crucial roles in the degradation of incompletely synthesized peptides and in the recycling of stalled ribosomes. Among mollicutes, tmRNA homolog was first found in *M. capricolum*[69]; the genome sequences of *M. genitalium, M. pneumoniae* and *U. urealyticum* also revealed the presence of the genes for tmRNA[13, 16, 21]. As tmRNAs have been found in all eubacteria so far analyzed[33], all mycoplasmas should contain tmRNA and trans-translation system. The mycoplasmal tmRNAs can be folded into the same secondary structures as that of *E. coli*[33]. One of the characteristic feature of mycolasmal tmRNAs is that the tag-coding sequences are much longer than those of the most eubacterial counterparts: *E coli* and *B. subtilis* tmRNA encode tags of 10 and 14 amino acids, respectively, while the presumed tags of *M. capricolum, M. genitalium, M. pneumoniae* and *U. urealyticum* are 27, 26, 26, and 28 amino acids long, respectively[33]. Accordingly, the full length of mycoplasmal tmRNA is about 410 nucleotides, which is more than 30 nucleotides longer than those of most other eubacteria. This RNA seems to be required for growth under extreme stress conditions[40].

# REFERENCES

1.    Andachi, Y., Yamao, F., Iwami, M., Muto, A. and Osawa, S., 1987, Occurrence of unmodified adenine and uracil at the first position of anticodon in threonine tRNAs in *Mycoplasma carpicolum. Proc. Natl. Acad. Sci. USA*. **84**:7393-7402

2.    Andachi, Y., Muto, A., Yamao, F. and Osawa, S., 1989, Codon recognition patterns as deduced from sequences of the complete set of transfer RNA species in *Mycoplasma capricolum*: Resemblance to mitochondria. *J. Mol. Biol.* **209**:37-54

3.    Andersson,S.G., Zomorodipour, A., Andersson, J.O., Sicheritz-Ponten, T., Alsmark, U.C., Podowski, R.M., Naslund,A.K., Eriksson, A.S., Winkler, H.H. and Kurland, C.G., 1998, The genome sequence of *Rickettsia prowazekii* and the origin of mitochondria. *Nature* **396**:133-140

4.    Aoki, H., Dekany, K., Adams, S.-L. and Ganoza, C., 1997, The gene encoding the elongation factor P protein is essential for viability and is required for protein synthesis. *J. Biol. Chem.* **272**:32254-32259

5.  Bjork, G.R., Jacobsson, K., Nilsson, K., Johansson, M.J.O., Bystrom, A.S. and Persson, O.P., 2001, A primordial tRNA modification required for the evolution of life? *EMBO J.* **20**: 231-239

6.  Blattner, F.R., Plunkett III, G., Bloch, C.A., Perna, N.T., Burland,V., Riley, M., Collado-Vides, J., Glasner,J.D., Rode, C.K., Mayhew, G.F., Gregor, J., Davis, N.W., Kirkpatrick, H.A., Goeden, M.A., Rose, D.J., Mau B. and Shao, Y., 1997, The complete genome sequence of *Escherichia coli* K-12. *Science* **277**:1453-1474

7.  Bork, P., Ouzounis, C., Casari, C., Schneider, R., Sander, C., Dolan, M., Gilbert, W. and Gillevet, P.M., 1995, Exploring the *Mycoplasma capricolum* genome: a minimal cell reveals its physiology. *Molec. Microbiol.* **16**: 955-967

8.  Bove, J.-M., 1993, Molecular features of mollicutes. *Clin. Infect. Dis.*, **17** (suppl. 1): S10-31

9.  Cashel, M., Gentry, D.R. Hernandez, V.J. and Vinella, D., 1996, The stringent response. In *"Escherichia coli* and *Salmonella typhimurium"*. (*ed.* Neidhardt, F.C.), Second edition, *American Society for Microbiology, Wshington DC*, pp1458-1496.

9a. Chambaud, I., Heilig, R., Ferris, S., Barbe, V., Samson, D., Galisson, F., Moszer, I., Dybvig, K., Wroblewski, H., Viari, A., Eduardo, P.C. Rocha, E. and Blanchard, A., 2001, The complete genome sequence of the murine respiratory pathogen *Mycoplasma pulmonis. Nucleic Acids Res.* **29**: 2145-2153.

10. Dandekar, T., Huynen, M.,Regula, J.T., Ueberle, B., Zimmermann, C.U., Andrade, M.A., Doerks, T., Sanchez-Pulido, L., Snel, B., Suyama, M., Yuan, Y.P., Herrmann, R. and Bork, P., 2000, Reannotating the *Mycoplasma pneumoniae* genome requence: adding value, function and reading frames. *Nucleic Acids Res.* **28**:3278-3288

11. Dybvig, K., 1990, Mycoplasmal genetics. *Annu. Rev. Microbiol.* **44**: 81-104

12. Dybvig, K. and Voelker, L.L., 1995, Molecular biology of mycoplasmas. *Annu. Rev. Microbiol.* **50**: 25-57

13. Fraser, C.M., Gocayne, J.D., White, O., Adams, M.D., Clayton, R.A., Fleischmann, R.D., Bult, C.J., Kerlavage, A.R., Sutton, G.G., Kelley,J .M., Fritchman, J.L., Weidman, J.F., Small, K.V., Sandusky, M., Fuhrmann, J.L., Nguyen, D.T., Utterback, T., Saudek, D.M., Phillips, C.A., Merrick, J.M., Tomb, J., Dougherty, B.A., Bott, K.F., Hu, P.C., Lucier, T.S., Peterson, S.N., Smith, H.O. and Venter, J.C., 1995, The minimal gene complement of *Mycoplasma genitalium. Science* **270**:397-403

14. Gaurivaud, P., Laigret, F. and Bove, J.-M., 1996, Insusceptibility of members of the class *Mollicutes* to rifampin: studies of the *Spiroplasma citri* RNA polymerase β-subunit gene. *Antimicrob. Agents Chemother.* **40**:858-862

15. Glaser, G., Amikam, D. and Razin, S., 1981, Stable RNA synthesis and its control in *Mycoplasma capricolum. Nucleic Acids Res.* **9**:3641-3646

16. Glass, J.I., Lefkowitz, E.J., Glass, S.C., Heiner, R., Chen, E.Y. and Cassell, G.H., 2000, The complete sequence of the mucosal pathogen *Ureaplasma urealyticum. Nature* **407**: 757-762

17. Gohlmann, H.M.H., Weiner III, J., Schon, A. and Herrmann, R., 2000, Identification of a small RNA within the *pdh* gene cluster of *Mycoplasma pneumoniae* and *Mycoplasma genitalium. J. Bacteriol.* **182**:3281-3284

18. Gourse, R.L., Gaal, T., Bartlett, M.S., Appleman, J.A. and Ross, W., 1996, rRNA transcription and growth rate-dependent regulation of ribosome synthesis in *Escherichia coli. Annu. Rev. Microbiol.* **50**: 645-677

19. Grill, S., Gualerzi, C.O., Londei, P. and Blasi, U., 2000, Selective stimulation of translation of leaderless mRNA by initiation factor 2: evolutionary implication for translation. *EMBO J.* **19**:4101-4110

20. Himeno, H., Sato, M., Tadaki, T., Fukushima, M., Ushida, C. and Muto, A., 1997, *In vitro* trans-translation mediated by alanine charged 10Sa RNA *J. Mol. Biol.* **268**:803-808.

21. Himmelreich, R., Hilbert, H., Plagens, H., Pirkl, E., Li, B.C. and Herrmann, R., 1996, Complete sequence analysis of the genome of the bacterium *Mycoplasma pneumoniae. Nucleic Acids Res.* **24**:4420-4449

22. Himmelreich, R., Plagens, E., Hilbert, H., Reiner, B. and Herrmann, R., 1997, Comparative analysis of the genomes of the bacteria *Mycoplasma pneumoniae* and *Mycoplasma genitalium. Nucleic Acids Res.* **25**:701-712

23. Huang, Y., Robertson, J.A. and Stemke, G.W., 1995, An unusual rRNA gene organization in *Mycoplasma fermentans* (incognitus strain). *Can. J. Microbiol.* **41**:425-427

24. Hutchison III, C.A., Peterson, S.N., Gill, S.R., Cline, R.T., White, O., Fraser, C.M., Smith, H.O. and Venter, J.C., 1999, Global transposon mutagenesis and a minimal mycoplasma genome. *Science* **286**: 2165-2169

25. Inagaki, Y., Bessho, Y. and Osawa, S., 1993, Lack of peptide-release activity responding to codon UGA in *Mycoplasma capricolum. Nucleic Acids Res.* **21**:1335-1338

26. Inagaki, Y., Bessho, Y., Hori, H. and Osawa, S., 1996, Cloning of the *Mycoplasma capricolum* gene encoding peptide-chain release factor. *Gene* **169**:101-103

27. Inagaki, Y., Kojima, A., Bessho, Y., Hori, H., Ohama, T. and Osawa, S., 1995, Translation of synonymous codons in family boxes by *Mycoplasma capricolum* tRNAs with unmodified uridine or adenosine at the first anticodon position. *J. Mol. Biol.* **251**:486-492

28. Jukes, T.H., Osawa, S., Muto, A. and Lehman, N., 1987, Evolution of anticodons: variation in the genetic code. *Cold Spr. Harb. Symp. Quant. Biol.* **52**:769-776

29. Kannan, T.R. and Baseman, J.B., 2000, Expression of UGA-containing *Mycoplasma* genes in *Bacillus subtilis. J. Bacteriol.* **182**:2664-2667

30. Karzai, A.W., Roch, E.D. and Sauer, R.T., 2000, The SsrA-SmpB system for protein tagging, directed degradation and ribosome rescue. Nature Struct. Biol. 7:449-455

31. Keiler, K.C., Waller, P.R. H. and Sauer, R.T., 1996, Role of peptide tagging system in degradation of protein synthesized from messenger RNA. *Science* **271**, 990-993.

32. Kochel, H.G., Lazarus, C.M., Basak, N. and Kuntzel, H., 1981, Mitochondrial tRNA gene clusters in *Aspergillus nidulans*: organization and nucleotide sequence. *Cell* **23**:625-633

33. Knudsen, J., Wower, J., Zwieb, C. and Gorodkin, J., 2001, tmRDB (tmRNA database). *Nucleic Acids Res.* **29**:171-172

34. Kunst, F., Ogasawara, N., Moszer, I., Albertini, A.M., Alloni, G., Azevedo, V., Bertero, M.G., Bessieres, P., Bolotin, A., Borchert, S., Borriss, R., Boursier, L., Brans, A., Braun, M., Brignell, S.C., Bron, S., Brouillet, S., Bruschi, C.V., Caldwell, B., Capuano, V., Carter, N.M., Choi, S.K., Codani, J.J., Connerton, I.F., Danchin, A. *et al.*, 1997, The complete genome sequence of the gram-positive bacterium *Bacillus subtilis. Nature* **390**: 249-256

35. Lim, O.P. and Sears, B.B., 1992, Evolutionary relationships of a plant-pathogenic mycoplasmalike organism and *Acholeplasma laidlawii* deduced from two ribosomal protein gene sequences. *J. Bacteriol.* **174**:2606-2611

36. Lovett, P.S., Ambulos, P., Mulbry, Jr., W., Noguchi, N. and Rogers, R.J., 1991, UGA can be decoded as tryptophan at low efficiency in *Bacillus subtilis. J. Bacteriol.* **173**:1810-1812

37. Luneberg, E., Kamla, V., Hadding, U. and Frosch, M., 1991, Sequence and expression in *Escherichia coli* of a *Mycoplasma hominis* gene encoding elongation factor Tu. *Gene* **102**:123-127

38. Muto, A., 1987, The genome structure of *Mycoplasma capricolum*. *Isr. J. Med. Sci.* **23**:334-341

39. Muto, A., Andachi, Y., Yuzawa, H., Yamao, F. and Osawa, S., 1990, The organization and evolution of transfer RNA genes in *Mycoplasma capricolum*. *Nucleic Acids Res.* **8**:5037-5043

40. Muto, A., Fujihara, A., Ito, K., Matsuno, J., Ushida, C.and Himeno, H., 2000, Requirement of transfer-messenger RNA for the growth of *Bacillus subtilis* under stresses. *Genes to Cells* **5**:627-636

41. Muto, A. and Osawa, S., 1987, The guanine and cytosine content of genomic DNA and bacterial evolution. *Proc. Natl. Acad. Sci. USA* **89**:166-169

42. Muto, A., Ushida, C. and Himeno, H., 1998, A bacterial RNA which functions both as a tRNA and an mRNA. *Trends Biochem. Sci.* **23**:25-29

43. Muto, A., Yamao, F. and Osawa, S., 1987, The genome of *Mycoplasma capricolum*. *Prog. Nucleic Acid Res. & Mol. Biol.* **34**:29-58

44. Nakagawa, T., Uemori, T., Asada, K., Kato, I. And Harasawa, R., 1992, *Acholeplasma laidlawii* has tRNA genes in the 16S-23S spacer of the rRNA operon. *J. Bacteriol.* **174**:8163-8165

45. Neyrolles, O., Ferris, S., Behbahani, N., Montagnier, L. and Blanchard, A., 1996, Organization of *Ureaplasma urealyticum* urease gene cluster and expression in a suppressor strain of *Eshcerichia coli*. *J. Bacteriol.* **178**:5853-3859

46. Oba, T., Andachi, Y., Muto, A. and Osawa, S., 1991, Translation *in vitro* of codon UGA as tryptophan in *Mycoplasma capricolum*. *Biochimie* **73**:1109-1112

47. Oba, T., Andachi, Y., Muto, A. and Osawa, S., 1991, CGG: an unassigned or nonsense codon in *Mycoplasma capricolum*. *Proc. Natl. Acad. Sci. USA* **88**:921-925

48. Ohkubo, S., Muto, A., Kawauchi, Y., Yamao, F. and Osawa, S., 1987, The ribosomal protein gene cluster of *Mycoplasma capricolum*. *Molec. Gen. Genet.* **210**:314-322

49. Osawa, S., Jukes, T.H., Muto, A., Yamao, F., Ohama, T. and Andachi, Y., 1987, Divergence of genetic code and directional mutation pressure. *Cold Spr. Harb. Symp. Quant. Biol.* **52**:777-789

50. Osawa, S., Jukes, T.H., Watanabe, K. and Muto, A., 1992, Recent evidence for evolution of the genetic code. *Microbiol. Rev.* **56**:229-264

51. Pellegrin, J.-L., Maugain, J., Clerc, M.-T., Leng, B., Bove, J.-M. and Bebear, C., 1990, Activity of rifampin against mollicutes, clostridia and L-forms. *Zentbl. Bakteriol. Suppl.* **20**:810-812

52. Razin, S., 1985, Molecular biology and genetics of mycoplasmas (*Mollicutes*). *Microbiol. Rev.* **49**:419-455

53. Razin, S., Yogev, D. and Naot, Y., 1998, Molecular Biology and pathogenicity of mycoplasmas. *Microbiol. Mol. Biol. Rev.* **62**:1094-1156

54. Renbaum, P., Abrahamove, D., Feinsod, A., Wilson, G.G., Rottem, S. and Razin, A., 1990, Cloning, characterization, and expression in *Escherichia coli* of the gene for the CpG DNA methylase from *Spiroplasma sp.* Strain MQ1 (M.SssI). *Nucleic Acids Res.* **18**:1145-1152

55. Rogers, M.A., Steinmetz, A.A. and Walker, R.T., 1986, The nucleotide sequence of a tRNA gene cluster from *Spiroplasma meliferum*. *Nucleic Acids Res.* **14**:3145

56. Samuelsson, T., Elias, P., Lustig, F. and Guidy, Y.S., 1985, Cloning and nucleotide sequence analysis of transfer RNA genes from *Mycoplasma mycoides*. *Biochem. J.* **232**:223-228

57.    Sibler, A.-P., Dirheimer, G. and Martin, R.P., 1986, Codon reading patterns in
       *Saccharomyces cerevisiae* mitochondria based on sequences of mitochondrial tRNAs.
       *FEBS Lett.* **194**:131-138

58.    Simoneau, P., Li, C.-M., Loechel, S., Wenzel, R., Herrmann, R. and Hu, R.-C., 1993,
       Codon reading scheme in *Mycoplasma pneumoniae* revealed by the analysis of the
       complete set of tRNA genes. *Nucleic Acids Res.* **21**:4967-4974

59.    Skamrov, A., Goldman, M., Klasove, J. and Beabealashvilli, 1995, *Mycoplasma
       gallisepticum* 16S rRNA genes. *FEMS Microbiol. Lett.* **129**:321-325

60.    Smart, C.D., Schneider, B., Blomquist, C.L., Guerra, L.J., Harrison, N.A., Ahrens, U.,
       Lorenz, K.-H., Seemuller, E. and Kirpatrick, B.C., 1996, Phytoplasma-specific PCR
       primers based on sequences of the 16S-23S rRNA spacer region. *Appl. Environ.
       Microbiol.* **62**:2988-2993

61.    Smiley, B.K. and Minion, F.C., 1993, Enhanced readthrough of opal (UGA) stop
       codons and production of *Mycoplasma pneumoniae* P1 epitopes in *Escherichia coli.*
       *Gene* **134**:33-40

61a.   Sprengart, M.L., Fuchs, E. and Porter, A.G., 1996, The downstream box: an efficient
       and independent translation initiation signal in Escherichia coli. *EMBO J.*, **15**: 665-
       674.

62.    Sprinzl, M., Horn, C., Brown, M., Ioudovitch, A. and Steinberg, S., 1998,
       Compilation of tRNA sequences and sequences of tRNA genes. *Nucleic Acids Res.*
       **26**:148-153

63.    Stemke, G.W., Huang, Y., Laigret, F. and Bove, J.-M., 1994, Cloning the ribosomal
       RNA operons of *Mycoplasma flocculare* and comparison with those of *Mycoplasma
       hypopneumoniae. Microbiology* **140**:857-860

64.    Tanaka, R., Muto, A. and Osawa, S., 1989, Nucleotide sequence of tryptophan tRNA
       gene in *Acholeplasma laidlawii. Nucleic Acids Res.* **17**:5843

65.    Tanaka, R., Andachi, Y., and Muto, A., 1991, Evolution of tRNAs and tRNA genes in
       *Acholeplasma laidlawii. Nucleic Acids Res.* **19**:6787-6792

66.    Taschke, C. and Herrmann R., 1988, Analysis of transcription and processing signals
       in the 5' regions of two *Mycoplasma capricolum* rRNA operons. *Mol. Gen. Genet.*
       **212**:522-530

67.    Taschke, C.,Klinkert, M.-Q., Wolters, J. and Herrmann R., 1986, Organization of
       ribosomal RNA genes in *Mycoplasma hyopneumoniae*: the 5S rRNA gene is
       separated from the 16S and 23S rRNA genes. *Mol. Gen. Genet.* **205**:428-433

68.    Toth, K.F., Harrison, N. and Sears, B.B., 1994, Phylogenetic relationships among
       members of the class *Mollicutes* deduced from *rps3* gene sequences. *Int. J. Syst.
       Bacteriol.* **44**:119-124

69.    Ushida, C., Himeno, H., Watanabe, T. and Muto, A., 1994, tRNA-like structures in
       10Sa RNAs of *Mycoplasma capricolum* and *Bacillus subtilis. Nucleic Acids Res.*
       **22**:3392-3396

70.    Ushida, C. and Muto, A., 1995, Novel small stable RNAs of *Mycoplasma capricolum.*
       *DNA Res.* **2**:229-230

71.    Waldo, R.H., Popham, P.L., Romero-Arroyo, C. E., Mothershed, E.A., Lee, K.K. and
       Krause, D.C., 1999, Transcriptional analysis of the *hmw* gene cluster of *Mycoplasma
       pneumoniae. J. Bacteriol.* **181**:4978-4985

72.    Weiner, J., Herrmann R. and Browning, G.F., 2000, Transcription in *Mycoplasma
       pneumoniae. Nucleic Acids Res.* **28**:4488-4496

73.    Weisburg, M.G., Tully, J.G., Rose, D.L., Retzel, J.P., Oyaizu, H., Yang, D.,
       Mandelco, L., Sechrest, J., Lawrence, T.G., Van Etten, J., Maniloff, J. and Woese,

C.R., 1989, Phylogenetic analysis of the mycoplasmas: basis for their classification. *J. Bacteriol.* **171**:6455-6467

74. Yamao, F., Muto, A., Kawauchi, Y., Iwami, M., Iwagami, S., Azumi, Y. and Osawa, S., 1985, UGA is read as tryptophan in *Mycoplasma capricolum. Proc. Natl. Acad. Sci. USA* **82**:2306-2309

75. Yamao, F., Iwagami, S., Azumi, Y., Muto, A. and Osawa, S., 1988, Evolutionary dynamics of tryptophan tRNAs in *Mycoplasma capricolum. Molec. Gen. Genet.* **212**:364-369

76. Zreik, L., Carle, P., Bove, J.-M. and Garnier, M., 1995, Characteriztion of the mycoplasmalike organism associated with witches'-bloom disease of lime and proposition of a *Candidatus* taxon for the organism, "*Candidatus* Phytoplasma aurantifolia" *Int. J. Syst. Bacteriol.* **45**:449-453

Chapter 15

# Extrachromosomal Elements and Gene Transfer

JOËL RENAUDIN
*Laboratoire de Biologie Cellulaire et Moléculaire, UMR GDPP, IBVM, INRA and Université Victor Segalen Bordeaux2, 71 avenue Edouard Bourleaux, BP 81, 33883 Villenave d'Ornon Cedex, France.*

## 1.    INTRODUCTION

Molecular dissection of plasmid or phage DNA provided many important clues to central biological processes such as DNA replication and gene expression in bacteria. In recent years, plasmid research has taken a new turn, with the advent of gene cloning, and the use of plasmids as gene vectors. Indeed, the availability of gene transfer systems is an absolute prerequisite to genetic studies. In mollicutes, the engineering of endogenous plasmid and viral DNA to cloning vectors has encountered limited success. However, during the past five years, significant progress in the development of new genetic tools has been accomplished, opening the way to genetic studies in mollicutes. Gene vectors based on artificial *oriC* plasmids (plasmids carrying the replication origin, *oriC* of *Spiroplasma citri*) and on transposons have been constructed, and methods such as transposon mutagenesis and gene disruption through homologous recombination have been used to generate mutants. Characterization of these mutants has contributed to the identification of genes associated with biological function such as motility, cytadherence, and pathogenicity. Exhaustive reviews on plasmids, viruses, and molecular genetics of mollicutes have been published previously[7,9,22,23,33,71,85,86,91]. Therefore this chapter will focus primarily on the recent data with special emphasis on viruses and plasmids used as tools for gene transfer.

*Molecular Biology and Pathogenicity of Mycoplasmas*, Edited by Razin and Herrmann, Kluwer Academic/Plenum Publishers, New York, 2002

## 2.        EXTRACHROSOMAL ELEMENTS

Both plasmids and viruses have been found in mollicutes. Spiroplasmas and acholeplasmas are the mollicutes most frequently infected by a variety of viruses whereas four viruses only have been isolated from Mycoplasma species. Plasmids have been frequently described in spiroplasmas and in phytoplasmas. In contrast, *Mycoplasma mycoides* subsp. *mycoides* is the only Mycoplasma species known to harbor plasmids.

### 2.1      Plasmids

Within the mollicutes, plasmids were first detected in spiroplasmas as extrachromosomal, covalently-closed-circular DNA and some of them were mapped and/or cloned in *E. coli*. However, the distinction of plasmid DNA from replicative forms (RF) of viral DNA requires serious consideration as spiroplasmas are infected by a variety of viruses[9,33,39,85]. Spiroplasma plasmids are still cryptic. No specific phenotype has been associated with any of them and none of them has been sequenced. More recently, two distinct plasmids, pADB201 and pKMK1, have been isolated from *Mycoplasma mycoides* and characterized. These small plasmids are related to Gram-positive bacterial plasmids that replicate via single-stranded DNA intermediates and do not code for factors other than those required for replication and maintenance[5,6,30,33,57].

The non-culturable plant pathogenic phytoplasmas have also been found to carry extrachromosomal DNAs (EC-DNAs)[12,16,17,48,62,65,79,80,102]. The phytoplasma EC-DNAs have been postulated to be plasmids because no virus particles were observed by electron microscopy[44,62,63]. Cloned fragments of phytoplasma plasmids have been used as probes to study their relatedness and to improve pathogen detection in plants and insect vectors[12,44,63,80,81,101]. The EC-DNAs associated with phytoplasmas range within at least two distinct groups. One includes the EC-DNAs of the AY (aster yellow)- and EY (elm yellow)- clade phytoplasmas and the other is associated with the X (western X)- clade phytoplasmas[88]. Sequencing a 3.6-kbp EC-DNA fragment of the onion yellow phytoplasma (OY-W) has revealed five ORFs two of which showed similarities to known proteins[61]. One (ORF4), encodes a putative protein showing similarities with single-strand binding proteins (SSB). SSBs specifically bind single-stranded DNA and are mainly involved in DNA replication and repair. The ORF4-encoded protein shares 26.1% identity with a SSB protein of *Bacillus subtilis* lacking the C-terminal glycine-rich region and 30% identity with the SSB protein of the conjugative plasmid RK2. The other (ORF5) encodes a putative replication (Rep) protein similar (22.8% identity) to that of the *M. mycoides* plasmid pKMK1[30], suggesting that the phytoplasma plasmid also belongs to the pE194 family of rolling-circle replication (RCR)-type plasmids[54]. Hybridization studies revealed heterogeneity between the plasmids of the wild-type OY-W and those of a mutant of lower pathogenicity OY-M. However, the link between the plasmid content and the pathogenicity of the phytoplasma has not been clearly established[61]. Similarly, no difference was found in the infected plant's symptom expression for X-clade phytoplasmas carrying EC-DNAs of different sizes[88]. Interestingly, the EC-DNA associated with the X-clade *Vaccinium* witches' broom phytoplasma, as well as a new EC-DNA from the OY-W phytoplasma were found to encode a protein similar to the geminivirus Rep protein[88]. In the case of the OY-W phytoplasma, the Rep protein homologue was shown to be expressed in OY-W infected plants, suggesting that this EC-DNA replicates via a geminivirus-like rolling circle replication mechanism[82]. Geminiviruses are a group of plant viruses that differ from all other plant viruses in possessing circular, single-stranded DNA genomes. The close phylogenetic relationship between phytoplasma extrachromosomal DNA and geminiviruses provides new evidence for the theory previously proposed that these circular single-stranded DNA viruses might have evolved from prokaryotic episomal replicons[97].

## 2.2 Viruses

Among the mollicutes, acholeplasmas and spiroplasmas are the most frequently infected by a variety of viruses whereas very few viruses are known to infect mycoplasmas. All mollicute viruses carry DNA genomes either circular or linear, and single- or double-stranded. With the exception of acholeplasma viruses L2 and L172 which are unique among viruses of eubacteria in that they possess an envelope[71], most of the mollicute viruses have their morphological counterparts among phages of bacteria with a cell wall. Classification, structure, physicochemical properties and molecular biology of mollicute virus infection have been extensively reviewed[9,71,85,91].

### 2.2.1 *Spiroplasma citri* virus SpV1

SpV1 is a naked, rod-shaped virus with a circular single-stranded DNA genome of approximately 8 kb, depending on the isolate. Nucleotide sequences and gene organization of two SpV1 isolates (SpV1-R8A2B and SpV1-C74) have been determined[89,91]. While most of the ORFs have not been identified, ORF3 of SpV1-R8A2B shares striking homologies with transposases of insertion sequences of the IS*30* family. Interestingly, the highest homology score was with the transposase of the *Staphylococcus* *aureus* prophage phiPV83. Similarly, ORF3 of SpV1-C74 shares significant homology with transposases of the IS*3* family, and in particular with a putative transposase of the archaeon *Archaeoglobus fulgidus*. In contrast, the genome of SVTS2, a SpV1-type virus primarily isolated from *S. melliferum*, contains no recognizable transposase[104]. A peculiar feature of *S. citri* SpV1 is the presence of viral sequences in the spiroplasmal host chromosome[90]. Viral sequences consisting of full-length or parts of the viral genome are distributed all along the spiroplasmal chromosome and could account for up to 10% of the *S. citri* genome[4,9,91,114]. In the case of *S. citri* strain R8A2, the presence of SpV1 viral sequences does not confer immunity to infection by viruses SpV1-R8A2B and SpV1-C74. In contrast, resistance to SVTS2 superinfection of *S. citri* lines MR2 and MR3, has been associated with integration of viral sequences into the host chromosomal and extrachromosomal DNAs[103]. Actually, the function of the SpV1-related sequences integrated into the *S. citri* chromosome is unknown. However, considering the number and the restriction fragment length polymorphism of SpV1-like sequences among various strains of *S. citri*, it was suggested that the viral sequences, considered as repeated elements, might play a role in genomic rearrangements and gene expression[114,115]. For example, in *S. citri* GII-3, the transcription of the fructose operon stops at a terminator structure located within SpV1-related sequences, immediately downstream of the 1-phosphofructokinase (*fruK*) gene[42]. Further comparisons of integrated and nonintegrated viral sequences also suggested that SpV1 sequences may have altered spiroplasma genomes by insertion within active genes, by mediating deletions of sequences adjacent to their integration sites, and by providing targets for site-specific or homologous recombination[76].

### 2.2.2 *Spiroplasma melliferum* virus SpV4

SpV4 is a naked isometric virus with a circular single-stranded DNA genome. It belongs to the *Microviridae* family. SpV4 was the first mollicute virus to be sequenced[94]. Sequence analyses revealed that gene organization, transcription, and translation were similar to those in eubacteria with the exception that in spiroplasmas UGA codes for tryptophan[7,91,96]. The SpV4 capsid protein (product of ORF1) and the translation product of ORF2 have their counterpart in Chlamydia phages Chp1 and Chp2. In addition the SpV4 capsid protein shares limited, but significant, homology with the F protein, the major coat protein of coliphage $\Phi$X174. The three-dimensional structure of SpV4 has been determined[13]. The detailed description of the SpV4 capsid structure was based on a three-dimensional cryo-electron microscopy image reconstruction of SpV4 samples and structural alignment with the atomic model of the major capsid protein (protein F) of the related coliphage

ΦX174. A striking feature of the SpV4 structure, which was not previously seen[95], is the presence of 20 protrusions that extend radially outwards from the capsid along the threefold icosahedral axes. These protrusions might function as the receptor-recognition site during host infection in an analogous manner to the G protein spikes of the coliphages. Interestingly, the amino acid sequence that folds to form this protrusion was identified as a 71-residue insertion in the SpV4 coat protein sequence, as compared to the ΦX174 F capsid protein. Sequence alignments of the F capsid proteins of the coliphages α3, G4 and ΦX174, and the capsid proteins of Chlamydia phage Chp1 and SpV4, suggest that distinct genera of the *Microviridae* family might have evolved from a common primordial ancestor, with capsid surface variations that have enabled them to explore diverse host ranges[13].

### 2.2.3 *Mycoplasma pulmonis* virus P1

The *M. pulmonis* virus P1 is a tailed, polyhedral virus[31]. The P1 genome consists of a linear, double-stranded DNA of 11.3 kbp with 350 bp inverted terminal repeats, and a putative terminal protein was found to be attached to the 5' termini of the P1 DNA[118]. The P1 virus shares the characteristic properties of the phage family *Podoviridae* and is most similar to phages Φ29 and Cp-1, the hosts of which are Gram-positive bacteria, phylogenetically related to mollicutes. In *M. pulmonis*, the *hsd1* element encodes a type I restriction and modification system which is regulated by DNA inversion[35]. Interestingly, virus P1 is currently used for assessing restriction and modification in *M. pulmonis*, as the DNA inversion that changes the restriction and modification properties of the organism can be monitored by changes in the plating efficiency of P1 on lawns of *M. pulmonis*[32].

### 2.2.4 *Mycoplasma arthritidis* MAV1

MAV1 is the most recently discovered mycoplasma virus, and it has been associated with virulence of *M. arthritidis*[112]. Highly arthritogenic strains were shown to carry MAV1 DNA integrated at various sites of the mycoplasmal chromosome, whereas low-virulence strains did not. Furthermore, lysogenization with MAV1 of a low-virulence strain considerably enhanced its arthritogenic potential. Determination of sequences flanking the integrated prophage and the ends of native MAV1 DNA led to the identification of the phage DNA (*attP*) and bacterial DNA (*attB*) recombination sites[110]. The MAV1 virion is highly resistant to proteinase K, and contains a linear, double-stranded DNA genome of 16 kbp[110]. Sequencing the MAV1 genome has identified 15 ORFs. Putative protein products from these ORFs comprise DNA replication, restriction-modification, structural, regulatory, and integration/excision proteins[111]. On the basis of significant matches with known proteins, ORFs *repB*, *marMP*, and *int* were predicted to encode the primary protein involved in phage DNA replication, a cytosine-5-specific DNA methyltransferase that is part of a phage encoded restriction-modification system, and a site specific recombinase of the Int family, respectively. ORF *exiS* overlapping the *int* gene has been anticipated to be an excisase. The MAV1 *imm* protein product was hypothesized to be the phage repressor as it shares the characteristics of phage repressors that are not inducible by UV-irradiation or mitomycin C, namely a small size, an acidic pI, and a leucine-zipper motif immediately preceded by a very basic motif. The predicted protein products of ORFs *htpN*, *htpT*, *htpH*, *htpE*, *htpA*, and *htpB* were thought likely to encode the MAV1 structural proteins. Based on the observations that MAV1 sequences are required for *M. arthritidis* to cause arthritis, and that the *vir* transcript is one of the only two MAV1 transcripts detected in MAV1 lysogens, it was hypothesized that the *vir* protein product, a lipoprotein of 25.4 kDa, is the MAV1 virulence determinant[111]. However, further experimental evidence is required to clearly establish the causal relationship between the MAV1 encoded lipoprotein and *M. arthritidis* pathogenicity.

## 3.     GENE TRANSFER

### 3.1     Transformation

Transformation of *Acholeplasma laidlawii* with mycoplasma virus (L2) DNA, i. e., transfection, was the first successful transfer of DNA into a member of the class mollicutes. The procedure required the presence of polyethylene glycol (PEG) during the transfection process. Subsequently, the so-called PEG method has been used to transfect *Spiroplasma melliferum* with SpV4 DNA, and *A. laidlawii* with DNA from mycoplasma viruses L1 and L172. However, transfection with L3 DNA could only be achieved by electroporation[23]. Electroporation has also proved to be more efficient in promoting penetration of SpV1 DNA into *S. citri* cells[40,75,106]. Currently, there is no general method, and both the PEG-method and electroporation are routinely used to transform mollicutes[29,77]. PEG-mediated transformation has been demonstrated for *A. laidlawii*[27], *M. pulmonis*[26,27], *M. mycoides*[56], *Mycoplasma capricolum*[59], *Mycoplasma gallisepticum*[10], *M. arthritidis*[28,108], *S. citri*[40], and *S. melliferum*[84], and electroporation has been used for transformation of *M. gallisepticum*[60], *Mycoplasma hominis*[1], *M. mycoides*[113], *Mycoplasma pneumoniae*[50], *Mycoplasma genitalium*[87], and *S. citri*[40,106].

Transformation of mollicutes to tetracycline resistance has also been achieved through conjugal transfer of Tn*916* from *Enterococcus faecalis* by a spontaneous mating process. It has been first demonstrated in *M. hominis*[98] and, more recently, in *M. arthritidis*[108] and in *M. gallisepticum*[100]. Detailed protocols for transformation and for conjugal transfer of Tn*916* in mycoplasmas have been described[77,109]. However, as was previously mentioned[33], the method of choice for transformation of mollicutes depends on the particular species being studied and on the particular DNA molecule being employed. For example, electroporation, but not the PEG-method, has been successful for transformation of *M. pneumoniae*[50], whereas PEG-mediated transformation, but not electroporation, has been successful in *M. pulmonis*[29]. *S. citri* strain R8A2 was efficiently transfected by electroporation with SpV1 viral RF[106] but was very poorly transformed with the *oriC* plasmid pBOT1[93]. Moreover, in a particular species, successful transformation may also depend on the strain being used. While *S. citri* ASP1 was readily transformed ($\sim 10^{-3}$ transformants per CFU) with plasmid pBOT1, *S. citri* R8A2 was very poorly transformed[93] (less than $10^{-8}$ transformants per CFU), and strain BR3G was not transformed at all (J. Fletcher, personal communication; J. Renaudin unpublished results). To explain these results, it was hypothesized that strains would possess different restriction-modification systems, or would differ in the expression of the corresponding genes. For example, in *M. pulmonis*, the sensitivity to infection by virus P1 was shown to correlate with the restriction-modification properties which are controlled by a DNA inversion at the *hsd1* locus[25,35]. Also, transformation of *M. mycoides* and *M. capricolum* with plasmid pIKΔ showed that the plasmid was introduced back into its original host (*M. mycoides* or *M. capricolum*) at a higher frequency than it was into the reciprocal host, suggesting that restriction and modification systems which are prevalent in mollicutes may represent barriers in gene transfer[59]. This has been further illustrated in *M. arthritidis*[108]. PEG-mediated transformation, as well as conjugal transfer of Tn*916*, were successful with strain H6061 but not with other strains of *M. arthritidis*, including strain 1581. Interestingly, this strain carries an *Alu*I-like endonuclease which cleaved the unmethylated AGCT recognition sites whereas strain H6061 does not. In this particular case, the role of restriction and modification systems as barriers in gene transfer has been further demonstrated by the fact that strain 1581 could not be transformed with Tn*916* unless the DNA was first modified with the *Alu*I methylase.

## 3.2      Selectable markers

Selection of bacterial cells transformed with exogenous DNA has been made possible by the use of selectable markers. In mollicutes, markers include resistance to toxic compounds such as arsenic acid, vanadium oxide, and xylitol, which have been used to isolate *S. citri* mutants[2,3,43], and resistance to antibiotics such as erythromycin, chloramphenicol, gentamycin, and tetracycline. The corresponding resistance genes have been used to construct gene vectors. The erythromycin resistance gene (*erm$^r$*) of the streptococcal plasmid pAMβ1 functions as a selection marker in *A. laidlawii* and *M. mycoides*[21,58], and a kanamycin-neomycin resistance gene from *S. aureus* plasmid pUB110 confers resistance to kanamycin in *A. laidlawii*[107]. However, the most commonly used markers in genetic studies of mollicutes are the tetracycline resistance gene (*tetM*) of Tn*916*[14] or the gentamycin resistance gene (*aacA-aphD*) of Tn*4001*[99]. With the exception of some spiroplasma species such as *S. melliferum and S. floricola*[15], mollicutes are generally sensitive to tetracycline. The *tetM* gene has conferred high level tetracycline resistance in all species into which it has been introduced including *A. laidlawii*[27], *M arthritidis*[108], *M. gallisepticum*[10], *M. hominis*[98], *M. mycoides*[113], *M. pulmonis* and *M. hyorhinis*[26], and *S. citri*[106]. The gentamycin resistance gene of Tn*4001* has been used for selection of insertional mutants in *Acholeplasma oculi*[24,69], *M. gallisepticum*[10], *M. genitalium*[87], *M. pneumoniae*[50], and *S. citri*[37], but is probably useless in *M. mycoides* which is rather insensitive to gentamycin[33]. The inability to transform *M. arthritidis* and *M. pulmonis* with Tn*4001* was recently correlated with the failure of the *aacA-aphD* gene of Tn*4001* to confer a selectable level of resistance in these organisms[28]. To expand the versatility of Tn*4001* for genetic manipulations and for use in species resistant to gentamycin, the chloramphenicol acetyltransferase (*cat*) gene from *S. aureus* was used to construct Tn*4001* derivatives which have been successfully transferred into *M. pneumoniae*[47]. Similarly, the *cat* gene of the *E. coli* transposon Tn*9* driven by a mycoplasmal promoter has been used as a selection marker in *M. pulmonis*[28]. The *cat* gene of Tn*9* has also been used for selection in *S. citri*. However, *oriC* plasmids, carrying the *cat* gene driven by the *S. citri* fibril protein promoter, transformed spiroplasma cells at low frequency as compared to plasmids carrying the gentamycin resistance gene of Tn*4001* or the *tetM* gene of Tn*916* (S. Duret and J. Renaudin, unpublished results).

## 3.3     Cloning and expression vectors

Many species of mollicutes have been transformed to antibiotic resistance by delivery of transposons with suicide plasmids, however, there is no replicative, plasmid vector of general use[33,86]. Attempts to develop cloning and shuttle vectors made use of plasmids such as streptococcal plasmids[21,107], plasmid pMH1 from *S. citri*[105], and plasmid pKMK1 from *M. mycoides*[58], as well as viral DNA such as the replicative form of the *S. citri* virus SpV1[72,92,106]. However the more successful approach was the construction of artificial plasmids containing the replication origin (*oriC*) of the *S. citri* chromosome[92,93,116]. In mycoplasmas, the lack of plasmid vectors has been partially overcome by using a transposon-based, integrative vector as an alternative approach to the introduction of foreign genes into mycoplasma cells[36,60]. Vectors for gene expression and insertional mutagenesis in mollicutes have been listed in Table 1.

*Table 1.* Vectors for gene expression and insertional mutagenesis in mollicutes

| Vector | Size (kbp) | Replicon (a) | Selectable phenotype (b) | Mollicute host and/or relevant properties | Ref |
|---|---|---|---|---|---|
| SpV1-RF | 8.3 | SpV1-R8A2B | Plaque | *S. citri* strains | 106 |
| pNZ18 | 5.7 | *S. lactis* pSH71 | Kan[r] | shuttle *A laidlawii/E. coli* | 107 |
| p2D4 | 10.6 | *M. mycoides* pKMK1 | Tet[r] | shuttle *M. mycoides/E. coli* (c) | 58 |
| pIKΔ | 5.9 | *M. mycoides* pKMK1 | Tet[r] | *M. mycoides, M.capricolum* | 58 |
| pBOT1 | 9.35 | *S. citri oriC* | Tet[r] | shuttle *S. citri/E. coli* (c) | 93 |
| pBOG | 7.7 | *S. citri oriC* | Gm[r] | shuttle *S. citri/E. coli* (c) | (d) |
| pSD3/4 | 7.35 | *S. citri oriC* | Tet[r*] | shuttle *S. citri/E. coli* (c) | (e) |
| pC1/2 | 5.7 | *S. citri oriC* (*) | Tet[r*] | shuttle *S. citri/E. coli* | (f) |
| pAM120 | 21.4 | - | Tet[r] | Tn*916* delivery | 27 |
| pISM1001 | 13.45 | - | Gm[r] | Tn*4001* delivery | 69 |
| pMUT | 8.6 | - | Gm[r] | Tn*4001* delivery | 37 |
| pISM2062 | 8.7 | - | Gm[r] | Tn*4001*mod delivery | 60 |
| pKV98/9 | 10.3 | - | Gm[r], Cm[r] | Tn*4001*mod delivery | 47 |
| pKV103/4 | 8.7 | - | Cm[r] | Tn*4001*mod delivery | 47 |
| pIVC-1 | 9.9 | - | Gm[r], Cm[r*] | Tn*4001*mod delivery | 28 |
| pIVT-1 | 13.2 | - | Gm[r], Tet[r] | Tn*4001*mod delivery | 28 |

(*) in these plasmids the *oriC* region is restricted to the *dnaA-dnaN* intergenic region.

(a)   replication origin functioning in the mollicute host.

(b)   selectable phenotype of mollicute transformants. Markers in the plasmids are: Cm[r], *cat* gene from *S. aureus* pC194; Cm[r*], *cat* gene from *E. coli* Tn9 driven by the *M. pulmonis vsa* promoter; Gm[r], gene *aacA-aphD* from *S. aureus* Tn*4001*; Kan[r], *kan/neo* from *S. aureus* pUB110; Tet[r], *tetM* gene from *E. faecalis* Tn*916;* Tet[r*], *tetM* gene from Tn*916* driven by the *S. citri* spiralin gene promoter.

(c)   may integrate into the host chromosome during passaging of the mollicute transformant.

(d)   J. Renaudin, unpublished.

(e)   S. Duret and J. Renaudin, unpublished.

(f)   C. Lartigue and J. Renaudin, unpublished.

### 3.3.1    Replicative vectors

Various streptococcal plasmids were found to transform *A. laidlawii* but failed to transform mycoplasmas[33,86]. When inserted into pNZ18, a *Streptococcus lactis* replicon containing the kanamycin/neomycin and chloramphenicol resistance genes from *S. aureus* plasmids pUB110 and pC194, respectively, the α-amylase gene of *Bacillus licheniformis* as well as the *S. citri* spiralin gene were expressed into *A. laidlawii*[55]. The only plasmids known to replicate in mycoplasmas are plasmids p2D4 (shuttle) and pIKΔ (without the *E. coli* replicon), both derived from the *M. mycoides* plasmid pKMK1. By using these plasmids as vectors, the erythromycin resistance determinant (*erm'*) could be introduced into *M. mycoides* and *M. capricolum*[58,59] conferring resistance to erythromycin, but not in other mycoplasma species[33]. However the stability of inserts during plasmid replication in a mycoplasmal host requires further examination as the small plasmids of Gram-positive bacteria, which replicate via single-stranded intermediates, are known to display structural instability[45,54]. In spiroplasmas, transfection of *S. citri* with the SpV1 replicative form (RF) carrying the *cat* gene or an adhesin P1 gene fragment resulted in the synthesis of the relevant proteins in the transfected cells[72,106]. However, loss of the cloned DNA insert was observed during propagation of the transfected cells. The SpV1 genome also replicates by the "rolling circle model", a mechanism known to favour illegitimate recombination. However in this case, deletions in the viral RF were shown to occur not only by illegitimate recombination through a copy choice process but also by homologous recombination between the circular, free viral RF and the SpV1 viral sequences present in the *S. citri* host chromosome[74].

### 3.3.2    Integrative vectors

Transformation with integrative plasmid vectors (which do not replicate) rely on their ability to integrate into the mycoplasmal host chromosome. Two types of integrative vectors have been described in mycoplasmas. One, consists of plasmids carrying sequences homologous to the mycoplasma chromosome, has been used for transformation of *A. laidlawii*, *A. oculi*, *M. gallisepticum*, and *M. pulmonis*[10,24,34,70]. In this case the plasmid integrates through homologous recombination. The other type of integrative vectors consists of suicide plasmids carrying a transposon, namely the Tn*4001* derivative Tn*4001*mod, as the cloning vehicle[60]. Such a transposon-based vector has been used for functional complementation of a *M. pneumoniae* cytadherence-negative mutant[36].

### 3.3.3    *S. citri oriC* plasmids

Originally, *oriC* plasmids were developed to isolate chromosomal fragments containing autonomous replication activity and to study the initiation of chromosome replication. In *E. coli* multiple copies of *oriC* can be maintained extrachromosomally. In contrast to *E. coli*, the Gram-positive bacterium *Bacillus subtilis* strictly regulates the number of *oriC*[83]. In this organism, *oriC* plasmids cannot be maintained and have a tendency to integrate into the host chromosome[117].

In *S. citri*, gene organization of the chromosomal replication origin (*oriC*) region was determined and a 2-kbp chromosome fragment containing the *dnaA* gene with the flanking intergenic regions in which *dnaA* boxes were identified, was characterized as an autonomously replicating sequence (ars)[116]. Later on it was shown that sequences required for plasmid replication were carried on a 163-bp fragment of the *dnaA-dnaN* intergenic region (C. Lartigue and J. Remaudin, unpublished results). This fragment shares common features with *oriC* of other bacteria. In particular, it contains three putative *dnaA* boxes and three AT clusters. Shuttle plasmid pBOT1 was constructed by combining the 2-kbp *oriC* region of the spiroplasmal chromosome with the *tetM* gene of Tn916 and a colE1-derived *E. coli* replicon. When transformed in *S. citri*, the shuttle plasmid pBOT1 replicates first as a free extrachromosomal element. However, probably due to the incompatibility of *oriC* plasmids in Gram-positive bacteria, the plasmid integrates into the spiroplasmal chromosome during passaging of the spiroplasmal transformants. Plasmid integration was shown to occur by homologous recombination involving one crossover at the *oriC* region. Once integrated into the host chromosome the whole plasmid is stably maintained regardless of the presence or absence of selection pressure. This is also true for pBOT1-derived, recombinant plasmids carrying heterologous DNA inserts[93]. For these reasons, the *S. citri oriC* plasmids proved to be very useful in expression of cloned genes in *S. citri*[92,93]. In particular, pBOT1-derived recombinant plasmids were used for functional complementation of *S. citri* mutants[41,53]. The so-called nonpathogenic mutant GMT553, in which the transposon Tn4001 was inserted within the fructose operon, is unable to use fructose as the only carbon source and, when experimentally transmitted to periwinkle plants by the insect vector, this mutant does not induce symptoms[38]. However, both the ability to use fructose in vitro and the ability to produce severe symptoms in the host plant were restored when this mutant was transformed with the wild-type fructose operon carried on the pBOT1 plasmid[41]. Similarly, functional complementation of the *scm1*-disrupted *S. citri* motility mutant G540 has been achieved. Transformation with the wild-type *scm1* gene restored spiroplasmal motility[53]. In such studies, the stability of the integrated plasmid holds a considerable advantage in monitoring the behavior of spiroplasmal transformants in vivo, i. e. in the insect vector or in the host plant, in which no selection pressure can be applied.

The *S. citri oriC* plasmid pBOT1 failed to transform mycoplasmas such as *M. hominis* PG21 (C. Bébéar and J. Renaudin, unpublished results), and *M. pulmonis* UAB CTIP (A. Blanchard and J. Renaudin, unpublished results). The reason for this specificity is not known. However, whether the *oriC* plasmid approach can be applied to other mollicute species should be investigated, in particular with mycoplasmas such as *M. capricolum* and *M. pulmonis* in which gene organization at the origin of replication is similar to that of *S. citri*, due the presence of putative *dnaA* boxes in the vicinity of the *dnaA* gene[11,78].

In *S. citri* the *oriC* plasmid were also used to produce mutants by gene disruption through homologous recombination (see below).

## 3.4     Transposon mutagenesis

Transposon mutagenesis is generated by random insertion of transposons into genes causing their inactivation. The advantages of gene inactivation by transposon insertion include production of null alleles, minimal damage to other genes and easy cloning of the interrupted locus. Successful transformation of *A. laidlawii* and *M. pulmonis* with plasmid pAM120, which carries Tn916, has shown that both the *tetM* gene and the genes required for transposition were expressed, opening the way to insertional mutagenesis in mollicutes[27]. Following this pioneering work, a variety of mollicute species have been transformed to tetracycline resistance with Tn916 or to gentamycin, tetracycline or chloramphenicol resistance with Tn4001 and Tn4001-derivatives. Tn916 has been introduced into mollicutes by conjugal transfer from *E. faecalis* donor cells[98,100,108], as well as by PEG-mediated transformation[10,26,98,108,113] and electroporation[10,113] with the *E. coli* plasmid pAM120[14] whereas the *S. aureus* transposon Tn4001 has been delivered by PEG- or electro- transformation with plasmids derived from pIS1001[10,28,37,50,69]. Transformation frequencies ranged from $10^{-9}$ per CFU in *S. citri*[106] to $10^{-5}$ per CFU in *M. gallisepticum*[10] with Tn916, and from $10^{-8}$ per CFU in *S. citri*[37] to $10^{-4}$ in *M. pneumoniae*[50] with Tn4001.

Transposon mutagenesis has been applied to *M. genitalium*, *M. pneumoniae*, and *S. citri*, in order to identify genes that are involved in pathogenicity[8,38,49,53,87]. *M. pneumoniae* and *M. genitalium* mutants deficient in cytadherence were screened for their inability to hemadsorb or to attach to human T-cells. For example, in *M. pneumoniae*, characterization of such mutants has identified a cytadherence regulatory locus, the P65 operon[64]. Tn4001 insertion in the *orf*p216 encoding the HMW2 protein not only affected the level of all products of the P65 operon but also the products HMW1 and HMW3 from the *hmw* operon which maps 160 kbp apart from the P65 operon. It was thought that HMW2 stabilizes HMW1 and HMW3, integrating them into the mycoplasma cytoskeleton. The role of HMW2 in cytadherence has been further demonstrated through functional complementation of a spontaneous cytadherence mutant lacking HMW2. To overcome the absence of an adequate plasmid vector, the modified transposon Tn*4001*mod[60] was used as a gene vector to express cloned genes in *M. pneumoniae*[46]. Transposon delivery of the wild-type *hmw2* allele (cloned into the IS256L element of Tn*4001*mod) into the deficient mutant resulted in production of HMW2 and rescued cytadherence[36]. Also, a *M. pneumoniae* mutant with an aberrant morphology has been obtained by insertion of Tn*4001* into the *hmw3* gene. Characterization of this mutant led to the conclusion that although HMW3 is not required for production/stability of tip-associated proteins, this protein is necessary in order for a functional tip structure (M. Willby and D. Krause, unpublished results) (see "Cytadherence and Cytoskeletal Elements" chapter). Similarly, Tn*4001*mod was used as a vector with the *E. coli lacZ* gene as the reporter system to examine the role of the GAA repeats in M9/pMGA gene expression in *M. gallisepticum*. These in vivo studies strongly indicate that M9/pMGA gene expression is regulated by the length of the GAA repeat region located upstream of the putative promoter[66] (see "Antigenic Variation" chapter). The signature-tagged mutagenesis by transposition provides a powerful tool to identify genes encoding proteins essential for in vivo survival and infection[51]. Using a similar approach, a library of *M. gallisepticum* mutants was generated by random insertion of Tn*4001*mod in which non-redundant tag elements were inserted to allow selection. However this study revealed a critical limitation of the mutagenesis approach for the further identification of virulence determinants of mollicutes such as *M. gallisepticum*, displaying a highly versatile surface architecture due to spontaneous antigenic variation (P. Much, J. A. Aguero Fernandez, F. Winner, A. Lugmair, C. Citti, and Rosengarten, unpublished results).

Tn*4001* mutagenesis in *S. citri* has generated mutants deficient in motility and in pathogenicity to plants[38,53]. In the motility mutant G540, which was isolated on the basis of its sharp-edged colonies, characterization of the transposon insertion site has led to the identification of the motility gene *scm1*. The highly hydrophobic, *scm1*-encoded polypeptide is predicted to contain a signal peptide, ten transmembrane α-helices, and a leucine zipper motif. It does not show homology with known proteins and its function is still unknown. However, functional complementation of the motility mutant by transformation with a *oriC* plasmid carrying the wild-type *scm1* gene proved this gene to be involved, directly or indirectly, in spiroplasmal motility[53]. The so-called nonpathogenic mutant GMT553 has been isolated on the basis of its inability to produce severe symptoms in periwinkle plants through transmission by the insect vector[38]. In this mutant, insertion of Tn*4001* into the first gene *fruR* of the fructose operon abolishes transcription of all three genes *fruR, fruA,* and *fruK* of the fructose operon, and results in the inability of this mutant to use fructose[41]. The correlation between fructose utilization and pathogenicity was shown by functional complementation. When mutant GMT553 was complemented with the fructose operon genes that restore fructose utilization, severe pathogenicity similar to that of the wild-type strain was restored[41]. The *S. citri* mutant G76 was isolated on the basis of its low ability to be transmitted by the insect vector. In this mutant the transposon is inserted into a gene encoding a lipoprotein which shares limited homology with the MG040 encoded lipoprotein of *M. genitalium* (A. Boutareaud and C. Saillard, unpublished results).

Because mollicutes are considered as the smallest self-replicating organisms, transposon mutagenesis with Tn*4001* has also been used, in *M. genitalium* and *M. pneumoniae*, to address the question "what is a minimal set of essential cellular genes?"[52]. In agreement with the fact that all the ORFs of the smaller *M. genitalium* genome are contained in the *M. pneumoniae* genome, the data indicate that most of the nonessential genes (indicated by transposon insertions) were in the species-specific portion of the *M. pneumoniae* genome. In *M. genitalium*, the analysis suggests that 265 to 350 of the 480 protein-coding genes are essential under laboratory growth conditions. Surprisingly, a few mutations were found in genes generally believed to be essential. For example, putatively disruptive insertions have been found in the DNA replication gene *dnaA* as well as in the unique sigma factor gene. However, as indicated by the authors, conclusive proof of the dispensability of specific genes requires cloning and detailed characterization of a pure population carrying the disrupted gene and, in addition, genes that are individually dispensable may not be simultaneously dispensable (see "Minimal Cell Concept" chapter).

## 3.5      Gene disruption through homologous recombination

Production of mutants by gene disruption or allelic exchange, both of which are dependent on homologous recombination, is crucial in assessment of the role of individual genes in various processes, such as pathogenesis. In addition to gene disruption or allelic exchange, homologous recombination is also used to construct gene fusions for in vivo studies, in order to identify genes that are specifically expressed upon infection.

The process of homologous genetic recombination is based on the ability of enzymes such as the RecA protein to bring together homologous DNA molecules and promote the reciprocal exchange of DNA strands. In mollicutes, homologous recombination was first demonstrated to occur in *A. oculi*[68,70]. However, the first description of gene inactivation through homologous recombination was in *A. laidlawii*[34]. In this study, an internal fragment of the *recA* gene from *A. laidlawii* was cloned into a plasmid that does not replicate in this organism. When introduced into *A. laidlawii*, the plasmid integrated into the chromosome, disrupting the *recA* gene. The loss of RecA activity resulted in cells deficient in DNA repair. The occurrence of homologous recombination was also reported in *M. gallisepticum*[10] and *S. citri*[73,93]. Mutants generated by specific gene inactivation through homologous recombination have been obtained recently in *M. genitalium*[19] and in *S. citri*[20]. In *M. genitalium*, a 2-kbp fragment encompassing the 5' end of *orfmg218*, in which the gentamycin resistance gene had been inserted, was cloned into a plasmid with no mycoplasmal replication origin so that the selection of gentamycin-resistant transformants could result only from integration of the plasmid into the mycoplasma chromosome. Transformation of *M. genitalium* with this plasmid yielded transformants with hemadsorption-negative or reduced hemadsorption phenotypes that correlated with the absence or presence of a truncated gene product. Genotypic characterization of three transformants revealed that in two of these the mg218 locus was disrupted by integration of the whole plasmid at two different sites by single-crossover events whereas in the third one, the disruption of the *mg218* locus appeared to be a double-crossover event[19]. To investigate the role of the peptide methionine sulfoxide reductase (MsrA) in oxidative stress and virulence of *M. genitalium*, a similar strategy was used to isolate a *M. genitalium* mutant lacking MsrA activity. The *msrA*-dirupted mutant was found to exhibit a reduced ability to resist oxidative stress, bind erythrocytes, and elicit cytotoxicity[18].

In *S. citri*, the efficiency of recombination is expected to be very low because of the absence of RecA protein[73,74]. Indeed, attempts to inactivate the motility gene *scm1* by transformation with a suicide plasmid carrying homologous sequences were unsuccessful[20]. However, the low frequency of recombination could be overcome by using the replicative, *oriC* plasmid pBOT1 as the vector. Spiroplasma cells were transformed with an internal fragment of the *scm1* gene carried on the pBOT1 vector. During passaging of the transformants the plasmid was found to integrate into the chromosome by homologous recombination involving one crossover, either at the *oriC* or at the *scm1* gene. Plasmid integration at the *scm1* gene did lead, to a nonmotile phenotype due to inactivation of the target gene[20]. The stability of the *scm1*-disrupted mutant, even in the absence of selection pressure, was exploited to investigate the role of spiroplasmal motility in transmission by the vector and phytopathogenicity, leading to the conclusion that motility might not be essential for pathogenicity of *S. citri*[20]. The use of replicative, *oriC* plasmids has also been successful in producing a nonpathogenic mutant of *S. citri* by inactivation of the fructose operon[43]. As compared to suicide plasmids, the use of replicative, *oriC* plasmids increases the time over which recombination can occur. In addition, the incompatibility of *oriC* plasmids can be used as a selection pressure for plasmid integration into the chromosome. However, the results also showed that, in most of the transformants, plasmid integration occurred at the *oriC* rather than at the target gene. Therefore, considering that recombination frequency is in part a function of the length of homologous sequences, the use of an *oriC* plasmid vector in which the *oriC* fragment was reduced to the minimal sequences required for plasmid replication was expected to decrease the frequency of plasmid integration at the *oriC* region and, in turn, to increase the frequency of recombination at the target gene. In fact, by using such a plasmid to inactivate the *S. citri* motility gene *scm1* it was shown that plasmid integration did occur at the target gene rather than at the *oriC* region (C. Lartigue and J. Renaudin, unpublished results). These data highlight the potential of *oriC* plasmids to promote homologous recombination in *S. citri*, an organism which lacks the RecA protein.

## 3.6     Mycoplasma mating

Intraspecies chromosomal gene transfer was first shown to occur in *S. citri*[2] and then in *A. oculi*[24,70]. However, the mechanism of gene transfer, conjugation-like or membrane fusion, has still to be determined. Very recently, chromosomal gene transfer between cells of *M. pulmonis* strains harboring distinct antibiotic resistance determinants (*tetM* on Tn*916* and *cat* on Tn*4001*) was reported (K. Dybvig, personal communication). Gene transfer was detected at a frequency of $10^{-8}$ to $10^{-9}$ events per CFU. Mycoplasmal colonies exhibiting dual antibiotic resistance were shown to contain both the *tetM* and *cat* markers in their chromosome, as well as genes unique to each parent strain. The fact that the use of the fusing agent polyethylene glycol dramatically increased (up to $10^{-5}$) the frequency of the appearance of dual-antibiotic resistant clones, suggests a mechanism involving cell membrane fusion.

## 4.    CONCLUSIONS

It is likely that more and more complete genome sequences of mollicutes will become available in the next few years. The powerful technologies of comparative genomics, such as transcriptome and proteome analyses, will help to handle this tremendous amount of data, leading to the identification of "candidate" genes, i. e. genes or groups of genes putatively involved in a given function. However, experimental proof is required for ultimate validation of biological significance. In this respect, methodologies for specific gene targeting and complementation analyses are of paramount importance to assess the role of individual genes in various processes such as pathogenesis. As reported in this chapter, significant advances are being made toward the development of gene transfer systems in mollicutes, and the few genetic studies that have been conducted so far have already yielded new insights to the understanding of pathogenicity mechanisms. One good example is the finding of the role of sugar metabolism in pathogenicity of the plant mollicute *S. citri*[41]. However, such genetic studies are still limited to the few mollicute species for which genetic tools have been made available. It is hoped that in the near future these new approaches will be improved and extended to other mollicute species.

## ACKNOWLEDGMENTS

I thank the colleagues and students from the BCM laboratory who contributed to the work reported here. I also acknowledge the cooperation of many colleagues in sharing unpublished results. Research from the BCM laboratory was supported by grants from Ministère de la Recherche and Conseil Régional d'Aquitaine.

## REFERENCES

1.    **Aleksandrova, N. M., M. R. Bevova, and V. M. Govorum.** 2000. Transformation of *Mycoplasma hominis* by plasmid pAM120 using electroporation. Genetika **36**:309-313.
2.    **Barroso, G., and J. Labarère.** 1988. Chromosomal gene transfer in *Spiroplasma citri*. Science **241**:959-961.
3.    **Barroso, G., J-C. Salvado, and J. Labarère.** 1990. Influence of genetic markers and of the fusing agent polyethylene glycol on chromosomal transfer in *Spiroplasma citri*. Curr. Microbiol. **20**:53-56.
4.    **Bébéar C.-M., P. Aullo P., J. M. Bové, and J. Renaudin.** 1996. *Spiroplasma citri* virus SpV1. Characterization of viral sequences present in the spiroplasmal host chromosome. Curr. Microbiol. 32:134-140.

5.    **Bergemann, A. D., and L. R. Finch**. 1988. Isolation and restriction endonuclease analysis of a mycoplasma plasmid. Plasmid **19**:68-70.

6.    **Bergemann A. D., J. C. Whitley, and L. R. Finch**. 1989. Homology of mycoplasma plasmid pADB201 and staphylococcal plasmid pE194. J. Bacteriol. **171**:593-595.

7.    **Bové, J. M.** 1993. Molecular features of mollicutes. Clin. Infect. Dis. **17**(Suppl. 1):S10-S31.

8.    **Bové, J. M.** 1997. Spiroplasmas: infectious agents of plants, arthropods, and vertebrates. Wien. Klin. Wochenschr. **109**:604-612.

9.    **Bové, J. M., P. Carle, M. Garnier, F. Laigret, J. Renaudin, and C. Saillard**. 1989. Molecular and cellular biology of spiroplasmas, p. 243-364. *In* R. F. Whitcomb and J. G. Tully (ed.), The Mycoplasmas, vol. V. Academic Press, Inc., New York.

10.   **Cao, J., P. A. Kapke, and F. C. Minion**. 1994. Transformation of *Mycoplasma gallisepticum* with Tn*916*, Tn*4001* and integrative plasmid vectors. J. Bacteriol. **176**:4459-4462.

11.   **Chambaud, I., R. Heilig, S. Ferris, V. Barbe, D. Samson, F. Galisson, I. Moszer, K. Dybvig, H. Wroblewski, A. Viari, E. Rocha, and A Blanchard**. 2001. The complete genome sequence of the murine respiratory pathogen *Mycoplasma pulmonis*. Nucl. Acids Res. **29**:2145-2153.

12.   **Chen, T-A., J. D. Lei, and C. P. Lin**. 1992. Detection and identification of plant and insect mollicutes, p393-424. *In* R. F. Whitcomb and J. G. Tully (ed.), The mycoplasmas, vol. V. Academic Press, Inc., New York.

13.   **Chipman, P. R., M. Agbandje-McKenna, J. Renaudin, T. S. Baker, and R. McKenna**. 1998. Structural analysis of the spiroplasma virus, SpV4: implications for evolutionary variation to obtain host diversity among the *Microviridae*. Structure **6**:135-145.

14.   **Clewell, D. B., and C. Gawron-Burke**. 1986. Conjugative transposons and the dissemination of antibiotic resistance in streptococci. Ann. Rev. Microbiol. **40**:635-639.

15.   **Davis, R. E.** 1981. Antibiotic sensitivities in vitro of diverse spiroplasma strains associated with plants and insects. Appl. Environ. Microbiol. **41**:329-333.

16.   **Davies, M. J., J. H. Tsai, R. L. Cox, L. L. McDaniel, and N. A. Harrison**. 1988. Cloning of chromosomal and extrachromosomal DNA of the mycoplasma-like organism that causes maize bushy stunt disease. Mol. Plant-Microbe Interact. **1**:295-302.

17.   **Denes, A. S., and R. C. Sinha**. 1991. Extrachromosomal DNA elements of plant pathogenic mycoplasma-like organisms. Can. J. Plant Pathol. **13**:26-32.

18.   **Dhandayuthapani, S., M. Blaylock, C. M. Bébéar, W. G. Rasmussen, and J.B. Baseman**. 2001. Peptide methionine sulfoxide reductase (MsrA) is a virulence determinant in *Mycoplasma genitalium*. J. Bacteriol. **183**:5645-5650.

19.   **Dhandayuthapani, S., W. G. Rasmussen, and J.B. Baseman**. 1999. Disruption of gene *mg218* of *Mycoplasma genitalium* through homologous recombination leads to an adherence-deficient phenotype. Proc. Natl. Acad. Sci. USA **96**:5227-5232.

20.   **Duret, S., J. L. Danet, M. Garnier, and J. Renaudin**. 1999. Gene disruption through homologous recombination in *Spiroplasma citri*: an *scm1*-disrupted motility mutant is pathogenic. J. Bacteriol. **181**:7449-7456.

21.   **Dybvig, K.** 1989.Transformation of *Acholeplasma laidlawii* with streptococcal plasmids pVA868 and pVA920. Plasmid **21**:155-160.

22.   **Dybvig, K.** 1990. Mycoplamal genetics. Annu. Rev. Microbiol. **44**:81-104.

23.   **Dybvig, K.** 1992. Gene transfer, p 355-361. *In* J. Maniloff, R. N. McElhaney, L. R. Finch, and J. B. Baseman (ed.), Mycoplasmas: molecular biology and pathogenesis. American Society for Microbiology, Washington.

24.   **Dybvig, K.** 1993. The genetics and basic biology of *Mycoplasma pulmonis*: How much is actually *Acholeplasma*? Plasmid **30**:176-178.

25. **Dybvig, K.** 1993. DNA rearrangements and phenotypic switching in prokaryotes. Mol. Microbiol. **10**:465-471.
26. **Dybvig, K., and J. Alderete.** 1988. Transformation of *Mycoplasma pulmonis* ans *Mycoplasma hyorhinis*: transposition of Tn*916* and formation of cointegrate structures. Plasmid **20**:33-41.
27. **Dybvig, K., and G. H. Cassell.** 1987. Transposition of Gram-positive transposon Tn916 in *Acholeplasma laidlawii* and *Mycoplasma pulmonis*. Science **235**:1392-1394.
28. **Dybvig, K., C. T. French, and L. L. Voelker.** 2000. Construction and use of derivatives of transposon Tn*4001* that function in *Mycoplasma pulmonis* and *Mycoplasma arthritidis*. J. Bacteriol. 182: 4343-4347.
29. **Dybvig, K., G. E. Gasparich, and K. W. King.** 1995. Artificial transformation of mollicutes via polyethylene glycol- and electroporation-mediated methods, p. 179-184, *In* S. Razin and J. G. Tully (ed.), Molecular and diagnostic procedures in mycoplasmology, vol. I. Academic Press, Orlando, Fla.
30. **Dybvig K., and M. Kaled.** 1990. Isolation of a second cryptic plasmid from *Mycoplasma mycoides* subsp. *mycoides*. Plasmid **24**:153-155.
31. **Dybvig K., A. Liss, J. Alderete, R. M. Cole, and G. H. Cassell.** 1987. Isolation of a virus from *Mycoplasma pulmonis*. Isr. J. Med. Sci. **23**:418-422.
32. **Dybvig, K., R. Sitaraman, and C. T. French.** 1998. A family of phase-variable restriction enzymes with differing specificities generated by high-frequency gene rearrangements. Proc. Natl. Acad. Sci. USA **95**:13923-13928.
33. **Dybvig K., and L. L. Voelker.** 1996. Molecular biology of mycoplasmas. Annu. Rev. Microbiol. **50**:25-57.
34. **Dybvig, K., and A. Woodward.** 1992. Construction of *recA* mutants of *Acholeplasma laidlawii* by insertional inactivation with a homologous DNA fragment. Plasmid **28**:262-266.
35. **Dybvig, K., and H. Yu.** 1994. Regulation of a restriction and modification system via DNA inversion in *Mycoplasma pulmonis*. Mol. Microbiol. **12**:547-560.
36. **Fisseha, M., H. W. Gohlmann, R. Herrmann, and D. C. Krause.** 1999. Identification and complementation of frameshift mutations associated with loss of cytadherence in *Mycoplasma pneumoniae*. J. Bacteriol. **181**:4404-4410.
37. **Foissac, X., C. Saillard, and J. M. Bové.** 1997. Random insertion of Tn*4001* in the genome of *Spiroplasma citri* strain GII3. Plasmid **37**:80-86.
38. **Foissac, X., C. Saillard, J. L. Danet, P. Gaurivaud, C. Paré, F. Laigret, and J. M. Bové.** 1997. Mutagenesis by insertion of transposon Tn*4001* into the genome of *Spiroplasma citri*: Characterization of mutants affected in plant pathogenicity and transmission to the plant by the leafhopper vector *Circulifer haematoceps*. Mol. Plant-Microbe Interact. **10**:454-461.
39. **Gasparich, G. E., K. J. Hackett, E. A. Clark, J. Renaudin, and R. F. Whitcomb.** 1993. Occurrence of extrachromosomal deoxyribonucleic acids in spiroplasmas associated with plants, insects, and ticks. Plasmid **29**:81-93.
40. **Gasparich, G. E., K. J. Hackett, C. Stamburski, J. Renaudin, and J. M. Bove.** 1993. Optimization of methods for transfecting *Spiroplasma citri* strain R8A2 HP with the spiroplasma virus SpV1 replicative form. Plasmid **29**:193-205.
41. **Gaurivaud, P., J. L. Danet, F. Laigret, M. Garnier, and J. M. Bové.** 2000. Fructose utilization and pathogenicity of *Spiroplasma citri*. Mol. Plant-Microbe Interact. **13**:1145-1155.
42. **Gaurivaud, P., F. Laigret, M. Garnier, and J. M. Bové.** 2000. Fructose utilization and pathogenicity of *Spiroplasma citri*: characterization of the fructose operon. Gene **252**:61-69.
43. **Gaurivaud, P., F. Laigret, E. Verdin, M. Garnier, and J. M. Bové.** 2000. Fructose operon mutants of *Spiroplasma citri*. Microbiology **146**:2229-2236.

44.  **Goodwin, P. H., B. G. Xue, C. R. Kuske, and M. K. Sears**. 1994. Amplification of plasmid DNA to detect plant pathogenic mycoplasma-like organisms. Ann. Appl. Biol. **124**:27-36.
45.  **Gruss, A., and S. D. Ehrlich**. 1989. The family of highly interelated single-stranded deoxyribonucleic acid plasmids. Microbiol. Rev. **53**:231-241.
46.  **Hahn, T-W., K. A. Krebes, and D. C. Krause**. 1996. Expression in *Mycoplasma pneumoniae* of the recombinant gene encoding the cytadherence-associated protein HMW1 and identification of HMW4 as a product. Mol. Microbiol. **19**:1085-1093.
47.  **Hahn, T-W., E. A. Mothershed, R. H. Waldo, and D. C. Krause**. 1999. Construction and analysis of a modified Tn*4001* conferring chloramphenicol resistance in *Mycoplasma pneumoniae*. Plasmid **41**:120-124.
48.  **Harrison, N. A., J. H. Tsai, C. M. Bourne, and P. A. Richardson**. 1991. Molecular cloning and detection of chromosomal and extrachromosomal DNA of mycoplasma-like organisms associated with witches' broom disease of pigeon pea in Florida. Mol. Plant-Microbe Interact. **4**:300-307.
49.  **Hedreyda, C. T., and D. C. Krause**. 1995. Identification of a possible cytadherence regulatory locus in *Mycoplasma pneumoniae*. Infect. Immun. **63**:3479-3483.
50.  **Hedreyda, C. T., K. K. Lee, and D. C. Krause**. 1993. Transformation of *Mycoplasma pneumoniae* with Tn*4001* by electroporation. Plasmid **30**:170-175.
51.  **Hensel, M., J. E. Shea, C. Gleeson, M. D. Jones, E. Dalton, and D. W. Holden**. 1995. Simultaneous identification of bacterial virulence genes by negative selection. Science **269**:400-403.
52.  **Hutchison, C. A., S. N. Peterson, S. R. Gill, R. T. Cline, O. White, C. M. Fraser, H. O. Smith, and J. C. Venter**. 1999. Global transposon mutagenesis and a minimal mycoplasma genome. Science **286**:2165-2169.
53.  **Jacob, C., F. Nouzières, S. Duret, J. M. Bové, and J. Renaudin**. 1997. Isolation, characterization, and complementation of a motility mutant of *Spiroplasma citri*. J. Bacteriol. **179**:4802-4810.
54.  **Jannière, L., A. Gruss, and S. D. Ehrlich**. 1993. Plasmids, p. 625-644. *In* A. L. Sonenshein, J. A. Hoch, and R. Losick (ed.), *Bacillus subtilis* and other Gram-positive bacteria. American Society for Microbiology, Washington.
55.  **Jarhede, T. K., M. Le Henaff, and A. Wieslander**. 1995. Expression of foreign genes and selection of promoter sequences in *Acholeplasma laidlawii*. Microbiology **141**:2071-2079.
56.  **King, K. W., and K. Dybvig**. 1991. Plasmid transformation of *Mycoplasma mycoides* subspecies *mycoides* is promoted by high concentrations of polyethylene glycol. Plasmid **26**:108-115.
57.  **King, K. W., and K. Dybvig**. 1992. Nucleotide sequence of *Mycoplasma mycoides* subsp. *mycoides* plasmid pKMK1. Plasmid **28**:86-91.
58.  **King, K. W., and K. Dybvig**. 1994. Mycoplasmal cloning vectors derived from plasmid pKMK1. Plasmid **31**:49-59.
59.  **King, K. W., and K. Dybvig**. 1994. Transformation of *Mycoplasma capricolum* and examination of DNA restriction modification in *M. capricolum* and *Mycoplasma mycoides* subsp. *mycoides*. Plasmid **31**:308-311.
60.  **Knudtson, K. L., and F. C. Minion**. 1993. Construction of Tn*4001lac* derivatives to be used as promoter probe vectors in mycoplasmas. Gene **137**:217-222.
61.  **Kuboyama, T., C-C. Huang, X. Lu, T. Sawayanagi, T. Kanazawa, T. Kagami, I. Matsuda, T. Tsuchisaki, and S. Namba**. 1998. A plasmid isolated from phytopathogenic onion yellows phytoplasma and its heterogeneity in the phytopathogenic mutant. Mol. Plant-Microbe Interact. **11**:1031-1037.
62.  **Kuske, C. R., and B. C Kirkpatrick**. 1990. Identification and characterization of plasmids from western aster yellows mycoplasma-like organism. J. Bacteriol. **172**:1628-1633.

63. **Kuske, C. R., B. C. Kirkpatrick, M. J. Davies, and E. Seemüller.** 1991. DNA hybridization between western aster yellows mycoplasma-like organism plasmids and extrachromosomal DNA from other plant pathogenic mycoplasma-like organisms. Mol. Plant-Microbe Interact. **4**:75-80.

64. **Krause, D. C., T. Proft, C. T. Hedreyda, H. Hilbert, H. Plagens, and R. Herrmann.** 1997. Transposon mutagenesis reinforces the correlation between *Mycoplasma pneumoniae* cytoskeletal protein HMW2 and cytadherence. J. Bacteriol. **179**:2668-2677.

65. **Lee, I., R. E. Davis, and D. E. Gundersen-Rindal.** 2000. Phytoplasma: phytopathogenic mollicutes. Annu. Rev. Microbiol. **54**:221-255.

66. **Liu, L., K. Dybvig, V. S. Panangala, V. L. van Santen, and C. T. French.** 2000. GAA trinucleotide repeat region regulates M9/pMGA gene expression in *Mycoplasma gallisepticum.* Infect. Immun. **68**:871-876.

67. **Mahairas, G. G., C. Jian, and F. C. Minion.** 1990. Genetic exchange of transposon and integrative plasmid markers in *Mycoplasma pulmonis.* J. Bacteriol. **172**:2267-2272.

68. **Mahairas, G. G., C. Jian, and F. C. Minion.** 1993. Transformation of *Mycoplasma pulmonis*: demonstration of homologous recombination, introduction of cloned genes, and preliminary description of an integrating shuttle system-author's correction. J. Bacteriol. **175**:3692.

69. **Mahairas, G. G., and F. C. Minion.** 1989. Random insertion of the gentamycin resistance transposon Tn*4001* in *M. pulmonis.* Plasmid **21**:43-47.

70. **Mahairas, G. G., and F. C. Minion.** 1989. Transformation of *Mycoplasma pulmonis*: demonstration of homologous recombination, introduction of cloned genes, and preliminary description of an integrating shuttle system. J. Bacteriol. **171**:1775-1780.

71. **Maniloff, J.** 1992. Mycoplasma viruses, p. 41-59. *In* J. Maniloff, R. N. McElhaney, L. R. Finch, and J. B. Baseman (ed.), Mycoplasmas: molecular biology and pathogenesis. American Society for Microbiology, Washington.

72. **Marais, A., J. M. Bové, S. F. Dallo, J. B. Baseman, and J. Renaudin.** 1993. Expression in *Spiroplasma citri* of an epitope carried on the G fragment of cytadhesin P1 gene from *Mycoplasma pneumoniae.* J. Bacteriol. **175**:2783-2787.

73. **Marais, A., J. M. Bové, and J. Renaudin.** 1996. *Spiroplasma citri* virus SpV1-derived cloning vector: Deletion formation by illegitimate and homologous recombination in a spiroplasmal host strain which probably lacks a functional *recA* gene. J. Bacteriol. **176**:862-870.

74. **Marais, A., J. M. Bové, and J. Renaudin.** 1996. Characterization of the *recA* gene regions of *Spiroplasma citri* and *Spiroplasma melliferum.* J. Bacteriol. **178**:7003-7009.

75. **McCammon, S. L., E. L. Dally, and R. E. Davis.** 1990. Electroporation and DNA methylation effects on the transfection of spiroplasma, p. 60-65. *In* G. Stanek, G. H. Cassell, J. G. Tully, and J. G. Whitcomb (ed.), Recent advances in mycoplasmology. Fisher Verlag, Stuttgart, Germany.

76. **Melcher, U., Y. Sha, F. Ye, and J. Fletcher.** 1999. Mechanisms of spiroplasma genome variation associated with SpV1-like viral DNA inferred from sequence comparisons. Microb. Comp. Genomics **4**:29-46.

77. **Minion, F. C., and P. A. Kapke.** 1998. Transformation of mycoplasmas. Methods Mol. Biol. **104**:227-234.

78. **Miyata, M., K-I. Sano, R. Okada, and T. Fukumura.** 1993. Mapping of replication initiation site in Mycoplasma capricolum genome by two-dimensional gel-electrophoretic analysis. Nucl. Acids Res. **21**:4816-4823.

79. **Nakashima, K., and T. Hayashi.** 1997. Sequence analysis of extrachromosomal DNA of sugarcane white leaf phytoplasma. Ann. Phytopathol. Soc. Jpn. **63**:21-25.

80.  **Nakashima, K., S. Kato, S. Iwanami, and N. Murata.** 1991. Cloning and detection of chromosomal and extrachromosomal DNA from mycoplasma-like organisms that cause yellow dwarf disease of rice. Appl. Environ. Microbiol. **57**:3570-3575.

81.  **Nakashima, K., S. Kato, S. Iwanami, and N. Murata.** 1993. DNA probes reveal relatedness of rice yellow dwarf mycoplasma-like organisms (MLOs) and distinguish them from other MLOs. Appl. Environ. Microbiol. **59**:1206-1212.

82.  **Nishigawa, H., S. Miyata, K. Oshima, T. Sawayanagi, A. Komoto, T. Kuboyama, I. Matsuda, T. Tsuchisaki, and S. Namba.** 2001. In planta expression of a protein encoded by the extrachromosomal DNA of a phytoplasma and related to geminivirus replication proteins. Microbiology **147**:507-513.

83.  **Ogasawara, N., S. Moriya, and H. Yoshikawa.** 1991. Initiation of chromosome replication: structure and function of *oriC* and DnaA protein in eubacteria. Res. Microbiol. **142**:851-859.

84.  **Pascarel-Devilder, M. C., J. Renaudin, and J. M. Bové.** 1986. The spiroplasma virus 4 replicative form cloned in *Escherichia coli* transfects spiroplasma. Virology **151**:390-393.

85.  **Razin, S.** 1985. Molecular biology and genetics of mycoplasmas (Mollicutes). Microbiol. Rev. **49**:419-455.

86.  **Razin, S., D. Yogev, and Y. Naot.** 1998. Molecular biology and pathogenicity of mycoplasmas. Microbiol. Mol. Biol. Rev. **62**:1094-1156.

87.  **Reddy, S. K., W. G. Rasmussen, and J. B. Baseman.** 1996 Isolation and characterization of transposon Tn*4001*-generated, cytadherence-deficient transformants of *Mycoplasma pneumoniae and Mycoplasma genitalium*. FEMS Immunol. Med. Mic. **15**:199-211.

88.  **Rekab, D., L. Carraro, B. Schneider, E. Seemüller, J. Chen, C. J. Chang, R. Locci, and G. Firrao.** 1999. Geminivirus-related extrachromosomal DNAs of the X-clade phytoplasmas share high sequence similarity. Microbiology **145**:1453-1459.

89.  **Renaudin, J., P. Aullo, J. C. Vignault, and J. M. Bové.** 1990. Complete nucleotide sequence of the genome of *Spiroplasma citri* virus SpV1-R8A2B. Nucleic Acids Res. **18**:1293-1294.

90.  **Renaudin J., Bodin-Ramiro C., Vignault J.C. & Bové J.M.** (1990). Spiroplasmavirus 1: presence of viral DNA sequences in the spiroplasma genome, p125-130. *In* G. Stanek, G. H. Cassell, J. G. Tully, and J. G. Whitcomb (ed.), Recent advances in mycoplasmology. Fisher Verlag, Stuttgart, Germany.

91.  **Renaudin, J., and J. M. Bové.** 1994. SpV1 and SpV4, spiroplasma viruses with circular, single-stranded DNA genomes, and their contribution to the molecular biology of spiroplasmas. Adv. Virus Res. **44**:429-463.

92.  **Renaudin, J., and J. M. Bové.** 1995. Plasmid and viral vectors for gene cloning and expression in *Spiroplasma citri*, p 167-178. *In* S. Razin and J. G. Tully (ed.), Molecular and diagnostic procedures in mycoplasmology. Academic Press, San Diego, CA.

93.  **Renaudin, J., A. Marais, E. Verdin, S. Duret, X. Foissac, F. Laigret, and J. M. Bové.** 1995. Integrative and free *Spiroplasma citri oriC* plasmids: expression of the *Spiroplasma phoeniceum* spiralin in *Spiroplasma citri*. J. Bacteriol. **177**:2800-2877.

94.  **Renaudin, J., M-C. Pascarel, and J. M. Bové.** 1987. Spiroplasma virus 4: nucleotide sequence of the viral DNA, regulatory signals, and proposed genome organization. J. Bacteriol. **169**:4950-4961.

95.  **Renaudin, J., M-C. Pascarel, M. Garnier, P. Junca-Carle, and J. M. Bové.** 1984. SpV4, a new spiroplasma virus with circular, single-stranded DNA. Ann. Virol. **135E**:343-361.

96.  **Renaudin J., M-C. Pascarel-Devilder, C. Saillard, C. Chevalier, and J. M. Bové.** 1986. Chez les spiroplasmes le codon UGA n'est pas non-sens et semble coder pour le tryptophane. C. R. Acad. Sci. Paris Série III, **303**:539-540.

97.   **Rigden, J. E., I. B. Dry, L. R. Krake, and M. A. Rezaian**. 1996. Plant virus DNA replication processes in Agrobacterium: insight into the origins of geminiviruses. Proc. Natl. Acad. Sci. USA **93**:10280-10284.

98.   **Roberts, M. C., and G. E. Kenny**. 1987. Conjugal transfer of transposon Tn*916* from *Streptococcus faecalis* to *Mycoplasma hominis*. J. Bacteriol. **169**:3836-3839.

99.   **Rouch, D. A., M. E. Byrne, Y. C. Kong, and R. A. Skurray**. 1987. The *aac-aphD* gentamycin and kanamycin resistance determinant of Tn*4001* from *Staphylococcus aureus*: expression and nucleotide sequence analysis. J. Gen. Microbiol. **133**:3039-3052.

100.  **Ruffin, D. C., V. L. van Santen, Y. Zhang, L. L. Voelker, V. S. Panangala, and K. Dybvig**. 2000. Transposon mutagenesis of *Mycoplasma gallisepticum* by conjugation with *Enterococcus faecalis* and determination of insertion site by direct genomic sequencing. Plasmid **44**:191-195.

101.  **Schneider, B., R. Mäurer, C. Saillard, B. C. Kirkpatrick, and E. Seemüller**. 1992. Occurrence and relatedness of extrachromosomal DNAs in plant pathogenic mycoplasma-like organisms. Mol. Plant-Microbe Interact. **5**:489-495.

102.  **Sears, B. B., O. Lim, N. Holland, B. C. Kirkpatrick, K. L. Klomparens**. 1989. Isolation and characterization of DNA from a mycoplasma-like organism. Mol. Plant-Microbe Interact. **2**:175-180.

103.  **Sha, Y., U. Melcher, R. E. Davis, and J. Fletcher**. 1995. Resistance of *Spiroplasma citri* lines to the virus SVTS2 is associated with integration of viral DNA sequences into host chromosomal and extrachromosomal DNA. Appl. Environ. Microbiol. **61**:3950-3959.

104.  **Sha, Y., U. Melcher, R. E. Davis, and J. Fletcher**. 2000. Common elements of spiroplasma plectroviruses revealed by nucleotide sequence of SVTS2. Virus genes **20**:47-56.

105.  **Simoneau, P., and J. Labarère**. 1990. Construction of chimeric antibiotic resistance determinants and their use in the development of cloning vectors for spiroplasmas, p. 66-74. *In* G. Stanek, G. H. Cassell, J. G. Tully, and J. G. Whitcomb (ed.), Recent advances in mycoplasmology. Fisher Verlag, Stuttgart, Germany.

106.  **Stamburski, C., J. Renaudin, and J. M. Bové**. 1991. First step toward a virus-derived vector for gene cloning and expression in spiroplamas, organisms which read UGA as a tryptophan codon: synthesis of chloramphenicol acetyltranferase in *Spiroplasma citri*. J. Bacteriol. **173**:2225-2230.

107.  **Sundström, T. K., and A. Wieslander**. 1990. Plasmid transformation and replica filter plating of *Acholeplasma laidlawii*. FEMS Microbiol. Lett. **72**:147-152.

108.  **Voelker, L. L., and K. Dybvig**. 1996. Gene transfer in *Mycoplasma arthritidis*: transformation, conjugal transfer of Tn*916*, and evidence for a restriction system recognizing AGCT. J. Bacteriol. **178**:6078-6081.

109.  **Voelker, L. L., and K. Dybvig**. 1998. Transposon mutagenesis. Methods Mol. Biol. **104**:235-238.

110.  **Voelker, L. L., and K. Dybvig**. 1998. Characterization of the lysogenic bacteriophage MAV1 from *Mycoplasma arthritidis*. J. Bacteriol. **180**:5928-5931.

111.  **Voelker, L. L., and K. Dybvig**. 1999. Sequence analysis of the *Mycoplasma arthritidis* bacteriophage MAV1 genome identifies the putative virulence factor. Gene **233**:101-107.

112.  **Voelker, L. L., K. E. Weaver, L. J. Ehle, and L. R. Washburn**. 1995. Association of lysogenic bacteriophage MAV1 with virulence of *Mycoplasma arthritidis*. Infect. Immun. **63**:4016-4023.

113.  **Whitley, J. C., and L. R. Finch**. 1989. Location of sites of transposon Tn*916* insertion in the *Mycoplasma mycoides* genome. J. Bacteriol. **171**:6870-6872.

114. **Ye, F., F. Laigret, J. C. Whitley, C. Citti, L. R. Finch, P. Carle, J. Renaudin, and J. M. Bové.** 1992. A physical map of the *Spiroplasma citri* genome. Nucleic Acids Res. **20:**1559-1565.

115. **Ye, F., F. Laigret, P. Carle, and J. M. Bove.** 1995. Chromosomal heterogeneity among various strains of *Spiroplasma citri.* Int. J. Syst. Bacteriol. **45:**729-734.

116. **Ye, F., J. Renaudin, J. M. Bové, and F. Laigret.** 1994. Cloning and sequencing the replication origin (*oriC*) of the *Spiroplasma citri* chromosome and construction of autonomously replicating artificial plasmids. Curr. Microbiol., **28:**1-7.

117. **Yoshikawa, H., and R. G. Wake.** 1993. Initiation and termination of chromosome replication, p. 507-528. *In* A. L. Sonenshein, J. A. Hoch, and R. Losick (ed.), *Bacillus subtilis* and other Gram-positive bacteria. American Society for Microbiology, Washington.

118. **Zou, N. K. Park, and K. Dybvig.** 1995. Mycoplasma virus P1 has a linear double-stranded DNA genome with inverted terminal repeats. Plasmid **33:**41-49.

# Chapter 16

# Restriction-Modification Systems and Chromosomal Rearrangements in Mycoplasmas

RAMAKRISHNAN SITARAMAN and KEVIN DYBVIG
*Department of Genomics and Pathobiology University of Alabama at Birmingham, Birmingham, Alabama, USA*

## 1. TYPES OF RESTRICTION-MODIFICATION

The restriction-modification (R-M) systems are so named because of their ability to degrade foreign DNA (restrict viral infection) and to methylate (modify) unmethylated and hemimethylated DNA. They were discovered as genetic elements that determined the susceptibility of *E. coli* and *S. typhimurium* to phage infection in a strain-dependent manner, with bacteriophage plaquing efficiency being reduced upon infection of a heterologous host strain (2). At the time of writing, 3154 R-M enzymes, many with overlapping DNA recognition specificities, have been catalogued in the restriction enzyme database (REBASE) (65). R-M systems are ubiquitous and arose early in prokaryotic evolution

Restriction enzymes are classified into three main categories - types I, II and III, though there exist enzymes that do not fall into either of these major categories. The salient features of the three major types of restriction enzymes are shown in Table 1. The simplest (and commercially most valuable) ones are the type II systems which consist of a site-specific endonuclease and a cognate methylase, with each of these activities residing on independent and single polypeptides. Upon binding to unmethylated or hemimethylated DNA, the methylase converts adenine or cytosine residues in the recognition sequence to $N^6$-methyladenosine (m$^6$A) or 5-methylcytosine (m$^5$C), respectively. Methylation renders the recognition sequence refractory to endonucleolytic cleavage by the restriction enzyme, which cleaves unmethylated sequences. The host genome, being

*Molecular Biology and Pathogenicity of Mycoplasmas*, Edited by Razin and Herrmann, Kluwer Academic/Plenum Publishers, New York, 2002

methylated, is not affected by the restriction endonuclease but any incoming foreign DNA usually lacks characteristic methylation patterns at the recognition sequences and is efficiently cleaved by the endonuclease. In a minority of events, however, the incoming DNA sequences are methylated before recognition by the endonuclease, resulting in a productive phage infection or successful DNA uptake and maintenance.

*Table 1.* Characteristics of restriction and modification systems. Adapted from Bickle and Kruger (1993) (6) and Yuan (1981) (79).

|  | Type I | Type II | Type III |
|---|---|---|---|
| Structural features | Multifunctional, multisubunit enzyme | Independent polypeptides for methylation and restriction | Multifunctional, multisubunit enzyme |
| Co-factors | $Mg^{2+}$, SAM*, ATP | $Mg^{2+}$ | $Mg^{2+}$, SAM, ATP |
| Recognition sequence | Bipartite, asymmetric, separated by a nonspecific spacer of fixed length | Palindromic | Asymmetric |
| DNA cleavage | Occurs at sites far away from binding sites, accompanied by ATP hydrolysis | Occurs within the recognition sequence | Occurs at fixed distances from the recognition sequence |
| Enzyme activities | Methylation and restriction activities are mutually exclusive | Methylation and restriction activities are independent | Methylation and restriction activities are simultaneous |
| Enzyme turnover | No | Yes | Yes |

* - S-adenosyl methionine

The type III enzymes are of intermediate complexity, wherein the ability to recognize a specific DNA sequence resides in the same protein as the methylase activity. While restriction activity is completely dependent on the formation of a methylase-endonuclease complex, the methylase can function either independently or as part of a methylase-endonuclease complex. ATP is a cofactor for restriction activity but is not hydrolyzed during the reaction. S-adenosyl methionine is a stimulatory, allosteric effector for endonucleolysis. DNA cleavage usually occurs at a fixed distance from the recognition site (25-30 bp) but, owing to the competing methylation activity, does not proceed to completion.

Type I enzymes are the most complex of the known R-M enzymes, with the activities for DNA sequence recognition, methylation and cleavage residing on three different polypeptides. A complex of the sequence-recognition subunit (S) and methylase (M) is sufficient for methylation, while only the holoenzyme consisting of the endonuclease (R), M and S subunits is capable of carrying out the endonuclease reaction. DNA cleavage is preceded by extensive ATP-dependent translocation of the DNA and occurs at random sites as far as 7 kb from the target sequence recognized by the S subunit (5).

Of the three major types of R-M systems described above, the type I enzymes are singular in their ability to evolve new specificities. While mutations in the polypeptides of the other types could simultaneously affect their associated enzymatic activity, the separation of DNA sequence recognition from other enzymatic activities in type I enzymes facilitates the evolution of ever newer DNA specificities without significantly compromising the overall enzymatic activity (1, 27, 28, 31, 46, 50, 57, 58). A striking example of the plasticity of type I systems occurs in the *hsd* loci of *Mycoplasma pulmonis*, wherein site-specific recombination between *hsdS* genes results in the formation and induction of new *hsdS* genes (see below).

## 2. RESTRICTION-MODIFICATION SYSTEMS IN MYCOPLASMAS

R-M enzymes identified in *Spiroplasma citri, S. monobiae, Acholeplasma laidlawii, Mycoplasma fermentans,* and *Ureaplasma urealyticum* have been reviewed elsewhere (7, 18, 22, 42). No functional studies have been carried out on the putative R-M systems identified in the genome sequences of *M. pneumoniae, M genitalium,* and *U. urealyticum* (Table 2). We focus here on R-M systems in mycoplasma species for which relatively new information is available that has not been previously reviewed.

### 2.1 *Mycoplasma arthritidis*

The presence of a restriction system in *M. arthritidis* was deduced by an analysis of transformation efficiency of different strains (73). Strain H6061 is readily transformable but strain 1581 is not. However, strain 1581 can be transformed with DNA that has been modified *in vitro* using the *Alu*I methylase, which methylates cytosine in the sequence AGCT to $m^5C$. Genomic DNA of strain 1581 is resistant to cleavage by the *Alu*I restriction enzyme, indicating that the genome is methylated at the sequence AGCT. Therefore, *M. arthritidis* is predicted to have an R-M system that is an isochizomer of *Alu*I that serves as a barrier to transformation.

*Table 2.* Components of R-M systems detected by genomic sequencing of mycoplasmas.

| Organism and ORF | Predicted gene product |
| --- | --- |
| *M. pneumoniae* [a] | |
| MPN615 | HsdS[b] |
| MPN201 | HsdS |
| MPN290 | HsdS |
| MPN289 | HsdS |
| MPN365 | HsdS |
| MPN343 | HsdS |
| MPN507 | HsdS |
| MPN089 | HsdS |
| MPN342 | HsdM[c] |
| MPN198 | Adenine-specific methyltransferase |
| MPN345 | HsdR[d] (fragment) |
| MPN347 | HsdR (fragment) |
| MPN346 | HsdR (fragment) |
| *M. genitalium* | |
| MG438 | HsdS |
| MG184 | Adenine-specific methyltransferase |
| *U. urealyticum* | |
| UU096 | HsdS |
| UU097 | HsdS |
| UU099 | HsdS |
| UU098 | HsdM |
| UU100 | HsdM |
| UU477[e] | Adenine-specific methyltransferase |
| UU095 | HsdR |

[a]Gene numbers according to Dandekar *et al.* (2000) (11).
[b]Type I R-M enzyme subunit for DNA sequence recognition.
[c]Type I subunit for methyltransferase activity.
[d]Type 1 subunit for endonuclease activity.
[e]Probably the only functional gene in *U. urealyticum* listed in this table.

Interestingly, the *M. arthritidis* virus MAV1, which is integrated into the genome of virulent strains of *M. arthritidis*, harbors genes homologous to methylases and restriction endonucleases (74). The gene *marMP* is predicted to encode a C-5 DNA methyltransferase. Another gene, *marRP*, is predicted to encode a protein having a very low level of sequence similarity with the HsdR protein of *M. pulmonis*.

## 2.2      *Mycoplasma pulmonis*

The *M. pulmonis* strain KD735 *hsd1* and *hsd2* loci encode the most complex R-M systems described to date. The *hsdS* genes are arranged two to a locus and flank an *hsdR* and *hsdM* gene. One of the *hsdS* genes, *hsdM* and *hsdR* are organized as a single operon while the second *hsdS* is encoded on the other strand and oriented in the direction opposite to that of the other three. The *hsdS* genes undergo site-specific inversions that not only generate new *hsdS* polymorphs but also result in the transcriptional induction of the one of the two polymorphs, because the *hsd* promoter is located on only one side of each locus. The *hsdR* and *hsdM* genes of the *hsd1* locus are identical to their counterparts in *hsd2*. This ensures the production of a functional holoenzyme if the *hsdR* and *hsdM* genes of at least one of the two loci are transcribed. The production of chimeric *hsdS* genes via site-specific inversions resulting in altered restriction specificity of the holoenzyme is unique among all known R-M systems (24, 68). Presently, at least six distinct specificities have been discerned based on the relative plaquing efficiency of the mycoplasma virus P1 on various host strains derived from KD735 (21).

The genome of *M. pulmonis* strain CT has been completely sequenced and harbors at least 4 loci encoding R-M systems, three of type I and one of type III (10). Two of the three type I R-M loci are essentially identical in organization and sequence to the *hsd1* and *hsd2* loci described for KD735 (24, 68). The newly discovered *hsd3* locus is similar to *hsd1* and *hsd2* but does not contain *hsdR* and *hsdM* genes between the two *hsdS* genes.

## 3. RESTRICTION-MODIFICATION VIS-A-VIS MYCOPLASMA GENETICS

Classically, R-M systems have been viewed as cellular immune systems, protecting the bacterial host from invading DNA. The initial discovery of R-M systems was based on this defensive function and can explain the widespread occurrence of R-M systems in prokaryotes. In a majority of instances, R-M systems selectively destroy DNA that is unmethylated, methylate hemimethylated substrates, and spare fully methylated DNA. An additional hypothesis advanced in recent years to account for the ubiquity of R-M systems is that of R-M genes as "selfish" genetic elements. Studies of the maintenance of type II R-M systems in bacterial cells show that the loss of an R-M plasmid during cell division can lead to an imbalance between competing restriction and methylation activities owing to the dilution of the methylase. This increases the frequency of occurrence of unmethylated restriction sites in newly replicated host DNA, which is rendered vulnerable to endonuclease attack and consequently, cell death (51). Such a mechanism also promotes the stable maintenance of plasmids encoding type II R-M systems, allowing them to resist their displacement by incompatible plasmids through host killing (52). In some systems, the presence of linked regulatory genes (C genes) on incoming DNA allows for the establishment and maintenance of newly acquired type II R-M systems. A critical concentration of C protein is required for restriction, allowing methylation to precede restriction by an incoming R-M system whenever the level of C protein expressed *de novo* is sub-critical. However, the prior presence of an "incompatible" C gene in the host would result in the expression of restriction activity by any incoming R-M system with a different DNA sequence specificity, killing the invaded host (53). Thus, type II restriction enzymes may be classified into incompatibility groups based not only on their sequence specificity but also the previously observed ability of associated C genes to cross-complement (37).

The cellular defense and selfish gene hypotheses satisfactorily account for the presence and persistence of type II R-M systems in the minimal genome of mycoplasmas. However, type I R-M systems have additional features that are not readily explicable by these two hypotheses. In type I systems, restriction is not as efficient as that of the other types; 1 phage in 100 to 10000 escapes (57). The inefficiency of restriction could be because a complex of S, M and R subunits is necessary for restriction but can also methylate target DNA. Additionally, a complex of S and M without R can methylate target DNA in the absence of restriction. Secondly, the loss of type I R-M genes would not lead to host killing as seen in the type II systems because of the inseparability of endonuclease activity from the methylation complex, implying that type I genes are not selfish elements.

The phase-variable *hsd* systems of *M. pulmonis* generate new R-M specificities *de novo*, indicating that type I R-M systems could have functions in addition to protection against phage invasion. The change in DNA recognition specificity could result in an altered pattern of chromosomal methylation, perhaps leading to the regulation of various genes. The concomitant change in restriction specificity and the production of double-stranded breaks at random sites in the chromosome could promote recombination and lead to evolutionary diversification (17, 39). The cleavage of incoming DNA at random sites by type I enzymes could also facilitate its recombination and integration into the genome. Such additional functions could explain the persistence of the type I R-M systems in the reduced genomes of mycoplasmas as well as the evolution of the elaborate phase-variable *hsd* loci of *M. pulmonis*.

## 4.    CHROMOSOMAL REARRANGEMENTS IN MYCOPLASMAS

The genetic material of organisms is dynamic. It has the capacity to undergo recombination, duplications, deletions and transpositions. Organisms, especially ones with short generation times such as microbes, have harnessed these abilities of the genetic material to enhance their adaptability to biological and biochemical circumstances in a variety of ecological niches. All genetic material has an intrinsic property to mutate and is by no means an inflexible repository of pre-determined information. From studies of rearrangements in the chromosomes of bacteria, we seek to understand the mechanisms whereby the reversible and irreversible accumulation of variations in genetic content is reflected in the variety of strains and species.

## 4.1 Mycoplasmal speciation and chromosomal rearrangements

The *Mollicutes* are closely related to the gram-positive *Lactobacillus* group of bacteria, such as *Bacillus* and *Streptococcus*, with whom they are thought to have shared a common ancestor. It has been proposed that loss of the cell wall in this ancestor and fusion of the resulting L-forms led to recombination between genomes. This could have been followed by the deletion of unique regions between non-tandem repeats, leading to a net reduction of genetic content that is expected to have obligated the *Mollicutes* to existence as parasites with little biosynthetic capability (69). Mycoplasmas are the smallest autonomously replicating organisms known, and the analysis of their genomes could facilitate the delineation of a minimal set of essential genes (48). The theoretical compilation of a minimal gene set has been performed by a comparison of the genomes of *Haemophilus influenzae* and *M. genitalium*, with the assumption that conserved genes belonging to different lineages (gram-negative and gram-positive, respectively) are most likely essential (49). An attempt to experimentally define the minimal gene set by global transposon mutagenesis in *M. genitalium* and *M. pneumoniae* has lent credence to this hypothesis (36).

Comparisons among mycoplasmal genomes throw light on the role of gene rearrangements as evolutionary milestones in genetic history. *M. genitalium* has the smallest of bacterial genomes known (580070 bp) and displays a high level of genomic parsimony as compared with *H. influenzae* and *E. coli*. Comparison of the genomes of *M. genitalium* and *M. pneumoniae* (816394 bp) indicates that gene order is conserved among orthologs but that the order of occurrence of orthologous gene segments is not. These segments are bound by repetitive elements in *M. pneumoniae*. Vestiges of these elements are detectable in *M. genitalium*, leading to the suggestion that translocation via homologous recombination over the course of evolution is responsible for differences in genome organization between the two species (32).

A recent study of homologous overlapping genes in *M. genitalium* and *M. pneumoniae* has uncovered some interesting features (26). The compilation of gene pairs whose homologues do not overlap in one of the two species shows that the overlap is caused by deletion of the stop codon in 64% of overlapping gene sets.  In mycoplasmas, the occurrence of overlapping genes could presumably be explained as the result of evolutionary pressure to minimize genome size.  But, analysis of mycoplasmal genes whose 3' ends were elongated by more than 45 codons compared to homologues in other species shows that elongation of genes can occur even without the acquisition of any new functional motifs.  Therefore, incidental elongation of coding regions may also be important in the production of overlapping genes. That mycoplasmal genomes are subject to reorganization of genetic information has become increasingly clear from several studies on phase and antigenic variation that have been reviewed elsewhere (60). (see Antigenic Variation chapter)

## 4.2     Recombination

Genetic recombination results in rearrangements that can take the form of reciprocal or non-reciprocal exchanges, inversions, insertions, deletions and duplications. DNA recombination is further classified into three categories depending on the mechanism involved - homologous, site-specific or illegitimate. Homologous recombination occurs between two sequences with long stretches of homology and is predominantly mediated by the RecA protein. Site-specific recombination occurs at specific DNA sequences recognized and bound by a recombinase that carries out strand exchange, cleavage and religation, with or without the involvement of additional host factors. Insertion elements, transposons and the genomes of lysogenic viruses such as phage λ use this mechanism to insert into sites in the genome and are important catalysts for genetic change. By insertion into, and excision from, specific sites in the genome, they can activate or inactivate neighboring genes and also cause gene rearrangements in the process. Some gene families, such as the flagellar genes of *Salmonella typhimurium*, have their expression controlled by site-specific DNA inversion events. Illegitimate recombination occurs between sequences exhibiting little or no homology.  It can occur due to errors by enzymes capable of cleaving and religating DNA (nucleases and topoisomerases), copy-choice errors, or slipped-strand mispairing during DNA replication. Reiterated sequences are especially susceptible to expansion and contraction due to slipped-strand mispairing.

### 4.2.1    Homologous recombination

Some species of mycoplasma have the ability to undergo homologous recombination. All of the sequenced mollicutes genomes have *recA*. The ability of mollicutes to undergo homologous recombination has been exploited to inactivate particular genes in *A. laidlawii* (23), *S. citri* (16, 29) and *M. genitalium* (13). In *A. laidlawii,* it was additionally noted that no recombinants could be obtained upon transformation of a RecA-deficient strain (23) showing that mycoplasmas are capable of RecA-dependent homologous recombination. It is conceivable that homologous recombination occurs between repetitive elements in the mycoplasmal genome. At present, the available evidence on this issue points to gene conversion, rather than reciprocal exchange (38, 56). Surprisingly, *S. citri* apparently lacks a functional *recA* gene but is nevertheless capable of homologous recombination (43).

### 4.2.2    Site-specific recombination

#### 4.2.2.1    IS elements
All IS elements discovered in mycoplasmas display high A+T percentages and use the UGA codon to specify tryptophan, indicating that they are not recent acquisitions via horizontal transfer. IS*1138* (*M. pulmonis*), IS*1221* (*M. hyorhinis*, *M. hyopneumoniae*, and *M. flocculare*), and IS*1296* (*M. mycoides* subp. *mycoides*) have been previously reviewed (22, 60) (see also the chapter Extrachromosomal Elements and Gene Transfer). In this section, we will focus on interesting examples of more recently identified IS elements in *Mollicutes*.

#### 4.2.2.1.1    IS*1634*
IS*1634* is a novel insertion element specific for small-colony isolates of *M. mycoides* subp. *mycoides* (*Mmm*SC). It consists of 1874 bp including two 13-bp terminal inverted repeats (IRs). Based on transposase homology, it is most closely related to IS*1549* of the IS*4* family of insertion sequences. A unique feature of the IS*1634* is its ability to cause large duplications of variable length at its insertion sites. Duplications of 17, 21, 70, 75, 127 and 478 bp have been documented, the last being one of the largest duplications known to be associated with any insertion element (71).

#### 4.2.2.1.2    IS*1630*
*M. fermentans* strains PG18 and II-29/1 contain a minimum of six copies of IS*1630* in their genome. IS*1630* belongs to the IS*30* family on the basis of

transposase homology. IS*1630* contains 1377 bp and has 27-bp IRs at the termini. It causes duplications of variable length at its insertion site, ranging from 19 to 26 bp. Interestingly, the insertion of IS*1630* into target DNA occurs preferentially at sites that are *Rho*-independent transcription terminators (8).

### 4.2.2.1.3   IS*Mi*1 and related elements

IS*Mi1* (also termed ISLE – insertion sequence-like element and probably the same as IS*1550*) is a 1405-bp element in *M. fermentans* that has 29-bp terminal IRs with seven mismatches and causes a duplication of 3 bp at its site of insertion. The insertion and excision of the element in the mycoplasmal chromosome may be associated with gene rearrangements occurring at a high frequency (33). An indication of a possible role of IS*Mi1* transposition influencing bacterial phenotype comes from studies of the proteinase-resistance antigen (Pra), wherein there is a positive correlation between a specific IS*Mi1* genotype and the Pra$^+$ phenotype (9). There are also indications, based on PCR using IS*Mi1* primer sets, that *M. orale* might harbor a closely related insertion sequence (15). IS*Mi1* has two overlapping ORFs that would encode a protein with high homology to the IS*150* (included in the IS*3* family) transposase (8). From the nucleotide sequence and secondary structure predictions, a ribosomal frameshift of -1 likely results in the synthesis of the transposase as a fusion product of the two ORFs. IS*1550* has recently been shown to transpose in *E. coli* (35).

A PCR primer complementary to the terminal IRs of IS*Mi1* can amplify a 1542-bp fragment of the *M. fermentans* genome (34). Sequence analysis revealed that this fragment, though possessing terminal IRs similar to IS*Mi1*, carries a completely different set of genes. There are four ORFs, none of which encode a transposase. ORF1 has no homology to any known genes. ORFs 2, 3 and 4 form an operon. On the basis of homology, ORF2 (*infC*) encodes the translation initiation factor IF3. ORF3 (*rpmI*) and ORF4 (*rplT*) encode ribosomal proteins L35 and L20, respectively. This genetic element is amplifiable with a single primer to the terminal IRs only in the incognitus strain of *M. fermentans*. It is possible that this element is capable of mobility if the termini are recognized and acted upon by the IS*Mi1* transposase.

### 4.2.2.1.4   The SpV1 genome

*S. citri* harbors repetitive DNA elements related to the genome of spiroplasma virus SpV1. These elements have been suggested to be IS elements based on the observations that some of these elements have 31-bp inverted repeats, a gene potentially encoding a transposase, and appear to be associated with a variety of types of chromosomal rearrangements (3, 47).

#### 4.2.2.2    Transposons

No uniquely mycoplasmal transposons have been isolated, though some transposons derived from Gram-positive bacteria function in mycoplasmas. Tetracycline-resistant *U. urealyticum* strains harbor sequences similar to conjugative transposon Tn*916* (12). Both Tn*916* and the staphylococcal transposon Tn*4001* are capable of transposing in several species of mycoplasma (19, 22, 73). Transformation of mycoplasmas requires treatment with agents such as polyethylene glycol or electroporation and may not occur in nature. However, transformation is not the only way to introduce transposons into mycoplasmas. Tn*916* can be conjugally transferred from *Enterococcus faecalis* to several mycoplasmas including *M. hominis*, *M. pulmonis* and *M. arthritidis* (22). Conjugal transfer is a likely natural route for mycoplasmas to have acquired antibiotic resistance genes from other bacteria (63, 64).

#### 4.2.2.3    Lysogenic mycoplasma viruses

Mycoplasma viruses capable of lysogeny include the *A. laidlawii* virus L2 (20), *S. citri* virus *ai* (14), and the *M. arthritidis* virus MAV1 (72). It is not known if the SpV1-type *S. citri* viruses are capable of lysogeny *de novo*, though several SpV1-like sequences are found interspersed in the *S. citri* genome (62). One of the SpV1 strains, SpV1-R8A2B, contains an ORF whose putative translation product has limited homology with the phage P22 integrase (61).

*M. arthritidis* strains that are lysogenic for MAV1 are more pathogenic than non-lysogenic strains. MAV1 DNA can integrate at different sites the *M. arthritidis* genome and may also exhibit mobility after integration (75). The enhanced virulence of associated with MAV1 lysogenization is probably due to a cell-surface lipoprotein encoded by the MAV1 gene *vir* (74).

#### 4.2.2.4    Site-specific inversions at the *vsa* and *hsd* loci of *M. pulmonis*

The *vsa* (variable surface antigen) and *hsd* (host specificity determinant) loci of *M. pulmonis* contain the first discovered genes in any mycoplasma that undergo site-specific inversions. The *vsa* genes (at least 11 of them in strain KD735) are all transcriptionally silent except for the one allele that is located at the *vsa* expression site, which contains the *vsa* promoter and the first 714 nucleotides of the *vsa* coding region. Site-specific inversions at a 34-bp sequence (termed *vrs*) within the coding region result in the incorporation of formerly silent genes to the expression site (4, 67). There is also strain-related heterogeneity within the *vsa* locus. KD735 lacks some *vsa* genes found in strain CT and *vice versa*. It is possible that the strains have diverged, in part, due to the loss or translocation of specific *vsa* genes. Both strains undergo site-specific inversions at the *vsa* locus at high frequency. A gene encoding a putative site-specific recombinase is located adjacent to the *vsa* locus in CT (10).

The *hsd1* and *hsd2* loci of strain KD735, described above, encode phase-variable R-M systems. Site-specific recombination occurs at specific sites within the *hsdS* sequences designated *vip* (12 mer) (68) and *hrs* (20 bp) (21). The *vip* and *hrs* sites are also present in the *hsdS* genes of each of the *hsd* loci in strain CT (10).

## 4.3     Illegitimate recombination

### 4.3.1     Reiterated sequences in mycoplasmas

Repetitive sequences are ubiquitous in prokaryotic genomes and are found both within coding and intergenic regions. Tandemly reiterated sequences are subject to expansion and contraction over generations because of slipped-strand mispairing. Those of sufficient length are also likely substrates for homologous recombination and, if such sequences from widely separated loci recombine, drastic chromosomal rearrangements could ensue.

In general, prokaryotes have fewer long repeats than eukaryotes, possibly because of the compact nature of their genome. However, prokaryotic genomes show a significant excess of short (2 to 8 bp) mononucleotide repeats over expected values. These homopolymeric tracts are primarily composed of adenine residues (thymidine on the complementary strand). The frequency of their occurrence bears no correlation with the overall A+T content of the genome (25). There are no poly(A) tracts longer than 9 bp in the *M. genitalium* genome, except for one tract which is the longest A tract ($A_{19}$) found so far in mycoplasmas. The $A_{19}$ sequence occurs in *M. genitalium* in the vicinity of the *polC* gene. *M. pneumoniae* does not have poly(A) tracts longer than 9 bp, except for one $A_{15}$ and two $A_{16}$ tracts. Both of the $A_{16}$ tracts are located near putative lipoprotein genes, one in the 5' untranslated region and the other in the 3' untranslated region. *U. urealyticum*, by contrast, possesses the a large number of A-tracts 13 or more bases long. In the genome sequence of *U. urealyticum*, $A_{17}$ and $A_{16}$ occur thrice, and $A_{19}$, $A_{18}$, $A_{15}$, $A_{14}$ and $A_{13}$ once each. The locations of these poly(A)-tracts in these three mycoplasmal genomes are listed in Table 3. Their physiological significance is unknown.

*Table 3.* List of homopolymeric tracts in selected mycoplasma genomes

| Organism | Repetitive sequence[a] | Position in genome |
|---|---|---|
| *M. genitalium* | $A_{19}$ | 36790-36808 |
| *M. pneumoniae*[b] | $T_{16}$ | 528807-528822 |
| | $A_{16}$ | 195460-195475 |
| | $T_{15}$ | 622602-622616 |
| *U. urealyticum* | $T_{19}$ | 612779-612797 |
| | $A_{18}$ | 611416-611433 |
| | $A_{17}$ | 351828-351844 |
| | $T_{17}$ | 553393-553409 |
| | | 45200-45216 |
| | $A_{16}$ | 576483-576498 |
| | | 572952-572967 |
| | $T_{16}$ | 537553-537568 |
| | $T_{15}$ | 407508-407522 |
| | $A_{14}$ | 621463-621476 |

[a] As read on the + strand in the database.
[b] According to Dandekar *et al.* (2000) (11).

A recent study of polypurine (polyR) tracts 15 bases or longer in eighteen microbial genomes indicates that their frequency of occurrence bears no correlation to the genome size or A+T content (59). While the density of such tracts is very close to the expected value in *M. pneumoniae*, it is much higher than expected in *M. genitalium*. The longest polyR-tract known thus far in any bacterial genome is a 61-nucleotide sequence in *M. genitalium*. It is interesting to note that polyR sequences are most abundant in the coding strand of the majority of 18 microbial genomes analyzed.

Homopolymeric tracts are involved in several examples of prokaryotic gene regulation. The *vlp* genes of *M. hyorhinis* exhibit phase-variable expression dependent on the number of adenine residues in a poly(A) tract within the promoter (78). Transcription of the pMGA genes of *M. gallisepticum* depends on the number of GAA (polyR) repeats present in the 5' non-coding region (30, 41). GAA repeats also occur in the 5' non-coding region of the *M. imitans vlhA2* gene (45). Many mycoplasmal surface antigens are subject to size variation due to expansion and contraction of tandem repeats in the gene's coding region.

### 4.3.2   Illegitimate recombination in *S. citri*

The circular, double-stranded replicative form of spiroplasma virus SpV1 DNA has been used as a vector in *S. citri* to express an epitope of the P1 adhesin of *M. pneumoniae* (44). This construct was unstable in *S. citri*, and the cloned insert was sometimes deleted. The sequence of the construct was such that the insert was flanked by *Mbo*I and *Xba*I restriction sites at both ends, forming short direct repeats. The deletion mutants lacked the entire insert and 6 or 9 nucleotides of sequence upstream of the *Mbo*I site. The direct repeats formed by the restriction sites could be substrates for a copy-choice process of during replication, leading to deletion (43).

## 4.4     Other instances of chromosomal rearrangements in mycoplasmas

### 4.4.1     *Mycoplasma hominis*

A large gene rearrangement has been observed in *M. hominis*. The mapping of the genomes of various strains of *M. hominis* by pulsed-field gel electrophoresis revealed that strain 7488 had probably undergone a DNA inversion of 290 kb. Furthermore, the average size of the chromosome of the analyzed strains varied between 696 and 745 kb, indicating that insertions or deletions had occurred (40).

### 4.4.2     *Mycoplasma genitalium*

Duplication, divergence and domain rearrangements of gene sequences are manifested at the protein level as the evolution, acquisition and recombining of specific functional domains. Classification of *M. genitalium* genes into families based on the structural comparison of encoded proteins reveals that gene segments corresponding to particular domains are duplicated within each family. There are 72 families with just one member each (no evidence of duplication) and 43 superfamilies that contain from 2 to 51 members (duplications) (70).

### 4.4.3     *Spiroplasma citri*

The chromosome of *S. citri* contains many copies of SpV1-like sequences that could act as direct or inverted repeats, allowing the inversion, deletion and transposition to take place between widely separated genetic loci (62, 76). A strain designated BR3-3X was found to have undergone deletions and inversions in the process of repeated propagation. Upon repeated passaging in turnip plants by grafting, rather than plant-to-plant transmission by the natural insect vector, a strain designated BR3-G was obtained. BR3-G exhibited a loss of insect transmissibility. Another strain, BR3-T was derived after repeated cycles of insect-mediated transmission in turnip plants. The genomes of strain BR3-3X as well as that of its derivatives revealed several chromosomal rearrangements. The genome of BR3-G was 1870 kb, which is 270 kb larger than that of the parent strain (1600 kb). BR3-T has a genome of 1750 kb, which is 150 kb larger than that of BR3-3X. A segment of about 1000-kb was rearranged relative to the corresponding segment in the BR3-G genome and is the largest rearrangement found in the *Mollicutes* to date. Relative to the genomes of strain BR3-G, that of BR3-T has deletions of 10-20 kb and 5-10 kb in two different regions (77). It is not known if these rearrangements are causally linked to the observed phenotypic differences among the three strains.

### 4.4.4    *Mycoplasma synoviae*

The *M. synoviae* undergoes antigenic variation of a hemagglutinin encoded by *vlhA* (54). The *M. synoviae* genome has eight additional tandemly-repeated coding regions partially homologous to *vlhA*. These repeated sequences, or pseudogenes, all lack the promoter region and the 5' end of the *vlhA* gene. The pseudogenes apparently recombine with *vlhA*, leading to the observed antigenic variation. There are no sequence homologies in the *vlh* recombination sites indicative of site-specific recombination. The process seems to be non-reciprocal and the previously expressed sequence is displaced by the incoming pseudogene, indicative of gene conversion (55).

### 4.4.5    Recombination among dispersed repetitive sequences

Repetitive sequences related to the genes encoding the P1 cytadhesin of *M. pneumoniae* and the homologous MgPa adhesin of *M. genitalium* are involved in the generation of sequence diversity in those genes.  These dispersed repetitive sequences may act as reservoirs of sequence variation, which could be incorporated into the expressed gene sequence by homologous recombination or gene conversion (38, 56, 66).

## REFERENCES

1. **Abadjieva, A., J. Patel, M. Webb, V. Zinkevich, and K. Firman.** 1993. A deletion mutant of type IC restriction endonuclease *Eco*R124I expressing a novel DNA specificity. Nucleic Acids Res. **21:**4435-4443.
2. **Arber, W., and S. Linn.** 1969. DNA modification and restriction. Annu. Rev. Biochem. **38:**467-500.
3. **Bebear, C.-M., P. Aullo, J.-M. Bove, and J. Renaudin.** 1996. *Spiroplasma citri* virus SpV1: characterization of viral sequences present in the spiroplasma host chromosome. Curr.Microbiol. **32:**134-140.

4. **Bhugra, B., L. L. Voelker, N. Zou, H. Yu, and K. Dybvig.** 1995. Mechanism of antigenic variation in *Mycoplasma pulmonis*: interwoven site-specific DNA inversions. Mol. Microbiol. **18**:703-714.

5. **Bickle, T. A.** 1987. DNA restriction and modification systems, p. 692-696. *In* F. C. Neidhardt, J. L. Ingraham, K. B. Low, B. Magasanik, M. Schaechter, and H. E. Umbarger (ed.), *Escherichia coli* and *Salmonella typhimurium*: Cellular and Molecular Biology. American Society for Microbiology, Washington, D.C.

6. **Bickle, T. A., and D. H. Kruger.** 1993. Biology of DNA restriction. Microbiol. Rev. **57**:434-450.

7. **Bove, J. M.** 1993. Molecular features of mollicutes. Clin. Infect. Dis. **17(Suppl 1)**:S10-S31.

8. **Calcutt, M. J., J. L. Lavrrar, and K. S. Wise.** 1999. IS*1630* of *Mycoplasma fermentans*, a novel IS*30*-type insertion element that targets and duplicates inverted repeats of variable length and sequence during insertion. J. Bacteriol. **181**:7597-7607.

9. **Campo, L., P. Larocque, T. L. Malfa, W. D. Blackburn, and H. L. Watson.** 1998. Genotypic and phenotypic analysis of *Mycoplasma fermentans* strains isolated from different host tissues. J. Clin. Microbiol. **36**:1371-1377.

10. **Chambaud, I., R. Heilig, S. Ferris, V. Barbe, D. Samson, F. Galisson, I. Moszer, K. Dybvig, H. Wroblewski, A. Viari, E. P. C. Rocha, and A. Blanchard.** 2001. The complete genome sequence of the murine respiratory pathogen *Mycoplasma pulmonis*. Nucl. Acids Res. **29**:2145-2153.

11. **Dandekar, T., M. Huynen, J. T. Regula, B. Ueberle, C. U. Zimmermann, M. A. Andrade, T. Doerks, L. Sanchez-Pulido, B. Snel, M. Suyama, Y. P. Yuan, R. Herrmann, and P. Bork.** 2000. Re-annotating the *Mycoplasma pneumoniae* genome sequence: adding value, function and reading frames. Nucleic Acids Res. **28**:3278-3288.

12. **de Barbeyrac, B., M. Dupon, R. Rodriguez, H. Renaudin, and C. Bebear.** 1996. A Tn*1545*-like transposon carries the *tet*(M) gene in tetracycline resistant strains of *Bacteroides ureolyticus* as well as *Ureaplasma urealyticum*, but not *Neisseria gonorrhoeae*. J. Antimicrob. Chemother. **37**:223-232.

13. **Dhandayuthapani, S., W. G. Rasmussen, and J. B. Baseman.** 1999. Disruption of gene *mg218* of *Mycoplasma genitalium* through homologous recombination leads to an adherence-deficient phenotype. Proc. Natl. Acad. Sci. USA **96**:5227-5232.

14. **Dickinson, M. J., and R. Townsend.** 1985. Lysogenisation of *Spiroplasma citri* by a type 3 spiroplasmavirus. Virology **146**:102-110.

15. **Ditty, S. E., M. A. Connolly, B. J. Li, and S.-C. Lo.** 1999. *Mycoplasma orale* has a sequence similar to the insertion-like sequence of *M. fermentans*. Mol. Cell. Probes **13**:183-189.

16. **Duret, S., J. L. Danet, M. Garnier, and J. Renaudin.** 1999. Gene disruption through homologous recombination in *Spiroplasma citri*: an *scm1*-disrupted motility mutant is pathogenic. J. Bacteriol. **181**:7449-7456.

17. **Dybvig, K.** 1993. DNA rearrangements and phenotypic switching in prokaryotes. Mol. Microbiol. **10**:465-471.

18. **Dybvig, K.** 1990. Mycoplasmal genetics. Annu. Rev. Microbiol. **44**:81-104.

19. **Dybvig, K., C. T. French, and L. L. Voelker.** 2000. Construction and use of derivatives of transposon Tn*4001* that function in *Mycoplasma pulmonis* and *Mycoplasma arthritidis*. J. Bacteriol. **185**:4343-4347.

20. **Dybvig, K., and J. Maniloff.** 1983. Integration and lysogeny by an enveloped mycoplasma virus. J. Gen. Virol. **64**:1781-1785.

21.   **Dybvig, K., R. Sitaraman, and C. T. French.** 1998. A family of phase-variable restriction enzymes with differing specificities generated by high-frequency gene rearrangements. Proc. Natl. Acad. Sci. USA **95**:13923-13928.

22.   **Dybvig, K., and L. L. Voelker.** 1996. Molecular biology of mycoplasmas. Annu. Rev. Microbiol. **50**:25-57.

23.   **Dybvig, K., and A. Woodard.** 1992. Construction of *recA* mutants of *Acholeplasma laidlawii* by insertional inactivation with a homologous DNA fragment. Plasmid **28**:262-266.

24.   **Dybvig, K., and H. Yu.** 1994. Regulation of a restriction and modification system via DNA inversion in *Mycoplasma pulmonis*. Mol. Microbiol. **12**:547-560.

25.   **Field, D., and C. Wills.** 1998. Abundant microsatellite polymorphism in *Saccharomyces cerevisiae*, and the different distributions of microsatellites in eight prokaryotes and *S. cerevisiae*, result from strong mutation pressures and a variety of selective forces. Proc. Natl. Acad. Sci. USA **95**:1647-1652.

26.   **Fukuda, Y., T. Washio, and M. Tomita.** 1999. Comparative study of overlapping genes in the genomes of *Mycoplasma genitalium* and *Mycoplasma pneumoniae*. Nucleic Acids Res. **27**:1847-1853.

27.   **Fuller-Pace, F. V., L. R. Bullas, H. Delius, and N. E. Murray.** 1984. Genetic recombination can generate altered restriction specificity. Proc. Natl. Acad. Sci. USA **81**:6095-6099.

28.   **Gann, A. A. F., A. J. B. Campbell, J. F. Collins, A. F. W. Coulson, and N. E. Murray.** 1987. Reassortment of DNA recognition domains and the evolution of new specificities. Mol. Microbiol. **1**:13-22.

29.   **Gaurivaud, P., F. Laigret, E. Verdin, M. Garnier, and J. M. Bove.** 2000. Fructose operon mutants of *Spiroplasma citri*. Microbiol. **146**:2229-2236.

30.   **Glew, M. D., N. Baseggio, P. F. Markham, G. F. Browning, and I. D. Walker.** 1998. Expression of the pMGA genes of *Mycoplasma gallisepticum* is controlled by variation in the GAA trinucleotide repeat lengths within the 5' noncoding regions. Infect. Immun. **66**:5833-5841.

31.   **Gubler, M., D. Braguglia, J. Meyer, A. Piekarowicz, and T. A. Bickle.** 1992. Recombination of constant and variable modules alters DNA sequence recognition by type IC restriction-modification enzymes. EMBO J. **11**:233-240.

32.   **Himmelreich, R., H. Plagens, H. Hilbert, B. Reiner, and R. Herrmann.** 1997. Comparative analysis of the genomes of the bacteria *Mycoplasma pneumoniae* and *Mycoplasma genitalium*. Nucleic Acids Res. **25**:701-712.

33.   **Hu, W. S., M. M. Hayes, R. Y.-H. Wang, J. W.-K. Shih, and S.-C. Lo.** 1998. High-frequency DNA rearrangements in the chromosomes of clinically isolated *Mycoplasma fermentans*. Curr. Microbiol. **37**:1-5.

34.   **Hu, W. S., R. Y.-H. Wang, J. W.-K. Shih, and S.-C. Lo.** 1993. Identification of a putative *infC-rpmI-rplT* operon flanked by long inverted repeats in *Mycoplasma fermentans* (incognitus strain). Gene **127**:79-85.

35.   **Hu, W. S., and C.-C. Yang.** 2001. IS*1550* from *Mycoplasma fermentans* is transposable in *Escherichia coli*. FEMS Microbiol. Lett. **198**:159-164.

36.   **Hutchison, C. A., S. N. Peterson, S. R. Gill, R. T. Cline, O. White, C. M. Fraser, H. O. Smith, and J. C. Venter.** 1999. Global transposon mutagenesis of a minimal mycoplasma genome. Science **286**:2165-2169.

37.   **Ives, C. L., A. Sohail, and J. E. Brooks.** 1995. The regulatory C proteins from different restriction-modification systems can cross-complement. J. Bacteriol. **177**:6313-6315.

38.  **Kenri, T., R. Taniguchi, Y. Sasaki, N. Okazaki, M. Narita, K. Izumikawa, M. Umetsu, and T. Sasaki.** 1999. Identification of a new variable sequence in the P1 cytadhesin gene of *Mycoplasma pneumoniae*: Evidence for the generation of antigenic variation by DNA recombination between repetitive sequences. Infect. Immun. **67**:4557-4562.

39.  **Kusano, K., K. Sakagami, T. Yokochi, T. Naito, Y. Tokinaga, E. Ueda, and I. Kobayashi.** 1997. A new type of illegitimate recombination is dependent on restriction and homologous interaction. J. Bacteriol. **179**:5380-5390.

40.  **Ladefoged, S. A., and G. Christiansen.** 1992. Physical and genetic mapping of the genomes of five *Mycoplasma hominis* strains by pulsed-field gel electrophoresis. J. Bacteriol. **174**:2199-2207.

41.  **Liu, L., K. Dybvig, V. S. Panangala, V. L. van Santen, and C. T. French.** 2000. GAA trinucleotide repeat region regulates M9/pMGA gene expression in *Mycoplasma gallisepticum*. Infect. Immun. **68**:871-876.

42.  **Maniloff, J., K. Dybvig, and T. L. Sladek.** 1992. Mycoplasma DNA restriction and modification, p. 325-330. *In* J. Maniloff, R. N. McElhaney, L. R. Finch, and J. B. Baseman (ed.), Mycoplasmas: Molecular Biology and Pathogenesis. American Society for Microbiology, Washington, D.C.

43.  **Marais, A., J. M. Bove, and J. Renaudin.** 1996. *Spiroplasma citri* virus SpV1-derived cloning vector: deletion formation by illegitimate and homologous recombination in a spiroplasmal host strain which probably lacks a functional *recA* gene. J. Bacteriol. **178**:862-870.

44.  **Marais, A., J. M. Bove, J. Renaudin, S. F. Dallo, and J. B. Baseman.** 1993. Expression in *Spiroplasma citri* of an epitope carried on the G fragment of the cytadhesin P1 gene from *Mycoplasma pneumoniae*. J. Bacteriol. **175**:2783-2787.

45.  **Markham, P. F., M. F. Duffy, M. D. Glew, and G. F. Browning.** 1999. A gene family in *Mycoplasma imitans* closely related to the pMGA family of *Mycoplasma gallisepticum*. Microbiology **145**:2095-2103.

46.  **Meister, J., M. MacWilliams, P. Hubner, H. Jutte, E. Skrzypek, A. Piekarowicz, and T. A. Bickle.** 1993. Macroevolution by transposition: drastic modification of DNA recognition by the type I restriction enzyme following Tn5 transposition. EMBO J. **12**:4585-4591.

47.  **Melcher, U., Y. Sha, F. Ye, and J. Fletcher.** 1999. Mechanisms of spiroplasma genome variation associated with SpV1-like viral DNA inferred from sequence comparisons. Micro. Comp. Genomics **4**:29-46.

48.  **Morowitz, H. J.** 1984. The completeness of molecular biology. Isr. J. Med. Sci. **20**:750-753.

49.  **Mushegian, A. R., and E. V. Koonin.** 1996. A minimal gene set for cellular life derived by comparison of complete bacterial genomes. Proc. Natl. Acad. Sci. USA **93**:10268-10273.

50.  **Nagaraja, V., J. C. W. Shepherd, and T. A. Bickle.** 1985. A hybrid recognition sequence in a recombinant restriction enzyme and the evolution of DNA sequence specificity. Nature **316**:371-372.

51.  **Naito, T., K. Kusano, and I. Kobayashi.** 1995. Selfish behavior of restriction-modification systems. Science **267**:897-899.

52.  **Naito, Y., T. Naito, and I. Kobayashi.** 1998. Selfish restriction-modification genes: Resistance of a resident R/M plasmid to displacement by an incompatible plasmid mediated by host killing. Biol. Chem. **379**:429-436.

53. **Nakayama, Y., and I. Kobayashi.** 1998. Restriction-modification gene complexes as selfish gene entities: Roles of a regulatory system in their establishment, maintenance, and apoptotic mutual exclusion. Proc. Natl. Acad. Sci. USA **95:**6442-6447.

54. **Noormohammadi, A. H., P. F. Markham, M. F. Duffy, K. G. Whitear, and G. F. Browning.** 1998. Multigene families encoding the major hemagglutinins in phylogenetically distinct mycoplasmas. Infect. Immun. **65:**2542-2547.

55. **Noormohammadi, A. H., P. F. Markham, A. Kanci, K. G. Whitear, and G. F. Browning.** 2000. A novel mechanism for control of antigenic variation in the haemagglutinin gene family of *Mycoplasma synoviae.* Mol. Microbiol. **35:**911-923.

56. **Peterson, S. N., C. C. Bailey, J. S. Jensen, M. B. Borre, E. S. King, K. F. Bott, and C. A. Hutchison, III.** 1995. Characterization of repetitive DNA in the *Mycoplasma genitalium* genome: possible role in the generation of antigenic variation. Proc. Natl. Acad. Sci. USA **92:**11829-11833.

57. **Price, C., and T. A. Bickle.** 1986. A possible role for DNA restriction in bacterial evolution. Microbiol. Sci. **3:**296-299.

58. **Price, C., J. Lingner, T. A. Bickle, K. Firman, and S. W. Glover.** 1989. Basis for changes in DNA recognition by the *Eco*R124 and *Eco*R124/3 type I DNA restriction and modification enzymes. J. Mol. Biol. **205:**115-125.

59. **Raghavan, S., R. Hariharan, and S. K. Brahmachari.** 2000. Polypurine.polypyrimidine sequences in complete bacterial genomes: preference for polypurines in protein-coding regions. Gene **242:**275-283.

60. **Razin, S., D. Yogev, and Y. Naot.** 1998. Molecular biology and pathogenicity of mycoplasmas. Microbiol. Mol. Biol. Rev. **62:**1094-1156.

61. **Renaudin, J., P. Aullo, J. C. Vignault, and J. M. Bove.** 1990. Complete nucleotide sequence of the genome of *Spiroplasma citri* virus SpV1-R8A2 B. Nucleic Acids Res. **18:**1293.

62. **Renaudin, J., and J. M. Bove.** 1994. SpV1 and SpV4, spiroplasma viruses with circular, single-stranded DNA genomes and their contribution to the molecular biology of spiroplasmas. Adv. Virus Res. **44:**429-463.

63. **Roberts, M. C., and G. E. Kenny.** 1986. Dissemination of the *tet* M tetracycline resistance determinant to *Ureaplasma urealyticum.* Antimicrob. Agents Chemother. **29:**350-352.

64. **Roberts, M. C., L. A. Koutsky, K. K. Holmes, D. J. LeBlanc, and G. E. Kenny.** 1985. Tetracycline-resistant *Mycoplasma hominis* strains contain streptococcal *tet*M sequences. Antimicrob. Agents Chemother. **28:**141-143.

65. **Roberts, R. J., and D. Macelis.** 2000. REBASE - restriction enzymes and methylases. Nucleic Acids Res. **28:**306-307.

66. **Ruland, K., R. Himmelreich, and R. Herrmann.** 1994. Sequence divergence in the ORF6 gene of *Mycoplasma pneumoniae.* J. Bacteriol. **176:**5202-5209.

67. **Shen, X., J. Gumulak, H. Yu, C. T. French, N. Zou, and K. Dybvig.** 2000. Gene rearrangements in the *vsa* locus of *Mycoplasma pulmonis.* J. Bacteriol. **182:**2900-2908.

68. **Sitaraman, R., and K. Dybvig.** 1997. The *hsd* loci of *Mycoplasma pulmonis:* organization, rearrangements and expression of genes. Mol. Microbiol. **26:**109-120.

69. **Sladek, T. L.** 1986. A hypothesis for the mechanism of mycoplasma evolution. J. Theor. Biol. **120:**457-465.

70. **Teichmann, S. A., J. Park, and C. Chothia.** 1998. Structural assignments to the *Mycoplasma genitalium* proteins show extensive gene duplications and domain rearrangements. Proc. Natl. Acad. Sci. USA **95:**14658-14663.

71.  **Vilei, E. M., J. Nicolet, and J. Frey.** 1999. IS*1634*, a novel insertion element creating long, variable-length direct repeats which is specific for *Mycoplasma mycoides* subsp. *mycoides* small-colony type. J. Bacteriol. **181:**1319-1323.

72.  **Voelker, L. L., and K. Dybvig.** 1998. Characterization of the lysogenic bacteriophage MAV1 from *Mycoplasma arthritidis.* J. Bacteriol. **180:**5928-5931.

73.  **Voelker, L. L., and K. Dybvig.** 1996. Gene transfer in *Mycoplasma arthriditis*: transformation, conjugal transfer of Tn*916*, and evidence for a restriction system recognizing AGCT. J. Bacteriol. **178:**6078-6081.

74.  **Voelker, L. L., and K. Dybvig.** 1999. Sequence analysis of the *Mycoplasma arthritidis* bacteriophage MAV1 genome identifies the putative virulence factor. Gene **233:**101-107.

75.  **Voelker, L. L., K. E. Weaver, L. J. Ehle, and L. R. Washburn.** 1995. Association of lysogenic bacteriophage MAV1 with virulence of *Mycoplasma arthritidis.* Infect. Immun. **63:**4016-4023.

76.  **Ye, F., F. Laigret, J. Whitley, C. Citti, L. Finch, P. Carle, J. Renaudin, and J. M. Bove.** 1992. A physical and genetic map of the *Spiroplasma citri* genome. Nucleic Acids Res. **20:**1559-1565.

77.  **Ye, F., U. Melcher, J. E. Rascoe, and J. Fletcher.** 1996. Extensive chromosome aberrations in *Spiroplasma citri* strain BR3. Biochem. Genet. **34:**269-286.

78.  **Yogev, D., R. Rosengarten, R. Watson-McKown, and K. S. Wise.** 1991. Molecular basis of *Mycoplasma* surface antigenic variation: a novel set of divergent genes undergo spontaneous mutation of periodic coding regions and 5' regulatory sequences. EMBO J. **10:**4069-4079.

79.  **Yuan, R.** 1981. Structure and mechanism of multifunctional restriction endonucleases. Annu. Rev. Biochem. **50:**285-315.

# Chapter 17

# Invasion of Mycoplasmas into and Fusion with Host Cells

SHLOMO ROTTEM
*Department of Membrane and Ultrastructure Research, The Hebrew University Hadassah Medical School, Jerusalem 91010, Israel*

## 1. INVASION INTO HOST CELLS

Although it is believed that mycoplasmas remain attached to the surface of epithelial cells[23], some mycoplasmas have evolved mechanisms for entering host cells that are not naturally phagocytic. The ability of *Mycoplasma penetrans* which was isolated from the urogenital tract of AIDS patients[19,20] to invade and live within host cells has been intensively studied. This microorganism has invasive properties and localizes in the cytoplasm and perinuclear regions[1,5,15]. Other mycoplasmas, known to be surface parasites, such as *M. fermentans*[35,37], *M. pneumoniae*[2], *M. genitalium*[17], and *M. galliseplticum*[40] were also shown, under certain circumstances, to reside within non-phagocytic cells.

### 1.1 Experimental systems

In studying bacterial invasion, it is essential to differentiate between microorganisms adhering to a host cell and those which have been internalized. The early light microscopic and electron microscopic observations of mycoplasmas engulfed in membrane vesicles lead to conflicting interpretations. Are the mycoplasmas intracytoplasmatic, or are they at the bottom of crypts formed by the invagination of the cell membrane[42]? A more sophisticated ultrastructural study was based on a combined immunochemistry and electron microscopy approach using ruthenium red to stain surface polysaccharides of the host cell allowing

*Molecular Biology and Pathogenicity of Mycoplasmas*, Edited by Razin and
Herrmann, Kluwer Academic/Plenum Publishers, New York, 2002
391

better differentiation between intracellular and extracellular mycoplasmas[37]. At the present time, the most common assay used to differentiate intracellular from extracellular bacteria is the gentamicin resistance assay[11,31]. In this assay, the extracellular bacteria are killed by gentamicin, while the intracellular bacteria are shielded from the antibiotic due to its limited penetration of gentamicin into eucaryotic cells. The gentamicin procedure was successfully adapted to mycoplasma systems[1,40]. *M. penetrans* and *M. gallisepticum* are relatively susceptible to gentamicin. In the case of *M. penetrans* the susceptibility to the antibiotic can be markedly increased by adding low concentrations of Triton X-100 to the medium[1]. Thus, a combination of 200 μg/ml gentamicin and 0.01% Triton X-100 resulted in an 8 log decrease in CFU within 1 h of incubation at 37°C[1]. The low Triton X-100 concentrations affected neither the viability of the host cells nor their permeability to gentamicin. Low Triton X-100 concentrations have only a slight effect on the viability of *M. penetrans* or on the binding of *M. penetrans* to HeLa cells[1]. Usually the number of intracellular bacteria is determined by washing the host cells free of the antibiotic, lysing them with mild detergents to release the bacteria and counting the colonies[12]. Since mycoplasmas are as susceptible to detergent lysis as the host cells, dilutions of the mycoplasma-infected host cells should be plated directly onto solid mycoplasma media without lysing them beforehand. Each mycoplasma colony thus obtained represents one infected host cell rather than a single intracellular mycoplasma[11].

Differential immunofluorescent staining of internalized bacteria and of those remaining on the cell surface, combined with confocal laser scanning microscopy, has been also used to demonstrate that *M. penetrans* is capable of penetrating eucaryotic cells[2,5]. This nondestructive, high-resolution method allowed infected host cells to be optically sectioned, following fixation and immuno-fluorescent labeling and localization of the mycoplasmas within the host cell. Single-cell imaging of infected HeLa cells revealed that invasion is both time- and temperature-dependent. Penetration of HeLa cells has been observed as early as 20 min post infection[5], whereas invasion of cultured HEp-2 cells by *M. penetrans* has been shown to begin after 2 h of infection[2].

## 1.2    Invasins and receptors

Bacterial invasion of eucaryotic cells is a complex process that involves a variety of bacterial and host cell factors. Invasion is associated with adhesins as well as host cell receptors that mediate the specific interaction of the bacteria with the host cell[6,31]. It is also likely that surface molecules (proteins and lipids) that facilitate the adhesion process will have an effect on invasion. Nevertheless, adherence to the surface of host cells is not sufficient to trigger events that lead to invasion and the signals generated by the interaction of host cells with invasive mycoplasmas have not been investigated. It has been shown that bacterial invasion is based on the ability of several bacteria to bind fibronectin[10] or sulfated polysaccharides[9]. It was suggested that these compounds form a molecular bridge between the bacteria and different types of host cell surface proteins that enables invasion[6,9]. Fibronectin binding activity was detected in the case of *M. penetrans*. This organism binds selectively immobilized fibronectin and a 65-kDa fibronectin binding protein was identified in this mycoplasma[15]. An increase in the invasive capacity of *M. fermentans* which arises from the potential of this organism to bind plasminogen and activate it by the plasminogen activator urokinase to plasmin has been recently described[38]. Plasmin, a protease with broad substrate specificity, may alter *M. fermentans*-cell surface proteins and thereby enable its internalization. Proteolytic modification of bacterial and/or host cell surface protein(s) is an emerging theme among bacterial pathogens. Thus, the plasminogen activator of *Yersinia pestis* degrades bacterial outer membrane proteins and is associated with virulence[34] and a secreted protease was shown to stimulate the fibronectin-dependent uptake of *Streptococcus pyogenes* into eucaryotic cells[6].

## 1.3    Changes in the host cell cytoskeleton

Almost all invasive bacteria were shown upon contact with the host cell surface to trigger cytoskeletal rearrangements that result in bacterial internalization. The role of the cytoskeletal components, microtubules and microfilaments, in internalization can be tested biochemically employing specific inhibitors. If invasion is inhibited after depolymerization of microfilaments by cytochalasin D, the invasion process is considered to be microfilament-dependent. The requirement for microtubules for bacterial internalization can be tested by disrupting the microtubules with drugs such as colchicine, nocodazole, vincristine, or vinblastine, or by "freezing" the microtubule network by taxol[22]. Taxol crosslinks tubulin and thereby blocks the ability of the host cell to restructure the microtubular framework. By disrupting microtubules or microfilaments, the invagination and transport of the membrane-bound bacteria from the plasma membrane into the cell are inhibited either by preventing their movement along the microtubules or by inhibiting the required actin-mediated contractile forces[13,16]. *M. penetrans* invasion of HeLa cells depends on the capacity of the cells to assemble actin microfilaments, as treatment with cytochalasin D has a dramatic effect on the invasion of HeLa cells by *M. penetrans*[1]. Furthermore, as vinblastine, which disrupts microtubules as well as taxol which freezes microtubules, virtually abolish penetration of *M. penetrans*[5], it seems that alterations in the polymerization dynamics and stability of microtubules also has an inhibitory effect on the invasion of *M. penetrans* into HeLa cells. On the other hand, the entry of *M. gallisepticum* into chicken embryo fibroblasts is inhibited by the microtubule inhibitor nocodazole but not by cytochalasin D, suggesting that *M. gallisepticum* may use a different strategy from that of *M. penetrans* for reaching the intracellular space[40].

## 1.4    Signal transduction

Involvement of the host cell cytoskeleton in internalization is considered to be the result of a host cell signal transduction cascade induced by the invasive bacterium. As observed in many signal transduction processes initiated by bacteria, kinases and/or phosphatases are typically involved[22,24]. It has been suggested that the invading mycoplasmas generate uptake signals which cause the assembly of highly organized cytoskeletal structures in the host cells[15]. Yet, the nature of these signals and the mechanisms used to transduce them are not fully understood. The quantitative appearance or disappearance of corresponding phosphorylated and/or dephosphorylated host cell proteins during the internalization process can be detected by gel electrophoretic and immunological techniques. Specific protein kinase activation has been observed during the internalization of most of the bacteria which are taken up by microtubule-dependent mechanisms[25]. It has been shown that invasion of HeLa cells by *M. penetrans* is associated with tyrosine-phosphorylation of a 145 kDa host cell protein[1]. Tyrosine phosphorylation is known to activate phospholipase C to generate two second messengers: phosphatdylinositol metabolites and diacylglycerol. Indeed, changes in host cell lipid turnover as a result of *M. penetrans* binding and/or invasion of Molt-3 (a human lymphocytes cell line) have been observed[29]. These changes included the accumulation of diacylglycerol and the release of unsaturated fatty acids, predominantly long chain polyunsaturated ones such as docosahexanoic acid $(C_{22:6})$[29]. Nonetheless, phosphatidylinositol metabolites were not detected. These observations support the notion that *M. penetrans* stimulates host phospholipases to cleave membrane phospholipids, thereby initiating the signal transduction cascade. Since diacylglycerol is generated in HeLa cell invaded by *M. penetrans,* it is likely that the protein kinase C in host cells is activated. Indeed, transient protein kinase C activation was demonstrated in invaded HeLa cells by several methods, including translocation to the plasma membrane and enzymatic activity[5]. However, activation was weak and transient peaking at 20 min post infection. How any of these different signal transduction events lead to specific microtubule activity resulting in mycoplasmal internalization is unknown. The role of these signals in the penetration, survival and proliferation of mycoplasmas within host cells, as well as the involvement of the lipid intermediates in the pathobiological alterations taking place in the host cells, merit further investigation.

## 1.5    Survival and multiplication within host cells

The intracellular fate of invading bacteria can vary greatly. Most invasive bacteria appear to be able to survive intracellularly for extended periods of time, at least if they have reached a suitable host cell. Other engulfed bacteria are degraded intracellularly via phagosome-lysosome fusion. The invasive bacteria either remain and multiply within the endosomes after invasion or are released via exocytosis, and/or the lysis of the endosomes which may allow multiplication within the cytoplasm. Most ultrastructural studies performed with engulfed mycoplasmas revealed mycoplasmas within membrane bound vesicles[17,35,37]. Persistence of *M. penetrans* within NIH/3T3 cells, Vero cells, human endothelial cells, HeLa cells, WI-38 cells and HEp-2 cells has been observed over a 48-96 h postinfection[15,20]. *M. gallisepticum* remains viable within HeLa cells during 24-48 h of intracellular residence[40]. The observation of vesicles stuffed with *M. penetrans* in various host cells was taken as an indication that *M. penetrans* is able to divide within intracellular vesicles of the host cells[20]. Nonetheless, the intracellular multiplication of mycoplasmas remains to be convincingly demonstrated.

## 1.6    Cytopathic effects

Bacterial attachment to eucaryotic cells, even without subsequent entry, may lead to a pronounced cytopathic effect[7,22]. In the case of *M. penetrans*, cytopathic effects on HeLa cells were observed as early as 4 h postinfection[5], whereas on HEp-2 lymphocytes or fibroblastic mammalian cells, cytopathic effects were reported to occur within 1-5 days postinfection[15,20]. These discrepancies could be explained by the type of cell line used, the infecting dose of mycoplasmas or differences in the assays performed[15]. With HeLa cells infected by *M. penetrans* the most pronounced effect was the vacuolation of the host cells[5]. The vacuoles appeared to be empty, differing from the described membrane-bound vesicles containing clusters of bacteria[20]. The number and size of the vacuoles depended on duration of infection. As vacuolation is not obtained with *M. penetrans* cell fractions[5], it is unlikely that a necrotizing cytotoxin is involved in the generation of the cellular lesions. A possible mechanism leading to vacuolization may be associated with accumulation of organic peroxides upon invasion of HeLa cells by *M. penetrans*. Indeed, when HeLa cells were grown with the antioxidant α-tocopherol, the level of accumulated organic peroxides was extremely low and vacuolation was almost completely abolished[5].

## 2.    FUSION WITH HOST CELLS

The lack of a rigid cell wall allows direct and intimate contact of the mycoplasma membrane with the cytoplasmic membrane of the host cell. Under appropriate conditions, such contact may lead to cell fusion. Fusion of mycoplasmas with eukaryotic host cells has been first observed in electron microscopic studies[27]. The development of energy transfer and fluorescence dequenching methods has enabled investigation of the fusion process on a quantitative basis, and has also allowed the identification of fusogenic mycoplasmas.

### 2.1    Factors mediating fusion

In all the fusogenic mycoplasmas tested (*M. fermentans, M. capricolum* and *Spiroplasma floricola*), fusogenicity is dependent on the unesterified cholesterol content of the cell membrane[36]. Fusogenic activity can be found only among mollicutes requiring unesterified cholesterol for growth, whereas *Acholeplasma* species, which do not require cholesterol, are nonfusogenic. Furthermore, adapting of *M. capricolum* to grow in the absence of cholesterol, results in a marked reduction in membrane-cholesterol content and renders the organism nonfusogenic[36]. Fusogenicity of M. *fermentans* with Molt-3 lymphocytes is markedly stimulated by $Ca^{+2}$ ions and depends on the proton gradient across the mycoplasma cell membrane, decreasing markedly when the proton gradient is collapsed by proton ionophores[8].

### 2.2    Molecules implicated in fusion

Among *Mycoplasma* species, *Mycoplasma fermentans* is highly fusogenic, capable of fusing with a variety of cells[14]. Furthermore, It has been shown that the polar lipid fraction of this organism is capable of enhancing the fusion of small, unilamellar phosphatidylcholine-cholesterol (1:1 molar ratio) vesicles with Molt-3 lymphocytes in a dose-dependent manner suggesting that a lipid component is acting as a fusogen[28]. In an attempt to identify the fusogen, detailed lipid analysis of *M. fermentans* membranes were performed[28,39,41] revealing that the polar lipid fraction of this organism is dominated by the presence of unusual choline-containing phosphoglycolipids (Fig. 1). The major type MfGL-II *(M. fermentans* glycolipid II) has been identified as 6'-O-(3" phosphorylcholine-2"-amino-1"-phospho-1",3"-propanediol)-α-D-glucopyranosyl (1'→3)-1, 2-diacyl-glycerol with hexadecanoyl (16:0) and octadecanoyl (18:0) in a molar ratio of 3.6:1 constituting the major acyl residues[41]. Other choline-containing lipids identified are MfGL-I [21] which is similar to MfGL-II but without the 2-amino-1,3-propanediol-1,3-bisphosphate, and the ether lipid 1-O-alkyl-/alkenyl-2-O-acyl-glycero-3-phosphocholine and its lysoform 1-O-alkyl-/alkenyl-glycero-3-phosphocholine (MfEL and lyso-MfEL)[39].

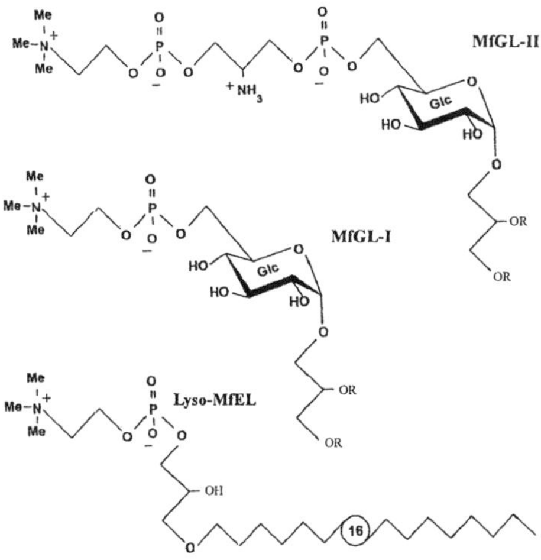

*Figure 1.* Structure of the choline containing lipids of *M. fermentans*.

It has been proposed that MfGL-II is the fusogenic component in *M. fermentans* strain PG18[28]. Nonetheless, a recent study showed that despite the fact that the respiratory isolates of *M. fermentans*, strain M39 and M52 have no MfGL-II, these strains fused with Molt-3 lymphocytes at almost the same rate and to about the same extent as PG18, suggesting that in these strains MfGL-II is not the fusogenic component[4]. It is widely accepted that the reorganization of the membrane structure which occurs during fusion requires that the lipid bilayer is broken up and that other inverted configurations, such as reversed nonbilayer aggregates, are being formed[7,30]. Nonetheless, analyses of the phase behaviour of MfGL-II/H$_2$O mixtures by solid state $^{31}$P and pulsed-field gradient diffusion NMR spectroscopy revealed that MfGL-II is a bilayer stabilizing lipid incapable of undergoing a phase transition from a lamellar to an inverted configuration[3]. This property of MfGL-II is difficult to reconcile with a role in membrane fusion. On the other hand, it is well established that lysolipids can substantially enhance the rate of fusion in model membranes as well as in biomembranes[7], and it is plausible that the lyso ether lipid found in all *M. fermentans* strains[39] may act as a fusogen. Very little is known about the role of membrane proteins in the fusion process. The observation that fusion of *M. fermentans* with Molt-3 lymphocytes was inhibited by pretreatment of intact *M. fermentans* with proteolytic enzymes[8] implies that this organism possess a proteinase sensitive adhesin(s) responsible for binding and/or the establishment of tight contact with the cell surface of the host cell involved in fusion.

## 2.3     Role in virulence

During the fusion process, mycoplasma components are apparently delivered into the host cell, and affect the normal functions of the cell. A whole array of potent hydrolytic enzymes has been identified in mycoplasmas, including, phospholipases, proteases and nucleases [23, 32, 33]. Recently, it has been shown that *M. fermentans* contains a potent phosphoprotein phosphatase[32]. Phosphorylation of cellular constituents by interacting cascades of serine/threonine and tyrosine protein kinases and phosphatases is a major mean by which a eucaryotic cell responds to exogenous stimuli[25]. The delivery of an active phosphoprotein phosphatase into the eurkaryotic cell upon fusion may interfere with the normal signal transduction cascade of the host cell. In addition to delivery of the mycoplasmal cell content into the host cell, fusion also allows insertion of mycoplasmal membrane components into the membrane of the eucaryotic host cell. This could alter receptor recognition sites, as well as affect the induction and expression of cytokines and alter the cross-talk between the various cells in an infected tissue.

## 3.     CONCLUSIONS

Although the ability of internalized mycoplasmas to multiply within the host cell remains to be convincingly demonstrated, the findings of invasive and fusogenic mycoplasmas offer new insights into the potential virulence strategies employed by mycoplasmas. Invasion of nonphagocytic host cells, if only for a short period of time, may provide mycoplasmas with the ability to cross mucosal barriers and gain access to internal tissues. The intracellular organisms are also resistant to the host defense mechanisms as well as to antibiotic treatment and may account for the difficulty to eradicate mycoplasmas from cell cultures[26]. Fusion or invasion of host cells with mycoplasmas is bringing up another emerging theme, the subversion by the mycoplasmas of host cell functions mainly signal-transduction pathways and cytoskeleton organization. Nonetheless, as all the studies of mycoplasma fusion and invasion of non-phagocytic host cells rely exclusively on in vitro experiments done with cultured mammalian cells rather than organ cultures of the actual target tissue or intact animals, the interpretation of the results should be made with caution.

# REFERENCES

1. **Andreev J., Borovsky, Z., Rosenshine, I. and Rottem, S.**, 1995, Invasion of Hela cells by *Mycoplasma penetrans* and the induction of tyrosine phosphorylation of a 145 kDa host cell protein. FEMS Lett. **132**: 189-194.

2. **Baseman, J.B., Lange, M., Criscimagna, N.L. and Thomas, C.A.**, 1995, Interplay between mycoplasmas and host target cells. Microb. Pathog. **19**: 105-116.

3. **Ben-Menachem, G., Byström, T., Rechnitzer, H., Rottem, S., Rilfors, L. and Lindblom, G.**, 2001, The physico-chemical characteristics of the phosphocholine containing glyco-glycerolipid MfGL-II govern the permeability properties of *Mycoplasma fermentans*. Eur. J. Biochem. (In press)

4. **Ben-Menachem, G., Zähringer, U. and Rottem, S.**, 2001, The phosphocholine motif in membranes of *Mycoplasma fermentans* strains. FEMS Letts. **199**: 137-141.

5. **Borovsky, Z., Tarshis, M., Zhang, P. and Rottem, S.**, 1998, *Mycoplasma penetrans* invasion of HeLa cells induces protein kinase C activation and vacuolation in the host cells. J. Med. Microbiol. **47**: 915-922.

6. **Chausee, M.S., Cole, R.L. and van Putten, J.P.M.**, 2000, Streptococcal erythrogenic toxin B abrogates fibronectin dependent internalization of *Streptococcus pyogenes* by cultured mammalian cells. Infect. Immun. **68**: 3226-3232.

7. **Cullis, P.R. and Hope, M.J.**, 1988, Lipid poymophism, lipid asymmetry and membrane fusion. *In* Molecular Mechanisms of Membrane Fusion (S. Ohki, D. Doyle, T.D. Flanagan, S.W. Hui, and E. Mayhew, eds.) Plenum Press, New York. pp. 37-51.

8. **Dimitrov, D.S., Franzoso, G., Salman, M., Blumenthal, R., Tarshis, M., Barile, M.F. and Rottem, S.**, 1993, *Mycoplasma fermentans*, incognitus strain, cells are able to fuse with T-lymphocytes. Clin. Infect. Dis. **17**:S305-S308.

9. **Duensing, T.D., Wing, J.S. and van Putten, J.P.M.**, 1999, Sulfated polysaccharide-directed recruitment of mammalian host proteins: a novel strategy in microbial pathogenesis. Infect. Immun. **67**: 4463-4468.

10. **Dziewanowska, K.J., Platt, M., Deobald, C.F.K., Bales, W., Trumble, W. R. and Bohach, G.A.**, 1999. Fibronectin binding protein and host cell tyrosine kinase are required for internalization of *Staphylococcus aurens* by epithelial cells. Infect. Immun. **67**: 4673-4678.

11. **Elsinghorst, E.A.**, 1994, Measurement of invasion by gentamicin resistance. Methods Enzymol. **236**: 405-420.

12. **Finlay, B.B. and Falkow, S.**, 1988, Comparison of the invasion strategies used by *Salmonella choleraesuis*, *Shigella flexnerii* and *Yersinia enterocolitica* to enter

cultured animal cells: endosomic acidification is not required for bacterial invasion or intracellular replication. Biochimie **80**: 248-254.

13.  **Finlay, B.B., Ruschkowski, S. and Dedhar, S.,** 1991, Cytoskeletal rearrangements accompanying *Salmonella* entry into epithelial cells. J. Cell Sci. **99**: 283-296.

14.  **Franzoso, G., Dimitrov, D.S., Blumenthal, R., Barile, M.F. and Rottem, S.,** 1992, Fusion of *M. fermentans*, strain incognitus, with T-lymphocytes. FEBS Lett. **303**: 251-254.

15.  **Girón, J.A., Lange, M. and Baseman, J.B.,** 1996, Adherence, fibronectin binding, and induction of cytoskeleton reorganization in cultured human cells by *Mycoplasma penetrans*. Infect. Immun. **64**: 197-208.

16.  **Guzman, C.A., Rohde, M. and Timmis, K.N.,** 1994, Mechanisms involved in uptake of *Bordetella bronchiseptica* by mouse dendritic cells. Infect. Immun. **62**: 5538-5544.

17.  **Jensen, J. S., Blom, J. and Lind, K.,** 1993, Intracellular location of *Mycoplasma genitalium* in cultured Vero cells as demonstrated by electron microscopy. Int. J. Path. **75**: 91-98.

18.  **Kelly, R.B.,** 1990. Microtubules, membrane traffic, and cell organization. Cell **61**: 5-7.

19.  **Lo, S.C.,** 1992, Mycoplasmas in AIDS, In *Mycoplasmas: Molecular Biology and Pathogenesis*. (Maniloff, J., McElhaney, R.N., Finch, L.R. and Baseman, J.B., eds.) American Society Microbiology, Washington, D.C. pp. 525-548.

20.  **Lo, S.C., Hayes, M.M. and Kotani, H. *et al.*,** 1993, Adhesion onto and invasion into mammalian cells by *Mycoplasma penetrans*: a newly isolated mycoplasma from patients with AIDS. Mod. Pathol. **6**: 276-280.

21.  **Matsuda, K., Kasama, T., Ishizuka, I., Handa, S., Yamamoto, N. and Taki, T.,** 1994, Structure of a novel phosphocholine-containing glycoglycerolipids from *Mycoplasma fermentans* J. Biol. Chem. **269**: 33123-33128.

22.  **Oelschlaeger, T.A. and Kopecko, D.J.,** 2000, Microtubule dependent invasion pathways to bacteria. In *Bacterial Invasion into Eukaryotic Cells*. Subcellular Biochemistry, Volume 33, (Oelschlaeger, T.A. and Hacker, J, eds.) Kluwer Academic/Plenum Publishers, New York pp. 3-19.

23.  **Razin, S., Yogev, D. and Naot, Y.,** 1998, Molecular biology and pathogenicity of mycoplasmas. Microbiol. Rev. **63**: 1094-1156.

24.  **Ring, A. and Tuomanen,** 2000, Host Cell Invasion by *Streptococcus pneumoniae*. In *Bacterial Invasion into Eukaryotic Cells*. Subcellular Biochemistry, Volume 33, (Oelschlaeger, T.A. and Hacker, J., eds.) Kluwer Academic/Plenum Publishers, New York pp. 125-135.

25.  **Rosenshine, I. and Finlay, B.B.,** 1993, Exploitation of host signal transduction pathways and cytoskeletal functions by invasive bacteria. BioEssays **15**: 17-24.

26.  **Rottem, S. and Barile, M.F.,** 1993, Beware of mycoplasmas.Trends Biotechnol. **11**: 143-151.

27.  **Rottem, S. and Naot, Y.,** 1998, Subversion and exploitation of host cells by mycoplasmas. Trends Microbiol. **6**: 436-440.

28.  **Salman, M., Deutsch, I. , Tarshis, M., Naot, Y. and Rottem, S.,** 1994, Membrane lipids of *Mycoplasma fermentans*. FEMS Microbiol. Lett. **123**: 255-260.

29.  **Salman, M., Borovsky, Z. and Rottem, S.,** 1998, *Mycoplasma penetrans* invasion of Molt-3 lymphocytes induces changes in the lipid composition of host cells. Microbiology **144**: 3447-3454.

30.  **Siegel, D. P.,** 1999, Energetics of intermediates in membrane fusion: comparison of stalk and inverted micellar intermediate structures. Biophys. J. **76**: 291-313.

31.  **Shaw, J.H. and Falkow, S.,** 1988, Mod
     by *Neisseria gonorrhoeae.* Infect. Immu

32.  **Shibata, K.-I., Mamoru, N. , Yoshihiko,**
     phosphatase purified from *Mycoplasma fern.*
     like activity. Infect. Immun. **62:** 313-315.

33.  **Shibata, K.-I., Sasaki, T. and Watanabe, T.,**
     possess phospholipases C in the membrane. Infe

34.  **Sodeinde, O.A., Subrahmanyam, Y.V.B.K., Sta**
     **Goguen, J.D.,** 1992, A surface protease and the in
     **258:** 1004-1007.

35.  **Stadtlander, C.T., Watson, H.L., Simecka, J.W. an**
     Cytopathogenicity of *Mycoplasma fermentans* (includi.
     Infect. Dis. **17:** 289-301.

36.  **Tarshis, M., Salman, M. and Rottem, S.,** 1993, Cholest
     fusion of single unilamellar vesicles with *M. capricolum.* ↓

37.  **Taylor-Robson, D., Davies, H.A., Sarathchandra, P. and**
     Intracellular location of mycoplasmas in cultured cells demon
     immunocytochemistry and electron microscopy. Int. J. Exp. Pa

38.  **Yavlovich, A., Higazi, A.A.-R., and Rottem, S.,** 2001, Plasmin
     activation by *Mycoplasma fermentans.* Infect. Immun. **69:** 1977-.

39.  **Wagner, F., Rottem, S., Held, H.-D., Uhlig, S. and Zähringer, U**
     lipids in the cell membrane of *Mycoplasma fermentans.* Eur. J. Bio.
     6286.

40.  **Winner, F., Rosengarten, R. and Citti, C.,** 2000, *In vitro* cell invasio.
     *Mycoplasma gallisepticum.* Infect. Immun. **68:** 4238-4244.

41.  **Zähringer, U., Wagner, F., Rietschel, E.Th., Ben-Menachem, G., Deu**
     **Rottem, S.,** 1997, Primary structure of a new phosphocholine-containing
     glycoglycerolipid of *Mycoplasma fermentans.* J. Biol. Chem. **272:** 26262-2

42.  **Zucker-Franklin, D., Davidson, M. and Thomas, L.,** 1966, The interaction
     mycoplasmas with mammalian cell. HeLa cells, neutrophiles, and eosinophils.
     Med. **124:** 521-532.

## Chapter 18

# Apoptotic, Antiapoptotic, Clastogenic and Oncogenic Effects

SHYH-CHING LO
*Department of Infectious and Parasitic Diseases Pathology, Armed Forces Institute of Pathology, Washington, DC, USA*

## 1. INTRODUCTION

Mycoplasmas are among the few prokaryotes that can grow almost "symbiotically" and have a close interaction with mammalian host cells for long periods of time without causing acute cytopathic effects. Their infections in cell culture are commonly unrecognized, and many persistent parasitic infections in humans and animals are often clinically silent. However, interactions on the surface of mammalian cells by a wide variety of ligands can trigger a cascade of growth-stimulating signals from membranes to nuclei and regulate expression of a broad assortment of genes (32, 64). Chronic and persistent infection with mycoplasmas may continually signal the infected host cells to proliferate or alter gene expression and significantly affect their crucial biological characteristics (29, 67). *In vitro* studies have demonstrated that chronic infection by some mycoplasmas can cause genomic instability, chromosomal changes and malignant transformation of the infected mammalian cells. This could represent a unique form of pathogenesis associated with infection by the seemingly low virulent mycoplasmas.

## 2. THE NATURE OF CANCERS

Normal cells are pre-programmed to have only a limited number of replication cell cycles. The growth of cells is tightly controlled by signals of

*Molecular Biology and Pathogenicity of Mycoplasmas*, Edited by Razin and
Herrmann, Kluwer Academic/Plenum Publishers, New York, 2002

various growth factors, cell-cell contact inhibition as well as telomere length. In addition, there are many cell cycle checkpoints and various DNA repair systems to ensure the fidelity of genetic material transmission to daughter cells (46). Failure to repair mutations or correct genetic changes would lead to tumor formation (51). In comparison, the unifying biological aspect of cancer cells is uncontrolled cell growth. Clonality studies have revealed that all of the cells in a tumor, which may number more than 100 billion, are the descendants of a single ancestor cell that underwent malignant transformation. The transformed cells differ from their normal counterparts in morphology, dependence upon growth-stimulating factors, metabolism and many other biological functions. The transformation process would clearly involve many changes. Development of cancers in nature requires a multistep progression often with distinct initiation and promotion stages (63). As mentioned, cancer is largely a somatic genetic disease. Cancer cells have inherent genetic instability and may acquire a highly mutable, plastic genome during the progression of malignancy.

## 3.    APOPTOTIC AND ANTIAPOPTOTIC EFFECTS OF MYCOPLASMAS

Extensive cell cycle studies in recent years have helped advance the important concepts of "checkpoints" and "rate-limiting steps" in the cell cycle (46). Cells possess internal sensors that can detect DNA damage or genetic changes. Damage to genetic materials would prompt cells to stop cell division for repairs or to undergo apoptosis, a programmed suicidal process, when the damage is deemed irreparable (52, 68). Hence, apoptosis is an important safety measure to prevent continued growth of damaged cells with genetic abnormality. Many chemical compounds, protein factors and irradiation cause cell cycle arrest and signal rapid cell apoptosis (52, 68). Mycoplasmas can apparently exert both apoptotic and antiapoptotic effects on infected mammalian cells.

### 3.1    Apoptotic effects

Table 1 summarizes studies of apoptotic effects associated with mycoplasmas in various cultured mammalian cells. Mycoplasmal infection of different lymphoid and epithelial tumor cell lines would lead to inhibition of proliferation, cell death accompanied by DNA fragmentation and the morphological changes characteristic of apoptosis (56). Many other studies showed various mycoplasmas could markedly enhance the sensitivity of cells to inducers of apoptosis. Mycoplasmal nucleases could apparently promote internucleosomal DNA fragmentation, one of the most commonly used biochemical hallmarks of apoptosis (47, 48).

*Table 1.* Apoptotic effects on mammalian cells associated with mycoplasmas

| Active Components | Mycoplasma | Mammalian cell type | Citation |
|---|---|---|---|
| Arginine deiminase* | *M. arginini* | Human T Lymphoblastoid cell line | 34 |
| | | Human leukemia cells | 23; 24 |
| Live mycoplasma | *M. bovis* | Human lymphoid cells | 56 |
| Endonucleases: | | | |
| P47 & P54 | *M. hyorhinis* | PaTu8902, NIH3T3 | 6; 47 |
| P50 | *M. penetrans* | Human T lymphocyte cell line | 48 |
| Membrane lipoprotein: | | | |
| P48 | *M. fermentans* | HL-60 human myeloid leukemia cell line | 27 |

* *M. arginini* arginine deiminase (AD) expressed in *E. coli* was an effective anti-tumor enzyme when injected I.V. into tumor implanted mice (39).

The mycoplasmal nucleases were most likely to penetrate cells when they were under stress or already in the process of apoptosis. However, endonuclease P40 of *M. penetrans* could be exerting its apoptosis-inducing cytotoxic effect in human T-lymphocytes through direct interaction with the cell surface receptor(s) (6). It was also reported that purified mycoplasmal protein arginine deiminase (AD) from *M. arginini* induced cell cycle arrest and apoptotic cell death in human lymphatic cell lines, human T cells, leukemia and T lymphoblastoid cells in culture (24, 34). However, AD could not induce similar apoptotic processes in B precursor cells and myeloid cell lines (34). Study results showed AD inhibited cell proliferation not only by depletion of arginine, but also by mechanisms involving cell cycle arrest and by directly providing a death signal (23). It is important to note that mycoplasmal AD expressed in *E. coli* induced rapid apoptosis of tumor cells when injected intravenously into tumor-implanted mice (39). In addition, a purified 48-kDa protein (P48), most likely a lipoprotein, and the recombinant P48 fusion protein derived from *M. fermentans* could modulate mammalian cell functions and induce leukemia cells to undergo differentiation and apoptosis (27). Paradoxically, mycoplasma AD also demonstrated a cytoprotective or an antiapoptotic effect on taxol-induced apoptosis in DU145 human prostate cancer cells (31).

## 3.2    Mitogenic effects and antiapoptotic effects

In contrast to the above-described effects of inhibiting cell proliferation and inducing apoptotic cell death, the potent mitogenic effects of mycoplasmas have long been documented (7, 9, 15, 43). Viable mycoplasmas, various membrane preparations as well as soluble fractions of washed mycoplasma pellets were found to activate resting murine or human lymphocytes and induce their proliferation in culture. These cells would otherwise quickly die of apoptosis (7, 9, 12, 15, 16, 43, 44, 58). Studies of different mycoplasmal preparations from different organ systems or animals could produce slightly different findings. However, it was evident that membrane preparations from many mycoplasmas, with only a few exceptions, were highly mitogenic to mammalian lymphoid cells (58). We reported that lipid-associated membrane protein (LAMP) preparations of *M. fermentans* and *M. penetrans,* using Triton X-114 extraction, induced murine splenic B-lymphocytes to proliferate, form colonies, mature and produce immunoglobulins (17). The finding was consistent with the earlier reports of other mycoplasmas, such as *M. neurolyticum, M. pneumoniae, M. pulmonis* and *A. laidlawii* that acted as a polyclonal B-cell activator (8, 43). The ability of mycoplasmas to cause normal lymphoid cells or macrophages to undergo proliferation is, at least in part, associated with their ability to stimulate production of many potent cytokines (22, 58) (see also chapter Immunomodulation by Mycoplasmas: Artifacts, Facts and Active Molecules).

*M. fermentans* and *M. penetrans,* and their membrane lipoproteins, were reported recently by our laboratory and others to rapidly activate nuclear factor-kappa B (NF-κB) and activator protein 1 (AP-1) in various mammalian cells (18, 21, 53, 54). Activation of the nuclear factors in macrophages markedly enhanced expression of the TNF-α and other cytokines. The preparation of membrane lipoproteins retained most of the nuclear factor activation activity following proteinase K treatment indicating the lipid nature of the active component (18). Studies showed that the lipopeptide with the structure of S-(2,3-dihydroxypropyl)-cysteine amino terminus was responsible for the majority of the activity (40, 41). The lipopeptide structure is believed to be present in many mycoplasmas, and may even be a general characteristic of the genus *Mycoplasma.* The activation pathway of the mycoplasmal lipopeptides was likely mediated through a Toll-like receptor (55).

Other than induction of TNF-α or other cytokine production in macrophages, activation of NF-κB in mammalian cells could apparently activate a large group of gene products that function cooperatively at various cell cycle checkpoints to suppress cell apoptosis. Numerous studies showed that activation of NF-κB had potent antiapoptotic effects (5, 61, 62). We recently studied 32D murine myeloid progenitor cells that were strictly growth factor interleukin (IL)-3 dependent. The 32D cells rapidly undergo apoptosis upon withdrawal of IL-3 supplement in culture. However, 32D cells can continue to grow in IL-3-depleted culture after infection by mycoplasmas that effectively activate NF-κB (19). Interestingly, infections by mycoplasmas that activated AP-1 but not NF-κB, failed to rescue 32D cells from undergoing apoptosis in IL-3-free culture. The results showed that mycoplasmas could exert their antiapoptotic effect through activation of NF-κB, rendering mammalian cells growth factor independent and preventing them from dying of apoptosis in the absence of growth factor support.

## 4.     CLASTOGENIC EFFECTS

Fogh and Fogh (20) first reported that infection by an unspeciated mycoplasma in the FL human amnion cell line produced chromosomal aberrations, or clastogenic changes (Table 2). They observed two phases of chromosomal changes in FL cells following mycoplasmal infection. Within days after mycoplasma was inoculated they noticed chromosomal loss in many cells (14% in one experiment). However, this initial phase of chromosome number loss was transient. Studies of cells infected with mycoplasma for 10-30 days revealed the chromosome numbers were essentially similar to the numbers in uninfected cells. The early-produced cells with low numbers of chromosomes had apparently lost their proliferative capacity or died of apoptosis. However, reduction of chromosome numbers (from 70-76 to 64) in the mycoplasma-infected cells was again noted in the following months. A concomitant increase of chromosomal aberration occurred, including open breaks and stable and unstable rearrangements. Many of the chromosomal changes, as Fogh and Fogh (20) pointed out, were strikingly similar to the reported virus-induced changes. Paton et al. reported that *M. orale* infection produced a significant increase of polyploid counts as well as breaks and rearrangement frequencies in human diploid fibroblasts *WI*-38 (50). The clastogenic effects were most prominent during a long period of mycoplasmal infection. However, as they reported even 5-6 days after infection, greater frequencies of polyploidy and numbers of breaks could already be observed in the infected cells more than in the control.

*Table 2.* Clastogenic effects of mycoplasmas

| Species of Mycoplasma | Mammalian cells infected | Citation |
|---|---|---|
| Unspeciated mycoplasma | FL human amnion cell line | 20 |
| *M. orale* | WI-38 human fibroblasts | 50 |
| *M. salivarium, M. hominis, M. fermentans* | Human leukocyte cultures | 2 |
| *M. orale, M. hominis, M. fermentans* | WI-38 human fibroblasts | 57 |
| *Ureaplasma urealyticum* | Human lymphocytes | 36 |
| *M. fermentans, M. penetrans* | C3H murine embryonic cells | 60 |
| *M. fermentans, M. penetrans* | 32D myeloid progenitor cells | 19 |

There were reports with conflicting results on clastogenic effects of various mycoplasmas. Aula and Nichols (2) reported that *M. salivalium*, but not *M. hominis* or *M. fermentans*, induced chromosomal aberrations in a 5-day human leukocyte culture. However, in a later study, McGarrity et al. reported no increase in chromosomal aberrations in continuous lymphocyte cultures similarly infected with *M. salivarium* (37). However, *Ureaplasma urealyticum*, *M. orale*, *M. hominis* and *M. fermentans* were subsequently all shown to induce chromosomal changes in human lymphocytes or human diploid fibroblasts WI-38 (36, 57). The actual mechanisms responsible for causing genetic instability and chromosomal changes in the mycoplasma-infected mammalian cells are still not clear. Depletion of arginine in the mycoplasma-infected culture could have produced chromosomal damage. Since ureaplasmas, acholeplasmas and other fermenting mycoplasmas could cause chromosomal aberrations as well, Stanbridge et al. suggested that mycoplasmal inhibition of DNA synthesis in mammalian cells somehow leads to chromosomal abnormality (57). However, inhibition of DNA synthesis or retardation of proliferation cycle may simply reflect damage of DNA or chromosomes in cells infected by mycoplasmas. The mechanisms could be closely related to the processes of mycoplasmal apoptotic effects described above.

More recently, we showed that chronic infections with *M. fermentans* and *M. penetrans* induced genomic instability with prominent chromosomal changes in murine C3H embryonic cells and 32D hematopoietic progenitor cells (19, 60). In our study, changes including chromosomal loss and translocations were identified after 18 weeks of persistent infection of C3H cells with *M. fermentans* or *M. penetrans*. We also showed that infections by *M. fermentans* and *M. penetrans* caused infidelity of genomic transmission and/or aberration of chromosome segregation in cell division of 32D cells. Following a few weeks of infections by either of the mycoplasmas, various kinds of chromosomal changes, especially with trisomy 19, began to appear and quickly accumulated in culture. Gaining an additional chromosome 19 in infected 32D cells appeared to confer significant advantage in cell growth. All the karyotypic changes in the mycoplasma-infected cells were evidently irreversible and did not require continued presence of the mycoplasmas in the culture.

## 5.     ONCOGENIC EFFECTS

Since cancer is largely a somatic genetic disease and the hallmark of cancer cells is their genetic instability, exposures to clastogenic agents, infectious or non-infectious, would likely induce malignant transformation of mammalian cells. Mycoplasmal infections, in addition to causing genomic instability with prominent chromosomal changes, have antiapoptotic effects that would allow continued growth or survival of cells that otherwise undergo apoptosis. In other words, mycoplasmas loosen the controls of cell cycle checkpoints and promote further proliferation of cells with apparent genetic damage or changes. The combination of these two potent biological effects of mycoplasmas could produce changes of an oncogenic nature in infected mammalian cells.

## 5.1     Acute form of induction

It was first reported by Macpherson and Russell in 1966 (38) that introduction of mycoplasmas into cultures of baby hamster kidney (BHK) cells produced immediate morphological transformation with a high soft agar cloning efficiency. The transforming process was readily identifiable shortly after the mycoplasmal infection and was considered similar to the virus-mediated cell transformation. No significant latency or distinct stages of progression were noted. Unfortunately, the BHK cells often had a high rate of spontaneous transformation. Twenty years later, another report was also made of rapid cell transformation involving an arthropod spiroplasma (35). The unusual transformation process involved transfecting spiroplasmal DNA into mammalian cells. However, most scientists thought mycoplasmas simply introduced confusing artifacts that mimic cell transformation. Few believed that infections with mycoplasmas would transform mammalian cells into tumor cells like some viruses.

## 5.2 Chronic form of induction

As opposed to earlier reports (35, 38), our study showed *M. fermentans* and *M. penetrans* did not induce rapidly identifiable acute cell transformation. Instead, persistent parasitic infection (for many months) with the mycoplasmas was necessary to transform normal mouse embryonic C3H cells into growth–unregulated tumor cells. More importantly, we found mycoplasma-mediated oncogenesis not only demonstrated a long latency but also had progressive phases with multiple stages characterized by reversibility and irreversibility in the course of malignant transformation (60). In the reversible state, the morphologically transformed cells became hyperchromatic in staining, lost cell to cell contact growth inhibition and formed foci with multiple cell layers. However, all of these malignant properties quickly disappeared if mycoplasmas were eradicated from the culture. In the later irreversible state following further persistent infection with the mycoplasma, the transformed cells gained the ability of anchorage-independent growth in soft agar and tumor formation in animals (Figure 1).

Over-expression of H-ras and c-myc proto-oncogenes were associated with both the initial reversible and the subsequent irreversible states of the mycoplasma-mediated transformation in C3H cells (65). The onset of irreversible transformation closely coincided with prominent chromosomal changes of the infected cells. Differing from tumorigenesis in animal cells induced by most oncogenic viruses or in plant cells induced by agrobacteria, mycoplasmas evidently did not cause malignant transformation by integrating their gene(s) into the mammalian cell genome (66). Our results suggested malignant transformation could be a very unique form of pathogenesis associated with the chronic and persistent nature of mycoplasmal infection in mammalian cells.

As described above, our study showed that 32D cells, a murine myeloid progenitor cell line whose survival and growth is strictly IL-3 dependent, would live and continue to grow in IL-3-depleted culture when infected with *M. fermentans* and *M. penetrans* (19). The ability of mycoplasmas to prevent 32D cells from dying of apoptosis and support continued cell growth in the absence of IL-3 depended upon the ability of mycoplasmas to rapidly activate NF-κB in the cells. The onset of malignant transformation and gaining tumorigenicity of 32D cells coincided with development of chromosomal changes and trisomy 19 following prolonged mycoplasmal infection. In this instance, obtaining an additional chromosome 19 appeared to provide better autonomous growth ability. 32D cells with trisomy 19 quickly prevailed in culture and showed high tumorigenicity in animals. Thus, infections by mycoplasmas could first render growth-factor-independence and subsequently induce genomic instability and ensuing malignant transformation of the normally IL-3 dependent hematopoietic cells *in vitro*.

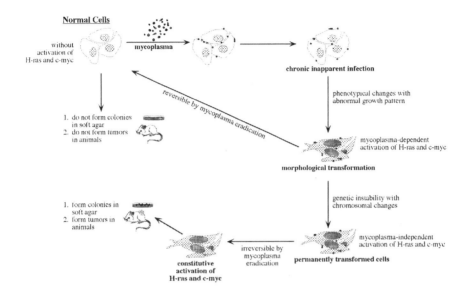

*Figure 1.* Schematic description of the multistage progression of mycoplasma-mediated malignant transformation in $C_3H$ cells. Reproduced from the article of Zhang *et al.* (ref. 65) with permission of the Society for Experimental Biology and Medicine.

## 6.     MYCOPLASMAS AS A MODEL SYSTEM FOR CANCER RESEARCH

During the early process of examining a potential role for viruses in leukemogenesis, a number of studies reported isolation of mycoplasmas from bone marrow or lymphomatous tissue of patients with leukemia (4, 25, 28, 42, 45). Barile reviewed the subject and concluded that the association did evidently exist (3), however an etiological association could not be confirmed. In addition to association with leukemia (30), there were occasional clinical observations suggesting possible association between infections by mycoplasmas and development of various tumor-like lesions as well as true malignancies in humans (1, 11, 49). Furthermore, findings of mycoplasmal genetic materials in benign tumors or malignant cancer were also reported recently (14, 26, 33). It is important to note that the high likelihood of mycoplasma isolation as a result of contamination from various laboratory procedures often complicated the assessment. The PCR detection of mycoplasmas in benign and malignant tumors would also need to be independently confirmed under stringent condition. Thus at present, it remains unclear if infections with any mycoplasmas could indeed induce or promote tumor formation in human or in animals. Further studies, including solid epidemiological studies and development of animal models, are clearly needed for better assessment of the possible role, if any, mycoplasmas may be playing in various malignancies.

However, we believe chronic persistent infection of mycoplasmas could be an excellent *in vitro* model system for cancer research. Studies of *in vitro* model systems, which are confined to a fixed condition and may not describe the full range of activities occurring in the infected host, could nevertheless provide valuable information on mechanisms of malignant cell transformation. The *in vitro* mycoplasma-oncogenesis model has a long latency with a multi-stage malignant progression, the paradigm in naturally occurring human cancers. In comparison, most virus model systems of cancer pathogenesis have not demonstrated these unique properties in culture. Moreover, differing from studying model systems with various viruses carrying acute transforming genes (10, 13, 59), the mycoplasmas are microbes that we naturally encounter. Successful dissection of the machinery promoting uncontrolled cell proliferation in the reversible and irreversible stages of transformation and the mechanism(s) producing clastogenic changes in mycoplasmal infections will certainly shed light on our understanding of the complex pathogenesis of cancers.

## 7.     CONCLUSION

*In vitro* studies have revealed that mycoplasmas can produce apoptotic, mitogenic as well as antiapoptotic effects on mammalian cells. In addition, mycoplasmal infections produce clastogenic effects that cause genomic instability and chromosomal changes. The chromosomal changes are apparently irreversible in the mammalian cells following chronic mycoplasmal infections in culture. Development of genetic instability and chromosomal aberrations in the mycoplasma-infected cells that fail to undergo apoptosis could lead to malignant transformation. A role for mycoplasmas in any human or animal cancers is still conjectural, however the mycoplasma-oncogenesis model can be an excellent *in vitro* tool for studying cancer cell transformation.

# REFERENCES

1. **Abo W., Y. Sawada, S. Ogawa, A. Toyoguchi, and S. Chiba.** 1988. Mesothelial cell proliferation and localized pleural pseudotumour associated with *Mycoplasma pneumoniae* infection. Eur. J. Pediatr. **148**:62-63.

2. **Aula, P. and W. W. Nichols.** 1967. The cytogenetic effects of Mycoplasma in human leukocyte cultures. J. Cell. Physiol. **70**:281-290.

3. **Barile, M. F.** 1967. Mycoplasma and leukemia. Ann. N.Y. Acad. Sci. **143**:557-571.

4. **Barile, M. F., G. P. Bodey, J. Snider, D. B. Riggs, and M. W. Grabowski.** 1966. Isolation of *Mycoplasma orale* from leukemic bone marrow and blood by direct culture. J. Natl. Cancer Inst. **36**:155-159.

5. **Beg, A. A., and D. Baltimore.** 1996. An essential role for NF-kappaB in preventing TNF-alpha-induced cell death [see comments]. Science. **274**:782-4.

6. **Bendjennat, M., A. Blanchard, M. Loutfi, L. Montagnier and E. Bahraoui.** 1999. Role of *Mycoplasma penetrans* endonuclease P40 as a potential pathogenic determinant. Infect. Immun. **67**:4456-4462.

7. **Biberfeld, G.** 1977. Activation of human lymphocytes subpopulations by *Mycoplasma pneumoniae*. Scand. J. Immunol. **6**:1145-1150.

8. **Biberfeld, G. and E. Gronowicz.** 1976. *Mycoplasma pneumoniae* is a polyclonal B-cell activator. Nature. **261**:238-239.

9. **Biberfeld, G. and Nilsson E.** 1978. Mitogenicity of *Mycoplasma fermentans* for human lymphocytes. Infect. Immun. **21**:48-54.

10. **Bishop J. M. and H. Varmus.** 1984. Functions and origins of retroviral transforming genes. In: *RNA Tumor Viruses: Molecular biology of tumor viruses*, second edition (R. Weiss, N. Teich, H. Varmus, and J. Coffin, eds). New York: Cold Spring Harbor Laboratory, p999.

11. **Boden W. E., R. A. Carleton, A. H. Khan, and M.T. McEnany.** 1983. Left ventricular hemangioma masquerading as mycoplasma pericarditis. Am. Heart J. **106**:771-74.

12. **Butler, G. H., U. Göbel and E. L. Stanbridge.** 1984. Blast transformation of mouse splenic B-lymphocytes with extracts of Mycoplasma hyorhinis. Isr. J. Med. Sci. **20**:891-894.

13. **Carmichael G.G. and T. L. Benjamin.** 1980. Identification of DNA sequence changes leading to loss of transforming ability in polyoma virus. J. Biol. Chem. **255**:230-235.

14. **Chan, P. J., I. M. Seraj, T. H. Kalugdan, and A. King.** 1996. Prevalence of mycoplasma conserved DNA in malignant ovarian cancer detected using sensitive PCR-ELISA. Gynecol. Oncol. **63**:258-260

15. **Cole, B. C., K. E. Aldridge and J. R. Ward.** 1977. Mycoplasma-dependent activation of normal lymphocytes: mitogenic potential of mycoplasmas for mouse lymphocytes. Infect. Immun. **18**:393-399.

16. **Cole, B. C., R. A. Danes and J. R. Ward.** 1981. Stimulation of mouse lymphocytes by a mitogen derived from *Mycoplasma arthritidis*. I. Transformation is associated with an H-2-linked gene that maps to the I-E/I-C subregion. J. Immunol. **127**:1931-1936.

17. **Feng, S. H. and S. C. Lo.** 1994. Induced mouse spleen B-cell proliferation and secretion of immunoglobulin by lipid-associated membrane proteins of *Mycoplasma fermentans* incognitus and *Mycoplasma penetrans*. Infect. Immun. **62**:3916-3921.

18.  **Feng, S.-H., and S.-C. Lo.** 1999. Lipid Extract of *Mycoplasma penetrans* proteinase K-digested lipid-associated membrane proteins rapidly activated NF-κB and activator protein 1. Infect. Immun. **67**:2951-2956.

19.  **Feng, S.-H., S. Tsai, J. Rodriguez and S.-C. Lo.** 1999. Mycoplasma infections prevent apoptosis and induce malignant transformation of Interleukin-3-dependent 32D hematopoietic cells. Infect. Mol. Cell. Biol. **19**:7995-8002.

20.  **Fogh, J. and H. Fogh.** 1965. Chromosome changes in PPLO-infected FL human amnion cells. Proc. Soc. Exp. Biol. Med. **119**:233-238.

21.  **Garcia, J., B. Lemercier, S. Roman-Roman, and G. Rawadi.** 1998. A *Mycoplasma fermentans*-derived synthetic lipopeptide induces AP-1 and NF-kappaB activity and cytokine secretion in macrophages via the activation of mitogen-activated protein kinase pathways. J. Biol. Chem. **273**:34391-8.

22.  **Gershon, H. and Y. Naot.** 1984. Induction of interleukin-2 and colony-stimulating factors in lymphoid cell cultures activated by mitogenic mycoplasmas. Isr. J. Med. Sci. **20**(9):882-885.

23.  **Gong, H., F. Zölzer, G. von Recklinghausen, J. Rossler, S. Breit, W. Havers, T. Fotsis and L. Schweigerer.** 1999. Arginine deiminase inhibits cell proliferation by arresting cell cycle and inducing apoptosis. Biochem. Biophys. Res. Commun. **261**:10-14.

24.  **Gong, H., F. Zölzer, G. von Recklinghausen, W. Havers and L. Schweigerer.** 2000. Arginine deiminase inhibits proliferation of human leukemia cells more potently than asparaginase by inducing cell cycle arrest and apoptosis. Leukemia. **14**:826-829.

25.  **Grace, J. T. Jr, J. S. Horoszewicz, T. B, Stim, E. A. Mirand, and C. James.** 1965. Mycoplasmas (PPLO) and human leukemia and lymphoma. Cancer **18**:1369-1376.

26.  **Gurr, A., A. Chakraverty, V. Callanan and S. J. Gurr.** 1996. The detection of *Mycoplasma pneumoniae* in nasal polyps. Clin. Otolaryngol. **21**:269-373.

27.  **Hall, R. E., A. Sujata and D. P. Kestler.** 2000. Induction of leukemia cell differentiation and apoptosis by recombinant P48, a modulin derived from *Mycoplasma fermentans*. Biochem Biophys. Res. Commun. **269**:284-289.

28.  **Hayflick, L and H. Koprowski.** 1965. Direct agar isolation of mycoplasmas from human leukaemic bone marrow. Nature. **13**:205:713-714.

29.  **Iyama, K., S. Zhang and S. C. Lo.** 2001. Effects of mycoplasmal LAMPs on receptor responses to steroid hormones in mammalian cells. Current Microbiol. **43**:163-169.

30.  **Jansson E, K. Wegelius, K. Hakkarainen, and A. Miettinen.** 1991. Mycoplasmas and leukaemia. Res. Virol. **142**:133.

31.  **Kang, S. W., H. Kang, I. S. Park, S. H. Choi, K. H. Shin, Y. S. Chun, B. G. Chun, and B. H. Min.** 2000. Cytoprotective effect of arginine deiminase on taxol-induced apoptosis in DU145 human prostate cancer cells. Mol. Cells. **10**:331-337.

32.  **Karp, J. E., S. Broder.** 1994. New directions in molecular medicine. Cancer Res. **54**:653-665.

33.  **Kidder, M., P. J. Chan, I. M. Seraj, W. C. Patton and A. King.** 1998. Assessment of archived paraffin-embedded cervical condyloma tissues for mycoplasma-conserved DNA using sensitive PCR-ELISA. Gynecol. Oncol. **71**:254-257.

34.  **Komada, Y., X. L. Zhang, Y. W. Zhou, M. Ido and E. Azuma.** 1997. Apoptotic cell death of human T lymphoblastoid cells induced by arginine deiminase. Int. J. Hematol. **65**:129-141.

35. **Kotani, H., D. Phillips and G.J. McGarrity.** 1986. Malignant transformation of NIH-3T3 and CV-1 cells by a helical mycoplasma, *Spiroplasma mirum*, strain SMCA. In Vitro Cell Dev. Biol. **22**:756-762.

36. **Kundsin, R. B., M. Ampola, S. Streeter and P. Neurath.** 1971. Chromosome aberrations induced by T strain mycoplasmas. J. Med. Genetics. **8**:181-187.

37. **McGarrity, G. J., D. Phillips, and A. B. Vaidya.** 1980. Mycoplasmal infection of lymphocyte cell cultures: infection with *M. salivarium*. In Vitro. **16**:346-356.

38. **McPherson, I. And W. Russell.** 1966. Transformation in hamster cells mediated by mycoplasmas. Nature **210**:1343-1354.

39. **Misawa S., M. Aoshima, H. Takaku, M. Matsumoto and H. Hayashi.** 1994. High-level expression of *Mycoplasma arginini* deiminase in *Escherichia coli* and its efficient renaturation as an anti-tumor enzyme. J. Biotechnol. **36**:145-155.

40. **Mühlradt, P., H. Meyer and R. Jansen.** 1996. Identification of S-(2,3-Dihydroxypropyl)cystein in a macrophage-activating lipopeptide from *Mycoplasma fermentans*. Biochemistry **35**:7781-7786.

41. **Mühlradt, P., M. Kiess, M. Holger, R. Süssmuth and J. Gunther.** 1998. Structure and specific activity of macrophage-stimulating lipopeptides from *Mycoplasma hyorhinis*. Infect. Immun. **66**:4804-4810.

42. **Murphy, W. H., D. Furtado and E. Plata.** 1965. Possible association between leukemia in children and virus-like agents. J. Amer. Med. Assoc. **191**:122-127.

43. **Naot, Y., J. G. Tully and H. Ginsburg.** 1977. Lymphocyte activation by various mycoplasma strains and species. Infect. Immun. **18**:310-317.

44. **Naot, Y., S. Merchav, E. Ben-David and H. Ginsburg.** 1979. Mitogenic activity of *Mycoplasma pulmonis*. I. Stimulation of rat B and T lymphocytes. Immunology. **36**: 399-406.

45. **Negroni, G.** 1964. Isolation of viruses from leukaemic patients. Br. Med. J. **1**:927-929.

46. **Nurse, P., Y. Masui, and L. Hartwell.** 1998. Understanding the cell cycle. Nat. Med. **4**:1103-1106.

47. **Paddenberg, R., A. Weber, S. Wulf, and H. G. Mannhertz.** 1998. Mycoplasma endonucleases able to induce internucleosomal DNA degradation in cultured cells possess many characteristics of eukaryotic apoptotic nucleases. Cell Death Diff. **5**:517-528.

48. **Paddenberg, R., S. Wulf, A. Weber, P. Heimann, L. A. Beck and H. G. Mannhertz.** 1996. Internucleosomal DNA fragmentation in cultured cells under conditions reported to induce apoptosis may be caused by mycoplasma endonucleases. Eur. J. Cell Biol. **71**:105-119.

49. **Park S. H., G. Y. Choe, C. W. Kim, J. G. Chi, and S. H. Sung.** 1990. Inflammatory pseudotumor of the lung in a child with *Mycoplasma pneumonia*. J. Korean Med. Sci. **5**:213-223.

50. **Paton, G. R., J. P. Jacobs and F. T. Perkins.** 1965. Chromosome changes in human diploid-cell cultures infected with Mycoplasma. Nature. **207**:43-45.

51. **Peltomäki, P.** 2001. DNA mismatch and repair. Mutation Research. **488**:77-85.

52. **Pucci, B., M. Kasten and A. Giordano.** 2000. Cell cycle and apoptosis. Neoplasia. **2**:291-299.

53. **Rawadi, G., J. Garcia, B. Lemercier, and S. Roman-Roman.** 1999. Signal transduction pathways involved in the activation of NF-κB, AP-1, and c-fos by *Mycoplasma fermentans* membrane lipoproteins in macrophages. J Immunol. **162**:2193-2203.

54.   **Sacht,G., A. Märten, U. Deiters, R. Süssmuth, G. Jung, E. Wingender, and P. F. Mulhradt.** 1998. Activation of nuclear factor-κB in macrophages by mycoplasmal lipopeptides. Eur. J. Immunol. **28**:4207-4212.

55.   **Sato, S., F. Nomura, T. Kawai, O. Takeuchi, P. F. Mühlradt, K. Takeda and S. Akira.** 2000. Synergy and cross-tolerance between Toll-like receptor (TLR) 2- and TLR4-mediated signaling pathways. J. Immunol. **165**:7096-7101.

56.   **Sokolova, I. A., A. T. Vaughan and N. N. Khodarev.** 1998. Mycoplasma infection can sensitize host cells to apoptosis through contribution of apoptotic-like endonuclease(s). Immunol. Cell Biol. **76**:526-534.

57.   **Stanbridge, E., M. Önen, F. T. Perkins and L. Hayflick.** 1969. Karyological and morphological characteristics of human diploid cell strain WI-38 infected with mycoplasmas. Exp. Cell Res. **57**:397-410.

58.   **Stuart, P. M., G. H. Cassell and J. G. Woodward.** 1990. Differential induction of bone marrow macrophage proliferation by Mycoplasmas involves granulocyte-macrophage colony stimulating factor. Infect. Immun. **58**:3558-3563.

59.   **Teich N, J. Wyke, T. Mac, A. Bernstein and W. Hardy.** 1984. Pathogenesis of retrovirus-induced disease. In: Weiss R, Teich N, Varmus H, Coffin J, Eds. RNA Tumor Viruses: Molecular biology of tumor viruses, second edition. New York: Cold Spring Harbor Laboratory, p785.

60.   **Tsai, S., D. J. Wear, J. W. Shih, and S. C. Lo.** 1995. Mycoplasmas and oncogenesis: persistent infection and multistage malignant transformation. Proc. Natl. Acad. Sci. U S A. **92**:10197-10201.

61.   **Van Antwerp, D. J., S. J. Martin, T. Kafri, D. R. Green, and I. M. Verma.** 1996. Suppression of TNF-alpha-induced apoptosis by NF-kappaB [see comments]. Science. **274**:787-9.

62.   **Wang, C. Y., M. W. Mayo, R. G. Korneluk, D. V. Goeddel, and A. S. Baldwin, Jr.** 1998. NF-kappaB antiapoptosis: induction of TRAF1 and TRAF2 and c-IAP1 and c-IAP2 to suppress caspase-8 activation. Science. **281**:1680-3.

63.   **Weinberg, R. A.** 1985. Oncogenes, antioncogenes, and molecular bases of multistep carcinogenesis. Cancer Res. **49**:3713-3721.

64.   **Weinberg, R. A.** 1985. The action of oncogenes in the cytoplasm and nucleus. Science **230**:770-776.

65.   **Zhang, B., J. W. Shih, D. J. Wear, S. Tsai and S. C. Lo.** 1997. High-level expression of H-ras and c-myc oncogenes in mycoplasma- mediated malignant cell transformation. Proc. Soc. Exp. Biol. Med. **214**:359-66.

66.   **Zhang, B., S. Tsai, J. W.-K. Shih, D. J. Wear and S.-C. Lo.** 1998. Absence of mycoplasmal gene in malignant mammalian cells transformed by chronic persistent infection of mycoplasmas. Proc. Soc. Exp. Biol. Med. **218**:82-88.

67.   **Zhang, S., D. J. Wear, and S. C. Lo.** 2000. Mycoplasmal infections alter gene expression in cultured human prostatic and cervical epithelial cells. FEMS Immunol. Med. Microbiol. **27**:43-50.

68.   **Zhou, B.-B., and S. J. Elledge.** 2000. The DNA damage response: putting checkpoints in perspective. Nature. **408**:433-439.

# Chapter 19

# Genetic Mechanisms of Surface Variation

DAVID YOGEV [1], GLENN F. BROWNING [2] and KIM S. WISE [3]

[1]*Department of Membrane and Ultrastructure Research, The Hebrew University-Hadassah Medical School, Jerusalem, Israel,* [2]*School of Veterinary Science, The University of Melbourne, Australia and* [3]*Department of Molecular Microbiology and Immunology, University of Missouri School of Medicine, Columbia, Missouri, USA*

## 1. INTRODUCTION

The mycoplasma cell surface is unique among prokaryotes and is highly evolved to accommodate the lifestyle of these diverse organisms during interaction with their respective vertebrate and plant hosts. Particularly in the absence of a cell wall, several adaptive features must be in place to support the survival of these organisms during transmission and residence in varied niches, both extracellular and in some cases within host cells (22, 24, 68). As with any pathogenic organism, the mycoplasma surface reflects the functional interaction of individual organisms with host compartments, including interactions with the innate and adaptive immune systems. Quite unlike other bacteria, however, all functions are expressed from remarkably limited gene sets that are rapidly becoming delineated through mycoplasma genome sequencing projects (11, 26, 28, 36). Insights from these projects as well as many experimental studies now reveal one striking aspect of mycoplasmal biology: the prevalence of mutation-based systems as an adaptive strategy for survival. In contrast to their eubacterial counterparts, mycoplasmal genomes so far have revealed few of the complex systems for classic gene regulation and environmental sensing that endow adaptive flexibility for individual bacterial cells. On the other hand, mycoplasmas are replete with systems providing variation in the expression and structure of specific gene products, including several examples affecting the critical cell surface. While other bacteria contain varying degrees of these adaptive

*Molecular Biology and Pathogenicity of Mycoplasmas,* Edited by Razin and Herrmann, Kluwer Academic/Plenum Publishers, New York, 2002

mutational systems, they abound in mycoplasmas. Acquisition and selection for such systems indeed may have played a critical evolutionary role that allowed host adaptation despite genomic reduction. Key characteristics of these systems include (i) the localized occurrence of mutable sequences strategically placed in specific genes or gene sets, (ii) the ability to revert to alternative phenotypes through reversible mutations, and (iii) the consequence that adaptation of variants within a population (rather than adaptation of individual cells to environments) is the primary strategy for survival. Through selective examples, this chapter reviews known mechanisms underlying these mutable gene systems (Table 1.). It also reviews recent data demonstrating the functional consequences of mutations in creating variable phenotypes either through the primary gene product, or through increasingly complex interactions of these products with other surface components. The clear conclusion is that mycoplasmas possess a richly diverse and variable surface structure that will be fundamental to our understanding of their biology and pathogenesis.

*Table 1.* Genetic mechanisms of mycoplasma surface variation

| Species | Genes involved | Mechanism | Type of surface component |
|---|---|---|---|
| *M. hyorhinis* | *vlp*/gene family | Strand slippage, (poly A) | Lipoprotein |
| *M. pulmonis* | *vsa*/gene family | Site-specific inversions | Lipoprotein |
| *M. bovis* | *vsp*/gene family | Site-specific inversions<br>Homologous recombination | Lipoprotein<br>Chimeric lipoprotein |
| *M. allisepticum* | *pMGA*/gene family<br>*pvpA*/single copy | Strand slippage, (GAA)<br>Point mutation, (GAA) | Lipoprotein<br>Putative adhesin |
| *M. hominis* | *vaa*/ single copy<br>P120/*single copy* | Strand slippage, (poly A)<br>Unknown, Phase-variable masking | Lipoprotein, adhesin |
| *M. fermentas* | P78/part of the ABC transporter operon<br>*malp*/single copy | Strand slippage, (poly A)<br><br>Posttranslational modification | Lipoprotein<br><br>Lipoprotein |
| *M. arthritidis* | *MAA2*/single copy | Strand slippage, (poly T) | Lipoprotein |
| *M. synoviae* | *vlh*/ pseudogene family | Gene conversions | Chimeric lipoproteins, Hemagglutinin |
| *M. penetrans* | *P35*/single copy | Unknown | Lipoprotein |

# 2. FREQUENT MUTATIONS AS GENETIC SWITCHES

The essence of this type of variation is the presence of small regions containing reiterated bases (homopolymeric repeats) or oligonucleotide repeats. These, so called "hot spots", provide favourite targets for frequent insertions or deletions of nucleotides in order to switch genes ON and OFF (56, 70, 76, 77). Loss or gain of nucleotides is thought to occur due to transient misalignment during DNA replication by a process termed slipped-strand mispairing (56, 70). In light of the under representation of classic gene regulation systems in mycoplasmas, including those affecting structural and functional features of surface-exposed components of eubacterial cells, the use of frequent mutations to alter gene function represents an efficient alternative to classical gene regulation and sensing systems adapted by the mycoplasmas as a survival strategy (68, 89, 90).

## 2.1 Mutations in regulatory elements

### 2.1.1 The *vlp* gene system of *Mycoplasma hyorhinis*

The first mechanism controlling variable expression of surface proteins in mycoplasmas to be characterised was that for the *vlp* (variable lipoprotein) genes of *M. hyorhinis* (72, 73, 91). There are up to seven *vlp* genes in the *M. hyorhinis* genome and each of these genes appears to be independently phase variable (15, 91). Each of the *vlp* genes is translationally competent and is preceded by very similar promoter regions. A polydeoxyadenosine tract lies 17 bases 5' to the transcriptional start point of each *vlp* gene, 5' to the predicted –10 box and overlapping with the –35 box of the promoter and this tract can vary in length. When the polydeoxyadenosine tract is exactly 17 residues long, the *vlp* gene is transcribed and expressed (13, 91). The most probable explanation for the influence of the length of the polydeoxyadenosine tract on transcription is that the capacity of RNA polymerase to bind to the –10 and –35 boxes is optimal when they are separated by exactly 17 bases. Changes in the number of bases between these two binding sites for RNA polymerase not only affect the distance between them, but also their relative positions around the DNA helix, and this change in orientation may have a greater influence on the efficiency of initiation of transcription than the distance *per se*. In addition to the influence the variation in the promoter has on transcription, the level of transcription is also affected by the length of the gene (13, 14). The shorter transcripts are more abundant, and hence there is a greater concentration of mature lipoprotein on the cell surface, when the *vlp* gene has less copies of the coding repeats (discussed later).

### 2.1.2 The *pMGA* gene system of *Mycoplasma gallisepticum*

The pMGA lipoprotein is the immunodominant antigen of the major avian pathogen *Mycoplasma gallisepticum* (50). This lipoprotein appears to be one of the haemagglutinins of this organism and has been shown to exhibit high frequency phase and antigenic variation during culture *in vitro*, particularly when growing in media containing antibodies against pMGA (53), and also during infections of the respiratory tract of the chicken (31). Observations on the variation during infection suggest that there are two stages. A switch off of expression of pMGA during the acute stages of infection, commencing from the first day after inoculation, then a switch to antigenic variants of pGMA during the chronic stage of infection, from two weeks after inoculation. This variation in expression has been shown to result from switches in expression of different members of a repertoire of over 30 translationally competent genes (2, 51, 52). In most cases only a single member of the repertoire appears to be transcribed in a cell at any one time (29). The control of transcription of each member of this gene family has been shown to reside in a short GAA trinucleotide repeat region that lies 18 bases 5' to the –35 box of the promoter for each gene (30, 45). Slipped strand mispairing resulting from loss or gain of trinucleotide units generates variability within the repeat region. The number of repeats varies between 8 and 24, but only when there are 12 repeats is the gene transcribed and the pMGA it encodes expressed. The mechanism by which this repeat motif influences the promoter of the gene has not yet been elucidated. A homologous gene family, members of which are also preceded by GAA repeat regions, has been identified in the closely related species *M. imitans* and presumably plays a similar role in this avian pathogen (54).

Unlike many other phase variable lipoproteins of mycoplasmas, the coding regions *pMGA* genes do not contain large regions of tandemly repeated sequence, although there is a proline rich region adjacent to the amino terminal end of the mature lipoprotein that is encoded by variable numbers of imperfect tandem repeats of a 12 base pair sequence.

### 2.1.3 The *MAA2* gene of *Mycoplasma arthritidis*

MAA2 is a phase and size variable lipoprotein of the rat pathogen *M. arthritidis*. It is a putative cytadhesin and a protective antigen. There appears to be a single gene for MAA2 in the genome, with the 3' end of the coding sequence composed of tandem repeats of a 264 base sequence (86). The mature protein consists of 8 amino terminal residues, then repeats of an 88 amino acid sequence, with size variation in MAA2 generated by variation in the number of these coding repeats. Different clones contain between four and six repeats. Some strains appear to contain distinct repeat regions, although the sequences of these are yet to be characterised.

The phase variation in the *MAA2* gene appears to be controlled by changes in the length of a polydeoxythymidine tract 105 bases 5' to the translational start codon of the lipoprotein gene. In a clone in which MAA2 was expressed, this polydeoxythymidine tract was 16 bases long, while in two clones in which it was not expressed, the tract was 14 bases long. No other differences were observed between the three genes in the region between base 441 5' to the start codon and base 38 of the coding sequence. The tract lies between putative −10 and −35 boxes and it is hypothesised that the length of the tract may thus influence promoter binding and transcription of the gene, in a manner established to function in control of the *vlp* genes of *M. hyorhinis*.

## 2.2 Mutations within coding regions

### 2.2.1 The *vaa* gene of *Mycoplasma hominis*

The Vaa (variable adherence-associated) protein of the human pathogen *Mycoplasma hominis* is an abundant surface lipoprotein and a putative adhesin molecule, expressed *in vivo* during chronic, active arthritis associated with *M. hominis* infection and is highly immunogenic in the human host (34, 63). The Vaa protein is subject to high-frequency phase variation in expression, which correlates precisely with the ability of *M. hominis* to adhere to cultured human cells. A hot spot region in the form of short homopolymeric tract of adenine residues (Poly-A) located in the N-terminal encoding region of the *vaa* gene was shown to undergo high-frequency, reversible insertion or deletion of adenine residues. These mutations create translational frameshifts in the gene open reading frame (ORF) that results in either a complete Vaa-ORF or a truncated Vaa-ORF due to an in-frame UAG stop codon immediately downstream of the poly-A tract (93, 94).

It should be noted, that the Vaa antigen is subject to another type of variation. Analysis of the *vaa* gene from 42 *M. hominis* strains revealed a modular composition of the *vaa* genes in the form of a variable number of interchangeable cassettes generating five distinct *vaa* gene categories (7). The presence of interchangeable cassettes displaying a high degree of homology and of short direct repeats within the *vaa* gene of the various isolates, provide favorite targets for homologous recombination events. An exchange of intact *vaa* gene cassettes by recombinative mechanisms, perhaps following intraspecies horizontal gene transfer, would generate an array of antigenically distinct Vaa proteins (7).

### 2.2.2    The *p78* gene of *Mycoplasma fermentans*

The *p78* gene is the 3' most distal gene of a four-gene ABC transporter operon of *M. fermentans* proposed as the substrate-binding lipoprotein of the transporter (84). Sequence analysis indicated that P78 contains a prokaryotic prolipoprotein signal peptide sequence. In addition, P78 was shown to be a surface-exposed antigen that undergoes high-frequency phase variation in expression. The molecular basis of P78 phenotypic switching lies in a high-frequency mutational event that independently dictates its expression at the translational level. A short homopolymeric tract of adenine residues (Poly-A) located in the N-terminal coding region of the mature product is subject to hypermutation. High-frequency, reversible insertion or deletion of adenine residues lead to frameshift mutations and to selective phase variation of P78 expression (84).

### 2.2.3    The *pvpA* gene of *Mycoplasma gallisepticum*

The *pvpA* gene (8), present as a single chromosomal copy, encodes a putative cytadhesin-related molecule of *M. gallisepticum* exhibiting high homology to the P30 and the P32 cytadhesin proteins of *M. pneumoniae* and *M. genitalium*, respectively (17, 69), and high homology to adhesin molecules of *M. gallisepticum* as well (32, 37, 40). The PvpA protein of strain R shown to undergo variation in its expression, possesses a proline-rich carboxy terminal region containing two identical directly-repeated sequences of 52 amino acids (8). The molecular basis of PvpA variation was revealed in a short tract of repeated GAA codons, encoding five successive glutamate resides, located at the N-terminal region and subject to frequent mutation generating an in-frame UAA stop codon. It should be noted, that in contrast to the frameshift mutations identified in the *vaa* gene of *M. hominis* (93, 94) and in the *p78* gene of *M. fermentans* (84), the mutation within the *pvpA* structural gene is a nonsense mutation in which the nucleotide guanine is replaced with the nucleotide thymidine. In several *M. gallisepticum* clonal isolates tested, the nucleotide substitution took place consistently within the fourth GAA codon of the GAA tract (8). This suggests that this site is preferred for the occurrence of such mutation. Although the mechanism is still unknown, such a mutation cannot be explained simply by slipped-strand mispairing during DNA replication and may rather reflect the inability to identify a reversible switching event. In several generations tested, colony immunoblot of the PvpA-positive phenotype allowed the identification of variants displaying the PvpA-negative phenotype, at a frequency of about $10^{-3}$ to $10^{-4}$ per cell per generation. However, plating of the negative phenotype did not allow the identification of progeny exhibiting the positive phenotype, suggesting that the occurrence of the nonsense mutation is either irreversible or occurs at a low frequency.

Interestingly, the PvpA protein exhibits another type of variation. Analysis of the *pvpA* gene of several *M. gallisepticum* strains differing in their adherence and pathogenicity capabilities, revealed a few types of deletions within the PvpA proline-rich carboxy-terminal region (8) suggesting that this domain may be under selective pressure in the natural host.

## 3.  SITE-SPECIFIC DNA INVERSIONS AND PHENOTYPIC SWITCHING

Targeted DNA rearrangements have been shown to mediate phenotypic switching of surface key antigens in a variety of bacterial pathogens. Homologous recombination, gene conversions, gene duplications, additions or deletions of tandem repetitive units, movement of transposable elements within the chromosome and DNA inversions are frequently mechanisms employed for regulating the expression of genes encoding surface antigens (9, 23, 66, 70, 76). In many systems, surface diversity is achieved by *recA*-independent conservative site-specific recombination in which an invertible DNA segment provides the ON/OFF genetic switch by disconnecting or connecting a gene to sequences required for its expression (27, 42). Although there are numerous examples of site-specific DNA inversion systems regulating the expression of variable surface antigens, most of the examples reflect relatively simple recombination systems consisting of two specific recombination core sites that function as DNA crossover sequences and a gene encoding a site-specific recombinase, which mediates the recombination between the specific sites (23, 27, 42).

Two site-specific DNA inversion systems associated with surface antigenic variation have so far been discovered in the mycoplasmas, the *vsa* system of the murine pathogen *Mycoplasma pulmonis* (6, 78) and the *vsp* gene system of the bovine pathogen *Mycoplasma bovis* (47, 48). Perhaps the most striking feature of these two inversion systems, when compared with other bacterial inversion systems, is their profound complexity: (i) both mycoplasmal systems utilize multiple recombination core sites; (ii) single or multiple DNA inversions occur, apparently in concert; and (iii) both systems regulate the phase-variable expression of a large gene family encoding a set of surface lipoproteins. In this respect, the mycoplasmal systems resemble most closely the R64-type shufflon system found in enteric bacterial plasmids within the IncI incompatibility group (42).

## 3.1 The *vsa* gene system of *Mycoplasma pulmonis*

The *vsa* locus of *M. pulmonis* contains at least 11 *vsa* genes (in strain KD735) encoding variable surface antigens that influence several properties of *M. pulmonis* such as hemadsorption capability, ligand binding, resistance to mycoplasma virus P1, and cultural properties (6, 20, 21, 78, 80). All Vsa proteins possess a highly conserved N-terminal end of 242 amino acids beginning with a consensus signal peptide sequence characteristic of prokaryotic lipoproteins (6, 22, 78, 88). The rest of the Vsa molecules display striking differences usually in the form of tandem repetitive domains.

The *vsa* genes represent the first documented example of genes encoding surface lipoproteins that undergo site-specific DNA inversions (6, 78). All *vsa* genes are transcriptionally silent except for the particular gene that is associated with an expression site. Each of the silent *vsa* genes lacks the first 714 nucleotides of the *vsa* coding region as well as the promoter region. The expression site is a single chromosomal sequence containing a single *vsa* promoter. Activation of silent *vsa* genes is achieved by site-specific DNA inversions that occur between oppositely oriented copies of specific recombination core sites of 34-bp designated *vrs* (*vsa* inversion sequence) that are present at the 5' end of each of the gene's ORF. DNA inversions within the *vsa* locus occur at a high frequency measured at a rate of about $10^{-2}$-$10^{-3}$ per cell per generation (5, 6). Each of the *vsa* genes has been found by PCR analysis to be sometimes in association with the expression site, indicating that the full spectrum of *vsa* genes is expressed in the heterogeneous population of cells that are present in a *M. pulmonis* culture.

An interesting feature of the *vsa* system was the finding that some DNA inversions within the *vsa* locus occur between silent genes and do not result in changes in Vsa expression. The occurrence of phenotypically silent DNA inversions, however, contribute significantly to the overall variability of the system by enabling other *vsa* genes to be repositioned within the *vsa* locus in the correct orientation with respect to the expression site. In addition, striking nucleotide similarities of almost 100% between the *vsa* region carrying the *vsaC2*-S and the *vsaE2*-S genes and the region containing the *vsaC1*-S and *vsaE1*-S genes, strongly argues in favour of gene duplication as another type of variation that may occur in the *vsa* locus (78). It should also be noted, that DNA inversions within the *vsa* locus apparently occur in concert with inversions of the *hsd1* (host specificity determinant) element, which regulates restriction and modification activity in *M. pulmonis* (see chapter: Restriction-modification systems and chromosomal rearrangements)

## 3.2 The *vsp* gene system of *Mycoplasma bovis*

### 3.2.1 Structural features of the *vsp* genes

Major lipoprotein antigens (Vsps) on the surface of *M. bovis* were shown to spontaneously undergo noncoordinate phase variation between ON and OFF expression states at a high frequency of $10^{-2}$-$10^{-3}$ per cell per generation (4, 46). The *vsp* genomic locus of strain PG45, contains 13 related single-copy *vsp* genes, each of which exists as a complete open reading frame encoding a putative surface lipoprotein (47). Each *vsp* gene is preceded by a conserved 5' noncoding sequence. All *vsp* genes encode highly conserved N-terminal domains for membrane insertion and lipoprotein processing, while the rest of the mature Vsp molecules display sequence divergence. A major portion of the *vsp* genes, in some *vsp* genes comprising about 80% of the coding region, is composed of in-frame tandem repeats creating a periodicity in the polypeptide structure. Eighteen distinct repetitive domains of different length and amino acid sequences were found within the various *vsp* genes (Figure 1) (47). The repeated domain is subjected to size variation by spontaneous expansion or contraction of these repeating units and was shown to possess immunogenic epitopes and adhesive structures, suggesting a possible role for the Vsps as complex adherence-mediating regions in pathogenesis (75).

### 3.2.2 Regulatory features of the *vsp* system

The *vsp* locus undergoes site-specific DNA inversions. These inversions occur at a high frequency as evidenced by the ease at which Vsp clonal variants were isolated (46, 48). The upstream region of each *vsp* gene contains two sequence cassettes. The first (cassette #1), a 71-bp region upstream of the ATG initiation codon exhibits 98% homology among all *vsp* genes, while the second (cassette #2) upstream of cassette #1, ranging in size from 50 to 180-bp, is more divergent. The specific recombination core sites, designated *vis* (*vsp* inversion sequence) that function as putative DNA crossover sequences, were identified within the highly conserved cassette #1 upstream of each *vsp* gene (48). In other words, 13 distinct recombination sites are present within the *vsp* locus. *vis* site-mediated DNA inversions can occur between *vis* copies that are oppositely oriented in the chromosome but not between copies that are oriented in the same direction. The fact that the *vsp* genes are not all similarly oriented within the chromosomal locus, allows for several DNA segments to be invertible. The essence of Vsp phase variation is the addition or removal of a promoter to the coding region of a *vsp* gene *via* site-specific DNA inversions. A specific cassette #2 present as a single chromosomal copy and carries the transcriptional start site for an expressed *vsp* gene was identified as an active promoter (48). In VspA-ON variants, all *vsp* genes were found to be transcriptionally silent except for the *vspA* allele that possesses the *vsp* promoter. Juxtaposition of this promoter to a silent *vsp* gene by DNA inversion events between two inverted *vis* sites, allows transcription initiation of the recipient gene (48).

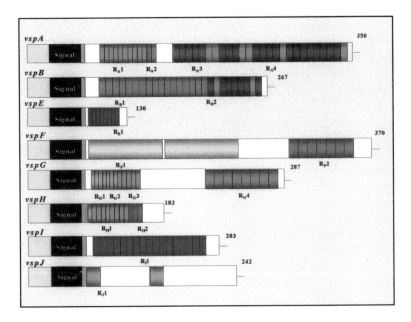

*Figure 1.* Structural features of some of the *M. bovis vsp* genes. The structure of each of the *vsp* genes and predicted Vsp proteins is schematically presented by aligned rectangles. A highly conserved upstream 5'-region extending about 80-bp 5' of each *vsp* gene is represented by the first light blue block. The second black block, represents a highly homologous 75-bp DNA sequence encoding a conserved prolipoprotein signal peptide. A sequence of 6 amino acids common to all Vsps is shown by the third pink block (47). In-frame reiterated coding sequences extending from the N-terminus to the C-terminus of the Vsp proteins and encoding periodic aa sequences are shown by different coloured blocks. Distinctive repetitive domains within each Vsp are labelled with R and the letter of the corresponding *vsp* gene. Repetitive units present in more than one Vsp molecule are similarly coloured. Numbers on the right end of each Vsp indicate the length of each Vsp polypeptide chain. (47, with permission of the ASM).

Among the most interesting features of the *vsp* system, were the findings that during phase transition (ON→OFF) of the VspA or of the VspC proteins, two site-specific DNA inversions were involved yielding three distinct *vsp* configurations (an example of VspA phase variation is given in Figure 2). Further studies revealed that the DNA inversions might occur concurrently, thus, a growing population contains apparently distinct subpopulations that have undergone single or multiple inversions (48). The population as a whole possesses, therefore, a wide spectrum of *vsp* configurations allowing an increased capability to diversify the antigenic repertoire of the mycoplasma cell surface and ensure the presence of a desirable variant needed for survival in the case of a change in the host environment.

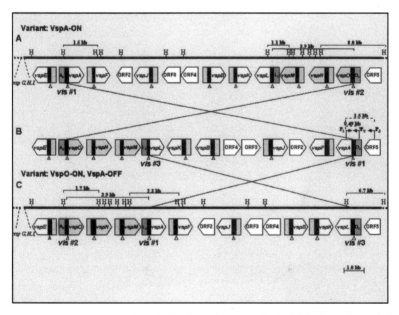

*Figure 2.* Schematic representation of DNA inversion events in the *M. bovis vsp* locus during VspA phase variation. (**A-C**) The solid line represents about 17-kb of the *vsp* locus that underwent inversion during the transition from VspA-ON to VspA-OFF (panels A and C, respectively). The positions of the *Hin*dIII (H) restriction sites are marked. Large yellow open arrows indicate the location and the direction of the *vsp* genes. The *vspM, vspN* and *vspO* genes are shown in darker yellow. The location of other *vsp* genes (*vspG, vspH* and *vspI*) is marked. The location of four non-*vsp* ORFs (ORFs 2-5) is shown by open-labelled arrows. A black block 5' to each *vsp* gene represents the homologous cassette #1, while cassettes #2 are shown by grey blocks. Cassette #2 of the *vspA, vspL* and *vspO* genes (A$_2$, L$_2$ and O$_2$, respectively) are differently coloured. Open triangles indicate the locations of the *vis* sites within the conserved cassette #1. The postulated DNA inversions yielding three distinct *vsp* configurations (A-C) are indicated by crossed lines and the involved *vis* sites (*vis* #1, *vis* #2 and *vis* #3) are indicated. (48, with permission of the ASM).

## 3.3    Catalysis by site-specific recombinases

In DNA rearrangements mediated by site-specific recombination, four DNA strands are broken, exchanged and resealed at specific positions of two separate recombination sites. These sites that are usually quite short (a few tens of base pairs) contain sequence elements that are recognized by one or more DNA-binding proteins and in many systems display identical nucleotide sequences (27, 57, 82). Most site-specific DNA inversions are catalysed by members of either the λ invertase family (Hin enzyme of *Salmonella*) or the bacteriophage integrases family of site-specific recombinases (23, 82).

In both systems, the *vsa* gene family of *M. pulmonis* and the *vsp* gene family of *M. bovis*, multiple specific recombination core sites (*vrs* and *vis*, respectively) that function as putative DNA crossover sequences were identified (48, 78). Comparison of the *vis* and the *vrs* nucleotide sequences did not reveal any homology, suggesting that different recombinases act in each of the mycoplasma species. Interestingly, two single-copy genes adjacent to the *vsp* locus and encoding putative recombinases were identified. The first, located 1 kb upstream of the *vsp* locus, exhibits high homology to the site-specific recombinase XerD of *E. coli* which belongs to the λ integrase family of site-specific recombinases (3, 16, 82). The second, located 3-kb downstream of the *vsp* locus, encodes a protein with a high identity to the aminopeptidase A (PepA) of *E. coli*. PepA is a multifunctional protein and was shown to play an important role as an accessory DNA-binding protein in Xer site-specific recombination *in vivo* and *in vitro* (1, 33).

In the *vsa* system, a crossover event involving a staggered cleavage reaction centering six nucleotides of the *vsr* site was shown to occur during recombination between *vsrA* and *vsrF* (78). Since the λ integrase and Tn3 resolvase families catalyze reactions involving breakage sites having six-to eight bases, it has been proposed that site-specific DNA inversions within the *vsa* locus are mediated by a member of the λ integrase family. Indeed, when the sequencing of the entire genome of *M. pulmonis* was completed (11), a gene, the product of which potentially encodes this type of recombinase, was located immediately downstream of the *vsa* locus. Further studies are needed to determine whether the high frequency site-specific DNA inversions observed in the *vsa* and in the *vsp* gene system are mediated by the newly discovered mycoplasmal recombinases.

# 4. EXTENDED REPERTOIRES OF HOMOLOGOUS GENES ENCODING VARIABLE SURFACE ANTIGENS

Several mycoplasma species contain in their small genome a cluster of related genes encoding surface lipoproteins (68). As discussed above, studies of some of these loci have revealed that these genes are subject to high frequency, reversible ON/OFF switching (phase variation) as well as structural variations (size variation) that generate alternate or composite expression profiles of corresponding gene products. However, the ability to generate and maintain surface diversity within populations even in a small, clonal population of bacteria, such as the limiting inoculum that initiates an infection, is also strongly dependent on the number of genes available, particularly with respect to the generation of combinatorial diversity. In other words, not only does the phase and size variation of surface antigens create an extensive potential for antigenic and structural diversification, but also extends the natural gene reservoir expressed in a particular propagating population. Extended reservoirs of genes encoding surface components are likely to contribute significantly to the successful persistence of chronic infections caused by pathogenic mycoplasmas in their natural host (79).

## 4.1 *vlp* gene repertoire of *M. hyorhinis*

*M. hyorhinis* was the first documented example of an extended gene family encoding variable lipoproteins (Vlps) (15, 92). Sequence analysis of the *vlp* genomic locus obtained from clonal isolates of the pathogenic strain SK76 revealed the presence of three or seven *vlp* genes in comparison to six *vlp* genes in the highly passaged, tissue culture-derived GDL strain. It has been proposed that the enlarged *vlp* gene family in the SK76 strain may be a prototype of *vlp* families in the natural host environment and that simple deletions from this configuration could yield the smaller gene families (15). Recent analysis of *vlp* gene families in geographically diverse field isolates confirms the predominance of larger families in the natural host environment (Droesse, M. and Wise K., unpublished). Moreover, despite a remarkable capacity for diversity in *vlp* genes due to the presence of multiple reading frames in each gene (90), these families show consistency in the reading frames actually utilized. The range and nature of functions associated with this fixed array of structurally diverse *vlp* genes is still an intriguing enigma. The strong conservation of gene organization and sequence prompts the speculation that the *vlp* gene families, in dynamic flux in propagating populations, may provide specific arrays of *vlp* gene products and endow selected functions needed for survival in the host.

## 4.2    *vsp* gene repertoire of *M. bovis*

Recent study has revealed an extended *vsp* gene repertoire in *M. bovis*. Different strains of *M. bovis* were shown to possess modified versions of the *vsp* gene complex (61). While the *vsp*-5' upstream regions and the N-terminal regions retained their highly homologous sequences, a substantial sequence divergence among individual *vsp* genes of different *M. bovis* isolates was observed, primarily in the reiterated coding sequences of the *vsp* structural genes. For example, when the *vsp* locus of the PG45 strain was compared with that of a field isolate designated 422, at least 33 distinct repetitive units, of different amino acids sequences and length (ranging between 6-84 aa) were identified in the two strains combined. Some of these repeats were strain-specific while others were common to both strains. Notably, however, the distribution of the common repeats within individual Vsps as well as their copy number vary considerably between the two-*vsp* gene families. It should be also noted, that unlike the *vlp* gene repertoire of *M. hyorhinis*, none of the *vsp* genes in *M. bovis* strain PG45 was identical to *vsp* genes of strain 422 (47, 61). Taken together, the presence of strain-specific *vsp*-repetitive coding sequences on the one hand and the profound distribution of the shared repeats on the other hand, may generate an amplified array of Vsp-phenotypes differing in their antigenic characteristics and thus, demonstrate a vastly expanded potential for antigenic variation within populations of this organism. The existence of *M. bovis* strains carrying modified versions of the *vsp* gene system, raises the intriguing speculation of natural gene transfer among cells that are in direct contact in propagating populations or even during infection within the natural host (67, 85). The presence of multiple regions of high sequence similarity upstream of each *vsp* gene or within the *vsp* coding regions and the ability of the *vsp* locus to undergo rearrangements, would then allow for the modulation of the *vsp* gene repertoire by recombination processes (66, 81).

## 5.    OTHER MECHANISMS OF SURFACE VARIATION

## 5.1    Generation of chimeric *vsp* genes by intragenic recombination in *M. bovis*

Generation and expression of chimeric genes might also provide an important element of genetic variation and an additional source of antigenic diversification within the mycoplasma population, increasing the microorganism's flexibility to deal with the immunologic attack imposed by the host. Chimeric genes can be generated as a result of recombination events among homologous sequences of related genes (41, 64, 66, 67). The *vsp* genomic locus provides favourable conditions for the occurrence of intergenic recombination. First, it has been shown that the *vsp* locus undergoes DNA inversions. Second, it displays highly homologous sequences 5' to all *vsp* genes, and third, the recurrence of similar reiterated sequences within the coding region of several *vsp* genes. These homologous sequences might serve as potential sites for intrachromosomal recombination events that could occur among *vsp* genes. Therefore, a large repertoire of chimeric *vsp* genes can potentially be generated, affecting the *vsp* genomic and antigenic repertoire. A unique example of chimeric gene formation is the *vspC* gene of *M. bovis* (49). Genetic analysis of *M. bovis* clonal isolates displaying distinct Vsp phenotypes has shown that the *vsp* gene repertoire is subject to changes. An intergenic recombination event between two closely related members of the *vsp* gene family, the formerly expressed *vspA* gene and the silent *vspO* gene, led to the formation of a new chimeric and functional *vsp* gene, namely the *vspC* gene (49). The *vspC* gene was shown to be an embodiment of the N-terminus region of the *vspO* gene and of the C-terminus of the *vspA* gene. Examination of the 5' and the 3' ends of the rearranged region revealed that homologous sequences were utilized. A 35-bp sequence (designated *vis*, see section 3.2.2) within the highly conserved *vsp*-5' upstream region was identified as the potential 5'-site for the recombination event, while identical reiterated sequences within the coding region of *vspA* and *vspO* genes (designated $R_A1$) (46, 47) served as the potential recombination site at the 3'-end. Recombination events within the *vsp* locus generating chimeric *vsp* genes may be initiated by cleavage within the *vis* sequence, while downstream recombination might then occur at any site bearing sufficient sequence similarity, including sites within the *vsp* genes themselves.

## 5.2 Generation of chimeric genes by multiple gene conversions: the *vlhA* system of *Mycoplasma synoviae*

The coding sequence of the *vlhA* gene for the haemagglutinin of the avian pathogen *M. synoviae* is homologous to those for the *pMGA* genes of *M. gallisepticum* and their homologues in *M. imitans*, even though *M. synoviae* is phylogenically distant from these two species (59). Like pMGA in *M. gallisepticum* the haemagglutinin is subject to high frequency phase and antigenic variation (58). However, in most species the product of the gene is cleaved to yield a lipoprotein and the haemagglutinin, both of which are membrane associated and exposed on the cell surface (59).

In spite of the coding similarities, the mechanism used to generate variation in the expressed *vlhA* gene is quite distinct from that controlling the pMGA gene repertoire (60). There is only one transcriptionally and translationally competent *vlhA* gene in the *M. synoviae* genome. However while there is a single copy of the promoter and the first 408 bases of the coding sequence, there are multiple, variant copies of the region encoding the 3' end of the gene. These pseudogenes appear to be clustered in the genome as tandem repeats. However, rather than encoding the full extent of the 3' end of the expressed *vlhA* gene, the pseudogenes extend between one of three 5' ends, corresponding to around base 408, 1068 or 1326 of the coding sequence of the expressed gene, and one of two 3' ends, corresponding to around base 1836 or the 3' end of the coding sequence of the expressed gene.

The specific start and endpoints of the pseudogenes suggest that variation in the expressed *vlhA* gene is likely to be achieved by recombination between the expressed gene and members of the pseudogene repertoire at one of three specific 5' sites and 2 specific 3' sites. Examination of sequential clones of *M. synoviae* has demonstrated that such recombination events do occur, with duplication of the pseudogene sequence and loss of the corresponding region in the expressed gene.

Multiple gene conversion events would result in a chimeric expressed gene, deriving different regions of its coding sequence from different pseudogenes. This model for generation of antigenic variation is supported by sequence comparisons of the coding region of different variants of the expressed gene. This has shown that there are highly conserved regions, extending from base 1 to base 720 and from base 2190 to the 3' end, a semi-conserved region between base 720 and base 1110, variable regions from base 1110 to 1326 and from base 1836 to base 2190 and a hypervariable region between base 1326 and base 1836. This distribution of variability correlates with the number of copies of each region in an 8 kb section of the genome that has been fully characterised. This section of the genome, which was entirely composed of 6 *vlh* pseudogenes of different lengths, contained 2 copies of the region from base 408 to 1068, 4 copies of the region from base 1068 to base 1326, 6 copies of the region from base 1068 to 1836, and 3 copies of the region from base 1836 to the 3' end of the gene. If all these regions of the pseudogenes are able to recombine independently with the expressed gene, there is the capacity to generate 144 variant genes from this 8 kb region. The full repertoire of pseudogenes in the genome is likely to be around 50 kb, suggesting the potential for generation of perhaps 180,000 variant *vlhA* genes.

## 5.3 Transcriptional effects determine variable P35 expression in *Mycoplasma penetrans*

An abundant surface lipoprotein antigen, P35 of *M. penetrans* (25) was shown to undergo high-frequency phase variation in its expression at a rate similar to that described for other mycoplasma variable lipoproteins (68). However, genetic analysis of P35 phase variants ruled out large scale chromosomal rearrangements, or translational frameshift mutations within the *p35* structural gene, or point mutations in its promoter. Further studies have suggested an interesting way of regulating P35 expression that involve changes at the level of *p35* transcription by yet an unknown mechanism. Recently, three additional lipoproteins (P38, P34A, P34B) have been shown to undergo phase variation in expression independently from P35 (71). These results suggest that *M. penetrans* employs systems for phase variation operating on multiple discrete genes, through mechanisms not yet fully revealed.

## 5.4 Repetitive DNA and antigenic variation in *Mycoplasma genitalium* and *Mycoplasma pneumoniae*

An interesting source for generating antigenic variation has been proposed for the proteins comprising the three-gene P1 operon of *M. pneumoniae* and its homologue, the MgPa operon of *M. genitalium* (39). The P1 and MgPa proteins are the major surface adhesin molecule of each of these species and have been shown to be highly immunogenic and necessary for attachment to red blood cells and to epithelial cells (55). A family of repetitive DNA elements consisting of segments of the P1 and MgPa operons are scattered in the *M. pneumoniae* and the *M. genitalium* genomes (65). Recombination events among the repetitive elements themselves and with regions of the P1 and MgPa operons have been suggested as a possible source of genetic variation. Such recombination events would give rise to variants of the P1 and MgPa genes, as well as ORF 6 and its homologue, contributing to antigenic variation in these surface proteins (65, 74).

In the *M. pneumoniae* genome there are 10 copies of RepMP2/3, corresponding to different extents of 37% of the P1 open reading frame, 8 copies of RepMP4, corresponding to different extents of a further 32% of the P1 open reading frame and 8 copies of the RepMP5, corresponding to different extents of 73% of the ORF6 open reading frame (36).

Comparison of the P1 gene sequences of five isolates of *M. pneumoniae* has suggested that intragenomic recombination can occur between RepMP2/3 elements and the expressed P1 gene (19), resulting in chimeric genes derived from multiple repetitive elements, although the frequency of these recombination events has not been determined, and studies have not yet confirmed that such gene conversion events result from intragenomic recombination with the other repetitive elements.

Notably, the homologous operon to the P1 operon in the avian pathogen *M. gallisepticum*, which is closely related phylogenically to *M. pneumoniae*, occurs as a single copy, suggesting that the acquisition of the repetitive elements has been a relatively recent evolutionary event (32, 37, 40).

## 6.        EFFECTS OF VARIABLE PRODUCTS: SURFACE ARCHITECTURE AND EXOPROTEIN MODIFICATION

### 6.1      Surface phenotypes affected by phase-variable masking

Primary gene products undergoing high-frequency phase variation in their expression, as well as variations in their size, generate extensive surface diversity on the mycoplasma cell. Interestingly, however, surface proteins that are constitutively expressed may also contribute to the plasticity of the cell surface through interaction with phase variable components. A recent study on *M. hominis* has shown that the exposure of surface epitopes on a constitutively expressed membrane protein P56, was subject to high-frequency phase variation (95). The variations in the accessibility of the P56 epitopes to cognate antibodies was due to high-frequency variation in the expression of another protein, the P120 protein, which is exposed on the mycoplasma surface and is highly immunodominant in the human host (12, 62). Noteworthy, whereas no changes in the expression profiles of other membrane proteins were detected, a striking inverse correlation was found between P56 accessibility and P120 expression, suggesting that the two antigens may specifically interact as nearest neighbours. These results demonstrate that variable expression of one surface protein can govern the phase-variable display of another surface molecule. Phase-variable masking provides another dimension of antigenic variation and may confer greater flexibility of mycoplasmas in adapting to host niches.

An intriguing example of phase variable masking has been observed in *M. fermentans*, where exposure to MAb-defined epitopes on the constituency expressed P29 surface lipoprotein is subject to high frequency variation (83). The recent demonstration that P29 is a major adhesin of *M. fermentans* (43, 44) raises the interesting prospect that phase-variable expression of some surface components could possibly affect the exposure of functional domains of another surface protein such as an adhesin. This notion has not been tested genetically, nor has the factor affecting P29 accessibility yet been identified.

## 6.2 Size variation

Several mycoplasma surface proteins, including some of those described in this chapter, contain regions of repeated sequences that range in size from a few to hundreds of amino acid residues and comprise from two to dozens of copies. Often these units are tandemly arranged, hence the corresponding genes are subject to in-frame insertions and deletions in the redundant coding regions. A consequence of this property is a resulting instability during propagation that allows spontaneous changes in the length of specific genes and their products, thereby endowing a population the capacity to diversify the surface features of its members. Despite the widespread occurrence of repeat structures in mycoplasmal surface proteins, the functional consequences of size variation remains for the most part unknown. One category of such products includes adhesins such as the Vaa protein of *M. hominis* (35, 93), the MAA1 protein of *M. arthritidis* (87) and the P97 products of *M. hyopneumoniae* (38). Although adaptive scaffolding, variable presentation of binding domains and redundancy of functional adherence domains have been postulated as possible consequences of variable repeat structures, experimental evidence fully supporting these possibilities is still limited.

Another group of products containing tandem repeats is the abundant and variable surface proteins encoded by gene families such as *pmga, vlp, vsa,* and *vsp*. The characteristic capacity for size variation among these gene families, as well as their widespread occurrence in diverse species of mycoplasma, suggests that size variation could underlie critical and general functional roles on the mycoplasma surface. These abundant products could, for example, provide properties that compensate for the absence of cell walls and other surface components found in Gram-positive bacteria. One consequence of size variation has been revealed in the Vlp system of *M. hyorhinis*, wherein increased length of Vlp coat proteins is shown to protect organisms from growth inhibiting host serum antibody (14). Vlp proteins are not the antigens targeted by these functional antibodies, but rather serve in a masking role. A notable feature of this system is that any of the distinctive Vlp proteins can confer protection, but only if its length is extended through multiple repeats (a capability embedded in the structure of all Vlp proteins). Analysis of escape variants surviving growth-inhibiting antibody revealed rapid selection for long variants. Survivors from selected susceptible populations appear to have elongated regions of expressed proteins, or have switched ON a transcriptionally silent gene encoding a longer alternative product. In contrast to other forms of immunological escape, this effect appears to be mediated through the protective structural properties of Vlp coat proteins rather than their antigenic diversification (which is in fact limited to only seven Vlp variants; see previous section). No obvious advantage of short forms is apparent, although the capacity to achieve both extremes seems to be the selected feature. In this regard it is also notable, and surprising, that the relative abundance on the cell of a long Vlp variant is actually much lower than that of its shorter counterpart (13). The intriguing structural and topological features of Vlps and other mycoplasmal coat proteins remain to be determined.

## 6.3    Posttranslational modification and variable release of exoproteins

The demonstrated spatial effects that a variable gene product may confer on the mycoplasma surface, either as a general property or through specific interaction with other surface components, are examples of phase variable phenotypes that involve more than the mere primary gene product. Extending this concept is the conjecture that some observed variable products i) may be capable of posttranslationally modifying another component or ii) may occur as a result of being subject to posttranslational modification. Recent evidence of variable phenotypes arising from posttranslational processing is therefore noteworthy in this regard. The MALP gene product of *M. fermentans* is encoded by the single copy *malp* gene and can occur as a membrane anchored lipoprotein of 404 amino acids, or as a lipopeptide comprising the N-terminal 14 residues of the protein (10). Interest in these products stems from the presence of a distinctive SLA (selective lipoprotein associated) motif and the mapping of other putative functions within the full-length form, and the potent Toll-like receptor 2 mediated macrophage stimulatory activity associated with the lipopeptide. Strains and isolates expressing these different forms have been observed, yet show no differences in genes or transcripts that would account for the phenomenon. Recently, the posttranslational processing of this full-length lipoprotein has been shown to occur through site specific cleavage between residues 14 and 15 of the mature lipoprotein. This results in a surface-bound lipopeptide and release of the entire remaining C-terminal portion as an abundant, stable, soluble product bearing the SLA motif (18). Initial observations of intraclonal variation of these phenotypes (Davis, K., Kim, M. and Wise, unpublished) raises the interesting possibility that variable processing of this component may confer the variable phenotype, through a mechanism not yet fully revealed. The presence of *malp* orthologs in several mycoplasmas (including all four whose genomes have been fully sequenced) further prompts speculation that "secretion" of these exoproteins, as well as corresponding surface phenotypes, may be subject to variation through posttranslational processing. It is becoming clear that posttranslational interactions among specific surface components has the potential to amplify the range of products and functions affected by phase or size variable gene products. The combined technologies for post-genomic analysis may be pivotal in unravelling these complex systems, using mycoplasmas as model organisms with the simplest gene sets.

## ACKNOWLEDGMENTS

This work was supported in part by: Research Grant Award No. IS-2540-95R from BARD, The United States-Israel Binational Agricultural Research and Development Fund (DY); the German-Israeli Foundation for Scientific Research and Development-GIF (DY); the Israel Academy of Sciences and Humanities (DY); the Rural Industries Research and Development Corporation, the Australian Research Council, Bioproperties (Australia) Pty Ltd and the National Health and Medical Research Council (GFB). PHHS grants AI32219 (KSW) and AI31656 (KSW) from the National Institute of Allergy and Infectious Diseases.

## REFERENCES

1. **Alen, C., D. J. Sherratt, and S. D. Colloms.** 1997. Direct interaction of aminopeptidase A with recombination site DNA in Xer site-specific recombination. EMBO J. **16**:5188-5197.
2. **Baseggio, N., M. D. Glew, P. F. Markham, K. G. Whithear, and G. F. Browning.** 1996. Size and genomic location of the pMGA multigene family of *Mycoplasma gallisepticum*. Microbiology **142**:1429–1435.
3. **Bath, J., D. J. Sherratt, and S. D. Colloms,** 1999. Topolgy of Xer recombination on catenanes produced by Lambda integrase. J. Mol. Biol. **289**:873-883.
4. **Behrens, A., M. Heller, H. Kirchhoff, D. Yogev, and R. Rosengarten.** 1994. A family of phase-and size-variant membrane surface lipoprotein antigens (Vsps) of *Mycoplasma bovis*. Infect. Immun. **62**:5075-5084.
5. **Bhugra, B., and K. Dybvig.** 1992. High-frequency rearrangements in the chromosome of *Mycoplasma pulmonis* correlate with   phenotypic switching. Mol. Microbiol. **6**:149-1154.
6. **Bhugra, B., L. L. Voelker, N. Zou, H. Yu, and K. Dybvig.** 1995. Mechanism of antigenic variation in *Mycoplasma pulmonis*: interwoven, site-specific DNA inversions. Mol. Microbiol. **18**:703-714.
7. **Boesen, T., J. Emmerson, L. T. Jensen, S. A. Ladefoged, P. Thorsen, S. Birkelund, and G. Christiansen.** 1998. The *Mycoplasma hominis vaa* gene displays a mosaic gene structure. Mol. Microbiol. **29**:97-110.

8.    Boguslavsky, S., D. Menaker, Lysnyansky, I. T. Liu, S. Levisohn, R. Rosengarten, M. Garcia, and D. Yogev. 2000. Molecular characterization of the *Mycoplasma gallisepticum pvpA* gene which encodes a putative variable cytadhesin protein. Infect. Immun. **68**: 3956-3964.

9.    Borst, P., and D. R. Greaves. 1987. Programmed gene rearrangements altering gene expression. Science. **235**:658-667.

10.   Calcutt, M. J., M. .F. Kim, A.B. Karpas, P.F.Mueand K.S. Wise. 1999. Differential posttranslational processing confers intraspecies variation of a major surface lipoprotein and a macrophage-activating lipopeptide of *Mycoplasma fermentans* Infect. Immun. **67**: 760-771.

11.   Chambaud, I., R. Heilig, S. Ferris, V. Barbe, D. Samson, F. Galisson, I. Moszer, K. Dybvig, H. Wroblewski, A. Viari, E. P. C. Rocha, and A. Blanchard. 2001. The complete genome sequence of the murine respiratory pathogen *Mycoplasma pulmonis*. Nucleic Acids Res. **29**:2145-2153.

12.   Christiansen, G., S. L. Mathiesen., C. Nyvold, and S. Birkelund. 1994. Analysis oF a *Mycoplasma hominis* membrane protein, P120. FEMS Microbiol. Lett. **121**:121-128.

13.   Citti, C., and K. S. Wise. 1995. *Mycoplasma hyorhinis vlp* gene transcription: critical role in phase variation and expression of surface lipoproteins. Mol. Microbiol. **18**:649–660.

14.   Citti, C., M. F. Kim, and K. S. Wise. 1997. Elongated versions of Vlp surface lipoproteins protect *Mycoplasma hyorhinis* escape variants from growth-inhibiting host antibodies. Infect.Immun. **65**:1773-1785.

15.   Citti, C., Watson-McKown, Droesse, M, and K. S. Wise. 2000. Gene families encoding phase- and size- variable surface lipoproteins of *Mycoplasma hyorhinis*. J. Bacteriol. **182**:1356-1363.

16.   Colloms, S. D., R. McCulloch., K. Grant., L.Neilson, and D. J. Sherratt. 1996. Xer-mediated site-specific recombination *in viro*. EMBO J. **15**:1172-1181.

17.   Dallo, S. F., A. Chavoya and J. B. Baseman. 1990. Characterization of the gene for a 30-kilodalton adhesin-related protein of *Mycoplasma pneumoniae*. Infect Immun. **58**:4163-4165.

18.   Davis, K. L. and K. S. Wise. 2002. Site-specific proteolysis of the MALP-404 lipoprotein determines the release of a soluble selective lipoprotein associated motif-containing fragment and alteration of the surface phenotype of *Mycoplasma fermentans*. Infect. Immun. **70 (3)**:1129-1135.

19.   Dorigo-Zetsma, J. W., B. Wilbrink, J. Dankert and S. A. J. Zaat. 2001. *Mycoplasma pneumoniae* P1 type 1- and type 2-specific sequences within the P1 cytadhesin gene of individual strains. Infect. Immun. **69**:5612-5618.

20.   Dybvig, K., J.Alderete, and G. H. Cassell. 1988. Adsorption of *Mycoplasma pulmonis* virus P1 to host cells. J. Bacteriol. **170**:4373-4375.

21.   Dybvig, K., J. K. Simecka, H. L.Watson, and G. H. Cassell. 1989. High-frequency variation in *Mycoplasma pulmonis* colony size. J. Bacteriol. **171**:5165-5168.

22.   Dybvig, K. 1990. Mycoplasmal genetics. Annu. Rev. Microbiol. **44**: 81-104.

23.   Dybvig, K. 1993. DNA rearrangements and phenotypic switching in prokaryotes. Mol. Microbiol. **10**:465-471.

24.   Dybvig, K., and L. L. Voelker. 1996. Molecular Biology of Mycoplasmas. Annu. Rev. Microbiol. **50**:25-57.

25.   Ferris, S., H. L. Watson, O. Neyrolles, L. Montagnier, and A. Blanchard. 1995. Characterization of a major *Mycoplasma penetrans* lipoprotein and its gene. FEMS Microbiol. Lett. **130**:313-320.

26.     **Fraser, C. M., J. D. Gocayne, O. White, M. D. Adams, R. A. Clayton, R. D. Fleischmann, C. J. Bult, A. R. Kerlavage, G. Sutton, J. M. Kelley, J. L. Fritchman, J. F. Weidman, K. V. Small, M. Sandusky, J. Fuhrmann, D. Nguyen, T. R. Utterback, D. M. Saudek, C. A. Phillips, J. M. Merrick, J. F. Tomb, B. A. Dougherty, K. F. Bott, P. C. Hu, T. S. Lucier, S. N. Peterson, H. O. Smith, C. A. Hutchison, III, and J. C. Venter.** 1995. The minimal gene complement of *Mycoplasma genitalium*. Science **270**:397-403.

27.     **Glasgow, A. C., K. T. Hughes, and M. I. Simon.** 1989. Bacterial DNA inversion systems, p. 637-659. *In* E. Berg and M. M. Howe (ed.), Mobile DNA. American Society for Microbiology, Washington, D.C.

28.     **Glass, J. I., E. J. Lefkowitz, J. S. Glass, C. R. Heiner, E. Y. Chen, and G. H. Cassell.** 2000. The complete sequence of the mucosal pathogen *Ureaplasma urealyticum*. Nature **407**:757-762.

29.     **Glew, M. D., P. F. Markham, G. F. Browning, and I. D. Walker.** 1995. Expression studies on four members of the pMGA multigene family in *Mycoplasma gallisepticum*. Microbiology 141:3005–3014.

30.     **Glew, M. D., P. F. Markham, G. F. Browning, and I. D. Walker.** 1998. Variation in the GAA trinucleotide repeat lengths within the non-coding regions of *Mycoplasma gallisepticum* pMGA genes controls their expression. Infect. Immun. **66**:5833–5841.

31.     **Glew, M. D., G. F. Browning, P. F. Markham, and I. D. Walker.** 2000. pMGA phenotypic variation in *Mycoplasma gallisepticum* occurs *in vivo* and is mediated by trinucleotide repeat length variation. Infect. Immun. **68**:6027–6033.

32.     **Goh, M. S., T. S. Gorton, M. H. Forsyth, K. E. Troy, and S. J. Geary.** 1998. Molecular and biochemical analysis of a 105 kDa *Mycoplasma gallisepticum* cytadhesin (GapA). Microbiology 144:2941-2950.

33.     **Guhathakurta, A., I. Viney, and D. Summers.** 1996. Accessory proteins impose site selectivity during ColE1 dimer resolution. Mol. Microbiol. **20**:613-620.

34.     **Henrich, B., Feldmann., and U. Hadding.** 1993. Cytoadhesins of *Mycoplasma hominis*. Infect. Immun. **61**:2945-2951.

35.     **Henrich, B., A. Kitzerow, R. C. Feldmann, H. Schaal, and U. Hadding.** 1996. Repetitive elements of the *Mycoplasma hominis* adhesin p50 can be differentiated by monoclonal antibodies. Infect.Immun. **64**:4027-4034.

36.     **Himmelreich, R., H. Hilbert, H. Plagens, E. Pirkl, B. C. Li, and R. Herrmann.** 1996. Complete sequence analysis of the genome of the bacterium *Mycoplasma pneumoniae*. Nucleic Acids Res. **24**:4420–4449.

37.     **Hnatow, L. L., C. L. Keeler Jr, L. L. Tessmer, K. Czymmek and J. E. Dohms.** 1998. Characterization of MGC2, a *Mycoplasma gallisepticum* cytadhesin with homology to the *Mycoplasma pneumoniae* 30-kilodalton protein P30 and *Mycoplasma genitalium* P32. Infect. Immun. **66**:3436-3442.

38.     **Hsu, T. and C. Minion.** 1998. Identification of the cilium binding epitope of the *Mycoplasma hyopneuminiae* P97 adhesin. Infect.Immun. **66**:4762-4766.

39.     **Inamine, J. M., S. Loechel., A. M. Collier., R.F. Barile, and P. C. Hu.** 1989. Nucleotide sequence of the *MgPa* (*mgp*) operon of *Mycoplasma* genitalium and comparison to the P1 (*mpp*) operon of *Mycoplasma pneumoniae*. Gene **82**:259-267.

40.     **Keeler, C. L. Jr, L. L. Hnatow, P. L. Whetzel and J. E. Dohms.** 1996. Cloning and characterization of a putative cytadhesin gene (mgc1) from *Mycoplasma gallisepticum*. Infect. Immun. 64:1541-7.

41.     **Kitten, T., A. V. Barrera, and A. G. Barbour.** 1993. Intragenic recombination and a chimeric outer membrane protein in the relapsing fever agent *Borrelia hermsii*. J. Bacteriol. **175**:2516-2522.

42.    **Komano Teruya.** 1999. Shufflons: Mutiple inversion systems and integrons. Ann. Rev. Gen. **33:** 171-191.

43.    **Leigh, S. A. and Wise, K. S.** *Mycoplasma fermentans* P29 surface lipoprotein is an adhesin. Abstr. 100th Ann. Meet. Amer. Soc. Microbiol. Abstract G-12, 344. 2000.

44.    **Leigh, S. A.** 2002. Functional analysis of the *Mycoplasma fermentans* P29 adhesin. Ph.D. Thesis. University of Missouri-Columbia, Columbia, MO. Submitted.

45.    **Liu, L., K. Dybvig, V. S. Panangala, V. L. van Santen and C. T. French.** 2000. GAA trinucleotide repeat region regulates M9/pMGA gene expression in *Mycoplasma gallisepticum.* Infect. Immun. **68:**871-876.

46.    **Lysnyansky, I., R. Rosengarten, and D. Yogev.** 1996. Phenotypic switching of variable surface lipoproteins in *Mycoplasma bovis* involves high-frequency chromosomal rearrangements. J. Bacteriol. **178:** 5395-5401.

47.    **Lysnyansky, I., K. Sachse, R. Rosenbusch, S. Levisohn, and D. Yogev.** 1999. The *vsp* locus of *Mycoplasma bovis*: gene organization and structural features. J. Bacteriol. **181:**5734-5741.

48.    **Lysnyansky, I., Y. Ron, and D. Yogev.** 2001. Juxtaposition of an active promoter to *vsp* genes via site-specific DNA inversions generates antigenic variation in *Mycoplasma bovis.* J. Bacteriol. **183:**5698-5708.

49.    **Lysnyansky, I., Y. Ron., K. Sachse, and D. Yogev.** 2001. Intrachromosomal recombination within the *vsp* locus of *Mycoplasma bovis* generates a chimeric variable surface lipoprotein antigen. Infect. Immun. **69:**703-3712.

50.    **Markham, P. F., M. D. Glew, M. R. Brandon, I. D. Walker, and K. G. Whithear.** 1992. Characterization of a major hemagglutinin protein from *Mycoplasma gallisepticum.* Infect. Immun. **60:**3885–3891.

51.    **Markham, P. F., M. D. Glew, K. G. Whithear, and I. D. Walker.** 1993. Molecular cloning of a member of the gene family that encodes pMGA, a hemagglutinin of *Mycoplasma gallisepticum.* Infect. Immun. **61:**903–909.

52.    **Markham, P. F., M. D. Glew, J. E. Sykes, T. R. Bowden, T. D. Pollocks, G. F. Browning, K. G. Whithear, and I. D. Walker.** 1994. The organisation of the multigene family which encodes the major cell surface protein, pMGA, of *Mycoplasma gallisepticum.* FEBS Lett. **352:**347–352.

53.    **Markham, P. F., M. D. Glew, G. F. Browning, K. G. Whithear, and I. D. Walker.** 1998. Expression of two members of the pMGA gene family of *Mycoplasma gallisepticum* oscillates and is influenced by pMGA-specific antibodies. Infect. Immun. **66:**2845–2853.

54.    **Markham, P. F, M. F. Duffy, M. D. Glew and G. F. Browning.** 1999. A gene family in *Mycoplasma imitans* closely related to the pMGA family of *Mycoplasma gallisepticum.* Microbiology. **145:**2095-2103.

55.    **Mernaugh, G. R., S.F. Dallo, S. C. Holt, and J. B. Baseman.** 1993. Properties of adhering and nonadhering populations of *Mycoplasma genitalium.* Clin. Infect. Dis. **17** (Suppl. 1) S69- S78.

56.    **Meyer, T. F., C. P. Gibbs, and R. Haas.** 1990. Variation and control of protein expression in *Neisseria.* Annu. Rev. Microbiol. **44:**451-477.

57.    **Moskowitz, I.P.G., K. A. Heichman, and R. C. Johnson.** 1991. Alignment of recombination sites in Hin-mediated site-specific DNA recombination. Genes & Development **5:**1635-1645.

58.    **Noormohammadi, A.H., Markham, P.F., Whithear, K.G., Walker, I.D., Gurevich, V.A., Ley, D.H., and G. F. Browning.** 1997 *Mycoplasma synoviae* has two distinct phase-variable major membrane antigens, one of which is a putative hemagglutinin. Infect. Immun. **65:**2542-2547.

59. **Noormohammadi, A.H., Markham, P.F., Duffy, M.F., Whithear, K.G., and Browning, G.F.** 1998. Multigene families encoding the major hemagglutinins in phylogenetically distinct mycoplasmas. Infect. Immun. **66**:3470-3475.

60. **Noormohammadi, A. H., P. F. Markham, A. Kanci, K. G. Whithear and G. F. Browning.** 2000. A novel mechanism for control of antigenic variation in the haemagglutinin gene family of *Mycoplasma synoviae*. Mol. Microbiol. **35**:911-923.

61. **Nussboum, S., I. Lysnyansky, K. Sachse, S. Levisohn and D. Yogev.** 2002. Extended Repertoire of Genes Encoding Variable Surface Lipoproteins (Vsps) in *Mycoplasma bovis* Strains Infect. Immun. **70: In Press.**

62. **Nyvold, C., S. Birkelund, and G. Christiansen.** 1997. The *Mycoplasma homonis* PG21 contains multiple directly repeated sequences. Infect. Immun. **63**:212-223.

63. **Olson, L.D., S.W. Shane, A.A. Karpas, T.M.. Cuningham, P.S, Probst and M.F. Barile.** 1991. Monoclonal antibodies to surface antigens of a pathogenic *Mycoplasma hominis* strain. Infect. Immun. **59**:1683-1689.

64. **Pays, E.** 1989. Pseudogenes, chimaeric genes and the timing of antigen variation in African trypanosomes. Trends. Genet. **5**:389-391.

65. **Peterson, S. N., C. C. Bailey, J. S. Jensen, M. B. Borre, E. S. King, K. F. Bott, and C. A. Hutchison, III.** 1995. Characterization of repetitive DNA in the *Mycoplasma genitalium* genome: possible role in the generation of antigenic variation. Proc. Natl. Acad. Sci. USA **92**:11829-11833.

66. **Petes, T. D., and C. W. Hill.** 1988. Recombination between repeated genes in microorganisms. Annu. Rev. Genet. **22**:147-168.

67. **Rainey, P. B., E. R. Moxon, and I. P. Thompson.** 1993. Intraclonal polymorphism in bacteria. Adv. Microb. Ecol. **13**:263-300.

68. **Razin, S., D. Yogev, and Y. Naot.** 1998. Molecular biology and pathogenicity of mycoplasmas. Microbiol. Mol. Biol. Rev. **62**:1094-1156.

69. **Reddy, S. P., W. G. Rasmussen, and J. B. Baseman.** 1995. Molecular cloning and characterization of an adherence-related operon of *Mycoplasma genitalium*. J. Bacteriol. **177**:5943-5951.

70. **Robertson, B. D., and T. F. Meyer.** 1992. Antigenic variation in bacterial pathogens, p. 61-73. *In* C. W. Hormaeche, C. W. Penn and C. J. Smyth (ed.), Molecular Biology of Bacterial Infection, Vol. 49. Cambridge University Press.

71. **Röske, K., A. Blanchard, I. Chambaud, C. Citti, J. H. Helbig, M. C. Prevost, R. Rosengarten, and E. Jacobs.** 2001. Phase variation among major surface antigens of *Mycoplasma penetrans*. Infect.Immun. **69**:7642-7651.

72. **Rosengarten, R. and K. S. Wise.** 1990. Phenotypic switching in mycoplasmas: Phase variation of diverse surface lipoproteins. Science. **247**:315-318.

73. **Rosengarten, R. and K. S. Wise.** 1991. The Vlp system of *Mycoplasma hyorhinis*: combinatorial expression of distinct size variant lipoproteins generating high-frequency surface antigenic variation. J. Bacteriol. **173**:4782-4793.

74. **Ruland, K., R. Himmelreich, and R. Herrmann.** 1994. Sequence divergence in the ORF6 gene of *Mycoplasma pneumoniae*. J. Bacteriol. **176**:5302-5209.

75. **Sachse, K., J. H. Helbig, I. Lysnyansky, C. Grajetzki, W. Muller, E. Jacobs, and D. Yogev.** 2000. Epitope mapping of immunogenic and adhesive structures in repetitive domains of *Mycoplasma bovis* variable surface lipoproteins. Infect. Immun. **68**:680-687.

76. **Saunders, J. R.** 1986. Genetic basis of phase and antigenic variation in bacteria, p. 57-76. *In* T. H. Birkbeck and C. W. Penn (ed.), Antigenic Variation in Infectious Diseases. Society for General Microbiology, IRL Press, Oxford.

77.    **Seifert, H. S., and M. So.** 1988. Genetic mechanisms of bacterial antigenic variation. Microbiol. Rev. **52**:327-336.
78.    **Shen, X., J. Gumulak, H, Yu, C. Todd French, N. Zou, and K. Dybvig.** 2000. Gene rearrangements in the *vsa* locus of *Mycoplasma pulmonis*. J. Bacteriol. **182**: 2900-2908.
79.    **Simecka, J. W., J. K. Davis, M. K. Davidson, S. E. Ross, C. T. K.-H. Stadtländer, and G. H. Cassell.** 1992. Mycoplasma diseases of animals, p. 391-416. *In* J. Maniloff, R. N. McElhaney, L. R. Finch and J. B. Baseman (ed.), Mycoplasmas: Molecular Biology and Pathogenesis. American Society for Microbiology, Washington, D.C.
80.    **Simmons, W. L., C. Zuhua, J.I. Glass, J.W. Simecka, G. H. Cassell, and H.L. Watson.** 1996. Sequence analysis of the chromosomal region around and within the V-1-encoding gene of *Mycoplasma pulmonis*: Evidence for DNA inversion as a mechanism for V-1 variation. Infect. Immun. **64**:472-479.
81.    **Smith, G. R.** 1994. Hotspots of homologous recombination. Experientia. **50**:234-241.
82.    **Stark, W. M., M. R. Boocock, and D. J. Sherratt.** 1992. Catalysis by site-specific recombinases. Trends Genet. **8**:432-439.
83.    **Theiss, P., A. Karpas, and K. S. Wise.** 1996. Antigenic topology of the P29 surface lipoprotein of *Mycoplasma fermentans*: differential display of epitopes results in high-frequency phase variation. Infect.Immun. **64**:1800-1809.
84.    **Theiss, P. and K. S. Wise.** 1997. Localized frameshift mutation generates selective, high-frequency phase variation of a surface lipoprotein encoded by a mycoplasma ABC transporter operon. J. Bacteriol. **179**:4013-4022.
85.    **Veal, A. D., H. W. Stokes and G. Daggard.** 1992. Genetic exchange in natural microbial communities. P. 383-430. *In* K. C. Marshall. (ed.), Advances in Microbial Ecology. Vol. 12. Plenum Press, New York.
86.    **Washburn, L. R., K. E. Weaver, E. J. Weaver, W. Donelan and S. Al-Sheboul** 1998. Molecular characterization of *Mycoplasma arthritidis* variable surface protein MAA2. Infect. Immun. **66**:2576–2586.
87.    **Washburn, L. R., E. J. Miller, and K. E. Weaver.** 2000. Molecular characterization of *Mycoplasma arthritidis* membrane lipoprotein MAA1. Infect. Immun. **68**:437-442.
88.    **Watson, H. L., K. Dybvig, D. K. Blalock, and G. H. Cassell.** 1989. Subunit structure of the variable V-1 antigen of *Mycoplasma pulmonis*. Infect. Immun. **57**:1684-1690.
89.    **Wise, K. S., D. Yogev, and R. Rosengarten.** 1992. Antigenic Variation. p. 473-489. *In* J. Maniloff, R. N. McElhaney, L. R. Finch and J. B. Baseman (ed.), Mycoplasmas: Molecular Biology and Pathogenesis. American Society for Microbiology, Washington, D.C.
90.    **Wise, K. S.** 1993. Adaptive surface variation in mycoplasmas. Trends Microbiol. **1**:59-63.
91.    **Yogev, D., R. Rosengarten, R. Watson-McKown, and K. S. Wise.** 1991. Molecular basis of Mycoplasma surface antigenic variation: a novel set of divergent genes undergo spontaneous mutation of periodic coding regions and regulatory sequences. EMBO J. **10**:4069–4079.
92.    **Yogev, D., Watson-McKown, R. Rosengarten, J. Im, and K. S. Wise.** 1995. Increased structural and combinatorial diversity in an extended family of genes encoding Vlp surface proteins of *Mycoplasma hyorhinis*. J. Bacteriol. **177**: 5636-5643.

93.    **Zhang, Q. and K. S. Wise**. 1996. Molecular basis of size and antigenic variation of a *Mycoplasma hominis* adhesin encoded by divergent *vaa* genes. Infect. Immun. **64**:2737-2744.

94.    **Zhang, Q. and K. S. Wise**. 1997. Localized reversible frameshift mutation in an adhesin gene confers a phase-variable adherence phenotype in mycoplasma. Mol. Microbiol. **25**:859-869.

95.    **Zhang, Q, and K. S. Wise**. 2001. Coupled phase-variable expression and epitope masking of selective surface lipoproteins increases surface phenotypic diversity in *Mycoplasma hominis*. Infect. Immun. **69**:5177-5181.

# Chapter 20

# Immunomodulation by Mycoplasmas: Artifacts, Facts and Active Molecules

PETER F. MÜHLRADT
*Immunobiology Research Group, Gesellschaft für Biotechnologische Forschung mbH, Mascheroderweg 1, D-38124 Braunschweig, Germany*

## 1.     MISHAPS AND MISINTERPRETATIONS

One of the earliest reports on mycoplasma-mediated immunomodulation was a study concerning the toxicity of heat-killed mycoplasmas[28]. As is now known, bacterial toxicity, if not caused by exotoxins, is the result of excessive stimulation of the innate immune system through the so-called bacterial modulins; these are mostly components of the cell wall[35] and, possibly, pieces of nucleic acids containing the motif CpG[51]. Cells of the innate immune system, in particular of the monocyte/macrophage lineage, react to these components by release of proinflammatory cytokines and IFN-$\gamma$, which in turn can lead to shock and death[116]. When the observations of Gabridge and Murphy[28] were made, knowledge about these potent mediators of various cellular responses was nonexistent and the data were uninterpretable in modern terms. However, their report is a good example of a study which intentionally dealt with mycoplasmas. Many more studies were published, and some until quite recently, in which immunomodulation due to the presence of mycoplasmas was unknowingly observed and reported.

It is curious that many researchers are aware of the immunomodulatory effects of the lipopolysaccharide (LPS) endotoxin, and therefore they take utmost care to avoid contamination with this compound. In contrast, the interference by mycoplasmas and their products with immunological studies is less commonly appreciated, although a warning came quite early (see review by Ruuth and Praz,[92]) and even effective remedies were offered[98].

*Molecular Biology and Pathogenicity of Mycoplasmas*, Edited by Razin and
Herrmann, Kluwer Academic/Plenum Publishers, New York, 2002

Institutions where cell lines are deposited estimate that about one third, if not more, of the deposited lines are mycoplasma-contaminated. This means that in any immunological studies with such lines, the "conditioned culture media", or other products from these lines, e.g. monoclonal antibodies, even if sterile, will lead to ill-defined results. Since the realization that many cells of the immune system interact with one another by means of soluble mediators and cell to cell contact, immunologists have tried to use well-characterized cell populations in order to define more explicitly their experimental system. The use of established cell lines is one comfortable way of doing this, but apart from the fact that such lines do not necessarily mirror the situation of a primary cell, there is always the danger of a contamination with mycoplasmas or their remnants, in particular if the lines have been passed among colleagues from one laboratory to another. Because both live mycoplasmas and their cellular components are immunologically highly active, it does not suffice to kill these microorganisms but it is advisable to use mycoplasma-free cell cultures to start with. A survey of the methods of detecting and eliminating mycoplasma contamination is beyond the scope of this chapter[49].

In this section I will select a few examples for studies with unintentional mycoplasma contaminations and the discovery of interesting effects because of them. My colleagues whose work I shall quote need not fear that I do this out of pure malicious glee. In fact my own interest in mycoplasmas began through studies with a mycoplasma-contaminated cell line and the resulting observations (see below). Out of pure luck, we published our data after having found out about this contamination. The point here is to show how various experimental systems can be influenced by mycoplasma contamination.

## 1.1    Mycoplasmas and malaria

A number of reports appeared (e.g. by Allan and Kwiatkowski[1]) which described the capacity of a particular, internationally widely-spread clone of *Plasmodium falciparum* (R29) to stimulate TNF in human monocytes. Biologically, these findings held promise, because malaria infections are associated with heavy attacks of fever, and TNF is one of the endogenous pyrogens[86]. However, early lines from clone R29 proved to be infected with *Mycoplasma orale*, and others of later origin, with *Mycoplasma arginini*. The earlier described ability of clone R29 to stimulate TNF thus proved to be due to the presence of these mycoplasmas[89].

## 1.2    Mycoplasmas and natural killer (NK) cells

When the possible involvement of soluble cytotoxic factors in the lytic mechanism of NK cell-mediated cytotoxicity was investigated, such factors were detected only when effector cells were stimulated with mycoplasma-contaminated target cells or with concanavalin A. Mycoplasma-free target cells were ineffective[39].

## 1.3    Mycoplasmas and T cell glycolipids

In a search of marker glycolipids for human helper T cells, the occurrence of novel phosphocholine-containing glycoglycerolipids in a human helper T-cell line was reported. These glycolipids turned out to be a lipid component of *M. fermentans* comprising a major immunological determinant[55].

## 1.4    Mycoplasmas and "lymphokines" which activate B cells to proliferate and produce immunoglobulins

Two groups observed at about the same time that T cell hybridomas or T cell lines which were contaminated with *M. hyorhinis* [78] or *M. arginini* [91,93], respectively, contained a lymphokine-like activity in their conditioned media which resembled "B cell stimulating factor", an activity which is probably a mixture of IL-4 and IL-6.

## 1.5    Mycoplasmas and a differentiation-inducing cytokine

P48 was an activity detected in the supernatant of the human null cell leukemia cell line Reh, which induced differentiation and cytolytic activity in HL-60 cells[4]. The cDNA for P48 was cloned from Reh cellular RNA. Subsequent studies using PCR and Southern analysis revealed P48 sequences in the DNA isolated from *M. fermentans* [33].

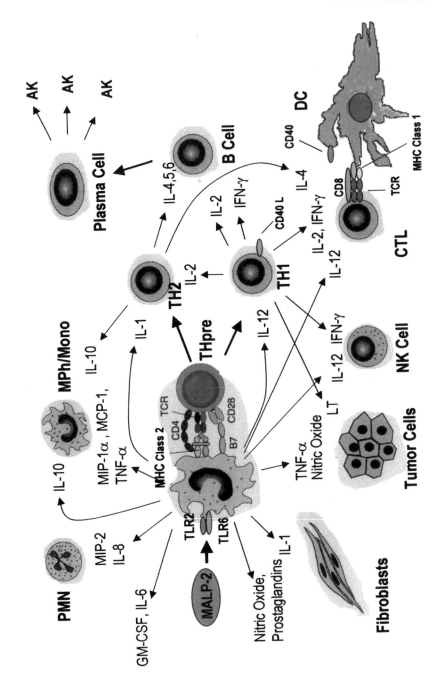

*Figure 1.* Interactions of immune cells. A macrophage activated by MALP-2 through its receptor pair TLR2 and TLR6 is shown in close contact with a naïve T cell at the left center. This T cell can differentiate into a T helper cell type 1 or 2. As indicated by the arrows, these cells influence each other as well as further bystander cells through soluble cytokines or other mediators. Abbreviations: CD = cluster of differentiation, CTL = cytolytic T lymphocyte, DC = dendritic cell, IL = interleukin, IFN = interferon, LT = lymphotoxin, MCP = monocyte chemoattractant protein, MHC = major histocompatibility antigen, MIP = macrophage inflammatory protein, MPh/Mono = macrophage or monocyte, NK = natural killer, PMN = polymorphonuclear phagocyte, TCR = T cell receptor, TLR = toll-like receptor, TNF = tumor necrosis factor.

## 1.6 Mycoplasmas and nonself discrimination of malignant cells

The recently described antigen M161, which activates human complement, is apparently expressed only by malignant cells and cell lines exposed to Fas or X-irradiation stimuli[56,57]. Further work showed that genomic and cDNA clones of M161Ag were highly homologous to a *M. fermentans* gene encoding P48[58].

## 1.7 Mycoplasmas and a cytotoxicity inducing activity

In a search for lymphokines from various cell lines a new "activity" was discovered which led to increased formation of cytotoxic T lymphocytes from Concanavalin A-stimulated thymocytes. This activity was found to be derived from contaminating mycoplasmas and was named MDHM for mycoplasma derived high mol. wt. material. MDHM was found to act on CTL precursors indirectly through stimulation of macrophages and T helper cells, involving the synthesis of IL-1, IL-6, IL-2, and IL-4[79,66,80].

Interestingly, the activities or effects mentioned in the three last examples were due to or derived from the same lipoprotein, the MALP-404, recently characterized and cloned from *M. fermentans*[15].

The preceding examples are only a few that show the effects of mycoplasmas and their components in very different experimental systems ranging from parasitology to tumor research and "pure" immunology. It may be appropriate to diverge here for a moment and look at a simplified model of cell interactions involved in immunological responses, and at the intricate interrelation between the innate and the acquired immune system, before we offer explanations for some of the more recent observations and comment on some of the pioneer work dealing with mycoplasmas and their interaction with immune cells in the widest sense.

## 2.    CAUSE AND EFFECTS

Fig. 1 shows a complicated looking but still simplified schematic drawing of the interactions of various cells of the innate and adaptive immune system. Thus, this model does for example not include long range effects such as those of inflammatory cytokines on the bone-marrow or central nervous system. Interactions take place through cell-cell contacts and soluble mediators. For example, antigen presenting cells like dendritic cells (DC) or macrophages (MPh) make contacts with T helper (TH) as well as cytolytic T cells (CTL). The molecules involved are not only the T cell receptor (TCR) and the antigen presenting molecules and the Class I and Class II major histocompatibility antigens (MHC) but also others such as CD4, CD8 and CD40 ligand (CD40 L) on the T cells and B7 and CD40 on the antigen presenting cells, respectively. These molecules impart a certain specificity of cell-cell interaction. Additionally a number of other cell surface components, not indicated in Fig. 1, provide less specifically stickiness and adhesion of various cells involved in immune responses or inflammation.

A second way of cell-cell communication is provided by soluble mediators, some of which, (e.g. TNF or IL-1), may be also cell surface bound. A certain amount of cell specificity is warranted by corresponding receptors for these mediators which may be specifically expressed by only the target cells. Many, but not all, of these mediators are peptides or proteins, so-called lymphokines, cytokines or chemokines. There are mediators that upregulate the function of other partner cells in the response and others that downregulate or shift the activity of their target cells. In addition chemokines, which are chemoattractant molecules, recruit cells to a certain site. Moreover, some of these mediators are indiscriminately produced by many different cells, others, such as IL-4 of IFN-γ, by only a few specialized ones. To make things really complicated, some cytokines produced by one particular cell induce secondary target cells to produce another lymphokine which in turn has effects on a third partner cell. Thus IL-12, produced by macrophages, may induce T helper 1 (TH1) cells to synthesize IFN-γ which in turn activates natural killer (NK) cells. Often there is redundancy in the system so that failure to produce one particular mediator, as it is the case in genetically deficient knock out animals, is not always easily noted, as other cytokines may take its place.

Non-peptide mediators or effector molecules are of mainly two types: the arachidonate metabolites (prostaglandins and leukotrienes) and very small molecules prone to high reactivity and rapid decay, such as active oxygen and nitrogen species. Prostaglandins are involved in the generation of fever and the sensation of pain. Leukotrienes have an effect on the chemotaxis of leukocytes. Formation of arachidonate metabolites is thus of relevance in the context of mycoplasma-generated inflammatory processes. Active oxygen and nitrogen species are effector molecules for the killing of extracellular and intracellular microorganisms and tumor cells respectively[7,19,63,90]. Additionally, nitric oxide is a mediator of blood pressure and a messenger in the central nervous system. Determination of the decay products of nitric oxide is also a very convenient way to measure activation of murine macrophage[90,106] (see e. g. Figs. 3 and 4 below).

Once the enormous complexity of the system is realized, one can see that modulation of the function of one cell will influence those of others. This may become particularly apparent when the cell whose function is modulated is central, for example a macrophage. Thus, stimulation of macrophages by mycoplasmal products, may have effects on cells rather distal in the system, such as plasma cells or cytolytic T cells.

## 2.1 Earlier reports on immunomodulation by mycoplasmas

A wealth of information on immunomodulation by preparations from quite different species of mycoplasmas in various experimental system is available. Some of the early work has been reviewed by Ruuth and Praz[92], and some of the more recent studies by Razin, Yogev and Naot[85]. Apart from the aforementioned studies with the *M. arthritidis* superantigen (see also chapter by Washburn and Cole), many, but not all, of these effects can be ascribed to macrophage activation (compare Fig. 1). This is self-evident in experimental systems in which the production of cytokines such as TNF-α, IL-1, IL-6 or GM-CSF was studied in response to mycoplasmal products[12,13,18,36,44,45,65,66,79,82,83,104,107,109,118,123]. Similarly, macrophage activation is very likely involved when tumoricidal or microbicidal activity is being measured which, depending on the experimental system, may be TNF-α or nitric oxide dependent[52,90,99,117,120]. An interpretation becomes difficult when effects on T or B lymphocytes or NK cells are measured[6,23,71,78,91,93,101,115], because indirect as well as direct effects of mycoplasmal substances on these cells are hypothetically possible (see Fig. 1). In such systems the presence of even a small percentage of macrophages or their precursors (which are poorly adherent but will mature in cell culture) has to be rigorously excluded before it can be claimed that the biologically active substance in question directly acts upon cells other than macrophages. As will be pointed out in the next sections, mycoplasmal lipoproteins are potent macrophage activators and ubiquitously found in mollicutes. They are comparable in activity and equally wide-spread in mollicutes as is LPS in Gram-negative bacteria. And they are as difficult to remove from other mycoplasmal material as is the LPS from components of Gram-negative bacteria. It is thus not surprising to find that many mycoplasmas or materials derived from them behave similarly in comparable immunological tests just as does LPS from different species of most Gram-negatives[9].

## 3.  MYCOPLASMAL LIPOPROTEINS AND LIPOPEPTIDES AS IMMUNOMODULATORY COMPONENTS

### 3.1    Pioneer work

Early results from Cole´s group[2] and that of Kirchner[43] had shown that culture media from *M. arthritidis* contained a material having potent immunostimulatory capacities. The substance responsible turned out to be a superantigen, a subject that will be covered elsewhere in this book (*Mycoplasma arthritidis* Pathogenicity: Membranes, MAM and MAV1 chapter). Other groups had shown in a number of early papers that immunomodulatory substances from mycoplasmas can be demonstrated in heat-killed organisms or in membrane preparations. Thus Rottem and Gallily reported in a series of publications that membranes from spiroplasmas and other mollicutes can stimulate in vitro TNF release, a classical indication of macrophage activation (see e.g. Sher[99]). It took many more years to isolate and characterize the active macrophage activator, and many groups were involved in this work. Thus, several studies demonstrated that immunomodulatory material could be extracted into the organic phase upon Triton X114 partition[23,45,82], a method that Kim Wise introduced into mycoplasmology, and which was used to enrich mycoplasmal lipoproteins[14]. Consequently, it was argued that the active substance, since extractable into the Triton phase, might be of lipoprotein nature[23,45]. This hypothesis was all the more convincing, as it was known from many reports of different groups that bacterial lipoproteins from various species have immunomodulatory, and in particular macrophage activating properties[5,34,60,72,81]. Moreover, such compounds had been synthesized and shown to be biologically active[25,61,62].

### 3.2    Mycoplasmal lipoproteins

That mycoplasmas were capable of synthesizing lipoproteins had been firmly established by biochemical methods, i.e. labelling with fatty acids and cleavage under defined conditions[41,75,122] (see also recent review by

Chambaud et al.[17]). Nyström et al. showed that acholeplasmas synthesize lipoproteins with the fatty acids in ester linkage only[75]. Likewise, *M. fermentans* and *M. hyorhinis* lipoproteins lack the N-acyl modification[67,68] (see Fig. 2). This type of lipoprotein with only two ester-bound fatty acids had previously only been discovered in *Rhodopseudomonas viridis* [121]. It is an unusual feature of bacterial lipoproteins which are commonly substituted by two ester-linked and a third amide-bound fatty acid at the lipid-modified N-terminus (see review by Braun and Wu[10]) to lack the amide-linked fatty acid, and this lack has great consequences for their biological activity (see below). It would be beyond the scope of this chapter to go into the biosynthesis of bacterial lipoproteins. It should be mentioned, though, that according to the studies of the late H. Wu, who investigated the biosynthesis of E. coli lipoproteins, a specific enzyme is required which transfers the third fatty acid to the free N terminus[95]. The gene for this fatty acid transferase was neither found in the genome of *M. pneumoniae* nor in that of *M. genitalium* [37]. The absence of the third fatty acid from the N-terminus may thus be common to many mycoplasma species and not restricted to *M. fermentans*.

However, reports by Jan et al., who studied lipoproteins from *M. gallisepticum*, showed that, like in lipoproteins from eubacteria, mycoplasmal lipoproteins can also be acylated at the N-terminus[41]. Similar findings were recently reported for *M. agalactiae* [47]. The issue, whether lipoproteins in mollicutes are generally carrying only ester-linked fatty acids or can be additionally N-substituted, is therefore still unresolved. There may be still surprises coming from another angle, as there is one report on isoprenylated proteins in acholeplasmas[74], but no information on isoprenylated proteins in other mycoplasma species nor their potential effects on immune cells is known.

## 3.3 MALP-2, the molecule

As mentioned above, we had discovered a "mycoplasma-derived high molecular weight material", MDHM, which was basically a macrophage activator indirectly leading to formation of cytolytic T cells[79]. Attempts to purify this substance were made more convenient by the introduction of the nitric oxide release assay, allowing many samples to be tested[90]. Material was isolated from HPLC which contained only traces of amino acids and displayed very high biological activity[64]. Had this substance been an "ordinary" bacterial lipoprotein, which requires μg/ml concentrations for stimulation, we expected to be able to detect much more protein than we did. Our initial interpretation was, therefore, that we were dealing with a lipid contaminated with some peptide. Only closer analysis of the HPLC-eluted

"MDHM" disclosed that fractions displaying macrophage activating properties indeed contained a glycerol-thioether characteristic for bacterial lipoproteins[67]. Since we still could not explain the extraordinarily high specific activity, we assumed that the presence of unsaturated fatty acids, which are not present in "ordinary" lipoproteins, were responsible for the comparatively much higher activity of the mycoplasmal product. Lack of sufficient material, so familiar to mycoplasma researchers dealing with poorly growing organisms, made it at first impossible to isolate enough material for a more detailed structure elucidation. This could be achieved only after the isolation of clones from *M. fermentans* with exceptionally high macrophage activating activity[68]. The macrophage activator turned out to be a lipopeptide with only two fatty acids and a free amino terminus and a molecular weight of about 2 kDa. The substance was therefore called MALP-2 for macrophage activating lipopeptide of 2 kDa mol. wt. Chemical synthesis by Jung and his group with their enormous experience in lipoprotein synthesis[61,62] confirmed the correct structure of MALP-2; it then became clear that the synthetic MALP-2 with saturated fatty acids was as active as the natural compound containing unsaturated acids[68]. The structure is depicted in Fig. 2.

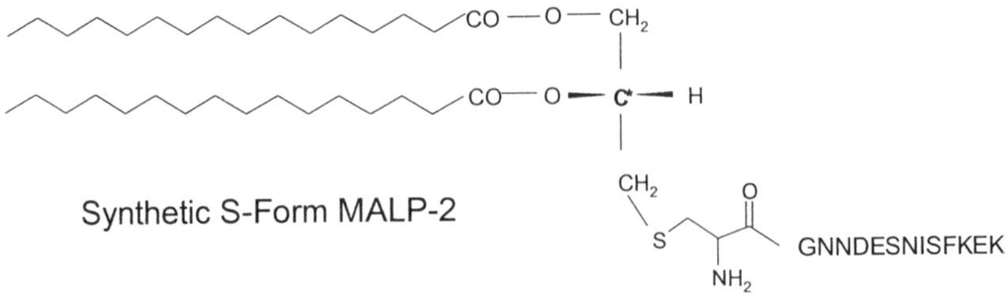

Synthetic S-Form MALP-2

*Figure 2.* The structure of synthetic MALP-2. Note the asymmetric C atom in the lipid moiety.

MALP-2 is derived from a larger lipoprotein, MALP-404, that has been previously described by Wise's laboratory as P41. It was genetically cloned[15] and shown to be very similar or identical to P48, cloned by Hall[33].

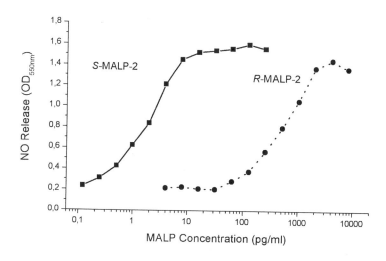

*Figure 3.* Comparison of the macrophage stimulatory activity of the two MALP-2 stereosiomers in the nitric oxide release assay.

## 3.4     Structural details that determine the biological activity of MALP-2

Valuable information on the importance of position and degree of fatty acid substitution of bacterial lipopeptides for their biological activity ensues from earlier studies of Jung et al. with synthetic lipoproteins and lipopeptides, such as that the lack of one ester-bound fatty acid destroys the biological activity[62]. As these data were collected at a time when quantitation of macrophage activation was difficult, we resumed and extended these experiments using the nitric oxide release assay for an accurate measurement of macrophage activation. The data are shown in Figs. 3 and 4. Several important conclusions can be derived from these studies: (i) the stereochemistry at the 2-C atom in the lipid moiety (compare Fig. 2) determines the biological activity. Mainly the *S*-form, derived from the natural phosphatidyl glycerol, is biologically active (Fig. 3) [111] (see erratum in the Nov. issue of J. Immunol. 2000). We think it is likely that a minor contamination (less than 1%) of the *R* with the *S*-form is responsible for the remaining activity of the *R*-form. Incidentally, the unsaturated fatty acids in the phospholipids of mycoplasmas are in the 1-position[88], and consequently also in the lipoproteins derived from this precursor[95], whereas in walled bacteria unsaturated fatty acids are more commonly found in the 2-position. Furthermore, as mentioned above, the lack of N-acylation makes at least some mycoplasmal lipoproteins so very active (Fig. 4). Whether the N-terminal amino group is substituted by a short or long chain fatty acid is of little consequence: the specific biological activity decreases by two orders of magnitude after acylation of the N-terminus (Fig. 4). However, removal of one ester-bound fatty acid diminishes the biological activity by a factor of 1000 (see also Metzger et al. [62]). Since macrophages are rich in esterases, this may be one way by which these cells deactivate this highly active molecule.

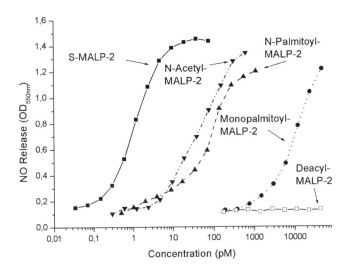

*Figure 4.* The influence of fatty acid substitution on the macrophage stimulatory activity in the nitric oxide release assay.

## 3.5    Structural similarities between macrophage activating lipopeptides and LPS endotoxins

MALP-2, as its name implies, is primarily a macrophage activator. Further macrophage activating lipopeptides were isolated from *M. hyorhinis*. They possess a comparable specific activity and lipid moiety as MALP-2 from *M. fermentans* but their peptide portions are of a different length and composition[69]. This finding and the data from the previous section show that it is the lipid portion which imparts the macrophage stimulatory activity to these molecules. We have thus a structural situation which is similar to that seen in LPS endototoxins from Gram-negative bacteria (reviewed in Brade et al.[9]). In both types of compounds the lipid moiety, in the case of LPS the Lipid A, for mycoplasmal lipoproteins the diacylated N-terminus, is responsible for the macrophage stimulatory activity, whereas the rest of the molecule provides antigenic variation (Fig. 5). As we shall see in the next section, there are not only structural but also great functional similarities between mycoplasma lipoproteins and LPS.

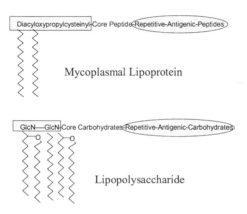

*Figure 5.* Structural similarities between mycoplasmal lipoproteins and LPS.

## 3.6    Biological activities of MALP-2 and other mycoplasmal lipoproteins

Macrophages are the source of many mediators and effector molecules, as indicated in Fig. 1. As appropriate for this volume dealing with mycoplasmas, mycoplasma-derived MALP-2 serves as stimulator in this model. However, for an immunologist mycoplasmal lipopeptides might as well be replaced by LPS (except for the respective receptors, see below). At least in vitro MALP-2 and LPS display much the same activities when added to macrophage/monocyte cultures, and both substances act at similarly low concentrations. Thus proinflammatory cytokines[31,42,111], chemokines[21,42] and prostaglandins are released in response to MALP-2 and other mycoplasmal lipoproteins and lipopeptides[30,65], and, in combination with IFN-γ, also the unstable gas nitric oxide is liberated. Nitric oxide is responsible for killing of tumor target cells or intracellular microorganisms[68,90].

Apart from inducing soluble mediators, mycoplasmal lipoproteins and MALP-2 may regulate the cell surface expression of molecules which are important for interactions between cells of the immune system. Thus several studies show the modulation of molecules involved in antigen presentation. Stuart et al. [102,103] have reported that heat-inactivated mycoplasmas from five species of mycoplasmas can stimulate the expression of MHC class I and class II antigens in the mouse myelomonocytic cell line WEHI-3. In a follow up study it was shown that neither IFN-γ, IL-4 nor GM-CSF were required for this effect[105]. Frisch et al.[27] reported that MALP-2 containing preparations were able first to stimulate and then suppress expression of MHC class II antigen on peritoneal macrophages, resulting in an overall suppression of antigen presentation. Also in this system no soluble mediators were found responsible for this regulation[27]. The expression of another cell surface molecule, the intercellular adhesion molecule-1 (ICAM-1), was also found to be upregulated by mycoplasmal lipoproteins, as reported by Dong et al. [22]. Lastly, MALP-2 appears to upregulate its own receptor (see also below) (Grote and Mühlradt, unpublished). In essence, mycoplasmal lipoproteins may exert quite a number of effects, apart from liberating cytokines, on immune and other cells.

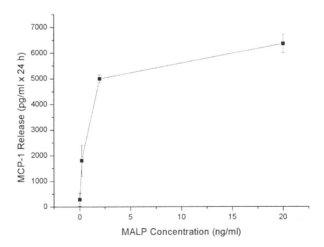

*Figure 6.* MALP-2-dependent release of MCP-1 from murine fibroblasts.

In vivo studies with MALP-2 are scarce, because until recently there was not enough material available. Deiters and Mühlradt[21] have shown that µg quantities of intraperitoneally injected MALP-2, as well as heat-killed mycoplasmas, will lead to an infiltration of leukocytes in mice and the release of proinflammatory cytokines into the serum. When MALP-2 was inserted into liposomes the effect was even stronger[21]. Further data by Deiters and Mühlradt[21] suggested that such infiltration, which is also the hallmark of a natural mycoplasmal infection[50,53,87,114,117], is caused by chemokines which could be identified in the serum of the animals[21]. The most likely source for these chemokines in this experimental model is the peritoneal macrophages. However, fibroblasts also are capable of reacting with MALP-2, releasing the chemokine MCP-1 (Fig. 6).

We had previously shown that membrane preparations of *M. fermentans* were pyrogenic[65]. It was a priori to be expected that MALP-2 would be similarly pyrogenic as LPS, being as potent a macrophage activator as LPS and generating proinflammatory cytokines such as the endogenous pyrogens IL-1 and TNF. Unexpectedly, µg/kg quantities of MALP-2 were required for a febrile response in rabbits (see Fig. 7), whereas less than 10 ng/kg of LPS are pyrogenic[26]. Several explanations are possible, among them higher adsorption rates to erythrocytes upon intravenous injection or, more likely, distinct receptors for MALP and LPS, respectively, which are differently expressed on various cells of the body (see below). Although not as pyrogenic as LPS, MALP-2 is quite toxic when given to mice. Lethal doses are between 100 and 200 µg/animal[29].

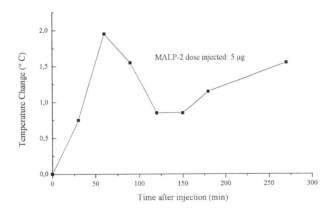

*Figure 7*. Febrile response to an intravenous MALP-2 injection in a 2.75 kg rabbit.

## 3.7    Receptors and signal transduction

Luckily for the experimenter, a mouse mutant, the C3H/HeJ strain, has been known for many years to exhibit exceptionally low susceptibility towards endotoxin[108]. Macrophages and other cells from this mouse do not respond to LPS, while being perfectly responsive to bacterial lipoproteins. Thus, it was feasible to study the effects of MALP-2 and other mycoplasma-components without the constant worry about LPS contamination. The exact nature of this LPS low-responsiveness in molecular terms was clarified only recently by the group of Beutler as being due to a defective LPS signaling receptor[77]. It belongs to a recently discovered novel family of receptors of the innate immune system, the so-called toll-like receptors (TLR) (see recent review[8]) and was defined as TLR4[77]. This finding was subsequently confirmed by the group of Akira[40] with TLR4 knock-out mice.

In general, the availability of knock-out animals with defined genetic deficiencies has made great advances possible in the field of immunology, and studies with TLR knock-out mice, in particular, helped to define the receptors for various bacterial modulins in addition to the LPS receptor. Because TLR4 defective C3H/HeJ mice do respond to MALP-2, it appeared evident that MALP-2 and LPS use different receptors. Indeed, a defect in TLR2 led to unresponsiveness to MALP-2 and other bacterial lipoproteins[38,48,111]. Further studies by Takeuchi et al. showed that a second toll-like receptor, TLR6, is equally important for MALP-2 signaling, though not for stimulation by lipoproteins with three fatty acid substitutents[112]. Consequently, TLR2, in combination with TLR6 and possibly as yet undefined other cell surface components, is vital to MALP-2-mediated signaling. TLR2 is expressed by many cells of the body, however, not necessarily by the same cells as TLR4[16,70]. The same may apply to the expression of TLR6. This implies that different cells may respond to LPS and mycoplasmal lipoproteins. In other words, the observed biological similarities between LPS and mycoplasmal lipoproteins, although very pronounced in cell cultures with monocytes or macrophages, may not be valid when these compounds are used in animal experiments, because the MALP-2 response will be restricted to cells expressing both TLR2 and TLR6, while the LPS response depends on the expression of TLR4. It is possible that, for example, the weaker pyrogenic response to MALP-2 as compared to LPS can be explained on this ground. In addition, different body distribution and clearance rates may contribute to these differences.

One of the endpoints of MALP signaling is the activation of the transcription factor NF-κB. This had been shown by several groups before the TLR-dependent signal pathway was known[24,31,94]. NF-κB binding sites are present on the promoters of many cytokine genes, including TNF-α and IL-6 (see reviews[3,59,100]). The simplified TLR-dependent signal pathway[110], depicted in Fig. 8, focuses on the MALP-2-mediated activation of the transcription factor NF-κB and the activation of genes with binding sites of this transcription factor in their promoters. However, there are other pathways involving several MAP kinases triggered by MALP-2. Thus, an involvement of small GTP binding proteins in MALP signaling were shown by Rawadi et al.[84].

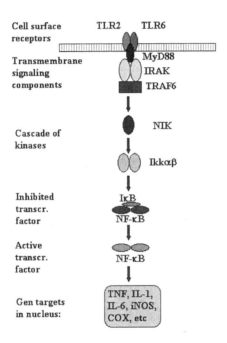

*Figure 8.* Simplified model of the toll-like receptor-mediated signal pathway leading to NF–κB activation. There are other parallel toll-like receptor-mediated MAP kinase pathways leading to additional transcription factors and target genes not shown here. MyD88 is an adaptor molecule that links the receptors to IL-1 receptor associated kinase (IRAK). TNF receptor associated factor (TRAF) 6 is a further adaptor molecule which is necessary for the initiation of a phosphorylation cascade including NF-κB inducing kinase (NIK) and the IκB kinases (Ikkαβ). This results in the phosphorylation and breakdown of IκB, an inhibitory component of NF-κB, thus activating the transcription factor NF-κB.

4.        IMMUNOMODULATION BY MYCOPLASMAL
          PRODUCTS NOT DUE TO MACROPHAGE
          ACTIVATION

## 4.1      Activation of complement

This section is not concerned with classical, antibody-mediated complement activation or fixation but rather with direct activation of complement by mycoplasmas or their products. Such activation and binding of C1 was reported long ago by Bredt et al. [11]. Later, Ziegler-Heitbrock and Burger observed that mycoplasma contaminated cell lines could be freed from mycoplasma infection by incubation with human, rabbit, or guinea pig sera and that this effect was independent of specific antibody but inhibitable by monoclonal antibodies against C3a and C3b. These authors also noted C3b fragments on the surface of mycoplasma-infected cells[124]. Also, Webster et al. reported activation of the first component of complement by mycoplasmas which then bound through complement receptors to neutrophiles[119]. Matsumoto et al. were the first to identify a mycoplasmal lipoprotein from *M. fermentans,* which they called M181Ag, and which is almost identical to MALP-404[15], that activates C3[58]. It is likely that other lipoproteins and lipopeptides show similar activity. Thus, Deiters found that MALP-2 activates complement when lipid-bound but not when in soluble form (doctoral thesis, Braunschweig 1999).

## 4.2      Interaction of mycoplasmal compounds with bone
          organ cultures

The effects of mycoplasmas and their components on bone metabolism is of particular interest since an increasing number of studies report the involvement of mycoplasmas in reactive arthritis[20,32,96,97]. It was reported by Novak et al. that high molecular weight material from *M. hyorhinis* and *M. arthritidis* causes bone resorption in bone organ cultures[73]. The authors postulated that damage to subchondral bone in arthritis associated with mycoplasma infection may be caused by a potent bone resorption inducing agent of mycoplasmal origin. Since this material resembled crude MALP-2 preparations, it was attempted by Piec at al.[76] to replace crude mycoplasma membrane material by pure MALP-2 in this system. Indeed, MALP-2 stimulated osteoclast-mediated bone resorption at sub nannomolar concentrations. MALP-2 could thus be identified as an active component in this system, and it was concluded that MALP-2 or other mycoplasmal lipoproteins might be able to stimulate bone resorption in the arthritogenic joint[76]. Strictly speaking, these effects can also be considered in the context of macrophage activation, as osteoclasts are macrophage related cells.

## 4.3    Interaction of mycoplasmal compounds with cells of the nervous system

Gallily et al., who were primarily interested in attenuation of inflammatory processes in the nervous system by various agents, exposed primary rat glial cells to heat-killed *M. fermentans* to initiate an inflammation. This procedure triggered the release of the proinflammatory cytokine TNF-α prostaglandin $E_2$, and nitric oxide[30]. This finding is also of interest in the context of a mycoplasma – or MALP-2 – induced febrile response: prostaglandins are local endogenous pyrogens.

## 4.4    Effects of MALP-2 and other mycoplasmal compounds on fibroblasts and alveolar type II lung epithelial cells

Fibroblasts, even in the widest sense, are not regarded as immune cells, but may all the same participate in immune reactions, be it that they present viral antigens or participate in inflammatory reactions. We have seen above (Fig. 6) that MALP-2 can induce fibroblasts to liberate MCP-1, a chemokine which is a chemoattractant for macrophages. In the same experiment, these fibroblasts released IL-6 (not shown). Thus, definitely cells other than macrophages can react to MALP-2, taken here as a prototype for mycoplasmal lipoproteins. Another example for reactivity of fibroblasts with mycoplasmal lipoprotein has been reported by Dong et al., who observed transcription of ICAM-1 mRNA in human primary fibroblasts in response to lipoproteins in membranes from *M. salivarium* and *M. fermentans* [22].

A relevant discovery explaining the effect of leukocyte infiltration in mycoplasma infected lungs was reported by Kruger and Baier. They investigated the ability of viable and heat-killed *M. hominis* to induce a type II epithelial cell line to produce chemokines chemotactic for neutrophiles. They were able to demonstrate that *M. hominis* preparations were potent stimulators of type II cell-derived IL-8 and epithelial cell-derived neutrophil activating peptide[46].

## 5.    SUMMARY AND CONCLUSIONS

This chapter starts with several examples of unintentionally discovered interesting effects of mycoplasmas and their products on cells of the innate or acquired immune system. Much has been learned from these observations, and valuable contributions to the understanding of mycoplasma-host interactions were made and could be added to results obtained from carefully planned studies with mycoplasmal products in well defined experimental systems. A great many effects could be explained on the ground of macrophage activation with all facets such as release of various soluble mediators and their sequelae. Again, macrophage activation could in many cases be assigned to the action of lipoproteins or lipopeptides ubiquitously occurring in mycoplasmas and ureaplasmas, and their particular structure which renders mycoplasmal lipopeptides about 100 times more active than lipoproteins from other bacteria. Lipoproteins may not be the only, but are certainly the most active components from mycoplasmas acting on the immune system, in particular, but not exclusively, on cells of the macrophage/monocyte lineage. These lipoproteins can, at least in vitro, be compared in extent and character of their biological activity with the endotoxin of Gram-negative bacteria. Even their recently identified receptors and the signal pathway are very similar to those of LPS. Not only because of personal bias, but also because of its synthetic availability the focus has been on MALP-2 which can be regarded as a prototype mycoplasmal lipopeptide. In the complex immunological world, it is helpful to have a well-defined tool. However, there may be other mycoplasmal compounds such as glycolipids playing a role in mycoplasma-host interactions[113], and more surprises may be waiting. Such fusogenic membrane components, as well as variable antigens, mycoplasmal superantigens, and oncogenic properties of mycoplasmas, will be covered elsewhere in this book. Influence of mycoplasmas on HIV in culture and disease are beyond the scope of this chapter, as well as interactions of mycoplasmas with neutrophiles. The reader is referred to an excellent review on this subject which has appeared relatively recently[54].

## ACKNOWLEDGMENTS

    I like to thank my wife Toni and all my colleagues, inside and outside of my laboratory, for helping me over the years of often exciting (but sometimes frustrating) research with discussions, helpful advice and good humour. The financial help of the Deutsche Forschungsgemeinschaft, grant DFG Mu 672/1-5, without which our work on mycoplasmas would not have been possible, is gratefully acknowledged. Last, I would like to thank my friends Joan and Wesley Volk for helpful advice with the manuscript.

# REFERENCES

1.   ALLAN, R. G., and KWIATKOWSKI, D., 1993, *Plasmodium falciparum* varies in its ability to induce tumor necrosis factor. *Infect. Immun.* 61:4772-4776

2.   ATKIN, C. L.; COLE, B. C.; SULLIVAN, G. J.; WASHBURN, L. R., and WILEY, B. B., 1986, Stimulation of mouse lymphocytes by a mitogen derived from *Mycoplasma arthritidis*. V. A small basic protein from culture supernatants is a potent T cell mitogen . *J. Immunol.* 137:1581-1589

3.   BAEUERLE, P. A., and HENKEL, T., 1994, Function and activation of NF-kB in the immune system. *Annu. Rev. Immunol.* 12:141-179

4.   BEEZHOLT, D. H.; LEFTWICH, J. A., and HALL. R. E., 1989, P48 induces tumor necrosis factor and IL-1 secretion by human monocytes. *J. Immunol.* 143:3217-3221

5.   BESSLER, W. G.; COX, M.; LEX, A.; SUHR, B.; WIESMÜLLER, K. H., and JUNG, G., 1985, Synthetic lipopeptide analogs of bacterial lipoprotein are potent polyclonal activators for murine B lymphocytes. *J. Immunol.* 135:1900-1905

6.   BIBERFELD, G., and NILSSON, E., 1978, Mitogenicity of *Mycoplasma fermentans* for human lymphocytes. *Infect. Immun.* 21:48-54

7.   BOGDAN, C.; ROLLINGHOFF, M., and DIEFENBACH, A., 2000, The role of nitric oxide in innate immunity. *Immunol. Rev.* 173:17-26

8.   BOWIE, A., and O'NEILL, L. A., 2000, The interleukin-1 receptor/Toll-like receptor superfamily: signal generators for pro-inflammatory interleukins and microbial products. *J. Leukoc. Biol.* 67:508-514

9.   BRADE, H.; OPAL, S. M.; VOGEL, S. N. und MORRISON, D.C.eds., 1999, Endotoxin in health and disease. Marcel Dekker, New York,

10.  BRAUN, V.; WU, H. C.,1994, Lipoproteins, structure, function, biosynthesis and model for protein export. Bacterial cell wall. ( GHUYSEN, J.-M. und HAKENBECK, R.eds.), Elsevier Science,,pp. 319-341.

11.  BREDT, W.; WELLEK, B.; BRUNNER, H., and LOOS, M., 1977, Interactions between *Mycoplasma pneumoniae* and the first component of complement. *Infect. Immun.* 15:7-12

12.  BRENNER, C.; WROBLEWSKI, H.; LE HENAFF, M.; MONTAGNIER, L., and BLANCHARD, A., 1997, Spiralin, a mycoplasmal membrane lipoprotein, induces T-cell-independent B-cell blastogenesis and secretion of proinflammatory cytokines . *Infect. Immun.* 65:4322-4329

13.  BRENNER, T.; YAMIN, A.; ABRAMSKY, O., and GALLILY, R., 1993, Stimulation of tumor necrosis factor-$\alpha$ production by mycoplasmas and inhibition by dexamethasone in cultured astrocytes. *Brain Research* 608:273-279

14.  BRICKER, T. M.; BOYER, M. J.; KEITH, J.; WATSON-MCKOWN, R., and WISE, K. S., 1988, Association of lipids with integral membrane surface membrane proteins of *Mycoplasma hyorhinis*. *Infect. Immun.* 56:295-301

15.  CALCUTT, M. J.; KIM, M. F.; KARPAS, A. B.; MÜHLRADT, P. F., and WISE, K. S., 1999, Differential post-translational processing confers intraspecies variation of a major lipopeptide (MALP-2) of *Mycoplasma fermentans*. *Infect. Immun.* 67:760-771

16.  CARIO, E.; ROSENBERG, I. M.; BRANDWEIN, S. L.; BECK, P. L.; REINECKER, H. C., and PODOLSKY, D. K., 2000, Lipopolysaccharide activates distinct signaling pathways in intestinal epithelial cell lines expressing Toll-like receptors. *J. Immunol.* 164:966-972

17.  CHAMBAUD, I.; WROBLEWSKI, H., and BLANCHARD, A., 1999, Interactions between mycoplasma lipoproteins and the host immune system. *Trends in Microbiology* 7:493-499

18.  CHELMONSKA-SOYTA, A.; MILLER, R. B., and ROSENDAL, S., 1994, Activation of murine macrophages and lymphocytes by *Ureaplasma diversum*. *Can. J. Vet. Res.* 58:275-280

19.  CIFONE, M. G.; CIRONI, L.; MECCIA, M. A.; RONCAIOLI, P.; FESTUCCIA, C.; DE.NUNTIIS, G.; DÁLO, S., and SANTONI, A., 1995, Role of nitric oxide in cell-mediated tumor cytotoxicity. *Adv. Neuroimmunol.* 5:443-461

20.  COLE, B. C., 1999, Mycoplasma-induced arthritis in animals: relevance to understanding the etiologies of the human rheumatic diseases. *Rev. Rhum. Engl. Ed.* 66(1 Suppl):45S-49S

21.  DEITERS, U., and MÜHLRADT, P. F., 1999, Mycoplasmal lipopeptide MALP-2 induces the chemoattractant proteins macrophage inflammatory protein 1a (MIP-1a), monocyte chemoattractant protein 1, and MIP-2 and promotes leukocyte infiltration in mice. *Infect. Immun.* 67:3390-3398

22.  DONG, L.; SHIBATA, K.; SAWA, Y.; HASEBE, A.; YAMAOKA, Y.; YOSHIDA, S., and WATANABE, T., 1999, Transcriptional activation of mRNA of intercellular adhesion molecule 1 and induction of its cell surface expression in normal human gingival fibroblasts by *Mycoplasma salivarium* and *Mycoplasma fermentans*. *Infect. Immun.* 67:3061-3065

23.  FENG, S.-H., and LO, S.-C., 1994, Induced mouse spleen B-cell proliferation and secretion of immunoglobulin by lipid-associated membrane proteins of *Mycoplasma fermentans* incognitus and *Mycoplasma penetrans*. *Infect. Immun.* 62:3916-3921

24.  FENG-SH, and LO-SC, 1999, Lipid extract of Mycoplasma penetrans proteinase K-digested lipid-associated membrane proteins rapidly activates NF-kappaB and activator protein 1. *Infect. Immun.* 67:2951-2956

25.  FIDLER, I. J.; NII, A.; UTSUGI, T.; BROWN, D.; BAKOUCHE, O., and KLEINERMAN, E. S., 1990, Differential release of TNF-α, IL 1, and PGE2 by human blood monocytes subsequent to interaction with different bacterial derived agents. *Lymphokine Res.* 9:449-463

26.  FREUDENBERG, M. A., and GALANOS, C., 1985, Alterations in rats in vivo of the chemical structure of lipopolysaccharide from *Salmonella equi*. *Eur. J. Biochem.* 152:353-359

27.  FRISCH, M.; GRADEHANDT, G., and MÜHLRADT, P. F., 1996, *Mycoplasma fermentans*-derived lipid inhibits class II major histocompatibility complex expression without mediation by interleukin-6, interleukin-10, tumor necrosis factor, transforming growth factor-ß, type I interferon, prostaglandins or nitric oxide. *Eur. J. Immunol.* 26:1050-1057

28.  GABRIDGE, M. G., and MURPHY, W. H., 1971, Toxic membrane fractions from *Mycoplasma fermentans*. *Infect. Immun.* 4:678-682

29.  GALANOS, C.; GUMENSCHEIMER, M.; MÜHLRADT, P. F.; JIRILLO, E., and FREUDENBERG, M. A., 2001, MALP-2, a mycoplasma lipopeptide with classical endotoxic properties: end of an era of LPS monopoly ?. *J. Endotoxin Res.* 6:471-476

30.  GALLILY, R.; KIPPER-GALPERIN, M., and BRENNER, T., 1999, *Mycoplasma fermentans*-induced inflammatory response of astrocytes: selective modulation by aminoguanidine, thalidomide, pentoxifylline and IL-10. *Inflammation* 23:495-505

31.  GARCIA, J.; LEMERCIER, B.; ROMAN-ROMAN, S., and RAWADI, G., 1998, A *Mycoplasma fermentans* -derived synthetic lipopeptide induces AP-1 and NF-kB activity and cytokine secretion in macrophages via the activation of mitogen-activated protein kinase pathway. *J. Biol.Chem.* 273:34391-34398

32.  HAIER, J.; NASRALLA, M.; FRANCO, A. R., and NICOLSON, G. L., 1999, Detection of mycoplasmal infections in blood of patients with rheumatoid arthritis. *Rheumatology Oxford* 38:504-509

33.  HALL, R. E.; AGARWAL, S.; KESTLER, D. P.; COBB, J. A.; GOLDSTEIN, K. M., and CHANG, N.-S., 1996, cDNA and genomic cloning and expression of the P48 monocytic differentiation/activation factor, a *Mycoplasma fermentans* gene product. *Biochem. J.* 319:919-927

34. HAUSCHILDT, S.; LÜCKHOFF, A.; MÜLSCH, A.; KOHLER, J.; BESSLER, W., and BUSSE, R., 1990, Induction and activity of NO synthase in bone-marrow -derived macrophages are independent of $Ca^{++}$. *Biochem. J.* 270:351-356

35. HENDERSON, B.; POOLE, S., and WILSON, M., 1996, Bacterial modulins: a novel class of virulence factors which cause host tissue pathology by inducing cytokine synthesis. *Microbiol. Rev.* 60:316-341

36. HERBELIN, A.; RUUTH, E.; DELORME, D.; MICHEL-HERBELIN, C., and PRAZ, F., 1994, *Mycoplasma arginini* TUH-14 membrane lipoproteins induce production of interleukin-1, interleukin-6, and tumor necrosis factor alpha by human monocytes. *Infect. Immun.* 62:4690-4694

37. HIMMELREICH, R.; PLAGENS, H.; HILBERT, H.; REINER, B., and HERRMAN, R., 1997, Comparative analysis of the genomes of the bacteria *Mycoplasma pneumoniae* and *Mycoplasma genitalium*. *Nucleic Acids Res.* 25:701-712

38. HIRSCHFELD, M.; KIRSCHNING, C. J.; SCHWANDNER, R.; WESCHE, H.; WEIS, J. H.; WOOTEN, R. M., and WEIS, J. J., 1999, Inflammatory signaling by *Borrelia burgdorferi* lipoproteins is mediated by toll-like receptor 2. *J. Immunol.* 163:2382-2386

39. HOMMEL-BERREY, G. A., and BRAHMI, Z., 1987, Relevance of soluble cytotoxic factors generated by mycoplasma-contaminated targets to natural killer cell-mediated killing. . *Hum. Immunol.* 20:33-46

40. HOSHINO, K.; TAKEUCHI, O.; KAWAI, T.; SANJO, H.; OGAWA, T.; TAKEDA, Y., and AKIRA, S., 1999, Toll-like receptor 4 (TLR4)-deficient mice are hyporesponsive to lipopolysaccharide: evidence for TLR4 as the Lps gene product. *J. Immunol.* 162:3749-3752

41. JAN, G.; FONTENELLE, C.; LE HÉNAFF, M., and WRÓBLEWSKI, H., 1995, Acylation and immunological properties of *Mycoplasma gallisepticum* membrane proteins. *Res. Microbiol.* 146:739-750

42. KAUFMANN, A.; MÜHLRADT, P. F.; GEMSA, D., and SPRENGER, H., 1999, Induction of cytokines and chemokines in human monocytes by *Mycoplasma fermentans*-derived lipoprotein MALP-2. *Infect. Immun.* 67:6303-6308

43. KIRCHNER, H.; BREHM, G.; NICKLAS, W.; BECK, R., and HERBST, F., 1986, Biochemical characterization of the T-cell mitogen derived from *Mycoplasma arthritidis*. *Scand. J. Immunol.* 24:245-249

44. KITA, M.; OHMOTO, Y.; HIRAI, Y.; YAMAGUCHI, N., and IMANISHI, J., 1992, Induction of cytokines in human peripheral blood mononuclear cells by mycoplasmas. *Microbiol. Immunol.* 36:507-516

45. KOSTYAL, D. A.; BUTLER, G. H., and BEEZHOLD, D. H., 1994, A 48-kilodalton *Mycoplasma fermentans* membrane protein induces cytokine secretion by human monocytes. *Infect. Immun.* 62:3793-3800

46. KRUGER, T., and BAIER, J., 1997, Induction of neutrophil chemoattractant cytokines by *Mycoplasma hominis* in alveolar type II cells. *Infect. Immun.* 65:5131-5136

47. LEHENNAFF, M.; GUEGEN, M. M., and FONTENELLE, C., 2000, Selective acylation of plasma membrane proteins of *Mycoplasma agalactiae*: the causal agent of agalactia. *Curr. Microbiol.* 40:23-28

48. LIEN-E; SELLATI-TJ; YOSHIMURA-A; FLO-TH; RAWADI-G; FINBERG-RW; CARROLL-JD; ESPEVIK-T; INGALLS-RR; RADOLF-JD, and GOLENBOCK-DT, 1999, Toll-like receptor 2 functions as a pattern recognition receptor for diverse bacterial products. *J-Biol-Chem.* 274:33419-33425

49. LINCOLN, C. K., and GABRIDGE, M. G., 1998, Cell culture contamination: sources, consequences, prevention, and elimination. *Methods Cell Biol.* 57:49-65

50. LINDSEY, J. R., and CASSELL, G. H., 1973, Experimental *Mycoplasma pulmonis* infection in pathogen-free mice. *Am. J. Pathol.* 72:63-90

51.    LIU, H. M.; NEWBROUGH, S. E.; BHATIA, S. K.; DAHLE, C. E.; KRIEG, A. M., and
       WEINER, G. J., 1998, Immunostimulatory CpG oligodeoxynucleotides enhance the
       immune response to vaccine strategies involving granulocyte-macrophage colony-
       stimulating factor. *Blood* 92:3730-3736

52.    LOEWENSTEIN, J.; ROTTEM, S., and GALLILY, R., 1983, Induction of macrophage-
       mediated cytolysis of neoplastic cells by mycoplasmas. *Cellular Immunology* 77:290-
       297

53.    LOPEZ, A.; MAXIE, M. G.; RUHNKE, L.; SAVAN, M., and THOMSON, R. G., 1986,
       Cellular inflammatory response in the lungs of calves exposed to bovine viral diarrhea
       virus, *Mycoplasma bovis*, and *Pasteurella haemolytica* . *Am. J. Vet. Res.* 47:1283-
       1286

54.    MARSHALL, A. J.; MILES, R. J., and RICHARDS, L., 1995, The phagocytosis of
       mycoplasmas. *J. Med. Microbiol.* 43:239-250

55.    MATSUDA, K.; KASAMARA, T.; ISHIZUKA, I.; HANDA, S.; YAMAMOTO, N., and
       TAKI, T., 1994, Structure of a novel phosphocholine-containing glycoglycerolipid
       from *Mycoplasma fermentans*. *J. Biol. Chem.* 269:33123-33128

56.    MATSUMOTO, M.; YAMASHITA, F.; IIDA, K.; TOMITA, M., and SEYA, T., 1995,
       Purification and characterization of a human membrane protein that activates the
       alternative complement pathway and allows the deposition of homologous
       complement C3. *J. Exp. Med.* 181:115-125

57.    MATSUMOTO, M.; TAKEDA, J.; INOUE, N.; HARA, T.; HATANAKA, M.; TAKAHASHI,
       K.; NAGASAWA, S.; AKEDO, H., and SEYA, T., 1997, A novel protein that
       participates in nonself discrimination of malignant cells by homologous complement.
       *Nature Medicine* 3:1266-1270

58.    MATSUMOTO, M.; NISHIGUCHI, M.; KIKKAWA, S.; NISHIMURA, H.; NAGASAWA, S.,
       and SEYA, T., 1998, Structural and functional properties of complement-activating
       protein M161Ag, a *Mycoplasma fermentans* gene product that induces cytokine
       production by human monocytes. *J. Biol. Chem.* 273:12407-12414

59.    MAY, M. J., and GHOSH, S., 1998, Signal transduction through NF-kappaB.
       *Immunol. today* 19:80-88

60.    MELCHERS, F.; BRAUN, V., and GALANOS, C., 1975, The lipoprotein of the outer
       membrane of *Escherichia coli*: a B-lymphocyte mitogen. *J. Exp. Med.* 142:473-482

61.    METZGER, J.; WIESMÜLLER, K.-H.; SCHAUDE, R.; BESSLER, W. G., and JUNG, G.,
       1991, Synthesis and novel immunologically active tripalmitoyl-S-glycerylcysteinyl
       lipopeptides as useful intermediates for immunogen preparations. *Int. J. Pept. Protein
       Res.* 37:46-57

62.    METZGER, J. W.; BECK-SICKINGER, A. G.; LOLEIT, M.; BESSLER, M. E. W. G.,
       and JUNG, G., 1995, Synthetic S-(2,3-dihydroxypropyl)-cysteinyl peptides derived
       from the N-terminus of the cytochrome subunit of the photoreaction centre of
       *Rhodopseudomonas viridis* enhance murine splenocyte proliferation. *J. Pept. Sci.*
       3:184-190

63.    MILLER, R. A., and BRITIGAN, B. E., 1997, Role of oxidants in microbial
       pathophysiology. *Clin. Microbiol. Rev.* 10:1-18

64.    MÜHLRADT, P. F., and FRISCH, M., 1994, Purification and partial biochemical
       characterization of a *Mycoplasma fermentans*-derived substance that activates
       macrophages to release nitric oxide, tumor necrosis factor, and interleukin-6. *Infect.
       Immun.* 62:3801-3807

65.    MÜHLRADT, P. F., and SCHADE, U., 1991, MDHM, a macrophage-stimulatory
       product of *Mycoplasma fermentans*, leads to in vitro interleukin-1 (IL-1), IL-6, tumor
       necrosis factor, and prostaglandin production and is pyrogenic in rabbits. *Infect.
       Immun.* 59:3969-3974

66. MÜHLRADT, P. F.; QUENTMEIER, H., and SCHMITT, E., 1991, Involvement of interleukin-1 (IL-1), IL-6, IL-2, and IL-4 in generation of cytolytic T cells from thymocytes stimulated by a *Mycoplasma fermentans*-derived product . *Infect. Immun.* 59:3962-3968

67. MÜHLRADT, P. F.; MEYER, H., and JANSEN, R., 1996, Identification of S-(2,3-dihydroxypropyl)cystein in a macrophage-activating lipopeptide from *Mycoplasma fermentans*. *Biochemistry* 35:7781-7786

68. MÜHLRADT, P. F.; KIESS, M.; MEYER, H.; SÜSSMUTH, R., and JUNG, G., 1997, Isolation, structure elucidation, and synthesis of a macrophage stimulatory lipopeptide from *Mycoplasma fermentans* acting at pico molar concentration. *J. Exp. Med.* 185:1951-1958

69. MÜHLRADT, P. F.; KIESS, M.; MEYER, H.; SÜSSMUTH, R., and JUNG, G., 1998, Structure and specific activity of macrophage stimulating lipopeptides from *Mycoplasma hyorhinis. Infect. Immun.* 66:4804-4810

70. MUZIO, M.; POLENTARUTTI, N.; BOSISIO, D.; PRAHLADAN, M. K., and MANTOVANI, A., 2000, Toll-like receptors: a growing family of immune receptors that are differentially expressed and regulated by different leukocytes. *J-Leukoc-Biol.* 67:450-456

71. NAOT, Y.; MERCHAV, S., and GINSBURG, H., 1979, *Mycoplasma neurolyticum*: a potent mitogen for rat B lymphocytes. *Eur. J. Immunol.* 9:185-189

72. NORGARD, M. V.; ARNDT, L. L.; AKINS, D. A.; CURETTY, L. L.; HARRICH, D. A., and RADOLF, J. D., 1996, Activation of human monocytic cells by *Treponema pallidum* and *Borrelia burgdorferi* lipoproteins and synthetic lipopeptides proceeds via a different pathway distinct from that of lipopolysaccharide but involves the transcriptional activator NF-kB. *Infect. Immun.* 64:3845-3852

73. NOVAK, J. F.; HAYES, J. D. J., and MCMASTER, J. H., 1995, Mycoplasma-mediated bone resorption in bone organ cultures. *Microbios* 81:241-260

74. NYSTRÖM, S., and WIESLANDER, A., 1992, Isoprenoid modification of proteins distinct from membrane acyl proteins in the procaryote *Acholeplasma laidlawii. Biochim. Biophys. Acta* 1107:39-43

75. NYSTRÖM, S.; WALLBRANDT, P., and WIESLANDER, A., 1992, Membrane protein acylation. Preference for exogenous myristic acid or endogenous saturated chains in *Acholeplasma laidlawii. Eur. J. Biochem.* 204:231-240

76. PIEC, G.; MIRKOVITCH, J.; PALACIO, S.; MÜHLRADT, P. F., and FELIX, R., 1999, Effect of MALP-2, a lipopeptide from *Mycoplasma fermentans*, on bone resorption in vitro . *Infect. Immun.* 67:6281-6285

77. POLTORAK, A.; HE, X.; SMIRNOVA, I.; LIU, M.-Y.; VANHUFFEL, C.; DU, X.; BIRDWELL, D.; ALEJOS, E.; SILVA, M.; GALANOS, C.; FREUDENBERG, M. R.-C. P.; LAYTON, B., and BEUTLER, B., 1998, Poltorak, A., X. He, I. Smirnova, M.-Y. Liu, C. V. Huffel, X. Du, D. Birdwell, E. Alejos, M. Silva, C. Galanos, M. Defective LPS signaling in C3H/HeJ and C57BL/10ScCr mice: mutations in Tlr4 gene. *Science* 282:2085-2088

78. PROUST, J. J.; BUCHHOLZ, M. A., and NORDIN, A. A., 1985, A "lymphokine-like" soluble product that induces proliferation and maturation of B cells appears in the serum-free supernatant of a T cell hybridoma as a consequence of mycoplasmal contamination. *J. Immunol.* 134:390-396

79. QUENTMEIER, H.; SCHMITT, E.; KIRCHHOFF, H.; GROTE, W., and MÜHLRADT, P. F., 1990, *Mycoplasma fermentans*-derived high molecular weight material induces interleukin-6 release in cultures of murine macrophages and human monocytes. *Infect. Immun.* 58:1273-1280

80. QUENTMEIER, H.; KLAUCKE, J.; MÜHLRADT, P. F., and DREXLER, H. G., 1992, Role of IL-6, IL-2, and IL-4 in the in vitro induction of cytotoxic T cells. *J. Immunol.* 149:3316-3320

81. RADOLF, J. D.; ARNDT, L. L.; AKINS, D. R.; CURETTY, L. L.; LEVI, M. E.; SHEN, Y.; DAVIS, L. S., and NORGARD, M. V., 1995, *Treponema pallidum* and *Borrelia burgdorferi* lipoproteins and synthetic peptides activate monocytes/macrophages. *J. Immunol.* 154:2866-2877

82. RAWADI, G., and ROMAN-ROMAN, S., 1996, Mycoplasma membrane lipoproteins induce proinflammatory cytokines by a mechanism distinct from that of lipopolysaccharide. *Infect. Immun.* 64:637-643

83. RAWADI, G.; ROMAN-ROMAN, S.; CASTEDO, M.; DUTILLEUL, V.; SUSIN, S.; MARCHETTI, P.; GEUSKENS, M., and KROEMER, G., 1996, Effects of *Mycoplasma fermentans* on the myelomonocytic lineage. *J. Immunol.* 156:670-678

84. RAWADI, G.; ZUGAZA, J. L.; LEMERCIER, B.; MARVAUD, J. C.; POPOFF, M.; BERTOGLIO, J., and ROMAN-ROMAN, S., 1999, Involvement of small GTPases in *Mycoplasma fermentans* membrane lipoproteins-mediated activation of macrophages. *J. Biol. Chem.* 274:30794-30798

85. RAZIN, S.; YOGEV, D., and NAOT, Y., 1998, Molecular biology and pathogenicity of mycoplasmas. *Microbiol. Mol. Biol. Rev.* 62:1094-1156

86. RICHARDS, A. L., 1997, Tumour necrosis factor and associated cytokines in the host′s response to malaria. *Int. J. Parasitol.* 27:1251-1263

87. ROLLINS, S.; COLBY, T., and CLAYTON, F., 1986, Open lung biopsy in *Mycoplasma pneumoniae* pneumonia. *Arch. Pathol. Lab. Med.* 110:34-41

88. ROTTEM, S., and MARKOWITZ, O., 1979, Unusual positional distribution of fatty acids in phosphatidylglycerol of sterol-requiring mycoplasmas. *FEBS Letters* 107:379-382

89. ROWE, J. A.; SCRAGG, I. G.; KWIATKOWSKI, D.; FERGUSON, D. J.; CARUCCI, D. J., and NEWBOLD, C. I., 1998, Implication of mycoplasma contamination in *Plasmodium falciparum* cultures and methods of its detection and eradication. *Mol. Biochem. Parasitol.* 92:177-180

90. RUSCHMEYER, D.; THUDE, H. J., and MÜHLRADT, P. F., 1993, MDHM, a macrophage-activating product from *Mycoplasma fermentans*, stimulates murine macrophages to synthesize nitric oxide and become tumoricidal. *FEMS Immunol. Medical Microbiol.* 7:223-230

91. RUUTH, E., and LUNDGREN, E., 1986, Enhancement of immunoglobulin secretion by the lymphokine-like activity of a *Mycoplasma arginini* strain. *Scand. J. Immunol.* 23:575-580

92. RUUTH, E., and PRAZ, F., 1989, Interactions between mycoplasmas and the immune system. *Immunol. Rev.* 112:133-160

93. RUUTH, E.; RANBY, M.; FRIEDRICH, B.; PERSSON, H.; GOUSTIN, A.; LEANDERSON, T.; COUTINHO, A., and LUNDGREN, E., 1985, Mycoplasma mimicry of lymphokine activity in T-cell lines. *Scand. J. Immunol.* 21:593-600

94. SACHT, G.; MÄRTEN, A.; DEITERS, U.; SÜßMUTH, R.; JUNG, G.; WINGENDER, E., and MÜHLRADT, P. F., 1998, Activation of nuclear factor-kappaB in macrophages by mycoplasmal lipopeptides. *Eur. J. Immunol.* 28:4207-4212

95. SANKARAN, K., and WU, H. C., 1994, Lipid modification of bacterial prolipoprotein. *J. Biol.Chem.* 269:19701-19706

96. SCHAEVERBEKE, T.; GILROY, C. B.; BÉBÉAR, C.; DEHAIS, J., and TAYLOR-ROBINSON, D., 1996, *Mycoplasma fermentans* in joints of patients with rheumatic arthritis and other joint disorders. *Lancet* 347:1418

97. SCHAEVERBEKE, T.; RENAUDIN, H.; CLERC, M.; LEQUEN, L.; VERNHES, J. P.; DE BARBEYRAC, B.; BANNWARTH, B.; BÉBÉAR, C., and DEHAIS, J., 1997, Systematic detection of mycoplasmas by culture and polymerase chain reaction (PCR) procedures in 209 synovial fluid samples. *Br. J. Rheumatol.* 36:310-314

98. SCHIMMELPFENG, L.; LANGENBERG, U., and PETERS, J. H., 1980, Macrophages overcome mycoplasma infections of cells in-vitro. *Nature* 285:661-662

99.  SHER, T.; YAMIN, A.; ROTTEM, S., and GALLILY, R., 1990, In vitro induction of tumor necrosis factor a, tumor cytolysis and blast transformation by *Spiroplasma* membranes . *J. Natl. Cancer Inst.* 82:1142-1145
100. SIEBENLIST, U.; FRANZOSO, G., and BROWN, K., 1994, Structure, regulation and function of NF-kB. *Annu. Rev. Cell Biol.* 10:405-455
101. SITIA, R.; RUBARTELLI, A.; DEAMBROSIS, S.; POZZI, D., and HÄMMERLING, U., 1985, Differentiation of the murine B cell lymphoma I.29: inductive capacities of lipopolysaccharide and *Mycoplasma fermentans* products. *Eur. J. Immunol.* 15:570
102. STUART, P. M., 1993, Mycoplasmal induction of cytokine production and major histocompatibility complex expression. *Clinical Infectious Dis.* 17(Suppl 1):S187-S191
103. STUART, P. M.; CASSELL, G. H., and WOODWARD, J. G., 1989, Induction of Class II MHC Antigen expression in macrophages by *Mycoplasma* species. *J. Immunol.* 142:3392-3399
104. STUART, P. M.; CASSELL, G. H., and WOODWARD, J. G., 1990, Differential induction of bone marrow macrophage proliferation by mycoplasmas involves granulocyte-macrophage colony-stimulating factor. *Infect. Immun.* 58:3558-3563
105. STUART, P. M.; EGAN, R. M., and WOODWARD, J. G., 1993, Characterization of MHC induction by *Mycoplasma fermentans* (incognitus strain). *Cell. Immunol.* 152:261-270
106. STUEHR, D. J.; GROSS, S. S.; SAKUMA, I.; LEVI, R., and NATHAN, C. F., 1989, Activated murine macrophages secrete a metabolite of arginine with the bioactivity of endothelium-derived relaxing factor and the chemical reactivity of nitric oxide. *J. Exp. Med.* 169:1011-1020
107. SUGAMA, K.; KUWANO, K.; FURUKAWA, M.; HIMENO, Y.; SATOH, T., and ARAI, S., 1990, Mycoplasma induce transcription and production of tumor necrosis factor in a monocytic cell line, THP1, by a protein kinase C-independent pathway. *Infect. Immun.* 58:3564-3567
108. SULTZER, B. M., 1968, Genetic control of leucocyte responses to Endotoxin. *Nature (London)* 219:1253-1254
109. TAKEMA, M.; OKA, S.; UNO, K.; NAKAMURA, S.; ARITA, H.; TAWARA, K.; INABA, K., and MURAMATSU, S., 1991, Macrophage-activating factor extracted from mycoplasmas. *Cancer Immunol. Immunother.* 33:39-44
110. TAKEUCHI, O.; TAKEDA, K.; HOSHINO, K.; ADACHI, O.; OGAWA, T., and AKIRA, S., 2000, Cellular responses to bacterial cell wall components are mediated through MyD88-dependent signaling cascades. *Int. Immunol.* 12:113-117
111. TAKEUCHI, O.; KAUFMANN, A.; GROTE, K.; KAWAI, T.; HOSHINO, K.; MORR, M.; MÜHLRADT, P. F., and AKIRA, S., 2000, Preferentially the R-stereoisomer of the mycoplasmal lipopeptide MALP-2 activates immune cells through a TLR2- and MyD88-dependent signaling pathway. *J. Immunol.* 164:558-561
112. TAKEUCHI, O.; KAWAI, T.; MÜHLRADT, P. F.; MORR, M.; RADOLF, J. D. ,. Z. A.; TAKEDA, Y., and AKIRA, S., 2001, Discrimination of microbial lipoproteins by Toll-like receptor (TLR) 6. *Int. Immunol.* 13:933-940
113. TARSHIS, M.; SALTMAN, M., and ROTTEM, S., 1991, Fusion of Mycoplasmas: the formation of cell hybrids. *FEMS Microbiol. Letters* 82:67-72
114. TAYLOR, G., and HOWARD, C. J., 1980, Interaction of *Mycoplasma pulmonis* with mouse peritoneal macrophages and polymorphonuclear leucocytes. *J. Med. Microbiol.* 13:19-30
115. TEH, H. S.; HO, M., and WILLIAMS, L. D., 1988, Suppression of cytotoxic responses by a supernatant factor derived from *Mycoplasma hyorhinis*-infected mammalian cell lines. *Infect. Immun.* 56:197-203
116. TRACEY, K. J., and CERAMI, A., 1994, Tumor necrosis factor: a pleiotropic cytokine and therapeutic agent. *Annu. Rev. Med.* 45:491-503

117.  TUFFREY, M. A.; FURR, P. M.; FALDER, P., and TAYLOR-ROBINSON, D., 1984, The anti-Chlamydial effect of experimental *Mycoplasma pulmonis* infection in the murine genital tract. *J. Med. Microbiol.* 17:357-362

118.  UNO, K.; TAKEMA, M.; HIDAKA, S.; TANAKA, R.; KONISHI, T.; KATO, T.; NAKAMURA, S., and MURAMATSU, S., 1990, Induction of antitumor activity in macrophages by mycoplasmas in concert with interferon. *Cancer Immunol. Immunother.* 32:22-28

119.  WEBSTER-AD; FURR-PM; HUGHES-JONES-NC; GORICK-BD, and TAYLOR-ROBINSON-D, 1988, Critical dependence on antibody for defence against mycoplasmas. *Clin-Exp-Immunol.* 71:383-387

120.  WEINBERG, J. B.; SMITH, P. F., and KAHANE, I., 1980, Bacterial lipopolysaccharides and mycoplasmal lipoglycans: a comparison between their abilities to induce macrophage-mediated tumor cell killing and limulus amebocyte lysate clotting. *Biochem. Biophys. Res. Com.* 97:493-499

121.  WEYER, K. A.; SCHÄFER, W.; LOTTSPEICH, F., and MICHEL, H., 1987, The cytochrome subunit of the photosynthetic reaction center from Rhodopseudomonas viridis is a lipoprotein. *Biochemistry* 26:2909-2914

122.  WISE, K. S.; KIM, M. F.; THEISS, P. M., and LO, S.-C., 1993, A family of strain-variant surface lipoproteins of *Mycoplasma fermentans*. *Infect. Immun.* 61:3327-3333

123.  YANG, G.; COFFMAN, F. D., and WHEELOCK, E. F., 1994, Characterization and purification of a macrophage-triggering factor produced in *Mycoplasma arginini*-infected L5178Y cell cultures. *J. Immunol.* 153:2579-2591

124.  ZIEGLER-HEITBROCK, H. W., and BURGER, R., 1987, Rapid removal of mycoplasma from cell lines mediated by a direct effect of complement. *Exp. Cell. Res.* 173:388-394

Chapter 21

## *Mycoplasma arthritidis* Pathogenicity: Membranes, MAM, and MAV1

LEIGH R. WASHBURN[1] and BARRY C. COLE[2]
*[1]Division of Basic Biomedical Sciences, University of South Dakota School of Medicine, Vermillion, SD, USA; [2]Division of Rheumatology, University of Utah School of Medicine, Salt Lake City, UT, USA*

## 1. INTRODUCTION

In 1937, W. A. Collier, working in Java, came into possession of a wild rat with unusual swellings in its hind legs. When injected into laboratory rats, material from the lesions induced a migratory, inflammatory polyarthritis (29). In 1938, two researchers from Columbia University discovered a "pyogenic filterable agent," possibly a virus, that induced a severe systemic infection with joint swelling, hind limb paralysis, and rapid weight loss on intravenous (iv) injection (91). At the same time, Emmy Klieneberger and her colleagues at the Lister Institute in London were systematically characterizing and classifying an unusual group of micro-organisms called "pleuropneumonia-like organisms" (PPLO), which were turning up with remarkable regularity in a wide variety of animals. They had seen phenomena similar to those reported by the other two groups in both wild and laboratory rats, and, using Klieneberger's "special medium," succeeded in isolating the responsible agents and identifying them as PPLOs. These agents were designated L4 and L7 (37, 53). The Columbia University strain was shown to be identical to L4 (52, 90), although Collier's strain was lost at the beginning of World War II before it could be thoroughly characterized (52). Over the next few years, similar isolates turned up in a variety of settings (37, 53, 65). Transplanted rat lymphosarcoma tissue proved to be an especially prolific source (43-45, 47, 91).

*Molecular Biology and Pathogenicity of Mycoplasmas*, Edited by Razin and Herrmann, Kluwer Academic/Plenum Publishers, New York, 2002

However, perhaps the most intriguing isolations were not from rats at all, but from humans (62, 70). These isolates were referred to as *Mycoplasma hominis* type 2. Whether they were, in fact, human isolates or laboratory contaminants will probably never be known. Aside from a report by Jansson *et al.* in 1971 on the apparent isolation of *M. arthritidis* from synovial tissues of rheumatoid arthritis (RA) patients (46), no other reports of *M. arthritidis* isolations from humans have appeared in the literature since the 1950s.

In 1956 Edward and Freundt proposed species status for the rat isolates, recommending the name *Mycoplasma arthritidis* (35). Before long, it became clear *M. hominis* type 2 was essentially identical to *M. arthritidis* (58, 74) and in 1965, it was recommended that these strains, including Campo, O7, H39, and PG27 (57, 62), be reclassified as members of that species (36). Shortly thereafter, Cole *et al.* showed that both human and rat isolates were equally arthritogenic for rats, providing further evidence that they were the same (18).

## 2.    DISEASE

*Mycoplasma arthritidis*-induced arthritis has been studied in three different experimental animals, rats, mice, and rabbits (27). The rat and mouse diseases more closely approximate "natural" conditions, while the rabbit disease is strictly experimental. However, it remains of interest because immune complex deposition is clearly an important factor in perpetuating chronic inflammation. Current information on each of these models is summarized below.

### 2.1    Rats

Naturally occurring *M. arthritidis* arthritis is rarely seen today under modern housing conditions, although *M. arthritidis* has been isolated from apparently healthy rats (30) and may be part of the normal microflora of these animals. In the past, *M. arthritidis* has also been associated with pathologic conditions other than arthritis, including respiratory, middle ear (55, 65, 71), and skin infections. Aside from a single comment by Woglom and Warren in 1938 that they had "seen no evidence that this filterable agent can be passed spontaneously from rat to rat," (91) there is essentially no information in the literature about contagion under natural conditions, and that question remains open.

A number of variables contribute to experimentally-induced disease, and for many years investigators struggled with problems of reproducibility and strain attenuation (91). These variables include dose [$10^8$-$10^{10}$ CFU are usually required for consistent induction of arthritis (25, 39, 55)], route of infection [the iv route is more reliable than intraperitoneal, subcutaneous, or footpad injection (39, 65)], strain of animal (5), and even its social status within a colony (38).

M. *arthritidis* arthritis of rats is usually described as acute and self-limiting, although prolonged persistence and even recurrences have been reported with strain ISR-1 (50). The earliest signs include redness and swelling in the feet and ankles two to five days after injection. This spreads progressively to other peripheral joints within seven to 14 days. Abscesses that rupture and heal spontaneously may form over affected joints. Spinal deformities and hind limb paralysis are common and are attributed to infection of the diarthrodial interspinal joints and involvement of the nerve roots within the spinal foramina. Within the joints, bone and cartilage erosions, pannus formation, periosteal proliferation, and new bone formation occur routinely. Healing is accompanied by fibrous and cartilaginous ankylosis and bone remodeling (27, 77). Extra-articular manifestations include rhinitis, conjunctivitis, pneumonia, inflammation of the brain and testes, splenomegaly, swollen lymph nodes, diarrhea, and fecal impaction (40, 41, 77, 91). Mycoplasmas can be recovered from actively inflamed joints at all stages and from a variety of other organs, including blood, brain, liver, kidney, lymph nodes, spleen, and thymus, early on (23, 40, 41, 50, 77).

Persistence of mycoplasmas in the face of the characteristically early and vigorous antibody response is intriguing. Although rats produce no or only minimal levels of growth-inhibiting and opsonizing antibodies against M. *arthritidis* (11, 14, 41, 81, 84, 91), they are highly resistant to reinfection and that resistance is antibody-mediated (23, 27). A possible explanation lies in the presence of an antibody in rat serum that can effectively kill M. *arthritidis* cells that are not actively replicating (21). This antibody may also be responsible for the low levels of metabolism-inhibiting activity seen by Kirchhoff *et al.* in sera from rats infected with M. *arthritidis* strain ISR-1 (51). Immune complexes can persist in rat sera for prolonged periods (51), and arthritis severity correlates directly and significantly with IgG and IgM antibody titers (84), suggesting a role for hypersensitivity in tissue damage, although this has not been definitively shown. Rats also mount a cell-mediated immune response against M. *arthritidis* (13), although unlike some other mycoplasmal infections, delayed-type hypersensitivity does not appear to participate in this disease (6, 48).

## 2.2 Mice

*Mycoplasma arthritidis* also induces experimental arthritis in mice. This disease is quite different from that seen in rats. Described as a chronic, proliferative polyarthritis, onset is more gradual and more variable in mice than in rats. Recovery is more prolonged, and disease may recur months after apparent recovery (25). There are fewer abscesses and less suppuration in mice than in rats and less involvement of bone marrow. Pannus appears but articular cartilage erosions are less pronounced. During the chronic stage, massive villous hypertrophy can sometimes occlude the entire joint space. Foci of mononuclear cells resembling lymphoid follicles appear, along with "punched out" areas at the cartilage-periosteal junction. All of these bear a more striking resemblance to RA than do changes in the rat (25). The antibody response to *M. arthritidis* develops more slowly in mice. Previously infected mice are less resistant to reinfection, and passive immunization with convalescent mouse serum is less protective (24). Mice mount a vigorous cell-mediated immune response to *M. arthritidis*, although there is no correlation between its intensity and disease severity (12). Cole *et al.* have suggested that the interaction of mycoplasmas with the immune systems of mice and rats may in part explain the differences in disease (14). In support of this, Kaklamani *et al.* reported that after experimental infection, macrophage phagocytic activity remained high considerably longer in rats than in mice (49).

Another important difference is that mice are susceptible to toxic shock and death after injection of large doses of viable organisms while rats are not. In addition, subcutaneous injection of *M. arthritidis* induces rapidly spreading necrotic lesions in mice but only localized abscesses in rats (25, 91). Both are now known to be caused by the soluble superantigen MAM, which may interact differently with the immune systems of these two hosts (see below).

## 2.3    Rabbits

Because it was difficult to obtain enough synovial tissue from rats and mice for detailed analyses of mycoplasmal interactions with host tissues, a new model was established in 1977. Cole *et al.* showed that *M. arthritidis* could induce a chronic inflammatory synovitis in rabbits that persisted for at least 12 weeks after a single intra-articular (ia) injection of viable organisms (16). Washburn *et al.* extended these observations through 52 weeks and found that although neither viable mycoplasmas nor mycoplasmal antigens could be detected in injected joints after seven weeks, chronic inflammation persisted throughout the entire observation period (78). Disease remains confined to the injected joint and is characterized by pronounced villous hypertrophy, pannus formation, and destruction of articular cartilage. Microscopically, an early predominance of heterophils gives way to mononuclear cell infiltration during the chronic stages, although heterophils persist in small numbers. The synovial membrane becomes thickened and lymphocytes organize into follicle-like structures (78). Large numbers of IgG- and IgA-containing plasma cells persist throughout the entire observation period, and foci of IgG- and complement-containing immune complexes are seen routinely in synovium and articular cartilage (79).

Thus, it appears as if this is an immune complex-mediated disease, although whether these immune complexes contain mycoplasmal or "self" antigens is not known. The immune complex hypothesis was supported by the finding that rabbits sensitized by immunization with killed *M. arthritidis* developed a severe chronic inflammatory arthritis on ia injection of nonviable mycoplasmas (80). This indicates that an immune response to persisting mycoplasmal antigens might be capable of sustaining chronic inflammation in rabbits injected with viable organisms. This is quite different from the rat and mouse diseases, in which active inflammation is associated with persistence of viable organisms. Unfortunately, highly sensitive detection methods such as polymerase chain reaction that could confirm persistence of organisms in rabbit joints were not available at the time these studies were done, and this model has not been further investigated.

## 3. PATHOGENIC MECHANISMS IN MURINE ARTHRITIS

As in most infectious diseases, pathogenesis of *M. arthritidis*-induced arthritis is a multifactorial process. Recent work in this area has focused on the rat and mouse models and has progressed on three fronts - 1) the role of *M. arthritidis* membrane components in adherence to host tissues and in biological mimicry, 2) the effects of potent soluble superantigen MAM on the host immune system, and 3) the role of factor(s) carried by temperate bacteriophage MAV1. Recent advances in each of these areas - Membranes, MAM, and MAV1 - are summarized below and a model of pathogenesis for the rat is proposed.

### 3.1 Membranes

Mycoplasmal lipoproteins are characteristically highly variable in both size and expression state [reviewed in (66)], and, regardless of their exact function, are highly antigenic, inducing vigorous antibody responses in host animals (4, 59, 63, 68). The fact that pathogenic mycoplasmas have evolved elaborate ways to avoid host responses indicates that these proteins are important in host-parasite interaction and that antibodies against them may play critical roles in recovery from and protection against infection. Although membrane lipoproteins of *M. arthritidis* have not been characterized as extensively as those of some other arthritogenic species, they, too, are highly immunogenic for rats (87) and at least some of them vary in both size and expression state *in vitro* (34, 86, 88) and *in vivo* (Droesse, M., *et al.*, 9th Congress of the International Organization for Mycoplasmology, 1992).

### 3.1.1    Adhesins

Since adherence to host tissues is the first and arguably the most critical step in most infectious diseases, recent work with *M. arthritidis* surface-exposed proteins has focused on that aspect. Washburn *et al.* have identified two surface-exposed lipoproteins, MAA1 and MAA2, in the virulent *M. arthritidis* strain 158p10p9 that may contribute to this process (82).

An anti-MAA1 monoclonal antibody (MAb) recognizes an epitope found on 14 of 20 strains tested (86). On the basis of DNA sequence analysis, the MAA1 prolipoprotein is predicted to be ~86 kDa in size, basic (p$I$ 8.14) and, except for the signal peptide, largely hydrophilic (83). It is neither size nor phase variable, contains no repetitive elements, and shows no significant sequence similarities to other known proteins. (83, 86) A spontaneous mutant that adheres poorly to rat L2 lung cells in culture compared to the wild-type contains a point mutation within the *maa1* coding region that generates a stop codon at that site. This results in expression of a highly truncated product (26 kDa) that is processed and inserted into the membrane but is apparently nonfunctional (83, 86).

MAA2 is a major antigenic marker for a group of strains serologically related to 158p10p9 (86) and is subject to both size and phase variation (86, 88). The bulk of MAA2 from strain 158p10p9 consists of six 88-amino acid tandem direct repeats, one of which is missing from strain H606, which expresses a proportionately smaller product. The mature lipoprotein is predicted to be highly basic (p$I$ = 9.43) and largely hydrophilic (87). Secondary structure analysis predicts a coiled-coil motif reminiscent of other bacterial adherence-associated proteins (54, 64). Expression correlates with the length of a poly(T) tract beginning 106 base-pairs upstream of the ATG start codon, situated between putative -10 and -35 sites (88).

Both MAA1 and MAA2 can elicit protective immunity against *M. arthritidis* in rats. Interestingly, passive immunization of rats with a MAb against MAA1 resulted in almost complete protection against challenge with virulent strain 158p10p9. Rats passively immunized with this MAb developed little or no antibody against *M. arthritidis* on iv challenge, suggesting that mycoplasmas were rapidly cleared (87). Although these experiments do not provide direct evidence for a role in disease pathogenesis, protective immunity in bacterial infections is often directed against major virulence factors.

### 3.1.2 Biological mimicry

The failure of rats to produce growth- or metabolism-inhibiting antibodies against *M. arthritidis* prompted Cahill *et al.* to look for a shared antigen that might permit critical *M. arthritidis* cell components to go unrecognized (7). Using complement-fixation, immunofluorescence, and gel diffusion, they found evidence of cross-reactivity between *M. arthritidis* and rat muscle tissue and lymphocytes (7). However, Washburn *et al.*, using an enzyme immunoassay (EIA), later found that serum components contaminating antigen preparations could account for most if not all of this apparent cross-reactivity (85). Meanwhile, Kirchhoff *et al.*, using immunofluorescence and EIA, observed that antiserum against *M. arthritidis* strain ISR-1 reacted with rat brain cells, chondrocytes, kidney, heart muscle, and serum (50, 72, 73). Since precautions were taken, it is unlikely that serum contamination of antigens and immunogens is entirely responsible for these observations. Interestingly, some regions of MAA2 show amino acid sequence similarities to certain rat cytoskeletal proteins (Washburn, unpublished observation). Thus, questions of biological mimicry as well as a role for cross-reacting antigens in induction of autoimmunity remain open.

## 3.2 MAM

The ability of *M. arthritidis* to activate splenocytes from immunologically naïve mice was first reported in 1977 (2). The key observations were that, unlike other known mitogens, the M. arthritidis mitogen (MAM) activated spleen cells from only certain strains of mice and that these responses were under control of the H2-E region of the murine major histocompatibility complex (MHC) [reviewed in (19)]. Subsequently, MAM was also shown to activate rat and human lymphocytes, and human lymphocyte activation is restricted to those cells bearing HLA.DRα, which is very similar to the murine H-2Eα. In 1989, it was recognized that MAM belonged to a newly described category of immunomodulating agents called "superantigens" (SAgs) (89). Each SAg binds to a large but characteristically different set of Vβ chains on T lymphocytes and activates these lymphocytes without being processed by accessory cells. Thus, SAgs can activate up to 25% of all T-cells within an individual, sometimes resulting in life-threatening illnesses such as toxic shock and necrotizing fasciitis (31, 33).

### 3.2.1    Molecular and functional analysis

MAM is released into culture supernatants by senescent cells of all *M. arthritidis* strains tested, regardless of their virulence (3). Its DNA sequence predicts a small (25 kDa), highly basic (p*I* 10.1), hydrophobic protein with a 39-amino acid signal peptide and no significant sequence similarities to other known proteins or phylogenetic relationships to other SAgs. However, MAM contains a functional domain near its amino-terminus that does exhibit some similarity to short sequences of staphylococcal and streptococcal SAgs. These regions may influence critical three-dimensional structures of these molecules. Another region, more centrally located, contains the legume lectin β motif, which is important in T-cell interaction (17). This peptide partially blocks proliferative responses to Concanavalin A, implying that it may function as an alternate binding mechanism for MAM. Interestingly, this peptide also possesses an adjacent sequence similar to that of an epitope present in the hypovariable region 3 of the RA-associated HLA-DRβ molecule. Recently, evidence for a third region at the carboxy-terminus that is involved in MAM/T-cell receptor (TCR) binding was also reported (56).

MAM is presented to T-cells by the α chains of H-2E molecules on accessory cells such as macrophages, B-cells, and dendritic cells. Human leukocytes expressing HLA-DRα can also present MAM, and recently Cole *et al.* showed that certain murine H-2A haplotypes can do so as well (19). Stimulation of mouse lymphocytes with MAM results in a clonal expansion of T-cells bearing primarily the Vβ 8.1, 8.2, 8.3, 5, and 6 chains of the α/β TCR, with a hierarchical preference for Vβ8. However, other Vβ chains can also be used, especially in strains of mice in which the preferred chains are absent due to genomic deletions (10). The most highly MAM-responsive human T-cells express counterparts to these murine molecules (19).

Recently it was shown that, in contrast to other bacterial SAgs, MAM also binds to the complementarity-determining region 3 (CDR3) of TCR-β. This additional contact point is used by conventional peptide antigens and may serve to further stabilize the MAM-T-cell bond. Thus, MAM may represent a new class of TCR ligand, distinct from both conventional antigens and other bacterial SAgs (42).

Bacterial SAgs, including MAM, induce B-cell differentiation and polyclonal immunoglobulin secretion by means of a "superantigen bridge," which links Vβ TCRs on T-helper cells to MHC molecules on B cells. MAM activates both murine and human B-cells in this manner, and T-cell IL-4, IL-6 and IL-10 production in response to MAM probably contributes to this activity (19). MAM enhances the murine B-cell response to T-cell-dependent antigens *in vitro* and was recently shown to do so *in vivo* as well (Stohl and Cole, unpublished observation). Although these studies were done in mice, similar activity in rats could explain the extremely vigorous antibody response that characteristically appears within 1-3 days after injection of viable *M. arthritidis* into immunologically naïve animals (McKenzie, M. and L.R. Washburn, Annual Meeting of the American Society for Microbiology, 1989).

That MAM also activates murine macrophages was suggested as early as 1982 (32). Induction of cytokine release by human monocytes was confirmed during the next decade (1, 67) as was the ability of MAM to activate human NK-cells *in vitro* (19). Direct activation of NK cells by MAM implies another receptor since this subpopulation is generally not considered to express MHC molecules. However NK cells can also be activated indirectly by cytokines from MAM-activated T-cells. Recently, it was shown that MAM also induces nitric oxide (NO) production in murine peritoneal macrophages and murine and human monocytic tumor cell lines (Mu and Cole, unpublished observation).

### 3.2.2  *In vivo* effects

All of these activities imply that MAM should have a powerful immunomodulating effect in *M. arthritidis* infections, and evidence is mounting in favor of such a role, particularly in mice. MAM is responsible for toxic death and dermal necrosis in mice injected intravenously and subcutaneously, respectively, with viable *M. arthritidis* (25). This effect is mouse strain-specific and correlates directly with lymphocyte responsiveness to MAM *in vitro*. In addition, MAM causes transient joint inflammation after ia injection into rats (8) and will induce anergy (28), prolong skin grafts, suppress contact sensitivity (9), and exacerbate collagen-induced arthritis in mice (15). The latter is a well-characterized model of autoimmune arthritis [reviewed in (26)], and the ability of MAM not only to exacerbate and reactivate the disease, but also to trigger it, is especially intriguing (see below).

Although early studies suggested that there was no correlation between the susceptibility of mice to *M. arthritidis* arthritis and the ability of their T-cells to respond to MAM (20), Mu *et al.* recently showed that MAM induces profoundly different cytokine profiles *in vivo*, although not *in vitro*, in arthritis-susceptible and arthritis-resistant mice. Thus, a protective type 2 response predominates in the more highly resistant BALB/c strain and a proinflammatory type 1 response predominates in the arthritis-susceptible C3H/HeJ strain (61). In addition, C3H/HeJ mice are more susceptible to *M. arthritidis*-induced toxic death than are BALB/c mice. These data suggest that responses to MAM *in vivo* may at least predict and at most be responsible for susceptibility to both arthritogenic and toxic properties of *M. arthritidis.*

C3H/HeJ mice, but not C3H/HeSnJ mice, are hyporesponsive to lipopolysaccharide (LPS) because of a spontaneous mutation, $lps^d$, that inactivates the toll-like receptor 4 (TLR4), a member of the IL-1 receptor family. The additional finding by Mu *et al.* (60) that, unlike the C3H/HeJ mouse, the C3H/HeSnJ mouse elaborates a type 2 protective cytokine profile in response to MAM and is more resistant to the toxic effects of MAM is noteworthy. The $lps^d$ mutation and *in vivo* responsiveness to MAM and to *M. arthritidis*-induced toxic shock are both inherited in a dominant negative fashion (19). These observations suggest that MAM may interact with or be influenced by the LPS-signaling pathway and that it may exert some of its immunomodulatory effects through innate pathways early in the course of infection.      However, there are additional complexities, since other determinants can control cytokine profiles induced by MAM. Susceptibility to collagen-induced arthritis is dependent upon MHC haplotype, and the permissive Vβ chains include those used by MAM.   MAM induces a proinflammatory type 1 cytokine profile in collagen-susceptible B10.RIII mice (H-2$^r$) but a type 2 protective response in collagen-resistant BALB/c (H-2$^d$) and C57BL/10 (B.10, H-2$^b$) mice. All three of these strains possess an intact, functional TLR4 molecule [(19) and Mu *et al.*, unpublished observation].

Although the role of MAM in rat arthritis remains largely unexplored, MAM may contribute to the early and vigorous antibody response, the intensity of inflammation that characterizes this disease, the absence of growth-inhibiting and opsonizing antibodies, and the autoimmune phenomena proposed by some investigators. In recent years, MAM has also become a valuable tool for studying the role of SAgs in human autoimmune diseases, including RA. The observation that antibodies to MAM are present in the sera of human RA patients is particularly interesting (69).   Space limitations preclude pursuing that topic here, but interested readers are referred to a recent review of this work by Cole *et al.* (19).

## 3.3    MAV1

Extrachromosomal DNA elements are rare in the genus *Mycoplasma*, so the discovery of one in *M. arthritidis* was of considerable interest.  This element, designated MAV1, is a 16-kb, double-stranded DNA, temperate bacteriophage. Even more interesting was the observation that only the most highly virulent *M. arthritidis* strains are lysogenized by MAV1 (76). Voelker and Dybvig sequenced the complete MAV1 genome and found 15 open reading frames, 14 of which are oriented in the same direction and appear to encode phage functions such as integration, transposition, excision, and immunity (75). The 15th ORF appears to encode a membrane lipoprotein designated Vir. The gene encoding Vir is transcribed in lysogenized *M. arthritidis* strains (75). The protein product partitions into the membrane fraction on TX-114 extraction and migrates in the 30-kDa range by polyacrylamide gel electrophoresis. Interestingly, some nonlysogenized strains produce a cross-reacting protein that migrates at about 70 kDa. This protein disappears and is apparently replaced by Vir on lysogenization (Tu, A-H.T. *et al.*, General Meeting of the American Society for Microbiology, 2001). Voelker and Dybvig proposed that Vir confers resistance to killing during the early stages of infection (75). This is consistent with the suggestion put forth earlier by Washburn *et al.* that virulence differences are based largely upon differences in resistance to nonspecific clearance mechanisms (84) and with the observation by Cole *et al.* that virulent, *i.e.*, lysogenized, strains persist much longer in peripheral blood of rats and mice than do avirulent strains (22). Elucidation of the role of Vir and/or other MAV1 factors in pathogenesis of *M. arthritidis* infections of both rats and mice are still under investigation.

## 4. MODEL OF PATHOGENESIS FOR RATS

Based on the observations summarized above, the following model for pathogenesis of *M. arthritidis*-induced arthritis in the natural host is proposed: The very large numbers of viable organisms required to induce arthritis in rats and mice suggest i) that only a small number of cells within a population of virulent *M. arthritidis* is able to avoid early clearance due to expression of the MAV1-encoded virulence determinant Vir on their surfaces; ii) that only a small number of cells within a population expresses the appropriate combination of adhesins allowing entry into and colonization of joint tissues, and/or iii) that large numbers of organisms are required to overwhelm innate immune pathways in otherwise healthy animals. Very likely all of these factors come into play. Once in the joints, mycoplasmas attach to host tissues by adhesins such as MAA1 and MAA2 and begin to replicate. Actively replicating mycoplasmas stimulate an early and vigorous antibody response. However, they persist because rats are unable to produce either growth-inhibiting or opsonizing antibodies against them, and virulent *M. arthritidis* is highly resistant to phagocytosis by murine macrophages in the absence of opsonization. The *M. arthritidis* superantigen MAM, released from dead or dying cells, nonspecifically activates significant numbers of T lymphocytes, possibly including those that would ordinarily have responded to the targets of neutralizing or opsonizing antibodies. Activated lymphocytes and macrophages release large amounts of proinflammatory cytokines that exacerbate the inflammatory process, resulting in the massive joint destruction characteristic of this disease. Serum from convalescent rats is able to protect passively immunized rats against infection. Although the antibody(ies) responsible for this and its target antigen have not been identified, two candidates are the "defective" mycoplasmacidal antibody active only against nonreplicating mycoplasmas and the antibody response against putative adhesin MAA1, which may prevent newly replicating cells from attaching to each other and to their target tissues. Antibodies against MAM and Vir may also neutralize these factors, reducing inflammation and allowing at least limited phagocytosis. Eventually the organisms are cleared and the damage is repaired. Clearly this model is incomplete, and work on all three fronts - Membranes, MAM, and MAV1 - continues.

# REFERENCES

1. **al-Daccak, R., K. Mehindate, J. Herbert, L. Rink, S. Mecheri, and W. Mourad.** 1994. *Mycoplasma arthritidis*-derived superantigen induces proinflammatory monokine gene expression in the THP-1 human monocytic cell line. Infect. Immun. **62**:2409-2416.

2. **Aldridge, K.E., B.C. Cole, and J.R. Ward.** 1977. Mycoplasma-dependent activation of normal lymphocytes: Induction of a lymphocyte-mediated cytotoxicity for allogeneic and syngeneic mouse target cells. Infect. Immun. **18**:377-385.

3. **Atkin, C.L., S. Wei, and B.C. Cole.** 1994. The *Mycoplasma arthritidis* superantigen MAM: Purification and identification of an active peptide. Infect. Immun. **62**:5367-5375.

4. **Behrens, A., M. Heller, H. Kirchhoff, D. Yogev, and R. Rosengarten.** 1994. A family of phase- and size-variant membrane surface lipoprotein antigens (Vsps) of *Mycoplasma bovis*. Infect. Immun. **62**:5075-5084.

5. **Binder, A., K. Gärtner, H.J. Hedrich, W. Hermanns, H. Kirchhoff, and K. Wonigeit.** 1990. Strain differences in sensitivity of rats to *Mycoplasma arthritidis* ISR 1 infection are under multiple gene control. Infect. Immun. **58**:1584-1590.

6. **Binder, A., H.J. Hedrich, K. Wonigeit, and H. Kirchhoff.** 1993. The *Mycoplasma arthritidis* infection in congenitally athymic nude rats. J. Exp. Anim. Sci. **35**:177-185.

7. **Cahill, J.F., B.C. Cole, B.B. Wiley, and J.R. Ward.** 1971. Role of biological mimicry in the pathogenesis of rat arthritis induced by *Mycoplasma arthritidis*. Infect. Immun. **3**:24-35.

8. **Cannon, G.W., B.C. Cole, J.R. Ward, J.L. Smith, and E.J. Eichwald.** 1988. Arthritogenic effects of *Mycoplasma arthritidis* T cell mitogen in rats. J. Rheumatol. **15**:735-741.

9. **Cole, B.C.** 1991. The immunobiology of *Mycoplasma arthritidis* and its superantigen MAM. Curr. Top. Microbiol. Immunol. **174**:107-119.

10. **Cole, B.C., R.A. Balderas, E.A. Ahmed, D. Kono, and A.N. Theofilopoulos.** 1993. Genomic composition and allelic polymorphisms influence Vβ usage by the *Mycoplasma arthritidis* superantigen. J. Immunol. **150**:3291-3299.

11. **Cole, B.C., J.F. Cahill, B.B. Wiley, and J.R. Ward.** 1969. Immunological responses of the rat to *Mycoplasma arthritidis.* J. Bacteriol. **98:**930-937.

12. **Cole, B.C., L. Golightly-Rowland, and J.R. Ward.** 1976. Arthritis of mice induced by *Mycoplasma arthritidis.* Humoral and lymphocyte responses of CBA mice. Ann. Rheum. Dis. **35:**14-22.

13. **Cole, B.C., L. Golightly-Rowland, and J.R. Ward.** 1975. Chronic proliferative arthritis of mice induced by *Mycoplasma arthritidis:* Demonstration of a cell-mediated immune response to mycoplasma antigens in vitro. Infect. Immun. **11:**1159-1161.

14. **Cole, B.C., L. Golightly-Rowland, J.R. Ward, and B.B. Wiley.** 1970. Immunological response of rodents to murine mycoplasmas. Infect. Immun. **2:**419-425.

15. **Cole, B.C., and M.M. Griffiths.** 1993. Triggering and exacerbation of autoimmune arthritis by the *Mycoplasma arthritidis* superantigen MAM. Arthritis Rheum. **36:**994-1002.

16. **Cole, B.C., M.M. Griffiths, E.J. Eichwald, and J.R. Ward.** 1977. New models of rabbit synovitis induced by mycoplasmas. I. Microbiological, histopathological, and immunological observations on rabbits injected with *M. arthritidis* and *M. pulmonis.* Infect. Immun. **16:**382-396.

17. **Cole, B.C., K.L. Knudtson, A. Oliphant, A.D. Sawitzke, A. Pole, M. Manohar, L.S. Benson, E. Ahmed, and C.L. Atkin.** 1996. The sequence of the *Mycoplasma arthritidis* superantigen, MAM: Identification of functional domains and comparison with microbial superantigens and plant lectin mitogens. J. Immunol. **183:**1105-1110.

18. **Cole, B.C., M.L. Miller, and J.R. Ward.** 1967. A comparative study on the virulence of *Mycoplasma arthritidis* and "*Mycoplasma hominis* type 2" strains in rats. Proc. Soc. Exp. Biol. Med. **124:**103-107.

19. **Cole, B.C., A. Sawitzke, and H.-H. Mu.** 2000. *Mycoplasma arthritidis* and its superantigen *M. arthritidis* mitogen as a model for inflammatory and autoimmune disease, p. 93-107. *In* M. W. Cunningham and R. S. Fujinami (ed.), Effects of microbes on the immune system. Lippincott Williams & Wilkins, Philadelphia, PA.

20. **Cole, B.C., R.N. Thorpe, L.A. Hassell, L.R. Washburn, and J.R. Ward.** 1983. Toxicity but not arthritogenicity of *Mycoplasma arthritidis* associates with the haplotype expressed at the major histocompatibility complex. Infect. Immun. **41:**1010-1015.

21. **Cole, B.C., and J.R. Ward.** 1973. Detection and characterization of defective mycoplasmacidal antibody produced by rodents against *Mycoplasma arthritidis.* Infect. Immun. **8:**199-207.

22. **Cole, B.C., and J.R. Ward.** 1973. Fate of intravenously injected *Mycoplasma arthritidis* in rodents and effect of vaccines. Infect. Immun. **7:**416-425.

23. **Cole, B.C., and J.R. Ward.** 1979. Mycoplasmas as arthritogenic agents, p. 367-398. *In* J. G. Tully and R. F. Whitcomb (ed.), The Mycoplasmas, Vol. II, Human and Animal Mycoplasmas. Academic Press, New York, NY.

24. **Cole, B.C., J.R. Ward, L. Golightly-Rowland, and G.A. Trapp.** 1971. Chronic proliferative arthritis of mice induced by *Mycoplasma arthritidis.* II. Serological responses of the host and effect of vaccines. Infect. Immun. **4:**431-440.

25. **Cole, B.C., J.R. Ward, R.S. Jones, and J.F. Cahill.** 1971. Chronic proliferative arthritis of mice induced by *Mycoplasma arthritidis.* I. Induction of disease and histopathological characteristics. Infect. Immun. **4:**344-355.

26. **Cole, B.C., L.R. Washburn, C.O. Samuelson, and J.R. Ward.** 1982. Experimental models of rheumatoid arthritis, systemic lupus erythematosus, and scleroderma, p. 22-52. *In* G. S. Panayi (ed.), Scientific Basis of Rheumatology. Churchill Livingston, Edinburgh.

27. **Cole, B.C., L.R. Washburn, and D. Taylor-Robinson.** 1985. Mycoplasma-induced arthritis, p. 108-160. *In* S. Razin and M. F. Barile (ed.), The Mycoplasmas, Vol. IV, Mycoplasma Pathogenicity. Academic Press, New York.

28. **Cole, B.C., and D.J. Wells.** 1990. Immunosuppressive properties of the *Mycoplasma arthritidis* T-cell mitogen in vivo: Inhibition of proliferative responses to T-cell mitogens. Infect. Immun. **58:**228-236.

29. **Collier, W.A.** 1939. Infectious polyarthritis of rats. J. Pathol. Bacteriol. 48:579-589.

30. **Cox, N.R., M.K. Davidson, J.K. Davis, J.R. Lindsey, and G.H. Cassell.** 1988. Natural mycoplasmal infections in isolator-maintained LEW/Tru rats. Lab. Anim. Sci. **38:**381-388.

31. **Cunningham, M.E.** 2000. Pathogenesis of group A streptococcal infections. Clin. Microbiol. Rev. **13:**470-511.

32. **Dietz, J.N., and B.C. Cole.** 1982. Direct activation of the J774.1 murine macrophage cell line by *Mycoplasma arthritidis*. Infect. Immun. 37:811-819.

33. **Dinges, M.M., P.M. Orwin, and P.M. Schlievert.** 2000. Exotoxins of *Staphylococcus aureus*. Clin. Microbiol. Rev. **13:**16-34.

34. **Droesse, M., G. Tangen, I. Gummelt, H. Kirchhoff, L.R. Washburn, and R. Rosengarten.** 1995. Major membrane proteins and lipoproteins as highly variable immunogenic surface components and strain-specific components and strain-specific antigenic markers of *Mycoplasma arthritidis*. Microbiology **141:**3207-3219.

35. **Edward, D.G.ff., and E.A. Freundt.** 1956. The classification and nomenclature of organisms of the pleuropneumonia group. J. Gen. Microbiol. **14:**197-207.

36. **Edward, D.G.ff., and E.A. Freundt.** 1965. A note on the taxonomic status of strains like 'Campo', hitherto classified as *Mycoplasma hominis*, type 2. J. Gen. Microbiol. **41:**263-265.

37. **Findlay, G.M., R.D. MacKenzie, F.O. MacCullum, and E. Klieneberger.** 1939. The aetiology of polyarthritis in the rat. Lancet **237:**7-10.

38. **Gärtner, K., H. Kirchhoff, K. Mensing, and R. Velleuer.** 1989. The influence of social rank on the susceptibility of rats to *Mycoplasma arthritidis*. J. Behav. Med. **12:**487-502.

39. **Hannan, P.C.T., and B.O. Hughes.** 1971. Reproducible polyarthritis in rats caused by *Mycoplasma arthritidis*. Ann. Rheum. Dis. **30:**316-321.

40. **Hermanns, W., L.-C. Schulz, H. Kirchhoff, and J. Heitmann.** 1983. Studies of polyarthritis caused by *Mycoplasma arthritidis* in rats. III. Histological findings. Zentralbl. Bakteriol. Parasitenkd. Infektionskr. Hyg. Abt. 1 Orig. **254:**423-434.

41. **Hill, A., and G.J.R. Dagnall.** 1975. Experimental polyarthritis in rats produced by *Mycoplasma arthritidis*. J. Comp. Pathol. **85:**45-52.

42. **Hodtsev, A.S., Y. Choi, E. Spanopoulou, and D. Posnett.** 1998. Mycoplasma superantigen is a CDR3-dependent ligand for the T cell antigen receptor. J. Exp. Med. **187:**319-327.

43. **Howell, E.V., and R.S. Jones.** 1963. Factors influencing pathogenicity of *Mycoplasma arthritidis* (PPLO). Proc. Soc. Exp. Biol. Med. **112:**69-72.

44. **Howell, E.V., W.K. Otto, and R.S. Jones.** 1961. Mycoplasma (PPLO) and Murphy-Sturm Lymphosarcoma. Proc. Soc. Exp. Biol. Med. **106:**673-677.

45. **Howell, E.V., J.R. Ward, and R.S. Jones.** 1959. Mycoplasmal (PPLO) polyarthritis and tumor regression in rats. Proc. Soc. Exp. Biol. Med. **102:**210-212.

46. **Jansson, E., U. Vanio, O. Snellman, and S. Tuuri.** 1971. Search for mycoplasma in rheumatoid arthritis. Ann. Rheum. Dis. **30:**413-418.

47. **Jasmin, G.** 1967. Experimental arthritis in rats. A comprehensive review with special reference to mycoplasmas. Rheumatology **1:**197-231.

48. **Jouanneau, M., H. Brouilhet, A. Kahan, A. Charles, and F. Delbarre.** 1973. The effect of immunosuppression on polyarthritis induced by *Mycoplasma arthritidis* in rats. Biomed. Express **19:**156-159.

49. **Kaklamani, E., D. Karalis, P. Kaklamanis, Y. Koumandaki, K. Katsouyanni, C. Blackwell, L. Sparos, D. Weir, and D. Trichopoulos.** 1991. The effect of *Mycoplasma arthritidis* infection on the phagocytic activity of macrophages in rats and mice. FEMS Microbiol. Lett. **76:**151-158.

50. **Kirchhoff, H., J. Heitmann, A. Ammar, W. Hermanns, and L.-C. Schulz.** 1983. Studies of polyarthritis caused by *Mycoplasma arthritidis* in rats. I. Detection of the persisting mycoplasma antigen by the enzyme immune assay (EIA) and conventional culture technique. Zentralbl. Bakteriol. Parasitenkd. Infektionskr. Hyg. Abt. 1 Orig. **254A:**129-138.

51. **Kirchhoff, H., J. Heitmann, H. Mielke, H. Dubenkropp, and R. Schmidt.** 1983. Studies of polyarthritis caused by *Mycoplasma arthritidis* in rats. II. Serological investigation of rats experimentally infected with *M. arthritidis* ISR 1. Zentralbl. Bakteriol. Parasitenkd. Infektionskr. Hyg. Abt. 1 Orig. **254:**275-280.

52. **Klieneberger, E.** 1940. The pleuropneumonia-like organisms: Further comparative studies and a descriptive account of recently discovered types. J. Hyg. Camb. **40:**204-222.

53. **Klieneberger, E.** 1939. Studies on pleuropneumonia-like organisms: The L4 organisms as the cause of Woglom's "pyogenic virus". J. Hyg. Camb. **39:**260-265.

54. **Krause, D.C., T. Proft, C.T. Hedreyda, H. Hilbert, H. Plagens, and R. Herrmann.** 1997. Transposon mutagenesis reinforces the correlation between *Mycoplasma pneumoniae* cytoskeletal protein HMW2 and cytadherence. J. Bacteriol. **179:**2688-2677.

55. **Laber, G., H. Walzl, and E. Schütze.** 1975. Klinische und histopathologische Befunde bei der Mykoplasma-Polyarthritis der Ratte. I. Der Ablauf der Infektion in den ertsen 8 Tagen. Zentralbl. Bakteriol. Parasitenkd. Infektionskr. Hyg. Abt. 1 Orig. **230:**385-397.

56. **Langlois, M.A., P. Etongue-Mayer, M. Ouellette, and W. Mourad.** 2000. Binding of *Mycoplasma arthritidis*-derived mitogen to human MHC class II molecules via its N terminus is modulated by invariant chain expression and its C terminus is required for T cell activation. Eur. J. Immunol. **30:**1748-1756.

57. **Leberman, P.R., P.F. Smith, and H.E. Morton.** 1950. The susceptibility of pleuropneumonia-like organisms to the in vitro action of antibiotics: Aureomycin, chloramphenicol, dihydrostreptomycin, streptomycin, and sodium penicillin G. J. Urol. **64:**167-173.

58. **Lemcke, R.** 1964. The serological differentiation of *Mycoplasma* strains (pleuropneumonia-like organisms) from various sources. J. Hyg. Camb. **62:**199-219.

59. **Markham, P.F., M.D. Glew, M.R. Brandon, I.D. Walker, and K.G. Whithear.** 1992. Characterization of a major hemagglutinin protein from *Mycoplasma gallisepticum.* Infect. Immun. **60:**3885-3891.

60. **Mu, H.-H., A. Sawitzke, and B.C. Cole.** 2001. Presence of the *lps^d* mutation influences cytokine regulation *in vivo* by the *Mycoplasma arthritidis* superantigen MAM and lethal toxicity in mice given *M. arthritidis*. Infect. Immun. **69:**3837-3844.

61. **Mu, H.-H., A.D. Sawitzke, and B.C. Cole.** 2000. Modulation of cytokine profiles by the mycoplasma superantigen *Mycoplasma arthritidis* mitogen parallels susceptibility to arthritis induced by *M. arthritidis.* Infect. Immun. **68:**1142-1149.

62. **Nicol, C.S., and G.D.ff. Edward.** 1953. Role of organisms of the pleuropneumonia group in human genital infections. Br. J. Vener. Dis. **29:**141-150.

63.  **Olson, L.D., S.W. Shane, A.A. Karpas, T.M. Cunningham, P.S. Probst, and M.F. Barile.** 1991. Monoclonal antibodies to surface antigens of a pathogenic *Mycoplasma hominis* strain. Infect. Immun. **59**:1683-1689.

64.  **Phillips, G.N., P.F. Flicker, C. Cohen, B.N. Manjula, and V.A. Fischetti.** 1981. Streptococcal M protein: alpha-helical coiled-coil structure and arrangement on the cell surface. Proc. Natl. Acad. Sci. USA **78**:4689-4693.

65.  **Preston, W.S.** 1942. Arthritis in rats caused by pleuropneumonia-like micro-organisms and the relationship of similar organisms to human rheumatism. J. Infect. Dis. **70**:180-184.

66.  **Razin, S., D. Yogev, and Y. Naot.** 1998. Molecular biology and pathogenicity of mycoplasmas. Microbiol. Molec. Biol. Rev. **62**:1094-1156.

67.  **Rink, L., W. Nicklas, J. Luhm, R. Kruse, and H. Kirchner.** 1996. Induction of a proinflammatory cytokine network by *Mycoplasma arthritidis*-derived superantigen (MAS). J. Interferon Cytok. Res. **16**:861-868.

68.  **Rosengarten, R., and K.S. Wise.** 1991. The Vlp system of *Mycoplasma hyorhinis*: Combinatorial expression of distinct size variant lipoproteins generating high-frequency surface antigenic variation. J. Bacteriol. **173**:4782-4793.

69.  **Sawitzke, A., D. Joyner, K.L. Knudtson, H.-H. Mu, and B.C. Cole.** 2000. anti-MAM antibodies in rheumatic disease: evidence for a MAM-like superantigen in rheumatoid arthritis. J. Rheumatol. **27**:358-364.

70.  **Schaub, I.G., and J.A. Guilbeau.** 1949. The occurrence of pleuropneumonia-like organisms in material from the postpartum uterus; simplified methods for isolation and staining. Bull. Johns Hopkins Hosp. **84**:1-10.

71.  **Stewart, D.D., and G.E. Buck.** 1975. The occurrence of *Mycoplasma arthritidis* in the throat and middle ear of rats with chronic respiratory disease. Lab. Anim. Sci. **25**:769-773.

72.  **Stulle, K., A. Binder, G. Sommer, and H. Kirchhoff.** 1988. Immunological reaction of rat tissue cells with antiserum against *Mycoplasma arthritidis* membranes in indirect immunofluorescence test. J. Vet. Med. B **35**:713-715.

73.  **Tangen, G., and H. Kirchhoff.** 1995. Investigation into the origin of the antigens of *Mycoplasma arthritidis* cross-reacting with host tissue antigens. FEMS Microbiol. Lett. **12**:9-16.

74.  **Tully, J.G.** 1965. Biochemical, morphological, and serological characterization of mycoplasma of murine origin. J. Infect. Dis. **115**:171-185.

75.  **Voelker, L.L., and K. Dybvig.** 1999. Sequence analysis of the *Mycoplasma arthritidis* bacteriophage MAV1 genome identifies the putative virulence factor. Gene **233**:101-107.

76.  **Voelker, L.L., K.E. Weaver, L.J. Ehle, and L.R. Washburn.** 1995. Association of lysogenic bacteriophage MAV1 with virulence of *Mycoplasma arthritidis*. Infect. Immun. **63**:4016-4023.

77.  **Ward, J.R., and R.S. Jones.** 1962. The pathogenesis of mycoplasmal (PPLO) arthritis in rats. Arthritis Rheum. **5**:163-175.

78.  **Washburn, L.R., B.C. Cole, M.I. Gelman, and J.R. Ward.** 1980. Chronic arthritis of rabbits induced by mycoplasmas. I. Clinical, microbiologic, and histologic features. Arthritis Rheum. **23**:825-836.

79.  **Washburn, L.R., B.C. Cole, and J.R. Ward.** 1980. Chronic arthritis of rabbits induced by mycoplasmas. II. Antibody response and the deposition of immune complexes. Arthritis Rheum. **23**:837-845.

80. **Washburn, L.R., B.C. Cole, and J.R. Ward.** 1982. Chronic arthritis of rabbits induced by mycoplasmas. III. Induction with nonviable *Mycoplasma arthritidis* antigens. Arthritis Rheum. **25:**937-946.

81. **Washburn, L.R., B.C. Cole, and J.R. Ward.** 1988. Expression of metabolism-inhibition antibodies against *Mycoplasma arthritidis* in rats. Amer. J. Vet. Res. **49:**52-57.

82. **Washburn, L.R., S. Hirsch, and L.L. Voelker.** 1993. Mechanisms of attachment of *Mycoplasma arthritidis* to host cells in vitro. Infect. Immun. **61:**2670-2680.

83. **Washburn, L.R., E.J. Miller, and K.E. Weaver.** 2000. Molecular characterization of *Mycoplasma arthritidis* membrane lipoprotein MAA1. Infect. Immun. **68:**437-442.

84. **Washburn, L.R., and J.R. Ramsay.** 1989. Experimental induction of arthritis in LEW rats and antibody response to four *Mycoplasma arthritidis* strains. Vet. Microbiol. **21:**41-55.

85. **Washburn, L.R., J.R. Ramsay, and M.B. Andrews.** 1988. Recognition of *Mycoplasma arthritidis* membrane antigens by rats and rabbits: Comparison by immunoblotting and radioimmunoprecipitation. Vet. Microbiol. **17:**45-57.

86. **Washburn, L.R., L.L. Voelker, L.J. Ehle, S. Hirsch, C. Dutenhofer, K. Olson, and B. Beck.** 1995. Comparison of *Mycoplasma arthritidis* strains by enzyme-linked immunosorbent assay, immunoblotting, and DNA restriction analysis. J. Clin. Microbiol. **33:**2271-2279.

87. **Washburn, L.R., and E.J. Weaver.** 1997. Protection of rats against *Mycoplasma arthritidis*-induced arthritis by active and passive immunizations with two surface antigens. Clin. Diag. Lab. Immunol. **4:**321-327.

88. **Washburn, L.R., K.E. Weaver, E.J. Weaver, W. Donelan, and S. Al-Sheboul.** 1998. Molecular characterization of *Mycoplasma arthritidis* variable surface protein MAA2. Infect. Immun. **66:**2576-2586.

89. **White, J., A. Herman, A.M. Pullen, R. Kubo, J.W. Kappler, and P. Marrack.** 1989. The Vβ-specific superantigen staphylococcal enterotoxin B: stimulation of mature T cells and clonal deletion in neonatal mice. Cell **56:**27-35.

90. **Woglom, W.H., and J. Warren.** 1939. The nature of pyogenic filterable agent in the white rat. J. Hyg. Camb. **39:**266-267.

91. **Woglom, W.H., and J. Warren.** 1938. A pyogenic filterable agent in the albino rat. J. Exp. Med. **68:**513-528.

# Chapter 22

# Cytadherence and the Cytoskeleton

MITCHELL F. BALISH and DUNCAN C. KRAUSE
*Department of Microbiology, University of Georgia, Athens, GA, USA*

## 1. INTRODUCTION

### 1.1 The attachment organelle

In order to associate specifically with host cells, *Mycoplasma pneumoniae* and closely related species employ a specialized polar structure, the attachment organelle, which is assembled from a set of unique proteins (49-51). Many of these cytadherence-associated proteins exhibit insolubility in the nonionic detergent Triton X-100 (TX) and are associated with a structure, the triton shell, that remains after extraction of cells with TX. Because of its solubility properties, its appearance, and its association with various features of cell morphology, cell motility, cell division, and cell-cell adhesion, this triton shell is regarded as a novel bacterial cytoskeleton.

### 1.2 The cytoskeleton

#### 1.2.1 Comparison with the eukaryotic cytoskeleton

The cytoskeleton is commonly regarded as a hallmark of eukaryotic cells. Polymeric proteinaceous filaments of three varieties, microtubules, microfilaments, and intermediate filaments, constitute the central players in the eukaryotic cytoskeleton. Early investigations of the *M. pneumoniae* TX-insoluble structure revealed analogies with actin-based microfilaments (71), but efforts to identify analogous proteins in *M. pneumoniae* and related

*Molecular Biology and Pathogenicity of Mycoplasmas*, Edited by Razin and Herrmann, Kluwer Academic/Plenum Publishers, New York, 2002

organisms have failed, culminating in the demonstrated absence of such genes from mycoplasma genomes (9,23,27,35). Thus, it is clear that the mycoplasma cytoskeleton is composed of proteins different from those of eukaryotic cells. As a result, insight into the mycoplasma cytoskeleton has seldom derived from information gleaned from studies of eukaryotic cells, but requiring instead direct analysis. Among the mycoplasmas, *M. pneumoniae* has been the primary focus of these experiments.

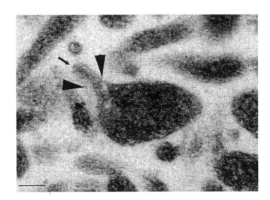

*Figure 1.* Transmission electron micrograph of a thin section of *M. pneumoniae* cells. Prominently featured in the terminal organelle is an electron-dense core (arrow) surrounded by electron-lucent space (arrowheads). Bar, 100 nm. (Courtesy of M. J. Willby.)

### 1.2.2    Distinguishing features of the *Mycoplasma* attachment organelle

Electron microscopic (EM) images of both the *M. pneumoniae* attachment organelle and the triton shell prominently feature an electron-dense rodlike core of uncertain composition, notably surrounded by electron-lucent space (Fig. 1) (28,68). The distal end of the core is wider than the core shaft and is referred to as the terminal button. The core itself is frequently observed to have either a banded or spiral pattern (68,82). The position of the core within the attachment organelle has led researchers to believe that it is critical for cytadherence, though the identities of all its components are unknown. In addition, other cytadherence-associated proteins are membrane-bound (3,59,77,80). Efforts to comprehend the components of the cytoskeleton in terms of their assembly and their dynamic interaction with the mycoplasma membrane have revealed a number of interesting similarities and differences with respect to eukaryotic cytoskeletons.

## 2. THE ATTACHMENT ORGANELLE AND *M. PNEUMONIAE* CELL DIVISION

Although the two processes have not been observed directly, there is strong evidence that *M. pneumoniae* cell division is temporally linked to attachment organelle duplication (6,8,89). Comparison between cellular DNA content and localization of the P1 adhesin, which is prominently clustered at the attachment organelle, suggests that a new attachment organelle is formed adjacent to the original one as the cellular DNA content begins to increase (89). The juxtaposition of the new and old tips raises the possibility that components from the original attachment organelle are used in creating the nascent one, providing a mechanism for the targeting of core components to the cell pole. Prior to doubling of DNA content and resolution of the daughter nucleoids, one tip structure can be observed alongside the cell body, suggesting that one attachment organelle migrates to the opposite pole during replication and before nucleoid separation. Indeed, within an asynchronous population one finds a subset of cells having an attachment organelle at each pole, presumably in the late stages of cell division. The mechanism of tip migration is not known but may be linked to that of nucleoid partitioning. The temporal linkage between attachment organelle biogenesis and cell division suggests that communication exists between this structure and the DNA replication and/or cell division machinery. Thus, regulation of cell division may be dependent upon proper attachment organelle assembly.(see Cell Division chapter)

## 3. ATTACHMENT ORGANELLE PROTEINS

### 3.1 Introduction

Insight into the process of *M. pneumoniae* attachment organelle assembly is critical to understanding the basic biology and pathogenesis of this organism. In recent years several laboratories have focused their research efforts on identifying organelle component proteins in *M. pneumoniae* and related mycoplasmas. These studies indicate that several species, including *M. pneumoniae, Mycoplasma genitalium, Mycoplasma gallisepticum,* and *Mycoplasma pirum,* produce at least some members of this novel set of proteins associated with attachment organelle formation. An hypothesis that drives much of this research is that these proteins assemble in a specific sequence to form a structure that is essential for mycoplasma colonization and virulence. Therefore, the identification of terminal organelle proteins is merely a first step to understanding the molecular basis for attachment organelle assembly and adherence to host cells. Analysis of noncytadhering mutants has led to identification of key components and characterization of the defects that ultimately cause a noncytadhering phenotype. These may include altered cell morphology, improper protein localization, and instability of specific proteins. Proteins involved in cytadherence were identified through the study of four classes of *M. pneumoniae* cytadherence mutants twenty years ago; these proteins are P1, B, C, P65, HMW1, HMW2, and HMW3 (Table 1) (50,52). Systematic evaluation of these mutants has begun to yield a better understanding of the attachment organelle assembly process and the roles of specific proteins therein.

*Table 1.* Levels of cytadherence-associated proteins in cytadherence mutants.

|         | P1  | B / C* | P30   | HMW1 | HMW2  | HMW3 | P65 |
|---------|-----|--------|-------|------|-------|------|-----|
| WT      | +++ | +++    | +++   | +++  | +++   | +++  | +++ |
| I-2     | +++ | +++    | ++    | +    | -     | +    | +   |
| H9      | +++ | +++    | +++   | +/-  | ++Δ   | +/-  | +/- |
| II-3    | +++ | +++    | -     | +++  | +++   | +++  | ++  |
| II-7/M7 | +++ | +++    | ++Δ   | +++  | +++   | +++  | ++  |
| III-4/M5| +++ | -      | +++   | +++  | +++   | +++  | +++ |
| IV-22   | -   | -      | +++   | +++  | +++   | +++  | +++ |
| M6      | +++ | +++    | ++ Δ  | -    | +     | +    | +   |
| *hmw3⁻* | +++ | +++    | +++   | +++  | +++   | -    | ++  |

\* The ORF6 gene products P90 and P40, respectively; +++, high levels; ++, intermediate levels; +, low levels; +/-, barely detectable; -, none detectable; Δ, protein is truncated. References in text.

## 3.2      The P1 adhesin

### 3.2.1     Structure, processing and localization of P1

Protein P1 (39), which is absent in the cytadherence mutant IV-22 due to a frameshift in its gene (52,93), is a major adhesin of *M. pneumoniae* (4,20,38). Its only known homologs are the adhesins of the closely related species *M. genitalium* (12,40), *M. gallisepticum* (29), and *M. pirum* (97), though its C-terminal domain is related to that of cytadherence-accessory protein B (42). Synthesized in *M. pneumoniae* strain M129 as a 1627-amino acid preprotein and subject to removal of an N-terminal signal peptide (41,

44,90,95), P1 is an integral membrane protein (43,77). In the FH strain of *M. pneumoniae*, the P1 gene contains a variant region (94); other P1 variants have also been identified in clinical isolates (18,47). In wild-type *M. pneumoniae* cells, P1 is present at the attachment organelle (Fig. 2)(20,38,89), where it binds host cell-surface molecules (4); however, according to both immunoelectron (IEM) and immunofluorescence (IFM) microscopy, lower concentrations of P1 are widespread on the cell surface. P1 clustering appears to be important for its function, as numerous cytadherence mutants share in common a failure to cluster P1 at the attachment organelle (4,33,59). An exception exists with mutants II-3 and II-7, in which the primary defect is in protein P30 (85). P1 is clustered in these mutants despite a significantly altered cell morphology, indicating that clustering of P1 is necessary but not sufficient for cytadherence.

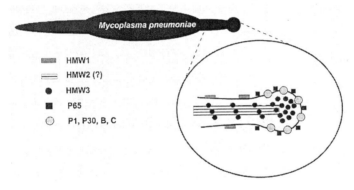

*Figure 2.* Schematic diagram of locations of cytadherence-accessory proteins in *M. pneumoniae*. P1, B, C, and P30 are grouped together based upon immunolocalization data for P1, B, and P30, and cross-linking data for P1, B, and C (58,61). Although HMW1 and P65 are indicated only on the cell surface subpopulation of these proteins may also exist in the cell interior.

### 3.2.2    TX partitioning of P1

Although P1 was the first protein identified as a component of the *M. pneumoniae* triton shell (46), reports of the degree of partitioning of P1 between soluble and insoluble fractions have varied. In fact, P1 can be entirely removed from the triton shell by extracting suspended cells multiple times in 2% TX, whereas no additional HMW1 is solubilized after the first extraction (R. H. Waldo III and D. C. K., unpublished results). The weaker association of P1 with the triton shell is likely to account for the variation in reports of its partitioning, especially in light of the different extraction protocols used. EM examination of a TX-insoluble *M. pneumoniae* fraction in which P1 was identified indicates that membranous material appeared to be associated with attachment organelle cores (82). Poor extraction of the membrane may depend upon its composition; for example, sterol-rich membrane is poorly solubilized by TX (62).

## 3.3    ORF6 gene products

### 3.3.1    Structures of the ORF6 gene and proteins B and C

Proteins A, B, and C are absent in some high passage *M. pneumoniae* strains (90) and mutants III-4 and IV-22 (52). At 90 and 40 kDa respectively, proteins B and C are thought to be identical to proteins P90 and P40, which are derived from cleavage of a 130-kDa precursor (57) encoded by a gene historically designated ORF6 (42) and known now as MPN142 (14). ORF6 is located just downstream of the gene encoding P1 and is probably translationally linked to it. The absence of proteins B and C in mutant IV-22 (Table I) is likely to result from failure to translate P1. Cellular localization of B and C is similar to that of P1 (Fig. 2) (22,57,89), with B predicted to be a transmembrane protein like P1 (57). The identity of the 72-kDa protein A is unknown. Like P1, the B/C precursor has homologs only in closely related mycoplasma species. Its homologs appear not to be subject to proteolytic cleavage (15,69,74), and the sequence predicted to be involved in recognition and cleavage is absent.

### 3.3.2    ORF6 mutants and implications for B/C function

Proteins B and C are absent, but P1 is present at normal levels, in *M. pneumoniae* mutants III-4 (52) and M5 (56) (Table 1). The identity of the genetic defects of mutants III-4 and M5 are unreported but may differ, as mutant III-4 cells fail to localize P1 at the tips of these cells (4), while mutant M5 reportedly does localize P1 properly (89). Whether these strains lack only proteins B and C can best be determined by complementation studies with the ORF6 gene. However, both the large size of this gene and the likely requirement of the P1 gene for translational coupling have hindered these studies. The adjacency of the genes encoding P1 and the B/C precursor, the likely translational coupling of their products, and the cross-linking of the three polypeptides *in vitro* (58) suggest that proteins B, C, and P1 act together. Furthermore, the presence of normal levels of other cytadherence-associated proteins in mutants III-4 and IV-22 (51) suggests that these proteins act late in the attachment organelle assembly pathway, subsequent to the formation of a macromolecular complex by upstream proteins. Nonetheless, the irregular, often branched morphology of mutant III-4 cells (45) belies a role for proteins B and C in attachment organelle assembly that may or may not be independent of their role in polar clustering of P1. It has been suggested that proteins B and C are involved in enabling association of P1 with the triton shell (56), but in light of the extractability of P1 from this fraction under rigorous conditions it is difficult to interpret these results. Seto et al. (89) suggested that in mutant M5, protein P65 is improperly localized, whereas Jordan et al. (45) reported normal polar localization of P65 in mutant III-4. This disparity, like the difference in P1 localization, suggests that there are other, unknown differences between the two mutants.

## 3.4     Protein P65

### 3.4.1     Anticipated involvement of protein P65 in cytadherence

No mutant lacking P65 has been identified, making it difficult to determine its cellular role. However, P65 is implicated indirectly in cytadherence by several means. First, it is present at reduced levels in mutants lacking cytadherence-accessory proteins HMW1, HMW2, HMW3, and P30 (Table I) (47; M. J. Willby and D. C. K., submitted); for at least the *hmw2* mutant this reduction occurs post-translationally (45). Second, its gene is the first in the operon of the same name, also known as the cytadherence regulatory locus, just upstream of the *hmw2* gene, to which it is transcriptionally, albeit not translationally, linked (35,52). Third, like HMW1 and HMW3, P65 contains an acidic, proline-rich (APR) domain (78). Fourth, it is present on the mycoplasma surface and in the triton shell (77,79). Finally, P65 is localized by IFM primarily to the attachment organelle in permeabilized cells (Fig. 2) (45, 86).

### 3.4.2     Structure and localization of P65

P65 (Fig. 3) is a 47-kDa protein in *M. pneumoniae* strain M129 (78) and, by virtue of a perfect repeat within the APR domain, it appears as a 49-kDa protein in strain FH (78). Nonetheless, it exhibits aberrant migration in sodium dodecyl sulfate-polyacrylamide gel electrophoresis (SDS-PAGE), by which P65 migrates at 65 and 68 kDa, respectively (78). This slowed migration is likely due to the APR domain, which constitutes residues 9-181 of P65 in strain M129. APR domains are defined not by primary structure but by compositional similarity (79). The APR domain of P65 is enriched in Pro, Asp, Asn, and Tyr, but is devoid of Glu and basic residues. It contains numerous repeats of the tripeptide Asp-Pro-Asn embedded within larger repeats. Interestingly, while the only known homolog of P65, encoded by *M. genitalium*, has an APR domain with repeats of the Asp-Pro-Asn tripeptide embedded, the sequences and layout of the repeats are distinct from those of *M. pneumoniae* P65 (23). Thus, repeated duplications of a segment including this tripeptide have likely occurred separately in the two species subsequent to their divergence. Following the APR domain of P65 is a predicted coiled coil. At least some P65 is present on the *M. pneumoniae* cell surface (45,78,89). In permeabilized wild-type cells P65 is concentrated at the attachment organelle, similar to P1 (Fig. 2) (45,89). P65 has no predicted transmembrane domains and behaves as a peripheral membrane protein (77). P65 also lacks a recognizable export signal; thus it requires a novel means of export to the cell surface.

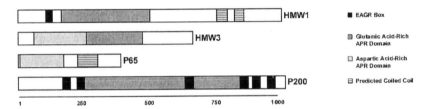

*Figure 3.* Schematic presentation of acidic proline-rich (APR) domain containing proteins in the *M. pneumoniae* cytoskeleton. EAGR box, enriched in glycine and aromatic residues (3). Domains are discussed in the text. Scale at bottom, amino acid residues.

### 3.4.3    P65 loss in cytadherence mutants

Steady-state P65 levels are reduced in the absence of any of the cytadherence-accessory proteins HMW1-HMW3 or P30, but not proteins A, B, and C or P1 (Table 1) (49,51). In mutant I-2, which lacks protein HMW2, this decrease occurs post-translationally, probably by proteolysis (45). This suggests that P65 stability is dependent upon HMW1-HMW3 and P30 but not A, B, C, or P1. Thus, P65 is likely to occupy a temporal position in the attachment organelle assembly process after the incorporation of HMW1-HMW3 and P30, and independent of that of B, C, and P1. When P65 is present at reduced levels, its localization to the attachment organelle is compromised, with the remaining P65 being visualized throughout the cell at low levels or not at all (45,89). The correlation between localization of P65 and its stability suggests that P65 is stabilized by proper assembly into a nascent attachment organelle containing HMW1-HMW3. Otherwise, uncomplexed P65 is a target of proteolysis.

## 3.5 Protein P30

### 3.5.1 Localization and structure of protein P30

The 30-kDa protein P30 (5) appears to be exclusively localized to the *M. pneumoniae* attachment organelle (Fig. 2). P30 behaves as an integral membrane protein consistent with the presence of two predicted transmembrane domains (11). Like P1, it is synthesized as a preprotein and subject to signal peptide cleavage following the first transmembrane domain (11). P30 and its homologs in *M. genitalium* (81) and *M. gallisepticum* (7,36) share a conserved cytoplasmic domain and transmembrane alpha-helix. The C-terminus of P30 is extracellular (13) and consists of sixteen 6-7-residue Pro-rich repeats; like the P65 repeats, the P30 repeats differ in sequence among organisms, suggesting that duplications occurred after speciation. The P30 gene is located near the 5'-end of the *hmw* operon, which also encodes HMW1 and HMW3 (35,98).

### 3.5.2 P30 mutants

#### 3.5.2.1 Morphology of P30 mutants

Numerous P30 mutants have been studied, including the null mutant II-3 (85). Especially for this mutant but also for mutants II-7 (52) and M7 (60), in which differing numbers of the C-terminal Pro-rich repeats are deleted (13,60), cell morphology is noticeably affected (85). Mutant cells exhibit multiple branches, some of which contain P1 clusters (85). The nucleoid occupies a proportionally greater volume of the cell than in wild-type cells (J. L. Jordan and D. C. K., unpublished results), suggesting that P30 may be required for proper nucleoid condensation in addition to attachment organelle formation. Thus, P30 may be involved in coordination of cell division with attachment organelle biogenesis.

#### 3.5.2.2 Partial complementation of a P30 mutant

A revertant strain of mutant II-3 was isolated based on its ability to adhere to plastic; its P30 gene was demonstrated to have a second-site mutation upstream of the original frameshift responsible for the mutant phenotype which restored the correct reading frame after a 17-amino acid stretch of novel sequence (85). Immunoblot analysis of the revertant protein P30 revealed dimers and tetramers, suggesting that the revertant P30 is subject to aggregation or disulfide linkage (J. L. Jordan and D. C. K., submitted). The revertant P30 restored hemadsorption to half that of the wild-type parent. Motility was not restored in this revertant as determined by both microcinematography and observation of colony morphology (J. L. Jordan and D. C. K., submitted). These findings suggest that formation of stable multimers by revertant P30 affect its function in gliding motility more overtly than its function in hemadsorption.

### 3.5.2.3    Effects of P30 loss on cytadherence-associated proteins

Levels of P65 are reduced in mutants II-3 and II-7, whereas other known cytadherence-related proteins are present at wild-type levels (Table 1) (51). This suggests that P30 is directly involved in some aspect of stabilization of P65, perhaps by contributing to its proper localization. The fact that in mutant II-7, in which only four Pro-rich repeats in the P30 molecule are deleted, P65 levels are as low as in mutant II-3, which has no P30, may indicate that these repeats interact with P65. However, as it is unknown whether the truncated P30 of mutant II-7 is localized to the attachment organelle, it is also possible that P65 localization is simply dependent upon P30 localization in general. In either case, the position of P30 in the attachment organelle assembly scheme is fairly early, but not prior to HMW1-HMW3, whose levels are normal in cells with mutated P30. P30 levels are slightly reduced in the *hmw2* mutant I-2 (5) but not in an *hmw3* mutant (Table 1) (M. J. Willby and D. C. K., submitted), suggesting that its optimum steady-state concentration is dependent upon HMW2 but not on HMW3.

### 3.5.3    TX partitioning of P30 and implications for P30 function

P30 exhibits a weak association with the triton shell, and like P1, P30 is completely solubilized by repeated extractions (R. H. Waldo III and D. C. K., unpublished results). Thus, the position of TX-soluble P30 in the assembly pathway between the HMW proteins and P65, all of which are TX-insoluble, is somewhat enigmatic. Protein P30 may not be directly involved in organelle assembly, instead serving as a regulatory link between attachment organelle duplication and cell division. Alternatively or in addition, P30 may play a direct but transient role in the folding or stabilization of other cytadherence-associated proteins. The observation that P1 is localized properly but is not functional in the P30 mutants (33,85) would appear to be consistent with such a role. Finally, antibodies against protein P30 have been reported to inhibit hemadsorption (70), raising the possibility that like P1, P30 is an adhesin. The absence of motility in P30 mutants suggests that motility itself is not dependent upon P1 clustering; however, motility may require proper cell morphology.

## 3.6 HMW3

### 3.6.1 Structure and localization of HMW3

Protein HMW3 (Fig. 3) is TX-insoluble (92), contains an APR domain (73,79), and exhibits retarded SDS-PAGE migration (73). Despite its predicted molecular mass of 74 kDa, HMW3 is observed primarily as a band at ~130 kDa. Its APR domain appears to be composed of two segments. From residues 88 to 270 its composition resembles the Asp-rich APR domain of P65, whereas residues 271-488 are more reminiscent of the Glu-rich HMW1 APR domain. Perhaps this bipartite structure confers modular interactions with other proteins or other HMW3 monomers. The N- and C-terminal domains of HMW3 are similar only to the corresponding regions of its *M. genitalium* homolog (81). IEM analysis indicates that HMW3 is associated with both the terminal button and the core shaft of the terminal organelle (Fig. 2) (92); to date it is the only protein definitively identified as a component of either structure. Anti-HMW3 immunogold labeling of cores frequently results in a linear pattern at oblique angles to the length of the core ("barbershop pole" pattern) (92), and the appearance of HMW3 as beads on a string in potassium iodide-treated triton shells suggests unwinding of this array. HMW3 that has been similarly stripped from membranes is largely insoluble and must subsequently be separated from the mycoplasma membrane by sucrose gradient centrifugation (M. F. B. and D. C. K., unpublished results). These properties are consistent with a polymeric organization for HMW3, as might be expected for proteins associated with the attachment organelle core. Although attachment organelles of whole cells were labeled by immunogold (92), perhaps as a consequence of damage to the cell membrane during fixation, the failure to immunoprecipitate HMW3 from whole cells (92) suggests that HMW3 is not surface-exposed in living mycoplasma cells.

### 3.6.2 Decreased HMW3 in cytadherence mutants

In mutant I-2, in which HMW2 is not produced (52,53), HMW3, like HMW1 and P65, is present at reduced steady-state levels (Table 1) (52,76). This is also true in mutant M6 (50), where the absence of HMW1 feeds back on HMW2 so as to cause reduced HMW2 levels (Table 1) (M. F. B. and D. C. K., unpublished results); this is independent of the P30 mutation in mutant M6, as complementation with the P30 gene has no effect on HMW3 levels (M. F. B. and D. C. K., unpublished results). When HMW2 is replaced by a variant lacking 1480 amino acid residues from the center of the protein, designated P38, HMW3 but not HMW1 is present at wild-type levels (M. Fisseha and D. C. K., unpublished results), indicating that stability of HMW3 is independent of HMW1.

### 3.6.3    HMW3 mutant and implications for HMW3 function

An *M. pneumoniae* mutant has been isolated in which *hmw3* is insertionally inactivated, resulting in loss of HMW3 (M. J. Willby and D. C. K., submitted). Cells of this mutant, while pleomorphic, are affected such that the tapered morphology characteristic of most cells in a wild-type population is absent. Colony morphology and microcinematography indicate that these cells are immotile (J. L. Jordan and D. C. K., unpublished results), and hemadsorption by the *hmw3* mutant is defective (M. J. Willby and D. C. K., submitted). Thin sections of mutant cells observed by EM reveal cells in which the core is bifurcated, resembling a V pointing toward the distal end of the tip (M. J. Willby and D. C. K., submitted). This suggests that HMW3 may have a role in core maintenance, consistent with its peripheral association with the core. The level of P65, but not those of other cytadherence-related proteins, is reduced in the *hmw3* mutant (M. J. Willby and D. C. K., submitted), suggesting that HMW3 acts ahead of P65 in the assembly pathway and stabilizes it. Thus, HMW3 appears to act between HMW1/HMW2 and P65 during attachment organelle assembly.

## 3.7    HMW1

### 3.7.1    HMW1 structure

HMW1 (Fig. 3) is one of the better-studied components of the *M. pneumoniae* cytoskeleton to date. Although its predicted size is 112 kDa (17), HMW1 migrates in SDS-PAGE at 210 kDa (33,52), probably due to the APR domain (Fig. 3) (79). This domain in HMW1 constitutes amino acid residues 170-522 and is enriched in Glu, Pro, Val, and Gln, and is devoid of Asp and basic residues. The N-terminal domain (domain I) of HMW1 is well conserved in the *M. genitalium* homolog and unremarkable in composition except for the presence of a 31-amino acid residue motif between amino acids 106 and 136, the EAGR box (enriched in aromatic and glycine residues) (3), which is unique to a few mycoplasma proteins. EAGR boxes have been identified only in two other proteins, P200 and MPN119/MG200, both of which also contain APR domains, suggesting that the functions of the two are linked. The APR domain and EAGR box are postulated to regulate interactions with other cytoskeletal proteins. Following the APR domain of HMW1 (domain II) is domain III, which extends to the C-terminus and is also conserved in *M. genitalium*. HMW1', a truncated variant of HMW1 lacking the C-terminal 113 amino acid residues, is relatively stable in cells in which HMW2 is not produced (33), raising the possibility that this region contains signals for targeting HMW1 for degradation. This C-terminal sequence has motifs that resemble the sequence at which both the *M. pneumoniae* B/C precursor and *Caulobacter crescentus* McpA are cleaved (76). Also present within domain III are two coiled coil motifs (only one in the *M. genitalium* homolog of HMW1 [23]). HMW1 and HMW2 both contain phospho-Ser and phospho-Thr (16). Phosphorylation, which might be regulatory or structural, is inhibited in extracts by fructose 1,6-bisphosphate (FBP) and stimulated by 3-phosphoglycerate (55). No kinase has been reported to respond likewise to those molecules, though FBP stimulates HPr (Ser) kinase (54,84), whose physiological target in other organisms is a transcriptional regulator, HPr, which is encoded in the *M. pneumoniae* genome (35).

### 3.7.2    Localization of HMW1 and the effects of HMW2 loss

HMW1 is detectable on wild-type *M. pneumoniae* cells by both IFM and whole cell radioimmunoprecipitation (3). HMW1 appears by IEM to be excluded from the cell body, localizing to the surface at both the attachment organelle and the trailing filament of the cell at the opposite pole (Fig. 2) (91). IFM analysis of unpermeabilized mutant I-2 colonies indicates little surface exposed HMW1 (3), suggesting that stabilization of HMW1 in the cytoskeleton is associated with its export to the cell surface. Like P65, HMW1 lacks recognizable export signals; thus, the means of its export must be novel. Cell fractionation studies reveal that *M. pneumoniae* HMW1 is a peripheral membrane protein, though at steady state a small but detectable portion is cytosolic (3). Pulse-chase studies indicate that after synthesis, HMW1 rapidly equilibrates between the TX-soluble and TX-insoluble fractions; subsequently, there is a slower movement of the remaining soluble HMW1 into the TX-insoluble fraction (3). HMW1, like HMW3 and P65, is present at reduced steady state levels in mutant I-2 cells (Table 1) (52). Pulse-chase experiments with this mutant indicate that although the initial synthesis and distribution of HMW1 are the same as in wild-type cells, the stabilization of HMW1 in the cytoskeleton is defective (3). Thus, HMW1 loss in mutant I-2 is attributed to proteolytic turnover of accumulated soluble HMW1 due to its failure to be stably incorporated into the triton shell. In mutant I-2 cells producing the truncated HMW2 variant P38, which lacks amino acid residues 98-1577, HMW1 but neither HMW3 nor P65 remains at reduced levels (M. Fisseha, M. F. B., and D. C. K., unpublished results), suggesting that HMW1 may be regulated differently by HMW2. It is possible that some determinant of the stability of HMW1 but not the other proteins lies in the region of HMW2 deleted in P38. Alternatively, the length of the coiled coil of HMW2 rather than its sequence may be the feature of HMW2 to which HMW1 is sensitive. These data demonstrate that HMW1 stability is dependent only upon HMW2 and not upon the other proteins whose levels are reduced in mutant I-2. Housekeeping proteases such as Lon and FtsH, both of which are encoded in the *M. pneumoniae* genome (35), are often employed to degrade protein molecules that are in stoichiometric excess over binding partners with which they form a more or less permanent complex (31,88). For example, *Escherichia coli* SecY is degraded by FtsH when overexpressed with regard to its binding partner, SecE (48).

### 3.7.3    HMW1 mutant and implications for function

The M6 mutant of *M. pneumoniae* harbors two mutations: one prevents HMW1 production, and the other results in a truncated P30 identical to the variant produced in mutant II-7 (13,59). Mutant M6 is defective in hemadsorption, motility, morphology, and P1 clustering (33,45,59). Restoration of P30 by transposon delivery results in near-wild-type levels of P30 (33), allowing study of the specific contribution of HMW1. There is little difference in phenotype between mutant M6 and M6/P30$^+$ cells. In both, HMW2 is present at approximately two-fold reduced levels, and HMW3 and P65 are also reduced (50), likely due to the reduction in HMW2 as in mutant I-2 (Table 1). The reduction of HMW2 indicates that there is reciprocal stabilization between HMW1 and HMW2. It is therefore difficult to distinguish the effects of HMW1 loss from the effects of diminution of the other proteins. Nonetheless, the instability of HMW2, HMW3, and P65 in the M6/P30 cells indicates that HMW1 is, along with HMW2, required close to the beginning of the attachment organelle assembly pathway.

## 3.8    HMW2 and P28

### 3.8.1    Structure of HMW2

HMW2 is a TX-insoluble 216-kDa protein that migrates at about 190 kDa by SDS-PAGE (52,53). Secondary structure predictions suggest that throughout its 1818-amino acid length it consists almost entirely of coiled coil structure (53). This coiled coil is predicted to be interrupted at many sites along the length of HMW2, raising the possibility that HMW2 is flexible. Most of the coiled coil sequences of HMW2 are predicted to be divalent, though two short stretches of trivalent coil are also predicted as well. Sequences near the termini of HMW2 have other features. Among the

N-terminal 72 amino acids of HMW2 are four Pro residues, and amino acids 1610 to 1631 and 1757 to 1783 constitute further Pro-rich regions.

### 3.8.2 Protein P28

Protein P28 is presumably derived from internal translation of *hmw2* near the 3'-end. Either of two candidate start codons, corresponding to amino acids 1578 and 1620 of HMW2, is likely to be the start site of translation of P28 (21). P28 is predicted to include most of the Pro-rich sequences near the C-terminus of HMW2 as well as both divalent coiled coil and one of the two stretches of trivalent coiled coil. If these structural features confer specific network interactions among HMW2 monomers, then P28 might serve to modulate cross-linking within this network. Although an HMW2 homolog, MG218, is expressed in *M. genitalium* (15), no protein corresponding to P28 has been reported in that organism. The C-terminal region of MG218 lacks a predicted trivalent coiled coil region, possibly obviating a P28-like molecule in this organism.

### 3.8.3 Localization of HMW2

Localizing HMW2 has been challenging because of the low *in situ* immunoreactivity of antibodies prepared against it. A fusion gene containing sequences encoding the inherently fluorescent protein from the jellyfish *Aequorea victoria*, enhanced green fluorescent protein (EGFP) (99), sandwiched between the codons for HMW2 amino acids 97 and 98, to be expressed in *M. pneumoniae* cells. EGFP-HMW2 was visualized at the leading pole of motile cells, corresponding to the attachment organelle (Fig. 2) (R. T. Santurri and D. C. K., unpublished results). Like HMW3, HMW2 is peripherally associated with the mycoplasma membrane. Its insolubility requires that it be separated from stripped membranes by sucrose gradient centrifugation (M. F. B. and D. C. K., unpublished results), suggesting that HMW2 forms higher-order structures. The combination of potential dimeric and trimeric tendencies in HMW2 may lead to interactions among HMW2 monomers that result in formation of a network that constitutes such structures. These structural predictions and localization data make HMW2 the best candidate yet for an integral component of the attachment organelle core.

### 3.8.4 HMW2 mutants

Cells of mutant I-2, which carry a mutation in *hmw2*, lack the tapered appearance of wild-type cells (4) and are nonmotile (J. L. Jordan and D. C.

K., submitted). HMW1, HMW3, and P65 are subject to enhanced loss in mutant I-2, presumably due to proteolytic degradation (45,52,76). For HMW1 this loss has been demonstrated to be a consequence of failure to be stably incorporated into the triton shell (3). In *M. genitalium*, the 114-kDa ORF6 homolog, like the P1 homolog, is present at lower steady-state levels in the absence of the HMW2 homolog (15); in contrast, in *M. pneumoniae*, levels of B, C, and P1 are unaffected by HMW2 loss (52). In mutants in which transposon Tn*4001* has inserted near or within the P28-encoding portion of *hmw2*, no P28 is made and truncated HMW2 is undetectable (34,53). However, in transposon mutant H9, in which HMW2 translation is interrupted and terminated after amino acid residues 1040 (34,53), both P28 and truncated HMW2 are produced. HMW2 may normally function to facilitate interactions between HMW1 and the unidentified HMW1 export machinery, and may have similar organizational roles in complexing P65 and HMW3 with their respective binding partners. In this regard HMW2 may be viewed as a possible cytoskeleton/membrane organizer, analogous to the function of spectrin in eukaryotic cells (19). These properties are also consistent with HMW2 as an essential component of the attachment organelle core, though it could perform these functions without being a core component.

### 3.8.5    Timing of HMW2 function and requisite interactions

The reduced levels of HMW1, HMW3, P65, and P30 in mutant I-2, as well as the apparent absence of cores, suggest that the action of HMW2 in *M. pneumoniae* attachment organelle assembly precedes those of the other proteins. However, the reduced levels of HMW2 in the absence of HMW1 (M6/P30 mutant) (M. F. B. and D. C. K., unpublished results) indicate that HMW1, which also appears to act early in assembly, may feed back on HMW2, such that the two proteins together constitute the earliest known step (although not necessarily the first step) in this process. However, it is not known whether levels of HMW2 in mutant M6 are reduced post-translationally or perhaps at the transcriptional level, as P65 is also reduced in this mutant (Table 1) (45,50). Whereas full-length HMW1 restores HMW2 to wild-type levels in the M6 mutant, C-terminally truncated HMW1 does not (M. J. Willby, M. F. B., and D. C. K., unpublished results), possibly indicating that the C-terminus of HMW1 is responsible for HMW2 stabilization.

## 3.9 Protein P200 as a candidate attachment organelle protein

Although HMW2 and HMW1 are the earliest known players in attachment organelle assembly, other TX-insoluble proteins are likely to be involved in this process. Little is known about *M. pneumoniae* protein P200 (Fig. 3), a component of the triton shell (77,79). Like certain other cytoskeletal proteins, P200 contains an APR domain (79); in addition, it contains six EAGR boxes (five in the *M. genitalium* homolog) (3). A 117-kDa protein, its SDS-PAGE migration is retarded such that it is observed at 200 kDa (79). A basic N-terminal domain of about 90 amino acids gives way to a region of mixed charge containing two EAGR boxes. The Glu-, Val-, and Gln-rich APR domain, similar to that of HMW1, stretches from amino acid 263 to amino acid 858, and is interrupted by an EAGR box. Three more EAGR boxes continue to amino acid 985, followed by a 52-amino acid C-terminal region of mixed charge. In *M. genitalium*, the homologous protein is much larger (1616 amino acids total) due to increased length of the APR domain, and lacks the EAGR box within the APR domain (23). The P200 gene is not linked to known cytadherence-associated genes (35). No mutant in P200 is known; P200 is present at wild-type levels in cytadherence mutants I-2, II-3, II-7, III-4, IV-22, and M6 (M. F. B. and D. C. K., unpublished results). Therefore, if P200 is involved in cytadherence, it is not associated with any of the phenotypes observed thus far. Nonetheless, both its structure and its TX-insolubility make P200 impossible to ignore as a candidate for function in cytadherence.

## 3.10 Order of attachment organelle assembly

The process of attachment organelle assembly is likely to involve early contributions by HMW2 in the formation of the core and direction of downstream proteins, followed consecutively by HMW1, HMW3, P30, and P65. Once these proteins have established a polar structure, an independently assembled complex of proteins B, C, and P1 is drawn to this structure (Fig. 4). Other proteins, including P200, may also be involved.

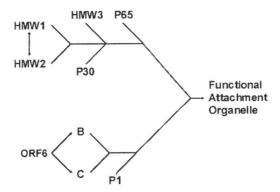

*Figure 4.* Order of assembly of components of the *M. pneumoniae* attachment organelle deduced from protein stability in cytadherence mutants.

## 4.    OTHER COMPONENTS OF THE MYCOPLASMA CYTOSKELETON

### 4.1    TX partitioning of *Mycoplasma* proteins

While some protein species, such as HMW2, are present almost exclusively in the TX-insoluble fraction (53,77), for others, like HMW1, a small but significant amount is TX-soluble (3), and for still others, like the protease FtsH, there is a considerable amount in both fractions (R. H. Waldo III and D. C. K., unpublished results). For HMW1, evidence suggests that the TX-soluble portion is awaiting conversion to an insoluble form (3). For housekeeping proteins like FtsH (1), the chaperone DnaK (30), and the redox enzyme thioredoxin (2), multimericity and/or transient interactions with aggregates or TX-insoluble proteins may contribute to partial insolubility. Finally, for proteins like pyruvate dehydrogenase (PDH), at least some component of their insolubility in *M. pneumoniae* is presumably a function of their size and complexity (66). The appearance of enzymes like PDH and lactate dehydrogenase in the TX-insoluble fraction of *M. pneumoniae* has been suggested to be concordant with the notion of the enzoskeleton, a hypothetical structure that results in part from interactions among enzymes of metabolic or other pathways (72). Finally, membranous structures have been identified in loose association with TX-insoluble cores (82). Perhaps the polar region of the cell membrane is enriched in sterols, which tend to be TX-insoluble. Thus, the mechanism of direction of tip-associated membrane proteins like P1 might be preferential association with certain membrane-resident lipids, perhaps analogous to lipid rafts in eukaryotic cells. Thus, a function of cytadherence-accessory proteins might be to create an environment enriched in these lipids near the attachment organelle; in the absence of such an environment, P1 would fail to cluster.

## 4.2 Identification of *Mycoplasma* cytoskeletal proteins

### 4.2.1 Successes and pitfalls

Two-dimensional gel electrophoresis coupled with mass spectrometry has constituted a powerful tool in proteomics (26). Both the *M. pneumoniae* total lysate (83) and TX-insoluble fraction (82) have been subjected to this analysis, revealing which of the predicted open reading frames of the genome are among those expressed and which are among the components of the TX-insoluble material. About 50 TX-insoluble *M. pneumoniae* proteins were identified by a combination of this method and immunoblotting (82). These include HMW1-HMW3, P65, P1, B, C, and P200, as well as others. Thus this work provides starting points for study of proteins that are potentially important for cytadherence and other cytoskeletal functions. A weakness of this technique is its underrepresentation of hydrophobic proteins, especially transmembrane ones. Thus, for example, FtsH, which is membrane-associated and partly TX-insoluble (R. H. Waldo III and D. C. K., unpublished observations), was not identified. Also, P1, which may be only loosely associated with the triton shell, was identified as TX-insoluble.

### 4.2.2 Cell division-associated proteins

FtsZ (86) was identified among the TX-insoluble proteins of *M. pneumoniae* (82). *ftsZ* genes have been identified in all prokaryotes with a few exceptions (64); in addition, FtsZ is present in some eukaryotic organelles derived from prokaryotes (64). FtsZ is predicted to form a linear homopolymer, the Z ring, on the inner face of the cytoplasmic membrane at the nascent constriction site during cell division, where it drives membrane constriction (63). It is likely that the polymerized state accounts for the insolubility of some *M. pneumoniae* FtsZ. In walled bacteria FtsZ activity is coordinated with that of membrane-inserted proteins that are involved in constructing a cell wall septum at the division site (64). Since *M. pneumoniae* cells have no wall, it is not surprising to find that these other division proteins are absent (35). However, the genomes of *M. pneumoniae* and *M. genitalium* have two genes of unknown function, *yabB* and *yabC*, the homologs of which are typically located in other organisms within the cluster of genes that is involved in division and cell wall biosynthesis, including *ftsZ* (64). In the mycoplasmas these genes also appear to be in the *ftsZ*-containing operon (21,32). YabB was identified as a component of the TX-insoluble fraction of *M. pneumoniae* (82), suggesting that it may interact with FtsZ. The identification of this protein as cytoskeletal underscores the potential utility of studying the mycoplasma cytoskeleton for revealing functions of uncharacterized genes and proteins in other organisms.

### 4.2.3    Other proteins

Several energy metabolism enzymes are present in the TX-insoluble fraction of *M. pneumoniae* (82), possibly analogous to the association of glycolytic enzymes with microfilaments in eukaryotic cells (65), where clustering of these enzymes might enhance metabolic efficiency. However, except for phosphotransacetylase and three of the polypeptide chains of pyruvate dehydrogenase (PDH), the specific metabolic enzymes that were identified as TX-insoluble are not anticipated to operate consecutively in metabolic pathways. Failure to identify other enyzmes might reflect low abundance or selective loss during preparation.    Alternatively, the association of some metabolic enzymes with the TX-insoluble fraction may result from factors independent of metabolic function, such as secondary roles, like those of the crystallins of vertebrate lens tissue, which are alternatively utilized forms of common enzymes that double as structural components in the lens (75). Other TX-insoluble proteins identified include those involved in protein folding, GroEL, DnaK, and trigger factor, which associate with other proteins; thus their presence in the TX-insoluble fraction may be a result of these interactions. It is also important to note that some of these are among the most abundant proteins in *M. pneumoniae* (82), so their identification in the triton shell may result from incomplete separation of soluble proteins. Ribosome-associated proteins were also identified as TX-insoluble; these proteins may be present in this fraction by virtue of the insolubility of this large macromolecular complex. Transcription elongation factor, uniquely among RNA polymerase components, was also found in the TX-insoluble fraction.

## 5.    THE NATURE OF THE MYCOPLASMA CYTOSKELETON

### 5.1    Biochemical definition of the cytoskeleton

Although the cytoskeleton is typically thought of as the body of structural elements of the cell, the definition of "cytoskeleton" depends upon who studies it. A primary, biochemical and rather operational definition of the cytoskeleton is simply the material that is not solubilized from intact cells by 0.5-2% TX (24). As observed by microscopic techniques, the most prominent feature of cytoskeletons of eukaryotic cells thus prepared is their filamentous appearance. The components of these filamentous systems fall into two broad groups: the inherently insoluble filamentous structures themselves, and the proteins rendered TX-insoluble through their association with these filaments. These filaments are involved in modification of cell shape, cell motility, cell division, intracellular transport, and positioning of organelles and other materials (87). The TX-insoluble fraction of eukaryotic cells also contains structures that are insoluble due to their association with membrane subdomains known as lipid rafts (37,62). Proteins present in large structures, such as oligomers of Golgi enzymes (10), ribosomes (25), or some components of other large multisubunit complexes like PDH (96), likewise exhibit variable TX-insolubility. Finally, specialized cells may have unique TX-insoluble structures, such as the post-synaptic density of neurons (67).

## 5.2    Physiological definition of the cytoskeleton

A second definition of the cytoskeleton is a functional one. To students of cell physiology, the cytoskeleton includes any proteins responsible for cell structure and dynamics (87). Thus, it is the cellular roles of the major filamentous TX-insoluble proteins that form the basis for this second definition, rather than the simple fact of insolubility. In this context, the salient defining property of the cytoskeleton is that it is a dynamic structure: its components polymerize and depolymerize in response to cellular conditions and signals, including post-translational modification, activity of accessory proteins that affect stability or length of filaments, and availability of certain small molecules, any of which may occur at either global or local levels. Furthermore, the proteins that modulate the polymerization state of the filaments or carry out functions intimately linked to them are regarded as cytoskeletal by those who employ this definition.

## 5.3    Defining the *Mycoplasma* cytoskeleton

While these biochemical and physiological definitions of the cytoskeleton overlap, they are distinct. As discussed above, there are TX-insoluble proteins that are not directly involved in cell structure and dynamics. Likewise, a protein need not be identified as a component of the TX-insoluble material for it to play a role in these functions. The imperfect overlap between detergent insolubility and function in cellular organization makes it difficult to classify novel proteins with regard to function based solely on biochemical fractionation properties. While TX-insoluble material constitutes only a small portion of most prokaryotic cells, this fraction of *M. pneumoniae* contains 25-33% of the total cellular protein (45; M. F. B. and D. C. K., unpublished results). Along with large protein complexes not thought to be directly involved in cell structure and dynamics, this fraction includes the proteins of the attachment organelle. Clearly the proteins of the *M. pneumoniae* terminal organelle are cytoskeletal in the biochemical sense, as would be expected for components of a large protein complex. While these proteins are involved in the same sorts of processes as eukaryotic cytoskeletal proteins (of the second definition), in the absence of specific knowledge of their mechanisms they may be hypothesized to carry out their functions either statically, by allowing assembly of other machinery that directly mediates the processes in question, or dynamically, by responding to signals as does the eukaryotic cytoskeleton. Thus, on one hand, the *M. pneumoniae* cytadherence proteins might merely comprise a scaffold for other molecules that carry out functions that result in adherence, motility, and normal shape. On the other hand, they may represent evolutionary convergence of function with the eukaryotic cytoskeleton, perhaps under the selective pressure of cell wall loss. Either way, the novel nature of the mycoplasma attachment organelle and its importance in pathogenesis warrant considerable attention from the cell biology and microbiology communities.

## 6.    CONCLUSION

The TX-insoluble fraction of mycoplasmas, though dominated by the components of the attachment organelle, consists of numerous proteins that may partition with this fraction by virtue of their physical characteristics, impurity of the fraction, or genuine association with the attachment organelle or other structural elements. The nature of the mycoplasma cytoskeleton, like the TX-insoluble fractions of other cells, is therefore eclectic, and parsing out which components are involved in which processes will require contributions from many areas of research, including cytadherence, cell division, metabolic regulation, protein folding, and physical biochemistry.

## ACKNOWLEDGMENTS

The authors are grateful to members of the Krause lab for sharing their unpublished results and their thoughts. This work was supported by the

Public Health Service research grant AI22362 (to D. C. K.) and National Research Service Award AI10500 (to M. F. B.) from the National Institute of Allergy and Infectious Diseases.

# REFERENCES

1. **Akiyama, Y., T. Yoshihisa, K. Ito.** 1995. FtsH, a membrane-bound ATPase, forms a complex in the cytoplasmic membrane of *Escherichia coli.* J. Biol Chem. **270:**23485-23490.
2. **Arner, E. S., and A. Holmgren.** 2000. Physiological functions of thioredoxin and thioredoxin reductase. Eur. J. Biochem. **267:**6102-6109.
3. **Balish, M. F., T.-W. Hahn, P. L. Popham, and D. C. Krause.** 2001. Stability of *Mycoplasma pneumoniae* cytadherence protein HMW1 correlates with its association with the triton shell. J. Bacteriol. **183:**3680-3688.
4. **Baseman, J. B., R. M. Cole, D. C. Krause, and D. K. Leith.** 1982. Molecular basis for cytadsorption of *Mycoplasma pneumoniae.* J. Bacteriol. **151:**1514-1522.
5. **Baseman, J. B., J. Morrison-Plummer, D. Drouillard, B. Puleo-Scheppke, V. V. Tryon, and S. C. Holt.** 1987. Identification of a 32-kilodalton protein of *Mycoplasma pneumoniae* associated with hemadsorption. Isr. J. Med. Sci. **23:**474-479.
6. **Boatman, E. S.** 1979. Morphology and ultrastructure of the mycoplasmatales, p. 63-102. *In* M. F. Barile and S. Razin (ed.), The Mycoplasmas, vol. I. Academic Press, New York, N. Y.
7. **Boguslavsky, S., D. Menaker, I. Lysnyansky, T. Liu, S. Levisohn, R. Rosengarten, M. García, and D. Yogev.** 2000. Molecular characterization of the *Mycoplasma gallisepticum pvpA* gene which encodes a putative variable cytadhesin protein. Infect. Immun. **68:**3956-3964.
8. **Bredt, W.** 1968. Motility and multiplication of *Mycoplasma pneumoniae*: a phase contrast study. Pathol. Microbiol. **32:**321-326.
9. **Chambaud, I., R. Heilig, S. Ferris, V. Barbe, D. Samson, F. Galisson, I. Moszer, K. Dybvig, H. Wrobleski, A. Viari, E. P. Rocha, and A. Blanchard.** 2001. The complete genome sequence of the murine respiratory pathogen *Mycoplasma pulmonis*. Nucleic Acids Res. **29:** 2145-2153.
10. **Chen, C., J. Ma, A. Lazic, M. Backovic, K. J. Colley.** 2000. Formation of insoluble oligomers correlates with ST6Gal I stable localization in the golgi. J. Biol. Chem. **275:**13819-13826.
11. **Dallo, S. F., A. Chavoya, and J. B. Baseman.** 1990. Characterization of the gene for a 30-kilodalton adhesin-related protein of *Mycoplasma pneumoniae*. Infect. Immun. **58:**4163-4165.
12. **Dallo, S. F., A. Chavoya, C.-J. Su, and J. B. Baseman.** 1989. DNA and protein sequence homologies between the adhesins of *Mycoplasma genitalium* and *Mycoplasma pneumoniae*. Infect. Immun. **57:**1059-1065.
13. **Dallo, S. F., A. L. Lazzell, A. Chavoya, S. P. Reddy, J. B. Baseman.** 1996. Biofunctional domains of the Mycoplasma pneumoniae P30 adhesin. Infect. Immun. **64:**2595-2601
14. **Dandekar, T., M. Huynen, J. T. Regula, B. Ueberle, C. U. Zimmermann, M. A. Andrade, T. Doerks, L. Sánchez-Pulido, B. Snel, M. Suyama, Y. P. Yuan, R. Herrmann, and P. Bork.** 2000. Re-annotating the *Mycoplasma pneumoniae* genome sequence: adding value, function and reading frames. Nucleic Acids Res. **28:**3278-3288.

15. **Dhandayuthapani, S., W. G. Rasmussen, and J. B. Baseman.** 1999. Disruption of gene *mg218* of *Mycoplasma genitalium* through homologous recombination leads to an adherence-deficient phenotype. Proc. Natl. Acad. Sci. USA **96:**5227-5232.

16. **Dirksen, L. B., K. A. Krebes, and D. C. Krause.** 1994. Phosphorylation of cytadherence-accessory proteins in *Mycoplasma pneumoniae*. J. Bacteriol. **176:**7499-7505.

17. **Dirksen, L. B., T. Proft, H. Hilbert, H. Plagens, R. Herrmann, and D. C. Krause.** 1996. Sequence analysis and characterization of the *hmw* gene cluster of *Mycoplasma pneumoniae*. Gene **171:**19-25.

18. **Dorigo-Zetsma, J. W., J. Dankert, and S. A. J. Zaat.** 2000. Genotyping of *M. pneumoniae* clinical isolates reveals eight P1 subtypes within two genomic groups. J. Clin. Microbiol. **38:**965-970.

19. **Dubreuil, R. R., and T. Grushko.** 1998. Genetic studies of spectrin: new life for a ghost protein. Bioessays **20:**875-878.

20. **Feldner, J., U. Göbel, and W. Bredt.** 1982. *Mycoplasma pneumoniae* adhesin localized to tip structure by monoclonal antibody. Nature (London) **298:**765-767.

21. **Fisseha, M., H. H. Göhlmann, R. Herrmann, and D. C. Krause.** 1999. Identification and complementation of frameshift mutations associated with loss of cytadherence in *Mycoplasma pneumoniae*. J. Bacteriol. **181:**4404-4410.

22. **Franzoso, G., P.-C. Hu, G. A. Meloni, and M. F. Barile.** 1993. The immunodominant 90-kilodalton protein is localized on the terminal tip structure of *Mycoplasma pneumoniae*. Infect. Immun. **61:**1523-1530.

23. **Fraser, C. M., J. D. Gocayne, O. White, M. D. Adams, R. A. Clayton, R. D. Fleischmann, C. J. Bult, A. R. Kerlavage, G. Sutton, J. M. Kelley, J. L. Fritchman, J. F. Weidman, K. V. Small, M. Sandusky, J. Fuhrmann, D. Nguyen, T. R. Utterback, D. M. Saudek, C. A. Phillips, J. M. Merrick, J.-F. Tomb, B. A. Dougherty, K. F. Bott, P.-C. Hu, T. S. Lucier, S. N. Peterson, H. O. Smith, C.A. Hutchison III, and J. C. Venter.** 1995. The minimal gene complement of *Mycoplasma genitalium*. Science **270:**397-403.

24. **Fulton, A.** 1984. The Cytoskeleton. p. 7-8. Chapman and Hall, New York, N. Y.

25. **Gavrilova, L. P., N. M. Rutkevitch, V. I. Gelfand, L. P. Motuz, J. Stahl, U. A. Bommer, and H. Bielka.** 1987. Immunofluorescent localization of protein synthesis components in mouse embryo fibroblasts. Cell Biol. Int. Rep. **11:**745-753.

26. **Gevaert, K., and J. Vandekerckhove.** 2000. Protein identification methods in proteomics. Electrophoresis **21:**1145-1154.

27. **Glass, J. I., E. J. Lefkowitz, J. S. Glass, C. R. Heiner, E. Y. Chen, and G. H. Cassell.** 2000. Nature **407:**757-762.

28. **Göbel, U., V. Speth, and W. Bredt.** 1981. Filamentous structures in adherent *Mycoplasma pneumoniae* cells treated with nonionic detergents. J. Cell Biol. **91:**537-543.

29. **Goh, M. S., T. S. Gorton, M. H. Forsyth, K. E. Troy, and S. J. Geary.** 1998. Molecular and biochemical analysis of a 105 kDa *Mycoplasma gallisepticum* cytadhesin (GapA). Microbiology (U. K.) **144:**2971-2978.

30. **Goloubinoff, P., A. Mogk, A. P. Zvi, T. Tomoyasu, and B. Bukau.** 1999. Sequential mechanism of solubilization and refolding of stable protein aggregates by a bichaperone network. Proc. Natl. Acad. Sci. USA **96:**13732-13737.

31. **Gottesman, S.** 1996. Proteases and their targets in *Escherichia coli*. Annu. Rev. Genet. **30:**465-506.

32. **Hahn, T.-W., K. A. Krebes, and D. C. Krause.** 1996. Expression in *Mycoplasma pneumoniae* of the recombinant gene encoding the cytadherence-accessory protein HMW1 and identification of HMW4 as a product. Mol. Microbiol. **19**:1085-1093.
33. **Hahn, T.-W., M. J. Willby, and D. C. Krause.** 1998. HMW1 is required for cytadhesin P1 trafficking to the attachment organelle in *Mycoplasma pneumoniae*. J. Bacteriol. **180**:1270-1276.
34. **Hedreyda, C. T., and D. C. Krause.** 1995. Identification of a possible cytadherence regulatory locus in *Mycoplasm pneumoniae*. Infect. Immun. **63**:3479-3483.
35. **Himmelreich, R., H. Hilbert, H. Plagens, E. Pirkl, B.-C. Li, and R. Herrmann.** 1996. Complete sequence analysis of the genome of the bacterium *Mycoplasma pneumoniae*. Nucleic Acids Res. **24**:4420-4449.
36. **Hnatow, L. L., C. L. Keeler, Jr., L. L. Tesmer, K. Czymmek, and J. E. Dohms.** 1998. Characterization of MGC2, a *Mycoplasma gallisepticum* cytadhesin with homology to the *Mycoplasma pneumoniae* 30-kilodalton protein and *Mycoplasma genitalium* P32. Infect. Immun. **66**:3436-3442.
37. **Hooper, N. M.** 1999. Detergent-insoluble glycosphingolipid/cholesterol-rich membrane domains, lipid rafts and caveolae. Mol. Membr. Biol. **16**:145-156.
38. **Hu, P.-C., R. M. Cole, Y. S. Huang, T. A. Graham, D. E. Gardner, A. M. Collier, and W. A. Clyde.** 1982. *Mycoplasma pneumoniae* infection: role of a surface protein in the attachment organelle. Science **216**:313-315.
39. **Hu, P.-C., A. M. Collier, and J. B. Baseman.** 1977. Surface parasitism by *Mycoplasma pneumoniae* of respiratory epithelium. J. Exp. Med. **145**:1328-1343.
40. **Hu, P.-C., U. Schaper, A. M. Collier, W. A. Clyde, Jr., M. Horikawa, Y. Huang, and M. F. Barile.** 1987. A *Mycoplasma genitalium* protein resembling the *Mycoplasma pneumoniae* attachment protein P1. Infect. Immun. **55**:1126-1131.
41. **Inamine, J. M., T. P. Denny, S. Loechel, U. Schaper, C.-H. Huang, K. F. Bott, and P.-C. Hu.** 1988a. Nucleotide sequence of the P1 attachment protein gene of *Mycoplasma pneumoniae*. Gene **64**: 217-219.
42. **Inamine, J. M., S. Loechel, and P.-C. Hu.** 1988b. Analysis of the nucleotide sequence of P1 operon of *Mycoplasma pneumoniae*. Gene **73**:175-183.
43. **Jacobs, E.** 1991. *Mycoplasma pneumoniae* virulence factors and the immune response. Rev. Med. Microbiol. **2**:83-90.
44. **Jacobs, E., K. Fuchte, and W. Bredt.** 1987. Amino acid sequence and antigenicity of the amino-terminus of the 168 kDa adherence protein of *Mycoplasma pneumoniae*. J. Gen. Microbiol. **133**:2233-2236.
45. **Jordan, J. L., K. M. Berry, M. F. Balish, and Duncan C. Krause.** 2001 Stability and subcellular localization of cytadherence – associated protein P65 in *Mycoplasma pneumoniae*. J. Bacteriol. **183**:7387-7391.
46. **Kahane, I., S. Tucker, D. K. Leith, J. Morrison-Plummer, and J. B. Baseman.** 1985. Detection of the major adhesin P1 in triton shells of virulent *Mycoplasma pneumoniae*. Infect. Immun. **50**:944-946.
47. **Kenri, T., R. Taniguchi, Y. Sasaki, N. Okazaki, M. Narita, K. Izumikawa, M. Umetsu, and T. Sasaki.** 1999. Identification of a new variable sequence in the P1 cytadhesin gene of *Mycoplasma pneumoniae*: evidence for the generation of antigenic variation by DNA recombination by repetitive sequences. Infect. Immun. **67**:4557-4562.
48. **Kihara, A., Y. Akiyama, and K. Ito.** 1995. FtsH is required for proteolytic elimination of uncomplexed forms of SecY, an essential protein translocase subunit. Proc. Natl. Acad. Sci. USA **92**:4532-4536.
49. **Krause, D. C.** 1996. *Mycoplasma pneumoniae* cytadherence: unravelling the tie that binds. Mol. Microbiol. **20**:247-253.

50.  **Krause, D. C.** 1998. *Mycoplasma pneumoniae* cytadherence: organization and assembly of the attachment organelle. Trends Microbiol. **6**:15-18.

51.  **Krause, D. C., and M. F. Balish.** 2001. Structure, function, and assembly of the terminal organelle of *Mycoplasma pneumoniae*. FEMS Microbiol. Lett. **198**:1-7.

52.  **Krause, D. C., D. K. Leith, R. M. Wilson, and J. B. Baseman.** 1982. Identification of *Mycoplasma pneumoniae* proteins associated with hemadsorption and virulence. Infect. Immun. **35**:809-817.

53.  **Krause, D. C., T. Proft, C. T. Hedreyda, H. Hilbert, H. Plagens, R. Herrmann.** 1997. Transposon mutagenesis reinforces the correlation between *Mycoplasma pneumoniae* cytoskeletal protein HMW2 and cytadherence. J. Bacteriol. **179**:2668-2677.

54.  **Kravanja, M., R. Engelmann, V. Dossonnet, M. Bluggel, H. E. Meyer, R. Frank, A. Galinier, J. Deutscher, N. Schnell, and W. Hengstenberg.** 1999. The *hprK* gene of *Enterococcus faecalis* encodes a novel bifunctional enzyme: the HPr kinase/phosphatase. Mol. Microbiol. **31**:59-66.

55.  **Krebes, K. A., L. B. Dirksen, and D. C. Krause.** 1995. Phosphorylation of *Mycoplasma pneumoniae* cytadherence-accessory proteins in cell extracts. J. Bacteriol. **177**:4571-4574.

56.  **Layh-Schmitt, G., and M. Harkenthal.** 1999. The 40- and 90-kDa membrane proteins (ORF6 gene product) of *Mycoplasma pneumoniae* are responsible for the tip structure formation and P1 (adhesin) associateion with the triton shell. FEMS Microbiol. Lett. **174**:143-149.

57.  **Layh-Schmitt, G., and R. Herrmann.** 1992. Localization and biochemical characterization of the ORF6 gene product of the *Mycoplasma pneumoniae* P1 operon. Infect. Immun. **60**:2904-2913.

58.  **Layh-Schmitt, G., and R. Herrmann.** 1994. Spatial arrangement of gene products of the P1 operon in the membrane of *Mycoplasma pneumoniae*. Infect. Immun. **62**:974-979.

59.  **Layh-Schmitt, G., H. Hilbert, and E. Pirkl.** 1995. A spontaneous hemadsorption-negative mutant of *Mycoplasma pneumoniae* exhibits a truncated adhesin-related 30-kilodalton protein and lacks the cytadherence-accessory protein HMW1. J. Bacteriol. **177**:843-846.

60.  **Layh-Schmitt, G., R. Himmelreich, and U. Leibfried.** 1997. The adhesin related 30-kDa protein of *Mycoplasma pneumoniae* exhibits size and antigen variability. FEMS Microbiol. Lett. **152**:101-8.

61.  **Layh-Schmitt, G., A. Podtelejnikov, and M. Mann.** 2000. Proteins complexed to the P1 adhesin of *Mycoplasma pneumoniae*. Microbiology (UK) **146**:741-747.

62.  **London, E., and D. A. Brown.** 2000. Insolubility of lipids in Triton X-100: physical origin and relationship to sphingolipid/cholesterol membrane domains (rafts). Biochim. Biophys. Acta **1508**:182-195.

63.  **Lu, C., M. Reedy, and H. P. Erickson.** 2000. Straight and curved conformations of FtsZ are regulated by GTP hydrolysis. J. Bacteriol. **182**:164-170.

64.  **Margolin, W.** 2000. Themes and variations in prokaryotic cell division. FEMS Microbiol. Rev. **24**:531-548.

65.  **Masters, C. J., S. Reid, and M. Don.** 1987. Glycolysis: new concepts in an old pathway. Mol. Cell Biochem. **76**:3-14.

66.  **Mattevi, A., G. Obmolova, E. Schulze, K. H. Kalk, A. H. Westphal, A. de Kok, and W. G. Hol.** 1992. Atomic structure of the cubic core of the pyruvate dehydrogenase multienzyme complex. Science. **255**:1544-1550.

67.  **Matus, A. I., and D. H. Taff-Jones.** 1978. Morphology and molecular composition of isolated postsynaptic junctional structures. Proc. R. Soc. Lond. B. Biol. Sci. **203**:135-151.

68. **Meng, K. E., and R. M. Pfister.** 1980. Intracellular structures of *Mycoplasma pneumoniae* revealed after membrane removal. J. Bacteriol. **144**:390-399.
69. **Mernaugh, G. R., S. F. Dallo, S. C. Holt, and J. B. Baseman.** 1993. Properties of adhering and nonadhering populations of *Mycoplasma genitalium*. Clin. Infect. Dis. **17**:S69-S78.
70. **Morrison-Plummer, J., D. K. Leith, and J. B. Baseman.** 1986. Biological effects of anti-lipid and anti-protein monoclonal antibodies on *Mycoplasma pneumoniae*. Infect. Immun. **53**:398-403.
71. **Neimark, H.** 1977. Extraction of an actin-like protein from the prokaryote *Mycoplasma pneumoniae*. Proc. Natl. Acad. Sci. USA **74**:4041-4045.
72. **Norris, V., G. Turnock, and D. Sigee.** 1996. The *Escherichia coli* enzoskeleton. Mol. Microbiol. **19**:197-204.
73. **Ogle, K. F., K. K. Lee, and D. C. Krause.** 1992. Nucleotide sequence analysis reveals novel features of the phase-variable cytadherence accessory protein HMW3 of *Mycoplasma pneumoniae*. Infect. Immun. **60**:1633-1641.
74. **Papazisi, L., K. E. Troy, T. S. Gorton, X. Liao, and S. J. Geary.** 2000. Analysis of cytadherence-deficient, GapA-negative *Mycoplasma gallisepticum* strain R. Infect. Immun. **12**:6643-6649.
75. **Piatigorsky, J.** 1998. Multifunctional lens crystallins and corneal enzymes: more than meets the eye. Ann. N. Y. Acad. Sci. **842**:7-15.
76. **Popham, P. L., T.-W. Hahn, K. A. Krebes, and D. C. Krause.** 1997. Loss of HMW1 and HMW3 in noncytadhering mutants of *Mycoplasma pneumoniae* occurs post-translationally. Proc. Natl. Acad. Sci. USA **94**:13979-13984.
77. **Proft, T., and R. Herrmann.** 1994. Identification and characterization of hitherto unknown *Mycoplasma pneumoniae* proteins. Mol. Microbiol. **13**: 337-348.
78. **Proft, T., H. Hilbert, G. Layh-Schmitt, and R. Herrmann.** 1995. The proline-rich P65 protein of *Mycoplasma pneumoniae* is a component of the Triton X-100-insoluble fraction and exhibits size polymorphism in the strains M129 and FH. J. Bacteriol. **177**:3370-3378.
79. **Proft, T., H. Hilbert, H. Plagens, and R. Herrmann.** 1996. The P200 protein of *Mycoplasma pneumoniae* shows common features with the cytadherence-associated proteins HMW1 and HMW3. Gene. **171**: 79-82.
80. **Razin, S., and E. Jacobs.** 1992. Mycoplasma adhesion. J. Gen. Microbiol. **138**: 407-422.
81. **Reddy, S. P., W. G. Rasmussen, and J. B. Baseman.** 1995. Molecular cloning and characterization of an adherence-related operon of *Mycoplasma genitalium*. J. Bacteriol. **177**:5943-5951.
82. **Regula, J. T., G. Boguth, A. Görg, J. Hegermann, F. Mayer, R. Frank, and R. Herrmann.** 2001. Defining the mycoplasma 'cytoskeleton': the protein composition of the Triton X-100 insoluble fraction of the bacterium *Mycoplasma pneumoniae* determined by 2-D gel electrophoresis and mass spectrometry. Microbiology (U. K.) **147**:1045-1057.
83. **Regula, J. T., B. Ueberle, G. Boguth, A. Görg, M. Schnolzer, R. Herrmann, and R. Frank.** 2000. Towards a two-dimensional proteome map of *Mycoplasma pneumoniae*. Electrophoresis **21**:3765-3780.
84. **Reizer, J., C. Hoischen, F. Titgemeyer, C. Rivolta, R. Rabus, J. Stulke, D. Karamata, M. H. Saier, Jr, and W. Hillen.** 1998. A novel protein kinase that controls carbon catabolite repression in bacteria. Mol. Microbiol. **27**:1157-1169.

518                                          *Mitchell Balish and Duncan Krause*

85. **Romero-Arroyo, C. E., J. Jordan, S. J. Peacock, M. J. Willby, M. A. Farmer, and D. C. Krause.** 1999. *Mycoplasma pneumoniae* protein P30 is required for cytadherence and associated with proper cell development. J. Bacteriol. **181**:1079-1087.
86. **Rothfield, L. I., S. Justice, J. García-Lara.** 1999. Bacterial cell division. Annu. Rev. Genet. **33**:423-448.
87. **Schliwa, M.** 1986. The cytoskeleton: an introductory survey, p. 1-2. *In* M. Alfert, W. Beermann, L. Goldstein, and K. R. Porter (ed.), Cell Biology Monographs, vol. XIII. Springer-Verlag, New York, N. Y.
88. **Schumann, W.** 1999. FtsH – a single-chain charonin? FEMS Microbiol. Rev. **23**:1-11.
89. **Seto, S., G. Layh-Schmitt, T. Kenri, and M. Miyata.** 2001. Visualization of the attachment organelle and cytadherence proteins of *Mycoplasma pneumoniae* by immunofluorescence microscopy. J. Bacteriol. **183**:1621-1630.
90. **Sperker, B., P.-C. Hu, and R. Herrmann.** 1991. Identification of gene products of the P1 operon of *Mycoplasma pneumoniae*. Mol. Microbiol. **5**:299-306.
91. **Stevens, M. K., and D. C. Krause.** 1991. Localization of the *Mycoplasma pneumoniae* cytadherence-accessory proteins HMW1 and HMW4 in the cytoskeletonlike triton shell. J. Bacteriol. **173**:1041-1050.
92. **Stevens, M. K., and D. C. Krause.** 1992. *Mycoplasma pneumoniae* cytadherence phase-variable protein HMW3 is a component of the attachment organelle. J. Bacteriol. **174**:4265-4274.
93. **Su, C.-J., A. Chavoya, and J. B. Baseman.** 1989. Spontaneous mutation results in loss of the cytadhesin (P1) of *Mycoplasma pneumoniae*. Infect. Immun. **57**:3237-3239.
94. **Su, C.-J., A. Chavoya, S. F. Dallo, and J. B. Baseman.** 1990. Sequence divergence of the cytadhesin gene of *Mycoplasma pneumoniae*. Infect. Immun. **58**:2669-2674.
95. **Su, C.-J., V. V. Tryon, and J. B. Baseman.** 1987. Cloning and sequence analysis of cytadhesin P1 gene from *Mycoplasma pneumoniae*. Infect. Immun. **55**:3023-3029.
96. **Sumegi, B., Z. Liposits, L. Inman, W. K. Paull, and P. A. Srere.** 1987. Electron microscopic study on the size of pyruvate dehydrogenase complex *in situ*. Eur. J. Biochem. **169**:223-230.
97. **Tham, T. N., S. Ferris, E. Bahraoui, S. Canarelli, L. Montagnier, and A. Blanchard.** 1994. Molecular characterization of the P1-like adhesin gene from *Mycoplasma pirum*. J. Bacteriol. **176**:781-788.
98. **Waldo, R. H. III, P. L. Popham, C. E. Romero-Arroyo, E. A. Mothershed, K. K. Lee, and D. C. Krause.** 1999. Transcriptional analysis of the *hmw* gene cluster of *Mycoplasma pneumoniae*. J. Bacteriol. **181**:4978-4985.
99. **Zhang, G., V. Gurtu, and S. R. Kain.** 1996. An enhanced green fluorescent protein allows sensitive detection of gene transfer in mammalian cells. Biochem. Biophys. Res. Comm. **227**:707-711

Chapter 23

# *Mycoplasma pneumoniae* Disease Manifestations and Epidemiology

ENNO JACOBS
*Institute for Medical Microbiology and Hygiene, Medical Faculty of the TU-University of Dresden, Fetscherstrasse 74, D-01307 Dresden, Germany*

## 1. INTRODUCTION

*Mycoplasma pneumoniae* represents one of the most common etiological agents of non-nosocomial acquired diseases of the ororespiratory tract causing different clinical signs, which include mild respiratory symptoms, tracheobronchitis, interstitial pneumonia and in rare cases various pulmonary and extra-pulmonary complications, including intravascular hemolysis[21, 28]. The incidence of *M. pneumoniae* disease in a community varies among major outbreaks reported at intervals of every 3 to 5 years[15, 29]. Disease manifestations appear in all age groups but *M. pneumoniae* infections are most prevalent in household- and school-contacts. The pathomechanisms of *M. pneumoniae* include the transmission by aerosols, the adhesion of *M. pneumoniae* to respiratory epithelium cells and motile activity of this pathogen to reach the ciliabasis for further colonization and growth. Hidden in this cell compartment from the immune system, the agent is causing delayed host responses[20,32]. Following an incubation time of about one to three weeks, the first clinical manifestations occur. They are mostly unspecific, including dry coughing and malaise. *M. pneumoniae* upper and lower respiratory tract acute diseases and the convalescence time are long lasting without resulting in protective immunity[20]. Differences in the amino acid sequences of the important functional sites of the major adhesin of *M. pneumoniae*[40] were proposed as possible reasons for incomplete host protection.

*Molecular Biology and Pathogenicity of Mycoplasmas*, Edited by Razin and Herrmann, Kluwer Academic/Plenum Publishers, New York, 2002

## 2.      THE AGENT'S HISTORY AND ITS RESERVOIR

Mild respiratory symptoms, disproportionate to massive infiltrations observed on chest roentgenograms, the failure of sulfonamide therapy and the appearance of hemagglutinins in patients' blood led the first search for an uncultivable microorganism, possibly a virus, causing the primary atypical pneumonia, which we nowadays know is due to the fastidious growing bacterium, *M. pneumoniae*. It was this 'walking pneumonia', a common term in military medical circles, which stimulated studies in 1944 by Eaton and his colleagues to define the infectious agent. Failing to culture the microorganism, which was not detectable by Gram-stain and passed filter membranes used to exclude bacteria from liquid medium led first to the naming the organism as Eaton's agent [30]. The study of respiratory diseases of active-duty soldiers was a main key because this agent is distributed in persons living in close contact. Even today military health services face problems handling respiratory infections associated mostly with prolonged nonproductive cough in afebrile soldiers lasting from two to six weeks before medical care is sought[43]. Neither the dry cough nor complications like a paroxysmal cough, posttussive emesis and/or displayed subconjunctival hemorrhages are not agent specific and may be due to other than *M. pneumoniae* infections, especially *Chlamydia pneumoniae* or *Bordetella pertussis*[37,38,43]. Almost 55 years after Eaton's description of this new human pathogen, the diagnostic tools to identify *M. pneumoniae* infections are still under development, although major efforts were undertaken to increase the sensitivity and specificity of serological and molecular tests for a "real" time diagnosis of *M. pneumoniae* (see Diagnosis chapter). Despite the latest breakthroughs in designing better test systems, physicians are generally not aware of the improvements of these diagnostic tools and stick to the perception of an agent which is almost undetectable during the acute phase of disease. Combined with a certain unwillingness to properly diagnose and treat this mild respiratory and in most cases self-limited disease, our knowledge of the epidemiology and reservoirs of this human pathogen is limited. Based on only a few studies, it appears that *M. pneumoniae* can be detected in the oral cavity of healthy persons or of recovered patients. The carrier state depends on the population screened for *M. pneumoniae*, on the season, but especially whether the investigation was carried out during a major outbreak or during the years between two outbreaks. Foy[15] detected *M. pneumoniae* up to 4 months in a recovered patients' throat culture. In most cases, the rate of *M. pneumoniae* isolation was low six weeks following the infection. In 1966, a year prior to an epidemic, the Seattle group cultured 2354 throat samples from school children and found only three to be positive for *M. pneumoniae*. In a prospective PCR study in the Netherlands from January 1997 to July 1999, 1172 patients showed acute respiratory infections in outpatient general practitioner settings. Among these patients 39 (3.9%)

were positive for *M. pneumoniae*. Dorigo-Zetsma et al.[13] followed the spread of *M. pneumoniae* in household contacts and found that 12 of 79 household contacts were positive for *M. pneumoniae*, nine of these positives were younger than 16 years and four of them did not develop respiratory signs but were obviously in a carrier state. During epidemic years, the rate of successful *M. pneumoniae* isolation rose up to 10% in military personal, presumably carriers[45] or up to 13.5% in a healthy non-selected population in Scandinavia[16]. Using PCR techniques, it was clearly demonstrated that sensitivity of tests could be increased by combining PCR with a hybridisation step or by using nested PCR approaches[1]. Application of more elaborate and sensitive techniques, i.e. real-time PCR or quantification of PCR-products[17,19] is expected to provide more definitive data on the carrier state of *M. pneumoniae* in a population. On the other hand it is obvious that it will be necessary to define a threshold of number of genomic DNA copies to differentiate between putative carriers and patients. Williamson et al.[46] first tried to develop such a threshold for this PCR-approach. Their suggested threshold was $10^4$ genomic DNA copies per ml of throat washing, beyond which no difference was observed between patients and carriers. Such a threshold might be discussed as a further academic "Limes"; it might help us during acute diseases to establish the association of a laboratory result with the ungoing respiratory infection of a patient if the result is above the threshold. Interestingly, in good laboratory practice this threshold is commonly used for all different conventional culturable respiratory pathogens i.e. *Streptococcus pneumoniae, Haemophilus influenzae, Moraxella catarrhalis*. Also, the latter three human pathogens can be frequently isolated from carriers out of the oral cavity in low numbers, therefore diagnosis of the latter three should always include the CFU/ml of the pathogen isolated from respiratory material. On the other hand, all microbiologists active in the diagnostic field are struggling with the fact that patient's material may be of quite a different quality, so that the cultured CFU or numbers of DNA copies in a patient sample alone are not too helpful to distinguish between a carrier or a patient. For this reason, the quality of a respiratory sample is evaluated by the additional finding of granulocytes, which are associated with disease, or of epithelial cells of the oral cavity, which are contaminated by oral flora. Concerning this common carrier status, there are major differences between the above mentioned three bacterial pathogens and *M. pneumoniae* infections: (i) Parameters of infection such as granulocytes are not found during the acute phase of *M. pneumoniae* infections. (ii) Due to the dry, non-productive cough, no appropriate sample can be obtained from *M. pneumoniae* patients, especially from children. (iii) Throat washings inevitably contain oral flora. (iv) It is not known whether the number of CFU of *M. pneumoniae* in a carrier varies during day-time probing or during epidemic and non-epidemic periods, and (v) Little is known of CFU content of samples taken from hospitalized and non-hospitalized patients before and after adequate therapy, and weeks later during convalescence[13]. Since an easy method, i.e., a Gram stain, to evaluate patients' respiratory material for granulocytes and to associate bacteriological findings with the respiratory disease, does not exist, we must be aware that with our new and more sensitive PCR methods one cannot exclude the detection of *M. pneumoniae* of a carrier and that another agent might be responsible for disease of this specific patient. Despite this carrier problem, especially when PCR techniques are used, the diagnosis of acute disease caused by *M. pneumoniae* can be performed within hours, which should increase the acceptance of this diagnostic tool to identify *M. pneumoniae* in patient's material, and help the physician to associate the laboratory result with the clinical and chest X-ray findings.

## 3.     EPIDEMIOLOGY

Because of the underreporting of *M. pneumoniae* diseases, only a few epidemiological studies exist to provide us with an insight of the epidemiological pattern of *M. pneumoniae* diseases. Two major studies covered the epidemiological aspects of *M. pneumoniae* diseases in Seattle[15] and in Denmark[29]. The Seattle study was based on an extensive long-term study in a prepaid care group (reviewed by Foy[15]) consisting of 10% of the Seattle population (about 125.000 persons) during the period from December 1962 to the spring of 1975. Based on culture techniques and documentation of a fourfold or greater increase in antibody titers tested by the complement fixation test, the Seattle group showed that *M. pneumoniae* diseases accounted for 15-20% of all cases of pneumonia or 2 cases per 1000 persons annually. In this study, the secular swings with a first epidemic in 1967 lasting for more than a year, followed by an increased number of cases in 1972 and a summer peak in 1974 when approximately half of the community-acquired pneumonias were due to *M. pneumoniae*, demonstrating that this agent belongs to the top five bacterial pathogens, which induce most of the lower respiratory tract diseases in a population. Regarding the pattern of outbreaks, the retrospective analysis of Lind et al.[29] in Denmark over a 50-year period from 1946 to1995 was made possible due to the very large serum and blood bank storage facilities at the central laboratory at The Statens Serum Institut in Copenhagen. Based on cold-agglutinins and the complement fixation tests, Lind et al. showed that up to the year 1973 *M. pneumoniae* occurred in a regular pattern of epidemics peaking every 4.5 years, followed by two premature epidemics in 1975 and 1977/78 and a hypoendemic period which ended by the big epidemic in the winter of 1991/2. This last outbreak led to a changing pattern during the four years of the study characterized by an unusual high number of cases of *M.*

*pneumoniae* pneumonia with no return to a low endemic phase. The recent introduction of diagnostic molecular methods as well as advancement of our knowledge of *M. pneumoniae* virulence factors i.e. the major P1-adhesin[32] and the deciphering of the complete sequence of the *M. pneumoniae* genome[33], may help to understand the changing pattern of the latest outbreaks. Collecting and testing sera and isolates, we confirmed a *M. pneumoniae* outbreak in Germany in 1992 and an unusual high number of cases in the following years[22]. Screening the isolates with a monoclonal antibody, which was able to inhibit the attachment to sheep erythrocytes of most *M. pneumoniae* isolates before 1992, we found no attachment inhibition of isolates of the year 1992 outbreak and no binding to the P1-adhesin in immunoblot analysis. Based on the knowledge of two different P1-sequences[40], various groups showed by different molecular techniques that an outbreak in a population might be due to a distinct P1-adhesin subtype and might be followed by another variant [8,22,35,41]. More recently a third P1-adhesin subtype was described by Kenri et al.[25]. It appears to be a variant of subtype 2 characterized by minor amino acid changes in the P1-protein, compared to the more extensive change of amino acid sequence between subtype one and two[25]. Recently, additional P1 cytadhesin gene divergences were described[8,14]. In contrast to subtypes one and two only a few isolates of the new subtype 2 variant were isolated over the last few years in different regions. The epidemiological pattern of subtypes one and two in Japan were linked to outbreaks. Sasaki et al.[35] tested two hundred and fifty *M. pneumoniae* clinical isolates by PCR-restriction fragment- length polymorphism and found that three of four old isolates of the year 1976 belonged to subtype 1 and one isolate to subtype two. In 1997 and 1980, twenty four isolates belonged exclusively to subtype 2 and from 1985 to 1991 almost all isolates were of subtype one. In Japan, the frequency of subtype one was declining in 1992 and in 1994 90% of the isolates belonged to subtype two again. Comparing this pattern with German isolates, the 1992 outbreak in Germany was due to subtype one and an increase in the number of isolates of subtype 2 was found in the following years[22]. These results pointing to an association of outbreaks with defined P1 subtypes may also explain the findings of Foy[15] who showed repeated episodes of *M. pneumoniae* pneumonia in six patients, one with humoral immunodeficiency and the other five in immunocompetent patients. Interestingly, the interval between both episodes was between 2.5 to 11 years, indicating that the second infection was due to the loss of the protecting antibodies or to contact with a different P1 subtype of a following outbreak. Antibody responses in respect to the activity of an inhibition of adherence of *M. pneumoniae* to host cells were shown to be specific either to subtype one or two, indicating an only subtype limited antibody protection in patients[22].

The seroepidemiological findings of Lind et al.[29] showing high numbers of cases after the 1991/1992 epidemic was confirmed in a study of 28 years of surveillance, from 1970 to 1997, in Poland on 25,932 serum specimen by Rastawiki et al.[31]. Whether this change of the epidemic pattern of *M. pneumoniae* in the last few years is due to different modifications or variations of the P1-adhesin and/or other virulence factors, or only due to an increasing number of fast, sensitive and specific diagnostic tools must await further studies. Without question, the development of modern molecular approaches has led to renewed interest in the epidemiology of *M. pneumoniae* diseases. Using the PCR technique in a prospective study of community-acquired pneumonia of bacterial etiology of adults in France during five successive winter periods (winter 1992/3 to winter 1996/7), Layani-Milon et al.[26] tested 3897 samples which were collected by 75 medical practitioners. Positive samples of *M. pneumoniae* were found in 7.3% of the patients. The percentage of positive samples declined from 10.1% in the winter of 1992/3, to 2 % in 1995/6, increasing again in the winter of 1996/7 to 4.2%. The calculated incidence of *M. pneumoniae* among these patient populations with acute respiratory symptoms was 190 to 1234 per 100.000 persons in France. In the Netherlands, the calculated incidence of *M. pneumoniae* among patients with respiratory disease approaching outpatient general practitioners was 587 per 100.000 persons per year. All these different calculations did not include patients who did not consult their physicians. Dorigo-Zetsma[12] followed the household contacts of *M. pneumoniae* positive patients who consulted general practitioners and found that only 25% of the *M. pneumoniae* positive household contacts went to see their physicians, reflecting that this disease is more often mild and does not lead to a visit to a physician. In these Dutch and French studies, all age groups were present and no significant difference associated with patient's age was found. Among the household contacts significantly more children up to 15 years were positive for *M. pneumoniae*. They had either no signs or symptoms of an acute respiratory infection or the symptoms were so mild that the children had no physician contact which led to the assumption that children especially serve as reservoirs. In the French study[26] and in an Indian[9] study, it was clearly demonstrated that *M. pneumoniae* behaves like other respiratory pathogens i.e. *Streptococcus pneumoniae* and *Haemophilus influenzae* in cases of a first viral respiratory infection or during viral epidemics. During the 1992/3 surveillance two peaks of *M. pneumoniae* infections were recorded coinciding with two viral epidemics of RSV and of influenza B virus, indicating that *M. pneumoniae* can cause superinfection and might cocirculate with different agents[26]. This finding may be the key to the changing pattern of *M. pneumoniae* to a high incidence in the years following the *M. pneumoniae* outbreak of 1991/2. *M. pneumoniae* was still there in a high number of carriers' oral cavities and because of the peaking viral epidemics *M. pneumoniae* got another chance to spread by aerosol to patients suffering from a viral infection and being to some extent immunocompromised. On the other hand because of a viral disease in a *M.*

*pneumoniae* carrier, the mycoplasma might get a better chance to spread from the oral cavity to the lower respiratory tract resembling the endogenous infections of S*treptococcus pneumoniae* and *Haemophilus influenzae* during influenza epidemics, which in both cases result in a superinfection with more severe manifestations than without the viral agent, and therefore resulting in a higher rate of consultation of the patients with practitioners. In addition, the more severe disease manifestations lead to more efforts to diagnose the respiratory disease, including *M. pneumoniae* as a probable pathogen. Beside this possibility of a prolonged spread of *M. pneumoniae* as a hitchhiker during virus outbreaks, the incubation time of *M. pneumoniae* of up to 20 days must result in quite different outbreak patterns from those seen during epidemics with viruses alone. The viral outbreaks show sharp increases and decreases in the number of cases during an epidemic. In contrast, *M. pneumoniae* outbreaks need months or even years to be recognized as an outbreak, followed by a peaking shoulder and a slow decline of case reports[31] after several months. The marked improvements in
*M. pneumoniae* diagnosis might help us to establish a sentinel system worldwild to follow up changes within *M. pneumoniae* subtypes and frequencies of  patient cases and provide the outpatient practitioners with fast diagnostic reports  for the benefit of patients suffering from *M. pneumoniae* diseases.

# 4.    MANIFESTATIONS OF *M. PNEUMONIAE* DISEASES

The main manifestations of *M. pneumoniae* diseases are related to upper respiratory tract infections and in a few cases to lower respiratory tract infections described as an interstitial pneumonia. Most of our knowledge of *M. pneumoniae* diseases published in the literature is based on hospitalized patients. In the Seattle study, Foy[15] showed that only 2% of all patients with *M. pneumoniae* pneumonia were hospitalized, whereas 16% of patients with pneumonia of all etiologies were admitted to the hospital, and of all patients hospitalized with pneumonia, 4.5-5% were infected with *M. pneumoniae*. Recently, in France, 6.7% of 150 episodes of infectious pneumonia were caused by *M. pneumoniae*[27]. In this study, all patients were adults of 27 years to 72 years old. Compared with former studies, mainly based on hospitalized children with *M. pneumoniae* diseases[21], the clinical pulmonary and extra-respiratory symptoms in both patient groups exhibited almost the same clinical signs. In all cases the leading clinical symptom was coughing and in most cases dyspnea. Regarding fever, children and adults showed a broad range from elevated temperatures, in most cases within the medium ranges and in rare cases high degrees of fever. Discrete scattered rhonchi were found in 15 of 18 children by auscultation[21]. Chest X-ray findings usually show involvement especially of lower lung lobes in children and adults.

Almost half of both groups of adults and children suffered from pharyngitis, all other clinical signs i.e. rhinitis, sinusitis, arthralgia and myalgia were only rarly found[21,27]. The delay in appearance of first symptoms and the first visit to a physician were between 4 and 24 days in adults and between 1 to 12 days in children. The fact that in 10 hospitalized adult patients eight antibiotic therapy regimens were ineffective in treating *M. pneumoniae* and of 18 hospitalized children six got only antipyretic treatment and four ineffective antibiotics prescribed by their general practitioners, and that in each group only two patients were treated with the drug of choice, erythromycin or another macrolide, reflect the underestimated weight given to *M. pneumoniae* infections accompanying the unspecific pathological signs, which are not different from those of viral respiratory diseases. One may speculate that most of these patients would not be hospitalized if treated with effective chemotherapy from the beginning. It is therefore of utmost importance that in community-acquired respiratory infections an adequate laboratory examination should be carried out and results should be transmitted to a sentinel system, which should inform physicians of increasing case numbers or outbreaks of *M. pneumoniae* in a population. Complications are rare and usually are not followed up in long-term surveillance studies; it is therefore not possible to estimate incidence rates for such complications[7,15,28]. Although the development of cold agglutinins[3,29], which are autoimmunoglobulines of the IgM class is a common finding during the acute phase of *M. pneumoniae* diseases and was used therefore as a diagnostic marker, the severe complication manifested as intravascular hemolysis is rarely found among *M. pneumoniae* patients. In my own experience during 15 years of working in the field of mycoplasmology I can report only two female cases with severe intravascular hemolysis requiring intensive care treatment for several days. The list of published respiratory and non-respiratory complications[9,28] is still increasing. The list includes the complications involving the skin i.e. skin rashes, erythema nodosum and the Steven-Johnson syndrome, or complications affecting different human organs i.e. glomerulitis, nephritis[34], pericarditis, arthritis, aplastic anemia[39] and a particularly long list of cases with neurological symptoms. However, comparing this list of complications with those of other infectious diseases one can find that this list of complications is not unique to *M. pneumoniae*. This might indicate that unknown and perhaps quite different host parameters i.e. cold agglutinins, cytokines and combinations of specific and unspecific, perhaps polyclonal stimulation of cellular immune responses, including defects in regulation or exacerbation of cellular mechanisms[22] might be induced or stimulated by different agents, including *M. pneumoniae*. This theory would implicate that because of the rarity of cases exhibiting complications only those patients who carry genetic predisposition are responding in a way leading to clinical symptoms characterizing these complications. Clinical symptoms follow different, distinct but limited tracks of host responses, which is obvious in the example of interstitial pneumonia that may be due to *M. pneumoniae* but also to a long list of quite different microorganisms. The interstitial pneumonia is not a specific clinical finding for *M. pneumoniae*, but when compared to other respiratory pathogens it is very frequently associated with a *M. pneumoniae* infection. This should be the answer to all different complications which were described in connection with serological, cultural and molecular test results indicating a *M. pneumoniae* acute or even postinfectious stage. Under this perception, one has to evaluate the numerous reports published recently concerning neurological symptoms, including encephalitis[11], meningoencephalitis[6], coma[24], mild forms of Guillain-Barre[42,44], pyramidal and extrapyramidal tract dysfunctions, seizures, cognitive abnormalities and cerebellar dysfunction[2,5,18,36,47]. It is still an enigma to explain the pathomechanisms of these symptoms or answer the questions of a possible invasion of *M. pneumoniae* from the local infected respiratory tract into the blood system and across the blood-brain barrier. Major questions concerning these neurological complications during or after *M. pneumoniae* infection are yet unanswered. Which, if any, factor or factors of the bacterium induce the different complications; is it possible to detect or culture *M. pneumoniae* from cerebrospinal fluid or detect intrathecal synthesis of specific antibodies[4] indicating that the organism or antigen is locally involved in this special situation. If so, it may enable us to start with an appropriate therapy to modify the outcome[36].

The application of molecular approaches to the field of infectious diseases facilitates studies of all the different complications and probable side effects, leading not only to answers but also to new questions concerning the relevance of a positive PCR result. Since the difficult culture of *M. pneumoniae* is no longer the gold standard, one has to think of a new standard which might be a combination of fast and easy molecular techniques for the diagnosis of *M. pneumoniae* in the acute phase of disease as well as serological tests based on specific antigens to demonstrate an active host immune response during pulmonary or extrapulmonary manifestations due to *M. pneumoniae*.

# REFERENCES

1.     Abele-Horn, M., Busch, U., Nitschko, H., Jacobs, E., Bax, R., Pfaff, F., Schaffer, B., and Heesemann, J., 1998, Molecular approaches to diagnosis of pulmonary diseases due to Mycoplasma pneumoniae. *J. Clin. Microbiol.* **36:** 548-551

2.   **Andrade, C., Munoz, D., Koukoulis, A., Martinez, C., and Gomez-Alonso, J.**, 1999, Encephalopathy with focal lesions due to Mycoplasma pneumoniae. *Neurologia* **14**:131-134

3.   **Bar, M., Amital, H., Levy, Y., Kneller, A., Bar-Dayan, Y., and Shoenfeld, Y.**, 2000, Mycoplasma-pneumoniae-induced thrombotic thrombocytopenic purpura. *Acta Haematol.* **103**:112-115

4.   **Bencina, D., Dovc, P., Mueller-Premru, M., Avsic-Zupanc, T., Socan, M., Beovic, B., Arnez, M. and Narat, M.**, 2000, Intrathecal synthesis of specific antibodies in patients with invasion of the central nervous system by Mycoplasma pneumoniae. *Eur.J.Clin.Microbiol.Infect.Dis.* **19**:521-530

5.   **Berger, R.P. and Wadowksy, R.M.**, 2000, Rhabdomyolysis associated with infection by Mycoplasma pneumoniae: a case report. *Pediatrics* **105**:433-436

6.   **Cambonie, G., Sarran, N., Leboucq, N., Luc, F., Bongrand, A.F., Slim, G., Lassus, P., Fournier-Favre, S., Montoya, F., Astruc, J., and Rieu, D.**, 1999, Mycoplasma pneumoniae meningoencephalitis. *Arch. Pediatr* **6**: 275-278

7.   **Chan, E.D., Kalayanamit, T., Lynch, D.A., Tuder, R., Arndt, P., Winn, R., and Schwarz, M.I.**, 1999, Mycoplasma pneumoniae-associated bronchiolitis causing severe restrictive lung disease in adults: report of three cases and literature review. *Chest* **115**:1188-1194

8.   **Cousin-Allery, A., Charron, A., de Barbeyrac, B., Fremy, G., Skov, J., Renaudin, H., and Bebear, C.**, 2000, Molecular typing of Mycoplasma pneumoniae strains by PCR-based methods and pulsed-field gel electrophoresis. Application to French and Danish isolates. *Epidemiol. Infect.* **124**:103-111

9.   **Daian, C.M., Wolff, A.H., and Bielory, L.**, 2000, The role of atypical organisms in asthma. *Allergy Asthma Proc.* **21**:107-111

10.  **Dey, A.B., Chaudhry, R., Kumar, P., Nisar, N., and Nagarkar, K.M.**, 2000, Mycoplasma pneumoniae and community-acquired pneumonia. *Natl. Med. J. India* **13**:66-70.

11.  **Dionisio, D., Valassina, M., Mata, S., Rossetti, R., Vivarelli, A., Esperti, F.C., Benvenuti, M., Catalani, C., and Uberti, M.**, 1999, Encephalitis caused directly by Mycoplasma pneumoniae. *Scand. J. Infect. Dis.* **31**:506-509.

12.  **Dorigo-Zetsma, J.W.**, 2000, Molecular detection of Mycoplasma pneumoniae among general practitioner patients with respiratory infection and their household contacts reveals children as human reservoir. In *Molecular diagnosis and epidemiology of Mycoplasma pneumoniae.* (J.Wendelien Dorigo-Zetsma ed.), Addix, Wijk bij Duurstede, Netherland, pp.69-81.

13.  **Dorigo-Zetsma, J.W., Zaat, S.A., Vriesema, A.J., and Dankert, J.**, 1999, Demonstration by a nested PCR for Mycoplasma pneumoniae that M. pneumoniae load in the throat is higher in patients hospitalised for M. pneumoniae infection than in non-hospitalised subjects. *J. Med. Microbiol.* **48**: 1115-1122

14.  **Dorigo-Zetsma, J.W., Dankert, J., and Zaat, S.A.**, 2000, Genotyping of Mycoplasma pneumoniae clinical isolates reveals eight P1 subtypes within two genomic groups. *J. Clin. Microbiol.* **38**:965-970

15.  **Foy, H.M.**, 1993, Infections caused by Mycoplasma pneumoniae and possible carrier state in different populations of patients. *Clin.Inf.Dis.* **17(Suppl 1)** :37-46

16.  **Gnarpe, J., Lundback, A., Sundelof, B., and Gnarpe, H.**, 1992, Prevalence of Mycoplasma pneumoniae in subjectively healthy individuals. *Scand. J. Infect. Dis.* **24**: 161-164

17.  **Hardegger, D., Nadal, D., Bossart, W., Altwegg, M., and Dutly, F.,** 2000, Rapid detection of Mycoplasma pneumoniae in clinical samples by real-time PCR. *J. Microbiol. Methods* **41**:45-51

18.  **Heckmann, J.G., Sommer, J.B., Druschky, A., Erbguth, F.J., Steck, A.J., and Neundorfer, B.,** 1999, Acute motor axonal neuropathy associated with IgM anti-GM1 following Mycoplasma pneumoniae infection. *Eur. Neurol.* **41**:175-176

19.  **Honda, J., Yano, T., Kusaba, M., Yonemitsu, J., Kitajima, H., Masuoka, M., Hamada, K., and Oizumi, K.,** 2000, Clinical use of capillary PCR to diagnose Mycoplasma pneumonia. *J.Clin. Microbiol.* **38**:1382-1384

20.  **Jacobs, E.,** 1991, *Mycoplasma pneumoniae* virulence factors and the immune response. *Rev. Med. Microbiol.* **2**: 83-90

21.  **Jacobs, E.,** 1997, Mycoplasma infections of the human respiratory tract. *Wien. Klin. Wochenschr.* **109**: 574-577

22.  **Jacobs, E., Vonski, M., Oberle, K., Opitz, O., Pietsch, K.,** 1996, Are outbreaks and sporadic    respiratory infections by *Mycoplasma pneumoniae* due to two distinct subtypes? *Eur. J. Clin. Microbiol. Infect. Dis.* **15**: 38-44

23.  **Kaneko, K., Fujinaga, S., Ohtomo, Y., Nagaoka, R., Obinata, K., and Yamashiro, Y.,** 1999, Mycoplasma pneumoniae-associated Henoch-Schonlein purpura nephritis. *Pediatr. Nephrol.* **13**: 1000-1001

24.  **Keegan, B.M., Lowry, N.J., and Yager, J.Y.,** 1999, Mycoplasma pneumoniae: a cause of coma in the absence of meningoencephalitis. *Pediatr. Neurol.* **21**: 822-825

25.  **Kenri, T., Taniguchi, R., Sasaki, Y., Okazaki, N., Narita, M., Izumikawa, K., Umetsu, M. , and Sasaki, T.,** 1999, Identification of a new variable sequence in the P1 cytadhesin gene of Mycoplasma pneumoniae: evidence for the generation of antigenic variation by DNA recombination between repetitive sequences. *Infect. Immun.* **67**: 4557-4562

26.  **Layani-Milon, M.P., Gras, I., Valette, M., Luciani, J., Stagnara, J., Aymard, M., and Lina, B.,** 1999, Incidence of upper respiratory tract Mycoplasma pneumoniae infections among outpatients in Rhone-Alpes, France, during five successive winter periods. *J. Clin. Microbiol.* **37**: 1721-1726

27.  **Lesobre, V., Azarian, R., Gagnadoux, F., Harzic, M., Pangon, B., and Petitpretz, P.,** 1999, Mycoplasma pneumoniae pneumopathy. 10 cases. *Presse Med.* **28**: 59-66.

28.  **Lind, K.,** 1983, Manifestations and complications of Mycoplasma pneumoniae disease: a review. *Yale J. Biol. Med.* **56**: 461-468

29.  **Lind, K., Benzon, M.W., Jensen, J.S., and Clyde, W.A.J.,** 1997, A seroepidemiological study of Mycoplasma pneumoniae infections in Denmark over the 50-year period 1946-1995. *Eur. J .Epidemiol.* **13** : 581-586

30.  **Marmion, B.P.,** 1990, Eaton agent-science and scientific acceptance: A historical commentary. *Rev. Infect. Diseases* **12**: 338-353

31.  **Rastawicki, W., Kaluzweski, S., Jagielski, M. and Gierczyski, R.,** 1998, Epidemiology of Mycoplasma pneumoniae infections in Poland: 28 years of surveillance in Warsow, 1970-1997. *Eurosurveillance* **3**:99-100

32.  **Razin, S. and Jacobs, E.,** 1992, Mycoplasma adhesion. *J. Gen. Microbiol.* **138**: 407-422

33.  **Razin, S., Yogev, D., and Naot, Y.,** 1998, Molecular biology and pathogenicity of mycoplasmas. *Microbiol. Mol. Biol. Rev.* **62**: 1094-1156

34.  **Said, M.H., Layani, M.P., Colon, S., Faraj, G., Glastre, C., and Cochat, P.,** 1999, Mycoplasma pneumoniae-associated nephritis in children. *Pediatr. Nephrol.* **13**: 39-44

35.  **Sasaki, T., Kenri, T., Okazaki, N., Iseki, M., Yamashita, R., Shintani, M., Sasaki, Y., and Yayoshi, M.,** 1996, Epidemiological study of Mycoplasma pneumoniae infections in japan based on PCR-restriction fragment length polymorphism of the P1 cytadhesin gene. *J. Clin. Microbiol.* **34:** 447-449

36.  **Smith, R. and Eviatar, L.,** 2000, Neurologic manifestations of Mycoplasma pneumoniae infections: diverse spectrum of diseases. A report of six cases and review of the literature. *Clin. Pediatr. (Phila)* **39:** 195-201

37.  **Socan, M., Marinic-Fiser, N., Kraigher, A., Kotnik, A., and Logar, M.,** 1999, Microbial aetiology of community-acquired pneumonia in hospitalised patients. *Eur. J. Clin. Microbiol. Infect. Dis.* **18** : 777-782

38.  **Sopena, N., Sabria, M., Pedro-Botet, M.L., Manterola, J.M., Matas, L., Dominguez, J., Modol, J.M., Tudela, P., Ausina, V., and Foz, M.,** 1999, Prospective study of community-acquired pneumonia of bacterial etiology in adults. *Eur. J. Clin. Microbiol. Infect. Dis.* **18** : 852-858

39.  **Stephan, J.L., Galambrun, C., Pozzetto, B., Grattard, F., and Bordigoni, P.,** 1999, Aplastic anemia after Mycoplasma pneumoniae infection: a report of two cases. *J. Pediatr. Hematol. Oncol.* **21:** 299-302

40.  **Su, C.J., Chavoya, A., Dallo, S.F., Baseman, J.B.,** 1990, Sequence divergency of the   cytadhesin gene of Mycoplasma pneumoniae. *Infect. Immun.* **58:** 2669-2674

41.  **Ursi, D., Ieven, M., van Bever, H., Quint, W., Niesters, H.G., and Goossens, H.,** 1994, Typing of Mycoplasma pneumoniae by PCR-mediated DNA fingerprinting. *J. Clin. Microbiol.* **32:** 2873-2875

42.  **Van Koningsveld, R., Van Doorn, P.A., Schmitz, P.I., Ang, C.W., and Van der Meche, F.G.,** 2000, Mild forms of Guillain-Barre syndrome in an epidemiologic survey in The Netherlands. *Neurology* **54:** 620-625

43.  **Vincent, J.M., Cherry, J.D., Nauschuetz, W.F., Lipton, A., Ono, C.M., Costello, C.N., Sakaguchi, L.K., Hsue, G., Jackson, L.A., Tachdjian, R., Cotter, P.A., and Gornbein, J.A.,** 2000, Prolonged afebrile nonproductive cough illnesses in American soldiers in Korea: a serological search for causation. *Clin. Infect. Dis.* **30:** 534-539

44.  **Vinzio, S., Andres, E., Goichot, B., and Schlienger, J.L.,** 2000, Guillain-Barre syndrome and Mycoplasma pneumoniae infection. *Ann. Med. Interne (Paris)* **151**: 309-310

45.  **Wenzel, R.P., Craven, R.B., Davies, J.A., Hendley, J.O., Hamory, B.H., Gwaltney, J.M. Jr.,** 1977, Protective efficacy of an inactivated Mycoplasma pneumoniae vaccine. *J.Infect.Dis.* **136 (suppl)** : 204-207

46.  **Williamson, J., Marmion, B.P., Worswick, D.A., Kok, T.W., Tannock, G., Herd, R., and Harris, R.J.,** 1992, Laboratory diagnosis of Mycoplasma pneumoniae infection. 4. Antigen capture and PCR-gene amplification for detection of the Mycoplasma: problems of clinical correlation. *Epidemiol. Infect.* **109:** 519-537

47.  **Yamashita, S., Ueno, K., Hashimoto, Y., Teramoto, H., and Uchino, M.,** 1999, A case of acute disseminated encephalomyelitis accompanying Mycoplasma pneumoniae infection. *No To Shink ei* **51:** 799-803

# Chapter 24

# Diagnosis of Mycoplasmal Infections

SHMUEL RAZIN
*Department of Membrane and Ultrastructure Research, The Hebrew University-Hadassah Medical School, Jerusalem, Israel 91120*

## 1. INTRODUCTION

Mycoplasma identification and laboratory diagnosis of mycoplasmal infections has been based on the classical bacteriological tests, including morphology, cultural characteristics, physiological and serological properties (47). While these tests still occupy a major part in mycoplasmal diagnostics, new tests based on the molecular analysis of genomic DNA, ribosomal RNAs, cell proteins, and lipids appear to push aside the classical tests, and one may expect that the molecular tests will shortly become the prevailing tests in mycoplasma identification, particularly with regards to the slow-growing and/or extremely fastidious species. Systematic and detailed description and evaluation of the classical procedures used in mycoplasma identification can be found in the two volumes of Methods in Mycoplasmology (49; 64), while the newer molecular methods are included in the two more recent volumes of Molecular and Diagnostic Procedures in Mycoplasmology (50; 65). A succinct practical review of current methods applied in routine laboratory diagnosis of mycoplasmal infections, with emphasis on infections in humans has recently been published (69).

## 2. CULTURAL PROPERTIES AND BIOCHEMICAL TESTS

No single medium formulation is adequate for all mollicute species due to their different nutritional requirements, optimal pH, etc. Self-prepared as

*Molecular Biology and Pathogenicity of Mycoplasmas*, Edited by Razin and Herrmann, Kluwer Academic/Plenum Publishers, New York, 2002

well as commercially available culture media used in diagnosis of mycoplasmal infections in humans, as well as specimen collection and transport media, are described in some detail in Waites et al. (69). Liquid specimens and particularly minced tissues should be serially diluted in broth at least $10^{-3}$ with subculture of each dilution onto agar. This is an extremely important step since it may help to overcome possible interference of mycoplasmal growth by antibodies, antibiotics and other inhibitors, such as lysophospholipids, that may be present in clinical specimens. The wall-less mollicutes are particularly sensitive to lysophospholipids produced by the action of phospholipases released in minced tissues (61). In some cases of fastidious mycoplasmas, such as *M. genitalium,* successful cultivation could be achieved by cocultivation with Vero cell cultures (33; 62). This complex methodology is certainly inadequate for routine cultivation of fastidious mycoplasmas, leaving the door open for the application of molecular techniques, such as PCR (29; 62). It should be emphasised in this context that even in the case of *M. pneumoniae* the value of culture is far from being satisfactory as it may require up to 5 weeks and may be successful in approximately only 60% of the *M. pneumoniae* patients, so that for fast diagnosis required for selecting the right treatment one has to resort to PCR which is still inapplicable in many diagnostic labs (30).

The International Committee on the Taxonomy of *Mollicutes* issued recommended tests for mycoplasma identification as well as for description of new species of the class *Mollicutes* (31). The recommended tests include those required to define a new isolate at the higher taxonomic levels (class, order, family) as well as tests for genus and species determination. Thus, a test for sterol requirement, based on growth promotion by cholesterol in the presence of either 0.01% or 0.04% Tween 80 (48; 63) separates the sterol-nonrequiring *Acholeplasma* and *Mesoplasma* species from the members of *Mollicutes* requiring cholesterol for growth. *Ureaplasma* identification is based on urea hydrolysis tests. Other key biochemical tests applied in mollicute identification include those testing sugar fermentation and arginine hydrolysis (49). A biochemical test based on the ability to utilize glycerol has recently been applied to differentiate the African *M. mycoides* subsp. *mycoides* strains from the less-pathogenic European strains of this species which are incapable of utilizing glycerol (53).

Antimicrobial sensitivity testing of mycoplasmas constitutes most frequently an integral part of the routine diagnosis of mycoplasmal infections. The principles and methodology of these tests are discussed in detail in the "Antimycoplasmal Agents" chapter in this volume and in Waites et al. (69). The conventional antibiotic sensitivity testing may be supplemented in the near future by nucleic acid-based testing for antibiotic resistance genes, such as *tetM*, the tetracycline-resistance determinant (9). PCR amplification of resistance genes can be currently considered as the most sensitive technique for detecting resistance in bacteria.

## 3.    SEROLOGICAL TESTS

A great variety of serological tests have been employed, particularly in mycoplasma species and strain identification. The classical recommended tests include growth and metabolism inhibition by specific antisera as well as direct and indirect immunofluorescence tests applied to mycoplasma colonies. A combined deformation-metabolism inhibition test system has been used to provide both screening and refined analysis and definition of *Spiroplasma* species (see the relevant chapters in Methods in Mycoplasmology, reference 64). A variety of other, in some cases more sensitive, tests based on principles of enzyme-linked-immunosorbent assay (ELISA or EIA) (21; 22; 32), immunobinding (40), immunoblotting (13;43) and immunoperoxidase tests (7), employing polyclonal or monoclonal antibodies, have been introduced to mycoplasma diagnosis. Due to the higher sensitivity and specificity of these tests, they are capable of identifying strains within a species. One particular advantage of the colony immunofluorescence, immunoperoxidase, and immunobinding techniques is their ability to distinguish a certain mycoplasma species within a mixed culture, or on primary isolation plates. However, the finding that many immunodominant antigens exposed on the mycoplasmal cell surface undergo rapid phase- and size-variation (see Genetic mechanisms of antigenic variation chapter) puts severe restrictions on the use of monoclonal antibodies directed to such a variable antigen as a reagent in colony immunoblotting (54). The use of a polyclonal antiserum is, therefore, preferable in these tests. Another solution to this problem can be based on a search for highly immunogenic and species-specific antigens not subjected to antigenic variation. Thus, monospecific polyclonal antibodies to the antigenic stable P30, a membrane-associated protein of *M. agalactiae,* enabled easy and dependable serological identification of isolates and serological subtyping of this mycoplasma (23).

A great variety of diagnostic serological procedures are based on detection of specific antibodies in sera of humans and animals suffering from mycoplasma infections. The procedures as well as the evaluation of the results are discussed in detail in the relevant chapters of Molecular and Diagnostic Procedures in Mycoplasmology, (65) and in Waites et al. (69). Serodiagnosis consists of examining serum samples for antibodies that inhibit the growth and metabolism of the organism or fix complement with mycoplasmal antigens. Being the best established and important human pathogen thus far, laboratory diagnosis of *M. pneumoniae* infections has been the subject of numerous studies. Antibody response in mycoplasmal pneumonia is easily demonstrated by complement fixation (CF), reacting acute- and convalescent-phase sera with intact organisms or their lipid extract as antigen. A fourfold or greater antibody rise is considered indicative of recent infection (22), whereas a sustained high antibody titer may not be significant, because a relatively high level of antibody may persist for at least 1 year after infection.

In view of the many limitations of CF tests, they have recently been largly replaced by methods applying alternative technologies, based mostly on EIA. Efforts were also invested in the development of more defined and specific antigens based on immunodominant *M. pneumoniae* membrane proteins, such as P1 (25; 66) or the 116 kDa protein (21). A P1-enriched antigen preparation was not only the most sensitive in detecting antibodies at the early stage of infection but also detected antibodies to *M. pneumoniae* produced in the convalescent phase of infection (66). This antigen is suitable for IgG, IgM, and IgA EIA testing (59). The late mounting of specific IgG antibodies in the course of *M. pneumoniae* infections, the possible lack of IgM induction in adult patients and a considerable IgG seroprevalence in the community (32) imposes serious limitations on the interpretation of serological tests. Young patients tend to have higher IgM levels than adults, whereas IgA antibodies are seen at higher levels in elderly patients. A combination of *M. pneumoniae*-specific PCR in nasopharyngeal aspirates with an IgM-capture immunoassay in acute phase sera, increases the sensitivity of rapid laboratory diagnosis of *M. pneumoniae,* enabling the attainment of a laboratory result in one or two days (22).

A variety of rapid tests based on indirect hemagglutination of erythrocytes or latex particles coated with *M. pneumoniae* antigens, or an immunofluorescence assay detecting IgM antibodies to *M. pneumoniae* (20) have been developed, and some are commercially available. EIA test kits achieve presently the largest market share of commercial mycoplasma serologic tests (59; 66; 70). A rather comprehensive list of current commercial serologic test kits and reagents for diagnosis of *M. pneumoniae* can be found in Waites et al. (69).

The high genomic resemblance between *M. genitalium* and *M. pneumoniae* is reflected in marked serological cross-reactions between these two human pathogens. Thus, the conventional microimmunofluorescence (MIF) test, CF tests and indirect hemagglutination (IHA) tests fail to differentiate the two mycoplasmas. However, immunoblotting with the *M. genitalium* MgPa cytadhesin antigen was shown to be specific for *M. genitalium* despite the high sequence homology between the MgPa and the *M. pneumoniae* P1 in the C-terminal region of these two cytadhesins (14). Nevertheless, the use of species-specific PCR probes are preferable to serology in order to distinguish between *M. genitalium* and *M. pneumoniae* infections. A similar diagnostic problem due to shared antigenic determinants is that of the porcine pathogenic *M. hyopneumoniae* and the genetically-related nonpathogenic *M. flocculare*. Also in this case the best solution is to apply PCR which is more effective and species-specific, though the use of monoclonal antibodies and a blocking EIA test proved to be effective and specific enough to detect *M. hyopneumoniae* infected pigs under field conditions (68).

In summation, although PCR considerably extends the arsenal of reliable laboratory techniques, serology still retains importance in diagnosis of *M. pneumoniae* infections and of complications associated with *M. pneumoniae* infections. It is well known that IgM antibodies are frequently diagnostic in children because they are most likely experiencing an episode of a primary disease. In adults, on the contrary, IgM antibdies alone may not be diagnostic. Serology using IgG and IgA- antibody detection remains a method of choice for the diagnosis of acute infection in adults (58; 70).

# 4. MOLECULAR GENETIC METHODS

## 4.1 Restriction endonuclease analysis (REA)

The classic tests for mycoplasma identification and classification have been supplemented by a variety of tests based on genomic DNA analysis. Restriction enzyme analysis (REA) of the mycoplasma genome provides a convenient and cost-effective means of determining DNA sequence variations among strains of mollicutes species (51). The method involves comparison of the number and size of fragments produced by digestion of the chromosomal DNA with a restriction endonuclease that cuts DNA at a constant position within a specific recognition site, usually composed of 4 to 6 bp. Because of the high specificity of restriction endonucleases, complete digestion of a given DNA with a specific restriction endonuclease provides a reproducible array of fragments. Separation of the fragments by agarose gel electrophoresis and staining with ethidium bromide provides a restriction pattern that can be compared with that of related strains. Variations in the array of fragments generated by a specific restriction endonuclease are called restriction fragment length polymorphisms (RFLPs ). RFLPs can result from sequence rearrangements, insertion or deletion of DNA segments, or base substitutions within the restriction endonuclease cleavage sites. Restriction endonuclease analysis has become a most useful taxonomic tool, facilitating the identification and classification of mycoplasmal isolates as well as providing means for evaluating the degree of genotypic heterogeneity of strains within established species (18; 44; 45).

The great advantage of chromosomal restriction endonuclease analysis is that it is universally applicable and sensitive, because the entire genome is evaluated for RFLP, and in addition it is relatively easy to perform. Moreover, unlike SDS-PAGE of mycoplasmal cell proteins, restriction endonuclease analysis is not susceptible to contamination by culture medium contaminants, and requires very little (3 to 5µg) of unlabeled DNA per test (51). A more recently developed technique, named amplified-fragment length polymorphism (AFLP) is a whole genome fingerprinting method based on the selective amplification of restriction fragments. This multistep procedure combines in an elegant manner the power of PCR with the informativeness of REA. AFLP yields more complex banding patterns than other DNA fingerprinting methods, increasing the discrimination between strains. In this way AFLP can be used as a device in studies of epidemiology, pathogenicity and genetic variability in natural populations of mollicute species (37).

Southern blot analysis of genomic DNA digested by a restriction enzyme and hybridized with a cloned conserved gene or a specific genomic fragment as a probe, has become a most useful tool in the identification, classification and subtyping of mollicute species (45;71). The plasmid pMC5, carrying the entire 23S and 5S and most of the 16S rRNA genes of *M. capricolum* (4) was among the first plasmids to be used as probes in Southern blot analysis of mycoplasmal DNA. The fact that the rRNA operons in the various mollicutes differ in restriction sites within the operon and in the flanking sequences (5; 52) results in the production of hybridization patterns specific to different mollicute species or strains. A restriction enzyme having a 6 bp recognition site will usually cut the one or two mycoplasmal rRNA operons in a few sites, so that the hybridization patterns are usually simple and much easier to compare than the multi-band restriction patterns obtained by restriction endonuclease analysis of entire genomic DNA. Southern blot hybridization with cloned rRNA genes as probes has been named ribotyping. However, other conserved genes can be employed as probes, such as the *tuf* gene, encoding the elongation factor EF-Tu (72). The mollicute genome carries only one copy of this gene, so that the hybridization patterns obtained with restricted mollicute DNAs are also very simple and easy to compare (72).

## 4.2     PCR-based diagnostic procedures

The introduction of PCR to diagnostics in the late 1980's pushed aside many of the previously developed DNA probes and commercial kits (46). PCR tests are several orders of magnitude more sensitive than those based on direct hybridization with a DNA probe. Moreover, PCR is fast, copying a single DNA sequence over a billion times within three hours. Nucleic acid amplification techniques are not limited by the ability of an organism to grow in culture, a feature of paramount importance considering the fastidious nature of some mycoplasmas, such as *M. genitalium,* and the fact that not all mollicutes can be cultured *in vitro*, notably the phytoplasmas and the haemoplasmas (*Eperythrozoon* and *Haemobartonella* species, see

Taxonomy of *Mollicutes* chapter). Thus, PCR-amplified 16S rDNA subjected to RFLP analysis has become the method of choice for routine detection, differentiation and classification of phytoplasmas (3; 42; 56; 57; see also the chapter on Mycoplasmas of plants and insects).

PCR methodology is still at a phase of rapid development and new procedures, modifications etc. are reported almost daily. Detailed descriptions of methodology and evaluation of PCR-based tests in diagnostics can be found in the relevant chapters in Molecular and Diagnostic Procedures in Mycoplasmology (65). The first major step in the development of a PCR-based test concerns the selection of the appropriate target sequences for amplification. The sequence to be amplified can be chosen from a published mycoplasmal gene sequence or from a randomly cloned DNA fragment demonstrated to be specific for the mycoplasma to be detected (10; 38; 41). Complete or almost complete sequences of the 16S rRNA genes are now available for almost all the established mollicute species, and can be retrieved from data banks (see Taxonomy of *Mollicutes* chapter). This facilitates the selection of a variety of target sequences, starting with sequences in the highly conserved regions of the genes, producing primers of wide specificity ("universal primers") which will react with the DNA of any mycoplasma or even with the DNA of other prokaryotes. This may be satisfactory for detection of mycoplasma infection in cell cultures, where the goal is just to screen the cultures for contamination (67). In case of a need to differentiate the contaminating mycoplasmas, a multiplex PCR system can be employed, consisting of a universal set of primers along with primer sets specific for the mycoplasma species commonly infecting cell cultures (12; 39). The mycoplasmal 16S rRNA genes carry in addition to the conserved regions more specific variable regions, as well as specific 16S-23S intergenic spacer regions (60). Primers can be selected from these regions with various degrees of specificity, ranging from clusters of species, single species, down to the subspecies level (34; 39). Some mycoplasmal gene sequences used to construct specific primers are listed in Table 4 in Razin (46).

Specimen preparation for PCR testing is an important parameter that should be optimised. Since mycoplasmas have no cell walls, boiling of the sample, following its concentration by centrifugation, is often sufficient to make the organisms' DNA accessible. However, some clinical samples may contain undefined inhibitors of the PCR reaction, reducing the efficiency of amplification, so that DNA extraction has to be employed (10). Appropriate positive and negative controls should be included in each PCR test to rule out the presence of inhibitory substances (1; 17; 20). The methodology of target sequence amplification and optimization of the PCR test conditions, as well as the identification of the PCR products by electrophoretic analysis are detailed in de Barbeyrac and Bebear (16). Capillary PCR in which amplification is carried out in sealed capillary glass tubes has been more recently introduced providing a more sensitive, specific and rapid procedure (28). We also evidence the development of calorimetric procedures for identification of specifically amplified double-stranded DNA captured in wells of a microtiter plate. These systems, some of which are commercially available as kits, are cost effective alternatives to gel electrophoresis, particularly for screening of moderately large sample numbers (6).

Although conventional PCR tests should under optimal conditions detect the DNA of a single mycoplasma, this rarely happens in testing clinical material. Two-step (nested) PCR has been devised to increase sensitivity to a level enabling detection of a single mycoplasma in a clinical sample (1; 8; 20; 68). Another advantage of nested PCR is that the second round PCR serves to confirm the specificity of the first round PCR (1; 27). Another approach is that termed "Random Amplified Polymorphic DNA (RAPD)", or "Arbitrary Primer PCR (AP-PCR)". It involves PCR amplification with a single arbitrary primer at low stringency, resulting in strain-specific arrays of DNA fragments that can reproducibly distinguish even closely related strains of a species (24). The simplicity of this procedure, and the availability of commercial kits should facilitate the use of RAPD analysis as a routine procedure for identification of strains of a mollicute species, employing the RAPD patterns to investigate disease outbreaks for epidemiologic tracking (11; 15).

### 4.2.1    Comparing PCR results with those of culture and serology

Numerous publications compare the results of the PCR technique with culture and serological testing for *M. pneumoniae*. Most publications indicate the higher sensitivity and the much faster results obtained by PCR (for example see references 1; 22; 26; 28; 36). The sensitivity of PCR assays is usually in the order of magnitude of a few femtograms of mycoplasmal DNA. When translated into numbers of organisms, 1fg of mycoplasmal DNA is approximately equivalent to the genomic DNA of a single mycoplasma cell, based on a genome size of about 1000kb. The ability to detect by PCR a single mycoplasma cell makes PCR the most sensitive detection method available, even more than culture. In theory, a positive culture can be derived from a single mycoplasma cell. In practice, however, due to many reasons (including mycoplasma cell aggregates, multinuclear filamentous forms, defective or non-viable cells) a positive mycoplasma culture requires an inoculum equivalent to about 100 to 1000 cells. This is known for other bacteria as well and is the reason for using the expression colony-forming units (cfus) or color-changing units (ccus) rather than number of cells per milliliter (46).

Obviously, PCR should be applied in the case of patients with an impaired immune response where serological tests may be much less effective. Positive PCR results for *M. pneumoniae* in culture-negative persons without evidence of respiratory disease suggests either inadequate specificity of the test, persistence of the organisms after infection, or their existence in asymptomatic carriers, making interpretation of such results difficult (69). In other words, would such a sensitive technique as PCR lead to 'over-diagnosis' of *M. pneumoniae* infections partly due to its detection in healthy carriers? It is still difficult to answer this question in a definitive manner.

PCR tests are less useful in the diagnosis of urogenital infections caused by the rapidly growing mycoplasmas *M. hominis* and *U. urealyticum*. Nevertheless, PCR assay of amniotic fluid samples in case of pre-term infections by ureaplasmas resulted in a significantly higher rate of *U. urealyticum* detection than by the standard microbiologic culture (73). These data were taken to suggest that patients with a positive PCR assay for *U. urealyticum* have a worse pregnancy outcome than those with a sterile amniotic fluid. PCR may also be useful in cases where it is difficult to get positive cultures, such as in synovial fluids. The search for mycoplasmas in synovial fluids of arthritic patients has been initiated over 30 years ago, with controversial results. The recent application of PCR methodology revealed the presence of *M. fermentans* in a significant percentage of arthritic joint exudates of rheumatoid arthritis patients, supporting the possible association of this hard-to-culture mycoplasma with human arthritides (2; 35; 55).

Another pathogenic mycoplasma hard to cultivate on primary isolation is *M. hyopneumoniae*, the major agent of enzootic pneumonia in swine. Due to the fastidious nature of this mycoplasma its culture and serological identification may take up to one month. Moreover, other swine mycoplasmas present in the clinical material, especially *M. hyorhinis,* easily overgrow and contaminate *M. hyopneumoniae* cultures. Current serological detection methods are hampered by cross-reactions between *M. hyopneumoniae* and *M. hyorhinis,* but more so with *M. flocculare.* Pairs of oligonucleotide primers designed to permit PCR amplification of genes encoding *M. hyopneumoniae* specific protein antigens exhibited specificity to *M. hyopneumoniae* enabling demonstration of this mycoplasma in lungs and tracheobronchial swabs from diseased pigs (10).

In conclusion of this Section, PCR detection of mycoplasmas is still too labor-intensive, expensive, and complex to be carried out routinely in most clinical microbiology laboratories (69). Development of satisfactory commercial PCR kits may simplify the procedures and bring about better standardization of the technique, and if available at a reasonable cost, PCR could become a major method for diagnosis of mycoplasmal infections (70).

# REFERENCES

1.    **Abele-Horn, M., U. Busch, H. Nitschko, E. Jacobs, R. Bax, F. Pfaff, B. Schaffer, and J. Heesemann.** 1998. Molecular approaches to diagnosis of pulmonary diseases due to *Mycoplasma pneumoniae* . J. Clin. Microbiol. **36**: 548-551.

2.    **Ainsworth, J. G. , J. Clarke, M. Lipman, D. Mitchell, and D. Taylor-Robinson.** 2000. Detection of *Mycoplasma fermentans* in broncho-alveolar lavage fluid specimens from AIDS patients with lower respiratory tract infection. HIV Medicine **1**: 219-223.

3.    **Alma, A., D.Bosco, A. Danielli, A. Bertaccini, M. Vibio, and A.Arzone.** 1997. Identification of phytoplasmas in eggs, nymphs and adults of *Scaphoideus titanus* Ball reared on healthy plants. Insect Mol. Biol. **6**: 115-121.

4.    **Amikam, D., S. Razin, and G. Glaser.** 1982. Ribosomal RNA genes in mycoplasma. Nucl. Acids Res. **10**:4215-4222.

5.    **Amikam, D., G. Glaser, and S. Razin.** 1984. Mycoplasmas *(Mollicutes)* have a low number of rRNA genes. J. Bacteriol. **158** : 376-378.

6.    **Bashiruddin, J. B., P. De Santis, A. Vacciana, and F. G. Santini.** 1999. Detection of *M. mycoides* subspecies *mycoides* SC in clinical material by a rapid colorimetric PCR. Mol. Cell. Probes **13**: 23-28.

7.    **Bencina,,D.,and J. M. Bradbury.** 1992. Combination of immunofluorescence and immunoperoxidase for serotyping mixtures of *Mycoplasma* species. J. Clin. Microbiol. **30**:407-410.

8.    **Berges, R., M. Rott, and E. Seemuller.** 2000. Range of phytoplasma concentrations in various plant hosts as determined by competative polymerase chain reaction. Phytopathology **90**: 1145-1152.

9.    **Blanchard, A., D. M. Crabb, K. Dybvig, L. B. Duffy, and G. H. Cassell.** 1992. Rapid detection of *tetM* in *Mycoplasma hominis* and *Ureaplasma urealyticum* by PCR: *tetM* confers resistance to tetracycline but not necessarily to doxycycline. FEMS Microbiol. Lett. **95**: 277-282.

10.   **Caron, J., M. Ouardani, and S. Dea.** 2000. Diagnosis and differentiation of *Mycoplasma hyopneumoniae* and *Mycoplasma hyorhinis* infections in pigs by PCR amplification of the p36 and p46 genes. J. Clin. Microbiol. **38**: 1390-1396.

11.   **Charlton, B. R., A. A. Bickford, R. L. Walker, and R. Yamamoto.** 1999. Complementary randomly amplified polymorphic DNA (RAPD) analysis patterns and primer sets to differentiate *Mycoplasma gallisepticum* strains. J. Vet. Diagn. Invest. **11**: 158-161.

12.   **Choppa, P. C., A. Vojdani, C. Tagle, R. Andrin, and L. Magtoto.** 1998. Multiplex PCR for the detection of *Mycoplasma fermentans*, *M. hominis* and *M. penetrans* in cell cultures and blood samples of patients with chronic fatigue syndrom. Mol. Cell. Probes **12**: 301-308.

13.   **Cimolai,N., and A. C. H. Cheong.** 1992. IgM anti-P1 immunoblotting. A standard for the rapid serologic diagnosis of *Mycoplasma pneumoniae* infection in pediatric care. Chest **102**: 477-481.

14.   **Clausen, H. F., J. Fedder, M. Drasbek, P. K. Nielsen, B. Toft, H. J. Ingerslev, S. Birkelund, and G. Christiansen.** 2001. Serological investigation of *Mycoplasma genitalium* in infertile women. Human Reproduction **16**: 1866-1874.

15.   **Cousin-Allery, A., A. Charron, B. De Barbeyrac, G. Fremy, J. S. Jensen, H. Renaudin, and C. Bebear.** 2000. Molecular typing of *Mycoplasma pneumoniae* strains by PCR-based methods and pulse-field gel electrophoresis. Application to French and Danish isolates. Epidemiol. Infect. **124**: 103-111.

16.   **De Barbeyrac, B., and C. Bebear.** 1996. PCR: amplification and identification of products, p. 65-73. *In* : J. G. Tully and S. Razin (ed.), Molecular and diagnostic

procedures in mycoplasmology, vol.II. Diagnostic procedures. Academic Press, San Diego.

17. **De Barbeyrac,B., C. Bebear, and D. Taylor-Robinson.** 1996. PCR: preparation de DNA from clinical specimens, p. 61-64. *In* : J. G. Tully and S. Razin (ed.), Molecular and diagnostic procedures in mycoplasmology, vol. II, Diagnostic procedures. Academic Press, San Diego.

18. **Djordjevic, S. P., W. A. Forbes, J. Forbes-Faulkner, P. Kuhnert, S. Hum, M. A. Hornitzky, E. M. Vilei, and J. Frey.** 2001. Genetic diversity among *Mycoplasma* species bovine group 7: clonal isolates from an outbreak of polyarthritis, mastitis, and abortion in dairy cattle. Electrophoresis **22**: 3551-3561.

19. **Dorigo-Zetsma, J. W., S. A. J. Zaat, P. M. E. Wertheim- van Dillen, L. Spanjaard, J. Rijntjes, G. van Waveren, J. S. Jensen, A. F. Angulo, and J. Dankert.** 1999. Comparison of PCR, culture, and serological tests for diagnosis of *Mycoplasma pneumoniae* respiratory tract infection in children. J. Clin. Microbiol. **37**: 14-17.

20. **Dorigo-Zetsma, J. W., S. A. J. Zaat, A. J. M. Vriesema, and J. Dankert.** 1999. Demonstration by a nested PCR for *Mycoplasma pneumoniae* that *M. pneumoniae* load in the throat is higher in patients hospitalised for *M. pneumoniae* infection than in non-hospitalised subjects. J. Med. Microbiol. **48**: 1115-1122.

21. **Duffy, M. F., K. G. Whithear, A. H. Noormohammadi, P. F. Markham, M. Catton, and G. F. Browning.** 1999. Indirect enzyme-linked immunosorbent assay for detection of immunoglobulin G reactive with a recombinant protein expressed from the gene encoding the 116-kilodalton protein of *Mycoplasma pneumoniae*. J. Clin. Microbiol. **37**: 1024-1029.

22. **Ferwerda, A., H. A. Moll, and R. de Groot.** 2001. Respiratory tract infections by *Mycoplasma pneumoniae* in children: a review of diagnostic and therapeutic measures. Eur. J. Pediatr. **160**: 483-491.

23. **Fleury, B., D. Bergonier, X. Berthelot, Y. Schlatter, J. Frey, and E. M. Vilei.** 2001. Characterization and analysis of a stable serotype-associated membrane protein (P30) of *Mycoplasma agalactiae*. J. Clin. Microbiol. **39**: 2814-2822.

24. **Geary, S. J.,and M. H. Forsyth.** 1996. PCR: random amplified polymorphic DNA fingerprinting, p. 81-85. *In* : J. G. Tully and S. Razin (ed.), Molecular and diagnostic procedures in mycoplasmology, vol. II, Diagnostic procedures. Academic Press, San Diego.

25. **Gerstenecker, B., and E. Jacobs.** 1993. Development of a capture-ELISA for the specific detection of *Mycoplasma pneumoniae* in patient's material, p. 195-205. *In:* Kahane, I. and A. Adoni (ed.). 1993. Rapid diagnosis of mycoplasmas. Plenum Press, New York.

26. **Haaheim, H., L. Vorland, and T. J. Gutteberg.** 2001. Laboratory diagnosis of respiratory diseases: PCR versus serology. Nucleosides, Nucleotides & Nucleic Acids **20**:1255-1258.

27. **Harasawa, R.** 1996. PCR: application of nested PCR to detection of mycoplasmas, p. 75-79. *In* : J. G. Tully and S. Razin (ed.), Molecular and diagnostic procedures in mycoplasmology, vol. II, Diagnostic procedures. Academic Press, San Diego.

28. **Honda, J., T. Yano, M. Kusaba, J. Yonemitsu, H. Kitajima, M. Masuoka, K. Hamada, and K. Oizumi.** 2000. Clinical use of capillary PCR to diagnose *Mycoplasma pneumoniae*, J. Clin. Microbiol. **38**: 1382-1384.

29. **Horner, P., B. Thomas, C. B. Gilroy, M. Egger, and D. Taylor-Robinson.** 2001. Role of *Mycoplasma genitalium* and *Ureaplasma urealyticum* in acute and chronic urethritis. Clin. Infect. Dis. **32**: 995-1003.

30. **Ieven, M., D. Ursi, H. Van Bever, W. Quint, H. G. M. Niesters, and H. Goossens.** 1996. Detection of *Mycoplasma pneumoniae* by two polymerase chain reactions and

role of *M. pneumoniae* in acute respiratory tract infections in pediatric patients. J. Infect. Dis. **173**: 1445-1452.

31.   International Committee on Systematic Bacteriology- Subcommittee on the Taxonomy of Mollicutes. 1995. Revised minimum standards for description of new species of the Class *Mollicutes* ( Division *Tenericutes* ). Int. J. Syst. Bacteriol. **45**: 605-612.

32.   **Jacobs, E., K. Fuchte, and W. Bredt.** 1986. A 168-kilodalton protein of *Mycoplasma pneumoniae* used as antigen in a dot enzyme-linked immunosorbent assay. Eur. J. Clin. Microbiol. **5**: 435-440.

33.   **Jensen, J. S., H. T. Hansen, and K. Lind.** 1996. Isolation of *Mycoplasma genitalium* strains from the male urethra. J. Clin. Microbiol. **34**: 286-292.

34.   **Johansson, K.-E.** 1996. Oligonucleotide probes complementary to 16S rRNA, p. 29-46. *In* : J. G. Tully and S. Razin (ed.), Molecular and diagnostic procedures in mycoplasmology, vol. II, Diagnostic procedures. Academic Press, San Diego.

35.   **Johnson, S., D. Sidebottom, F. Bruckner, and D. Collins.** 2000. Identification of *Mycoplasma fermentans* in synovial fluid samples from arthritis patients with inflammatory disease. J. Clin. Microbiol. **38**: 90-93.

36.   **Kessler, H. H., D. E. Dodge, K. Pierer, K. K. Y. Young, Y. Liao, B. I. Santner, E. Eber, M. G. Roeger, D. Stuenzner, B. Sixl-Voigt, and E. Marth.** 1997. Rapid detection of *Mycoplasma pneumoniae* by an assay based on PCR and probe hybridization in a nonradiactive microwell plate formate. *J.* Clin. Microbiol. **35**: 1592-1594.

37.   **Kokotovic, B., N. F. Friis, J. S. Jensen, and P. Ahrens.** 1999. Amplified-fragment length polymorphism fingerprinting of *Mycoplasma* species. J. Clin. Microbiol. **37**: 3300-3307.

38.   **Kong, F., X. Zhu, W. Wang, X. Zhou, S. Gordon, and G. L. Gilbert.** 1999. Comparative analysis and serovar-specific identification of multiple-banded antigen genes of *Ureaplasma urealyticum* biovar 1. J. Clin. Microbiol. **37**: 538-543.

39.   **Kong, F., G. James, S. Gordon, A. Zelynski, and G. L. Gilbert.** 2001. Species-specfic PCR for identification of common contaminant mollicutes in cell culture. Appl. Environ. Microbiol. **67**: 3195-3200.

40.   **Kotani, H., and G. J. McGarrity.** 1986. Identification of mycoplasma colonies by immunobinding. J. Clin. Microbiol. **23**: 783-785.

41.   **Kovacic, R., O. Grau, and A. Blanchard.** 1996. PCR: selection of target sequences, p. 53-60. *In* : J. G. Tully and S. Razin (ed.), Molecular and diagnostic procedures in mycoplasmology, vol. II, Diagnostic procedures. Academic Press, San Diego.

42.   **Marcone, C., I-M. Lee, R. E. Davis, A. Ragozzino, and E. Seemuller.** 2000. Classification of aster yellows-group phytoplasmas based on combined analyses of rRNA and *tuf* gene sequences. Int. J. Syst. Evol. Microbiol. **50**: 1703-1713.

43.   **Rastawicki, W., R. Rati, and M. Kleemola.** 1996. Detection of antibodies to *Mycoplasma pneumoniae* adhesin P1 in serum samples from infected and noninfected subjects by immunoblotting. Diagn. Microbiol. Infect Dis. **26**: 141-143.

44.   **Razin, S.** 1989. The molecular approach to mycoplasma phylogeny, p. 33-69. *In* : R. F. Whitcomb and J. G. Tully (ed.). The mycoplasmas, vol. 5: spiroplasmas, acholeplasmas, and mycoplasmas of plants and arthropods. Academic Press, San Diego.

45.   **Razin, S.** 1992. Mycoplasma taxonomy and ecology, p. 3-22. *In* : J. Maniloff, R. N. McElhaney, L. R. Finch, and J. B. Baseman( ed.), Mycoplasmas: molecular biology and pathogenesis. American Society for Microbiology, Washington, D. C.

46.   **Razin, S.** 1994. DNA probes and PCR in diagnosis of mycoplasma infections. Mol. Cell. Probes **8**:497-511.

47.     **Razin, S.** 2000. The genus *Mycoplasma*, and related genera (class *Mollicutes*). *In* M. Dworkin, S. Falkow, E. Rosenberg, K.-H. Schleifer, and E. Stackebrandt (eds.), The Prokaryotes. An evolving electronic resource for the microbiolgical community, 3rd edition, Springer-verlag, New York. (URL: http//:www.prokaryotes.com).

48.     **Razin, S. and J. G. Tully.** 1970. Cholesterol requirement of mycoplasmas. J. Bacteriol. **102**: 306-310.

49.     **Razin, S. and J. G. Tully (ed.).** 1983. Methods in mycoplasmology, vol. I , Mycoplasma characterization. Academic Press, New York.

50.     **Razin, S. and J. G. Tully (ed.) .** 1995. Molecular and diagnostic procedures in mycoplasmology, Vol. I , Molecular characterization. Academic Press, San Diego.

51.     **Razin, S. and D. Yogev.** 1995. Restriction endonuclease analysis, p. 355-360. *In* : Razin, S. and J. G. Tully (ed.), Molecular and diagnostic procedures in mycoplasmology, Vol. I , Molecular characterization. Academic Press, San Diego.

52.     **Razin, S., M. Gross, M. Wormser, Y. Pollack, and G. Glaser.** 1984. Detection of mycoplasmas infecting cell cultures by DNA hybridization. In Vitro **20**: 404-408.

53.     **Rice, P., B. M. Houshayami, E. A. M. Abu-Groun, R. A. J. Nicholas, and R.J. Miles.** 2001. Rapid screening of $H_2O_2$ production by *Mycoplasma mycoides* and differentiation of European subsp. *mycoides* SC (small colony) isolates. Vet. Microbiol. **78**: 343-351.

54.     **Rosengarten, R., and D. Yogev.** 1996. Variant colony surface antigenic phenotypes within mycoplasma strain populations: implications for species identification and strain standardization. J. Clin. Microbiol. **34**: 149-158.

55.     **Schaeverbeke, T., M. Clerc, L. Lequen, A. Charron, C. Bebear, B. de Barbeyrac, B. Bannwarth, J. Dehais, and C. Bebear.** 1998. Genotypic characterization of seven strains of *Mycoplasma fermentans* isolated from synovial fluids of patients with arthritis. J. Clin. Microbiol. **36**: 1226-1231.

56.     **Seemuller, E., and B. C. Kirkpatrick.** 1996. Detection of phytoplasma infections in plants, p. 299-311. *In* J. G. Tully, and S. Razin (eds.). Molecular and diagnostic procedures in mycoplasmology, vol. II: Diagnostic procedures. Academic Press. San Diego.

57.     **Seemuller, E., C. Marcone, U. Lauer, A. Ragozzino, and M. Goschl.** 1998. Current status of molecular classification of the phytoplasmas. J. Plant Pathol. **80**: 3-26.

58.     **Sillis, M.** 1990. The limitations of IgM assays in the serological diagnosis of *Mycoplasma pneumoniae* infections. J. Med. Microbiol. **33**: 253-258.

59.     **Suni, J., R. Vainionpaa, and T. Tuuminen.** 2001. Multicenter evaluation of the novel enzyme immunoassay based on P1-enriched protein for the detection of *Mycoplasma pneumoniae* infection. J. Microbiol. Meth. **47**: 65-71.

60.     **Tang, J., M. Hu, S. Lee, and R. Roblin.** 2000. A polymerase chain reaction based method for detecting *Mycoplasma/Acholeplasma* contaminants in cell culture. J. Microbiol. Meth. **39**: 121-126.

61.     **Taylor-Robinson, D., and T. A. Chen.** 1983. Growth inhibitory factors in animal and plant tissues, p. 109-114. *In* : S. Razin and J. G. Tully (ed.), Methods in mycoplasmology, vol. I , Mycoplasma characterization. Academic Press, New York.

62.     **Totten, P. A., M. A. Schwartz, K. E. Sjostrom, G. E. Kenny, H. Hunter Handsfield, J. B. Weiss, and W. L. H. Whittington.** 2001. Association of *Mycoplasma genitalium* with nongonococcal urethritis. J. Infect. Dis. **183**: 269-276.

63.     **Tully, J. G.** 1995. Determination of cholesterol and polyoxyethylene sorbitan growth requirements of mollicutes, p. 381-389. *In* : S. Razin, and J. G. Tully (eds.) Molecular and diagnostic procedures in mycoplasmology, vol. I , Molecular characterization. Academic Press, San Diego.

64.     **Tully, J. G., and S. Razin (eds.).** 1983. Methods in mycoplasmology, vol. II. Diagnostic mycoplasmology. Academic Press, San Diego.

65. **Tully, J. G., and S. Razin (eds.).** 1996. Molecular and diagnostic procedures in mycoplasmology, vol. II: Diagnostic procedures. Academic Press. San Diego.

66. **Tuuminen, T., J. Suni, M. Kleemola, and E. Jacobs.** 2001. Improved sensitivity and specificity of enzyme immunoassays with P1-adhesin enriched antigen to detect acute *Mycoplasma pneumoniae* infection. J. Microbiol. Meth. **44**: 27-37.

67. **Veilleux, C., S. Razin, and L. H. May.** 1996. Detection of mycoplasma infection by PCR, p. 431-438. *In* J. G. Tully, and S. Razin (eds.). Molecular and diagnostic procedures in mycoplasmology, vol. II: Diagnostic procedures. Academic Press. San Diego.

68. **Verdin, E., C. Saillard, A. Labbe, J. M. Bove, and M. Kobisch.** 2000. A nested PCR assay for the detection of *Mycoplasma hyopneumoniae* in tracheobronchiolar washings from pigs. Vet. Microbiol. **76**: 31-40.

69. **Waites, K. B., C. M. Bebear, J. A. Robertson, D. F. Talkington, and G. E. Kenny.** 2001. Laboratory diagnosis of mycoplasmal infections. Cumulative techniques and procedures in clinical microbiology (Cumitech 34). American Society for Microbiology Press, Washington, D.C.

70. **Watkins-Riedel, T., G. Stanek, and F. Daxboeck.** 2001. Comparison of SeroMP IgA with four other commercial assays for serodiagnosis of *Mycoplasma pneumoniae* pneumonia. Diagn. Microbiol. Infect. Dis. **40**: 21-25.

71. **Yogev, D., and S. Razin.** 1995. Southern blot analysis and ribotyping, p. 361-368. *In* : S. Razin, and J. G. Tully (eds.) Molecular and diagnostic procedures in mycoplasmology, vol. I , Molecular characterization. Academic Press, San Diego.

72. **Yogev, D., S. Sela, H. Bercovier, and S. Razin.** 1988. Elongation factor (Ef-Tu) gene probe detects polymorphism in *Mycoplasma* strains.

73. **Yoon, B. H., R. Romero, M. Kim, T. Kim, J. S. Park, and J. K. Jun.** 2000. Clinical implications of detection of *Ureaplasma urealyticum* in the amniotic cavity with the polymerase chain reaction. Am. J. Obstet. Gynecol. **183**: 1130-1137.

# Chapter 25

# Antimycoplasmal Agents

CÉCILE M. BÉBÉAR and CHRISTIANE BÉBÉAR
*Université Victor Segalen Bordeaux 2, Laboratoire de Bactériologie, 146 rue Léo Saignat, 33076 Bordeaux cedex, France.*

## 1.    INTRODUCTION

This chapter briefly reviews the methods used for antibiotic susceptibility testing of human mycoplasmas, the antimicrobials available for the treatment of mycoplasmal infections, the innate resistance of these microorganisms, and the development of acquired resistance. Guidelines for antimicrobial susceptibility testing of animal mycoplasma species are not considered here, but have been recently reviewed (41).

Mycoplasma infections concern mainly the urogenital and respiratory tracts. Furthermore, mycoplasmas can cause systemic infections in neonates and immunocompromized patients. In this chapter, *Ureaplasma urealyticum* will be referred to as *Ureaplasma* spp. Indeed, recently several authors presented evidence that the species *U. urealyticum* should be separated into two new species, namely, *U. parvum* (previously *U. urealyticum* biovar 1) and *U. urealyticum* (previously *U. urealyticum* biovar 2), (for details see the chapters "Mycoplasmas of humans" and "Taxonomy of *Mollicutes*").

## 2.    ANTIMICROBIAL SUSCEPTIBILITY TESTING

Actually, no specific guidelines for susceptibility testing of mycoplasmas were issued by the National Committee for Clinical Laboratory Standards (NCCLS). However, the recommendations of the Mycoplasmal Chemotherapy Working Team of the International Research Program on Comparative Mycoplasmology have been recently summarized by Waites *et*

*Molecular Biology and Pathogenicity of Mycoplasmas,* Edited by Razin and
Herrmann, Kluwer Academic/Plenum Publishers, New York, 2002

*al.* in the last edition of the Cumitech concerning the laboratory diagnosis of mycoplasmal infections (85). Furthermore, step-by-step procedures with detailed indications for testing human mycoplasmas are available in references 7 and 80. Standard broth and agar dilution methods used for susceptibility testing of bacteria have been adapted for use with mycoplasmas. These methods, performed with appropriate controls, give useful results for the treatment of mycoplasmal infections.

Several features are common for both methods using solid and liquid media. No single standardized medium or pH can be indicated since growth requirements vary with the species. Thus, SP4 and Hayflick's modified media are recommended for human *Mycoplasma* spp. while Shepard's 10B or A8 media are recommended for *Ureaplasma* spp. Concerning the pH, *Mycoplasma* spp. can be tested at pH 7.0 to 7.6 while acidic pH (6.0 to 6.5) is necessary for *Ureaplasma* spp. growth (85). This acidic pH could affect the activities of macrolides in vitro, as these antibiotics are much more active at pH 7 than at lower pH (49). Incubation conditions are at 35° to 37°C, under 5 % $CO_2$ for agar testing, for a length of time depending on the species tested and on the technique applied. Broth-based tests can be incubated under atmospheric conditions without $CO_2$. An inoculum of $10^4$ to $10^5$ color-changing units per ml has been recommended. Details concerning inoculum preparation have been reviewed in reference 85. When available, a control strain with known MICs of the tested drugs should be included to validate the test. Finally, mandatory controls such as sterile growth medium, control strain in medium alone with no drug, and medium with the highest concentration of the solvent used to dissolve the antimicrobial, are required.

The choice of test procedure is influenced by several factors such as the number of strains tested, their growth titer in broth or agar and the generation time of the species. Thus broth-based tests are preferred when a small number of strains has to be tested, when bactericidal testing is required (see below), or when *Ureaplasma* spp., for which the growth titer is lower than for *Mycoplasma* spp. and colonies more difficult to count on agar, are studied. Comparison of results obtained with broth or agar dilution methods allows similar categorization of the strains, the MICs determined by microbroth dilution tending to be slightly lower (8, 49, 90).

## 2.1    Broth dilution methods

Broth dilution and especially microbroth dilution methods have been modified for mycoplasmas. This method is the most widely used, particularly for *Ureaplasma* spp. testing. The method employs mycoplasmal or ureaplasmal media with decreasing antibiotic concentrations, inoculated with a standardized number of microorganisms in a 96-well microtiter plate.

Mycoplasma growth results in degradation of the metabolizable substrate such as glucose, arginine or urea, present in the medium with the resulting pH change visible as a color change of a pH indicator. The MIC is defined as the lowest concentration of antibiotic that prevents a color change at the time when the color in the control without antibiotic has changed. MICs for ureaplasmas will be available after 16-24 hours, for *Mycoplasma hominis* after 36-48 hours, and for *M. pneumoniae* after 5 days or more. Turbidity or color change in broth control indicates bacterial contamination.

Broth macrodilutions in tubes are convenient for testing small numbers of strains while microdilution is efficient for testing large numbers of strains and multiple antimicrobials in the same plate. Furthermore, mycoplasmacidal testing can be performed in the same system. However, the endpoint tends to shift over time for some drugs and since the inoculum is usually uncloned, a mixture of resistant and susceptible strains could occur.

Susceptibility testing kits such as Mycoplasma IST (bioMérieux), Mycoplasma SIR (Bio-Rad), Mycofast Evolution 2 (International Microbio), and Mycokit-ATB (PBS Orgenics) are available in Europe. They consist of microwells containing dried antimicrobials, generally in 2 or more concentrations corresponding to the threshold proposed for conventional bacteria to classify a strain as susceptible, intermediate, or resistant. Some of these kits combine organism growth and identification with susceptibility testing in the same product. Procedures and kit description have been recently reviewed in reference (91). These kits give results comparable to those obtained by established MIC determination and are a reasonable choice for diagnostic laboratories in countries where they are available (1, 21, 68). However, they are adapted only for use with *Ureaplasma* spp. and *M. hominis* and can be applied only after a primary culture. When used directly with the specimen, the lack of a defined inoculum can contribute to error since inoculum size can influence MICs. Thus tetracycline resistance may be undetected if the number of organisms is low (68).

## 2.2 Agar dilution and gradient diffusion methods

The standard agar dilution method adapted for testing several mycoplasmal species and ureaplasmas, has been used frequently as a reference method (7, 90). The MIC is the lowest concentration of the agent that prevents colony formation when read at the time when the antimicrobial-free control plate demonstrates growth. This usually occurs in 1-2 days for *U. urealyticum*, 2-4 days for *M. hominis* and 5 or more days for *M. pneumoniae* and other human mycoplasmas.

This method is time-consuming, labor-intensive and is not practical for testing small numbers of strains or occasional isolates. However, the endpoint is stable over time and allows detection of mixed cultures readily.

Using an uncloned inoculum, resistance can be detected. Furthermore large numbers of strains can be tested simultaneously.

Generally, agar disk diffusion is not acceptable for mycoplasma susceptibility testing because of the relatively long time required for growth of the microorganisms. However, the agar gradient diffusion technique or E-tests (AB Biodisk) have been recently adapted to determine fluoroquinolone, tetracycline and macrolide susceptibilities for *M. hominis* and *Ureaplasma* spp. (26, 86, 87, 91). Generally the results obtained for MICs agree within one to two dilutions with the values obtained by agar or broth microdilution method. The MIC is defined as the number on the E-test strip corresponding to the intersection of mycoplasmal growth with the strip. Reading is done when colonies are visible on the plate with the appearance of an ellipse. This method is simple and can be easily adapted for occasional testing or testing single isolates. E-tests are commercialized and have a long shelf live.

## 2.3      Minimal bactericidal concentration (MBC) tests

Detailed descriptions of procedures for mycoplasmicidal tests are available (11, 71, 80, 85).

Bactericidal activity can be tested directly using the microbroth dilution assay, by subculturing the mixture from wells that do not show a color change. The mixture from the MIC assay should be diluted in a sufficient volume of medium to inactivate the antibiotic. The mixture could also be washed of the antibiotic by passing through a 220 nm pore-size filter, which retains mycoplasmas, and is then placed in growth medium to culture viable organisms. After incubation of the broth and subculturing onto agar, if a color change occurs the MBC is defined as the lowest concentration of antibiotic inhibiting a color change or colony formation on agar.

The mycoplasmicidal concentration of an antibiotic can be also determined by time-kill experiments adapted to these microorganisms (71, 85), the end point being the amount of antibiotic needed to produce 99% killing of the original inoculum. Few killing curves for mycoplasmas have been reported.

Among the antibiotics active against mycoplasmas only two classes are bactericidal, the fluoroquinolones (11, 12, 14, 15, 67) and the ketolides (10, 12).

## 2.4      Interpretation of results

No breakpoints specific for mycoplasmas are available actually, so it may be preferable to merely report MICs. Some laboratories have adopted the breakpoints used for the interpretation of MICs for other bacteria. MIC values of ≤1 µg/ml should be considered predictive of effective treatment (85).

## 3. INNATE ANTIBIOTIC RESISTANCE OF MYCOPLASMAS

Innate antibiotic resistance is defined as resistance that is found in all members of a group of bacteria. For mycoplasmas, two kinds of innate or intrinsic resistance have to be considered, one related to the class *Mollicutes*, and one related to the different mycoplasma and ureaplasma species. These differences in innate antibiotic susceptibility have been used to isolate mycoplasmas from specimens contaminated with other bacteria, and to differentiate one mycoplasma species from another in the same specimen (71).

### 3.1 Resistance related to the class

The lack of a cell wall in mollicutes makes them innately resistant to β-lactams and to all antimicrobials, like glycopeptides and fosfomycin, which target the cell wall. In addition, mycoplasmas are also resistant to polymyxins, sulfonamides, trimethoprim, nalidixic acid, and rifampin. Mollicutes share resistance to rifampin with some Clostridia, *Clostridium innocuum* and *C. ramosum*, which are anaerobic gram-positive bacteria with a low guanine plus cytosine content, phylogenetically related to mollicutes (94). The molecular mechanism of this insusceptibility to rifampin has been investigated for the plant mollicute *Spiroplasma citri* (36). It was shown to be a natural mutation in the *rpoB* gene encoding the β subunit of the DNA-dependent RNA polymerase, the target of rifampin, a mutation which prevents the binding of the antibiotic to its target. Indeed the amino acid 526 in RpoB, previously described as a rifampin resistance hot spot in *Escherichia coli* (48), is occupied by an arginine in some mycoplasmas like *S. citri*, *M. gallisepticum* and *M. genitalium*, but by a histidine residue in *E. coli*. Engineering the substitution H526→N in the *E. coli rpoB* gene led to the appearance of resistance to rifampin in *E. coli*, demonstrating the responsibility of this substitution for resistance to the drug (36).

### 3.2 Resistance related to the species

Innate resistance related to specific mycoplasma species concerns essentially the macrolide-lincosamide-streptogramin (MLS) antibiotic class.

Thus *M. pneumoniae* and *M. genitalium* are susceptible to all MLS antibiotics and to ketolides, a new class of antimicrobials derived from erythromycin (12, 51). *Ureaplasma* spp. are susceptible to macrolides and ketolides, but resistant to lincosamides. The reverse is true for *M. hominis* and *M. fermentans*, these two species being resistant to 14- and 15-membered ring macrolides and ketolides, but susceptible to the 16-membered ring macrolide josamycin, and to lincosamides (see Table 1). Indeed, macrolides are classified into three subgroups according to their lactone ring sizes (14-, 15-, and 16-membered lactone ring); ketolides derive from the 14-membered erythromycin. Recently, the genetic basis of this innate resistance to erythromycin has been elucidated in part for *M. hominis* (35). As we will see below, macrolides bind to the bacterial ribosome and affect its peptidyltransferase activity, in the domain V of 23S rRNA. Macrolide resistance by target modification could involve a mutation in the domain V with a decreased macrolide binding to ribosomes (84). Furneri *et al.* (35) comparing the sequence of the 23S rRNA of domain V from *M. hominis*, and *M. pneumoniae* and *M. genitalium*, showed a G→A transition at position 2057 in the central loop for *M. hominis*. This 2057 transition has been found associated with erythromycin resistance, but 16-member macrolide susceptibility, in *E. coli* and propionibacteria (84). Furthermore, Bébéar *et al.* (unpublished results) found a dramatic decrease in *M. hominis* ribosome affinity for [$^{14}$C] erythromycin compared to ribosomes from *M. pneumoniae* and *E. coli*. The same substitution at 2057 was also detected in the 23S rRNA of *M. fermentans* (Bébéar *et al.*, unpublished results), and, interestingly, in the 23S rRNA sequence of *M. hyopneumoniae* (60), *M. flocculare* (77), and *M. pulmonis* (19). These different animal mycoplasmas have been found to be resistant to erythromycin but susceptible to 16-member macrolides and lincosamides (82).

Recently, two new antimicrobial agents, linezolid, a member of the oxazolidinone class, and evernimicin, belonging to the everninomicin group of oligosaccharide antibiotics, have been evaluated against human mycoplasmas (51). Both act against the 50S ribosomal subunit, inhibiting protein synthesis. Linezolid is active against *M. hominis* but not against *M. pneumoniae* or *Ureaplasma* spp. (Table 1). Evernimicin seems to have an activity against the three human species, *Ureaplasma* spp. being susceptible only at 16 µg/ml (Table 1).

## 4. ANTIBIOTICS USED FOR TREATMENT OF MYCOPLASMAL INFECTIONS

Because mycoplasmas lack a cell wall, the number of antibiotics that can be used for treating mycoplasmal infections is limited. Potentially active antimicrobials include tetracyclines, MLS group and ketolides, fluoroquinolones, aminoglycosides and chloramphenicol. The first three classes are the most widely used, except for the ketolides which were not commercially available in 2001. Chloramphenicol and aminoglycosides are kept for the treatment of special cases (see below). Tetracyclines, MLS and fluoroquinolones share the advantages of being potentially active against the microorganisms which could be associated with mycoplasmas in respiratory and urogenital tract infections. They share also the ability to reach high intracellular concentrations and since human mycoplasmas could localize within the cells (24), this property can be considered an advantage of these antimycoplasmal agents. Only fluoroquinolones and ketolides have mycoplasmicidal qualities.

Table 1 shows a compilation of MIC data obtained with different antimicrobials against several human mycoplasmas. A lot of data are available for the three main pathogens *M. pneumoniae*, *M. hominis* and *Ureaplasma* spp. The results for the small number of strains of *M. genitalium* that are currently available, indicate that this species has an antibiotic susceptibility profile similar to that of *M. pneumoniae*. The susceptibility profile of *M. fermentans* is comparable to that of *M. hominis*, with MICs of macrolides being slightly lower than those for *M. hominis*.

*Table 1.* Minimal inhibitory concentration ranges (µg/ml) for various antimicrobials against *M. pneumoniae, M. genitalium, M. hominis, M. fermentans,* and *Ureaplasma* spp.[a].

| Antimicrobial | M. pneumoniae | M. genitalium | M. hominis | M. fermentans | Ureaplasma spp. |
|---|---|---|---|---|---|
| *Tetracyclines[b] glycylcyclines* | | | | | |
| Tetracycline | 0.63-0.25 | ND | 0.2-2 | 0.1-1 | 0.05-2 |
| Doxycycline | 0.02-0.5 | ≤0.01-0.3 | 0.03-2 | 0.05-1 | 0.02-1 |
| Minocycline | 0.06-0.5 | ≤0.01-0.2 | 0.03-1 | ND | 0.06-1 |
| GAR-936 | 0.06-0.25 | ND | 0.125-0.5 | ND | 1-16 |
| *MLS group* | | | | | |
| Erythromycin | ≤0.004-0.06 | ≤0.01 | 32->1,000 | 0.5->64 | 0.02-4 |
| Dirithromycin | ≤0.01-0.06 | ≤0.01-0.12 | >64 | ≥64 | 0.25-4 |
| Roxithromycin | ≤0.01-0.03 | ≤0.01 | >16->64 | 32-64 | 0.06-4 |
| Clarithromycin | ≤0.004-0.125 | ≤0.01-0.06 | 16->256 | 1-64 | ≤0.004-2 |
| Azithromycin | ≤0.004-0.01 | ≤0.01-0.03 | 4->64 | 0.05-8 | 0.06-0.5 |
| Josamycin | ≤0.01-0.02 | 0.01-0.02 | 0.05-2 | 0.12-0.5 | 0.03-4 |
| Spiramycin | ≤0.01-0.25 | 0.12-1 | 32->64 | 2-4 | 4-32 |
| Clindamycin | ≤0.008-2 | 0.2-1 | ≤0.008-2 | 0.01-0.25 | 0.2-64 |

| | | | | | |
|---|---|---|---|---|---|
| Lincomycin | 4-8 | 1-8 | 0.2-2 | ND | 8-256 |
| Pristinamycin | 0.02-0.05 | ND | 0.1-0.5 | ND | 0.1-1 |
| Quinupristin/ Dalfopristin | 0.008-0.12 | 0.05 | 0.03-2 | 0.12-0.5 | 0.03-0.5 |
| *Ketolides* | | | | | |
| Telithromycin | 0.0002-0.06 | ≤0.015 | 2-32 | 0.06-0.25 | ≤0.015-0.25 |
| ABT-773 | ≤0.005 | ND | ND | ND | ND |
| *Fluoroquinolones[b]* | | | | | |
| Pefloxacin | 2 | ND | 1-4 | ND | 1-8 |
| Norfloxacin | ND | ND | 4-16 | ND | 4-16 |
| Ciprofloxacin | 0.5-2 | 2 | 0.5-4 | 0.02-0.25 | 0.1–4 |
| Ofloxacin | 0.05–2 | 1–2 | 0.5-4 | 0.02-0.25 | 0.2–4 |
| Sparfloxacin | ≤0.008-0.5 | 0.05-0.1 | ≤0.008-0.1 | ≤0.01-0.05 | 0.003-1 |
| Levofloxacin | 0.5–1 | 0.5–1 | ≤0.008-0.5 | 0.05 | 0.12–1 |
| Grepafloxacin | 0.06–0.25 | 0.06–0.12 | ≤0.008-0.06 | 0.03–0.12 | 0.03–2 |
| Trovafloxacin | ≤0.008-0.5 | ND | ≤0.008-0.06 | ≤0.008-0.03 | ≤0.008-0.5 |
| Gatifloxacin | 0.06-1 | ND | 0.06-0.25 | ND | 0.5-4 |
| Moxifloxacin | 0.06-0.3 | 0.03 | 0.06 | ≤0.015-0.06 | 0.12-0.5 |
| Gemifloxacin | ≤0.008-0.12 | 0.05 | 0.0025-0.01 | 0.001-0.01 | ≤0.008-0.25 |
| BMS-284756 | 0.015-0.12 | ND | 0.008-0.25 | ND | 0.12-1 |
| *Chloramphenicol* | 2-10 | 0.5-25 | 4-25 | 0.5-10 | 0.4–8 |
| *Aminosides* | | | | | |
| Gentamicin | 4 | ND | 2-16 | 0.25->500 | 0.1–13 |
| *New agents* | | | | | |
| Linezolid | 64-256 | ND | 2-8 | ND | >64 |
| Evernimicin | 2-4 | ND | 1-16 | ND | 8-16 |

[a]Data were compiled from published studies in which different methods were used (references 4, 10-12, 14, 15, 22, 29, 32, 33, 38-40, 42, 43, 47, 50-56, 62, 67, 69, 70, 79, 83, 85, 89, 95).
[b]Susceptible strains. ND, not determined.

## 4.1    Tetracyclines

Tetracyclines and the related compounds glycylcyclines inhibit bacterial protein synthesis by preventing the association of aminoacyl-tRNA to the acceptor (A) site of the ribosome (20). They are broad-spectrum antimicrobials active against gram-positive and gram-negative bacteria, and atypical microorganisms such as mycoplasmas, chlamydiae and *Legionella*. They are characterized by their bacteriostatic activity against bacteria, including mycoplasmas.

MIC ranges of tetracycline, doxycycline, minocycline and the glycylcycline GAR-936 are given in Table 1 for the main pathogenic mycoplasmas. Generally, tetracyclines are the drug of first choice for the treatment of mycoplasmal urogenital infections like nongonococcal urethritis and pelvic inflammatory disease. They are also indicated in respiratory tract infections due to *M. pneumoniae* in adults (20). They cannot be used in children and during pregnancy. However several publications reported the efficient use of tetracyclines, especially doxycycline in the treatment of central nervous system (CNS) infections in neonates (76, 88). Indeed, besides the innate resistance of several species, macrolides which are the treatment of choice for mycoplasma infections in children, penetrate poorly across the blood-brain barrier and into the cerebrospinal fluid (CSF).

## 4.2 Macrolide-Lincosamide-Streptogramin (MLS) group

The antibiotics constituting the MLS group have a similar mode of action and share a common binding site on the ribosome (84, 92). Macrolides are divided into three subgroups according to their lactone ring sizes. Thus erythromycin belongs to the first-generation 14-membered ring macrolides, while josamycin belongs to the 16-membered ring subgroup. Azithromycin is the sole member of macrolides with ring size of 15 atoms (18). Like tetracyclines, they inhibit protein synthesis at the ribosome level and are bacteriostatic. They bind to the 50S subunit of the ribosome and block the growth of the nascent peptide chain, probably causing premature dissociation of the peptidyl-tRNA from the ribosome (84). Recent X-ray crystallographic studies of the small and large subunits of the ribosome have contributed consequently to the knowledge of the drug binding site for macrolides (2, 63, 93). Macrolides and other MLS drugs have been mapped by chemical footprinting to the central loop in domain V of 23S rRNA, which is associated with the peptidyl transferase activity (23). Additionally, erythromycin and its derivative ketolides have been mapped to hairpin 35 in domain II of the 23S rRNA (84).

MICs of the main antimicrobials belonging to the MLS group are presented in Table 1. As discussed above, these products show differential antimycoplasmal activity according to the species concerned. The new streptogramin quinupristin-dalfopristin and ketolides, show a very high activity against *M. pneumoniae*, but do not differ from those of the prototype antibiotics, pristinamycin and erythromycin, respectively (6). MLS are mainly used for the treatment of respiratory tract infections, or when tetracyclines or fluoroquinolones are contraindicated, for instance in newborns and children or during pregnancy (5). Thus, erythromycin is the first-choice drug for ureaplasmal infections of the respiratory tract in neonates (88).

An experimental model in hamsters has been used for assessing the efficacy of erythromycin, roxithromycin, and pristinamycin in the treatment of respiratory infection caused by intranasal inoculation of *M. pneumoniae*. Different clinical trials showed also the efficiency of such treatments for community-acquired pneumonia (6).

## 4.3     Fluoroquinolones

Fluoroquinolones are broad-spectrum antibiotics widely used in clinical medicine. The intracellular targets of fluoroquinolones in bacteria are considered to be the type II topoisomerases, DNA gyrase and topoisomerase IV, which are essential for bacterial DNA replication (28, 45). The DNA gyrase is composed of two A and two B subunits, encoded by the *gyrA* and *gyrB* genes, respectively. This tetrameric enzyme catalyzes ATP-dependent negative supercoiling of DNA. Topoisomerase IV, a $C_2E_2$ tetramer encoded by the *parC* and *parE* genes, is essential for chromosome partitioning. ParC is homologous to GyrA while ParE is homologous to GyrB.

Nalidixic acid, the original non-fluorinated molecule of the quinolone class, is inactive against gram-positive bacteria and mycoplasmas (71). In contrast, the derived fluoroquinolone class became active against gram-positive bacteria and atypical microorganisms. The first members of the fluoroquinolone class such as norfloxacin, pefloxacin, ofloxacin, and ciprofloxacin were followed by new agents, like sparfloxacin, levofloxacin, grepafloxacin, trovafloxacin, gatifloxacin, moxifloxacin, gemifloxacin, and BMS-284756. It should be noted that among these newer fluoroquinolones, some compounds like grepafloxacin or trovafloxacin have never been commercialized or were withdrawn, while others like gemifloxacin and BMS-284756, are still under development. The activities of all these compounds are summarized in Table 1. Newer fluoroquinolones show an enhanced activity against all the human mycoplasmas studied, compared to the older ones. This class could be a very interesting alternative to tetracyclines and macrolides in both respiratory and urogenital tract infections caused by mycoplasmas. Furthermore, they showed a cidal activity against these microorganisms (6, 11, 12, 14, 15, 81). They should not be given to children because of possible juvenile cartilage damage. Besides, the increasing occurrence of bacterial resistance to fluoroquinolones in many species including mycoplasmas (see below) could limit their use in clinical settings.

Several compounds of this class have been developed and commercialized to treat cell cultures infected by mycoplasmas (27).

## 4.4    Miscellaneous

### 4.4.1    Chloramphenicol

Chloramphenicol acts by inhibiting protein synthesis. Its binding to the 50S subunit of the ribosome leads to the inhibition of the peptidyltransferase step (23). Mycoplasmas show a susceptibility level similar to that of gram-positive bacteria like *Staphylococcus aureus* (Table 1). Because of its potential toxicity, it is not used for treatment of mycoplasmal diseases. Exceptions are ureaplasmal or mycoplasmal infections of CNS in neonates, for the treatment of which chloramphenicol has been used successfully because of its ability to penetrate into CSF (76, 88).

### 4.4.2    Aminoglycosides

Aminoglycosides are a widely used group of broad-spectrum antimicrobials which are bactericidal and which target the bacterial ribosome. They bind on different sites in the 16S rRNA of the 30S subunit and to some ribosomal proteins, decreasing the fidelity of translation (58, 66). These drugs are not used for the treatment of human mycoplasmal infections, but they have been employed to eliminate mycoplasma contamination from cell cultures (27). They are moderately to highly effective in elimination of mycoplasmas, but only at concentrations cytotoxic for the eukaryotic cells. An alternative treatment could be fluoroquinolones or an association combining macrolides and tetracyclines (27).

## 5.    ACQUIRED RESISTANCE OF MYCOPLASMAS

There are three different ways of acquiring bacterial resistance: active efflux mechanisms, enzymatic modification of the drug or alteration in the drug target site. Antibiotic resistance that results from altered cellular physiology and structure, can be acquired by genetic mutation or acquisition of new genes via gene transfer (see chapter "Extrachromosomal elements and gene transfer"). Both mechanisms have been documented in mycoplasmas.

Mycoplasmas are characterized by high mutation frequencies. It was first related to the lack of 3'-to-5' exonuclease activity of their DNA polymerase, proved later to be wrong (30). Recently the availability of the four sequenced mollicute genomes show the limited amount of genetic information dedicated to DNA repair systems (19, 31, 37, 44). It has been shown in other bacteria that the lack of some DNA repair systems like the *mut* gene is associated to a mutator phenotype (see chapter "Mycoplasmas of humans", section 3.1.2). Thus, a link could be hypothesized between high mutation rates and antibiotic resistance among mycoplasmas, as it has been found for other bacteria (64). This resistance mechanism concerns all the antibiotic classes used for the treatment of mycoplasmal infections. Selection for single- or multi-step mutations can be done in vitro by stepwise selection in broth or on agar.

Concerning the acquisition of new resistance genes from other bacteria, only cryptic plasmids have been found in mollicutes, and none has been described in human species (see chapter "Extrachromosomal elements and gene transfer"). However transposons carrying antibiotic resistance genes have been recognized in mycoplasmas (71). The main example remains the *tetM* determinant leading to resistance to tetracyclines (see below).

Until recently, relatively few in vitro and in vivo resistance studies were available in the literature on human mycoplasmas. They are increasing now in number, concerning more *M. hominis* and *Ureaplasma* spp. than *M. pneumoniae*. Indeed, as this mycoplasma is quite difficult to isolate and remains greatly susceptible to antibiotics, the number of clinical resistant strains of *M. pneumoniae* is limited and the occurrence of acquired resistance in this mycoplasma is not well documented.

## 5.1     Tetracyclines

Acquired resistance to tetracyclines has been well documented for almost two decades. It concerns only the mollicutes of the urogenital tract, *Ureaplasma* spp. and *M. hominis*, but not *M. pneumoniae*. High-level resistance to tetracyclines (MICs $\geq 8$ µg/ml) in *M. hominis* and *Ureaplasma* spp. has been associated with the presence of the *tetM* determinant (17, 74, 75). In mycoplasmas, this determinant is associated with a conjugative transposon member of the Tn*1545* family, named Tn*916*. This transposon is widely distributed among urogenital bacteria of human origin (25, 71). The *tetM* transposon, Tn*916*, was shown to be present entirely in both *M. hominis* and *Ureaplasma* spp. (72), and is the sole naturally acquired antibiotic resistance determinant in clinical strains of mycoplasmas. Sequencing the *tetM* gene from a *U. urealyticum* strain showed a 95 % homology at both DNA and peptidic sequence levels to *tetM* from the gram-positive genus, *Streptococcus* (71). The *tetM* determinant encodes the TetM protein which protects the ribosome from the action of tetracyclines and confers resistance to doxycycline and minocycline. The TetM protein shows homology to elongation factors EF-Tu and EF-G, overlapping binding sites on ribosomes with EF-G (20). Thus the ribosomal protection protein leads to a ribosomal conformational change preventing the tetracycline binding to ribosome, without altering protein synthesis (20).

The extent to which tetracycline resistance occurs in genital mycoplasmas varies geographically and according to prior antimicrobial exposure in different populations. For instance the frequency of resistant strains has been evaluated to about 10% in patients attending STD clinics in United Kingdom (81) and about 5% in France (3). Regulation of the expression of the *tetM* determinant is probably linked to the sequences directly upstream from the structural gene, as suggested by the high homology of the upstream sequences from different *tetM* genes (20).

In addition the *tetM* resistance gene is one of the commonly used markers in genetic studies of mollicutes, as it has been transferred by conjugation or transformation in several mycoplasma species including *M. hominis* (73, see chapter "Extrachromosomal elements and gene transfer").

High-level resistant strains of some mycoplasma species like *M. hominis* and *M. pneumoniae* have been obtained by stepwise selection in the presence of increasing concentrations of tetracyclines (71; Bébéar *et al.*, unpublished data). The mechanisms of such a resistance remain to be elucidated.

## 5.2   MLS group

Acquired resistance to macrolides is not well documented in human mycoplasmas, and appears to remain a rare phenomenon. Some cases have been reported during treatment with erythromycin of *M. pneumoniae* infections, without affecting the clinical course of the illness (6). Resistant strains of *M. pneumoniae* have also been obtained in vitro by selection in the presence of erythromycin (59, 61, 78). The described strains presented an $MLS_B$ resistance phenotype with an acquired resistance to macrolides, lincosamides and streptogramins B. Lucier *et al.* (59) showed that erythromycin resistance of in vitro mutants of *M. pneumoniae* was associated with point mutations in the peptidyl transferase loop of domain V in the 23S rRNA, leading to a target modification. Two A→G transitions have been identified in positions 2058 and 2059, as hot spots of macrolide resistance in other bacteria (20). The ribosome binding studies showed a lower affinity to [$^{14}$C] erythromycin for ribosomes of resistant strains than those of the wild-type strain (59). The frequency of macrolide resistance among strains of *M. pneumoniae* is probably low, since such resistance was not detected among 150 strains isolated from France and Denmark since at least the last 10 years (Bébéar *et al.* unpublished data).

One clinical strain of *Ureaplasma* spp. resistant to macrolides has been isolated from a patient with nongonococcal urethritis treated with erythromycin (65). This isolate harbored an acquired resistance to 14- and 16-membered macrolides and natural resistance to lincosamides. Here again, the resistant isolate exhibited a decrease in antibiotic binding to ribosomes (65). The molecular mechanism was not investigated. It should be noted that *Ureaplasma* spp. could be classified incorrectly as intermediate in sensitivity to macrolides because of a technique bias in the susceptibility testing of this species (see section 2 of this chapter).

Two clinical isolates of *M. hominis* resistant to josamycin have been recently described, isolated from a patient with a chronic obstructive pulmonary disease repetitively exposed to antibiotic treatments (16). They showed a multi-resistance profile with additional resistance to fluoroquinolones due to target alterations, and to tetracyclines due to the presence of the *tetM* gene. The mechanism of josamycin resistance remains to be studied. Two mutants of *M. hominis* selected in vitro on josamycin have been very recently described (34). Their acquired resistance phenotype (resistance to 16-member macrolides but susceptibility to lincosamides) was associated with a mutation at position 2062 within domain V of the 23S rRNA. This position was found previously to be associated with macrolide resistance in other bacteria (20). The frequency of acquired macrolide resistance in clinical strains of *M. hominis* and *Ureaplasma* spp. is not known but, like for *M. pneumoniae*, is probably very low.

Neither the methytransferase enzyme (Erm family) that methylates position 2058 nor the enzyme modifying the antibiotic have been identified so far in mycoplasmas.

## 5.3    Fluoroquinolones

The quinolone resistance mechanisms that have been described, form two classes: (i) mutations in the four target genes, *gyrA*, *gyrB*, *parC* and *parE*, and (ii) reduction in the level of quinolone accumulation inside the cells either by active efflux or lack of penetration. Concerning the target-related resistance for each gene, mutations cluster within a conserved region referred to as the quinolone resistance-determining region (QRDR). QRDRs were first described for GyrA and GyrB, and homologous regions were identified later in the ParC and ParE enzymes (45, 46).

To date, acquired resistance to fluoroquinolones has been reported only for genital mycoplasmas. Recently Bébéar *et al.* have described fluoroquinolone resistance in clinical isolates of *M. hominis* (16) and *Ureaplasma* spp. (unpublished data), obtained from patients treated previously with fluoroquinolones. Most of these patients were immunocompromized. For both microorganisms, high-level resistance was encountered in vivo, and was associated with alterations in both DNA gyrase and topoisomerase IV. These alterations were located in the QRDRs of the gyrase *gyrA* gene and of the topoisomerase IV *parC* and *parE* genes. The mechanism of quinolone resistance has been studied also for in vitro-selected mutants of *M. hominis,* confirming the results obtained in vivo (8, 9, 13, 57).

Although all fluoroquinolones are active against both enzymes, they differ in their relative activities (45, 46). In *E. coli*, a gram-negative bacterium, DNA gyrase and topoisomerase IV were found to be the targets for fluoroquinolones primarily and secondarily, respectively. In contrast, in the gram-positive species *S. aureus*, topoisomerase IV was the primary quinolone target. However, some notable exceptions have been seen with *Streptococcus pneumoniae*, another gram-positive bacterium. In this microorganism genetic studies have shown that some fluoroquinolones like ciprofloxacin and levofloxacin targeted topoisomerase IV, while sparfloxacin and gatifloxacin targeted DNA gyrase (45). In *M. hominis*, a gram-positive related organism, the primary target was different depending on the fluoroquinolone used, resembling in this respect *S. pneumoniae*. In view of the in vitro genetic studies (9, 13, 57), topoisomerase IV was the primary target of pefloxacin, ofloxacin, ciprofloxacin, and trovafloxacin whereas DNA gyrase was the primary target of sparfloxacin.

The other resistance mechanism related to the expression or over-expression of energy-dependent efflux pumps that can actively remove antibacterials and other compounds from the cell, has not yet been described in mycoplasmas. However, recent data (Bébéar *et al.*, unpublished results) suggest the presence of an active efflux system implicated in the resistance to ciprofloxacin and unrelated compounds like ethidium bromide in the human mycoplasma *M. hominis*.

## 5.4    Aminoglycosides

Acquired resistance to aminoglycosides in bacteria can occur either by mutations in the ribosome target, presence of efflux mechanisms, or by the presence of most common enzymes which modify the antibiotic (58, 66). None of these enzymes has been found to occur naturally in mycoplasmas. However, the resistance gene *aacA-aphD* found in the transposon Tn*4001* which confers resistance to gentamicin, has been commonly used as a marker in genetic studies of mollicutes (see chapter "Extrachromosomal elements and gene transfer").

High-level resistant strains of different mycoplasma species to several aminoglycosides were previously described in vitro by stepwise selection (39, 71, 81), but the resistance mechanism was not elucidated. High-level resistance to aminoglycosides has been described in strains of *M. fermentans* isolated from cell cultures, but not from human sources (39). Indeed aminoglycosides have been used to cure cell cultures from mycoplasmal contamination, but not to treat human mycoplasmal infections.

## 5.5    Chloramphenicol

Since chloramphenicol is not used for treatment of mycoplasmal infections except for some rare cases of neonate infections, resistance mechanisms to this antibiotic have not been investigated in mycoplasmas.

The *cat* gene encoding the chloramphenicol acetyltransferase which modifies and inactivates the antibiotic, has been used as a selection marker in two non-human mollicutes and in *M. pneumoniae* (see chapter "Extrachromosomal elements and gene transfer").

## 6.    CONCLUSION

Standardization of susceptibility testing of mycoplasmas should be recommended and has to be conducted.

Considering the profile of acquired resistance in human mycoplasmas, *M. pneumoniae* and *M. genitalium* are predictably susceptible to antibiotics. Thus, their susceptibility testing is not necessary except for the in vitro evaluation of new antimicrobials. In contrast, the susceptibility of urogenital mycoplasmas needs to be evaluated, mainly in view of antibiotic resistance documented for these species.

Extragenital or respiratory infections caused by mycoplasmas occur often in immunocompromized patients, submitted frequently to an antibiotic selection pressure. So, in these patients the risk of multidrug-resistant mycoplasmas is higher, leading to the necessity of in vitro susceptibility testing of these microorganisms.

The appearance now of acquired resistance to the three main antibiotic classes used in mycoplasmal infections prompts us to keep watch over the susceptibility of human mycoplasmas, especially in immunocompromized patients. Furthermore, it is recommended to test recent isolates of mycoplasmas when studying a new compound, to be sure to include in the study some potential new cases of acquired resistance.

## REFERENCES

1.    **Abele-Horn, M., C. Blendinger, C. Becher, P. Emmerling, and G. Ruckdeschel.** 1996. Evaluation of commercial kits for quantitative identification and tests on antibiotic susceptibility of genital mycoplasmas. Zentralbl. Bakteriol. **284**:540-549.

2. **Ban, N., P. Nissen, J. Hansen, P. B. Moore, and T. A. Steitz.** 2000. The complete atomic structure of the large ribosomal subunit at 2.4 A resolution. Science. **289:**905-920.

3. **Bébéar, C.** 1996. Antibiotic sensitivity testing: introducing remarks, p. 181-183. *In* J. G. Tully, and S. Razin (eds), Molecular and diagnostic procedures in mycoplasmology, vol. II. Academic Press, San Diego.

4. **Bébéar, C., and D. H. Bouanchaud.** 1997. A review of the in-vitro activity of quinupristin/dalfopristin against intracellular pathogens and mycoplasmas. J. Antimicrob. Chemother. **39:**59-62.

5. **Bébéar, C., B. de Barbeyrac, C. M. Bébéar, H. Renaudin, and A. Allery.** 1997. New developments in diagnostic and treatment of mycoplasma infections in humans. Wien. Klin. Wochenschr. **109:**594-599.

6. **Bébéar, C., M. Dupon, H. Renaudin, and B. de Barbeyrac.** 1993. Potential improvements in therapeutic options for mycoplasmal respiratory infections. Clin. Infect. Dis. **17 Suppl 1:**S202-S207.

7. **Bébéar, C., and J. Robertson.** 1996. Determination of minimal inhibitory concentration, p. 189-199. *In* J. G. Tully, and S. Razin (eds), Molecular and diagnostic procedures in mycoplasmology, vol. II. Academic Press, San Diego.

8. **Bébéar, C. M., J. M. Bové, C. Bébéar, and J. Renaudin.** 1997. Characterization of *Mycoplasma hominis* mutations involved in resistance to fluoroquinolones. Antimicrob. Agents Chemother. **41:**269-273.

9. **Bébéar, C. M., O. Grau, A. Charron, H. Renaudin, D. Gruson, and C. Bébéar.** 2000. Cloning and nucleotide sequence of the DNA gyrase (*gyrA*) gene from *Mycoplasma hominis* and characterization of quinolone-resistant mutants selected in vitro with trovafloxacin. Antimicrob. Agents Chemother. **44:**2719-2727.

10. **Bébéar, C. M., H. Renaudin, M. D. Aydin, J. F. Chantot , and C. Bébéar.** 1997. In-vitro activity of ketolides against mycoplasmas. J. Antimicrob. Chemother. **39:**669-670.

11. **Bébéar, C. M., H. Renaudin, A. Boudjadja, and C. Bébéar.** 1998. In vitro activity of BAY 12-8039, a new fluoroquinolone, against mycoplasmas. Antimicrob. Agents Chemother. **42:**703-704.

12. **Bébéar, C. M., H. Renaudin, A. Bryskier, and C. Bébéar.** 2000. Comparative activities of telithromycin, levofloxacin, and other antimicrobials against human mycoplasmas. Antimicrob. Agents Chemother. **44:**1980-1982.

13. **Bébéar, C. M., H. Renaudin, A. Charron, J. M. Bové, C. Bébéar, and J. Renaudin.** 1998. Alterations in topoisomerase IV and DNA gyrase in quinolone-resistant mutants of *Mycoplasma hominis* obtained in vitro. Antimicrob. Agents Chemother. **42:**2304-2311.

14. **Bébéar, C. M., H. Renaudin, A. Charron, D. Gruson, M. Lefrançois, and C. Bébéar.** 2000. In vitro activity of trovafloxacin compared to those of five antimicrobials against *Mycoplasma hominis* and *Ureaplasma urealyticum* fluoroquinolone-resistant isolates that have been genetically characterized. Antimicrob. Agents Chemother. **44:**2557-2560.

15. **Bébéar, C. M., H. Renaudin, T. Schaeverbeke, F. Leblanc, and C. Bébéar.** 1999. In-vitro activity of grepafloxacin, a new fluoroquinolone, against mycoplasmas. J. Antimicrob. Chemother. **43:**711-714.

16. **Bébéar, C. M., J. Renaudin, A. Charron, H. Renaudin, B. de Barbeyrac, T. Schaeverbeke, and C. Bébéar.** 1999. Mutations in the *gyrA*, *parC*, and *parE* genes associated with fluoroquinolone resistance in clinical isolates of *Mycoplasma hominis*. Antimicrob. Agents Chemother. **43:**954-956.

17.   **Brunet, B., B. de Barbeyrac, H. Renaudin, and C. Bébéar.** 1989. Detection of tetracycline-resistant strains of *Ureaplasma urealyticum* by hybridization assays. Eur. J. Clin. Microbiol. Infect. Dis. **8:**636-638.
18.   **Bryskier, A. J., J. P. Butzler, H. C. Neu, and P. M. Tulkens** (eds.). 1993. Macrolides: chemistry, pharmacology and clnical uses. Arnette Blackwell, Paris, France.
19.   **Chambaud, I., R. Heilig, S. Ferris, V. Barbe, D. Samson, F. Galisson, I. Moszer, K. Dybvig, H. Wroblewski, A. Viari, E. P. Rocha, and A. Blanchard.** 2001. The complete genome sequence of the murine respiratory pathogen *Mycoplasma pulmonis*. Nucleic Acids Res. **29:**2145-2153.
20.   **Chopra, I., and M. Roberts.** 2001. Tetracycline antibiotics: mode of action, applications, molecular biology, and epidemiology of bacterial resistance. Microbiol. Mol. Biol. Rev. **65:**232-260.
21.   **Clegg, A., M. Passey, M. Yoannes, and A. Michael.** 1997. High rates of genital mycoplasma infection in the highlands of Papua New Guinea determined both by culture and by a commercial detection kit. J. Clin. Microbiol. **35:**197-200.
22.   **Cohen, M. A., and M. D. Huband.** 1997. In-vitro susceptibilities of *Mycoplasma pneumoniae*, *Mycoplasma hominis* and *Ureaplasma urealyticum* to clinafloxacin, PD 131628, ciprofloxacin and comparator drugs. J. Antimicrob. Chemother. **40:**308-309.
23.   **Cundliffe, E.** 1990. Recognition sites for antibiotics within rRNA, p. 479-490. *In* W. E. Hill, A. Dahlberg, R. A. Garrett, M. P. B., D. Schlessinger, and J. R. Waener (eds), The ribosome: structure, functions, and evolution. American society for Microbiology, Washington, D. C.
24.   **Dallo, S. F., and J. B. Baseman.** 2000. Intracellular DNA replication and long-term survival of pathogenic mycoplasmas. Microb. Pathogen. **29:**301-309.
25.   **de Barbeyrac, B., M. Dupon, P. Rodriguez, H. Renaudin, and C. Bébéar.** 1996. A Tn*1545*-like transposon carries the *tet(M)* gene in tetracycline resistant strains of *Bacteroides ureolyticus* as well as *Ureaplasma urealyticum* but not *Neisseria gonorrhoeae*. J. Antimicrob. Chemother. **37:**223-232.
26.   **Dosa, E., E. Nagy, W. Falk, I. Szoke, and U. Ballies.** 1999. Evaluation of the Etest for susceptibility testing of *Mycoplasma hominis* and *Ureaplasma urealyticum*. J. Antimicrob. Chemother. **43:**575-578.
27.   **Drexler, H. G., and C. C. Uphoff.** 2000. Contamination of cell cultures, *Mycoplasma*, p. 609-627. *In* H. G. Drexler (ed.), The leukemia lymphoma cell lines Factsbook. Academic Press, San Diego.
28.   **Drlica, K., and X. L. Zhao.** 1997. DNA gyrase, topoisomerase IV, and the 4-quinolones. Microbiol. Mol. Biol. Rev. **61:**377-392.
29.   **Duffy, L. B., D. Crabb, K. Searcey, and M. C. Kempf.** 2000. Comparative potency of gemifloxacin, new quinolones, macrolides, tetracycline and clindamycin against *Mycoplasma* spp. J. Antimicrob. Chemother. **45:**29-33.
30.   **Dybvig, K, and L. L. Voelker.** 1996. Molecular biology of mycoplasmas. Annu. Rev. Microbiol. **50:**25-57.
31.   **Fraser, C. M., J. D. Gocayne, O. White, M. D. Adams, R. A. Clayton, R. D. Fleischmann, C. J. Bult, A. R. Kerlavage, G. Sutton, J. M. Kelley, and et al.** 1995. The minimal gene complement of *Mycoplasma genitalium*. Science. **270:**397-403.
32.   **Fung-Tomc, J., B. Minassian, B. Kolek, T. Washo, E. Huczko, and D. Bonner.** 2000. In vitro antibacterial spectrum of a new broad-spectrum 8-methoxy fluoroquinolone, gatifloxacin. J. Antimicrob. Chemother. **45:**437-446.
33.   **Fung-Tomc, J. C., B. Minassian, B. Kolek, E. Huczko, L. Aleksunes, T. Stickle, T. Washo, E. Gradelski, L. Valera, and D. P. Bonner.** 2000. Antibacterial spectrum

of a novel des-fluoro(6) quinolone, BMS-284756. Antimicrob. Agents Chemother. **44:**3351-3356.

34. **Furneri, P. M., G. Rappazzo, M. P. Musumarra, P. Di Pietro, L. S. Catania, and L. S. Roccasalva.** 2001. Two New Point Mutations at A2062 Associated with Resistance to 16-Membered Macrolide Antibiotics in Mutant Strains of *Mycoplasma hominis.* Antimicrob. Agents Chemother. **45:**2958-2960.

35. **Furneri, P. M., G. Rappazzo, M. P. Musumarra, G. Tempera, and L. S. Roccasalva.** 2000. Genetic basis of natural resistance to erythromycin in *Mycoplasma hominis.* J. Antimicrob. Chemother. **45:**547-548.

36. **Gaurivaud, P., F. Laigret, and J.-M. Bové.** 1996. Insusceptibility of members of the class Mollicutes to rifampin: studies of the *Spiroplasma citri* RNA polymerase β-subunit gene. Antimicrob. Agents Chemother. **40:**858-862.

37. **Glass, J. I., E. J. Lefkowitz, J. S. Glass, C. R. Heiner, E. Y. Chen, and G. H. Cassell.** 2000. The complete sequence of the mucosal pathogen *Ureaplasma urealyticum.* Nature. **407:**757-762.

38. **Hamamoto, K., T. Shimizu, N. Fujimoto, Y. Zhang, and S. Arai.** 2001. In vitro activities of moxifloxacin and other fluoroquinolones against *Mycoplasma pneumoniae.* Antimicrob. Agents Chemother. **45:**1908-1910.

39. **Hannan, P. C.** 1995. Antibiotic susceptibility of *Mycoplasma fermentans* strains from various sources and the development of resistance to aminoglycosides in vitro. J. Med. Microbiol. **42:**421-428.

40. **Hannan, P. C.** 1998. Comparative susceptibilities of various AIDS-associated and human urogenital tract mycoplasmas and strains of *Mycoplasma pneumoniae* to 10 classes of antimicrobial agents in vitro. J. Med. Microbiol. **47:**1115-1122.

41. **Hannan, P. C.** 2000. Guidelines and recommendations for antimicrobial minimum inhibitory concentration (MIC) testing against veterinary mycoplasma species. International Research Programme on Comparative Mycoplasmology. Vet. Res. **31:**373-395.

42. **Hannan, P. C. T., and G. Woodnutt** 2000. In vitro activity of gemifloxacin (SB 265805; LB20304a) against human mycoplasmas. J. Antimicrob. Chemother. **45:**367-369.

43. **Hayes, M. M., H. H. Foo, H. Kotani, D. J. Wear, and S. C. Lo.** 1993. In vitro antibiotic susceptibility testing of different strains of *Mycoplasma fermentans* isolated from a variety of sources. Antimicrob. Agents Chemother. **37:**2500-2503.

44. **Himmelreich, R., H. Hilbert, H. Plagens, E. Pirkl, B. C. Li, and R. Herrmann.** 1996. Complete sequence analysis of the genome of the bacterium *Mycoplasma pneumoniae.* Nucleic Acids Res. **24:**4420-4449.

45. **Hooper, D. C.** 2000. Mechanisms of action and resistance of older and newer fluoroquinolones. Clin. Infect. Dis. **31:**S24-S28.

46. **Hooper, D. C.** 1999. Mechanisms of fluoroquinolone resistance. Drug Resist. Updat. **2:**38-55.

47. **Ishida, K., M. Kaku, K. Irifune, R. Mizukane, H. Takemura, R. Yoshida, H. Tanaka, T. Usui, K. Tomono, N. Suyama, H. Koga, S. Kohno, and K. Hara.** 1994. In-vitro and in-vivo activity of a new quinolone AM-1155 against *Mycoplasma pneumoniae.* J. Antimicrob. Chemother. **34:**875-83.

48. **Jin, D. J., and C. A. Gross.** 1988. Mapping and sequencing of mutations in the *Escherichia coli rpoB* gene that lead to rifampicin resistance. J. Mol. Biol. **202:**45-58.

49. **Kenny, G. E., and F. D. Cartwright.** 1993. Effect of pH, inoculum size, and incubation time in the susceptibility of *Ureaplasma urealyticum* to erythromycin in vitro. Clin. Infect. Dis. **17:**S215-S218.

50.    **Kenny, G. E., and F. D. Cartwright.** 1991. Susceptibilities of *Mycoplasma hominis*
       and *Ureaplasma urealyticum* to two new quinolones, sparfloxacin and WIN 57273.
       Antimicrob. Agents Chemother. **35:**1515-1516.
51.    **Kenny, G. E., and F. D. Cartwright.** 2001. Susceptibilities of *Mycoplasma hominis*,
       *M. pneumoniae*, and *Ureaplasma urealyticum* to GAR-936, dalfopristin,
       dirithromycin, evernimicin, gatifloxacin, linezolid, moxifloxacin, quinupristin-
       dalfopristin, and telithromycin compared to their susceptibilities to reference
       macrolides, tetracyclines and fluoroquinolones. Antimicrob. Agents Chemother.
       **45:**2604-2608.
52.    **Kenny, G. E., and F. D. Cartwright.** 1993. Susceptibilities of *Mycoplasma hominis*,
       *Mycoplasma pneumoniae*, and *Ureaplasma urealyticum* to a new quinolone, OPC
       17116. Antimicrob. Agents Chemother. **37:**1726-1727.
53.    **Kenny, G. E., and F. D. Cartwright.** 1994. Susceptibilities of *Mycoplasma hominis*,
       *Mycoplasma pneumoniae*, and *Ureaplasma urealyticum* to new glycylcyclines in
       comparison with those to older tetracyclines. Antimicrob. Agents Chemother.
       **38:**2628-2632.
54.    **Kenny, G. E., and F. D. Cartwright.** 1996. Susceptibilities of *Mycoplasma
       pneumoniae*, *Mycoplasma hominis* and *Ureaplasma urealyticum* to a new quinolone,
       trovafloxacin (CP-99,219). Antimicrob. Agents Chemother. **40:**1048-1049.
55.    **Kenny, G. E., and F. D. Cartwright.** 1991. Susceptibility of *Mycoplasma
       pneumoniae* to several new quinolones, tetracycline, and erythromycin. Antimicrob.
       Agents Chemother. **35:**587-589.
56.    **Kenny, G. E., T. M. Hooton, M. C. Roberts, F. D. Cartwright, and J. Hoyt.** 1989.
       Susceptibilities of genital mycoplasmas to the newer quinolones as determined by the
       agar dilution method. Antimicrob. Agents Chemother. **33:**103-107.
57.    **Kenny, G. E., P. A. Young, F. D. Cartwright, K. E. Sjostrom, and W. M. Huang.**
       1999. Sparfloxacin selects gyrase mutations in first-step *Mycoplasma hominis*
       mutants, whereas ofloxacin selects topoisomerase IV mutations. Antimicrob. Agents
       Chemother. **43:**2493-2496.
58.    **Kotra, L. P., J. Haddad, and S. Mobashery.** 2000. Aminoglycosides: perspectives
       on mechanisms of action and resistance and strategies to counter resistance.
       Antimicrob. Agents Chemother. **44:**3249-3256.
59.    **Lucier, T. S., K. Heitzman, S.-K. Liu, and P.-C. Hu.** 1995. Transition mutations in
       the 23S rRNA of erythromycin-resistant isolates of *Mycoplasma pneumoniae*.
       Antimicrob. Agents Chemother. **39:**2770-2773.
60.    **Ludwig, W., G. Kirchof, N. Klugbauer, M. Weizenegger, D. Betzl, M. Ehrmann,
       C. Hertel, S. Jilg, R. Tatzel, H. Zitzelsberger, S. Liebl, M. Hochberger, J. Shah,
       D. Lane, and P. R. Wallnoef.** 1992. Complete 23S ribosomal RNA sequences of
       Gram-positive bacteria with a low DNA G+C content. Syst. Appl. Microbiol. **15:**487-
       501.
61.    **Niitu, Y., S. Hasegawa, and H. Kubota.** 1974. Usefulness of an erythromycin-
       resistant strain of *Mycoplasma pneumoniae* for the fermentation-inhibition test.
       Antimicrob. Agents Chemother. **5:**111-113.
62.    **Nilius, A. M., M. H. Bui, L. Almer, D. Hensey-Rudloff, J. Beyer, Z. Ma, Y. S. Or,
       and R. K. Flamm.** 2001. Comparative in vitro activity of ABT-773, a novel
       antibacterial ketolide. Antimicrob. Agents Chemother. **45:**2163-2168.
63.    **Nissen, P., J. Hansen, N. Ban, P. B. Moore, and T. A. Steitz.** 2000. The structural
       basis of ribosome activity in peptide bond synthesis. Science. **289:**920-930.

64. **Oliver, A., R. Canton, P. Campo, F. Baquero, and J. Blazquez.** 2000. High frequency of hypermutable *Pseudomonas aeruginosa* in cystic fibrosis lung infection. Science. **288**:1251-1254.

65. **Palu, G., S. Valisena, M. F. Barile, and G. A. Meloni.** 1989. Mechanisms of macrolide resistance in *Ureaplasma urealyticum*: a study on collection and clinical strains. Eur. J. Epidemiol. **5**:146-153.

66. **Puglisi, J. D., S. C. Blanchard, K. D. Dahlquist, R. G. Eason, D. Fourmy, S. R. Lynch, M. I. Recht, and S. Yoshizawa.** 2000. Aminoglycoside antibiotics and decoding, p. 419-430. *In* R. A. Garrett, S. R. Douthwaite, A. Liljas, A. T. Matheson, P. B. Moore, and H. F. Noller (eds), The ribosome: structure, functions, antibiotics and cellular interactions. American society for Microbiology, Washington, D. C.

67. **Renaudin, H., and C. Bébéar.** 1992. Activité in vitro de la sparfloxacine sur les mycoplasmes. Pathol. Biol. **40**:450-454.

68. **Renaudin, H., and C. Bébéar.** 1990. Evaluation of the Mycoplasma Plus and the SIR Mycoplasma kits for quantitative detection and antibiotic susceptibility testing of genital mycoplasmas. Pathol Biol (Paris). **38**:431-435.

69. **Renaudin, H., J. G. Tully, and C. Bébéar.** 1992. In vitro susceptibilities of *Mycoplasma genitalium* to antibiotics. Antimicrob. Agents Chemother. **36**:870-872.

70. **Ridgway, G. L., H. Salman, M. J. Robbins, C. Dencer, and D. Felmingham.** 1997. The in-vitro activity of grepafloxacin against *Chlamydia* spp., *Mycoplasma* spp., *Ureaplasma urealyticum* and *Legionella* spp. J. Antimicrob. Chemother. **40 (Suppl. A):**31-34.

71. **Roberts, M. C.** 1992. Antibiotic resistance, p. 513-524. *In* J. Maniloff, R. N. McElhaney, L. R. Finch, and J. B. Baseman (eds), Mycoplasmas: molecular biology and pathogenesis. American Society for Microbiology, Washington, D. C.

72. **Roberts, M. C.** 1990. Characterization of the *tetM* determinants in urogenital and respiratory bacteria. Antimicrob. Agents Chemother. **34**:476-478.

73. **Roberts, M. C., and G. E. Kenny.** 1987. Conjugal transfer of transposon Tn916 from *Streptococcus faecalis* to *Mycoplasma hominis*. J. Bacteriol. **169**:3836-3839.

74. **Roberts, M. C., and G. E. Kenny.** 1986. Dissemination of the *tetM* tetracycline resistant determinant to *Ureaplasma urealyticum*. Antimicrob. Agents Chemother. **29**:350-352.

75. **Roberts, M. C., L. A. Koutsky, K. K. Holmes, D. J. LeBlanc, and G. E. Kenny** 1985. Tetracycline-resistant *Mycoplasma hominis* strains contain streptococcal *tetM* sequences Antimicrob. Agents Chemother. **28**:141-143.

76. **Sarlangue, J., and C. Bébéar.** 1999. Infections néonatales à mycoplasmes. Médecine Thérapeutique Pédiatrie. **2**:105-109.

77. **Stemke, G. W., Y. Huang, F. Laigret, and J. M. Bové.** 1994. Cloning the ribosomal RNA operons of *Mycoplasma flocculare* and comparison with those of *Mycoplasma hyopneumoniae*. Microbiology. **140**:857-860.

78. **Stopler, T., and D. Branski.** 1986. Resistance of *Mycoplasma pneumoniae* to macrolides, lincomycin and streptogramin B. J. Antimicrob. Chemother. **18**:359-364.

79. **Takahata, M., M. Shimakura, R. Hori, K. Kizawa, Y. Todo, S. Minami, Y. Watanabe, and H. Narita.** 2001. In vitro and in vivo efficacies of T-3811ME (BMS-284756) against *Mycoplasma pneumoniae*. Antimicrob. Agents Chemother. **45**:312-315.

80. **Taylor-Robinson, D.** 1996. Cidal activity testing, p. 199-205. *In* J. G. Tully, and S. Razin (eds), Molecular and diagnostic procedures in mycoplasmology, vol. II. Academic Press, San Diego.

81.  **Taylor-Robinson, D., and C. Bébéar.** 1997. Antibiotic susceptibilities of mycoplasmas and treatment of mycoplasmal infections. J. Antimicrob. Chemother. **40:**622-630.

82.  **Ter Laak, E. A., A. Pijpers, J. H. Noordergraaf, E. C. Schoevers, and J. H. Verheijden.** 1991. Comparison of methods for in vitro testing of susceptibility of porcine *Mycoplasma* species to antimicrobial agents. Antimicrob. Agents Chemother. **35:**228-233.

83.  **Ullmann, U., S. Schubert, and R. Krausse.** 1999. Comparative in vitro activity of levofloxacin, other fluoroquinolones, doxycycline and erythromycin against *Ureaplasma urealyticum* and *Mycoplasma hominis.* J. Antimicrob. Chemother. **43 (Suppl. C):**33-36.

84.  **Vester, B., and S. Douthwaite.** 2001. Macrolide resistance conferred by base substitutions in 23S rRNA. Antimicrob. Agents Chemother. **45:**1-12.

85.  **Waites, K. B., C. M. Bébéar, J. A. Roberston, D. F. Talkington, and G. E. Kenny.** 2001. Cumitech 34, Laboratory diagnosis of mycoplasmal infections. Coordinating ed., F. S. Nolte. American Society for Microbiology, Washington D. C.

86.  **Waites, K. B., K. C. Canupp, and G. E. Kenny.** 1999. In vitro susceptibilities of *Mycoplasma hominis* to six fluoroquinolones as determined by E test. Antimicrob. Agents Chemother. **43:**2571-2573.

87.  **Waites, K. B., D. M. Crabb, L. B. Duffy, and G. H. Cassell** 1997. Evaluation of the E test for detection of tetracycline resistance in *Mycoplasma hominis.* Diagn. Microbiol. Infect. Dis. **27:**117-122.

88.  **Waites, K. B., D. T. Crouse, and G. H. Cassell.** 1993. Therapeutic considerations for *Ureaplasma urealyticum* infections in neonates. Clin. Infect. Dis. **17:**S208-S214.

89.  **Waites, K. B., L. B. Duffy, T. Schmid, D. Crabb, M. S. Pate, and G. H. Cassell.** 1991. In vitro susceptibilities of *Mycoplasma pneumoniae, Mycoplasma hominis,* and *Ureaplasma urealyticum* to sparfloxacin and PD127391. Antimicrob. Agents Chemother. **35:**1181-1185.

90.  **Waites, K. B., T. A. Figarola, T. Schmid, D. M. Crabb, L. B. Duffy, and J. W. Simecka.** 1991. Comparison of agar versus broth dilution techniques for determining antibiotic susceptibilities of *Ureaplasma urealyticum.* Diagn. Microbiol. Infect. Dis. **14:**265-271.

91.  **Waites, K. B., D. F. Talkington, and C. M. Bébéar.** 2001. Mycoplasmas p.201-224. *In* A. L. Truant (ed.), Manual of commercial methods in clinical microbiology. American society for Microbiology, Washington D. C.

92.  **Weisblum, B.** 1998. Macrolide resistance. Drug Resist. Updat. **1:**29-41.

93.  **Wimberly, B. T., D. E. Brodersen, W. M. Clemons, Jr., R. J. Morgan-Warren, A. P. Carter, C. Vonrhein, T. Hartsch, and V. Ramakrishnan.** 2000. Structure of the 30S ribosomal subunit. Nature. **407:**327-339.

94.  **Woese, C. R.** 1987. Bacterial evolution. Microbiol. Rev. **51:**221-271.

95.  **Yamaguchi, T., Y. Hirakata, K. Izumikawa, Y. Miyazaki, S. Maesaki, K. Tomono, Y. Yamada, S. Kamihira, and S. Kohno.** 2000. In vitro activity of telithromycin (HMR3647), a new ketolide, against clinical isolates of *Mycoplasma pneumoniae* in Japan. Antimicrob. Agents. Chemother. **44:**1381-1382.

# Index